2008+
Solved Problems in
ELECTROMAGNETICS

Syed A. Nasar, Ph.D.

SciTECH
PUBLISHING, INC.

Raleigh, North Carolina
www.scitechpub.com

SciTech Publishing, Inc.
911 Paverstone Drive, Suite B
Raleigh, NC 27615
Phone: 919-847-2434   Fax: 919-847-2568
www.scitechpub.com

Editor: Dudley R. Kay
Production Director: Susan Manning
Production Assistant: Robert Lawless
Cover design by Kathy Gagne

Cover image courtesy of Greg Shirah of NASA,
http://pwg.gsfc.nasa.gov/istp/outreach/images/Aurora/71300.jpg

First edition published as *2000 Solved Problems in Electromagnetics*
copyright 1992 by McGraw-Hill.

ISBN: 1891121464
ISBN 13: 9781891121463

CONTENTS

1.1   Quantities having magnitude only are called *scalars,* and those having magnitudes *and* directions are called *vectors.* Classify: **(a)** 20 °C temperature in a room; **(b)** 50 N centripetal force; **(c)** 60 km/h speed of an automobile; **(d)** 60 km/h horizontal velocity of an automobile; **(e)** $100 in $1 bills.

▐   **(a)** scalar; **(b)** vector; **(c)** scalar; **(d)** vector; **(e)** scalar.

1.2   A vector may be represented by a directed line segment. Show how the two vectors given in Fig. 1-1(*a*) may be added together.

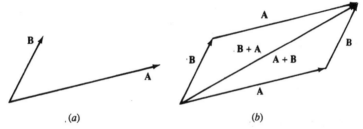

Fig. 1-1

▐   See Fig. 1-1(*b*).

1.3   For the vector **A** of Fig. 1-2(*a*), draw the vector *k***A** for $k > 0$ and $k < 0$.

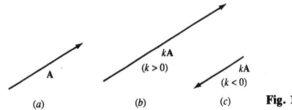

Fig. 1-2

▐   See Fig. 1-2(*b*) and (*c*).

1.4   Obtain the vector **A** − **B** for the vectors **A** and **B** given in Fig. 1-1.

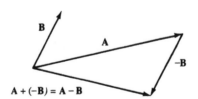

Fig. 1-3

▐   See Fig. 1-3.

1.5   We define *unit vectors* $\mathbf{a}_x$, $\mathbf{a}_y$, and $\mathbf{a}_z$ in a rectangular coordinate system as vectors of length 1 and directed along the x, y, and z axes, respectively. Given A_x, A_y, and A_z, the scalar components (projections) of a vector **A** along the three axes, express **A** in terms of these components and unit vectors.

▐   The *vector* components of **A** are $\mathbf{A}_x = A_x\mathbf{a}_x$, $\mathbf{A}_y = A_y\mathbf{a}_y$, $\mathbf{A}_z = A_z\mathbf{a}_z$; hence
$$\mathbf{A} = \mathbf{A}_x + \mathbf{A}_y + \mathbf{A}_z = A_x\mathbf{a}_x + A_y\mathbf{a}_y + A_z\mathbf{a}_z$$

1.6   A vector **A** has beginning rectangular coordinates (x_1, y_1, z_1) and end rectangular coordinates (x_2, y_2, z_2). Show **A** and its vector components.

▐   See Fig. 1-4.

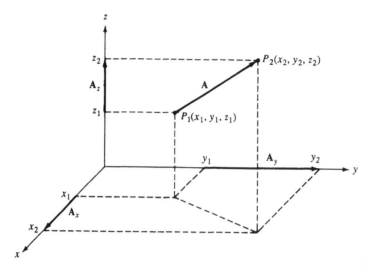

Fig. 1-4

1.7 Find the magnitude of the vector **A** of Problem 1.6 in terms of (*a*) the given coordinates, and (*b*) the scalar components of **A**.

▌ From Fig. 1-4 it follows that $|\mathbf{A}| = \sqrt{(x_2 - x_1)^2 + (y_2 - y_1)^2 + (z_2 - z_1)^2)} = \sqrt{A_x^2 + A_y^2 + A_z^2}$.

1.8 Obtain a unit vector in the direction of a vector **A**.

▌ The desired vector must have the form $\mathbf{a}_A = k\mathbf{A}$, with $k > 0$. Thus we require
$$|\mathbf{a}_A| = k\,|\mathbf{A}| = 1 \quad \text{or} \quad k = 1/|\mathbf{A}|$$
whence $\mathbf{a}_A = \mathbf{A}/|\mathbf{A}|$.

1.9 Given two vectors $\mathbf{A} = A_x\mathbf{a}_x + A_y\mathbf{a}_y$ and $\mathbf{B} = B_x\mathbf{a}_x + B_y\mathbf{a}_y$, find the components of $\mathbf{C} = \mathbf{A} + \mathbf{B}$.

▌ $\mathbf{C} = (\mathbf{A}_x + \mathbf{A}_y) + (\mathbf{B}_x + \mathbf{B}_y) = (\mathbf{A}_x + \mathbf{B}_x) + (\mathbf{A}_y + \mathbf{B}_y) = (A_x + B_x)\mathbf{a}_x + (A_y + B_y)\mathbf{a}_y$
whence $C_x = A_x + B_x$ and $C_y = A_y + B_y$.

1.10 Let *d***s** represent a directed element of a path in *xyz* space. Show the scalar components of *d***s**.

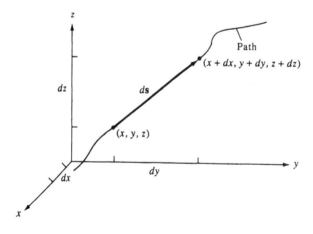

Fig. 1-5

▌ See Fig. 1-5.

1.11 Give analytical expressions for the vector *d***s** of Problem 1.10 and for the length.

▌ $$d\mathbf{s} = dx\,\mathbf{a}_x + dy\,\mathbf{a}_y + dz\,\mathbf{a}_z \qquad |d\mathbf{s}| = ds = \sqrt{dx^2 + dy^2 + dz^2}$$

1.12 Given two points $P_1 = (3, 1, 3)$ and $P_2 = (1, 3, 2)$, draw the vectors $\mathbf{R}_1 = \overrightarrow{OP_1}$, $\mathbf{R}_2 = \overrightarrow{OP_2}$, and $\mathbf{D} = \overrightarrow{P_1P_2}$.

▌ See Fig.1-6.

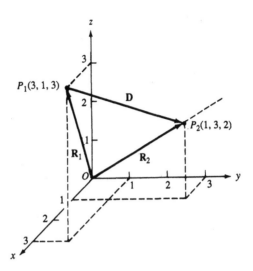

Fig. 1-6

1.13   Give mathematical expressions for the vectors $\mathbf{R}_1$, $\mathbf{R}_2$, and $\mathbf{D}$ of Problem 1.12.

▌ $\qquad \mathbf{R}_1 = 3\mathbf{a}_x + 1\mathbf{a}_y + 3\mathbf{a}_z \qquad \mathbf{R}_2 = 1\mathbf{a}_x + 3\mathbf{a}_y + 2\mathbf{a}_z \qquad \mathbf{D} = \mathbf{R}_2 - \mathbf{R}_1 = -2\mathbf{a}_x + 2\mathbf{a}_y - 1\mathbf{a}_z$

1.14   Obtain a unit vector in the direction of the vector $\mathbf{D}$ of Problem 1.12.

▌ By Problem 1.8,

$$\mathbf{a}_D = \frac{\mathbf{D}}{|\mathbf{D}|} = -\frac{2}{3}\mathbf{a}_x + \frac{2}{3}\mathbf{a}_y - \frac{1}{3}\mathbf{a}_z$$

1.15   An element of volume in rectangular coordinates is shown in Fig. 1-7. Give vector expressions for the surface areas of the faces.

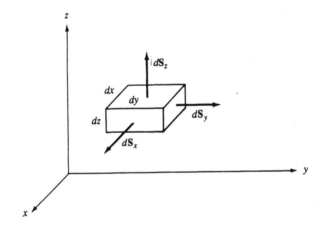

Fig. 1-7

▌ $\qquad\qquad d\mathbf{S}_x = dy\,dz\,\mathbf{a}_x \qquad d\mathbf{S}_y = dx\,dz\,\mathbf{a}_y \qquad d\mathbf{S}_z = dx\,dy\,\mathbf{a}_z$

Notice that the direction of a vectorial surface element is defined as the direction of the local outer normal to the surface.

1.16   A vector $\mathbf{A}$ is drawn from point $(0, -1, 3)$ to $(5, 1, -2)$. Find a unit vector in the direction of $\mathbf{A}$.

▌ $$\mathbf{A} = (5-0)\mathbf{a}_x + [1-(-1)]\mathbf{a}_y + (-2-3)\mathbf{a}_z = 5\mathbf{a}_x + 2\mathbf{a}_y - 5\mathbf{a}_z$$
$$|\mathbf{A}| = \sqrt{5^2 + 2^2 + 5^2} \approx 7.35$$
$$\mathbf{a}_A = \frac{\mathbf{A}}{|\mathbf{A}|} \approx \frac{1}{7.35}(5\mathbf{a}_x + 2\mathbf{a}_y - 5\mathbf{a}_z) = 0.68\mathbf{a}_x + 0.27\mathbf{a}_y - 0.68\mathbf{a}_z$$

1.17   Given $\mathbf{A} = 2\mathbf{a}_x + 3\mathbf{a}_y - \mathbf{a}_z$, $\mathbf{B} = \mathbf{a}_x + \mathbf{a}_y - 2\mathbf{a}_z$, and $\mathbf{C} = 3\mathbf{a}_x - \mathbf{a}_y + \mathbf{a}_z$, calculate: (*a*) $\mathbf{A} + \mathbf{B}$, (*b*) $\mathbf{B} - \mathbf{C}$, (*c*) $\mathbf{A} + 3\mathbf{B} - 2\mathbf{C}$, (*d*) $|\mathbf{A}|$, (*e*) $\mathbf{a}_B$.

■   **(a)** $3\mathbf{a}_x + 4\mathbf{a}_y - 3\mathbf{a}_z$ **(d)** $\sqrt{2^2 + 3^2 + 1^2} \approx 3.74$

   (b) $-2\mathbf{a}_x + 2\mathbf{a}_y - 3\mathbf{a}_z$ **(e)** $\dfrac{\mathbf{B}}{|\mathbf{B}|} = \dfrac{1}{\sqrt{6}}(\mathbf{a}_x + \mathbf{a}_y - 2\mathbf{a}_z) \approx 0.41\mathbf{a}_x + 0.41\mathbf{a}_y - 0.82\mathbf{a}_z$

   (c) $-\mathbf{a}_x + 8\mathbf{a}_y - 9\mathbf{a}_z$

1.18   Show how to represent an arbitrary vector **A** in terms of scalar components in a cylindrical coordinate system.

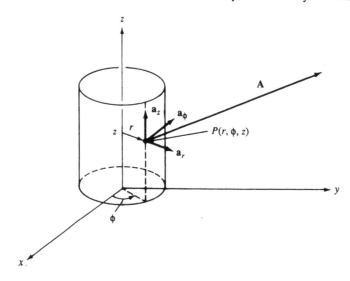

Fig. 1-8

■   Unit vectors $\mathbf{a}_r$, $\mathbf{a}_\phi$, and $\mathbf{a}_z$ at $P(r, \phi, z)$ are shown in Fig. 1-8. Then,

$$\mathbf{A} = A_r\mathbf{a}_r + A_\phi\mathbf{a}_\phi + A_z\mathbf{a}_z$$

and (as in any orthogonal system) $|\mathbf{A}|^2 = A_r^2 + A_\phi^2 + A_z^2$.

1.19   Figure 1-9 shows a volume element in cylindrical coordinates. Give expressions for **(a)** the differential volume, **(b)** the vectorial areas of the faces, **(c)** the differential arc length.

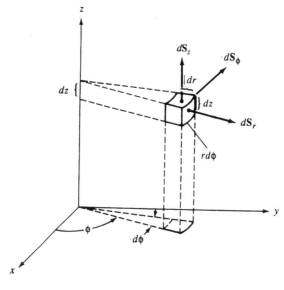

Fig. 1-9

■   **(a)** $dV = r\,d\phi\,dr\,dz$. **(b)** $d\mathbf{S}_r = r\,d\phi\,dz\,\mathbf{a}_r$, $d\mathbf{S}_\phi = dr\,dz\,\mathbf{a}_\phi$, $d\mathbf{S}_z = r\,d\phi\,dr\,\mathbf{a}_z$. **(c)** $d\mathbf{s}$ is the directed diagonal of the volume element: $d\mathbf{s} = dr\,\mathbf{a}_r + r\,d\phi\,\mathbf{a}_\phi + dz\,\mathbf{a}_z$.

1.20   Relate the rectangular coordinates (x_p, y_p, z_p) of a point P to its cylindrical coordinates (r_p, ϕ_p, z_p).

■   The z_p-coordinate is, of course, the same in both coordinate systems. Projecting P onto the xy plane (Fig. 1-10), we see that

$$r_p = \sqrt{x_p^2 + y_p^2} \qquad \text{and} \qquad \phi_p = \tan^{-1}\frac{y_p}{x_p}$$

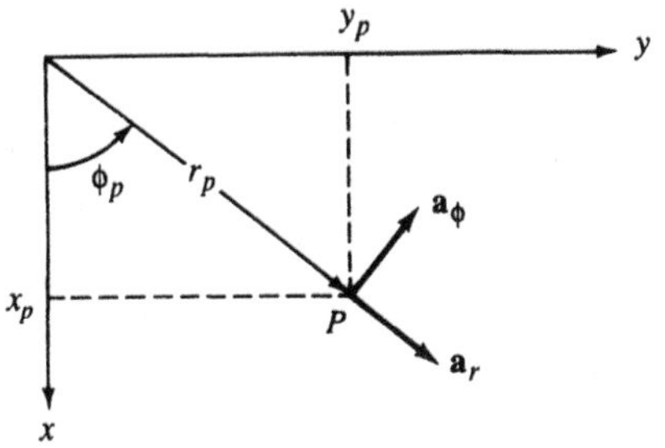

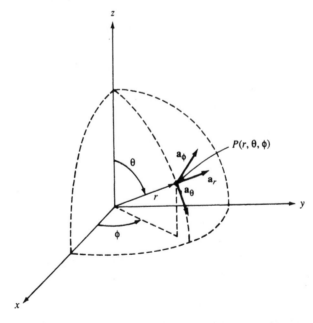

Fig. 1-10

1.21   Invert the relations found in Problem 1.20.

❚ $x_p = r_p \cos \phi_p$, $y_p = r_p \sin \phi_p$.

1.22   By integration in cylindrical coordinates, determine the volume enclosed by a cylinder of radius R and length L.

❚ Make the axis of the cylinder the z axis, from $z = 0$ to $z = L$. Then, using Problem 1.19(a),

$$V = \int dV = \int_{r=0}^{R} \int_{\phi=0}^{2\pi} \int_{z=0}^{L} r \, d\phi \, dr \, dz = \int_{r=0}^{R} \int_{\phi=0}^{2\pi} Lr \, d\phi \, dr = \int_{r=0}^{R} 2\pi Lr \, dr = \pi R^2 L$$

1.23   By integration, determine the surface area of the closed cylinder of Problem 1.22.

❚ Using Problem 1.19(b),

$$S = \int_{\text{lateral}} dS_r + \int_{\text{top cap}} dS_z + \int_{\text{bottom cap}} dS_z = \int_{z=0}^{L} \int_{\phi=0}^{2\pi} R \, d\phi \, dz + 2 \int_{r=0}^{R} \int_{\phi=0}^{2\pi} r \, d\phi \, dr = 2\pi RL + 2\pi R^2$$

1.24   Define the spherical-coordinate unit vectors at a point $P(r, \theta, \phi)$.

Fig. 1-11

❚ See Fig. 1-11. $\mathbf{a}_r$ is normal to the sphere $r = $ const. through P and points in the direction of increasing r; $\mathbf{a}_\theta$ is normal to the cone $\theta = $ const. through P and points in the direction of increasing θ; $\mathbf{a}_\phi$ is normal to the half-plane $\phi = $ const. through P and points in the direction of increasing ϕ.

1.25   Repeat Problem 1.19 for spherical coordination (Fig. 1-12).

❚ (**a**) $dV = r^2 \sin \theta \, dr \, d\theta \, d\phi$. (**b**) $d\mathbf{S}_r = r^2 \sin \theta \, d\phi \, d\theta \, \mathbf{a}_r$, $d\mathbf{S}_\phi = r \, d\theta \, dr \, \mathbf{a}_\phi$, $d\mathbf{S}_\theta = r \sin \theta \, dr \, d\phi \, \mathbf{a}_\theta$. (**c**) $d\mathbf{s} = dr \, \mathbf{a}_r + r \, d\theta \, \mathbf{a}_\theta + r \sin \theta \, d\phi \, \mathbf{a}_\phi$.

1.26   Convert between rectangular and spherical coordinates.

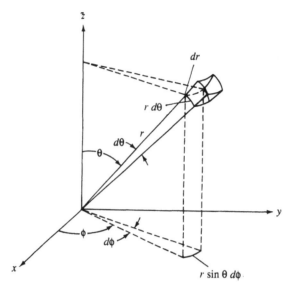

Fig. 1-12

▌ From Fig. 1-12, it follows that

$$x = r \sin \theta \cos \phi \qquad\qquad r = \sqrt{x^2 + y^2 + z^2}$$

$$y = r \sin \theta \sin \phi \qquad \text{or} \qquad \theta = \tan^{-1} \frac{\sqrt{x^2 + y^2}}{z}$$

$$z = r \cos \theta \qquad\qquad \phi = \tan^{-1} \frac{y}{x}$$

1.27 Find the surface area of a sphere of radius R, by integration in spherical coordinates.

▌
$$S = \int dS_r = \int_{\phi=0}^{2\pi} \int_{\theta=0}^{\pi} R^2 \sin \theta \, d\theta \, d\phi = \int_0^{2\pi} 2R^2 \, d\phi = 4\pi R^2$$

1.28 Determine the volume of the sphere of Problem 1.27, by integration.

▌
$$V = \int dV = \int_{r=0}^{R} \int_{\phi=0}^{2\pi} \int_{\theta=0}^{\pi} r^2 \sin \theta \, dr \, d\theta \, d\phi = \frac{R^3}{3} \cdot 2 \cdot 2\pi = \frac{4}{3} \pi R^3$$

1.29 Given the two vector functions $\mathbf{A} = y\mathbf{a}_x + 3x\mathbf{a}_y - \mathbf{a}_z$ and $\mathbf{B} = 2\mathbf{a}_y - xy\mathbf{a}_z$, determine at $x = 1$, $y = 2$, and $z = 4$, the vector $\mathbf{C} = \mathbf{A} + \mathbf{B}$.

▌ At the given point,
$$\mathbf{A} = 2\mathbf{a}_x + 3\mathbf{a}_y - \mathbf{a}_z \qquad \mathbf{B} = 2\mathbf{a}_y - 2\mathbf{a}_z \qquad \mathbf{C} = \mathbf{A} + \mathbf{B} = 2\mathbf{a}_x + 5\mathbf{a}_y - 3\mathbf{a}_z$$

1.30 Given two vectors expressed in cylindrical coordinates as $\mathbf{A} = 2\mathbf{a}_r + \pi\mathbf{a}_\phi + \mathbf{a}_z$ and $\mathbf{B} = -\mathbf{a}_r + \frac{3\pi}{2}\mathbf{a}_\phi - 2\mathbf{a}_z$, determine $\mathbf{C} = \mathbf{A} + 2\mathbf{B}$.

▌
$$\mathbf{C} = [2 + 2(-1)]\mathbf{a}_r + [\pi + 2(3\pi/2)]\mathbf{a}_\phi + [1 + 2(-2)]\mathbf{a}_z = 4\pi\mathbf{a}_\phi - 3\mathbf{a}_z$$
Warning: The expression for $\mathbf{C}$ makes sense only if $\mathbf{A}$ and $\mathbf{B}$ emanate from the same point (for $\mathbf{a}_r$ and $\mathbf{a}_\phi$ vary from point to point).

1.31 Let P and Q be two diametrically opposite points of the sphere $r = 1$. How are the unit vectors at the two points related?

▌ $\mathbf{a}_r(Q) = -\mathbf{a}_r(P)$, $\mathbf{a}_\theta(Q) = \mathbf{a}_\theta(P)$, $\mathbf{a}_\phi(Q) = -\mathbf{a}_\phi(P)$.

1.32 Plot the following scalar fields by sketching the contours of constant value $f = 0$, $f = 1$, and $f = 2$:

    (a) $f(x, y, z) = x + y$ (d) $f(r, \phi, z) = r$

    (b) $f(x, y, z) = x^2 + y^2$ (e) $f(r, \theta, \phi) = \phi/\pi$

    (c) $f(x, y, z) = x + y + z$

▌ See Fig. 1-13.

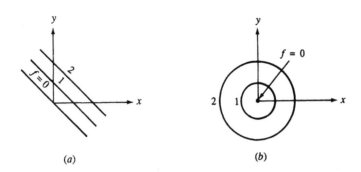

(a)   (b)

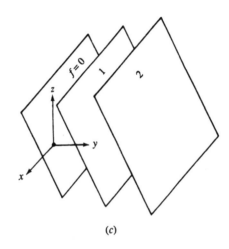

(c)

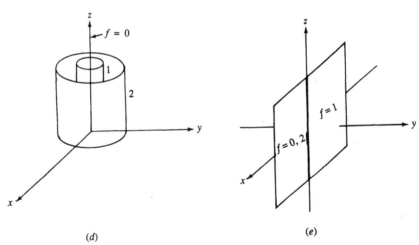

(d)   (e)   **Fig. 1-13**

1.33   Determine by integration the volume V of a region defined in a cylindrical coordinate system as $1 \le r \le 2$ m, $0 \le \phi \le \pi/3$ rad, and $0 \le z \le 1$ m.

▌ See Fig. 1-14.

$$V = \int_{z=0}^{1} \int_{\phi=0}^{\pi/3} \int_{r=1}^{2} r \, dr \, d\phi \, dz = \frac{2^2 - 1^2}{2} \cdot \frac{\pi}{3} \cdot 1 = \frac{\pi}{2} \, \text{m}^3$$

Check: $V = \frac{1}{6}$(difference of cylindrical volumes).

1.34   Determine by integration the area S of a surface defined in a spherical coordinate system as $r = 2$ m and $\pi/4 \le \theta \le \pi/3$ rad (Fig. 1-15).

▌

$$S = \int_{\theta=\pi/4}^{\pi/3} \int_{\phi=0}^{2\pi} r^2 \sin \theta \, d\theta \, d\phi = 4(2\pi) \int_{\theta=\pi/4}^{\pi/3} \sin \theta \, d\theta = 8\pi \left(\cos \frac{\pi}{4} - \cos \frac{\pi}{3} \right) = 5.21 \, \text{m}^2$$

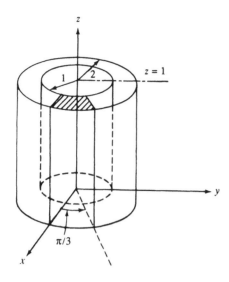

Fig. 1-14

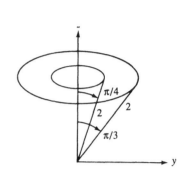

Fig. 1-15

1.35 Sketch the following vector fields:

(a) $\mathbf{F}(x, y, z) = x\mathbf{a}_x$ (d) $\mathbf{F}(r, \theta, \phi) = \mathbf{a}_\phi$ (g) $\mathbf{F}(x, y, z) = x\mathbf{a}_x - y\mathbf{a}_y$

(b) $\mathbf{F}(x, y, z) = y\mathbf{a}_x$ (e) $\mathbf{F}(r, \theta, \phi) = r\mathbf{a}_\theta$ (h) $\mathbf{F}(x, y, z) = -x\mathbf{a}_x - y\mathbf{a}_y$

(c) $\mathbf{F}(x, y, z) = z\mathbf{a}_y$ (f) $\mathbf{F}(x, y, z) = x\mathbf{a}_x + y\mathbf{a}_y + z\mathbf{a}_z$

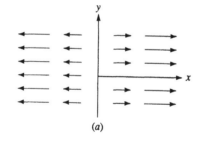

(a)

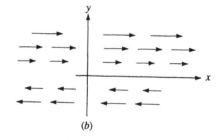

(b)

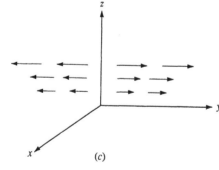

(c)

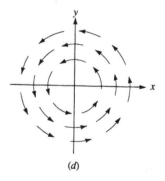

(d)

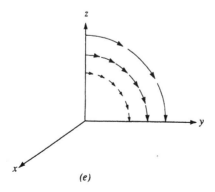

(e)

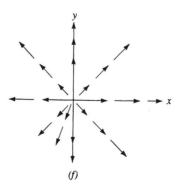

(f)

Fig. 1-16

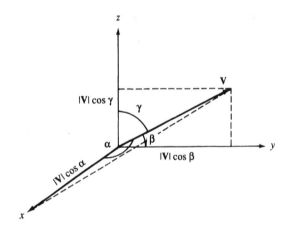

(g)   (h)   **Fig. 1-16** (cont'd).

▌ See Fig. 1-16.

1.36   The *dot product* (or *scalar product*) of two vectors **A** and **B** is defined by
$$\mathbf{A} \cdot \mathbf{B} = |\mathbf{A}|\,|\mathbf{B}|\cos\theta_{AB} = |\mathbf{A}|\,(\text{scalar proj. } \mathbf{B} \text{ on } \mathbf{A}) = |\mathbf{B}|\,(\text{scalar proj. } \mathbf{A} \text{ on } \mathbf{B}) \qquad (1)$$
Show that in rectangular coordinates an arbitrary vector **V** may be represented as
$$\mathbf{V} = (\mathbf{V} \cdot \mathbf{a}_x)\mathbf{a}_x + (\mathbf{V} \cdot \mathbf{a}_y)\mathbf{a}_y + (\mathbf{V} \cdot \mathbf{a}_z)\mathbf{a}_z \qquad (2)$$

Fig. 1-17

▌ From Fig. 1-17, $\mathbf{V} = (|\mathbf{V}|\cos\alpha)\mathbf{a}_x + (|\mathbf{V}|\cos\beta)\mathbf{a}_y + (|\mathbf{V}|\cos\gamma)\mathbf{a}_z$. But $\mathbf{V} \cdot \mathbf{a}_x = |\mathbf{V}|\,1\cos\alpha$, etc. It is clear that a representation of form (2) holds in any orthogonal coordinate system.

1.37   On the basis of the definition of the dot product given in Problem 1.36, determine the dot products of the unit vectors, in rectangular coordinates.

▌ The three unit vectors $\mathbf{a}_x$, $\mathbf{a}_y$, $\mathbf{a}_z$ are mutually perpendicular; therefore,
$$\mathbf{a}_x \cdot \mathbf{a}_x = \mathbf{a}_y \cdot \mathbf{a}_y = \mathbf{a}_z \cdot \mathbf{a}_z = 1$$
and
$$\mathbf{a}_x \cdot \mathbf{a}_y = \mathbf{a}_y \cdot \mathbf{a}_x = \mathbf{a}_x \cdot \mathbf{a}_z = \mathbf{a}_z \cdot \mathbf{a}_x = \mathbf{a}_y \cdot \mathbf{a}_z = \mathbf{a}_z \cdot \mathbf{a}_y = 0$$
Likewise for any other orthogonal coordinate system.

1.38   Demonstrate graphically that $\mathbf{A} \cdot (\mathbf{B} + \mathbf{C}) = \mathbf{A} \cdot \mathbf{B} + \mathbf{A} \cdot \mathbf{C}$.

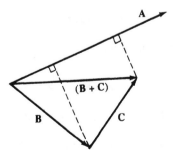

Fig. 1-18

▮ See Fig. 1-18, which shows that the scalar projection of **B** + **C** on **A** is the sum of the scalar projections of **B** and **C** on **A**.

1.39 Let **A** and **B** be two vectors defined at a point P in an orthogonal coordinate system; the unit vectors at P are $\mathbf{a}_1, \mathbf{a}_2, \mathbf{a}_3$. Show that $\mathbf{A} \cdot \mathbf{B} = A_1B_1 + A_2B_2 + A_3B_3$.

▮ By (2) of Problem 1.36,
$$\mathbf{A} \cdot \mathbf{B} = (A_1\mathbf{a}_1 + A_2\mathbf{a}_2 + A_3\mathbf{a}_3) \cdot (B_1\mathbf{a}_1 + B_2\mathbf{a}_2 + B_3\mathbf{a}_3)$$
Using the distributive property established in Problem 1.38, expand the right-hand side and apply Problem 1.37:
$$\mathbf{A} \cdot \mathbf{B} = A_1B_1 + A_2B_2 + A_3B_3$$

1.40 Given $\mathbf{A} = \mathbf{a}_x + 2\mathbf{a}_y - 3\mathbf{a}_z$ and $\mathbf{B} = 2\mathbf{a}_x - \mathbf{a}_y + \mathbf{a}_z$, determine
  (a) the scalar projection of **B** on **A**;
  (b) the smaller angle between **A** and **B**.

▮ (a)
$$\text{proj. } \mathbf{B} \text{ on } \mathbf{A} = \frac{\mathbf{A} \cdot \mathbf{B}}{|\mathbf{A}|} = \frac{(1)(2) + (2)(-1) + (-3)(1)}{\sqrt{1^2 + 2^2 + 3^2}} = \frac{-3}{\sqrt{14}}$$

  (b)
$$\cos\theta = \frac{\mathbf{A} \cdot \mathbf{B}}{|\mathbf{A}||\mathbf{B}|} = \frac{-3}{\sqrt{14}\sqrt{6}} \qquad \text{or} \qquad \theta = 109.1°$$

1.41 For the two vectors of Problem 1.30, obtain $\mathbf{A} \cdot \mathbf{B}$.

▮ $\mathbf{A} \cdot \mathbf{B} = (2)(-1) + (\pi)(3\pi/2) + (1)(-2) \approx 10.8$.

1.42 The *cross product* (or *vector product*) of two vectors **A** and **B** is defined as
$$\mathbf{A} \times \mathbf{B} = |\mathbf{A}||\mathbf{B}|\sin\theta_{AB}\,\mathbf{a}_n = \text{directed area spanned by } \mathbf{A} \text{ and } \mathbf{B}$$
Here, $\mathbf{a}_n$ is a unit vector normal to the plane that contains **A** and **B**, as determined by the right-hand rule (Fig. 1-19). Verify that $\mathbf{A} \times (\mathbf{B} + \mathbf{C}) = \mathbf{A} \times \mathbf{B} + \mathbf{A} \times \mathbf{C}$, in the special case that **A**, **B**, and **C** are coplanar. (The relation holds in general.)

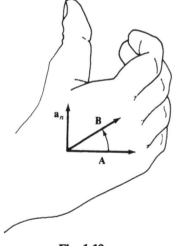

Fig. 1-19

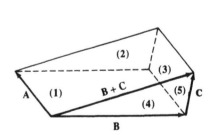

Fig. 1-20

▮ By congruent triangles in the planar diagram Fig. 1-20, area (2) = area (4) + area (5). Thus, dealing with magnitudes (as we can),
$$\mathbf{A} \times (\mathbf{B} + \mathbf{C}) = (1) + (2) + (3) = [(1) + (4)] + [(5) + (3)] = \mathbf{A} \times \mathbf{B} + \mathbf{A} \times \mathbf{C}$$

1.43 From the definition of the cross product in Problem 1.42, determine the cross products of the unit vectors, in rectangular coordinates.

▮
$$\mathbf{a}_x \times \mathbf{a}_x = \mathbf{a}_y \times \mathbf{a}_y = \mathbf{a}_z \times \mathbf{a}_z = 0$$
and
$$\mathbf{a}_x \times \mathbf{a}_y = -\mathbf{a}_y \times \mathbf{a}_x = \mathbf{a}_z \qquad \mathbf{a}_x \times \mathbf{a}_z = -\mathbf{a}_z \times \mathbf{a}_x = -\mathbf{a}_y \qquad \mathbf{a}_y \times \mathbf{a}_z = -\mathbf{a}_z \times \mathbf{a}_y = \mathbf{a}_x$$

1.44 Evaluate $\mathbf{A} \times \mathbf{B}$ in terms of rectangular components.

▌ By the associative property (Problem 1.42) and the relations of Problem 1.43,

$$\mathbf{A} \times \mathbf{B} = (A_x \mathbf{a}_x + A_y \mathbf{a}_y + A_z \mathbf{a}_z) \times (B_x \mathbf{a}_x + B_y \mathbf{a}_y + B_a \mathbf{a}_z)$$
$$= A_x B_x (\mathbf{a}_x \times \mathbf{a}_x) + A_x B_y (\mathbf{a}_x \times \mathbf{a}_y) + A_x B_z (\mathbf{a}_x \times \mathbf{a}_z) + A_y B_x (\mathbf{a}_y \times \mathbf{a}_x) + A_y B_y (\mathbf{a}_y \times \mathbf{a}_y)$$
$$+ A_y B_z (\mathbf{a}_y \times \mathbf{a}_z) + A_z B_x (\mathbf{a}_z \times \mathbf{a}_x) + A_z B_y (\mathbf{a}_z \times \mathbf{a}_y) + A_z B_z (\mathbf{a}_z \times \mathbf{a}_z)$$
$$= (A_y B_z - A_z B_y) \mathbf{a}_x + (A_z B_x - A_x B_z) \mathbf{a}_y + (A_x B_y - A_y B_x) \mathbf{a}_z$$

A handy form for this result is

$$\mathbf{A} \times \mathbf{B} = \begin{vmatrix} \mathbf{a}_x & \mathbf{a}_y & \mathbf{a}_z \\ A_x & A_y & A_z \\ B_x & B_y & B_z \end{vmatrix}$$

1.45 Show that

$$(\mathbf{A} \times \mathbf{B}) \cdot \mathbf{C} = \begin{vmatrix} A_x & A_y & A_z \\ B_x & B_y & B_z \\ C_x & C_y & C_z \end{vmatrix}$$

▌ By Problems 1.39 and 1.44,

$$(\mathbf{A} \times \mathbf{B}) \cdot \mathbf{C} = (A_y B_z - A_z B_y) C_x + (A_z B_x - A_x B_z) C_y + (A_x B_y - A_y B_x) C_z$$
$$= \begin{vmatrix} C_x & C_y & C_z \\ A_x & A_y & A_z \\ B_x & B_y & B_z \end{vmatrix} = \begin{vmatrix} A_x & A_y & A_z \\ B_x & B_y & B_z \\ C_x & C_y & C_z \end{vmatrix}$$

1.46 Interpret the determinant of Problem 1.45 geometrically.

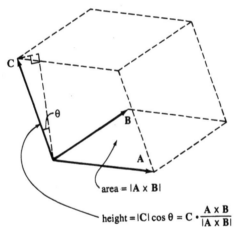

area = |A × B|

height = |C| cos θ = C · $\dfrac{\mathbf{A} \times \mathbf{B}}{|\mathbf{A} \times \mathbf{B}|}$

volume = area × height = C · (A × B) = (A × B) · C   **Fig. 1-21**

▌ See Fig. 1-21.

1.47 Express **A** × **B** in (*a*) cylindrical and (*b*) spherical coordinates.

▌ In either system the unit vectors compose a right-handed triad, just like $\{\mathbf{a}_x, \mathbf{a}_y, \mathbf{a}_z\}$.

(*a*)
$$\mathbf{A} \times \mathbf{B} = \begin{vmatrix} \mathbf{a}_r & \mathbf{a}_\phi & \mathbf{a}_z \\ A_r & A_\phi & A_z \\ B_r & B_\phi & B_z \end{vmatrix}$$

(*b*)
$$\mathbf{A} \times \mathbf{B} = \begin{vmatrix} \mathbf{a}_r & \mathbf{a}_\theta & \mathbf{a}_\phi \\ A_r & A_\theta & A_\phi \\ B_r & B_\theta & B_\phi \end{vmatrix}$$

1.48 Infer the smaller angle between $\mathbf{A} = 1\mathbf{a}_x - 3\mathbf{a}_y + 2\mathbf{a}_z$ and $\mathbf{B} = -3\mathbf{a}_x + 4\mathbf{a}_y - \mathbf{a}_z$ from their cross product.

▌
$$\sin \theta_{AB} = \frac{|\mathbf{A} \times \mathbf{B}|}{|\mathbf{A}| \, |\mathbf{B}|} = \frac{\sqrt{(-5)^2 + (-5)^2 + (-5)^2}}{\sqrt{(1)^2 + (-3)^2 + (2)^2} \, \sqrt{(-3)^2 + (4)^2 + (-1)^2}} \approx 0.45$$

so that $\theta = 27°, 153°$. Note that there are *two* angles compatible with a given cross product, whereas [Problem

1.40(b)] there is only one for a given dot product. By sketching the vectors we see that $\theta = 153°$ is the correct answer.

1.49 Determine a unit vector perpendicular to the plane containing the vectors **A** and **B** of Problem 1.48.

▮ Either the vector

$$\mathbf{a}_n = \frac{\mathbf{A} \times \mathbf{B}}{|\mathbf{A} \times \mathbf{B}|} = \frac{-5a_x - 5a_y - 5a_z}{\sqrt{(-5)^2 + (-5)^2 + (-5)^2}} = -\frac{1}{\sqrt{3}}\mathbf{a}_x - \frac{1}{\sqrt{3}}\mathbf{a}_y - \frac{1}{\sqrt{3}}\mathbf{a}_z$$

or the vector $-\mathbf{a}_n$ will do.

1.50 Prove that $\mathbf{A} \cdot (\mathbf{B} \times \mathbf{C}) = (\mathbf{A} \times \mathbf{B}) \cdot \mathbf{C}$ (*exchange of dot and cross*).

▮ By Problem 1.45,

$$\mathbf{A} \cdot (\mathbf{B} \times \mathbf{C}) = (\mathbf{B} \times \mathbf{C}) \cdot \mathbf{A} = \begin{vmatrix} B_x & B_y & B_z \\ C_x & C_y & C_z \\ A_x & A_y & A_z \end{vmatrix} = \begin{vmatrix} A_x & A_y & A_z \\ B_x & B_y & B_z \\ C_x & C_y & C_z \end{vmatrix} = (\mathbf{A} \times \mathbf{B}) \cdot \mathbf{C}$$

1.51 Given

$$\mathbf{A} = 1\mathbf{a}_x + 2\mathbf{a}_y + 3\mathbf{a}_z$$
$$\mathbf{B} = 4\mathbf{a}_x + 5\mathbf{a}_y + 6\mathbf{a}_z$$
$$\mathbf{C} = 7\mathbf{a}_x + 8\mathbf{a}_y + 9\mathbf{a}_z$$

show, without computation, that $\mathbf{A} \cdot (\mathbf{B} \times \mathbf{C}) = 0$.

▮ By inspection, $\mathbf{B} = \frac{1}{2}(\mathbf{A} + \mathbf{C})$. Thus the three vectors are coplanar, and determine a parallelopiped (Problem 1.46) of zero volume.

1.52 Given $\mathbf{A} = 2\mathbf{a}_y - xy\mathbf{a}_z$, determine $\mathbf{a}_y \times \mathbf{A}$ at $x = 1$, $y = 2$, $z = 4$.

▮ At the given point, $\mathbf{A} = 2\mathbf{a}_y - 2\mathbf{a}_z$; hence $\mathbf{a}_y \times \mathbf{A} = -2\mathbf{a}_x$.

1.53 Two vectors in cylindrical coordinates are $\mathbf{A} = 2\mathbf{a}_r + \pi\mathbf{a}_\phi + \mathbf{a}_z$ and $\mathbf{B} = -\mathbf{a}_r + (3\pi/2)\mathbf{a}_\phi - 2\mathbf{a}_z$. On the assumption that **A** and **B** are defined at the same point, evaluate $\mathbf{A} \times \mathbf{B}$ and find a unit vector perpendicular to the plane containing **A** and **B**.

▮
$$\mathbf{A} \times \mathbf{B} = (-2\pi - 3\pi/2)\mathbf{a}_r + (-1 + 4)\mathbf{a}_\phi + (3\pi + \pi)\mathbf{a}_z = -(7\pi/2)\mathbf{a}_r + 3\mathbf{a}_\phi + 4\pi\mathbf{a}_z$$
$$\mathbf{a}_n = \frac{\mathbf{A} \times \mathbf{B}}{|\mathbf{A} \times \mathbf{B}|} = \frac{-(7\pi/2)\mathbf{a}_r + 3\mathbf{a}_\phi + 4\pi\mathbf{a}_z}{\sqrt{(7\pi/2)^2 + 9 + (4\pi)^2}} = -0.65\mathbf{a}_r + 0.18\mathbf{a}_\phi + 0.74\mathbf{a}_z$$

1.54 Repeat Problem 1.53 for spherical coordinates, with
$$\mathbf{A} = 2\mathbf{a}_r + \pi\mathbf{a}_\theta - (\pi/2)\mathbf{a}_\phi \qquad \text{and} \qquad \mathbf{B} = \mathbf{a}_r - (\pi/3)\mathbf{a}_\theta$$

▮
$$\mathbf{A} \times \mathbf{B} = -(\pi^2/6)\mathbf{a}_r - (\pi/2)\mathbf{a}_\theta + (-2\pi/3 - \pi)\mathbf{a}_\phi = -(\pi^2/6)\mathbf{a}_r - (\pi/2)\mathbf{a}_\theta - (5\pi/3)\mathbf{a}_\phi$$
$$\mathbf{a}_n = \frac{\mathbf{A} \times \mathbf{B}}{|\mathbf{A} \times \mathbf{B}|} = \frac{-(\pi^2/6)\mathbf{a}_r - (\pi/2)\mathbf{a}_\theta - (5\pi/3)\mathbf{a}_\phi}{\sqrt{32.59}} = -0.288\mathbf{a}_r - 0.275\mathbf{a}_\theta - 0.918\mathbf{a}_\phi$$

1.55 For what value of α are $\mathbf{A} = \mathbf{a}_x + 2\mathbf{a} - \mathbf{a}_z$ and $\mathbf{B} = \alpha\mathbf{a}_x + \mathbf{a}_y + 3\mathbf{a}_z$ perpendicular?

▮ $\mathbf{A} \cdot \mathbf{B} = 0 = \alpha + 2 - 3$; thus $\alpha = 1$.

1.56 For what values of α and β are $\mathbf{A} = \mathbf{a}_r + \pi\mathbf{a}_\phi + 3\mathbf{a}_z$ and $\mathbf{B} = \alpha\mathbf{a}_r + \beta\mathbf{a}_\phi - 6\mathbf{a}_z$ parallel (or antiparallel)?

▮ $\mathbf{A} \times \mathbf{B} = 0 = (-6\pi - 3\beta)\mathbf{a}_r + (3\alpha + 6)\mathbf{a}_\phi + (\beta - \pi\alpha)\mathbf{a}_z$, whence $\alpha = -2$, $\beta = -2\pi$; i.e., $\mathbf{B} = -2\mathbf{A}$ (antiparallel).

1.57 Corresponding to a scalar field $V = 2xy - 5z$, a vector field is given by $\mathbf{E} = (\partial V/\partial x)\mathbf{a}_x + (\partial V/\partial y)\mathbf{a}_y + (\partial V/\partial z)\mathbf{a}_z$. Find **E** at the point $(1, 2, 3)$.

▮ $\mathbf{E}(x, y, z) = 2y\mathbf{a}_x + 2x\mathbf{a}_y - 5\mathbf{a}_z$ whence $\mathbf{E}(1, 2, 3) = 4\mathbf{a}_x + 2\mathbf{a}_y - 5\mathbf{a}_z$

1.58 For the vector **E** of Problem 1.57, determine a unit vector $\mathbf{a}_E$.

▮
$$\mathbf{a}_E = \frac{4\mathbf{a}_x + 2\mathbf{a}_y - 5\mathbf{a}_z}{\sqrt{4^2 + 2^2 + 5^2}} = 0.596\mathbf{a}_x + 0.298\mathbf{a}_y - 0.745\mathbf{a}_z$$

1.59   Given $\mathbf{A} = \alpha \mathbf{a}_y - 7\mathbf{a}_y + 4\mathbf{a}_z$ and $\mathbf{B} = 5\mathbf{a}_x + 4\mathbf{a}_y - 3\mathbf{a}_z$, find α such that the angle between the two vectors is 62.1°.

    ❙ $\mathbf{A} \cdot \mathbf{B} = 5\alpha - 28 - 12 = (\sqrt{\alpha^2 + 49 + 16})(\sqrt{25 + 16 + 9}) \cos 62.1°$

whence $\alpha = 2.43$ or 26.0. Since $5\alpha - 40 > 0$, $\alpha = 26.0$.

1.60   Calculate the angle between $\mathbf{A} = 3\mathbf{a}_x - 4\mathbf{a}_y + 5\mathbf{a}_z$ and $\mathbf{B} = -\mathbf{a}_x + 2\mathbf{a}_y - 3\mathbf{a}_z$, using the definition of the (**a**) dot and (**b**) cross product.

    (**a**) $\theta = \cos^{-1}\left(\dfrac{-3 - 8 - 15}{\sqrt{50}\,\sqrt{14}}\right) = 169.3°$

    (**b**) $\theta = \sin^{-1}\left(\dfrac{|2\mathbf{a}_x + 4\mathbf{a}_y + 2\mathbf{a}_z|}{\sqrt{50}\,\sqrt{14}}\right) = 169.3°$ (see Problem 1.48).

1.61   In cylindrical coordinates, at a certain point two vectors are defined as $\mathbf{A} = 5\mathbf{a}_r - 8\mathbf{a}_\phi + 3\mathbf{a}_z$ and $\mathbf{B} = -4\mathbf{a}_r + 2\mathbf{a}_\phi + 10\mathbf{a}_z$. Obtain the scalar component of $\mathbf{A}$ in the direction of $\mathbf{B}$.

    ❙ $\dfrac{\mathbf{A} \cdot \mathbf{B}}{|\mathbf{B}|} = \dfrac{-20 - 16 + 30}{\sqrt{4^2 + 2^2 + 10^2}} = \dfrac{-6}{\sqrt{120}} = -0.548$

1.62   For the vectors of Problem 1.61, give the vector component of $\mathbf{A}$ in the direction of $\mathbf{B}$.

    ❙   Multiply the unit vector along $\mathbf{B}$ by the scalar component:

$$\left(\frac{\mathbf{A} \cdot \mathbf{B}}{|\mathbf{B}|}\right)\frac{\mathbf{B}}{|\mathbf{B}|} = \frac{(\mathbf{A} \cdot \mathbf{B})\mathbf{B}}{|\mathbf{B}|^2} = \frac{-6(-4\mathbf{a}_r + 2\mathbf{a}_\phi + 10\mathbf{a}_z)}{120} = 0.2\mathbf{a}_r - 0.1\mathbf{a}_\phi - 0.5\mathbf{a}_z$$

1.63   A vector field $\mathbf{F}$ is given as $\mathbf{F} = 25\mathbf{a}_r + 12\mathbf{a}_\phi - 20\mathbf{a}_z$ at the point $(8, 120°, 5)$. Find the vector component of $\mathbf{F}$ which is perpendicular to the cylinder $r = 8$.

    ❙ $(\mathbf{F} \cdot \mathbf{a}_r)\mathbf{a}_r = 25\mathbf{a}_r$.

1.64   For $\mathbf{F}$ of Problem 1.63, find the vector component of $\mathbf{F}$ tangent to the cylinder $r = 8$.

    ❙   The required component is $\mathbf{F} - 25\mathbf{a}_r = 12\mathbf{a}_\phi - 20\mathbf{a}_z$.

1.65   For $\mathbf{F}$ of Problem 1.63, find a unit vector that is perpendicular to $\mathbf{F}$ and tangent to the cylinder $r = 8$.

    ❙   Since $\mathbf{a}_r$ is normal to the cylinder at the given point, the unit vector may be chosen as

$$\mathbf{a} = \frac{\mathbf{a}_r \times \mathbf{F}}{|\mathbf{a}_r \times \mathbf{F}|} = \frac{20\mathbf{a}_\phi + 12\mathbf{a}_z}{\sqrt{20^2 + 12^2}} = 0.857\mathbf{a}_\phi + 0.514\mathbf{a}_z$$

1.66   A vector field is given by $\mathbf{E} = (50/r)\mathbf{a}_r - 4\mathbf{a}_z$. Represent in rectangular coordinates the unit vector $\mathbf{a}_E$ at $(10, 20°, 2)$.

    ❙   At the given point, $\mathbf{E} = 5\mathbf{a}_r - 4\mathbf{a}_z$. Now, $E_x = 5 \cos 20° = 4.7$ and $E_y = 5 \sin 20° = 1.71$; hence

      $\mathbf{E} = 4.7\mathbf{a}_x + 1.71\mathbf{a}_y - 4\mathbf{a}_z$ and $\mathbf{a}_E = \dfrac{4.7\mathbf{a}_x + 1.71\mathbf{a}_y - 4\mathbf{a}_z}{\sqrt{4.7^2 + 1.71^2 + 4^2}} = 0.734\mathbf{a}_x + 0.267\mathbf{a}_y - 0.625\mathbf{a}_z$

1.67   For the field of Problem 1.66, find the surface on which $|\mathbf{E}| = 10$.

    ❙ $|\mathbf{E}|^2 = \dfrac{2500}{r^2} + 16 = 10^2$ or $r = 5.46$ (a cylindrical surface)

1.68   At a point P, two vectors are defined in spherical coordinates as $\mathbf{A} = 10\mathbf{a}_r - 3\mathbf{a}_\theta + 5\mathbf{a}_\phi$ and $\mathbf{B} = 2\mathbf{a}_r + 5\mathbf{a}_\theta + 3\mathbf{a}_\phi$. At P, find the scalar component of $\mathbf{B}$ in the direction of $\mathbf{A}$.

    ❙ $\dfrac{\mathbf{B} \cdot \mathbf{A}}{|\mathbf{A}|} = \dfrac{20 - 15 + 15}{\sqrt{10^2 + 3^2 + 5^2}} = \dfrac{20}{\sqrt{134}} = 1.728$

1.69   For the vectors of Problem 1.68, find, at P, the vector component of $\mathbf{B}$ in the direction of $\mathbf{A}$.

    ❙   From Problem 1.62,

$$\frac{(\mathbf{B} \cdot \mathbf{A})\mathbf{A}}{|\mathbf{A}|^2} = \frac{20(10\mathbf{a}_r - 3\mathbf{a}_\theta + 5\mathbf{a}_\phi)}{(10^2 + 3^2 + 5^2)} = 1.493\mathbf{a}_r - 0.448\mathbf{a}_\theta + 0.746\mathbf{a}_\phi$$

1.70 For the vectors of Problem 1.68, find a unit vector perpendicular to both **A** and **B** at P.

▮ One unit vector is given by

$$\mathbf{a} = \frac{\mathbf{B} \times \mathbf{A}}{|\mathbf{B} \times \mathbf{A}|} = \frac{\begin{vmatrix} \mathbf{a}_r & \mathbf{a}_\theta & \mathbf{a}_\phi \\ 2 & 5 & 3 \\ 10 & -3 & 5 \end{vmatrix}}{|\mathbf{B} \times \mathbf{A}|} = \frac{34\mathbf{a}_r + 20\mathbf{a}_\theta - 56\mathbf{a}_\phi}{\sqrt{34^2 + 20^2 + 56^2}} = 0.496\mathbf{a}_r + 0.292\mathbf{a}_\theta - 0.818\mathbf{a}_\phi$$

1.71 Express the vector field $\mathbf{A} = (x^2 - y^2)\mathbf{a}_y + xz\mathbf{a}_z$ in cylindrical coordinates at P: $r = 6$, $\phi = 60°$, $z = -4$.

▮ At P, $x = r \cos \phi$, $y = r \sin \phi$, and $\mathbf{a}_y = (\mathbf{a}_y \cdot \mathbf{a}_r)\mathbf{a}_r + (\mathbf{a}_y \cdot \mathbf{a}_\phi)\mathbf{a}_\phi = \sin \phi \, \mathbf{a}_r + \cos \phi \, \mathbf{a}_\phi$. Thus

$$\mathbf{A} = r^2(\cos^2 \phi - \sin^2 \phi)(\sin \phi \, \mathbf{a}_r + \cos \phi \, \mathbf{a}_\phi) + zr \cos \phi \, \mathbf{a}_z$$

$$= 36\left(\frac{1}{4} - \frac{3}{4}\right)\left(\frac{\sqrt{3}}{2}\mathbf{a}_r + \frac{1}{2}\mathbf{a}_\phi\right) - 4(6)\left(\frac{1}{2}\right)\mathbf{a}_z$$

$$= -9\sqrt{3}\,\mathbf{a}_r - 9\mathbf{a}_\phi - 12\mathbf{a}_z$$

1.72 Express the vector field of Problem 1.71 in spherical coordinates at Q: $r = 4$, $\theta = 30°$, $\phi = 120°$.

▮ At Q, $x = r \sin \theta \cos \phi$, $y = r \sin \theta \sin \phi$, $z = r \cos \theta$; and

$$\mathbf{a}_y = \sin \theta \sin \phi \, \mathbf{a}_r + \cos \theta \sin \phi \, \mathbf{a}_\theta + \cos \phi \, \mathbf{a}_\phi \qquad \mathbf{a}_z = \cos \theta \, \mathbf{a}_r - \sin \theta \, \mathbf{a}_\theta$$

Then, substituting numerical values, we obtain

$$\mathbf{A} = (16)\left(\frac{1}{4}\right)\left(\frac{1}{4} - \frac{3}{4}\right)\left[\left(\frac{1}{2}\right)\left(\frac{\sqrt{3}}{2}\right)\mathbf{a}_r + \left(\frac{\sqrt{3}}{2}\right)\left(\frac{\sqrt{3}}{2}\right)\mathbf{a}_\theta - \frac{1}{2}\mathbf{a}_\phi\right] + (16)\left(\frac{1}{2}\right)\left(\frac{\sqrt{3}}{2}\right)\left(-\frac{1}{2}\right)\left[\left(\frac{\sqrt{3}}{2}\right)\mathbf{a}_r - \left(\frac{1}{2}\right)\mathbf{a}_\theta\right]$$

$$= -\frac{\sqrt{3} + 6}{2}\mathbf{a}_r + \frac{2\sqrt{3} - 3}{2}\mathbf{a}_\theta + \mathbf{a}_\phi$$

1.73 Given a vector field $\mathbf{F} = \dfrac{\cos \theta}{r^2}\mathbf{a}_r + \dfrac{\sin \theta}{r}\mathbf{a}_\theta$ (spherical coordinates), express $\mathbf{F}$ in rectangular coordinates.

▮

$$F_x = \frac{\cos \theta}{r^2}\mathbf{a}_r \cdot \mathbf{a}_x + \frac{\sin \theta}{r}\mathbf{a}_\theta \cdot \mathbf{a}_x = \frac{1}{r^2}\sin \theta \cos \theta \cos \phi + \frac{1}{r}\sin \theta \cos \theta \cos \phi$$

Similarly,

$$F_y = \frac{1}{r^2}\cos \theta \sin \theta \sin \phi + \frac{1}{r}\sin \theta \cos \theta \sin \phi \qquad F_z = \frac{1}{r^2}\cos^2 \theta - \frac{1}{r}\sin^2 \theta$$

These expressions, together with the relations

$$\cos \theta = \frac{z}{r} \qquad \sin \theta = \frac{\sqrt{x^2 + y^2}}{r} \qquad \cos \phi = \frac{x}{\sqrt{x^2 + y^2}} \qquad \sin \phi = \frac{y}{\sqrt{x^2 + y^2}}$$

give

$$\mathbf{F} = \frac{1}{(x^2 + y^2 + z^2)^2}\{xz(1 + \sqrt{x^2 + y^2 + z^2})\mathbf{a}_x + yz(1 + \sqrt{x^2 + y^2 + z^2})\mathbf{a}_y + [z^2 - (x^2 + y^2)\sqrt{x^2 + y^2 + z^2}]\mathbf{a}_z\}$$

1.74 Given two vectors **A** and **B**, obtain **C**, the vector component of **A** perpendicular to **B**.

▮

$$\mathbf{A} = \frac{\mathbf{A} \cdot \mathbf{B}}{\mathbf{B} \cdot \mathbf{B}}\mathbf{B} + \mathbf{C} \qquad \text{or} \qquad \mathbf{C} = \mathbf{A} - \frac{\mathbf{A} \cdot \mathbf{B}}{\mathbf{B} \cdot \mathbf{B}}\mathbf{B}$$

1.75 Given $\mathbf{A} = \mathbf{a}_x + \mathbf{a}_y$, $\mathbf{B} = \mathbf{a}_x + 2\mathbf{a}_z$, and $\mathbf{C} = 2\mathbf{a}_y + \mathbf{a}_z$, find $(\mathbf{A} \times \mathbf{B}) \times \mathbf{C}$ and compare it with $\mathbf{A} \times (\mathbf{B} \times \mathbf{C})$.

▮ $\mathbf{A} \times \mathbf{B} = \begin{vmatrix} \mathbf{a}_x & \mathbf{a}_y & \mathbf{a}_z \\ 1 & 1 & 0 \\ 1 & 0 & 2 \end{vmatrix} = 2\mathbf{a}_x - 2\mathbf{a}_y - \mathbf{a}_z$ whence $(\mathbf{A} \times \mathbf{B}) \times \mathbf{C} = \begin{vmatrix} \mathbf{a}_x & \mathbf{a}_y & \mathbf{a}_z \\ 2 & -2 & -1 \\ 0 & 2 & 1 \end{vmatrix} = -2\mathbf{a}_y + 4\mathbf{a}_z$

A similar calculation gives $\mathbf{A} \times (\mathbf{B} \times \mathbf{C}) = 2\mathbf{a}_x - 2\mathbf{a}_y + 3\mathbf{a}_z$. Thus the parentheses that indicate which cross product is to be taken first are essential in the vector triple product.

1.76 Show that

$$|\mathbf{A} \times \mathbf{B}|^2 = \begin{vmatrix} \mathbf{A} \cdot \mathbf{A} & \mathbf{A} \cdot \mathbf{B} \\ \mathbf{B} \cdot \mathbf{A} & \mathbf{B} \cdot \mathbf{B} \end{vmatrix}$$

(The determinant, which is seen to be positive unless **A** and **B** are linearly dependent, is known as the *Gram determinant* of the two vectors.)

/ By Problems 1.42 and 1.74, $|\mathbf{A} \times \mathbf{B}| = |\mathbf{B}| \, |\mathbf{C}|$, so that

$$|\mathbf{A} \times \mathbf{B}|^2 = (\mathbf{B} \cdot \mathbf{B})(\mathbf{C} \cdot \mathbf{C}) = (\mathbf{B} \cdot \mathbf{B})\left[\mathbf{A} \cdot \mathbf{A} - 2\frac{(\mathbf{A} \cdot \mathbf{B})^2}{\mathbf{B} \cdot \mathbf{B}} + \frac{(\mathbf{A} \cdot \mathbf{B})^2}{\mathbf{B} \cdot \mathbf{B}} \right]$$

$$= (\mathbf{A} \cdot \mathbf{A})(\mathbf{B} \cdot \mathbf{B}) - (\mathbf{A} \cdot \mathbf{B})^2 = (\mathbf{A} \cdot \mathbf{A})(\mathbf{B} \cdot \mathbf{B}) - (\mathbf{A} \cdot \mathbf{B})(\mathbf{B} \cdot \mathbf{A})$$

1.77 Express the unit vector which points from $z = h$ on the z axis of cylindrical coordinates toward $(r, \phi, 0)$. See Fig. 1-22.

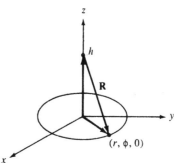

Fig. 1-22

/ $\qquad$ $\mathbf{R} = r\mathbf{a}_r - h\mathbf{a}_z$ $\qquad$ and so $\qquad$ $\mathbf{a}_R = \dfrac{\mathbf{R}}{|\mathbf{R}|} = \dfrac{r\mathbf{a}_r - h\mathbf{a}_z}{\sqrt{r^2 + h^2}}$

$\mathbf{a}_R$ varies implicitly with ϕ, through $\mathbf{a}_r$.

1.78 Express the unit vector which is directed toward the origin from an arbitrary point on the plane $z = -5$, as shown in Fig. 1-23.

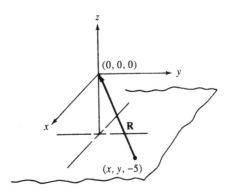

Fig. 1-23

/ $\qquad$ $\mathbf{R} = -x\mathbf{a}_x - y\mathbf{a}_y + 5\mathbf{a}_z$ $\qquad$ and $\qquad$ $\mathbf{a}_R = \dfrac{-x\mathbf{a}_x - y\mathbf{a}_y + 5\mathbf{a}_z}{\sqrt{x^2 + y^2 + 25}}$

1.79 Find the area of the strip $\alpha \le \theta \le \beta$ on a sphere of radius a.

/ In the natural spherical coordinates, the differential surface element is $dS = r^2 \sin \theta \, d\theta \, d\phi$; thus

$$A = \int_0^{2\pi} \int_\alpha^\beta a^2 \sin \theta \, d\theta \, d\phi = 2\pi a^2 (\cos \alpha - \cos \beta)$$

1.80 Transform

$$\mathbf{A} = y\mathbf{a}_x + x\mathbf{a}_y + \frac{x^2}{\sqrt{x^2 + y^2}} \mathbf{a}_z$$

from cartesian to cylindrical coordinates.

/ First transform the scalar components, obtaining $\mathbf{A} = r(\sin \phi \, \mathbf{a}_x + \cos \phi \, \mathbf{a}_y + \cos^2 \phi \, \mathbf{a}_z)$. Next, the scalar projections of the cartesian unit vectors on $\mathbf{a}_r$, $\mathbf{a}_\phi$, and $\mathbf{a}_z$ are obtained:

$$\begin{array}{lll} \mathbf{a}_x \cdot \mathbf{a}_r = \cos \phi & \mathbf{a}_x \cdot \mathbf{a}_\phi = -\sin \phi & \mathbf{a}_x \cdot \mathbf{a}_z = 0 \\ \mathbf{a}_y \cdot \mathbf{a}_r = \sin \phi & \mathbf{a}_y \cdot \mathbf{a}_\phi = \cos \phi & \mathbf{a}_y \cdot \mathbf{a}_z = 0 \\ \mathbf{a}_z \cdot \mathbf{a}_r = 0 & \mathbf{a}_z \cdot \mathbf{a}_\phi = 0 & \mathbf{a}_z \cdot \mathbf{a}_z = 1 \end{array}$$

Therefore $\mathbf{a}_x = \cos \phi \, \mathbf{a}_r - \sin \phi \, \mathbf{a}_\phi$, $\mathbf{a}_y = \sin \phi \, \mathbf{a}_r + \cos \phi \, \mathbf{a}_\phi$, $\mathbf{a}_z = \mathbf{a}_z$, giving

$$\mathbf{A} = r[2 \sin \phi \cos \phi \, \mathbf{a}_r + (\cos^2 \phi - \sin^2 \phi)\mathbf{a}_\phi + \cos^2 \phi \, \mathbf{a}_z]$$

1.81 A vector **A** of magnitude 10 points from $(5, 5\pi/4, 0)$ in cylindrical coordinates toward the origin (Fig. 1-24). Express the vector in cartesian coordinates.

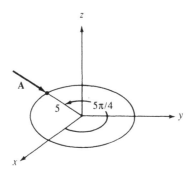

Fig. 1-24

▌ Since the vector lies in the xy plane,

$$\mathbf{A} = 10\left(\cos\frac{\pi}{4}\mathbf{a}_x + \sin\frac{\pi}{4}\mathbf{a}_y\right) = \frac{10}{\sqrt{2}}(\mathbf{a}_x + \mathbf{a}_y)$$

Notice that the value of the radial coordinate, 5, is immaterial.

1.82 Given the scalar field $f(x, y, z)$ in rectangular coordinates, define the *gradient* of f.

▌ $$\operatorname{grad} f \equiv \nabla f \equiv \left(\frac{\partial}{\partial x}\mathbf{a}_x + \frac{\partial}{\partial y}\mathbf{a}_y + \frac{\partial}{\partial z}\mathbf{a}_z\right)f = \frac{\partial f}{\partial x}\mathbf{a}_x + \frac{\partial f}{\partial y}\mathbf{a}_y + \frac{\partial f}{\partial z}\mathbf{a}_z$$

1.83 With reference to Problem 1.82, show that the change in f corresponding to an infinitesimal displacement $d\mathbf{s}$ is given by $df = \nabla f \cdot d\mathbf{s}$.

▌ The derivative of f in the direction of the unit vector

$$\mathbf{u} = \frac{d\mathbf{s}}{|d\mathbf{s}|} = \frac{dx}{ds}\mathbf{a}_x + \frac{dy}{ds}\mathbf{a}_y + \frac{dz}{ds}\mathbf{a}_z$$

is given by the chain rule:

$$\frac{df}{ds} = \frac{\partial f}{\partial x}\frac{dx}{ds} + \frac{\partial f}{\partial y}\frac{dy}{ds} + \frac{\partial f}{\partial z}\frac{dz}{ds} = \nabla f \cdot \mathbf{u} \tag{1}$$

Hence, $df = \nabla f \cdot (ds\,\mathbf{u}) = \nabla f \cdot d\mathbf{s}$. Note that this result, as an equation between two absolute scalars, holds in any coordinate system.

1.84 From Problem 1.83, verify that the maximum rate of change of f is the magnitude of its gradient.

▌ From (1) it is clear that df/ds will be greatest when $\mathbf{u}$ has the direction of ∇f:

$$\left(\frac{df}{ds}\right)_{\max} = \nabla f \cdot \frac{\nabla f}{|\nabla f|} = |\nabla f|$$

1.85 Obtain an expression for the gradient of the scalar field $f(r, \phi, z)$ in cylindrical coordinates.

▌ In cylindrical coordinates,

$$d\mathbf{s} = dr\,\mathbf{a}_r + r\,d\phi\,\mathbf{a}_\phi + dz\,\mathbf{a}_z \qquad \text{and} \qquad df = \frac{\partial f}{\partial r}dr + \frac{\partial f}{\partial \phi}d\phi + \frac{\partial f}{\partial z}dz$$

Thus $df = \nabla f \cdot d\mathbf{s}$ (Problem 1.83) only if

$$\nabla f = \frac{\partial f}{\partial r}\mathbf{a}_r + \frac{1}{r}\frac{\partial f}{\partial \phi}\mathbf{a}_\phi + \frac{\partial f}{\partial z}\mathbf{a}_z$$

1.86 In rectangular coordinates, the *Laplacian* of a scalar field $\psi(x, y, z)$ is defined as

$$\nabla^2 \psi \equiv \frac{\partial^2 \psi}{\partial x^2} + \frac{\partial^2 \psi}{\partial y^2} + \frac{\partial^2 \psi}{\partial z^2}$$

Show that, formally, $\nabla^2 = \nabla \cdot \nabla$.

▌ By Problem 1.82,

$$\nabla \cdot \nabla = \left(\frac{\partial}{\partial x}\mathbf{a}_x + \frac{\partial}{\partial y}\mathbf{a}_y + \frac{\partial}{\partial z}\mathbf{a}_z\right) \cdot \left(\frac{\partial}{\partial x}\mathbf{a}_x + \frac{\partial}{\partial y}\mathbf{a}_y + \frac{\partial}{\partial z}\mathbf{a}_z\right) = \frac{\partial^2}{\partial x^2} + \frac{\partial^2}{\partial y^2} + \frac{\partial^2}{\partial z^2} = \nabla^2$$

1.87 A scalar function V is independent of x and z, and the gradient of V is $20y$ in the y direction. Given $V = 10$ at $y = 0$, find V.

$$\frac{\partial V}{\partial y} = 20y \qquad \text{gives} \qquad V = 10y^2 + C$$

Since $V = 10$ at $y = 0$, $C = 10$; hence, $V = 10y^2 + 10$.

1.88 A sphere of radius 0.2 m, centered at the origin, contains electrical charge of density $2/\sqrt{x^2 + y^2}\ C/m^3$. Find the total charge Q contained within the sphere.

Integrate in spherical coordinates:

$$Q = \int \rho\, dV = \int_{r=0}^{0.2}\int_{\theta=0}^{\pi}\int_{\phi=0}^{2\pi} \left(\frac{2}{r\sin\theta}\right) r^2 \sin\theta\, dr\, d\theta\, d\phi = \int 2r\, dr \int d\theta \int d\phi = (0.2)^2(\pi)(2\pi) = 0.08\pi^2 \approx 0.789\ C$$

1.89 Find the gradient of (a) $f(x, y, z) = xy^2 + 2z$, (b) $f(r, \phi, z) = 2r\sin\phi$.

(a) $\nabla f = \dfrac{\partial f}{\partial x}\mathbf{a}_x + \dfrac{\partial f}{\partial y}\mathbf{a}_y + \dfrac{\partial f}{\partial z}\mathbf{a}_z = y^2\mathbf{a}_x + 2xy\mathbf{a}_y + 2\mathbf{a}_z$

(b) $\nabla f = \dfrac{\partial f}{\partial r}\mathbf{a}_r + \dfrac{1}{r}\dfrac{\partial f}{\partial \phi}\mathbf{a}_\phi + \dfrac{\partial f}{\partial z}\mathbf{a}_z = 2\sin\phi\,\mathbf{a}_r + 2\cos\phi\,\mathbf{a}_\phi$

1.90 Generalize Problem 1.85 to arbitrary orthogonal coordinates (x_1, x_2, x_3), in which $ds^2 = h_1^2(x_1, x_2, x_3)\, dx_1^2 + h_2^2(x_1, x_2, x_3)\, dx_2^2 + h_3^2(x_1, x_2, x_3)\, dx_3^2$.

$$df = \frac{\partial f}{\partial x_1}dx_1 + \frac{\partial f}{\partial x_2}dx_2 + \frac{\partial f}{\partial x_3}dx_3 \qquad \text{and} \qquad d\mathbf{s} = h_1\, dx_1\, \mathbf{a}_1 + h_2\, dx_2\, \mathbf{a}_2 + h_3\, dx_3\, \mathbf{a}_3$$

Compare $\nabla f \cdot d\mathbf{s} = (\nabla f)_1\, h_1\, dx_1 + (\nabla f)_2\, h_2\, dx_2 + (\nabla f)_3\, h_3\, dx_3$ with df to obtain

$$\nabla f = \frac{1}{h_1}\frac{\partial f}{\partial x_1}\mathbf{a}_1 + \frac{1}{h_2}\frac{\partial f}{\partial x_2}\mathbf{a}_2 + \frac{1}{h_3}\frac{\partial f}{\partial x_3}\mathbf{a}_3 \tag{1}$$

1.91 Obtain the gradient in spherical coordinates.

In (1) of Problem 1.90 make the identifications:

$$x_1 = r \qquad x_2 = \theta \qquad x_3 = \phi$$
$$h_1 = 1 \qquad h_2 = r \qquad h_3 = r\sin\theta$$

1.92 Show that the vector ∇f at the point P is normal to the surface $f = \text{const.} = f(P)$ through P.

By (1) of Problem 1.83, $\nabla f \cdot \mathbf{u} = 0$ whenever $\mathbf{u}$ lies in the surface $f = \text{const.}$ In the special case of two dimensions, the "surface" becomes the *contour line* or *level curve* $f(x, y) = \text{const.}$ in the xy plane. According to our result, the gradient vector $\nabla f = (\partial f/\partial x)\mathbf{a}_x + (\partial f/\partial y)\mathbf{a}_y$ lies along the curve normal at each point of the contour line.

1.93 By use of Problem 1.90 or other means, calculate the gradients of

(a) $f(x, y, z) = 5x + 10xz - xy + 6$ (c) $f(r, \theta, \phi) = 2r\cos\theta - 5\phi + 2$
(b) $f(r, \phi, z) = 2\sin\phi - rz + 4$

(a) $\nabla f = (5 + 10z - y)\mathbf{a}_x + (-x)\mathbf{a}_y + (10x)\mathbf{a}_z$

(b) $\nabla f = -z\mathbf{a}_r + \dfrac{2}{r}\cos\phi\,\mathbf{a}_\phi - r\mathbf{a}_z$

(c) $\nabla f = 2\cos\theta\,\mathbf{a}_r - 2\sin\theta\,\mathbf{a}_\theta - \dfrac{5}{r\sin\theta}\mathbf{a}_\phi$

1.94 A scalar field is given by $V = -(Q\cos\theta)/r^2$, where Q is a constant. Verify that ∇V makes a constant angle with the z axis on a cone $\theta = \text{const.}$

Problems 1.90 and 1.91 give

$$\nabla V = \frac{\partial V}{\partial r}\mathbf{a}_r + \frac{1}{r}\frac{\partial V}{\partial \theta}\mathbf{a}_\theta + \frac{1}{r\sin\theta}\frac{\partial V}{\partial \phi}\mathbf{a}_\phi = \frac{2Q}{r^3}\cos\theta\,\mathbf{a}_r + \frac{Q}{r^3}\sin\theta\,\mathbf{a}_\theta$$

The angle γ between ∇V and $\mathbf{a}_z$ is given by [use $\mathbf{a}_r \cdot \mathbf{a}_z = \cos \theta$, $\mathbf{a}_\theta \cdot \mathbf{a}_z = -\sin \theta$]:

$$\cos \gamma = \frac{\nabla V \cdot \mathbf{a}_z}{|\nabla V|} = \frac{(Q/r^3)(2 \cos^2 \theta - \sin^2 \theta)}{(Q/r^3)\sqrt{4 \cos^2 \theta + \sin^2 \theta}} = f(\theta) \qquad (1)$$

so that $\theta = \text{const.}$ implies $\gamma = \text{const.}$

1.95 The gradient field of Problem 1.94 is perpendicular to the xy plane at each point of that plane—infer this **(a)** from Problem 1.94, **(b)** from Problem 1.92(a).

❚ **(a)** Set $\theta = 90°$ in (1) of Problem 1.94, to obtain $\cos \gamma = -1$, or $\gamma = 180°$. **(b)** On the xy plane, $V = -(Q \cos 90°)/r^2 = 0 = \text{const.}$

1.96 The temperature field over a flat plate is given by $T = (10\,°\text{C/m})x + (20\,°\text{C/m})y$, for x and y in meters. Obtain a unit vector in the direction of maximum temperature change.

❚ $\qquad \nabla T = (10\,°\text{C/m})\mathbf{a}_x + (20\,°\text{C/m})\mathbf{a}_y \qquad$ and $\qquad |\nabla T| = \sqrt{100 + 400} = 22.36\,°\text{C/m}$

Thus, $\nabla T/|\nabla T| = 0.45\mathbf{a}_x + 0.89\mathbf{a}_y = $ required unit vector. Notice that the vector is dimensionless and, for this special field, independent of x and y.

1.97 Determine the rate of change of the scalar field $f(x, y, z) = xy + 2z^2$ at $(1, 1, 1)$ in the direction of the vector $\mathbf{a}_x - 2\mathbf{a}_y + \mathbf{a}_z$.

❚ Use (1) of the Problem 1.83.

$$\nabla f = y\mathbf{a}_x + x\mathbf{a}_y + 4z\mathbf{a}_z = \mathbf{a}_x + \mathbf{a}_y + 4\mathbf{a}_z \quad \text{at } (1, 1, 1) \qquad \mathbf{u} = \frac{\mathbf{a}_x - 2\mathbf{a}_y + \mathbf{a}_z}{\sqrt{6}}$$

$$\frac{df}{ds} = \nabla f \cdot \mathbf{u} = \frac{1 - 2 + 4}{\sqrt{6}} = \frac{3}{\sqrt{6}}$$

1.98 The *line integral* of a vector field $\mathbf{F}$ from point A to point B is defined by $\int_A^B \mathbf{F} \cdot d\mathbf{s}$, where a definite path connecting A and B is understood. (If $\mathbf{F}$ is a force field, the line integral represents the work done over the specified path.) Illustrate this definition with a diagram.

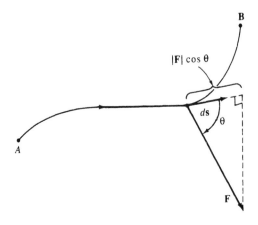

Fig. 1-25

❚ See Fig. 1-25.

1.99 Show that the line integral of the gradient of a scalar field around a closed path is identically zero.

❚ By Problem 1.83,

$$\oint \nabla f \cdot d\mathbf{s} = \oint df = 0$$

1.100 Verify Problem 1.99 for the scalar field $f(x, y) = 2xy + 3$ and the triangular path indicated in Fig. 1-26.

❚ For the given field, $\nabla f = 2y\mathbf{a}_x + 2x\mathbf{a}_y$ and $\nabla f \cdot d\mathbf{s} = 2y\,dx + 2x\,dy$.

$$\text{\textit{path } } P_1 \qquad \int_{(0,0)}^{(1,0)} 2y\,dx + 2x\,dy = \int_{x=0}^{x=1} 0\,dx = 0$$

$$\text{\textit{path } } P_2 \qquad \int_{(1,0)}^{(1,1)} 2y\,dx + 2x\,dy = \int_{y=0}^{y=1} 2\,dy = 2$$

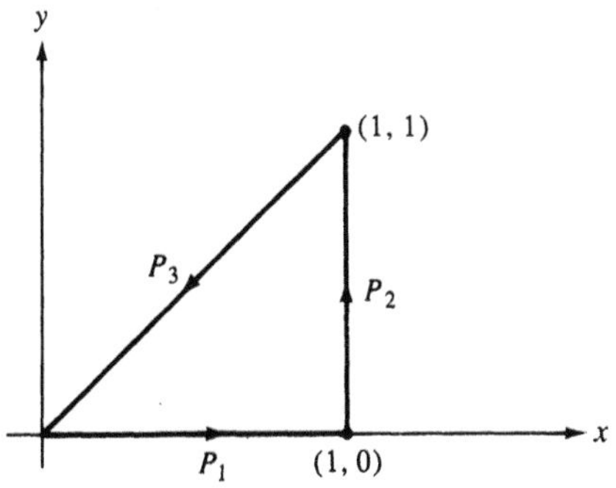

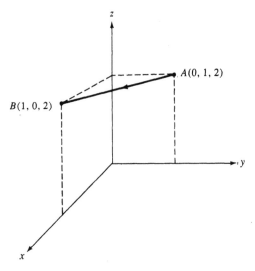

Fig. 1-26

path P_3 $\quad \int_{(1,1)}^{(0,0)} 2y\,dx + 2x\,dy = \int_{x=1}^{0} 2x\,dx + 2x\,dx = 2x^2 \Big]_1^0 = -2$

Consequently, $\oint = 0 + 2 + (-2) = 0$.

1.101 Refer to Problem 1.98. When the contour of integration is specified in terms of a parameter t, the line integral becomes an ordinary one-dimensional integral in t. Illustrate in cartesian coordinates.

▌ Let $\mathbf{F} = F_x\mathbf{a}_x + F_y\mathbf{a}_y + F_z\mathbf{a}_z$, where the components are functions of position; $A = (x_A, y_A, z_A)$, $B = (x_B, y_B, z_B)$; and let the path be given by

$$x = x(t) \qquad y = y(t) \qquad z = z(t)$$

where $x(t_A) = x_A, \ldots, z(t_B) = z_B$. Then

$$\int_A^B \mathbf{F}\cdot d\mathbf{s} = \int_{t_A}^{t_B}\left(\mathbf{F}\cdot\frac{d\mathbf{s}}{dt}\right)dt = \int_{t_A}^{t_B}(F_x x' + F_y y' + F_z z')\,dt$$

1.102 Evaluate the line integral of the vector field $\mathbf{F}(x, y, z) = (x + y)\mathbf{a}_x - x\mathbf{a}_y + z\mathbf{a}_z$ along the straight-line path shown in Fig. 1-27.

Fig. 1-27

▌ Parameterize the path as $x = t$, $y = 1 - t$, $z = 2$ $(0 \leq t \leq 1)$ and use Problem 1.101:

$$\int_A^B \mathbf{F}\cdot d\mathbf{s} = \int_0^1 [(1)(1) + (-t)(-1) + (2)(0)]\,dt = t + \frac{t^2}{2}\Big]_0^1 = \frac{3}{2}$$

1.103 The force exerted on an object is given by $\mathbf{F}(x, y, z) = 2x\mathbf{a}_x + 3z\mathbf{a}_y - 4\mathbf{a}_z$ N. Determine the work done on the object as it moves in a straight line from (**a**) $(0, 0, 1\,\text{m})$ to $(0, 0, -3\,\text{m})$; (**b**) $(1\,\text{m}, 1\,\text{m}, 0)$ to $(0, 1\,\text{m}, 0)$; (**c**) $(1\,\text{m}, 1\,\text{m}, 1\,\text{m})$ to $(0, 0, 1\,\text{m})$.

▌ With $\mathbf{F}\cdot d\mathbf{s} = 2x\,dx + 3z\,dy - 4\,dz$:

(**a**) $\quad W = \int_{z=1}^{-3} -4\,dz = -4(-4) = 16\,\text{J}$

(**b**) $\quad W = \int_{x=1}^{0} 2x\,dx = -1\,\text{J}$ (the object does 1 J of work against the force)

(c) $W = \int_{(1,1,1)}^{(0,0,1)} 2x\,dx + 3(1)\,dy = \int_{x=1}^{0} (2x+3)\,dx = x^2 + 3x \Big]_{1}^{0} = -1 - 3 = -4\,\text{J}$

1.104   Is the force field of Problem 1.103 a gradient field?

▮ A test for the general field $\mathbf{A} = A_x\mathbf{a}_x + A_y\mathbf{a}_y + A_z\mathbf{a}_z$ may be obtained as follows. If $\mathbf{A} = \text{grad}\ \psi$, then
$$d\psi \doteq A_x\,dx + A_y\,dy + A_z\,dx \qquad (1)$$
whence
$$A_x = \frac{\partial \psi}{\partial x} \qquad A_y = \frac{\partial \psi}{\partial y} \qquad A_z = \frac{\partial \psi}{\partial z}$$
Thus the components of $\mathbf{A}$ must satisfy the compatibility relations
$$\frac{\partial A_x}{\partial y} = \frac{\partial A_y}{\partial x} \qquad \frac{\partial A_x}{\partial z} = \frac{\partial A_z}{\partial x} \qquad \frac{\partial A_y}{\partial z} = \frac{\partial A_z}{\partial y} \qquad (2)$$
Conversely, if (2) holds, then (1) can be integrated to give the function ψ.

For the field of Problem 1.103,
$$\frac{\partial F_y}{\partial z} \neq \frac{\partial F_z}{\partial y}$$

so that $\mathbf{F}$ *is not* a gradient field.

1.105   Evaluate the line integral of the vector field $\mathbf{F} = \mathbf{a}_x + 2\mathbf{a}_y + \mathbf{a}_z$ along a circular arc of unit radius from $(1, 0, 1)$ to $(0, 1, 1)$.

▮ The path is given by $x = \cos \phi$, $y = \sin \phi$, $z = 1$, with $0 \le \phi \le \pi/2$; so
$$\int \mathbf{F} \cdot d\mathbf{s} = \int_0^{\pi/2} \left(F_x \frac{dx}{d\phi} + F_y \frac{dy}{d\phi} + F_z \frac{dz}{d\phi} \right) d\phi$$
$$= \int_0^{\pi/2} (-\sin \phi + 2\cos \phi)\,d\phi = \cos \phi + 2 \sin \phi \Big]_0^{\pi/2} = 1$$

1.106   A scalar field is given by $f(x, y, z) = 2x + yz - xy$; evaluate the line integral of $\mathbf{F} = \nabla f$ along a three-segment path consisting of segment 1 from $(1, -1, 1)$ to $(1, -1, 0)$, segment 2 from $(1, -1, 0)$ to $(1, 0, 0)$, and segment 3 from $(1, 0, 0)$ to $(0, 0, 0)$.

▮
$$\int \mathbf{F} \cdot d\mathbf{s} = f(0, 0, 0) - f(1, -1, 1) = 0 - 2 = -2$$

1.107   Let P be a point of space in the domain of definition of a vector field $\mathbf{F}$; let V be the volume enclosed by a surface S that contains P in its interior. The *divergence of* $\mathbf{F}$ at P is defined as
$$\text{div}\ \mathbf{F} \equiv \lim_{V \to 0} \frac{\int_S \mathbf{F} \cdot d\mathbf{S}}{V}$$

Thus div $\mathbf{F}$ is the *net outward flux of* $\mathbf{F}$ *per unit volume* as the volume shrinks to point P. In rectangular coordinates, show that
$$\text{div}\ \mathbf{F} = \frac{\partial F_x}{\partial x} + \frac{\partial F_y}{\partial y} + \frac{\partial F_z}{\partial z} \quad \text{(derivatives evaluated at } P\text{)}$$

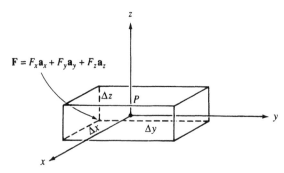

Fig. 1-28

▮ One is free to use the geometry of Fig. 1-28. Integrate over the faces of the box in the order back-front-left-right-bottom-top:

$$\frac{1}{V}\int_S \mathbf{F} \cdot d\mathbf{S} = \frac{1}{\Delta x \, \Delta y \, \Delta z}\left[-F_x \, \Delta y \, \Delta z + \left(F_x + \frac{\partial F_x}{\partial x}\Delta x + O(\overline{\Delta x^2})\right)\Delta y \, \Delta z\right.$$

$$-F_y \, \Delta x \, \Delta z + \left(F_y + \frac{\partial F_y}{\partial y}\Delta y + O(\overline{\Delta y^2})\right)\Delta x \, \Delta z$$

$$\left.-F_z \, \Delta x \, \Delta y + \left(F_z + \frac{\partial F_z}{\partial z}\Delta z + O(\overline{\Delta z^2})\right)\Delta x \, \Delta y\right]$$

$$= \frac{\partial F_x}{\partial x} + \frac{\partial F_y}{\partial y} + \frac{\partial F_z}{\partial z} + O(\Delta x) + O(\Delta y) + O(\Delta z)$$

Now let Δx, Δy, and Δz approach zero, to obtain the desired expression.

1.108 Generalize Problem 1.107 to obtain the form of the divergence in arbitrary orthogonal coordinates (Problem 1.90).

▎ In Fig. 1-28 make the replacements $x \to x_1$, $y \to x_2$, $z \to x_3$; $\Delta x \to h_1 \Delta x_1$, $\Delta y \to h_2 \Delta_2$, $\Delta z \to h_3 \Delta x_3$. Now follow Problem 1.107, keeping in mind that the h's are variables:

$$\frac{1}{V}\int_S \mathbf{F} \cdot d\mathbf{s} = \frac{1}{h_1 h_2 h_3 \, \Delta x_1 \, \Delta x_2 \, \Delta x_3}\left[-F_1 h_2 h_3 \, \Delta x_2 \, \Delta x_3 + \left(F_1 + \frac{\partial F_1}{\partial x_1}\Delta x_1 + \cdots\right)\left(h_2 + \frac{\partial h_2}{\partial x_1}\Delta x_1 + \cdots\right) \times\right.$$

$$\times \left(h_3 + \frac{\partial h_3}{\partial x_1}\Delta x_1 + \cdots\right)\Delta x_2 \, \Delta x_3$$

$$-\cdots$$

$$\left.-\cdots\right]$$

$$= \frac{1}{h_1 h_2 h_3 \, \Delta x_1 \, \Delta x_2 \, \Delta x_3}\left[\frac{\partial}{\partial x_1}(h_2 h_3 F_1) \, \Delta x_1 \, \Delta x_2 \, \Delta x_3 + \cdots\right.$$

$$-\cdots$$

$$\left.-\cdots\right]$$

Passage to the limit then gives

$$\text{div } \mathbf{F} = \frac{1}{h_1 h_2 h_3}\left[\frac{\partial}{\partial x_1}(h_2 h_3 F_1) + \frac{\partial}{\partial x_2}(h_3 h_1 F_2) + \frac{\partial}{\partial x_3}(h_1 h_2 F_3)\right] \qquad (1)$$

Warning: In rectangular coordinates $h_1 = h_2 = h_3 = 1$, and (1) gives

$$\text{div } \mathbf{F} = \frac{\partial F_1}{\partial x_1} + \frac{\partial F_2}{\partial x_2} + \frac{\partial F_3}{\partial x_3} = \mathbf{\nabla} \cdot \mathbf{F}$$

in agreement with Problem 1.107. However, Problem 1.90 gives, for a more general orthogonal system,

$$\mathbf{\nabla} \cdot \mathbf{F} = \frac{1}{h_1}\frac{\partial F_1}{\partial x_1} + \frac{1}{h_2}\frac{\partial F_2}{dx_2} + \frac{1}{h_3}\frac{dF_3}{\partial x_3}$$

which is clearly *not equal* to div $\mathbf{F}$. Unfortunately, the notation $\mathbf{\nabla} \cdot \mathbf{F}$ for the divergence of $\mathbf{F}$ in spherical coordinates, etc., is quite common; in this book it will generally be avoided.

1.109 Calculate div $\mathbf{F}$, assuming (a) $\mathbf{F} = r\mathbf{a}_r + z \sin \phi \mathbf{a}_\phi + 2\mathbf{a}$, (b) $\mathbf{F} = 2\mathbf{a}_r + r \cos \theta \mathbf{a}_\theta + r\mathbf{a}_\phi$.

▎ Using the appropriate scale factors h in (1) of Problem 1.108, we have

(a)
$$\text{div } \mathbf{F} = \frac{1}{r}\left[\frac{\partial}{\partial r}(r^2) + \frac{\partial}{\partial \phi}(z \sin \phi) + \frac{\partial}{\partial z}(2r)\right] = 2 + \frac{z}{r}\cos \phi$$

(b)
$$\text{div } \mathbf{F} = \frac{1}{r^2 \sin \theta}\left[\frac{\partial}{\partial r}(2r^2 \sin \theta) + \frac{\partial}{\partial \theta}(r^2 \sin \theta \cos \theta) + \frac{\partial}{\partial \phi}(r^2)\right]$$

$$= \frac{4}{r} + \frac{\cos 2\theta}{\sin \theta}$$

1.110 Compute the divergences of the following vector fields, and show by sketching the fields that divergence properly measures net flux "through a point": (a) $\mathbf{F}(x, y, z) = \mathbf{a}_y$, (b) $\mathbf{F}(x, y, z) = y\mathbf{a}_y$, (c) $\mathbf{F}(x, y, z) = x\mathbf{a}_y$.

▎ (a) $\mathbf{\nabla} \cdot \mathbf{F} = 0$; (b) $\mathbf{\nabla} \cdot \mathbf{F} = 1$; and (c) $\mathbf{\nabla} \cdot \mathbf{F} = 0$. The sketches are shown in Fig. 1-29.

1.111 Support Problem 1.110(a) by calculating the flux of the vector field out of a closed cylinder of radius 2 that is centered on the z axis and extends from $z = 0$ to $z = 2$.

▎ The net flux out of the ends of the cylinder is zero, owing to the direction of the field. On the lateral surface

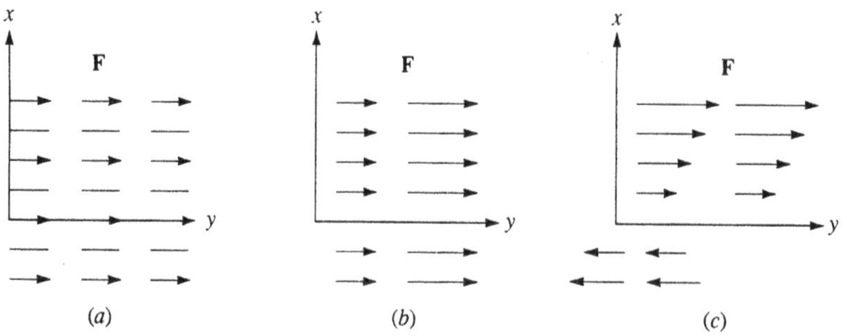

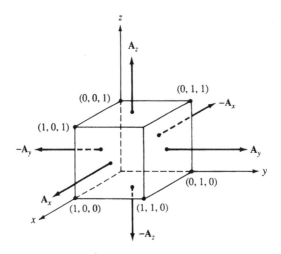

Fig. 1-29

we have, in cylindrical coordinates, $d\mathbf{S} = 2\,d\phi\,dz\,\mathbf{a}_r$; hence, since $\mathbf{a}_y \cdot \mathbf{a}_r = \sin\phi$,

$$\text{total flux} = \int_{\phi=0}^{2\pi}\int_{z=0}^{2} \mathbf{a}_y \cdot (2\,d\phi\,dz\,\mathbf{a}_r)$$
$$= \int_{\phi=0}^{2\pi}\int_{z=0}^{2} 2\sin\phi\,d\phi\,dz = \int_{\phi=0}^{2\pi} 4\sin\phi\,d\phi = -4\cos\phi\Big]_0^{2\pi} = 0$$

1.112 The definition of the divergence (Problem 1.107) was fashioned retrospectively from Gauss' *divergence theorem*:

$$\int_V \text{div }\mathbf{F}\,dV = \int_S \mathbf{F}\cdot d\mathbf{S} \tag{1}$$

Use the divergence theorem to confirm that the sphere of radius 3 has (units aside) equal volume and surface area.

▌ Make the center of the sphere the origin of spherical coordinates and choose $\mathbf{F} = r\mathbf{a}_r = x\mathbf{a}_x + y\mathbf{a}_y + z\mathbf{a}_z$. Since div $\mathbf{F} = 3$ everywhere, and since, on the spherical surface, $d\mathbf{S} = dS\,\mathbf{a}_r$, (1) gives

$$\int_V 3\,dV = \int_S 3\,dS \qquad \text{or} \qquad V = S$$

1.113 Verify the divergence theorem for the vector field $\mathbf{A} = 2xy\mathbf{a}_x + 3\mathbf{a}_y + z^2 y\mathbf{a}_z$ and the unit cube shown in Fig. 1-30.

Fig. 1-30

▌ The divergence of $\mathbf{A}$ is $\nabla\cdot\mathbf{A} = 2y + 0 + 2zy = 2y(z+1)$, whence

$$\int_V \nabla\cdot\mathbf{A}\,dV = \int_{z=0}^{1}\int_{y=0}^{1}\int_{x=0}^{1} 2y(z+1)\,dx\,dy\,dz = \frac{3}{2}$$

On the other hand, integration of the normal component of $\mathbf{A}$ over the faces—in the order top-bottom-right-left-front-back—yields

$$\int_S \mathbf{A}\cdot d\mathbf{S} = \int_{x=0}^{1}\int_{y=0}^{1} y\,dy\,dx + \int_{x=0}^{1}\int_{y=0}^{1} 0\,dx\,dy + \int_{z=0}^{1}\int_{x=0}^{1} 3\,dx\,dz + \int_{z=0}^{1}\int_{x=0}^{1} (-3)\,dx\,dz$$
$$+ \int_{y=0}^{1}\int_{z=0}^{1} 2y\,dy\,dz + \int_{y=0}^{1}\int_{z=0}^{1} 0\,dz\,dy = \frac{1}{2} + 1 = \frac{3}{2}$$

1.114 Determine the net flux of the vector field $\mathbf{F}(x, y, z) = 2x^2 y\mathbf{a}_x + z\mathbf{a}_y + y\mathbf{a}_z$ emerging from the unit cube $0 \le x, y, z \le 1$.

▌$\mathbf{F} \cdot d\mathbf{S} = 2x^2y\, dy\, dz + z\, dx\, dz + y\, dx\, dy$. Thus

$$\text{net } x\text{-flux} = \int_{y=0}^{1} \int_{z=0}^{1} (2x^2y)\, dy\, dz \bigg|_{x=1} - \text{same} \bigg|_{x=0} = 1 - 0 = 1$$

$$\text{net } y\text{-flux} = \int_{x=0}^{1} \int_{z=0}^{1} z\, dx\, dz \bigg|_{y=1} - \text{same} \bigg|_{y=0} = \frac{1}{2} - \frac{1}{2} = 0$$

$$\text{net } z\text{-flux} = \int_{x=0}^{1} \int_{y=0}^{1} y\, dx\, dy \bigg|_{z=1} - \text{same} \bigg|_{z=0} = \frac{1}{2} - \frac{1}{2} = 0$$

Hence, $\int_S \mathbf{F} \cdot d\mathbf{S} = 1 + 0 + 0 = 1$.

1.115 Solve Problem 1.114 by use of the divergence theorem.

▌$\text{div}\,\mathbf{F} = 4xy + 0 + 0 = 4xy$; hence

$$\int_S \mathbf{F} \cdot d\mathbf{S} = \int_V 4xy\, dx\, dy\, dz = 4\int_0^1 x\, dx \int_0^1 y\, dy \int_0^1 dz = 4\left(\frac{1}{2}\right)\left(\frac{1}{2}\right)(1) = 1$$

1.116 Water flows through a cylindrical pipe of radius R whose axis is the z axis; the flow vector is $\mathbf{F} = K[R/R_0 - r/(1+r)]\mathbf{a}_z$ (with K in, say, kg/m$^2 \cdot$ s). Verify the conservation of mass in this flow.

▌By Problem 1.108,

$$\text{div}\,\mathbf{F} = \frac{1}{r}\frac{\partial}{\partial z}\{K[R/R_0 - r/(1+r)]\} = 0$$

so that, for any closed surface S, $\int_S \mathbf{F} \cdot d\mathbf{S} = 0$ [zero rate of mass creation inside S].

1.117 Determine the net flux of the vector field $\mathbf{F}(r, \phi, z) = r\mathbf{a}_r + \mathbf{a}_\phi + z\mathbf{a}_z$ leaving the closed half-cylinder $r = 1$, $0 \le \phi \le \pi$, $0 \le z \le 1$.

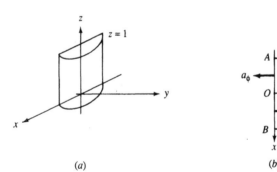

(a) (b) **Fig. 1-31**

▌$\mathbf{F} \cdot d\mathbf{S} = r^2\, d\phi\, dz + dr\, dz + zr\, dr\, d\phi$. For the curved side, Fig. 1-31(a) gives

$$\text{flux} = \int_{\phi=0}^{\pi} \int_{z=0}^{1} r^2\, d\phi\, dz \bigg|_{r=1} = (1)(\pi)(1) = \pi$$

Net flux through the flat side of the semicylinder is zero, since the "outward flux" from the surface OA [Fig. 1-31(b)] cancels the "inward flux" through the surface OB.

For the top:

$$\text{flux} = \int_{\phi=0}^{\pi} \int_{r=0}^{1} 1r\, dr\, d\phi = \frac{\pi}{2}$$

and, for the bottom, flux = 0. Hence, total flux = $3\pi/2$.

1.118 Check Problem 1.117 with the divergence theorem.

▌By Problem 1.108,

$$\text{div}\,\mathbf{F} = \frac{1}{r}\left[\frac{\partial}{\partial r}(r^2) + 0 + \frac{\partial}{\partial z}(rz)\right] = 3$$

whence

$$\int_S \mathbf{F} \cdot d\mathbf{S} = 3\int_0^1 r\, dr \int_0^\pi d\phi \int_0^1 dz = 3\pi/2$$

1.119 A vector field in spherical coordinates is

$$\mathbf{E} = \begin{cases} kr\mathbf{a}_r, & r \le a \\ k\dfrac{a^3}{r^2}\mathbf{a}_r, & r > a \end{cases}$$

where k and a are constants. Determine the divergence of this field.

▮ By Problem 1.108,

$$\text{div }\mathbf{E} = \frac{1}{r^2 \sin\theta}\frac{\partial}{\partial r}\begin{cases} kr^3 \sin\theta \\ ka^3 \sin\theta \end{cases} = \begin{cases} 3k \\ 0 \end{cases}$$

1.120 For the field of Problem 1.119, evaluate the outward flux through the sphere $r = b \le a$.

▮ Since the field is radial, flux $= |\mathbf{E}| \times$ area $= (kb)(4\pi b^2) = 4\pi kb^3$.

1.121 Obtain the result of Problem 1.120 by using the divergence theorem.

▮
$$\text{flux} = |\text{div }\mathbf{E}| \times \text{volume} = (3k)(\tfrac{4}{3}\pi b^3) = 4\pi kb^3$$

1.122 Verify that the divergence of a field directed radially and varying inversely with the radius within a cylinder is zero.

▮ From Problem 1.108,

$$\text{div}\left(\frac{k}{r}\mathbf{a}_r\right) = \frac{1}{r}\frac{\partial}{\partial r}\left(r\frac{k}{r}\right) = 0$$

1.123 A vector field within a sphere varies inversely with the square of the distance from the center and is radially directed. Find its divergence.

▮ From Problem 1.108,

$$\text{div}\left(\frac{k}{r^2}\mathbf{a}_r\right) = \frac{1}{r^2 \sin\theta}\frac{\partial}{\partial r}\left(r^2 \sin\theta \frac{k}{r^2}\right) = 0$$

except at $r = 0$.

1.124 Given $\mathbf{A} = e^{-\alpha y}(\cos\beta x\, \mathbf{a}_x - \sin\beta x\, \mathbf{a}_y)$, find $\nabla\cdot\mathbf{A}$.

▮ $\quad \nabla\cdot\mathbf{A} = \dfrac{\partial}{\partial x}(e^{-\alpha y}\cos\beta x) + \dfrac{\partial}{\partial y}(-e^{-\alpha y}\sin\beta x) = -\beta e^{-\alpha y}\sin\beta x + \alpha e^{-\alpha y}\sin\beta x = (\alpha - \beta)e^{-\alpha y}\sin\beta x$

1.125 Given $\mathbf{A} = xy\mathbf{a}_x + yz\mathbf{a}_y + zx\mathbf{a}_z$, find $\nabla\cdot\mathbf{A}$.

▮
$$\nabla\cdot\mathbf{A} = \frac{\partial}{\partial x}(xy) + \frac{\partial}{\partial y}(yz) + \frac{\partial}{\partial z}(zx) = y + z + x$$

1.126 Given $\mathbf{A} = 5x^2 \sin\pi x\, \mathbf{a}_x$, evaluate div $\mathbf{A}$ at $x = \tfrac{1}{2}$.

▮
$$\text{div }\mathbf{A}\big|_{x=1/2} = \frac{\partial}{\partial x}(5x^2 \sin\pi x)\bigg|_{x=1/2} = 5\pi x^2 \cos\pi x + 10x\sin\pi x\bigg|_{x=1/2} = 5$$

1.127 Given $\mathbf{A} = \sqrt{x^2 + y^2}\,\mathbf{a}_x$, obtain $\nabla\cdot\mathbf{A}$ at $(3, 4, 0)$.

▮
$$\nabla\cdot\mathbf{A}\big|_{(3,4,0)} = \frac{\partial}{\partial x}\sqrt{x^2 + y^2}\bigg|_{(3,4,0)} = \frac{x}{\sqrt{x^2 + y^2}}\bigg|_{(3,4,0)} = \frac{3}{5}$$

1.128 Calculate div $\mathbf{A}$, assuming $\mathbf{A} = r\sin\phi\, \mathbf{a}_r + r\cos 2\phi\, \mathbf{a}_\phi + z^2\, \mathbf{a}_z$.

▮ By Problem 1.108,

$$\text{div }\mathbf{A} = \frac{1}{r}\left[\frac{\partial}{\partial r}(r^2 \sin\phi) + \frac{\partial}{\partial\phi}(r\cos 2\phi) + \frac{\partial}{\partial z}(rz^2)\right] = 2z$$

1.129 Evaluate div $\mathbf{A}$ at $(100, \phi, 100)$, if $\mathbf{A} = r\sin^2\phi\, \mathbf{a}_r + (z^2/r)\cos^2\phi\, \mathbf{a}_z$.

▮ $\quad \text{div }\mathbf{A} = \dfrac{1}{r}\left[\dfrac{\partial}{\partial r}(r^2 \sin^2\phi) + 0 + \dfrac{\partial}{\partial z}(z^2 \cos^2\phi)\right] = 2\left(\sin^2\phi + \dfrac{z}{r}\cos^2\phi\right)$

When $z = r$, div $\mathbf{A} = 2$.

1.130 Show that the divergence of $\mathbf{A} = r^{-2}\sin\theta\,\mathbf{a}_r + r\cot\theta\,\mathbf{a}_\theta + r\sin\theta\cos\phi\,\mathbf{a}_\phi$ is nowhere positive.

▮ Using Problem 1.108, we obtain

$$\text{div }\mathbf{A} = \frac{1}{r^2\sin\theta}\left[\frac{\partial}{\partial r}(\sin^2\theta) + \frac{\partial}{\partial\theta}(r^2\cos\theta) + \frac{\partial}{\partial\phi}(r^2\sin\theta\cos\phi)\right]$$

$$= \frac{1}{r^2\sin\theta}[0 - r^2\sin\theta - r^2\sin\theta\sin\phi] = -(1+\sin\phi) \leq 0$$

1.131 Find div $\mathbf{A}$, given $\mathbf{A} = 5r^{-2}\mathbf{a}_r + 10\cos\theta\mathbf{a}_\theta + r^2\phi\sin\theta\,\mathbf{a}_\phi$.

▮ From Problem 1.108,

$$\text{div }\mathbf{A} = \frac{1}{r^2\sin\theta}\left[\frac{\partial}{\partial r}(5\sin\theta) + \frac{\partial}{\partial\theta}(10r\sin\theta\cos\theta) + \frac{\partial}{\partial\phi}(r^3\phi\sin\theta)\right] = r + \frac{10\cos 2\theta}{r\,\sin\theta}$$

1.132 Evaluate div $\mathbf{A}$ at $(1, \pi/2, \pi)$, for $\mathbf{A} = \sin\theta\,\mathbf{a}_\theta + \sin\phi\,\mathbf{a}_\phi$.

$$\text{div }\mathbf{A} = \frac{1}{r^2\sin\theta}\left[\frac{\partial}{\partial\theta}(r\sin^2\theta) + \frac{\partial}{\partial\phi}(r\sin\phi)\right] = \frac{2\cos\theta}{r} + \frac{\cos\phi}{r\sin\theta}$$

Thus, at $(1, \pi/2, \pi)$, div $\mathbf{A} = -1$.

1.133 Electrical charge density, ρ, is related to electrical flux density $\mathbf{D}$ through div $\mathbf{D} = \rho$. Determine ρ if

$$\mathbf{D} = \begin{cases} cz\mathbf{a}_z & |z| \leq 1 \\ (cz/|z|)\mathbf{a} & |z| > 1 \end{cases}$$

▮

$$\rho = \begin{cases} \dfrac{\partial}{\partial z}(cz) \\ \dfrac{\partial}{\partial z}(\pm c) \end{cases} = \begin{cases} c \\ 0 \end{cases}$$

1.134 In cylindrical coordinates, a vector field is given by $\mathbf{B} = b(r^2 + z^2)^{-3/2}(r\mathbf{a}_r + z\mathbf{a}_z)$. Show that its divergence is zero everywhere except at the origin.

▮
$$\text{div }\mathbf{B} = \frac{1}{r}\left\{\frac{\partial}{\partial r}[b(r^2 + z^2)^{-3/2}r^2] + \frac{\partial}{\partial z}[br(r^2 + z^2)^{+3/2}z]\right\}$$

$$= \frac{b}{r}\left[-\frac{3}{2}(r^2 + z^2)^{-5/2}(2r^3) + (r^2 + z^2)^{-3/2}(2r)\right] + b\left[-\frac{3}{2}(r^2 + z^2)^{-5/2}(2z^2) + (r^2 + z^2)^{-3/2}\right]$$

$$= b(r^2 + z^2)^{-5/2}[-3r^2 + (r^2 + z^2)(2) - 3z^2 + (r^2 + z^2)] = 0$$

unless $r = z = 0$.

1.135 Show that the result of Problem 1.134 is consistent with that of Problem 1.123.

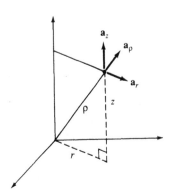

Fig. 1-32

▮ Transforming $\mathbf{B}$ to spherical coordinates (ρ, θ, ϕ) with the aid of Fig. 1-32, we obtain $\mathbf{B} = (b/\rho^2)\mathbf{a}_\rho$.

1.136 Determine the force field $\mathbf{F}$ with the following properties: (i) $\mathbf{F}$ is a central force and its magnitude depends only on the distance from the center; (ii) $\mathbf{F}$ vanishes at the center; (iii) div $\mathbf{F} = -3$, everywhere.

▮ Take the center as origin of spherical coordinates; by (i), $\mathbf{F} = f(r)\mathbf{a}_r$. By (iii),

$$\frac{1}{r^2}\frac{d}{dr}(r^2 f) = -3$$

Integrating and using (ii), $f = -r$.

1.137 Given $\mathbf{A} = r^2\mathbf{a}_r + 2\sin\theta\,\mathbf{a}_\theta$, evaluate div $\mathbf{A}$ at $(1, \pi, \phi)$.

▮ $$\text{div }\mathbf{A} = \frac{1}{r^2}\frac{\partial}{\partial r}(r^4) + \frac{1}{r\sin\theta}\frac{\partial}{\partial\theta}(2\sin^2\theta) = 4r + \frac{4\cos\theta}{r}$$

which vanishes at $(1, \pi, \phi)$.

1.138 Evaluate the divergence of $\mathbf{F} = (2r\sin\theta\cos\phi + \cos\theta)\mathbf{a}_r + (r\cos\theta\cos\phi - \sin\theta)\mathbf{a}_\theta - r\sin\phi\,\mathbf{a}_\phi$ at the point $(2, 30°, 90°)$.

▮ $$\text{div }\mathbf{F} = \frac{1}{r^2\sin\theta}\left[\frac{\partial}{\partial r}(2r^3\sin^2\theta\cos\phi + r^2\sin\theta\cos\theta) + \frac{\partial}{\partial\theta}(r^2\sin\theta\cos\theta\cos\phi - r\sin^2\theta) + \frac{\partial}{\partial\phi}(-r^2\sin\phi)\right]$$

$$= 6\sin\theta\cos\phi + \frac{2}{r}\cos\theta + \frac{r\cos\phi\cos 2\theta - \sin 2\theta}{r\sin\theta} - \frac{\cos\phi}{\sin\theta} \tag{1}$$

At $r = 2$, $\theta = 30°$, and $\phi = 90°$, (1) becomes div $\mathbf{F} = (\sqrt{3}/2) - (\sqrt{3}/2) = 0$.

1.139 Derive the useful identity

$$\text{div }g\mathbf{F} = g\,\text{div }\mathbf{F} + (\text{grad }g)\cdot\mathbf{F} \tag{1}$$

where $\mathbf{F}$ is any vector field and g is any scalar field.

▮ As an equation among scalars, (1) will hold generally if it holds in a particular coordinate system. Now, in rectangular coordinates,

$$\text{div }g\mathbf{F} = \frac{\partial}{\partial x}(gF_x) + \frac{\partial}{\partial y}(gF_y) + \frac{\partial}{\partial z}(gF_z)$$

$$= g\left(\frac{\partial F_x}{\partial x} + \frac{\partial F_y}{\partial y} + \frac{\partial F_z}{\partial z}\right) + \left(\frac{\partial g}{\partial x}F_x + \frac{\partial g}{\partial y}F_y + \frac{\partial g}{\partial z}F_z\right)$$

$$= g\,\text{div }\mathbf{F} + (\text{grad }g)\cdot\mathbf{F}$$

1.140 Let P be a point of space in the domain of definition of a vector field $\mathbf{F}$. In a fixed plane through P, with unit normal $\mathbf{a}_n$, consider a closed curve K that shrinks to the interior point P; the direction of $\mathbf{a}_n$ is made to correspond to description of K in the positive sense via the right-hand rule. By definition, the *curl* of $\mathbf{F}$ at P has the normal component

$$(\text{curl }\mathbf{F})\cdot\mathbf{a}_n \equiv \lim_{S\to 0}\frac{\oint\mathbf{F}\cdot d\mathbf{s}}{S} \tag{1}$$

where S is the area bounded by K. By three applications of (1), evaluate curl $\mathbf{F}$ in rectangular coordinates.

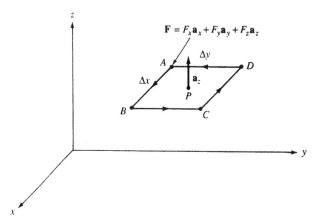

Fig. 1-33

▮ Let us first find the z component of curl $\mathbf{F}$. We choose as the curve K the infinitesimal rectangle $ABCD$

shown in Fig. 1-33. By (1),

$$(\text{curl } \mathbf{F})_z = \lim_{\Delta x, \Delta y \to 0} \frac{1}{\Delta x\, \Delta y}\left[F_x\, \Delta x + \left(F_y + \frac{\partial F_y}{\partial x}\Delta x + \cdots\right)\Delta y - \left(F_x + \frac{\partial F_x}{\partial y}\Delta y + \cdots\right)\Delta x - F_y\, \Delta y\right]$$

$$= \frac{\partial F_y}{\partial x} - \frac{\partial F_x}{\partial y} \quad \text{(derivatives evaluated at } P\text{)}$$

In like manner we find

$$(\text{curl } \mathbf{F})_x = \frac{\partial F_z}{\partial y} - \frac{\partial F_y}{\partial z} \qquad (\text{curl } \mathbf{F})_y = \frac{\partial F_x}{\partial z} - \frac{\partial F_z}{\partial x}$$

Consequently,

$$\text{curl } \mathbf{F} = \left(\frac{\partial F_z}{\partial y} - \frac{\partial F_y}{\partial z}\right)\mathbf{a}_x + \left(\frac{\partial F_x}{\partial z} - \frac{\partial F_z}{\partial x}\right)\mathbf{a}_y + \left(\frac{\partial F_y}{\partial x} - \frac{\partial F_x}{\partial y}\right)\mathbf{a}_z \qquad (2)$$

1.141 Comment on the common notation $\nabla \times \mathbf{F}$ for curl $\mathbf{F}$.

▌ The notation is sensible *in rectangular coordinates*, where indeed (Problem 1.44):

$$\nabla \times \mathbf{F} = \begin{vmatrix} \mathbf{a}_x & \mathbf{a}_y & \mathbf{a}_z \\ \dfrac{\partial}{\partial x} & \dfrac{\partial}{\partial y} & \dfrac{\partial}{\partial z} \\ F_x & F_y & F_z \end{vmatrix} = \text{curl } \mathbf{F}$$

However, a process analogous to that in Problem 1.108 gives, in arbitrary orthogonal coordinates

$$\text{curl } \mathbf{F} = \begin{vmatrix} \dfrac{\mathbf{a}_1}{h_2 h_3} & \dfrac{\mathbf{a}_2}{h_1 h_3} & \dfrac{\mathbf{a}_3}{h_1 h_2} \\ \dfrac{\partial}{\partial x_1} & \dfrac{\partial}{\partial x_2} & \dfrac{\partial}{\partial x_3} \\ h_1 F_1 & h_2 F_2 & h_3 F_3 \end{vmatrix} \qquad (1)$$

which is manifestly different from

$$\nabla \times \mathbf{F} = \begin{vmatrix} \mathbf{a}_1 & \mathbf{a}_2 & \mathbf{a}_3 \\ \dfrac{1}{h_1}\dfrac{\partial}{\partial x_1} & \dfrac{1}{h_2}\dfrac{\partial}{\partial x_2} & \dfrac{1}{h_3}\dfrac{\partial}{\partial x_3} \\ F_1 & F_2 & F_3 \end{vmatrix}$$

In this book the notation $\nabla \times \mathbf{F}$ will be used sparingly, and only in rectangular coordinates.

1.142 Show that a vector field $\mathbf{F}$ is the gradient of some scalar field f (called a *potential function* for $\mathbf{F}$) if and only if curl $\mathbf{F} \equiv \mathbf{0}$.

▌ Compare (2) of Problem 1.140 with (2) of Problem 1.104.

1.143 (a) Sketch the field given by $\mathbf{F}(x, y, z) = z\mathbf{a}_y$. (b) Calculate curl $\mathbf{F}$ and comment on the value obtained.

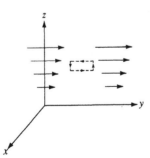

Fig. 1-34

▌ (a) See Fig. 1-34.

(b)
$$\text{curl } \mathbf{F} = \begin{vmatrix} \mathbf{a}_x & \mathbf{a}_y & \mathbf{a}_z \\ \dfrac{\partial}{\partial x} & \dfrac{\partial}{\partial y} & \dfrac{\partial}{\partial z} \\ 0 & z & 0 \end{vmatrix} = -1\mathbf{a}_x$$

It is obvious that the "force" $\mathbf{F}$ will perform more negative work along the top side of the rectangle in Fig. 1-34 than positive work along the bottom side. Thus, the net work, and with it $(\text{curl } \mathbf{F})_x$, will be negative.

1.144 Find the curl of: (**a**) $\mathbf{F} = xy\mathbf{a}_x + 2yz\mathbf{a}_y - \mathbf{a}_z$, (**b**) $\mathbf{F} = 2\mathbf{a}_r + \sin\phi\,\mathbf{a}_\phi - z\mathbf{a}_z$.

▮ (**a**)
$$\text{curl }\mathbf{F} = \left(\frac{\partial F_z}{\partial y} - \frac{\partial F_y}{\partial z}\right)\mathbf{a}_x + \left(\frac{\partial F_x}{\partial z} - \frac{\partial F_z}{\partial x}\right)\mathbf{a}_y + \left(\frac{\partial F_y}{\partial x} - \frac{\partial F_x}{\partial y}\right)\mathbf{a}_z = -2y\mathbf{a}_x - x\mathbf{a}_z$$

(**b**) By (*1*) of Problem 1.141

$$\text{curl }\mathbf{F} = \begin{vmatrix} \dfrac{1}{r}\mathbf{a}_r & \mathbf{a}_\phi & \dfrac{1}{r}\mathbf{a}_z \\[2mm] \dfrac{\partial}{\partial r} & \dfrac{\partial}{\partial \phi} & \dfrac{\partial}{\partial z} \\[2mm] F_r & rF_\phi & F_z \end{vmatrix}$$

$$= \left(\frac{1}{r}\frac{\partial F_z}{\partial \phi} - \frac{\partial F_\phi}{\partial z}\right)\mathbf{a}_r + \left(\frac{\partial F_r}{\partial z} - \frac{\partial F_z}{\partial r}\right)\mathbf{a}_\phi + \frac{1}{r}\left[\frac{\partial}{\partial r}(rF_\phi) - \left(\frac{\partial F_r}{\partial \phi}\right)\right]\mathbf{a}_z$$

$$= 0\mathbf{a}_r + 0\mathbf{a}_\phi + \frac{\sin\phi}{r}\mathbf{a}_z = \frac{1}{r}\sin\phi\,\mathbf{a}_z$$

1.145 Given $\mathbf{A} = (y\cos ax)\mathbf{a}_x + (y + e^x)\mathbf{a}_z$, find $\nabla \times \mathbf{A}$ at the origin.

▮
$$\nabla \times \mathbf{A} = \begin{vmatrix} \mathbf{a}_x & \mathbf{a}_y & \mathbf{a}_z \\[2mm] \dfrac{\partial}{\partial x} & \dfrac{\partial}{\partial y} & \dfrac{\partial}{\partial z} \\[2mm] y\cos ax & 0 & y + e^x \end{vmatrix} = \mathbf{a}_x - e^x\mathbf{a}_y - \cos ax\,\mathbf{a}_z \to \mathbf{a}_x - \mathbf{a}_y - \mathbf{a}_z$$

1.146 Show that the vector field $\mathbf{A} = \dfrac{-y\mathbf{a}_x + x\mathbf{a}_y}{x^2 + y^2}$ is curl-free at all points except the origin.

▮
$$\text{curl }\mathbf{A} = \begin{vmatrix} \mathbf{a}_x & \mathbf{a}_y & \mathbf{a}_z \\[2mm] \dfrac{\partial}{\partial x} & \dfrac{\partial}{\partial y} & \dfrac{\partial}{\partial z} \\[2mm] \dfrac{-y}{x^2+y^2} & \dfrac{x}{x^2+y^2} & 0 \end{vmatrix} = \left[\frac{\partial}{\partial x}\left(\frac{x}{x^2+y^2}\right) - \frac{\partial}{\partial y}\left(\frac{-y}{x^2+y^2}\right)\right]\mathbf{a}_z = \mathbf{0}$$

except at $x = y = 0$.

1.147 (**a**) Show that any central field of the form $\mathbf{F} = f(r)\mathbf{a}_r$ (spherical coordinates) is curl-free. (**b**) Give a potential function (Problem 1.142) for this field.

▮ (**a**)
$$\text{curl }\mathbf{F} = \begin{vmatrix} \dfrac{1}{r^2\sin\theta}\mathbf{a}_r & \dfrac{1}{r\sin\theta}\mathbf{a}_\theta & \dfrac{1}{r}\mathbf{a}_\phi \\[2mm] \dfrac{\partial}{\partial r} & \dfrac{\partial}{\partial \theta} & \dfrac{\partial}{\partial \phi} \\[2mm] f & 0 & 0 \end{vmatrix} = \mathbf{0}$$

(**b**) The potential ψ must vary with r only; thus

$$\mathbf{F} = \nabla\psi = \frac{d\psi}{dr}\mathbf{a}_r = f\mathbf{a}_r, \qquad \text{or} \qquad \psi = \int_{r_0}^{r} f(\rho)\,d\rho$$

1.148 Evaluate the curl of $\mathbf{A} = 5e^{-r}\cos\phi\,\mathbf{a}_r - 5\cos\phi\,\mathbf{a}_z$ over the xz plane.

▮
$$\text{curl }\mathbf{A} = \frac{1}{r}\frac{\partial}{\partial \phi}(-5\cos\phi)\,\mathbf{a}_r + \left[\frac{\partial}{\partial z}(5e^{-r}\cos\phi) - \frac{\partial}{\partial r}(-5\cos\phi)\right]\mathbf{a}_\phi - \frac{1}{r}\frac{\partial}{\partial \phi}(5e^{-r}\cos\phi)\,\mathbf{a}_z$$

$$= \left(\frac{5}{r}\sin\phi\right)\mathbf{a}_r + \left(\frac{5}{r}e^{-r}\sin\phi\right)\mathbf{a}_z$$

which vanishes for $\phi = 0$ and $\phi = \pi$.

1.149 By Problem 1.148, the field $\mathbf{A}$ has a potential function ψ over the half-plane $\phi = 0$; determine this function.

▮ In the zx plane (keeping to a right-handed system) one has, for $x > 0$, $\mathbf{A}(z, x) = -5\mathbf{a}_z + 5e^{-x}\mathbf{a}_x$. Thus

$$\frac{\partial \psi}{\partial z} = -5 \qquad \text{or} \qquad \psi = -5z + C(x)$$

and

$$\frac{\partial \psi}{\partial x} = C'(x) = 5e^{-x} \quad \text{or} \quad C(x) = -5e^{-x} \ (+ \text{ constant})$$

The required function is then $\psi = -5(z + e^{-x})$.

1.150 Given $\mathbf{H} = (Ir/2\pi a^2)\mathbf{a}_\phi$ (cylindrical coordinates), where I and a are constants, show that its curl is invariant in space.

▐ By (1) of Problem 1.141, we obtain

$$\text{curl } \mathbf{H} = \frac{I}{2\pi a^2} \text{curl } r\mathbf{a}_\phi = \frac{I}{2\pi a^2} \begin{vmatrix} \frac{1}{r}\mathbf{a}_r & \mathbf{a}_\phi & \frac{1}{r}\mathbf{a}_z \\ \frac{\partial}{\partial r} & \frac{\partial}{\partial \phi} & \frac{\partial}{\partial z} \\ 0 & r^2 & 0 \end{vmatrix} = \frac{I}{\pi a^2}\mathbf{a}_z$$

which is constant in magnitude and direction.

1.151 A vector field $\mathbf{B}$ is related to another vector field $\mathbf{A}$ by $\mathbf{B} = \text{curl } \mathbf{A}$. For $\mathbf{B} = (\mu_0 I/2\pi r)\mathbf{a}_\phi$, where μ_0 and I are constants, and for the boundary condition $\mathbf{A} = \mathbf{0}$ at $r = r_0$, determine an $\mathbf{A}$.

▐ $\mathbf{B} = \text{curl } \mathbf{A}$ is equivalent to the three component equations

$$\frac{1}{r}\left(\frac{\partial A_z}{\partial \phi} = -\frac{\partial A_\phi}{\partial z}\right) = 0 \qquad \frac{\partial A_r}{\partial z} - \frac{\partial A_z}{\partial r} = \frac{\mu_0 I}{2\pi r} \qquad \frac{1}{r}\left(\frac{\partial A_\phi}{\partial r} - \frac{\partial A_r}{\partial \phi}\right) = 0$$

which can evidently be satisfied by taking $A_r \equiv 0$, $A_\phi \equiv 0$, $A_z = A_z(r)$. Then

$$-\frac{dA_z}{dr} = \frac{\mu_0 I}{2\pi r} \quad \text{or} \quad A_z = -\frac{\mu_0 I}{2\pi}\ln r + C = -\frac{\mu_0 I}{2\pi}\ln r + \frac{\mu_0 I}{2\pi}\ln r_0$$

where the constant of integration was determined from the boundary condition. Consequently,

$$\mathbf{A} = \frac{\mu_0 I}{2\pi}\left(\ln \frac{r_0}{r}\right)\mathbf{a}_z$$

1.152 Confirm that the solution of Problem 1.151 is not unique.

▐ Let $f(\rho)$ be any function such that $f(0) = 0$ and let $\mathbf{A}' = f(r - r_0)\mathbf{a}_r$. One easily sees that curl $\mathbf{A}' = \mathbf{0}$, whence $\mathbf{A} + \mathbf{A}'$ also solves Problem 1.151.

1.153 Prove directly that the curl of a gradient is zero. (Problem 1.104 contains an indirect proof.)

▐ If $\mathbf{F} = \nabla f$, the definition (1) of Problem 1.140 yields

$$(\text{curl } \mathbf{F}) \cdot \mathbf{a}_n = \lim_{S \to 0} \frac{\oint df}{S} = \lim_{S \to 0} 0 = 0$$

whatever the orientation of $\mathbf{a}_n$. It follows that curl $\mathbf{F} = \mathbf{0}$.

1.154 Prove that the divergence of a curl is zero.

▐ Make point P the center of a sphere of radius ϵ. By Problem 1.107,

$$(\text{div curl } \mathbf{A})_P = \lim_{\epsilon \to 0} \frac{\int_S (\text{curl } \mathbf{A}) \cdot d\mathbf{S}}{\frac{4}{3}\pi\epsilon^3}$$

Evaluate the surface integral in spherical coordinates:

$$\int_{r=\epsilon} (\text{curl } \mathbf{A})_r \, dS = \epsilon \int_{\phi=0}^{2\pi} \int_{\theta=0}^{\pi} \left[\frac{\partial}{\partial \theta}(\sin \theta \, A_\phi) - \frac{\partial A_\theta}{\partial \phi}\right] d\theta \, d\phi$$

$$= \epsilon \int_0^{2\pi} (\sin \theta \, A_\phi)\bigg]_{\theta=0}^{\theta=\pi} d\phi - \epsilon \int_0^{\pi} A_\theta\bigg]_{\phi=0}^{\phi=2\pi} d\theta = 0$$

identically in ϵ, because $\sin 0 = \sin \pi = 0$ and $A_\theta(\epsilon, \theta, 0) = A_\theta(\epsilon, \theta, 2\pi)$. Hence (div curl $\mathbf{A})_P = 0$.

1.155 Confirm Problem 1.154 by an explicit calculation of div curl $\mathbf{A}$ in rectangular coordinates.

$$\text{div curl } \mathbf{A} = \frac{\partial}{\partial x}[(\text{curl } A)_x] + \frac{\partial}{\partial y}[(\text{curl } A)_y] + \frac{\partial}{\partial z}[(\text{curl } A)_z]$$

$$= \frac{\partial}{\partial x}\left(\frac{\partial A_z}{\partial y} - \frac{\partial A_y}{\partial z}\right) + \frac{\partial}{\partial y}\left(\frac{\partial A_x}{\partial z} - \frac{\partial A_z}{\partial x}\right) + \frac{\partial}{\partial z}\left(\frac{\partial A_y}{\partial x} - \frac{\partial A_x}{\partial y}\right)$$

$$= \left(\frac{\partial^2 A_x}{\partial y\,\partial z} - \frac{\partial^2 A_x}{\partial z\,\partial y}\right) + \left(\frac{\partial^2 A_y}{\partial z\,\partial x} - \frac{\partial^2 A_y}{\partial x\,\partial z}\right) + \left(\frac{\partial^2 A_z}{\partial x\,\partial y} - \frac{\partial^2 A_z}{\partial y\,\partial x}\right)$$

$$= 0 + 0 + 0 = 0$$

1.156 *In rectangular coordinate systems only*, one allows the Laplacian (Problem 1.86) to operate on vector fields, according to

$$\nabla^2 \mathbf{F} \equiv (\nabla^2 F_x)\mathbf{a}_x + (\nabla^2 F_y)\mathbf{a}_y + (\nabla^2 F_z)\mathbf{a}_z$$

Establish the important identity

$$\nabla^2 \mathbf{F} = \text{grad div } \mathbf{F} - \text{curl curl } \mathbf{F} \qquad (1)$$

$\blacksquare$ It suffices to show that (1) obtains among the three x components.

$$(\text{grad div } \mathbf{F})_x = \frac{\partial}{\partial x}(\text{div } \mathbf{F}) = \frac{\partial}{\partial x}\left(\frac{\partial F_x}{\partial x} + \frac{\partial F_y}{\partial y} + \frac{\partial F_z}{\partial z}\right) = \frac{\partial^2 F}{\partial x^2} + \left(\frac{\partial^2 F_y}{\partial x\,\partial y} + \frac{\partial^2 F_z}{\partial x\,\partial z}\right)$$

$$(\text{curl curl } \mathbf{F})_x = \frac{\partial}{\partial y}[(\text{curl } \mathbf{F})_z] - \frac{\partial}{\partial z}[(\text{curl } \mathbf{F})_y] = \frac{\partial}{\partial y}\left[\frac{\partial F_y}{\partial x} - \frac{\partial F_x}{\partial y}\right] - \frac{\partial}{\partial z}\left[\frac{\partial F_x}{\partial z} - \frac{\partial F_z}{\partial x}\right]$$

$$= \frac{\partial^2 F_y}{\partial y\,\partial x} - \frac{\partial^2 F_x}{\partial y^2} - \frac{\partial^2 F_x}{\partial z^2} + \frac{\partial^2 F_z}{\partial z\,\partial x} = -\frac{\partial^2 F_x}{\partial y^2} - \frac{\partial^2 F_x}{\partial z^2} + \left(\frac{\partial^2 F_y}{\partial x\,\partial y} + \frac{\partial^2 F}{\partial x\,\partial z}\right)$$

Hence, subtracting, we obtain $(\text{grad div } \mathbf{F})_x - (\text{curl curl } \mathbf{F})_x = \nabla^2 F_x = (\nabla^2 \mathbf{F})_x$.

1.157 A vector field $\mathbf{F}$ is *conservative* (in a given region of space) if $\oint \mathbf{F} \cdot d\mathbf{s} = 0$ for every closed contour (within the region). Show that a conservative field is curl-free, and vice versa.

$\blacksquare$ If $\mathbf{F}$ is conservative, the line integral $\int_{P_0}^{P} \mathbf{F} \cdot d\mathbf{s}$ is independent of the path from the fixed point P_0 to the variable point P. It thus defines a single-valued function of the coordinates of P; i.e., in cartesian coordinates,

$$\int_{(x_0, y_0, z_0)}^{(x,y,z)} F_x\,dx + F_y\,dy + F_z\,dz = f(x, y, z)$$

But then

$$\frac{\partial f}{\partial x} = F_x \qquad \frac{\partial f}{\partial y} = F_y \qquad \frac{\partial f}{\partial z} = F_z$$

which is to say, $\boldsymbol{\nabla} f = \mathbf{F}$; by Problem 1.153, curl $\mathbf{F} = \mathbf{0}$.

Conversely, if curl $\mathbf{F} = \mathbf{0}$, then (Problem 1.142) $\mathbf{F} = \boldsymbol{\nabla}\psi$, for some potential function ψ. Consequently, $\oint \mathbf{F} \cdot d\mathbf{s} = \oint d\psi = 0$.

From the above it is seen that a "curl-free field," a "gradient or potential field," and a "conservative field" mean one and the same thing. Another synonym is "irrotational field"—used mainly in reference to fluid flow.

1.158 State *Stokes' theorem* and give its main application.

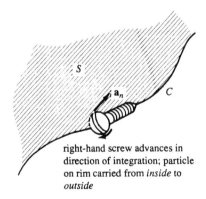

right-hand screw advances in direction of integration; particle on rim carried from *inside* to *outside*

Fig. 1-35

$\blacksquare$ Stokes' theorem is related to the definition of the curl exactly as the divergence theorem is related to that of the divergence. It reads

$$\int_S (\text{curl } \mathbf{F}) \cdot d\mathbf{S} = \oint_C \mathbf{F} \cdot d\mathbf{s} \qquad (1)$$

Here, S is an *open* surface spanning the closed curve C. For a given sense of description of C, the *outside* of S (the local *outer* normal gives the direction of dS) is determined by the rule of Fig. 1-35.

The theorem is useful in evaluating line integrals of nonconservative fields. [For a conservative field, (1) reduces to the trivial $\int_S 0 \, dS = 0$.]

1.159 Show how Stokes' theorem (Problem 1.158) predicts the computational result of Problem 1.154:

$$\int_{\text{spherical surface}} (\text{curl } \mathbf{A}) \cdot d\mathbf{S} = 0$$

▮ Let C denote the equator of the sphere; then

$$\int_{\text{northern hemisphere}} (\text{curl } \mathbf{A}) \cdot d\mathbf{S} = \oint_C \mathbf{A} \cdot d\mathbf{s} = -\int_{\text{southern hemisphere}} (\text{curl } \mathbf{A}) \cdot d\mathbf{S}$$

1.160 Verify Stokes' theorem for the vector field $\mathbf{F} = \mathbf{a}_x + zy^2 \mathbf{a}_y$ and the flat surface in the yz plane shown in Fig. 1-36. Describe the contour C in the clockwise direction, as shown.

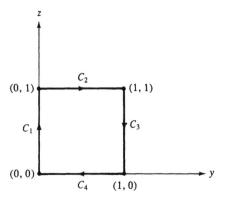

Fig. 1-36

▮
$$\oint \mathbf{F} \cdot d\mathbf{s} = \int_{z=0}^{z=1} 0 \, dz + \int_{y=0}^{y=1} 1y^2 \, dy + \int_{z=1}^{z=0} 0 \, dz + \int_{y=1}^{0} 0 \, dy = \frac{1}{3}$$

Also, since curl $\mathbf{F} = -y^2 \mathbf{a}_x$ and $d\mathbf{S} = dy \, dz \, (-\mathbf{a}_x)$,

$$\int_S (\text{curl } \mathbf{F}) \cdot d\mathbf{S} = \int_{y=0}^{1} \int_{z=0}^{1} y^2 \, dy \, dz = \frac{1}{3}$$

1.161 Verify Stokes' theorem for the vector field $\mathbf{F}(r, \theta, \phi) = \mathbf{a}_r + \mathbf{a}_\theta + \mathbf{a}_\phi$, the surface being the octant of a sphere: $r = 2$, $0 \le \theta \le \pi/2$, $0 \le \phi \le \pi/2$.

▮ By (1) of Problem 1.141,

$$\text{curl } \mathbf{F} = \begin{vmatrix} \dfrac{1}{r^2 \sin \theta} \mathbf{a}_r & \dfrac{1}{r \sin \theta} \mathbf{a}_\theta & \dfrac{1}{r} \mathbf{a}_\phi \\[2mm] \dfrac{\partial}{\partial r} & \dfrac{\partial}{\partial \theta} & \dfrac{\partial}{\partial \phi} \\[2mm] 1 & r & r \sin \theta \end{vmatrix}$$

$$= \frac{\cos \theta}{r \sin \theta} \mathbf{a}_r - \frac{1}{r} \mathbf{a}_\theta + \frac{1}{r} \mathbf{a}_\phi$$

so that, with $d\mathbf{S} = 4 \sin \theta \, d\theta \, d\phi \, \mathbf{a}_r$,

$$\int_S (\text{curl } \mathbf{F}) \cdot d\mathbf{S} = \int_{\theta=0}^{\pi/2} \int_{\phi=0}^{\pi/2} \left(\frac{\cos \theta}{2 \sin \theta} \right) (4 \sin \theta) \, d\theta \, d\phi = \pi \sin \theta \Big]_0^{\pi/2} = \pi$$

Correspondingly, with $\mathbf{F} \cdot d\mathbf{s} = dr + 2 \, d\theta + 2 \sin \theta \, d\phi$ and the proper direction of integration,

$$\oint \mathbf{F} \cdot d\mathbf{s} = \int_{\phi=0}^{\pi/2} 2 \sin \frac{\pi}{2} \, d\phi + \int_{\theta=\pi/2}^{0} 2 \, d\theta + \int_{\theta=0}^{\pi/2} 2 \, d\theta = \pi - \pi + \pi = \pi$$

1.162 Show that, in cylindrical coordinates, Stokes' theorem *does not hold* for $\mathbf{F} = r^{-1/2} \mathbf{a}_\phi$, if S is the circle $r \le 1$, $z = 0$.

▌ We have $\oint_C \mathbf{F} \cdot d\mathbf{s} = \int_0^{2\pi} 1^{1/2} \, d\phi = 2\pi$. However, curl $\mathbf{F} = -\frac{1}{2} r^{-5/2} \mathbf{a}_z$, so that

$$\int_S (\text{curl } \mathbf{F}) \cdot d\mathbf{S} = \int_0^{2\pi} d\phi \int_0^{1\cdot} -\frac{1}{2} r^{-3/2} \, dr \to \infty$$

The r-integral diverges because of the singularity of curl $\mathbf{F}$ at $r = 0$.

1.163 A scalar field varies with x only, in a rectangular coordinate system. The gradient of the field is $30x^2 \mathbf{a}_x$ and the field is zero at the origin. Determine the field.

▌ Let the field be $\psi = \psi(x)$; then

$$\frac{d\psi}{dx} = 30x^2 \qquad \psi = \int_0^x 30u^2 \, du = 10x^3$$

1.164 Determine the *(a)* divergence and *(b)* curl of $\mathbf{A} = x^2 \mathbf{a}_x - y^2 \mathbf{a}_y$.

▌ *(a)* $\quad \nabla \cdot \mathbf{A} = \dfrac{\partial}{\partial x}(x^2) + \dfrac{\partial}{\partial y}(-y^2) = 2x - 2y$

$\quad$ *(b)* $\quad \nabla \times \mathbf{A} = \begin{vmatrix} \mathbf{a}_x & \mathbf{a}_y & \mathbf{a}_z \\ \dfrac{\partial}{\partial x} & \dfrac{\partial}{\partial y} & \dfrac{\partial}{\partial z} \\ x^2 & -y^2 & 0 \end{vmatrix} = 0\mathbf{a}_x - 0\mathbf{a}_y + 0\mathbf{a}_z = \mathbf{0}$

1.165 The curl and the divergence of a two-dimensional vector field $\mathbf{B} = B_x(x, y)\mathbf{a}_x + B_y(x, y)\mathbf{a}_y$ are both zero. Obtain a differential equation satisfied by B_x and B_y.

▌ $\nabla^2 B_x = \nabla^2 B_y = 0$, by *(1)* of Problem 1.56.

1.166 Generalize Problem 1.165 to an arbitrary coordinate system.

▌ curl $\mathbf{B} = \mathbf{0}$ implies (Problem 1.142) $\mathbf{B} = \text{grad } \psi$, for some potential function ψ. Then div $\mathbf{B} = 0$ gives
$$\nabla^2 \psi \equiv \text{div grad } \psi = 0 \qquad\qquad (1)$$
which is the general Laplace's equation for ψ.

1.167 A vector field is obtained from a scalar field $f(r, \phi, z) = r \sin \phi$ as $\mathbf{F} = \nabla f$; evaluate the line integral of $\mathbf{F}$ between the origin and $x = 0$, $y = 1$, $z = 1$.

▌ The field is conservative and
$$\int_{(0,\phi,0)}^{(1,\pi/2,1)} \mathbf{F} \cdot d\mathbf{s} = f(1, \pi/2, 1) - f(0, \phi, 0) = 1 - 0 = 1$$

1.168 Evaluate the line integral of the vector field $\mathbf{F}(x, y, z) = y\mathbf{a}_x + (x + z)\mathbf{a}_y + 3yz\mathbf{a}_z$ along the broken-line contour of Fig. 1-37.

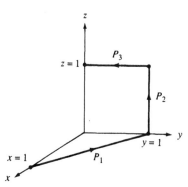

Fig. 1-37

▌
$$\int_{P_1+P_2+P_3} y \, dx + (x + z) \, dy + 3yz \, dz = \int_{y=0}^1 y(-dy) + (1 - y) \, dy + \int_{z=0}^1 3(1)z \, dz + \int_{y=1}^0 (1) \, dy$$
$$= 0 + \frac{3}{2} - 1 = \frac{1}{2}$$

1.169 Evaluate the line integral of the vector field $\mathbf{F}(x, y, z) = xy\mathbf{a}_x + 2\mathbf{a}_y$ around the closed path of Fig. 1-38.

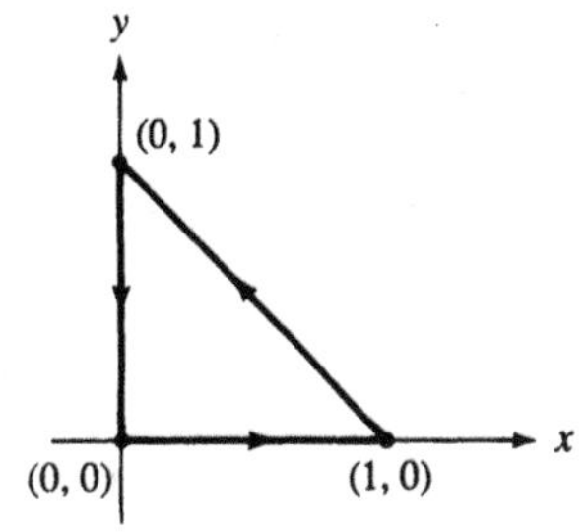

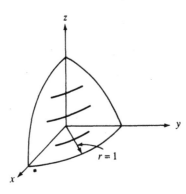

Fig. 1-38

▐ $$\oint xy\,dx + 2\,dy = \oint xy\,dx = 0 + \int_{1}^{0} x(1-x)\,dx + 0 = -\frac{1}{6}$$

1.170 Evaluate $\int (\nabla \times \mathbf{F}) \cdot d\mathbf{S}$ for the field of Problem 1.169 and hence verify Stokes' theorem.

▐ $(\nabla \times \mathbf{F}) \cdot d\mathbf{S} = (-x\mathbf{a}_z) \cdot (1-x)\,dx\,\mathbf{a}_z = -x(1-x)\,dx$, and this is integrated from $x = 0$ to $x = 1$.

1.171 Verify the divergence theorem for the vector field $\mathbf{F} = x\mathbf{a}_x + y\mathbf{a}_y + z\mathbf{a}_z$ and the volume $x \geq 0$, $y \geq 0$, $z \geq 0$, $x^2 + y^2 + z^2 \leq 1$ (Fig. 1-39).

Fig. 1-39

▐ Since div $\mathbf{F} = 3$,

$$\int_V \text{div } \mathbf{F}\,dV = 3\left(\frac{1}{8}\right)\left(\frac{4}{3}\pi\right) = \frac{\pi}{2}$$

As for the surface integral, the normal component of $\mathbf{F}$ vanishes on each flat face, leaving

$$\int_{\text{octant}} \mathbf{F} \cdot d\mathbf{S} = |\mathbf{F}| \times \text{area} = 1 \times \frac{4\pi}{8} = \frac{\pi}{2}$$

1.172 Verify Stokes' theorem for the vector field $\mathbf{V}(r, \theta, \phi) = \cos \theta\, \mathbf{a}_r$ on the hemisphere of unit radius defined by $r = 1$, $0 \leq \phi \leq 2\pi$, and $0 \leq \theta \leq \pi/2$.

▐ On $r = 1$, curl $\mathbf{V} = \sin \theta\, \mathbf{a}_\phi$ is orthogonal to $d\mathbf{S} = dS\, \mathbf{a}_r$; so, $\int_S (\text{curl } \mathbf{V}) \cdot d\mathbf{S} = 0$. Likewise, along the equator, $\mathbf{V}$ is orthogonal to $d\mathbf{s} = d\phi\, \mathbf{a}_\phi$; therefore, $\oint \mathbf{V} \cdot d\mathbf{s} = 0$.

1.173 Evaluate the line integral of

$$\mathbf{H} = \frac{10^4}{r}\left(\frac{1}{a^2}\sin ar - \frac{r}{a}\cos ar\right)\mathbf{a}_\phi \quad \text{(A/m)} \qquad \text{where } a \equiv \frac{\pi}{2r_0} \quad \text{(m}^{-1}\text{)}$$

around the circle $r = r_0$, $z = 0$.

▐ $$\oint_{r=r_0} \mathbf{H} \cdot d\mathbf{s} = H_\phi(r_0) \times 2\pi r_0 = \frac{8 \times 10^4 r_0^2}{\pi} \quad \text{(A)}$$

1.174 Use Stokes' theorem to duplicate the result of Problem 1.173.

$$\text{curl } \mathbf{H} = \begin{vmatrix} \dfrac{1}{r}\mathbf{a}_r & \mathbf{a}_\phi & \dfrac{1}{r}\mathbf{a}_z \\ \dfrac{\partial}{\partial r} & \dfrac{\partial}{\partial \phi} & \dfrac{\partial}{\partial z} \\ 0 & rH_\phi & 0 \end{vmatrix} = \frac{1}{r}\frac{d}{dr}(rH_\phi)\mathbf{a}_z$$

$$\int_{r \le r_0} (\text{curl } \mathbf{H}) \cdot d\mathbf{S} = \int_{\phi=0}^{2\pi} \int_{r=0}^{r_0} \frac{1}{r}\frac{d}{dr}(rH_\phi) r \, dr \, d\phi = 2\pi r_0 H_\phi(r_0)$$

1.175 Does the scalar field $H_\phi(r)$ of Problems 1.173 and 1.174 satisfy Laplace's equation, (1) of Problem 1.166?

⬛ No: If $\psi = \psi(r)$ (cylindrical coordinates) then

$$\text{grad } \psi = \frac{d\psi}{dr}\mathbf{a}_r, \qquad \nabla^2\psi = \text{div grad }\psi = \frac{1}{r}\frac{d}{dr}\left(r\frac{d\psi}{dr}\right)$$

Thus $\nabla^2\psi = 0$ only if

$$r\frac{d\psi}{dr} = C \qquad \psi = C \ln r + D$$

It is clear that no specialization of the constants C and D can result in $\psi = H_\phi$.

CHAPTER 2
Electrostatics

2.1 Sources of electric (**E**-) fields are electric charges, measured in coulombs (C). Various space distributions of charge are: *line, surface,* and *volume* distributions; these are respectively described by density functions ρ_ℓ (C/m), ρ_s (C/m²), and ρ_v (C/m³). Find the total charge on the x axis from 0 to 10 m if $\rho_\ell = 3x^2$ (μC/m).

❚
$$Q = \int \rho_\ell \, ds = \int_{x=0}^{10} 3x^2 \, dx = x^3 \Big]_0^{10} = 1000 \, \mu C = 1 \text{ mC}$$

2.2 Given $\rho_\ell(x, y, z) = 2x + 3y - 4z$ (C/m), find the charge on the line segment extending from $(2, 1, 5)$ to $(4, 3, 6)$.

❚ Representing the line segment by
$$x = 2 + t(4 - 2) \qquad y = 1 + t(3 - 1) \qquad z = 5 + t(6 - 5)$$
with $0 \le t \le 1$, one has $\rho(t) = -13 + 6t$ and
$$ds = \sqrt{\left(\frac{dx}{dt}\right)^2 + \left(\frac{dy}{dt}\right)^2 + \left(\frac{dz}{dt}\right)^2} \, dt = 3 \, dt$$

Hence
$$Q = \int_0^1 (-13 + 6t)(3) \, dt = -30 \text{ C}$$

2.3 A circular disk of radius R has a surface charge density that increases linearly away from the center, the constant of proportionality being k. Determine the total charge on the disk.

❚
$$\int_S \rho_s \, dS = \int_0^R (kr)(2\pi r) \, dr = \frac{2\pi k R^3}{3}$$

2.4 Find the total charge on the triangle of Fig. 2-1, given surface density $\rho_s = 6xy$ (C/m²).

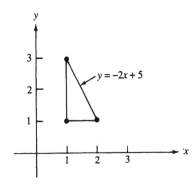

Fig. 2-1

❚
$$Q = \int_{x=1}^2 \int_{y=1}^{-2x+5} 6xy \, dy \, dx = 13 \text{ C}$$

2.5 A sphere of 200-mm radius contains electrical charge of density $2/(r \sin \theta)$ (C/m³). What is the total charge contained within the sphere?
$$Q = \int_V \rho_v \, dV = \int_{r=0}^{0.200} \int_{\theta=0}^{\pi} \int_{\phi=0}^{2\pi} \left(\frac{2}{r \sin \theta}\right) r^2 \sin \theta \, dr \, d\theta \, d\phi = \int_0^{0.200} 2r \, dr \int_0^{\pi} d\theta \int_0^{2\pi} d\phi = 0.08\pi^2 \approx 0.789 \text{ C}$$

2.6 A certain electron beam may be approximated by a right circular cylinder of radius R that contains a volume charge density $\rho_v = k/(c + r^2)$ (C/m³), where $k < 0$ and $c > 0$ are constants. Evaluate the total charge per unit length of the beam.

❚ In the natural cylindrical coordinates, one has, per meter of length, $dV = 2\pi r \, dr$ and
$$Q = \int_0^R \frac{k}{c + r^2} 2\pi r \, dr = \pi k[\ln (c + r^2)]_0^R = \pi k \ln \frac{c + R^2}{c} \quad \text{(C/m)}$$

2.7 A charge distribution $\rho_v = 2z$ (C/m³) is present in a region defined in cylindrical coordinates as $0 \le z \le 2$ m, $0 \le r \le 1$ m, and $45° \le \phi \le 90°$. Determine the total charge contained in the region.

$$Q = \int_{z=0}^{2} \int_{\phi=\pi/4}^{\pi/2} \int_{r=0}^{1} 2z \, r \, dr \, d\phi \, dz = \frac{\pi}{2} \text{ C}$$

2.8 In free space, *Coulomb's law* gives the force on a point of charge q due to the **E**-field of a point charge Q as

$$\mathbf{F}_Q = \frac{qQ}{4\pi\epsilon_0 R^2} \mathbf{a}_R \quad \text{(N)} \tag{1}$$

The unit vector $\mathbf{a}_R$ is directed along the line joining the two charges and points in the direction from Q to q, as shown in Fig. 2-2(a). In (1) the constant ϵ_0 is known as the *permittivity of free space* and is given (in farads per meter) by

$$\epsilon_0 = 8.854 \times 10^{-12} \approx \frac{10^{-9}}{36\pi} \text{ F/m} \tag{2}$$

Assuming that q is located at a point (x, y, z) and Q at (X, Y, Z), express (1) in terms of these coordinates.

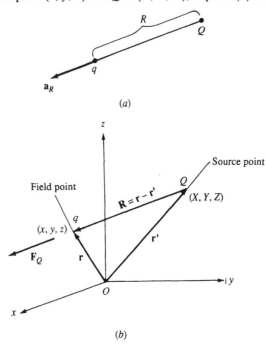

(a)

(b)   **Fig. 2-2**

From Fig. 2-2(b), $\mathbf{R} = (x - X)\mathbf{a}_x + (y - Y)\mathbf{a}_y + (z - Z)\mathbf{a}_z$, and $\mathbf{a}_R = \mathbf{R}/R$, where
$$R = |\mathbf{R}| = \sqrt{(x - X)^2 + (y - Y)^2 + (z - Z)^2}$$
Consequently (1) becomes
$$\mathbf{F}_Q = \frac{qQ[(x - X)\mathbf{a}_x + (y - Y)\mathbf{a}_y + (z - Z)\mathbf{a}_z]}{4\pi\epsilon_0[(x - X)^2 + (y - Y)^2 + (z - Z)^2]^{3/2}} \quad \text{(N)}$$
$\mathbf{F}_Q$ will have the direction indicated in Fig. 2-2(b) if q and Q have the same algebraic sign.

2.9 Express (1) of Problem 2.8 in terms of the vectors $\mathbf{r}$ and $\mathbf{r}'$ of Fig. 2-2(b).

$$\mathbf{F}_Q = \frac{qQ}{4\pi\epsilon_0} \frac{\mathbf{r} - \mathbf{r}'}{|\mathbf{r} - \mathbf{r}'|^3} \quad \text{(N)}$$

2.10 Consider a pair of point charges in free space: charge $q = -300 \, \mu\text{C}$ is located at $(2, 4, 5)$ m and charge $Q = 10 \, \mu\text{C}$ is located at $(1, 1, 3)$ m. What is the force (magnitude and direction) on q?

$$|\mathbf{F}_Q| = \frac{|q| \, |Q|}{4\pi\epsilon_0 R^2} \approx \frac{(300 \times 10^{-6})(10 \times 10^{-6})}{4\pi(10^{-9}/36\pi)[(1 - 2)^2 + (1 - 4)^2 + (3 - 5)^2]} = 1.93 \text{ N}$$

The force is directed from q toward Q.

2.11 A 100 μC point charge (Q_1), is located in a rectangular coordinate system at $(1, 1, 1)$ m, and another point charge (Q_2), of 50 μC, is at $(-1, 0, -2)$ m. Find the vector force on the first charge.

❚ The vector from source point to field point is $\mathbf{R}_{21} = 2\mathbf{a}_x + \mathbf{a}_y + 3\mathbf{a}_z$ (m); thus

$$\mathbf{F}_{21} = \frac{Q_1 Q_2}{4\pi\epsilon_0 R_{21}^3} \mathbf{R}_{21} \approx \frac{(100 \times 10^{-6})(50 \times 10^{-6})}{(10^{-9}/9)(2^2 + 1^2 + 3^2)^{3/2}} (2\mathbf{a}_x + \mathbf{a}_y + 3\mathbf{a}_z)$$

$$= 1.718\mathbf{a}_x + 0.859\mathbf{a}_y + 2.577\mathbf{a}_z \quad \text{N}$$

2.12 A positive charge and a negative charge are respectively located at $(0, 1, 3)$ m and $(3, 0, 0)$ m. Determine the direction cosines of the force on the positive change.

❚ The force, which is attractive, is parallel to the vector $3\mathbf{a}_x - 1\mathbf{a}_y - 3\mathbf{a}_z$. Hence, the desired direction cosines are

$$\alpha = \frac{3}{\sqrt{19}} \qquad \beta = \frac{-1}{\sqrt{19}} \qquad \gamma = \frac{-3}{\sqrt{19}}$$

2.13 Find the magnitude of the electrostatic force between a $100\ \mu\text{C}$ charge at $(-1, 1, -3)$ m and a $20\ \mu\text{C}$ charge at $(3, 1, 0)$ m.

❚
$$|\mathbf{F}| = \frac{(100 \times 10^{-6})(20 \times 10^{-6})}{4\pi(10^{-9}/36\pi)(4^2 + 0^2 + 3^2)} = 0.72\ \text{N}$$

2.14 Four like charges of $30\ \mu\text{C}$ each are located at the four corners of a square, the diagonal of which measures 8 m. Find the force on a $150\ \mu\text{C}$ charge located 3 m above the center of the square (Fig. 2-3).

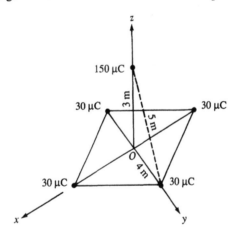

Fig. 2-3

❚ From symmetry, the force is z-directed and is given by

$$\mathbf{F} = \frac{4(150 \times 10^{-6})(30 \times 10^{-6})}{4\pi(10^{-9}/36\pi)(5)^2} \left(\frac{3}{5}\right)\mathbf{a}_z = 3.888\mathbf{a}_z\ \text{N}$$

2.15 In the xy plane, $Q_1 = 200\ \mu\text{C}$, at $(2, 4)$ m, experiences a repulsive force of 7.2 N because of Q_2, at $(10, 7)$ m. Determine Q_2.

❚
$$F = 7.2 = \frac{(200 \times 10^{-6})Q_2}{4\pi(10^{-9}/36\pi)[(10 - 2)^2 + (7 - 4)^2]} \qquad \text{or} \qquad Q_2 = 292\ \mu\text{C}$$

2.16 A circular disk of radius 3 m carries a uniformly distributed charge of $450\pi\ \mu\text{C}$. Calculate the force on a $75\text{-}\mu\text{C}$ charge located on the axis of the disk and 4 m from its center.

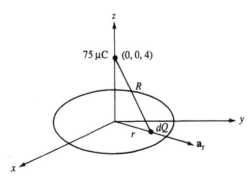

Fig. 2-4

❚ Referring to Fig. 2-4, we see that the resultant force has the direction $+\mathbf{a}_z$. Since $R^2 = r^2 + 4^2$ and

$$dQ = \frac{dS}{S}(450\pi \times 10^{-6}) = 50 \times 10^{-6} r\, dr\, d\phi$$

we obtain, by projection on the z axis,

$$F = \int_{\phi=0}^{2\pi} \int_{r=0}^{3} \frac{(75 \times 10^{-6})(50 \times 10^{-6})r\, dr\, d\phi}{4\pi(10^{-9}/36\pi)(r^2 + 16)} \cdot \frac{4}{\sqrt{r^2 + 16}} = 270\pi \frac{-1}{\sqrt{r^2 + 16}}\Big]_0^3 = 13.5\pi \text{ N}$$

2.17   Replace the disk of Problem 2.16 by a concentrated charge at the origin that gives the same **E**-field at $(0, 0, 4)$.

❚

$$13.5\pi = \frac{(75 \times 10^{-6})Q}{(10^{-9}/9)(16)} \quad \text{or} \quad Q = 320\pi\,\mu\text{C}$$

2.18   Replace the disk of Problem 2.16 by a uniformly distributed charge on the rim ($r = 3$ m) that gives the same **E**-field at $(0, 0, 4)$.

❚

$$13.5\pi = \frac{(75 \times 10^{-6})Q}{10^{-9}/9)5^2} \cdot \frac{4}{5} \quad \text{or} \quad Q = 625\pi\,\mu\text{C}$$

2.19   Write the superposition law for the force exerted on a point charge q by point charges $Q_1, Q_2, \ldots Q_N$.

❚

$$\mathbf{F} = \frac{qQ_1}{4\pi\epsilon_0 R_1^2}\mathbf{a}_{R_1} + \frac{qQ_2}{4\pi\epsilon_0 R_2^2}\mathbf{a}_{R_2} + \cdots + \frac{qQ_N}{4\pi\epsilon_0 R_N^2}\mathbf{a}_{R_N} = \frac{q}{4\pi\epsilon_0}\sum_{i=1}^{N}\frac{Q_i}{R_i^2}\mathbf{a}_{R_i} \qquad (1)$$

where R_i is the distance between q and Q_i and $\mathbf{a}_{R_i}$ is the unit vector pointing from Q_i to q.

2.20   Generalize (1) of Problem 2.19 to a continuous volume distribution of charge.

❚

$$\mathbf{F} = \frac{q}{4\pi\epsilon_0}\int_V \mathbf{a}_R \frac{\rho_v\, dV}{R^2} \qquad (1)$$

2.21   In Fig. 2-5, the segment $-a \leq y \leq a$ carries charge of uniform density ρ_ℓ. Calculate the force on the charge q.

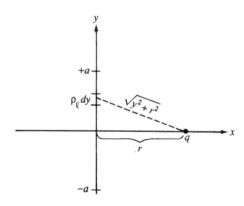

Fig. 2-5

❚ By symmetry, the integrated y component will be zero, leaving

$$F_x = 2\int_0^a \frac{\rho_\ell q}{4\pi\epsilon_0(y^2 + r^2)} \cdot \frac{r}{\sqrt{y^2 + r^2}}\, dy$$

$$= \frac{\rho_\ell q}{2\pi\epsilon_0 r}\frac{y}{\sqrt{y^2 + r^2}}\Big]_0^a = \frac{\rho_\ell q}{2\pi\epsilon_0 r\sqrt{1 + (r/a)^2}}$$

2.22   Generalize the result of Problem 2.21 to the case of an infinite line charge.

❚

$$\lim_{a \to \infty} F_x = \frac{\rho_\ell q}{2\pi\epsilon_0 r} \quad [\text{a ``}1/r\text{'' force}]$$

2.23   Four 100-μC point charges are located at $(1, 0, 0)$ m, $(0, 1, 0)$ m, $(-1, 0, 0)$ m, and $(0, -1, 0)$ m in the rectangular coordinate system of Fig. 2-6. Determine the vector force exerted on another 100-μC charge that is located at **(a)** $(0, 0, 0)$ m, **(b)** $(0, 0, 1)$ m, **(c)** $(1, 1, 0)$ m. [Use $\epsilon_0 = (10^{-9}/36\pi)$ F/m.]

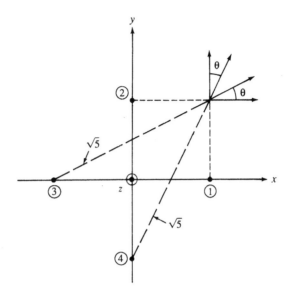

Fig. 2-6

▮ (a) $\mathbf{F} = 0$ (by symmetry)

(b) $\mathbf{F} = 4 \times \dfrac{(100 \times 10^{-6})^2}{4\pi\epsilon_0(\sqrt{2})^2} \cos 45° \, \mathbf{a}_z = \dfrac{36 \times 10^{-8} \times 10^9}{(\sqrt{2})^3} \, \mathbf{a}_z = 127.28\mathbf{a}_z \text{ N}$

(c) $\mathbf{F}_1 = \dfrac{(100 \times 10^{-6})^2}{4\pi\epsilon_0(1)^2} \, \mathbf{a}_y = 90\mathbf{a}_y \text{ N}$ $\mathbf{F}_3 = \dfrac{90}{5}(\cos\theta\,\mathbf{a}_x + \sin\theta\,\mathbf{a}_y) \text{ N}$

$\mathbf{F}_2 = 90\mathbf{a}_x \text{ N}$ $\mathbf{F}_4 = \dfrac{90}{5}(\sin\theta\,\mathbf{a}_x + \cos\theta\,\mathbf{a}_y) \text{ N}$

Now, $\cos\theta = 2/\sqrt{5}$ and $\sin\theta = 1/\sqrt{5}$. Consequently,

$$\mathbf{F} = \mathbf{F}_1 + \mathbf{F}_2 + \mathbf{F}_3 + \mathbf{F}_4 = 90\left(1 + \frac{3}{5\sqrt{5}}\right)(\mathbf{a}_x + \mathbf{a}_y) = 114.15(\mathbf{a}_x + \mathbf{a}_y) \text{ N}$$

2.24 Two positive charges, Q and $4Q$, are separated by a distance d, in air. A third charge is so placed that the entire system is in equilibrium. Determine the location, the magnitude, and the sign of the third charge.

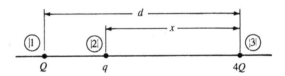

▮ The unknown charge q has to be negative and located between Q and $4Q$, as shown in Fig. 2-7. For equilibrium, $\mathbf{F}_{12} = \mathbf{F}_{32} = \mathbf{F}_{13}$. Hence,

$$\frac{Qq}{(d-x)^2} = \frac{4Qq}{x^2} = \frac{(4Q)(Q)}{d^2}$$

which yield $q = -\frac{4}{9}Q$, $x = \frac{2}{3}d$.

2.25 In Fig. 2-8, for what value of x is the system of charges in equilibrium?

Fig. 2-8

▮ For the pairwise interaction forces to be equal,

$$\frac{(18)(8)}{x^2} = \frac{(18)(72)}{30^2} = \frac{(8)(7)}{(30-x)^2}$$

whence $x = 10$ mm. Note that this system is overdetermined; thus, equilibrium is impossible for, say, $B = -8.001 \ \mu C$.

2.26 Two equal and opposite 35 μC charges are separated in air by a distance of 20 cm, and an electron ($e = 1.6 \times 10^{-19}$ C) is located midway between these charges. Calculate the force experienced by the electron.

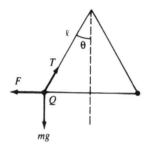

+35 μC   ⟵—— 10 cm ——⟶   e   ⟵—— 10 cm ——⟶   −35 μC   **Fig. 2-9**

▮ From Fig. 2-9, the force on e^- is

$$F_{\text{net}} = 2F = 2\frac{(35 \times 10^{-6})(1.6 \times 10^{-19})}{4\pi(10^{-9}/36\pi)(0.10)^2} = 1.0 \times 10^{-11}\,\text{N}$$

2.27 Two point charges Q, each having mass m, are suspended by identical, massless strings of length ℓ from a common point. Determine the charge Q, given the angle θ between each string and a vertical line.

Fig. 2-10

▮ Figure 2-10 shows the geometry of the problem. Isolating the left-hand charge, we have

horizontal equilibrium $F = \dfrac{Q^2}{4\pi\epsilon_0(2\ell \sin \theta)^2} = T \sin \theta$

vertical equilibrium $mg = T \cos \theta$

Elimination of T yields $Q = \pm(4\ell \sin \theta)\sqrt{\pi\epsilon_0 mg \tan \theta}$.

2.28 Electric field is produced by electrical charges and is measured by the force experienced by a unit positive test charge located in the field. Explicitly, the electric field **E** is given by

$$\mathbf{E} = \lim_{q \to 0} \frac{\mathbf{F}}{q} \qquad (1)$$

We use the limiting operation in (1) so that the test charge will not disturb the original charge distribution. Using this definition, find the **E**-field due to a point charge Q.

▮ By (1) of Problem 2.8,

$$\mathbf{E} = \frac{\mathbf{F}_Q}{q} = \frac{Q}{4\pi\epsilon_0 R^2}\mathbf{a}_R = \frac{Q}{4\pi\epsilon_0 |\mathbf{R}|^3}\mathbf{R} \qquad (2)$$

where **R** is the vector from Q to the observation point.

2.29 What is the unit of the electric field **E**?

▮ The definition (1) of Problem 2.28 implies the unit N/C. However, the equivalent unit V/m is more commonly used: as was shown in Problem 1.147, **E** is the gradient of a potential function, and the unit of electrical potential is the volt (V).

2.30 Find the **E**-field due to a volume distribution of charge.

▮ By (1) of Problem 2.20,

$$\mathbf{E} = \frac{\mathbf{F}}{q} = \frac{1}{4\pi\epsilon_0}\int_v \mathbf{a}_R \frac{\rho_v\,dV}{R^2} = \frac{1}{4\pi\epsilon_0}\int_v \mathbf{R}\frac{\rho_v\,dV}{|\mathbf{R}|^3} \qquad (1)$$

which specializes in the usual fashion to line and surface charge distributions. In (1), **R** is the vector from the source element $\rho_v\,dV$ to the observation point.

2.31 Express the electric field at (x, y, z) due to a point charge Q at (x_1, y_1, z_1).

▮ By (2) of Problem 2.28, with $\mathbf{R} = (x - x_1)\mathbf{a}_x + (y - y_1)\mathbf{a}_y + (z - z_1)\mathbf{a}_z$,

$$\mathbf{E} = \frac{Q}{4\pi\epsilon_0} \frac{(x - x_1)\mathbf{a}_x + (y - y_1)\mathbf{a}_y + (z - z_1)\mathbf{a}_z}{[(x - x_1)^2 + (y - y_1)^2 + (z - z_1)^2]^{3/2}} \qquad (1)$$

In spherical coordinates with origin at Q, (1) reduces to the familiar $\mathbf{E} = (Q/4\pi\epsilon_0 r^2)\mathbf{a}_r$.

2.32 A 25 μC charge is located at the point $(3, 4, 0)$ m. Evaluate the resulting electric field at the origin (*a*) in spherical coordinates and (*b*) in cartesian coordinates.

▮ (*a*) The vector from the charge to the origin is $\mathbf{R} = \sqrt{3^2 + 4^2}\,(-\mathbf{a}_r) = -5\mathbf{a}_r$; hence

$$\mathbf{E} = \frac{25 \times 10^{-6}}{4\pi(10^{-9}/36\pi)5^2}(-\mathbf{a}_r) = -9\mathbf{a}_r \text{ kV/m}$$

(*b*) $\mathbf{R} = -3\mathbf{a}_x - 4\mathbf{a}_y$, and

$$\mathbf{E} = \frac{25 \times 10^{-6}}{4\pi(10^{-9}/36\pi)5^2}\left(\frac{-3\mathbf{a}_x - 4\mathbf{a}_y}{5}\right) = 9\left(-\frac{3}{5}\mathbf{a}_x - \frac{4}{5}\mathbf{a}_y\right) \text{ kV/m}$$

2.33 A 2 μC charge is located at $(0, 3, 0)$ m and a 4 μC charge is at $(4, 0, 0)$ m. Determine $\mathbf{E}$ at $(0, 0, 5)$ m.

▮ $\mathbf{R}_1 = -3\mathbf{a}_y + 5\mathbf{a}_z, \quad \mathbf{R}_2 = -4\mathbf{a}_x + 5\mathbf{a}_z$

$$\mathbf{E}_1 = \frac{2 \times 10^{-6}}{4\pi(10^{-9}/36\pi)(34)}\left(\frac{-3\mathbf{a}_y + 5\mathbf{a}_z}{\sqrt{34}}\right) = -272.38\mathbf{a}_y + 453.97\mathbf{a}_z$$

$$\mathbf{E}_2 = \frac{4 \times 10^{-6}}{4\pi(10^{-9}/36\pi)(41)}\left(\frac{-4\mathbf{a}_x + 5\mathbf{a}_z}{\sqrt{41}}\right) = -548.51\mathbf{a}_x + 685.64\mathbf{a}_z$$

$$\mathbf{E} = \mathbf{E}_1 + \mathbf{E}_2 = -548.51\mathbf{a}_x - 273.38\mathbf{a}_y + 1139.61\mathbf{a}_z \quad \text{V/m}$$

2.34 An infinitely long, straight, nonconductive wire carries a uniform line charge of density 3 μC/m. Determine the electric field 2 m away from the wire.

▮ From Problem 2.22,

$$\mathbf{E} = \frac{\rho_\ell}{2\pi\epsilon_0 r}\mathbf{a}_r = \frac{3 \times 10^{-6}}{2\pi(10^{-9}/36\pi)(2)}\mathbf{a}_r = 27\mathbf{a}_r \text{ kV/m}$$

2.35 A straight nonconductive wire is parallel to the z axis and passes through the point $(3, -3, 0)$ m. The wire carries a uniform line charge of density 0.4 μC/m. Evaluate the E-field at $(-3, 0, 5)$ m.

▮ With $\mathbf{R} = (-3\mathbf{a}_x + 0\mathbf{a}_y + 5\mathbf{a}_z) - (3\mathbf{a}_x - 3\mathbf{a}_y + 5\mathbf{a}_z) = -6\mathbf{a}_x + 3\mathbf{a}_y$,

$$\mathbf{E} = \frac{\rho_\ell}{2\pi\epsilon_0}\frac{\mathbf{R}}{|\mathbf{R}|^2} = \frac{0.4 \times 10^{-6}}{2\pi(10^{-9}/36\pi)}\left(\frac{-6\mathbf{a}_x + 3\mathbf{a}_y}{45}\right) = -960\mathbf{a}_x + 480\mathbf{a}_y \quad \text{V/m}$$

2.36 Two straight nonconductive wires, parallel to the z axis, pass through points O and A, as shown in Fig. 2-11. The wires carry equal and uniform charges of density 0.4 μC/m. Determine the $\mathbf{E}$ field at point P.

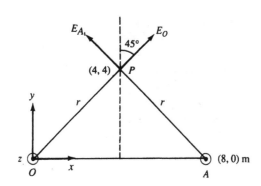

Fig. 2-11

▮ The magnitude of the E-field due to either of the two line charges is $|\mathbf{E}| = \rho_\ell/2\pi\epsilon_0 r$. The resultant E-field is then

$$\mathbf{E}_r = 2\,|\mathbf{E}|\cos 45° \,\mathbf{a}_y = 2\frac{0.4 \times 10^{-6}}{2\pi(10^{-9}/36\pi)(4\sqrt{2})}\cdot\frac{1}{\sqrt{2}}\mathbf{a}_y = 1800\mathbf{a}_y \text{ V/m}$$

2.37 The upper semicircle in Fig. 2-12 carries a line charge of constant density $+\rho_\ell$, while the lower semicircle carries $-\rho_\ell$. Determine **E** at the center.

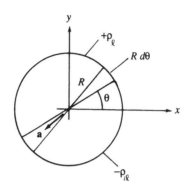

Fig. 2-12

▌ The double line element shown in Fig. 2-12 produces at the origin the field

$$d\mathbf{E} = \frac{\rho_\ell\, d\theta}{2\pi\epsilon_0 R}\,\mathbf{a}$$

By the symmetry of the charge distribution, the x component of $d\mathbf{E}$ will integrate to zero; hence

$$\mathbf{E} = (-\mathbf{a}_y)\frac{\rho_\ell}{2\pi\epsilon_0 R}\int_0^\pi \sin\theta\, d\theta = -\frac{\rho_\ell}{\pi\epsilon_0 R}\,\mathbf{a}_y$$

2.38 Obtain an expression for **E** everywhere off an infinite plane having uniform charge density ρ_s.

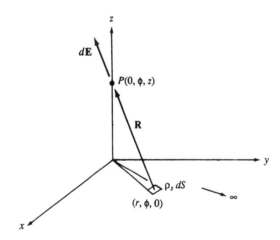

Fig. 2-13

▌ In the cylindrical coordinate system of Fig. 2-13,

$$d\mathbf{E} = \frac{\rho_s r\, dr\, d\phi}{4\pi\epsilon_0(r^2+z^2)}\left(\frac{-r\mathbf{a}_r + z\mathbf{a}_z}{\sqrt{r^2+z^2}}\right) \qquad (z>0)$$

Symmetry about the z axis results in cancellation of the radial components:

$$\mathbf{E} = \int_0^{2\pi}\int_0^\infty \frac{\rho_s rz\, dr\, d\phi}{4\pi\epsilon_0(r^2+z^2)^{3/2}}\,\mathbf{a}_z = \left(\frac{\rho_s z}{2\epsilon_0}\right)\frac{-1}{\sqrt{r^2+z^2}}\Big]_0^\infty \mathbf{a}_z = \frac{\rho_s}{2\epsilon_0}\,\mathbf{a}_z$$

For $z<0$, replace $\mathbf{a}_z$ by $-\mathbf{a}_z$. The electric field is everywhere normal to the plane of the charge and its magnitude is independent of the distance from the plane.

2.39 The plane $y=3$ m carries surface charge of density $(1/600\pi)\ \mu C/m^2$. Determine the E-field at all points.

▌ By Problem 2.38,

$$\mathbf{E} = \frac{\rho_s}{2\epsilon_0}\,\mathbf{a}_n = \begin{cases} 30\mathbf{a}_y\ V/m & y>3\ m \\ -30\mathbf{a}_y\ V/m & y<3\ m \end{cases}$$

2.40 Two infinite uniform sheets of charge, each with density ρ_s, are located at $x=\pm1$ (Fig. 2-14). Determine **E** in all regions.

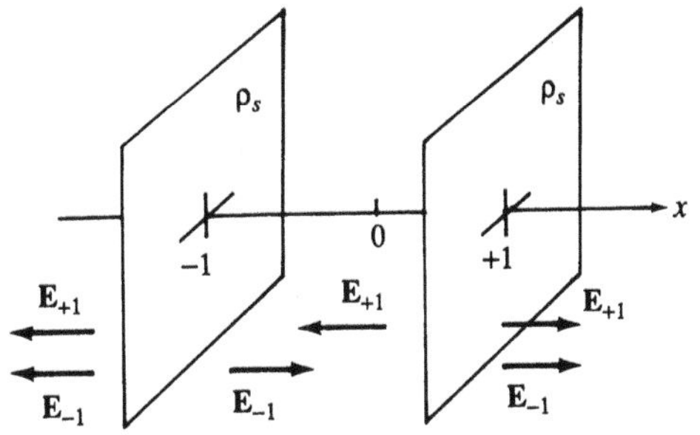

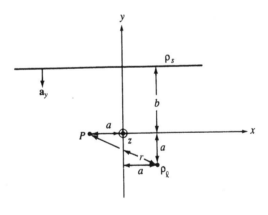

Fig. 2-14

▮ Both sheets result in **E**-fields that are directed along x, independent of the distance. Then,

$$\mathbf{E} = \mathbf{E}_{+1} + \mathbf{E}_{-1} = \begin{cases} -(\rho_s/\epsilon_0)\mathbf{a}_x & x < -1 \\ 0 & -1 < x < 1 \\ (\rho_s/\epsilon_0)\mathbf{a}_x & x > 1 \end{cases}$$

2.41 In Problem 2.40, $\mathbf{E} = \mathbf{0}$ everywhere between the two charged sheets. Does this violate the general principle that no location within an **E**-field can be a point of stable equilibrium?

▮ If a test charge anywhere in $-1 < x < 1$ is given a small displacement, it simply remains where it is put, there being no restoring force. Therefore, the principle is not violated.

2.42 Figure 2-15 shows the charged plane $y = b$ and, parallel to it, the charged line $x = -y = a$. Find the resultant **E**-field at point P.

Fig. 2-15

▮ At P, the plane produces the field $\mathbf{E}_s = (\rho_s/2\epsilon_0)(-\mathbf{a}_y)$ (Problem 2.38), and the line produces the field

$$\mathbf{E}_\ell = \frac{\rho_\ell}{2\pi\epsilon_0 r}\mathbf{a}_r = \frac{\rho_\ell}{2\pi\epsilon_0\sqrt{(2a)^2 + a^2}}\left(\frac{-2a\mathbf{a}_x + a\mathbf{a}_y}{\sqrt{(2a)^2 + a^2}}\right) = \frac{\rho_\ell}{10\pi\epsilon_0 a}(-2\mathbf{a}_x + \mathbf{a}_y)$$

Then

$$\mathbf{E} = \mathbf{E}_s + \mathbf{E}_\ell = -\frac{\rho_\ell}{5\pi\epsilon_0 a}\mathbf{a}_x - \frac{1}{2\epsilon_0}\left(\rho_s - \frac{\rho_\ell}{5\pi a}\right)\mathbf{a}_y$$

2.43 Specialize Problem 2.42 for the following numerical data: $\rho_s = (1/3\pi)$ nC/m^3; $\rho_\ell = (25/9)$ nC/m, located at $(3, -3)$ m; and P located at $(-3, 0, 0)$ m.

▮ $$\mathbf{E} = -\frac{(25/9) \times 10^{-9}}{5\pi(10^{-9}/36\pi)(3)}\mathbf{a}_x - \frac{1}{2(10^{-9}/36\pi)}\left[\frac{10^{-9}}{3\pi} - \frac{(25/9) \times 10^{-9}}{5\pi(3)}\right]\mathbf{a}_y = -\frac{20}{3}\mathbf{a}_x - \frac{8}{3}\mathbf{a}_y \quad \text{V/m}$$

2.44 For the surface and line charge distributions shown in Fig. 2-16, determine the **E**-field at $P(0, 4\text{ m}, 0)$. The line charge is parallel to the z axis.

▮ By superposition we have

$$\mathbf{E}_p = \frac{\rho_{s1}}{2\epsilon_0}\mathbf{a}_{n1} + \frac{\rho_{s2}}{2\epsilon_0}\mathbf{a}_{n2} + \frac{\rho_\ell}{2\pi\epsilon_0 r}\mathbf{a}_r$$

$$= \frac{(1/2\pi)10^{-9}}{2(10^{-9}/36\pi)}\mathbf{a}_y + \frac{(-1/2\pi)10^{-9}}{2(10^{-9}/36\pi)}(-\mathbf{a}_y) + \frac{(-2)10^{-9}}{2\pi(10^{-9}/36\pi)8}(-\mathbf{a}_y) = 22.5\mathbf{a}_y \quad \text{V/m}$$

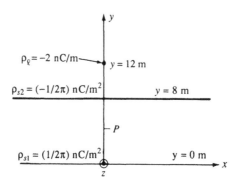

Fig. 2-16

2.45 A charge of 160 nC is distributed uniformly along the z axis between $z = \pm 4$ m. Evaluate $\mathbf{E}$ at $(3, 0, 0)$ m.

By Problem 2.21 (with the coordinates renamed),

$$\mathbf{E} = \frac{\rho_\ell}{2\pi\epsilon_0 x \sqrt{1 + (x/a)^2}} \mathbf{a}_x = \frac{(160/8) \times 10^{-9}}{2\pi(10^{-9}/36\pi)(3)\sqrt{1 + (3/4)^2}} \mathbf{a}_x = 96\mathbf{a}_x \text{ V/m}$$

2.46 Find $\mathbf{E}$ at P in Fig. 2-17.

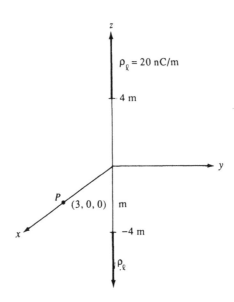

Fig. 2-17

Note that ρ_ℓ has the same numerical value as in Problem 2.45. If the entire z axis were charged, the field at P would be

$$\mathbf{E}'_P = \frac{\rho_\ell}{2\pi\epsilon_0 x} \mathbf{a}_x = \frac{20 \times 10^{-9}}{2\pi(10^{-9}/36\pi)(3)} \mathbf{a}_x = 120\mathbf{a}_x \text{ V/m}$$

Subtracting from this the field due to $-4 < z < +4$ m, we obtain $\mathbf{E} = 120\mathbf{a}_x - 96\mathbf{a}_x = 24\mathbf{a}_x$ V/m.

2.47 Figure 2-18 shows a "kite" shape in the xy plane, having line charges parallel to the z axis through three of its corners. Show that the electric field at the fourth corner is zero.

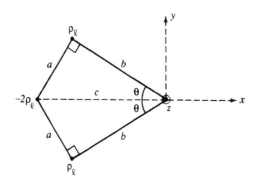

Fig. 2-18

$$\mathbf{E} = \left[\frac{\rho_\ell}{2\pi\epsilon_0 b}(2\cos\theta) + \frac{(-2\rho_\ell)}{2\pi\epsilon_0 c} \right]\mathbf{a}_x = \frac{\rho_\ell}{\pi\epsilon_0 b}\left(\cos\theta - \frac{b}{c}\right)\mathbf{a}_x = \mathbf{0}$$

since $\cos\theta = b/c$.

2.48 A circular disk of radius a carries a surface charge of uniform density ρ_s (C/m^2). Determine the **E**-field on the axis of the disk at a distance z from the center of the disk.

▌ This is just like Problem 2.38, except that the r-integration is now from 0 to a; thus, $\mathbf{E} = E_z\mathbf{a}_z$, where

$$E_z = \frac{\rho_s z}{z\epsilon_0}\frac{-1}{\sqrt{r^2+z^2}}\Bigg]_0^a = \frac{\rho_s}{2\epsilon_0}\left(1 - \frac{z}{\sqrt{a^2+z^2}}\right) = \frac{\rho_s}{2\epsilon_0}(1 - \cos\alpha) \quad \text{(V/m)}$$

Here, α is the half-angle subtended at the observation point by the disk.

2.49 Use Problem 2.48 to establish the following basic result: At any point outside a uniformly charged solid sphere, the **E**-field is as though produced by the total charge concentrated at the center.

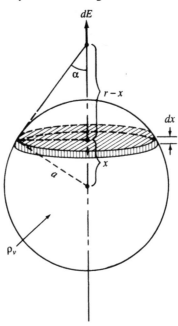

Fig. 2-19

▌ Section the sphere into disks of thickness dx (Fig. 2-19). The effective surface charge density on a disk is $\rho_s = \rho_v\,dx$, and, from the right triangles,

$$\cos\alpha = \frac{r-x}{\sqrt{a^2+r^2-2rx}}$$

Hence, by Problem 2.48, we obtain

$$\begin{aligned}
E &= \frac{\rho_v}{2\epsilon_0}\int_{-a}^{a}\left(1 - \frac{r-x}{\sqrt{a^2+r^2-2rx}}\right)dx \\
&= \frac{\rho_v}{2\epsilon_0}\left[\int_{-a}^{a} 1\,dx - \int_{-a}^{a}\frac{r\,dx}{\sqrt{a^2+r^2-2rx}} + \int_{-a}^{a}\frac{x\,dx}{\sqrt{a^2+r^2-2rx}}\right] \\
&= \frac{\rho_v}{2\epsilon_0}\left[2a - 2a + \frac{2a^3}{3r^2}\right] = \frac{\rho_v(\frac{4}{3}\pi a^3)}{4\pi\epsilon_0 r^2} = \frac{Q_{\text{tot}}}{4\pi\epsilon_0 r^2}
\end{aligned}$$

wherein the integrals were evaluated under the condition $r > a$.

To solve this very problem (for the gravitational field) Newton invented the calculus! A simpler solution is given in Problem 2.76.

2.50 Extend the result of Problem 2.49 to a uniformly charged spherical shell (Fig. 2-20) of arbitrary thickness.

▌ By superposition and Problem 2.49,

$$E_{s+\Delta s} = E_s + E_{\Delta s}$$
$$\frac{Q_s + Q_{\Delta s}}{4\pi\epsilon_0 r^2} = \frac{Q_s}{4\pi\epsilon_0 r^2} + E_{\Delta s}$$
$$\frac{Q_{\Delta s}}{4\pi\epsilon_0 r^2} = E_{\Delta s}$$

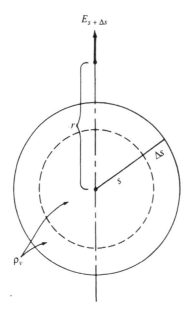

Fig. 2-20

2.51   Extend the result of Problem 2.49 to an arbitrary spherically symmetric distribution of charge.

❚ Decompose the distribution into spherical shells of infinitesimal thickness. In each shell the charge density is constant, so Problem 2.50 applies. The desired result now obtains by integration (superposition).
   Problems 2.48 through 2.51 illustrate the power of the superposition principle in electromagnetics.

2.52   A charged square of side 4 m is oriented as in Fig. 2-21. Given $\rho_s = 2(x^2 + y^2 + z^2)^{3/2}$ nC/m^2, find **E** at O.

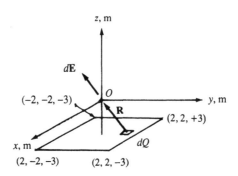

Fig. 2-21

❚

$$\mathbf{R} = -x\mathbf{a}_x - y\mathbf{a}_y + 3\mathbf{a}_z \quad \text{(m)} \qquad dQ = \rho_s\, dS = (2 \times 10^{-9})\,|\mathbf{R}|^3\, dS \quad \text{(C)}$$

and so

$$d\mathbf{E} = \frac{dQ}{4\pi\epsilon_0 |\mathbf{R}|^3}\mathbf{R} = \frac{10^{-9}\, dS}{2\pi\epsilon_0}(-x\mathbf{a}_x - y\mathbf{a}_y + 3\mathbf{a}_z) \quad \text{(V/m)}$$

As a result of symmetry, only the z component of **E** exists:

$$\mathbf{E} = \left(\frac{3 \times 10^{-9}}{2\pi\epsilon_0}\mathbf{a}_z\, \frac{\text{V}}{\text{m}^3}\right)(4\text{ m})^2 = 864\mathbf{a}_z \text{ V/m}$$

2.53   An infinitely long, line charge of uniform density ρ_ℓ produces an electric field. Show that the divergence of this field is zero everywhere (except at the line).

❚ For a line charge $\mathbf{E} = (\rho_\ell/2\pi\epsilon_0 r)\mathbf{a}_r$ so that, in cylindrical coordinates,

$$\text{div}\,\mathbf{E} = \frac{1}{r}\frac{\partial}{\partial r}(rE_r) = \frac{1}{r}\frac{\partial}{\partial r}(\text{const.}) = 0 \qquad (r > 0)$$

2.54   Two point charges $A = 20$ nC and $B = 10$ nC are separated from each other by a distance of 25 cm in free space. Calculate the electric field at a point P that is 15 cm away from A and 20 cm from B.

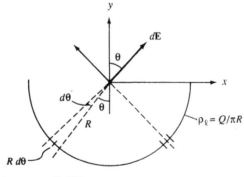

Fig. 2-22

❚ $\mathbf{E}_A$ and $\mathbf{E}_B$ are perpendicular (3-4-5 triangle; Fig. 2-22).

$$|\mathbf{E}_A| = \frac{20 \times 10^{-9}}{4\pi(10^{-9}/36\pi)(0.15)^2} = 8 \text{ kV/m}$$

$$|\mathbf{E}_B| = \frac{10 \times 10^{-9}}{4\pi(10^{-9}/36\pi)(0.20)^2} = 2.25 \text{ kV/m}$$

$$|\mathbf{E}| = \sqrt{(8)^2 + (2.25)^2} = 8.31 \text{ kV/m}$$

$$\theta = \tan^{-1}\frac{2.25}{8} \approx 15.7°$$

2.55 Charge is uniformly distributed throughout a sphere of radius a at unit density. A redistribution of the charge results in the density function $\rho_v(r) = k(3 - r^2/a^2)$. Evaluate k.

❚
$$Q = (1)\tfrac{4}{3}\pi a^3 = \int_0^a k\left(3 - \frac{r^2}{a^2}\right) 4\pi r^2 \, dr = 4\pi k\left[r^3 - \frac{r^5}{5a^2}\right]_0^a = \frac{16\pi k a^3}{5}$$

Hence $k = 5/12$.

2.56 Determine the **E**-field at a point $r = 2a$ due to the redistributed charge in Problem 2.55.

❚ For the symmetric distribution, Problem 2.51 gives
$$\mathbf{E} = \frac{Q}{4\pi\epsilon_0 r^2}\,\mathbf{a}_r = \frac{4\pi a^3/3}{4\pi\epsilon_0(2a)^2}\,\mathbf{a}_r = \frac{a}{12\epsilon_0}\,\mathbf{a}_r$$

2.57 A charge Q is uniformly distributed over a wire in the form of a semicircle of radius R, as shown in Fig. 2-23. Determine the electric field at the origin.

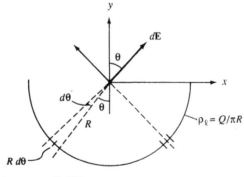

Fig. 2-23

❚ The **E** field due to a charge element $\rho_\ell R \, d\theta$ is
$$d\mathbf{E} = \frac{(Q/\pi R)R \, d\theta}{4\pi\epsilon_0 R^2}\,\mathbf{a}_r = \frac{Q \, d\theta}{4\pi^2\epsilon_0 R^2}(\cos\theta\,\mathbf{a}_y + \sin\theta\,\mathbf{a}_x)$$

By symmetry, x components of the field cancel. Thus
$$\mathbf{E} = \frac{Q}{4\pi^2\epsilon_0 R^2}\,\mathbf{a}_y \int_{-\pi/2}^{\pi/2} \cos\theta \, d\theta = \frac{Q}{2\pi^2\epsilon_0 R^2}\,\mathbf{a}_y$$

2.58 Two infinitely long straight wires running in the x direction are separated by a distance D and lie in the xy plane. One wire holds a positive charge per unit length $+\alpha/D$ (C/m) and the other a negative charge $-\alpha/D$ (C/m), as shown in Fig. 2-24. Derive an expression for the electric field at point P.

❚ At P, as a result of the two line charges,
$$E_y = \frac{\alpha/D}{2\pi\epsilon_0 y} - \frac{\alpha/D}{2\pi\epsilon_0(y + D)} = \frac{\alpha}{2\pi\epsilon_0 y(y + D)} \text{ V/m}$$

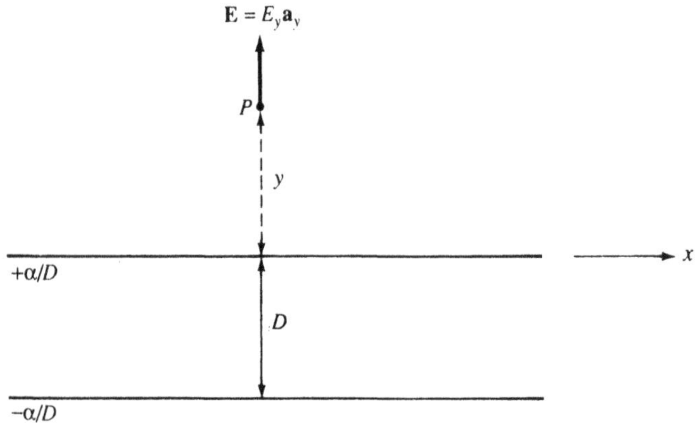

$$\mathbf{E} = E_y \mathbf{a}_y$$

Fig. 2-24

2.59 For the configuration of Fig. 2-24, what happens when the positive wire is held fast and the negative wire is brought up to it?

❚ As $D \to 0$, $E_y \to \alpha / 2\pi\epsilon_0 y^2$. From P, the dipolar line "looks like" a point charge of magnitude 2α located at the origin.

2.60 Within a sphere of radius a the charge density is $\rho_v = 1 - (r/a)^3$; determine the **E**-field at $r = 4a$.

❚

$$Q = \int_0^a \left(1 - \frac{r^3}{a^3}\right) 4\pi r^2 \, dr = \frac{2}{3}\pi a^3$$

By Problem 2.51, at $r = 4a$,

$$E_r = \frac{Q}{4\pi\epsilon_0 r^2} = \frac{2\pi a^3/3}{4\pi\epsilon_0(16a^2)} = \frac{a}{96\epsilon_0}$$

2.61 A hollow spherical shell of radius a is charged to uniform density ρ_s. Following Problem 2.49, show that the **E**-field vanishes everywhere inside the shell.

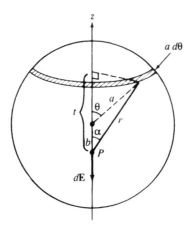

Fig. 2-25

❚ In Fig. 2-25 the arbitrary interior point P, a distance $b < a$ from the center of the shell, and the center determine the z axis of a spherical coordinate system. Let $t = b + a \cos \theta$ be the distance from P to the plane of a ring-shaped element of charge. By symmetry, the contribution, $d\mathbf{E}$, of the ring to the field at P must lie along the z axis; we have

$$dE_z = -\frac{\rho_s(2\pi a \sin \theta)(a \, d\theta)}{4\pi\epsilon_0 r^2} \cos \alpha$$

Now, by trigonometry,

$$r^2 = a^2 + b^2 + 2ab \cos \theta = a^2 - b^2 + 2bt$$

$$\cos \alpha = \frac{t}{r} = \frac{t}{(a^2 - b^2 + 2bt)^{1/2}}$$

and $dt = -a \sin\theta \, d\theta$. Hence (dropping multiplicative constants as we go),

$$E_z = \int_{b-a}^{b+a} \frac{t \, dt}{(a^2 - b^2 + 2bt)^{3/2}} = * \frac{a^2 - b^2 + bt}{(a^2 - b^2 + 2bt)^{1/2}} \Bigg]_{b-a}^{b+a}$$

$$= \frac{a(a+b)}{a+b} - \frac{a(a-b)}{a-b} = 0$$

2.62 An electric field is given by $\mathbf{E} = xy^2\mathbf{a}_x - xz^2\mathbf{a}_y + xyz\mathbf{a}_z$. Calculate the x component of its curl at $(2, 1, 3)$.

▌ In rectangular coordinates,

$$(\text{curl } \mathbf{E})_x = (\nabla \times \mathbf{E})_x = \frac{\partial E_z}{\partial y} - \frac{\partial E_y}{\partial z} = xz - (-2zx) = 3xz = 3(2)(3) = 18$$

2.63 Evaluate the divergence of the E-field of Problem 2.62 at $(1, 2, 3)$.

▌ In rectangular coordinates,

$$\text{div } \mathbf{E} = \nabla \cdot \mathbf{E} = \frac{\partial}{\partial x}(xy^2) + \frac{\partial}{\partial y}(-xz^2) + \frac{\partial}{\partial z}(xyz) = y^2 + xy = (2)^2 + (1)(2) = 6$$

2.64 Three infinitely long charged lines run parallel to the z axis, as shown in Fig. 2-26. The lines have uniform charge densities as shown. Determine the E-field at P.

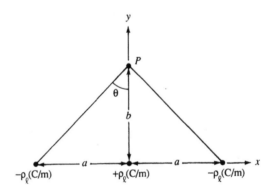

Fig. 2-26

▌ For an infinitely long charged line we have

$$\mathbf{E} = \frac{\rho_\ell}{2\pi\epsilon_0 r}\mathbf{a}_r \quad (\text{V/m})$$

Hence, by superposition of components,

$$E_y = \frac{1}{2\pi\epsilon_0}\left(-\frac{\rho_\ell \cos\theta}{\sqrt{a^2 + b^2}} - \frac{\rho_\ell \cos\theta}{\sqrt{a^2 + b^2}} + \frac{\rho_\ell}{b}\right) \quad (\text{V/m})$$

Substituting $\cos\theta = b/\sqrt{a^2 + b^2}$ yields

$$E_y = \frac{\rho_\ell}{2\pi\epsilon_0}\left(\frac{1}{b} - \frac{2b}{a^2 + b^2}\right) = \frac{\rho_\ell}{2\pi\epsilon_0 b}\left(\frac{a^2 - b^2}{a^2 + b^2}\right) \quad (\text{V/m}) \tag{1}$$

2.65 Show that, in Problem 2.64, point P "sees" a single line with charge density $-\rho_\ell \cos\alpha$, where $\alpha = 2\theta$ is the angle subtended at P.

▌ Substitute $\cos\alpha = \cos^2\theta - \sin^2\theta = (b^2 - a^2)/(a^2 + b^2)$ in (1).

2.66 Verify that the field of Problems 2.64 and 2.65 possesses the expected limiting values on the y axis.

▌ Near the origin ($\alpha \to \pi$), the effective charge density becomes $-\rho_\ell \cos\pi = +\rho_\ell$ (the charge density on the central line). At large distances ($x \to 0$), it becomes $-\rho_\ell \cos\theta = -\rho_\ell$ (the net charge density of the three-line system).

---

* Integrate by parts. All of this, of course, is unnecessary with Gauss' law.

2.67   Evaluate the divergence of $\mathbf{E} = r^2\mathbf{a}_r - r^2\sin\theta\,\mathbf{a}_\theta + 10\mathbf{a}_\phi$ at $(2, \pi, 0)$.

▮ In spherical coordinates (see Problem 1.108),

$$\text{div } \mathbf{E} = \frac{1}{r^2}\frac{\partial}{\partial r}(r^2 E_r) + \frac{1}{r\sin\theta}\frac{\partial}{\partial\theta}(E_\theta\sin\theta) + \frac{1}{r\sin\theta}\frac{\partial E_\phi}{\partial\phi}$$

$$= \frac{1}{r^2}\frac{\partial}{\partial r}(r^4) + \frac{1}{r\sin\theta}\frac{\partial}{\partial\theta}(-r^2\sin^2\theta) + \frac{1}{r\sin\theta}\frac{\partial}{\partial\phi}(10) \quad (10)$$

$$= 4r - 2r\cos\theta = 4(2) - 2(2)\cos\pi = 12$$

2.68   A *dipole* consists of a pair of equal and opposite charges $(\pm Q)$ separated by a distance $2a$, as shown in Fig. 2-27. Find the E-field at P.

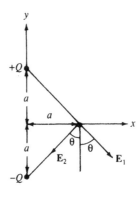

Fig. 2-27

▮
$$|\mathbf{E}_1| = |\mathbf{E}_2| = \frac{Q}{4\pi\epsilon_0(a^2 + a^2)} \qquad \text{and} \qquad \cos\theta = \frac{1}{\sqrt{2}}$$

$$\text{Resultant } \mathbf{E} = -(|\mathbf{E}_1| + |\mathbf{E}_2|)\cos\theta\,\mathbf{a}_y = \frac{-Q}{(4\sqrt{2})\pi\epsilon_0 a^2}\mathbf{a}_y$$

2.69   *Electric flux density* $\mathbf{D}$ (C/m^2) is related to the electric field $\mathbf{E}$ (V/m) in free space by $\mathbf{D} = \epsilon_0\mathbf{E}$, with ϵ_0 in farads per meter. At the surface of a cylinder of radius a, the E-field is given by $\mathbf{E} = a\sin\phi\,\mathbf{a}_r$. Determine the total outward flux ψ for the cylinder, the axial length of the cylinder being ℓ.

▮ Since $\mathbf{D}$ is radial and $\mathbf{D}(a, \phi + \pi, z) = \mathbf{D}(a, \phi, z)$ (both $\sin\phi$ and $\mathbf{a}_r$ change sign),

$$\psi = \int_S \mathbf{D}\cdot d\mathbf{S} = 0$$

2.70   Evaluate the integral $\int \mathbf{E}\cdot d\mathbf{S}$ over a closed surface S, where the E-field is due to a point charge Q inside the surface. Show that the result is consistent with the units for $\mathbf{D}$ given Problem 2.69.

▮ If $\mathbf{R}$ is the vector from Q to the surface element $d\mathbf{S}$,

$$\mathbf{E}\cdot d\mathbf{S} = \frac{Q}{4\pi\epsilon_0 R^2}\mathbf{a}_R \cdot d\mathbf{S}$$

But, from solid geometry, the solid angle at Q subtended by dS is given by

$$d\Omega = \frac{dS\cos\overset{\mathbf{a}_R}{<}d\mathbf{S}}{R^2} = \frac{\mathbf{a}_R\cdot d\mathbf{S}}{R^2}$$

Hence,

$$\int_S \mathbf{E}\cdot d\mathbf{S} = \frac{Q}{4\pi\epsilon_0}\int_S d\Omega = \frac{Q}{4\pi\epsilon_0}\Omega = \frac{Q}{\epsilon_0} \quad (1)$$

since the total solid angle Ω at Q is 4π (steradians). From *(1)*, $\int \mathbf{D}\cdot d\mathbf{S} = Q$, so that D must have the dimensions of charge per area.

2.71   Repeat Problem 2.70 if Q is *outside* the surface.

▮ The flux of $\mathbf{E}$ is still given by $(Q/4\pi\epsilon_0)\,\Omega$, but now $\Omega = 0$. [See Fig. 2-28: The projections of S_1 and S_2 on the sphere of radius R (area A) are equal in magnitude but opposite in sign; hence $\Omega = (A/R^2) + (-A/R^2) = 0$.]

2.72   Generalize the result of Problems 2.70 and 2.71 to an arbitrary (volume) distribution of charge ρ and state the conclusion in words.

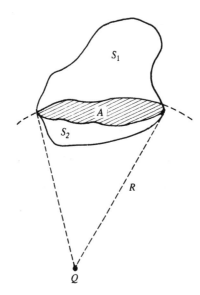

Fig. 2-28

∎ Superposition of the results for point charges $\rho\, dV$ yields

$$\int_S \mathbf{D} \cdot d\mathbf{S} = \int_V \rho\, dV = Q \tag{1}$$

In words: total *outward* electric flux from a *closed* surface equals the total electric charge *within* the surface. This statement is known as *Gauss' law* (in integral form).

2.73 Express *Gauss' law in point form.*

∎ By Problem 1.107, at a fixed point P,

$$\operatorname{div} \mathbf{D} = \lim_{V \to 0} \frac{\displaystyle\int_S \mathbf{D} \cdot d\mathbf{S}}{V} = \lim_{V \to 0} \left(\frac{1}{V} \int_V \rho\, dV \right) = \rho$$

2.74 Use Gauss' law to show that no charge can be present in a uniform electric field.

∎ At each point, $\operatorname{div} \mathbf{D} = 0$ (uniform field); hence $\rho = 0$ (no charge).

2.75 Two infinite line charges, one having line charge density ρ_ℓ and the other $-\rho_\ell$, are separated by a distance a. Use Gauss' law to find the electric intensity at point P of Fig. 2-29.

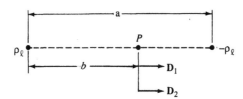

Fig. 2-29

∎ By Gauss' law, for cylindrical surfaces of unit height going through P, $2\pi b D_1 = \rho_\ell$ and $-2\pi(a-b)D_2 = -\rho_\ell$, whence

$$D_1 = \frac{\rho_\ell}{2\pi b} \quad \text{and} \quad D_2 = \frac{\rho_\ell}{2\pi(a-b)}$$

Thus

$$D = \frac{\rho_\ell}{2\pi} \left(\frac{1}{b} + \frac{1}{a-b} \right) = \frac{\rho_\ell a}{2\pi b(a-b)} \quad \text{and} \quad E = \frac{D}{\epsilon_0} = \frac{a\rho_\ell}{2\pi\epsilon_0 b(a-b)}$$

2.76 Charge is uniformly distributed throughout a spherical volume of radius R with density ρ_v. Determine the electric field intensity at all points. Obtain your result by using Gauss' law.

▮ By symmetry, the field at distance r from the center is purely radial.

$$D(4\pi r^2) = \begin{cases} \frac{4}{3}\pi r^3 \rho_v & r \leq R \\ \frac{4}{3}\pi R^3 \rho_v & r > R \end{cases}$$

or

$$E = \begin{cases} \dfrac{\rho_v r}{3\epsilon_0} & r \leq R \\ \dfrac{\rho_v R^3}{3\epsilon_0 r^2} & r > R \end{cases}$$

Note how Gauss' law gets around the lengthy integrations of Problems 2.49 and 2.61.

2.77 What is the total outward electric flux from the cube $0 < x, y, z < 1$ m, containing a volume charge density $\rho_v = 16xyz \ \mu C/m^3$?

▮ By Gauss' law,

$$\psi_e = Q = \int_V \rho \, dV = 16 \int_0^1 \int_0^1 \int_0^1 xyz \, dx \, dy \, dz = 2 \ \mu C$$

2.78 Charge is uniformly distributed (density ρ_s) over an infinite (in the x-direction) strip of width w. By direct integration, determine the electric field at a point P which is on a line perpendicular to the strip and at a distance z from the center of the strip.

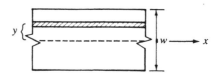

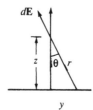

Fig. 2-30

▮ Decompose the strip into infinite line charges (Fig. 2-30). The effective linear charge density is $\rho_\ell = \rho_s \, dy$; thus $|d\mathbf{E}| = \rho_\ell/2\pi\epsilon_0 r = (\rho_s \cos\theta \, dy)/2\pi\epsilon_0 z$ and $(y = z \tan\theta)$

$$E_z = \frac{\rho_s}{2\pi\epsilon_0 z} \int_{-w/2}^{w/2} \cos^2\theta \, dy = \frac{\rho_s}{2\pi\epsilon_0} \int_{-\tan^{-1}(w/2z)}^{+\tan^{-1}(w/2z)} d\theta = \frac{\rho_s}{\pi\epsilon_0} \tan^{-1}\frac{w}{2z}$$

2.79 Show that Gauss' law is equivalent to the divergence theorem (Problem 1.112).

▮ Substitute $\rho = \operatorname{div} \mathbf{D}$ (Problem 2.73) into (1) of Problem 2.72:

$$\int_S \mathbf{D} \cdot d\mathbf{S} = \int_V (\operatorname{div} \mathbf{D}) \, dV$$

which is the divergence theorem for $\mathbf{D}$. A sufficiently regular vector field $\mathbf{F}$ can be interpreted as an electric flux density $\mathbf{D}$.

2.80 A volume charge density $\rho_s = k/r$ $(r \neq 0, k = \text{const.})$ exists within a sphere of radius a. Determine the magnitude of a point charge placed at the origin which will produce the same electric field at $r > a$.

▮ By Gauss' law,

$$\int \mathbf{D} \cdot d\mathbf{S} = \int \rho_v \, dV \qquad \text{or} \qquad D_r(4\pi r^2) = \int_0^a \frac{k}{r} (4\pi r^2) \, dr = 2\pi k a^2$$

Thus

$$D_r = \frac{2\pi k a^2}{4\pi r^2} \qquad \text{or} \qquad Q_{\text{equiv}} = 2\pi k a^2$$

2.81 A circular flat ring of inner radius 1 m and outer radius 2 m has surface charge density $\rho_s = 100/r$ $(\mu C/m^2)$. Determine the resulting $\mathbf{E}$-field on the axis of the ring, 10 m away from the center.

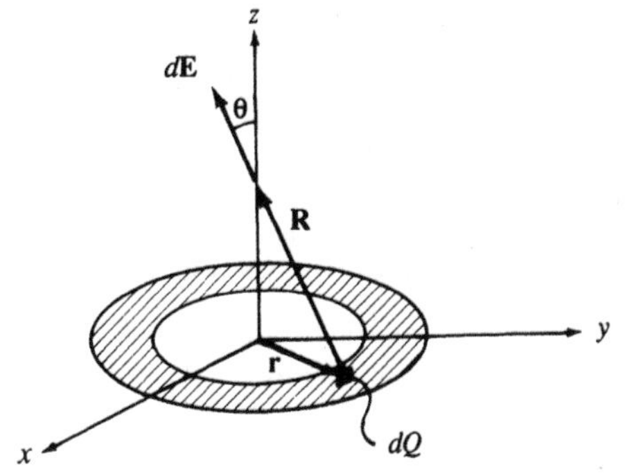

Fig. 2-31

I From Fig. 2-31, by symmetry, only the z component of the **E**-field exists along the z axis. Now

$$dQ = \rho_s r\, dr\, d\phi \qquad \cos\theta = \frac{z}{R} \qquad R = \sqrt{r^2 + z^2}$$

whence

$$dE_z = \frac{dQ}{4\pi\epsilon_0 R^2}\cos\theta = 9 \times 10^5 z \frac{dr\, d\phi}{(r^2 + z^2)^{3/2}} \quad \text{(V/m)}$$

and, at $z = 10$ m,

$$E_z = (9 \times 10^6)\int_{\phi=0}^{\pi}\int_{r=1}^{2}\frac{dr\, d\phi}{(r^2 + 100)^{3/2}} = (18\pi \times 10^6)\frac{r}{100\sqrt{r^2 + 100}}\Bigg]_1^2 = 54.63 \text{ kV/m}$$

2.82 A 100 μC point charge is located at the origin of a cartesian coordinate system. Determine the electric flux passing through the square $|x| < 1$ m, $|y| < 1$ m, $z = 2$ m.

I The square may be taken as one face of a 2-m cube enclosing the point charge. By Gauss' law and symmetry,

$$\text{flux through square} = \tfrac{1}{6}\,(\text{flux through cube}) = \tfrac{1}{6}Q = 16.67 \ \mu\text{C}$$

2.83 Rework Problem 2.62 for the hemisphere $r = 78$ m, $0 \le \theta \le \pi/2$.

I $\qquad\qquad$ flux through hemisphere $= \tfrac{1}{2}\,(\text{flux through sphere } r = 78 \text{ m}) = \tfrac{1}{2}Q = 50 \ \mu\text{C}$

2.84 Repeat Problem 2.82 for the spherical band $r = 11$ m, $\pi/4 \le \theta \le \pi/3$.

I This time we have to integrate.

$$\psi_e = \int \mathbf{D}\cdot d\mathbf{S} = \int_{\theta=\pi/4}^{\pi/3}\int_{\phi=0}^{2\pi}\frac{Q}{4\pi r^2}r^2\sin\theta\, d\phi\, d\theta = \frac{Q}{2}\int_{\pi/4}^{\pi/3}\sin\theta\, d\theta$$

$$= \frac{Q}{2}\left(\cos\frac{\pi}{4} - \cos\frac{\pi}{3}\right) = \frac{100}{2}\left(\frac{1}{\sqrt{2}} - \frac{1}{2}\right) = 10.355 \ \mu\text{C}$$

2.85 A hollow sphere of radius R has a surface charge of constant density ρ_s. Without recourse to Gauss' law, determine the resulting **E**-field at a distance $r > R$ from the center of the sphere.

I By Problem 2.50,

$$\mathbf{E} = \frac{4\pi R^2 \rho_s}{4\pi\epsilon_0 r^2}\mathbf{a}_r = \frac{\rho_s}{\epsilon_0}\left(\frac{R}{r}\right)^2\mathbf{a}_r$$

Notice that E_r jumps from 0 just inside the sphere (by Problem 2.61) to ρ_s/ϵ_0 just outside the sphere. Thus the discontinuity in $D_r = D_n$ is precisely ρ_s, in accord with the general result of Problem 2.167.

2.86 Show that the results of Problems 2.38 and 2.78 are consistent.

I $$\lim_{w\to\infty}\left(\frac{\rho_s}{\pi\epsilon_0}\tan^{-1}\frac{w}{2z}\right) = \frac{\rho_s}{\pi\epsilon_0}\frac{\pi}{2} = \frac{\rho_s}{2\epsilon_0}$$

2.87 Consider a spherical *conducting* shell with inner radius a and outer radius b and assume that the shell is initially uncharged and a point charge $+Q$ is placed at the center of the shell. Show by Gauss' law that a charge $-Q$ must be induced on the interior of the shell.

▮ At $r = a + \epsilon$ (inside the conductor), $\mathbf{D} = \mathbf{0}$ and so

$$\int_{r=a+\epsilon} \mathbf{D} \cdot d\mathbf{S} = 0 = Q_{net}$$

To make the net charge vanish, additional charge in the amount $-Q$ must be present inside $r = a + \epsilon$; the only place it can be is the inner surface $r = a$.

2.88 Refer to Problem 2.87. Is charge also induced on the outer surface $r = b$? If so, how much?

▮ Yes; overall electrical neutrality of the shell requires that $+Q$ be induced on the outer surface. At an external point the **E**-field is exactly as if the conductor were absent.

2.89 A sphere of 5-m radius contains an electric charge whose density is $\rho = 1.2(5 - 2r)$ $(\mu C/m^3)$. Evaluate the electric flux density 100 m away from the center of the sphere.

▮ Total charge within the sphere is

$$Q = \int \rho \, dV = \int_0^5 (1.2 \times 10^{-6})(5 - 2r) \, 4\pi r^2 \, dr = -5\pi \times 10^{-4} \, \text{C}$$

By Gauss' law,

$$D_r(4\pi r^2) = Q \qquad \text{or} \qquad D_r = \frac{Q}{4\pi r^2} = \frac{-5\pi \times 10^{-4}}{4\pi(100)^2} = -12.5 \, \text{nC/m}^2$$

2.90 The electrical charge density within a sphere of radius a is given by $\rho = kr^2/a^2$, where k is a constant. Determine the resulting electric field everywhere.

▮ By symmetry the **E**-field is everywhere radial. For $r < a$ (within the sphere),

$$Q = \int_0^r \frac{k\lambda^2}{a^5} 4\pi\lambda^2 \, d\lambda = \frac{4\pi k r^5}{5a^2}$$

and, for $r \geq a$, $Q = 4\pi k/5$. Then, by Gauss' law (or Problems 2.51 and 2.61),

$$E_r = \frac{Q}{4\pi\epsilon_0 r^2} = \begin{cases} kr^3/5\epsilon_0 a^5 & r < a \\ k/5\epsilon_0 r^2 & r \geq a \end{cases}$$

2.91 A continuous, spherically symmetric distribution of electric flux density is given by

$$\mathbf{D} = \begin{cases} (5r^2/4)\mathbf{a}_r & r \leq 2 \\ (20/r^2)\mathbf{a}_r & r > 2 \end{cases}$$

Find the charge density in each region.

▮ By Gauss' law in point form,

$$\rho = \text{div } \mathbf{D} = \frac{1}{r^2} \frac{\partial}{\partial r} (r^2 D_r) = \begin{cases} 5r & r < 2 \\ 0 & r > 2 \end{cases}$$

Note that the charge distribution is spherically symmetric, but is not continuous.

2.92 Determine the total charge contained within a cube of side L, oriented as in Fig. 2-32, if the electric field within and on the cube is given by

$$\mathbf{E} = \frac{1}{L^4}(\alpha x^2 yz \mathbf{a}_x + \beta xy^2 z \mathbf{a}_y + \gamma xyz^2 \mathbf{a}_z)$$

where α, β, and γ are constants (with units V/m).

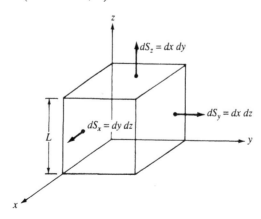

Fig. 2-32

▌ By Gauss' law, $Q/\epsilon_0 = \int\int \mathbf{E} \cdot d\mathbf{S}$. The given field vanishes on three faces of the cube (those in the coordinate planes), and the fluxes through the other three faces are in the ratio $\alpha : \beta : \gamma$. Hence

$$\frac{Q}{\epsilon_0} = \frac{\alpha + \beta + \gamma}{L^4} \int_0^L \int_0^L (L^2 yz)\, dy\, dz = \frac{\alpha + \beta + \gamma}{L^2} \int_0^L y\, dy \int_0^L z\, dz = (\alpha + \beta + \gamma)L^2/4$$

2.93 Suppose that, in Problem 2.92, $\alpha + \beta + \gamma = 0$; the cube would then contain no *net* charge. Could this be the result of a 50-50 mixture of positive and negative charge?

▌ No: Gauss' law in point form gives, inside the cube,

$$\rho = \epsilon_0 \nabla \cdot \mathbf{E} = \frac{2\epsilon_0(\alpha + \beta + \gamma)}{L^4} xyz \equiv 0$$

2.94 Granting the result of Problem 2.93, is it still possible that the **E**-field of Problem 2.92 is due to a stationary distribution of charge located outside of the cube?

▌ No; the electrostatic field is conservative (Problem 1.157), which requires in particular that

$$\frac{\partial E_x}{\partial y} = \frac{\partial E_y}{\partial x}$$

But here

$$\frac{\partial E_x}{\partial y} = \frac{\alpha}{L^4} x^2 z \quad \text{and} \quad \frac{\partial E_y}{\partial x} = \frac{\beta}{L^4} y^2 z$$

2.95 The **E**-field in the cubical region $|x| \le a$, $|y| \le a$, $|z| \le a$ is given in cylindrical coordinates by

$$\mathbf{E} = E_0\left(\frac{r^2}{a^2}\mathbf{a}_r + \sin\frac{\pi z}{2a}\mathbf{a}_z\right) \quad (\text{V/m})$$

Determine the total charge within the cylinder bounded by $r = a$, $z = a$, and $z = -a$.

▌ The cylinder is contained in the cube; hence the flux through it is

$$Q = \epsilon_0 E_0\left[\int_{-a}^{a} (1)^2\, 2\pi a\, dz + \int_0^a (1)\, 2\pi r\, dr + \int_0^a (1)\, 2\pi r\, dr\right] = 6\pi a^2 \epsilon_0 E_0 \quad (\text{C})$$

2.96 A charge distribution produces the field $\mathbf{E} = r^2\mathbf{a}_r - r^2 \sin\theta\, \mathbf{a}_\theta + 10\mathbf{a}_\phi$. Determine the charge density at $(2, \pi, 0)$.

▌ By Gauss' law, $\rho = \epsilon_0 \operatorname{div} \mathbf{E}$; and the divergence in spherical coordinates may be obtained from Problem 1.108.

$$\rho = \epsilon_0\left[\frac{1}{r^2}\frac{\partial}{\partial r}(r^2 E_r) + \frac{1}{r\sin\theta}\frac{\partial}{\partial\theta}(E_\theta \sin\theta) + \frac{1}{r\sin\theta}\frac{\partial}{\partial\phi}(E_\phi)\right]_{(2,\pi,0)}$$

$$= \epsilon_0[4r - 2r\cos\theta + 0]_{(2,\pi,0)} = 12\epsilon_0 \quad (\text{C/m}^3)$$

2.97 The infinite annular volume between $r = 2$ and $r = 3$ (cylindrical coordinates) contains a uniform charge density ρ. Determine the corresponding **E**-field at all points.

▌ Choose as Gaussian surfaces coaxial cylinders of radius r and height 1. For $0 < r < 2$, $Q = 0$. Thus $\mathbf{D} = 0$ and $\mathbf{E} = 0$. For $2 < r < 3$,

$$Q = \pi(r^2 - 2^2)(1)\rho = D(2\pi r)(1)$$

whence

$$E = \frac{D}{\epsilon_0} = \frac{\rho}{2\epsilon_0}\left(r - \frac{4}{r}\right)$$

For $3 < r$,

$$Q = \pi(3^2 - 2^2)(1)\rho = D(2\pi r)(1) \quad \text{or} \quad E = \frac{5\rho}{2\epsilon_0 r}$$

2.98 A 10 μC charge is at the origin of a spherical coordinate system. Calculate the electric flux crossing the portion of a spherical shell described by $0 \le \theta \le \pi/4$.

▌ By Gauss' law, flux Q crosses the area $4\pi r^2$. The area of the shell is

$$A = \int_{\phi=0}^{2\pi} \int_{\theta=0}^{\pi/4} r^2 \sin\theta\, d\theta\, d\phi = 2\pi r^2(-\cos\pi/4 + \cos 0) = \pi(2 - \sqrt{2})r^2$$

Thus, the flux through the shell is

$$\psi_{\text{shell}} = \left(\frac{A}{4\pi r^2}\right)Q = 5\left(1 - \frac{1}{\sqrt{2}}\right)\mu\text{C}$$

2.99 Find the total outward flux from the spherical shell $1 \le r \le 2$ m, if the charge density in the volume is $\rho = (5 \cos^2 \phi)/r^4$ (C/m^3).

▮ By Gauss' law,

$$\psi_e = Q = \int \rho \, dV = \int_{\phi=0}^{2\pi} \int_{\theta=0}^{\pi} \int_{r=1}^{2} \left(\frac{5 \cos^2 \phi}{r^4} \right) r^2 \sin \theta \, dr \, d\theta \, d\phi = 5\pi \text{ C}$$

2.100 In a certain region the electrical flux density is given by $\mathbf{D} = \sqrt{2} \, x\mathbf{a}_x + \sqrt{2}(\sqrt{2} - y)\mathbf{a}_y + 2z\mathbf{a}_z$ (C/m^2). Approximate the flux crossing a 1-mm^2 area on the surface of cylindrical shell at $r = 10$ m, $\phi = 45°$, and $z = 2$ m.

▮ At the point in question,

$$x = 10 \cos 45° = (10/\sqrt{2}) \text{ m} \qquad y = 10 \sin 45° = (10/\sqrt{2}) \text{ m} \qquad \mathbf{D} = 10\mathbf{a}_x - 8\mathbf{a}_y + 4\mathbf{a}_z \text{ C/m}^2$$

and $\qquad d\mathbf{S} \approx 10^{-6} \mathbf{a}_r = 10^{-6}(\cos 45° \, \mathbf{a}_x + \sin 45° \, \mathbf{a}_y) = (10^{-6}/\sqrt{2})(\mathbf{a}_x + \mathbf{a}_y) \text{ m}^2$

Thus $\qquad \psi_e = \mathbf{D} \cdot d\mathbf{S} \approx (10^{-6}/\sqrt{2})[(10)(1) - (8)(1) + (4)(0)] = \sqrt{2} \, \mu\text{C}$

2.101 Within the cylindrical region $r \le 4$ m,

$$\mathbf{E} = \frac{5}{2\epsilon_0 r} \left[\frac{1}{2} - e^{-2r} \left(r^2 + r + \frac{1}{2} \right) \right] \mathbf{a}_r \quad \text{(V/m)}$$

Evaluate the charge density at $r = 2$ m.

▮ By the point form of Gauss' law,

$$\rho = \epsilon_0 \, \text{div } \mathbf{E} = \frac{\epsilon_0}{r} \frac{\partial}{\partial r} (rE_r)$$

$$= \frac{5}{2r} \frac{d}{dr} \left[\frac{1}{2} - e^{-2r} \left(r^2 + r + \frac{1}{2} \right) \right] = 5re^{-2r} = 10e^{-4} \text{ C/m}^3$$

2.102 If the lower side of the upper plate and the upper side of the lower plate of a parallel-plate capacitor have ρ_s (C/m^2) and $-\rho_s$ (C/m^2) surface charge densities, respectively, what is the **E**-field within the capacitor?

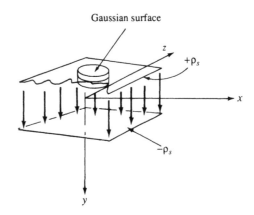

Fig. 2-33

▮ Consider a Gaussian surface enclosing a portion of the upper plate, as shown in Fig. 2-33. The field must be in the y direction (except at the edges), and Gauss' law gives

$$Q = \int_{\text{top}} \mathbf{D} \cdot d\mathbf{S} + \int_{\text{side}} \mathbf{D} \cdot d\mathbf{S} + \int_{\text{bottom}} \mathbf{D} \cdot d\mathbf{S} \qquad \text{or} \qquad \rho_s A = 0 + 0 + D_y A$$

Thus, $\mathbf{D} = \rho_s \mathbf{a}_y$ (C/m^2) and $\mathbf{E} = (\rho_s/\epsilon_0)\mathbf{a}_y$ (V/m).

2.103 A uniform line charge ρ_ℓ lies along the axis of an infinitely long cylinder of radius a, the surface of which has a uniform surface charge ρ_s. Determine the electrical flux density everywhere.

▮ By Gauss' law, for $0 < r < a$,

$$2\pi r L D = \rho_\ell L \qquad \text{or} \qquad \mathbf{D} = (\rho_\ell/2\pi r)\mathbf{a}_r$$

For $r > a$,

$$\rho_\ell L + \rho_s(2\pi r L) = D(2\pi r L) \qquad \text{or} \qquad \mathbf{D} = \frac{1}{2\pi r}(\rho_\ell + 2\pi r \rho_s)\mathbf{a}_r$$

2.104 Verify that the **E**-field of Problem 2.101 is consistent with the charge density obtained in that problem.

▮ Apply Gauss' law in integral form to a length L of a cylinder of radius $r > 4$ m:

$$D_r(2\pi rL) = \epsilon_0 E_r(2\pi rL) = \int_0^L \int_0^{2\pi} \int_0^r 5re^{-2r}\, r\, dr\, d\phi\, dz = 5\pi L\left[e^{-2r}\left(-r^2 - r - \frac{1}{2}\right) + \frac{1}{2}\right]$$

which leads to the given $\mathbf{E}$-field.

2.105 Find the total charge enclosed within the sphere $r \leq a$ if the $\mathbf{E}$-field there is given by $\mathbf{E} = (\rho r/2\epsilon_0)\mathbf{a}_r$.

▮
$$Q = D_r|_{r=a} \times 4\pi a^2 = \frac{\rho a}{2} \times 4\pi a^2 = 2\pi\rho a^3$$

using Gauss' law.

2.106 Show that the divergence of $\mathbf{D} = e^{-y}(\cos x\, \mathbf{a}_x - \sin x\, \mathbf{a}_y)$ is zero everywhere.

▮ In cartesian coordinates,

$$\nabla \cdot \mathbf{D} = \frac{\partial}{\partial x}(e^{-y}\cos x) - \frac{\partial}{\partial y}(e^{-y}\sin x) = e^{-y}(-\sin x) + e^{-y}(\sin x) = 0$$

2.107 Relate Problem 2.106 to Problem 1.154.

▮ $\mathbf{D} = \text{curl}(-e^{-y}\cos x\, \mathbf{a}_z)$.

2.108 Show that the $\mathbf{E}$-field given by $\epsilon_0\mathbf{E} = r\sin\phi\, \mathbf{a}_r + 2r\cos\phi\mathbf{a}_\phi + 2z^2\mathbf{a}_z$ can be realized by a charge density increasing linearly with z.

▮ By Gauss' law,

$$\rho = \text{div }\epsilon_0\mathbf{E} = \frac{1}{r}\frac{\partial}{\partial r}(r^2\sin\phi) + \frac{1}{r}\frac{\partial}{\partial \phi}(2r\cos\phi) + \frac{\partial}{\partial z}(2z^2) = 2\sin\phi - 2\sin\phi + 4z = 4z$$

2.109 Find the charge density at $(1/2, \pi/2, 0)$ corresponding to $\mathbf{D} = r\sin\phi\, \mathbf{a}_r + r^2\cos\phi\, \mathbf{a}_\phi + 2re^{-5z}\mathbf{a}_z$.

▮
$$\rho = \text{div }\mathbf{D} = \frac{1}{r}\frac{\partial}{\partial r}(r^2\sin\phi) + \frac{1}{r}\frac{\partial}{\partial\phi}(r^2\cos\phi) + \frac{\partial}{\partial z}(2re^{-5z})$$

$$= 2\sin\phi - r\sin\phi - 10re^{-5z} = 2\sin\frac{\pi}{2} - \frac{1}{2} - 10\left(\frac{1}{2}\right)e^0 = -3.5$$

2.110 Show that the volume charge density corresponding to $\mathbf{D} = 10\sin^2\phi\, \mathbf{a}_r + ra_\phi + (10z/r)\cos^2\phi\, \mathbf{a}_z$ is independent of ϕ and z.

▮ By Gauss' law,

$$\rho = \text{div }\mathbf{D} = \frac{1}{r}\frac{\partial}{\partial r}(10r\sin^2\phi) + \frac{1}{r}\frac{\partial}{\partial\phi}(r) + \frac{\partial}{\partial z}\left(\frac{10z}{r}\cos^2\phi\right)$$

$$= \frac{10}{r}\sin^2\phi + 0 + \frac{10}{r}\cos^2\phi = \frac{10}{r}$$

2.111 Verify the divergence theorem for the $\mathbf{D}$-field of Problem 2.110 and the cylindrical volume of Fig. 2-34.

▮
$$\int_V \text{div }\mathbf{D}\, dV \overset{?}{=} \int_{\text{lateral}} \mathbf{D}\cdot d\mathbf{S} + \int_{\text{top}} + \int_{\text{bottom}}$$

$$\int_0^b \frac{10}{r}(2\pi rh)\, dr \overset{?}{=} \int_0^{2\pi}(10\sin^2\phi)hb\, d\phi + 2\int_0^{2\pi}\int_0^b\left(\frac{5h}{r}\cos^2\phi\right)r\, dr\, d\phi$$

$$20\pi hb \overset{?}{=} 10hb\int_0^{2\pi}(\sin^2\phi + \cos^2\phi)\, d\phi$$

$$20\pi hb = 20\pi hb$$

2.112 Determine the average charge density within the cylinder of Problem 2.111.

▮
$$\rho_{\text{av}} = \frac{\text{total charge}}{\text{volume}} = \frac{20\pi hb}{\pi hb^2} = \frac{20}{b}$$

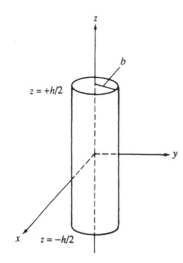

Fig. 2-34

2.113 If the electric flux density within a sphere of radius a is given by

$$\mathbf{D} = \frac{5}{r^2}\mathbf{a}_r - r^2\phi \sin\theta\, \mathbf{a}_\phi \quad (\text{C/m}^2)$$

what is the volume charge density in the sphere?

▮ Assuming the validity of Gauss' law (point form), we have, for $r > 0$,

$$\rho = \operatorname{div}\mathbf{D} = \frac{1}{r^2}\frac{\partial}{\partial r}(5) + \frac{1}{r\sin\theta}\frac{\partial}{\partial\phi}(-r^2\phi\sin\theta) = 0 - r = -r \quad (\text{C/m}^3)$$

2.114 How much charge is contained within the sphere of Problem 2.113?

▮ The answer is: there is no answer. The "field" $\mathbf{D}$ of Problem 2.113 is physically impossible, and Gauss' law does not apply to it.

To clarify the situation, write $\mathbf{D} = \mathbf{D}_r + \mathbf{D}_\phi$ (vector components) and note that Gauss' law is linear in $\mathbf{D}$. Now, $\mathbf{D}_r = (20\pi/4\pi r^2)\mathbf{a}_r$ is obviously a true electric field; it can be attributed to a point charge of 20π C located at the origin. Sure enough, Gauss' law gives for it:

$$\iint_{r=a}\mathbf{D}_r \cdot d\mathbf{S} = \frac{20\pi}{4\pi a^2}(4\pi a^2) = 20\pi \text{ C}$$

On the other hand, $\mathbf{D}_\phi$, which is nonperiodic in ϕ, cannot represent an electric field. If we try to apply Gauss' law to $\mathbf{D}_\phi$, we obtain

$$\iiint_{r<a}(\operatorname{div}\mathbf{D}_\phi)\,dV = \int_0^{2\pi}\int_0^\pi\int_0^a(-r)r^2\sin\theta\,dr\,d\theta\,d\phi = -\pi a^4 \text{ C}$$

whereas, $\mathbf{D}_\phi$ being tangential,

$$\iint_{r=a}\mathbf{D}_\phi \cdot d\mathbf{S} = 0 \text{ C}$$

2.115 Postscript to Problem 2.114:   Compute the work done by $(1/\epsilon_0)\mathbf{D}_\phi$ around the circle $r = a/2$, $\theta = \pi/2$.

▮

$$\frac{1}{\epsilon_0}\oint\mathbf{D}_\phi \cdot d\mathbf{s} = \frac{1}{\epsilon_0}\int_0^{2\pi}\left[\left(-\frac{a^2}{4}\right)\phi(1)\right]\frac{a}{2}\,d\phi = -\frac{\pi^2 a^3}{4\epsilon_0} \neq 0$$

2.116 A cube of 2-m side is centered at the origin with the edges parallel to the axes. Within the cube, $\mathbf{D} = (10x^3/3)\mathbf{a}_x$ (C/m^3). What is the total charge contained in the cube?

▮

$$\rho = \nabla\cdot\mathbf{D} = \frac{10}{3}(3x^2) = 10x^2 \quad (\text{C/m}^3)$$

$$Q = \int\rho\,dV = \int_{-1}^1\int_{-1}^1\int_{-1}^1 10x^2\,dx\,dy\,dz = \frac{80}{3}\text{ C}$$

2.117 Verify the result of Problem 2.116 by applying Gauss' law in integral form.

▮

$$Q = \int\mathbf{D}\cdot d\mathbf{S} = \int_{-1}^1\int_{-1}^1\frac{10(1)}{3}\,dy\,dz + \int_{-1}^1\int_{-1}^1\frac{10(-1)}{3}\,dy\,dz = \frac{40}{3} + \frac{40}{3} = \frac{80}{3}\text{ C}$$

2.118 The flux density within the cylindrical volume bounded by $r = 2$ m, $z = 0$ and $z = 5$ m is given by $\mathbf{D} = 30e^{-r}\mathbf{a}_r - 2z\mathbf{a}_z$ (C/m²). What is the total outward flux crossing the surface of the cylinder?

$$\psi_e = \int_{\text{side}} + \int_{\text{top}} + \int_{\text{bottom}} \mathbf{D} \cdot d\mathbf{S} = \int_0^5 \int_0^{2\pi} 30e^{-2}\mathbf{a}_r \cdot 2\, d\phi\, dz\, \mathbf{a}_r + \int_0^{2\pi} \int_0^2 (-2)(5)\mathbf{a}_z \cdot r\, dr\, d\phi\, \mathbf{a}_z + 0$$
$$= 60e^{-2}(2\pi)5 - 10(2\pi)^2 = 129.44 \text{ C}$$

2.119 In spherical coordinates, the region $r \le 4$ m, $\theta \le \pi/4$ contains a charge distribution $\rho_v = 5r$ (C/m³). Determine the total outward flux through the boundary of the region (the lateral surface of a cone plus a spherical cap).

$$\psi_e = Q = \int \rho_v\, dV = \int_0^{2\pi} \int_0^{\pi/4} \int_0^4 (5r)\, r^2 \sin\theta\, dr\, d\theta\, d\phi$$
$$= \left.\frac{5r^4}{4}\right]_0^4 \left.(-\cos\theta)\right]_0^{\pi/4} \left.\phi\right]_0^{2\pi} = 640\pi\left(1 - \frac{1}{\sqrt{2}}\right) \text{ C}$$

2.120 Determine the charge density function corresponding to the field

$$\mathbf{D} = \begin{cases} \rho_0\left(\dfrac{r^2 - a^2}{2r}\right)\mathbf{a}_r, & a < r < b \\[2mm] \rho_0\left(\dfrac{b^2 - a^2}{2r}\right)\mathbf{a}_r, & b < r \\[2mm] 0 & r < a \end{cases}$$

$$\rho = \text{div } \mathbf{D} = \frac{1}{r}\frac{\partial}{\partial r}(rD_r) = \begin{cases} \rho_0 & a < r < b \\ 0 & b < r \\ 0 & r < a \end{cases}$$

2.121 Under the action of a mechanical force $\mathbf{F}_m$, a point charge q is moved at constant speed along a contour C that runs through an electric field $\mathbf{E}$ (Fig. 2-35). Express (a) the mechanical work required to move q from a to b; (b) the electrical work done on q by the $\mathbf{E}$-field in the course of the motion.

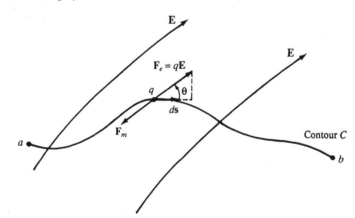

Fig. 2-35

(b)
$$W_e = \int_a^b \mathbf{F}_e \cdot d\mathbf{s} = \int_a^b q\, |\mathbf{E}|\cos\theta\, ds \tag{1}$$

(a) For motion at constant speed, the mechanical and electrical forces must be in equilibrium. Thus
$$W_m = \int_a^b \mathbf{F}_m \cdot d\mathbf{s} = \int_a^b (-\mathbf{F}_e) \cdot d\mathbf{s} = -W_e \tag{2}$$

2.122 Suppose that, in Problem 2.121, the field $\mathbf{E}$ is that due to a point charge Q located off the contour C. Verify that in this case W_e, and therefore W_m, is independent of the form of C, for fixed endpoints a and b.

To indicate the direction of motion, relabel W_m as $W_{a \to b}$. By Problem 2.121 and Fig. 2-36,
$$W_{a \to b} = -\int_a^b \mathbf{F}_e \cdot d\mathbf{s} = -\frac{qQ}{4\pi\varepsilon_0} \int_a^b \frac{\mathbf{R} \cdot d\mathbf{R}}{r^3}$$

Recalling that $2R\, dR = d(R^2) = d(\mathbf{R} \cdot \mathbf{R}) = d\mathbf{R} \cdot \mathbf{R} + \mathbf{R} \cdot d\mathbf{R} = 2\mathbf{R} \cdot d\mathbf{R}$, we obtain
$$W_{a \to b} = -\frac{qQ}{4\pi\epsilon_0} \int_{R_a}^{R_b} \frac{dR}{R^2} = \frac{qQ}{4\pi\epsilon_0}\left(\frac{1}{R_b} - \frac{1}{R_a}\right) \tag{1}$$

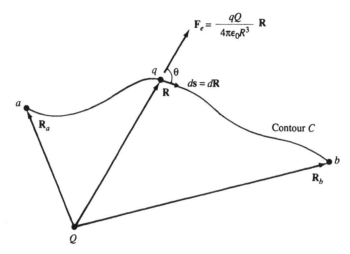

Fig. 2-36

As asserted, $W_{a \to b}$ depends only on the coordinates of the endpoints a and b. This result generalizes at once to the **E**-field of an arbitrary static distribution of charge.

2.123 Show that an electrostatic field obeys curl $\mathbf{E} = \mathbf{0}$.

▌ By Problem 2.122, $\oint \mathbf{E} \cdot d\mathbf{s} = 0$ for any closed contour. Thus, **E** is conservative and curl-free (Problem 1.157).

2.124 Define a potential-energy function for a point charge q in the field of another point charge Q.

▌ According to (1) of Problem 2.122, the work $W_{a \to b}$ needed to move q from a to b equals the increase from a to b of the point function $\phi(R) = qQ/4\pi\epsilon_0 R$. Hence, $\phi(R)$ may be taken as the desired potential energy. [The most general choice would be $\phi(R) + c$, with c an arbitrary constant. Normally one sets $c = 0$, to make the potential energy vanish at infinity.]

2.125 Generalize the result of Problem 2.124 to the field of an arbitrary static distribution of charge.

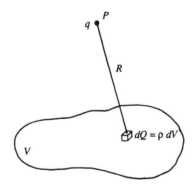

Fig. 2-37

▌ See Fig. 2-37. By Problem 2.124, the charge element dQ creates at P the potential energy

$$d\phi(P) = \frac{q(\rho \, dV)}{4\pi\epsilon_0 R} \quad \text{or} \quad \phi(P) = \frac{q}{4\pi\epsilon_0} \int_V \frac{\rho \, dV}{R}$$

The integral defining $\phi(P)$ may fail to exist if P lies within the charge cloud V, or if V extends to infinity.

2.126 Define the *electrostatic potential* or *voltage* (a scalar field) associated with an electric field.

▌ At each point of space the voltage is defined as the potential energy of a unit positive charge at that point. Thus, from Problems 2.124 and 2.125,

$$\textbf{\textit{point charge }} Q \qquad V(P) = \frac{Q}{4\pi\epsilon_0 R} \quad (R \equiv \text{distance from } Q \text{ to } P)$$

$$\textbf{\textit{volume distribution }} \rho \qquad V(P) = \frac{1}{4\pi\epsilon_0} \int_V \frac{\rho \, dV}{R} \quad (R \equiv \text{distance from } \rho \, dV \text{ to } P)$$

2.127   Give an integral relation between an **E**-field and the associated V-field.

▌ By Problems 2.124 and 2.126,

$$\frac{W_{a \to b}}{q} = V_b - V_a \tag{1}$$

But

$$\frac{W_{a \to b}}{q} = -\int_a^b \frac{\mathbf{F}_e}{q} \cdot d\mathbf{s} = -\int_a^b \mathbf{E} \cdot d\mathbf{s}$$

so that

$$V_b - V_a = -\int_a^b \mathbf{E} \cdot d\mathbf{s} \tag{2}$$

In (2), choose point a at infinity ($V_a = 0$) and relabel b as P. Then

$$V(P) = -\int_\infty^P \mathbf{E} \cdot d\mathbf{s} = \int_P^\infty \mathbf{E} \cdot d\mathbf{s} \tag{3}$$

is the desired relation. Because **E** is conservative, the path of integration may be chosen arbitrarily. To change the reference level from ∞ to point A, replace ∞ by A in (3).

2.128   Express the **E**-V relation (Problem 2.127) in differential form.

▌ Equation (2) or (3) of Problem 2.127 shows that the differential of V in an arbitrary direction $d\mathbf{s}$ is given by $dV = -\mathbf{E} \cdot d\mathbf{s}$. Thus (see Problem 1.83),

$$-\mathbf{E} = \nabla V \qquad \text{or} \qquad \mathbf{E} = -\nabla V \tag{1}$$

where **E** and the partial derivatives of V are evaluated at the same point P.

Equation (1) above and equation (3) of Problem 2.127 are equivalent, just as are the differential and integral forms of Gauss' law.

2.129   Figure 2-38 shows a cylindrical surface of radius λ that is infinite in length and supports a uniform surface charge distribution ρ_s (C/m^2). Determine the potential difference between two points at radial distances r_a and r_b from the axis, with $r_b > r_a > \lambda$.

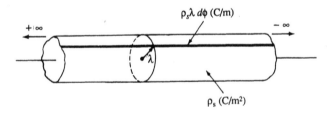

$\rho_s \lambda \, d\phi$ (C/m)

$+\infty$ $-\infty$

λ

ρ_s (C/m^2)   **Fig. 2-38**

▌ Using symmetry and Gauss' law, we find that the **E**-field is directed radially away from the cylinder and is given by $\mathbf{E} = (\rho_s \lambda / \epsilon_0 r)\mathbf{a}_r$, for points outside the cylinder. Then, by (2) of Problem 2.127, choosing a radial path, we obtain

$$V_b - V_a = -\int_{r_a}^{r_b} E_r \, dr = \frac{\rho_s \lambda}{\epsilon_0} \ln \frac{r_a}{r_b} \tag{1}$$

2.130   Refer to Problem 2.129. For $\rho_s > 0$, (1) shows that $V_b < V_a$. Is this a reasonable result?

▌ Yes. The *positive* cylinder repels (does positive work on) a unit *positive* test charge. Thus, *negative* external work is needed to move the test charge, which means a *decrease* in potential.

2.131   Verify the relation $\mathbf{E} = -\nabla V$ for the field of an arbitrary charge cloud.

▌ From Problems 2.20 and 2.126,

$$\mathbf{E}(P) = \frac{1}{4\pi\epsilon_0} \iiint \frac{\rho \, dV}{R^2} \mathbf{a}_R \qquad \text{and} \qquad V(P) = \frac{1}{4\pi\epsilon_0} \iiint \frac{\rho \, dV}{R}$$

For old times' sake let us work in cartesian coordinates, with (x, y, z) as the coordinates of P and (x', y', z') as those of the charge element $\rho \, dV$. Then

$$V(x, y, z) = \frac{1}{4\pi\epsilon_0} \iiint \frac{\rho(x', y', z') \, dx' \, dy' \, dz'}{\sqrt{(x - x')^2 + (y - y')^2 + (z - z')^2}}$$

Since $\mathbf{R} = (x - x')\mathbf{a}_x + (y - y')\mathbf{a}_y + (z - z')\mathbf{a}_z$, we have

$$\frac{\partial V(x, y, z)}{\partial x} = \frac{1}{4\pi\epsilon_0} \iiint \frac{\partial}{\partial x}\left[\frac{1}{\sqrt{(x - x')^2 + (y - y')^2 + (z - z')^2}}\right]\rho(x', y', z')\, dx'\, dy'\, dz'$$

$$= -\frac{1}{4\pi\epsilon_0} \iiint \frac{x - x'}{R^3}\rho(x', y', z')\, dx'\, dy'\, dz'$$

with symmetric expressions for $\partial V/\partial y$ and $\partial V/\partial z$. Multiply by the unit vectors (which are fixed vectors in cartesian coordinates) and add:

$$\nabla V = \frac{\partial V}{\partial x}\mathbf{a}_x + \frac{\partial V}{\partial y}\mathbf{a}_y + \frac{\partial V}{\partial z}\mathbf{a}_z$$

$$= -\frac{1}{4\pi\epsilon_0}\iiint \frac{(x - x')\mathbf{a}_x + (y - y')\mathbf{a}_y + (z - z')\mathbf{a}_z}{R^3}\rho(x', y', z')\, dx'\, dy'\, dz'$$

$$= -\frac{1}{4\pi\epsilon_0}\iiint \frac{R\mathbf{a}_R}{R^3}\rho\, dV = -\mathbf{E}$$

2.132 A point charge is moved in an **E**-field. What path must be followed for a maximum rate of expenditure of energy?

❙ Maximum work means maximum voltage increase, which occurs (Problem 1.84) in the direction of ∇V. But this is the direction of $-\mathbf{E}$; so the path must be a field line, but heading toward positive charge.

2.133 An optical fiber of length $4a$ carries the line charge distribution graphed in Fig. 2-39. Calculate the potential on the circle $y^2 + z^2 = h^2$ of the yz plane.

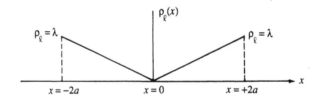

Fig. 2-39

❙

$$dV = \frac{\rho_\ell\, dx}{4\pi\epsilon_0\sqrt{h^2 + x^2}} = \frac{\lambda(|x|/2a)\, dx}{4\pi\epsilon_0\sqrt{h^2 + x^2}}$$

or (since the integrand is an even function),

$$V = \frac{2}{4\pi\epsilon_0}\int_0^{2a}\left(\frac{\lambda}{2a}\right)\frac{x\, dx}{\sqrt{h^2 + x^2}} = \frac{\lambda}{4\pi\epsilon_0 a}\left.\sqrt{h^2 + x^2}\right]_0^{2a} = \frac{\lambda}{4\pi\varepsilon_0 a}\left(\sqrt{h^2 + 4a^2} - h\right)$$

2.134 Find **E** on the circle of Problem 2.133.

❙ In a cylindrical system with h as the radial coordinate,

$$\mathbf{E} = -\nabla V = -\frac{\partial V}{\partial h}\mathbf{a}_h = \frac{\lambda}{4\pi\epsilon_0 a}\left(1 - \frac{h}{\sqrt{h^2 + 4a^2}}\right)\mathbf{a}_h$$

2.135 Four point charges are located at the corners of a square, of length $2L$ on each side, symmetrically located relative to the origin, as shown in Fig. 2-40. Derive an expression for the potential V at a position x along the axis.

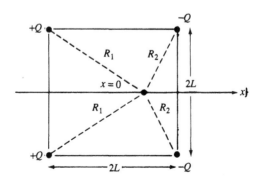

Fig. 2-40

$$V(x) = \frac{1}{4\pi\epsilon_0}\left(\frac{Q}{R_1} + \frac{Q}{R_1} - \frac{Q}{R_2} - \frac{Q}{R_2}\right) = \frac{2Q}{4\pi\epsilon_0}\left(\frac{1}{R_1} - \frac{1}{R_2}\right)$$

$$= \frac{Q}{2\pi\epsilon_0}\left[\frac{1}{\sqrt{L^2 + (L+x)^2}} - \frac{1}{\sqrt{L^2 + (L-x)^2}}\right] \qquad (1)$$

2.136 For the charges of Problem 2.135, evaluate the potential at (*a*) $x = 0$ and (*b*) $x = L/3$.

    ❙ (*a*) $V = 0$ (by symmetry). (*b*) From (*1*) of Problem 2.135,

$$V = \frac{Q}{2\pi\epsilon_0}\left[\frac{1}{\sqrt{L^2 + (L+L/3)^2}} - \frac{1}{\sqrt{L^2 + (L-L/3)^2}}\right] = \frac{Q}{2\pi\epsilon_0 L}(1.5)(0.2 - 0.278) = -\frac{0.117Q}{\pi\epsilon_0 L}$$

2.137 For the charged circle of Problem 2.37 and Fig. 2-12, show that the potential is zero at the center.

    ❙

$$V = \frac{\rho_\ell(\pi R)}{4\pi\epsilon_0 R} + \frac{(-\rho_\ell)(\pi R)}{4\pi\epsilon_0 R} = 0$$

2.138 An E-field is given by $\mathbf{E} = 4x\mathbf{a}_x + 2\mathbf{a}_y$ (V/m). Determine by direct integration the work required to move a unit positive charge along the curve $xy = 4$ from $(2, 2)$ to $(4, 1)$.

    ❙ In the notation of Problem 2.122, $W_{(2,2)\to(4,1)} = -\int_{(2,2)}^{(4,1)} \mathbf{E}\cdot d\mathbf{s}$.

$$\mathbf{E}\cdot d\mathbf{s} = 4x\,dx + 2\,dy = 4x\,dx + 2\,d(4/x) = \left(4x - \frac{8}{x^2}\right)dx$$

whence

$$W_{(2,2)\to(4,1)} = -\int_2^4 \left(4x - \frac{8}{x^2}\right)dx = -22\text{ J}$$

that is, the field does 22 J of work on the charge.

2.139 Obtain a potential function for the E-field of Problem 2.138 and use it to check the answer to that problem.

    ❙ $E_x = 4x = -\partial V/\partial x$; so $V = -2x^2 + f(y)$. $E_y = 2 = -\partial V/\partial y = -f'(y)$; so $f(y) = -2y + c$. Choosing $V(x, y) = -2(x^2 + y)$ (V), we obtain

$$W_{(2,2)\to(4,1)} = V_{(4,1)} - V_{(2,2)} = -2(17) + 2(6) = -22\text{ J}$$

2.140 In a certain region the electric potential distribution is as shown in Fig. 2-41(*a*). Sketch the corresponding E-field.

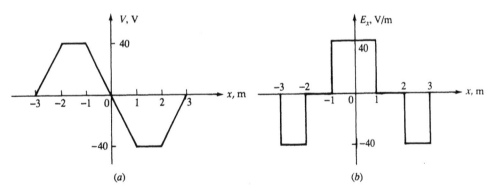

Fig. 2-41

    ❙ Since $\mathbf{E} = -\nabla V$, $E_x = -\partial V/\partial x$, we obtain the sketch shown in Fig. 2-41(*b*). Note that an antisymmetric potential corresponds to a symmetric field.

2.141 Determine the potential distribution for the hollow sphere of Problem 2.85.

    ❙ From Problem 2.85,

$$\mathbf{E} = \begin{cases} \mathbf{0} & r < R \\ (\rho_s R^2/\epsilon_0 r^2)\mathbf{a} & r > R \end{cases}$$

Therefore, for $r > R$,

$$V(r) = \int_r^\infty \frac{\rho_s R^2}{\epsilon_0 u^2}\,du = \frac{\rho_s R^2}{\epsilon_0 r}$$

and, for $r < R$,

$$V(r) = 0 + \frac{\rho_s R^2}{\epsilon_0 R} = \frac{\rho_s R}{\epsilon_0}$$

2.142 A circular plate of radius a has a uniform charge density ρ_s (C/m²) distributed over its surface. Evaluate the potential at at a point on the axis of the plate h meters from the center.

❚ In the natural cylindrical coordinates,

$$V(h) = \frac{1}{4\pi\epsilon_0} \int_0^a \frac{\rho_s(2\pi r)\,dr}{\sqrt{r^2 + h^2}} = \frac{\rho_s}{2\epsilon_0}(\sqrt{a^2 + h^2} - h) \quad (V)$$

2.143 A slab of charge (Fig. 2-42) has a uniform charge density ρ. At $x = x_1$, $V = V_1$ and $E_x = E_1$. There is no variation of E_x in the y and z directions. Find the potential $V(x)$ within the slab.

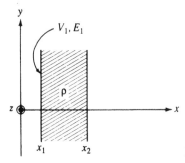

Fig. 2-42

❚ Since $\nabla \cdot \mathbf{D} = \rho$,

$$\frac{dE_x}{dx} = \frac{\rho}{\epsilon_0} \quad \text{or} \quad \int_{E_1}^{E_x} dE_x = \frac{\rho}{\epsilon_0} \int_{x_1}^x dx \quad \text{or} \quad E_x = \frac{\rho}{\epsilon_0}(x - x_1) + E_1$$

for $x_1 < x < x_2$. Now

$$-\frac{dV}{dx} = E_x = \frac{\rho}{\epsilon_0}(x - x_1) + E_1 \quad \text{or} \quad \int_{V_1}^V dV = -\int_{x_1}^x \left[\frac{\rho}{\epsilon_0}(x - x_1) + E_1\right] dx$$

or

$$V(x) = -\frac{\rho}{2\epsilon_0}(x - x_1)^2 - E_1(x - x_1) + V_1 \qquad (1)$$

For $x_1 < x < x_2$.

2.144 If the slab of charge of Problem 2.143 is made infinitely thin, show that the potential is continuous through the thin layer of charge.

❚ From (1) of Problem 2.143, as $x_2 - x_1 \to 0$,

$$V_2 - V_1 = -\frac{\rho}{2\epsilon_0}(x_2 - x_1)^2 - E_1(x_2 - x_1) \to -E_1(x_2 - x_1) \to 0$$

More generally, because V is differentiable ($\nabla V = -\mathbf{E}$), it is necessarily continuous.

2.145 An infinitely long cylinder of radius a is filled with a charge of uniform density ρ. If the potential on the surface of the cylinder is V_0, what is the potential within the cylinder?

❚ Apply Gauss' law to a coaxial cylinder of radius $r < A$ and length 1:

$$D_r(2\pi r)(1) = (\pi r^2)(1)\rho$$

$$D_r = \frac{\rho}{2}r$$

By (2) of Problem 2.127, for $0 < r < a$,

$$V = V_0 - \int_a^r \frac{\rho}{2\epsilon_0} r\,dr = V_0 + \frac{\rho}{4\epsilon_0}(a^2 - r^2)$$

2.146 Given a line charge density $\rho_\ell = 5\,\mu\text{C/m}$ along the z axis, determine the potential difference between the points $A(1\,\text{m}, \pi, 0)$ and $B(3\,\text{m}, \pi, 4\,\text{m})$.

❚

$$V(B) - V(A) = -\int_A^B E_r\,dr = -\int_1^3 \frac{5 \times 10^{-6}}{2\pi\epsilon_0 r}\,dr = -\frac{5 \times 10^{-6}}{2\pi\epsilon_0}\ln 3 \quad \text{V}$$

2.147 In Fig. 2-43, P is a fixed observation point outside the fixed circle; unit positive charges Q and Q' are placed where a ray from P hits the circle. Show that the potential at P depends only on the distance $R_{av} \equiv \overline{PM}$.

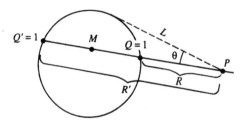

Fig. 2-43

$\blacksquare$ Elementary geometry (try it!) shows that $RR' = $ const. $= L^2$, independent of the ray angle θ. Hence

$$V(P) = \frac{1}{4\pi\epsilon_0}\left(\frac{1}{R} + \frac{1}{R'}\right) = \frac{1}{4\pi\epsilon_0}\frac{R+R'}{RR'} = \frac{R_{av}}{2\pi\epsilon_0 L^2}$$

2.148 For Problem 2.147, prove that, for all θ, $V(P) \geq 1/2\pi\epsilon_0 L$.

$\blacksquare$ By the "inequality of the means,"

$$R_{av} = \frac{R+R'}{2} \geq \sqrt{RR'} = L$$

and the result follows at once. Observe that the minimum potential is attained when Q and Q' coalesce into a single charge of 2 at the point of tangency.

2.149 A point charge of $1\ \mu C$ is located at the origin of a cylindrical coordinate system. Determine the potential difference between two points 1 m and 2 m radially away from the origin.

$\blacksquare$

$$V_1 - V_2 = \frac{Q}{4\pi\epsilon_0}\left(\frac{1}{r_1} - \frac{1}{r_2}\right) = \frac{10^{-6}}{10^{-9}/9}\left(1 - \frac{1}{2}\right) = 4500\ \text{V}$$

2.150 A circular loop 6 m in diameter carries a uniform charge density of $1\ \mu C/m$. What is the potential at a point on its axis 4 m away from the center of the loop?

$\blacksquare$ Every point of the loop is 5 m from the observation point; hence

$$V = \frac{(10^{-6}\ \text{C/m})\pi(6\ \text{m})}{4\pi\left(\dfrac{10^{-9}}{36\pi}\ \text{F/m}\right)(5\ \text{m})} = 34\ \text{kV}$$

2.151 The potential outside of the conducting cone $\theta = \alpha$ (Fig. 2-44) is given by $V = \ln\cot(\theta/2)$. Determine the surface charge per unit length in the z direction.

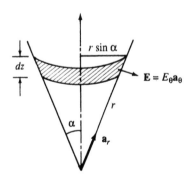

Fig. 2-44

$\blacksquare$ For a conductor, the surface charge density at any point is equal to the outer normal component of **D** at that point (see Problem 2.167). Thus

$$\rho_s = D_\theta|_{\theta=\alpha} = \epsilon_0 E_\theta|_{\theta=\alpha} = \epsilon_0\left(-\frac{1}{r}\frac{\partial V}{\partial \theta}\right)\bigg|_{\theta=x} = \frac{\epsilon_0}{r\sin\alpha}$$

$$\rho_\ell = \frac{dQ}{dz} = \frac{\rho_s(2\pi r\sin\alpha)\,dr}{dr\cos\alpha} = 2\pi\epsilon_0\sec\alpha$$

2.152 Two infinite line charges, $\pm\rho_\ell$, run parallel to the z axis, a distance b apart. Find the equipotential curves in the xy plane.

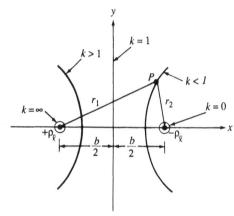

Fig. 2-45

▮ For a line charge, $V(r) = (\rho/2\pi\epsilon_0)\ln(1/r)$; therefore (see Fig. 2-45)

$$V(P) = \frac{\rho_\ell}{2\pi\epsilon_0}\ln\frac{r_2}{r_1} \tag{1}$$

so that the equipotentials are given by $r_2/r_1 = k$, where k assumes all positive values. By symmetry, we need only elucidate the curves for $0 < k < 1$. Going over to rectangular coordinates, we have

$$r_1^2 = \left(x + \frac{b}{2}\right)^2 + y^2 \qquad r_2^2 = \left(x - \frac{b}{2}\right)^2 + y^2 \tag{2}$$

whence

$$\left(x - \frac{1+k^2}{1-k^2}\frac{b}{2}\right)^2 + y^2 = \left(\frac{kb}{1-k^2}\right)^2 \tag{3}$$

This is a family of eccentric circles, clustering on the negative line as $k \to 0$ and approaching the y axis as $k \to 1$.

2.153 Consider two concentric spheres. The inner sphere has radius a, and a charge $+Q$ is distributed uniformly over its outer surface. The outer sphere has radius b, and a charge $-Q$ is distributed uniformly over its inner surface. Determine the potential difference between the two spheres.

▮ Between the spheres the field is that of a point charge $+Q$.

$$V_b - V_a = -\int_a^b E_r\,dr = -\int_a^b \frac{Q}{4\pi\epsilon_0 r^2}\,dr = \frac{Q}{4\pi\epsilon_0}\left(\frac{1}{b} - \frac{1}{a}\right)$$

Note that the outer sphere is at lower potential, consistent with the fact that a positive test charge would "fall" from the inner to the outer sphere.

2.154 A cylindrical conductor and its resistive analog are respectively shown in Fig. 2-46(a) and (b). Ohm's law for the two representations is stated as $\mathbf{J} = \sigma\mathbf{E}$ and $I = V/R$, where $\mathbf{J}$ is the current density vector in the conductor and σ is its conductivity. Assuming $\mathbf{J}$ constant in magnitude and direction across the conductor cross section, obtain an expression for R in terms of L, A, and σ.

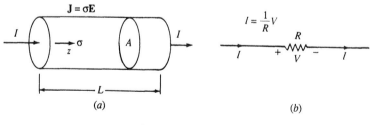

(a) (b) Fig. 2-46

▮

$$I = \int_A \mathbf{J}\cdot d\mathbf{S} = J_z A = \sigma E_z A$$

But

$$V = \int_0^L E_z\,dz = E_z L$$

Hence

$$R = \frac{V}{I} = \frac{L}{\sigma A}$$

2.155 In Fig. 2-47, two point charges $+q$ and $-q$ separated by a small distance $|\mathbf{s}| = s$ constitute an (extended) *electric dipole*. Determine the potential at a point P far away from the dipole.

▮
$$V_P = \frac{q}{4\pi\epsilon_0 r_+} - \frac{q}{4\pi\epsilon_0 r_-} \qquad (1)$$

Substitute $r_+ = r - (s/2)\cos\theta$ and $r_- = r + (s/2)\cos\theta$ to obtain

$$V_P = \frac{q}{4\pi\epsilon_0}\left(\frac{s\cos\theta}{r^2 - \frac{s^2}{4}\cos^2\theta}\right) \qquad (2)$$

For $r \gg s/2$, (2) becomes

$$V_P = \frac{p\cos\theta}{4\pi\epsilon_0 r^2} \quad (p \equiv qs) \qquad (3)$$

The expression (3) is valid for all r for a *point dipole*, obtained by letting $s \to 0$ and $q \to \infty$ in such fashion that qs remains fixed at p.

2.156 Determine the **E**-field of a point dipole.

▮
$$\mathbf{E} = -\nabla V = -\left(\mathbf{a}_r \frac{\partial V}{\partial r} + \mathbf{a}_\theta \frac{1}{r}\frac{\partial V}{\partial \theta}\right)$$

Using (3) of Problem 2.155 yields

$$\mathbf{E} = \frac{p}{4\pi\epsilon_0 r^3}(2\cos\theta\, \mathbf{a}_r + \sin\theta\, \mathbf{a}_\theta) \qquad (1)$$

2.157 The *dipole moment* of a point dipole is defined as the vector $\mathbf{p} = p\mathbf{a}_p$, where (see Fig. 2-47) $\mathbf{a}_p$ is the limiting direction from $-q$ to $+q$. Express the potential function for a point dipole in terms of its dipole moment.

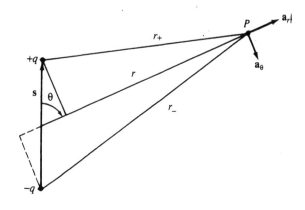

Fig. 2-47

▮ Since $\mathbf{p} \cdot \mathbf{a}_r = (p)(1)\cos\theta$, (3) of Problem 2.155 yields
$$V_P = \frac{\mathbf{p} \cdot \mathbf{a}_r}{4\pi\epsilon_0 r^2} \qquad (1)$$

Here, $\mathbf{a}_r$ is the unit vector directed from the dipole toward point P.

2.158 In contrast to conductors, certain materials have bound charges rather than free charges. These *dielectrics*, when immersed in an **E**-field, become *polarized*. The *polarization vector* **P** is defined as the dipole moment per unit volume; that is,

$$\mathbf{P} = \frac{d\mathbf{p}}{dV} \qquad (1)$$

Express the potential outside a polarized dielectric in terms of the internal **P**-field and its derivatives.

▮ Consider an elemental volume dV' of the polarized dielectric (Fig. 2-48). The vector $\mathbf{r}' = x'\mathbf{a}_x + y'\mathbf{a}_y + z'\mathbf{a}_z$ is directed from the origin to dV' (the "source point"). From (1) of Problem 2.157, we have, at point A,

$$dV_A = \frac{d\mathbf{p} \cdot \mathbf{a}_R}{4\pi\epsilon_0 R^2} = \frac{\mathbf{P} \cdot \mathbf{a}_R\, dV'}{4\pi\epsilon_0 R^2} \qquad (1)$$

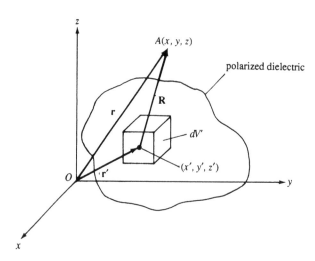

Fig. 2-48

Since $\mathbf{R} = \mathbf{r} - \mathbf{r}'$, we can show that $\mathbf{\nabla}'(1/R) = \mathbf{a}_R/R^2$, where $\mathbf{\nabla}'$ is the gradient with respect to the primed coordinates. Therefore, (1) becomes

$$dV_A = \frac{1}{4\pi\epsilon_0}\left[\mathbf{P}\cdot\mathbf{\nabla}'\left(\frac{1}{R}\right)dV'\right] = \frac{1}{4\pi\epsilon_0}\left[\text{div}'\left(\frac{1}{R}\mathbf{P}\right) - \frac{1}{R}\text{div}'\,\mathbf{P}\right]dV'$$

where the last equality follows from (1) of Problem 1.139. Hence, applying the divergence theorem,

$$V_A = \frac{1}{4\pi\epsilon_0}\int_V \text{div}'\left(\frac{1}{R}\mathbf{P}\right)dV' - \frac{1}{4\pi\epsilon_0}\int_V \frac{1}{R}\text{div}'\,\mathbf{P}\,dV'$$

$$= \frac{1}{4\pi\epsilon_0}\oint_S \frac{\mathbf{P}'\cdot\mathbf{n}'}{R}dS' - \frac{1}{4\pi\epsilon_0}\int_V \frac{\text{div}'\,\mathbf{P}}{R}dV' \qquad (2)$$

2.159 Exhibit bound charge distributions to which can be attributed the external field of a polarized dielectric.

▮ According to (2) of Problem 2.158, the voltage at an external point is that due to surface charge of density $\rho_{sb} = \mathbf{P}\cdot\mathbf{n}$ and volume charge of density $\rho_{vb} = -\text{div}\,\mathbf{P}$.

2.160 Using the results of Problems 2.158 and 2.159 and Gauss' law, obtain an expression for the electrical flux density vector $\mathbf{D}$ in a region containing free charges of volume density ρ_{vf} and bound charges of volume density ρ_{vb}, the latter replacing an infinite dielectric.

▮ The $\mathbf{E}$-field is that arising from a net charge density $\rho_t = \rho_{vf} + \rho_{vb}$, in free space. By Gauss' law,
$$\text{div}\,(\epsilon_0\mathbf{E}) = \rho_t = \rho_{vf} + \rho_{vb} = \rho_{vf} - \text{div}\,\mathbf{P}$$
or
$$\text{div}\,(\epsilon_0\mathbf{E} + \mathbf{P}) = \rho_{vf}$$
Hence, if we define $\mathbf{D} \equiv \epsilon_0\mathbf{E} + \mathbf{P}$, the $\mathbf{D}$-field *arises from free charge only*.

2.161 If the polarization $\mathbf{P}$ and the electric intensity $\mathbf{E}$ are related via $\mathbf{P} = \chi_e\epsilon_0\mathbf{E}$, where χ_e is known as the *electric susceptibility*, relate $\mathbf{E}$ and $\mathbf{D}$ in a polarized dielectric.

▮ By Problem 2.160,
$$\mathbf{D} = \epsilon_0\mathbf{E} + \mathbf{P} = \epsilon_0(1 + \chi_e)\mathbf{E} \equiv \epsilon\mathbf{E} \qquad (1)$$
where $\epsilon \equiv \epsilon_0(1 + \chi_e)$ is the *permittivity* of the dielectric.

2.162 A dielectric sphere of permittivity ϵ and radius R is centered at the origin of a spherical coordinate system and polarized radially such that $\mathbf{P} = kr\mathbf{a}_r$, where k is a constant. Evaluate the electric potential at the center of the sphere.

▮ Since there are no free charges,
$$\rho_v = -\text{div}\,\mathbf{P} = -\frac{1}{r^2}\frac{\partial}{\partial r}(r^2 kr) = -3k$$

From Gauss' law, within the sphere, we have
$$D_r(4\pi r^2) = (-3k)\left(\frac{4}{3}\pi r^2\right) \qquad \text{or} \qquad E_r = \frac{D_r}{\epsilon} = -\frac{kr}{\epsilon}$$

Hence, V at the center is given by
$$V_0 = -\int_\infty^0 E_r\,dr = -\int_\infty^R 0\,dr + \int_R^0 \frac{kr}{\epsilon}\,dr = -\frac{kR^2}{2\epsilon}$$

2.163 *Relative permittivity* ϵ_r is defined as the ratio of the permittivity of a dielectric to the permittivity of free space. If ϵ_r for distilled water is 80, what is its permittivity?

$$\epsilon = 80\epsilon_0 \approx 80\left(\frac{10^{-9}}{36\pi}\right) = 0.708 \text{ nF/m}$$

2.164 Express the *conservation of electric charge* in integral and in point form.

▮ The outflux of charge through any closed surface must equal the rate of decrease of charge within the surface. Thus,

$$\int_S \mathbf{J} \cdot d\mathbf{S} = -\frac{d}{dt}\int_V \rho_v \, dV = -\int_V \frac{\partial \rho_v}{\partial t} \, dV \qquad (1)$$

By the divergence theorem, the left member can be replaced by $\int_V \text{div } \mathbf{J} \, dV$. The resulting equation can hold for arbitrary V only if

$$\text{div } \mathbf{J} = -\frac{\partial \rho_v}{\partial t} \qquad (2)$$

which is the relationship in point form, and is also known as the *continuity equation*.

2.165 At time $t = 0$, excess free charge of volume density ρ_0 is injected into the interior of a conductor. Find the time for the charge density to decrease to $1/e$ of its initial value. Assume uniform conductor properties.

▮ The governing differential equation is (2) of Problem 2.164, which under the substitution $\mathbf{J} = \sigma\mathbf{E} = (\sigma/\epsilon)\mathbf{D}$ becomes

$$\frac{\sigma}{\epsilon}\text{div } \mathbf{D} = -\frac{\partial \rho_v}{\partial t} \qquad \text{or} \qquad \frac{\partial \rho_v}{\partial t} + \frac{\sigma}{\epsilon}\rho_v = 0$$

Solving, $\rho_v = \rho_0 e^{-t/(\epsilon/\sigma)}$, which shows that the required time is $\tau \equiv \epsilon/\sigma$. (The thinning of the charge cloud is known as *charge relaxation*, and τ is called the *relaxation time* or *time constant*.)

2.166 An interface between two dielectrics is shown in Fig. 2-49. Obtain the relationship between the tangential components of the **E**-field at either side of the interface.

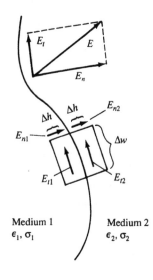

Fig. 2-49

▮ *Convention for interface problems:* At any point of the interface the unit normal vector **n** points out of medium 1 and into medium 2.

Since the electrostatic field is conservative, $\oint \mathbf{E} \cdot d\mathbf{s} = 0$ around the rectangular contour of Fig. 2-49:

$$E_{t2}\,\Delta w - E_{n2}\,\Delta h - E_{n1}\,\Delta h - E_{t1}\,\Delta w + E_{n1}\,\Delta h + E_{n2}\,\Delta h = 0$$

In the limit as the rectangular path approaches the surface, i.e., as $\Delta h \to 0$, this becomes

$$(E_{t2} - E_{t1})\,\Delta w = 0 \qquad \text{or} \qquad E_{t2} = E_{t1} \qquad (1)$$

Thus, *the tangential component of* **E** *is continuous across an interface.*

2.167 If a surface charge density ρ_s exists at the interface of two material media as shown in Fig. 2-50, obtain the relationship between the normal components of the **D**-vector at either side of the interface.

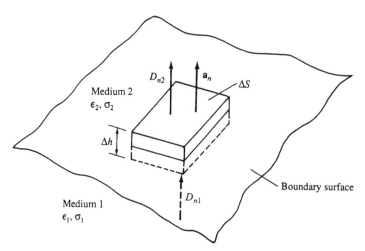

Fig. 2-50

▮ Apply Gauss' law, $\int_S \mathbf{D} \cdot d\mathbf{S} = \int_V \rho_v \, dV$, to the infinitesimal box shown in Fig. 2-50. As the height of this box approaches zero, i.e., as $\Delta h \to 0$, only the component of $\mathbf{D}$ normal to the boundary contributes to Gauss' law:

$$D_{n2}\,\Delta S - D_{n1}\,\Delta S = \lim_{\Delta h \to 0} \rho_v(\Delta h\,\Delta S) = \rho_s\,\Delta S \qquad \text{or} \qquad D_{n2} - D_{n1} = \rho_s \qquad (1)$$

Thus, at a point of an interface, *the jump in the normal component of* $\mathbf{D}$ *equals the local free surface charge density*.

2.168 Express the results of Problems 2.166 and 2.167 in terms of the tangential components of the $\mathbf{D}$-vector and the normal components of the $\mathbf{E}$-vector, respectively.

▮ $$\frac{D_{t2}}{\epsilon_2} = \frac{D_{t1}}{\epsilon_1} \qquad \text{and} \qquad \epsilon_2 E_{n2} - \epsilon_1 E_{n1} = \rho_s$$

2.169 Given medium 1 a perfect conductor ($\sigma_1 = \infty$) and medium 2 a perfect dielectric ($\sigma_2 = 0$), obtain the conditions on the normal and tangential components of $\mathbf{D}$ and $\mathbf{E}$ at the interface of the media.

▮ $\mathbf{E} = \mathbf{D} = 0$ in medium 1. Then Problems 2.166 through 2.168 give

$$E_{t2} = 0 \qquad D_{n2} = \rho_s \qquad D_{t2} = 0 \qquad E_{n2} = \frac{\rho_s}{\epsilon_2}$$

2.170 Repeat Problem 2.169 given both media perfect dielectrics, with $\rho_s = 0$.

▮ $$E_{t2} = E_{t1} \qquad D_{n2} = D_{n1} \qquad \frac{D_{t2}}{\epsilon_2} = \frac{D_{t1}}{\epsilon_1} \qquad \epsilon_2 E_{n2} = \epsilon_1 E_{n1}$$

2.171 Two cubes of dielectric materials have a common face in the xy plane of rectangular coordinates. An electric field $\mathbf{E}_2 = 3\mathbf{a}_x + 4\mathbf{a}_y - 12\mathbf{a}_z$ V/m exists in cube 2 ($z \geq 0$), the material of which has relative permittivity 3. Find $\mathbf{D}_1$ in cube 1, where the relative permittivity is 1.5.

▮ The normal to the interface is $+\mathbf{a}_z$. By Problem 2.170,

$$\mathbf{E}_1 = 3\mathbf{a}_x + 4\mathbf{a}_y + \frac{\epsilon_2}{\epsilon_1}(-12\mathbf{a}_z) = 3\mathbf{a}_x + 4\mathbf{a}_y - 24\mathbf{a}_z \quad \text{V/m}$$

whence $$\mathbf{D}_1 = \epsilon_1 \mathbf{E}_1 = 1.5\epsilon_0 \mathbf{E}_1 = \epsilon_0(4.5\mathbf{a}_x + 6.0\mathbf{a}_y - 36.90\mathbf{a}_z) \quad \text{C/m}^2$$

2.172 Write a set of boundary conditions on the components of $\mathbf{D}$ and $\mathbf{E}$ for the layers of dielectrics shown in Fig. 2-51.

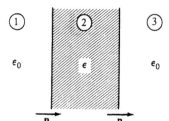

Fig. 2-51

▐ From Problem 2.170,

$$D_{n1} = D_{n2} \qquad \epsilon_0 E_{n1} = \epsilon E_{n2} \qquad D_{n2} = D_{n3} \qquad \epsilon E_{n2} = \epsilon_0 E_{n3}$$

2.173 For the **E**-fields shown at the interface of two dielectrics in Fig. 2-52, relate the angles θ_1 and θ_2.

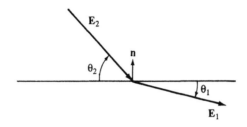

Fig. 2-52

▐ From Problem 2.170,

$$E_2 \cos \theta_2 = E_1 \cos \theta_1 \qquad \text{and} \qquad \epsilon_2(-E_2 \sin \theta_2) = \epsilon_1(-E_1 \sin \theta_1)$$

whence, by division, $\epsilon_2 \tan \theta_2 = \epsilon_1 \tan \theta_1$.

2.174 Write a set of boundary conditions for the normal and tangential components of the current density vector at the interface of two conducting media having conductivities σ_1 and σ_2.

▐ Problem 2.164 implies, for $\rho_v \equiv 0$, $\int \mathbf{J} \cdot d\mathbf{S} = 0$, which in turn implies

$$J_{n2} = J_{n1}$$

Since $\mathbf{J} = \sigma \mathbf{E}$ and the tangential component **E** is continuous,

$$\frac{J_{t2}}{\sigma_2} = \frac{J_{t1}}{\sigma_1}$$

2.175 Express the surface charge density ρ_s at the interface of Problem 2.174 in terms of E_{n1} or E_{n2} and the media constants.

▐ Problems 2.168 and 2.174 give $\epsilon_2 E_{n2} - \epsilon_1 E_{n1} = \rho_s$ and $\sigma_2 E_{n2} = \sigma_1 E_{n1}$. Hence

$$\rho_s = \left(\epsilon_2 - \epsilon_1 \frac{\sigma_2}{\sigma_1}\right) E_{n2} = \left(\epsilon_2 \frac{\sigma_1}{\sigma_2} - \epsilon_1\right) E_{n1}$$

2.176 Three point charges, Q_1, Q_2, and Q_3, are initially at infinity; they are brought to the respective locations P_1, P_2, and P_3, all other charge being absent. What is the potential energy of the three-charge configuration?

▐ By the work-energy theorem, the potential energy is equal to the work W required to assemble the configuration. Let the absolute potentials at point P due to Q_1, Q_2, and Q_3 (each in its final location) be $V_1(P)$, $V_2(P)$, and $V_3(P)$. If the charges are brought in from infinity in the order 1-2-3,

$$W = 0 + Q_2 V_1(P_2) + Q_3[V_1(P_3) + V_2(P_3)] \qquad (1)$$

Because the value of W is the independent of the order of assemblage, we obtain $3! - 1 = 5$ other equations from (1) by permuting the subscripts. Adding all six equations gives

$$6W = 3Q_1[V_2(P_1) + V_3(P_1)] + 3Q_2[V_1(P_2) + V_2(P_2)] + 3Q_3[V_1(P_3) + V_2(P_3)] \qquad (2)$$

Now, $[V_2(P_1) + V_3(P_1)] = V_{-1}(P_1)$, the absolute potential at P_1 due to all charges except Q_1; similarly for the other two sums. Thus (2) may be rewritten as

$$W = \tfrac{1}{2}[Q_1 V_{-1}(P_1) + Q_2 V_{-2}(P_2) + Q_3 V_3(P_3)] \qquad (3)$$

2.177 Generalize (2) of Problem 2.176 to a configuration of $n > 3$ point charges.

▐ $W = \dfrac{1}{2} \displaystyle\sum_{i=1}^{n} Q_i V_i(P_i)$.

2.178 Generalize the result of Problem 2.177 to a volume distribution of charge of (variable) density ρ.

▐ Writing the volume element as $d\tau = dx\, dy\, dz$, we have

$$W = \frac{1}{2} \iiint \rho V\, d\tau \qquad (1)$$

where the intergration is over the entire charge cloud.

2.179 Rewrite the energy formula (1) of Problem 2.178 in terms of **D**- and **E**-fields.

▌ By (1) of Problem 1.139,
$$\rho V = (\text{div } \mathbf{D})V = \text{div }(V\mathbf{D}) - (\text{grad } V) \cdot \mathbf{D} = \text{div }(V\mathbf{D}) + \mathbf{E} \cdot \mathbf{D}$$
so that, applying the divergence theorem

$$W = \frac{1}{2} \iint V\mathbf{D} \cdot d\mathbf{S} + \frac{1}{2} \iiint \mathbf{E} \cdot \mathbf{D} \, d\tau \qquad (1)$$

If we allow the volume of integration to expand indefinitely, the surface integral in (1) vanishes as $1/r$ (because $V \sim 1/r$, $D \sim 1/r^2$, and $S \sim r^2$). Thus,

$$W = \frac{1}{2} \iiint_{\text{all space}} \mathbf{E} \cdot \mathbf{D} \, d\tau \qquad (2)$$

2.180 Determine the energy density of an **E**-field within an infinite, homogenous, and isotropic medium having permittivity ϵ.

▌ By (2) of Problem 2.179,

$$W = \frac{1}{2} \iiint_{\text{all space}} \mathbf{E} \cdot \epsilon \mathbf{E} \, d\tau = \frac{\epsilon}{2} \iiint_{\text{all space}} E^2 \, d\tau \qquad (1)$$

which implies an energy density

$$w = \frac{dW}{d\tau} = \frac{\epsilon E^2}{2} \qquad (2)$$

2.181 The cubes in Problem 2.171 are so oriented that the points $(0, 0, 0)$ m and $(0, 0, -2)$ m both belong to cube 1. Find the potential difference between these two points.

▌ Integrate along the z axis:

$$V(0, 0, 0) - V(0, 0, -2) = -\int_{-2}^{0} E_{1z} \, dz = -\int_{-2}^{0} (-24) \, dz = 48 \text{ V}$$

2.182 Obtain the energy density in cube 2 of Problem 2.171.

▌
$$w_2 = \frac{\epsilon_2 E_2^2}{2} = \frac{3\epsilon_0 (3^2 + 4^2 + 12^2)}{2} = 253.5\epsilon_0 \ (\text{J/m}^3)$$

2.183 (a) Sketch the field lines for Problem 2.152 and (b) derive a differential equation for them.

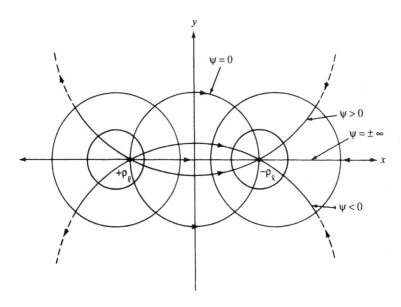

Fig. 2-53

▌ (a) The field lines are the orthogonal trajectories of the equipotentials; see Fig. 2-53. (b) First obtain a differential equation for the equipotentials, of the form $dy/dx = f(x, y)$; the field lines will then satisfy

$dy/dx = -1/f(x, y)$. Now, writing the equation of the equipotentials as

$$\frac{r_2^2}{r_1^2} = k^2 \qquad \text{or} \qquad \frac{\left(x - \frac{d}{2}\right)^2 + y^2}{\left(x + \frac{d}{2}\right)^2 + y^2} = k^2$$

differentiation by the quotient rule gives

$$\left[\left(x + \frac{d}{2}\right)^2 + y^2\right]\left[2\left(x - \frac{d}{2}\right)dx + 2y\,dy\right] - \left[\left(x - \frac{d}{2}\right)^2 + y^2\right]\left[2\left(x + \frac{d}{2}\right)dx + 2y\,dy\right] = 0$$

or

$$\frac{dy}{dx} = -\frac{x^2 - y^2 - (b^2/4)}{2xy}$$

Thus the required differential equation is

$$\frac{dy}{dx} = -\frac{2xy}{x^2 - y^2 - (b^2/4)} \tag{1}$$

Note that the particular field line $y = 0$ is a solution of (1).

2.184 Solve (1) of Problem 2.183 and thereby show that the field lines, too, are circles (strictly, arcs of circles).

❚ Rewrite (1) as

$$2xy\,dx + (y^2 - x^2 + b^2/4)\,dy = 0$$

Potential theory guarantees the existence of an integrating factor; it turns out to be $1/y^2$. Thus, we have

$$d\psi = \frac{2x}{y}\,dx + \frac{y^2 - x^2 + b^2/4}{y^2}\,dy = 0$$

Integrating, we obtain

$$\psi = \frac{1}{y}\int 2x\,dx = \frac{x^2}{y} + \omega(y)$$

$$\frac{\partial\psi}{\partial y} = \frac{y^2 - x^2 + b^2/4}{y^2} = -\frac{x^2}{y^2} + \frac{d\omega}{dy}$$

$$\frac{d\omega}{dy} = 1 + \frac{b^2}{4y^2}$$

$$\omega = y - \frac{b^2}{4y}$$

so that, finally,

$$\frac{x^2}{y} + y - \frac{b^2}{4y} = \psi \tag{1}$$

is the equation of the field lines, with one field line for each fixed value of ψ. Rewriting (1) as

$$x^2 + \left(y - \frac{\psi}{2}\right)^2 = \frac{b^2 + \psi^2}{4} \tag{2}$$

we recognize the field lines to be circles, centered on the y axis and passing through the two line charges. Several ψ-values are indicated in Fig. 2-53.

2.185 In a Millikan oil-drop experiment, the weight of a 1.6×10^{-14}-kg drop is exactly balanced by the electric force in a vertically directed 200 kV/m field. Calculate the charge on the drop in units of the electronic charge ($e = 1.6 \times 10^{-19}$ C).

❚ For equilibrium, $qE = mg$, or

$$\frac{q}{e} = \frac{mg}{Ee} = \frac{(1.6 \times 10^{-14}\text{ kg})(9.8\text{ m/s}^2)}{(200 \times 10^3\text{ N/C})(1.6 \times 10^{-19}\text{ C})} = 4.9$$

i.e., the drop had been stripped of five electrons.

2.186 A certain subatomic particle is heavier when charged to 1.6×10^{-19} C than when in the neutral state. The mass increase, translated into an energy increase via Einstein's relation $\Delta\mathscr{E} = c^2\,\Delta m$, amounts to 4.6 MeV. Identifying this energy with the electrostatic energy stored in the field of a uniformly charged sphere, calculate the radius of the particle.

❚ From Gauss' law, for the total charge Q and a sphere of radius R,

$$E_r = \begin{cases} \dfrac{Qr}{4\pi\epsilon_0 R^3} & 0 < r < R \\[2mm] \dfrac{Q}{4\pi\epsilon_0 r^2} & R < r < \infty \end{cases}$$

Energy stored in the electric field becomes

$$W_E = \frac{1}{2}\epsilon_0 \int_0^\infty E_r^2 (4\pi r^2)\, dr$$

$$= \frac{1}{2}\epsilon_0 \left[\int_0^R \left(\frac{Qr}{4\pi\epsilon_0 R^3}\right)^2 (4\pi r^2)\, dr + \int_R^\infty \left(\frac{Q}{4\pi\epsilon_0 r^2}\right)^2 (4\pi r^2)\, dr \right] = \frac{3Q^2}{20\pi\epsilon_0 R}$$

Therefore,

$$R = \frac{3Q^2}{20\pi\epsilon_0 W_E} = \frac{3(1.6 \times 10^{-19}\,\text{C})^2}{20\pi\left(\dfrac{10^{-9}}{36\pi}\,\text{F/m}\right)[(4.6 \times 10^6\,\text{V})(1.6 \times 10^{-19}\,\text{C})]} = 1.88 \times 10^{-16}\,\text{m}$$

2.187 Check Problem 2.186 by means of (1) of Problem 2.178.

❙ Within the sphere, the charge density and the *absolute* potential are given by

$$\rho = \frac{Q}{\frac{4}{3}\pi R^3} \qquad V(r) = c - \frac{Qr^2}{8\pi\epsilon_0 R^3} = \frac{Q}{8\pi\epsilon_0 R}\left(3 - \frac{r^2}{R^2}\right)$$

Here the value of c was fixed by the requirement that $V(R)$ equal the absolute potential of a point charge Q located at the origin. Then

$$W = \frac{\rho}{2}\int_0^R V(r)\, 4\pi r^2\, dr = \frac{3Q}{16\pi\epsilon_0 R^4}\int_0^R \left(3r^2 - \frac{r^4}{R^2}\right) dr = \frac{3Q^2}{20\pi\epsilon_0 R}$$

2.188 Check the result of Problems 2.186 and 2.187 by bringing in charge from infinity dq at a time (cf. Problem 2.176).

❙ Suppose that, shell by shell, a homogeneous sphere of charge, of radius r, has already been built up. All shells are of the same density ρ, so that the amount of charge in the sphere is $q = \rho(\frac{4}{3}\pi r^3)$ and the potential at the surface of the sphere is

$$V = \frac{q}{4\pi\epsilon_0 r} = \frac{q}{4\pi\epsilon_0 (3q/4\pi\rho)^{1/3}} = \frac{\rho^{1/3}}{(4\pi)^{2/3}\epsilon_0 3^{1/3}}q^{2/3}$$

The work needed to bring up the next shell dq is $dW = V\, dq$; hence

$$W = \int_0^Q V\, dq = \frac{\rho^{1/3}}{(4\pi)^{2/3}\epsilon_0 3^{1/3}}\left(\frac{3}{5}Q^{5/3}\right) \tag{1}$$

But $\rho = Q/(\frac{4}{3}\pi R^3)$, and when this is substituted in (1) the result is

$$W = \frac{3Q^2}{20\pi\epsilon_0 R}$$

as before.

2.189 Consider an electrically neutral two-conductor system. The *capacitance* of the system is defined as the ratio of the magnitude of the surface charge on either conductor to the magnitude of the potential difference between the conductors; that is,

$$C = \frac{Q}{V} \quad \text{(F)} \tag{1}$$

Determine the capacitance of a parallel-plate capacitor, where A is the surface area of each plate and d is the separation between them.

❙ Problem 2.102 gives $E = \rho_s/\epsilon_0$. But $V = Ed$ and $\rho_s = Q/A$; so

$$V = \frac{Qd}{\epsilon_0 A} \qquad \text{or} \qquad C = \frac{Q}{V} = \frac{\epsilon_0 A}{d}$$

2.190 By the method of Problem 2.188, show that the energy stored in a capacitor is given by

$$W = \frac{Q^2}{2C} = \frac{1}{2}CV^2 \tag{1}$$

❙ Starting with zero charge on either conductor, imagine the capacitor charged by transporting increments dq from the negative to the positive conductor. Figure 2-54 shows the process at an arbitrary stage; the work required to transport "the next" dq is $dW = v\, dq$. But $C = q/v$; that is, the capacitance, as a *geometric* property of the system, is the same at every stage. Thus,

$$W = \int_0^Q v\, dq = \frac{1}{C}\int_0^Q q\, dq = \frac{Q^2}{2C}$$

2.191 How much work is needed to charge the capacitor of Problem 2.189 to a voltage V_0?

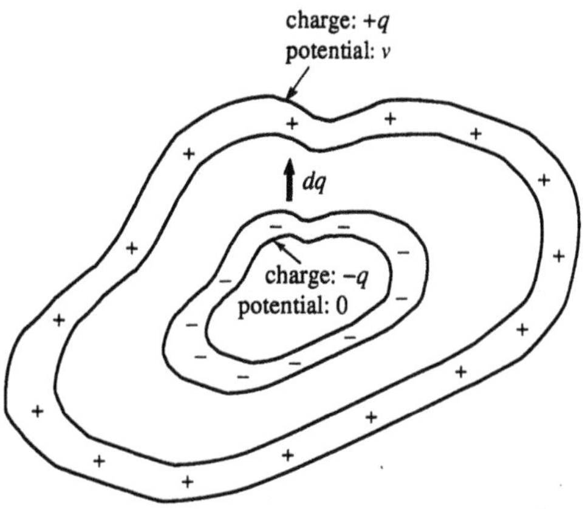

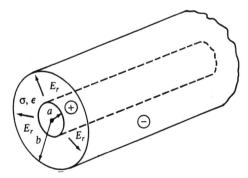

Fig. 2-54

▮ By Problems 2.189 and 2.190,

$$W = \frac{1}{2} CV_0^2 = \frac{\epsilon_0 A V_0^2}{2d}$$

2.192 Find the capacitance per unit length of the (infinite) coaxial cable indicated in Fig. 2-55.

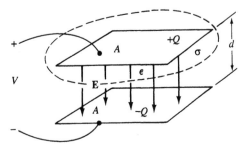

Fig. 2-55

▮ If $+q$ is the charge per unit length on the inner element, then, according to Gauss' law,

$$E_r = \frac{q}{2\pi\epsilon r} \quad (a < r < b)$$

Then the voltage difference between the inner and outer elements is

$$V = -\int_b^a E_r \, dr = \frac{q}{2\pi\epsilon} \ln \frac{b}{a}$$

and

$$c = \frac{q}{V} = \frac{2\pi\epsilon}{\ln(b/a)} \quad \text{(F/m)} \tag{1}$$

2.193 Express the time constant (Problem 2.165) of the capacitor of Fig. 2-56 in terms of its resistance R and capacitance C.

Fig. 2-56

▮ We have, as in Problem 2.154, $R = d/\sigma A$; and from Problem 2.189, $C = \epsilon A/d$. Therefore,

$$RC = \frac{\epsilon}{\sigma} \equiv \tau \tag{1}$$

Relation (1) holds for any capacitative system.

2.194   Determine the conductance per unit length g of the coaxial cable of Fig. 2-55.

▌ By Problems 2.192 and 2.193,

$$g = \frac{1}{r} = \frac{\sigma}{\epsilon} c = \frac{2\pi\sigma}{\ln(b/a)} \quad \text{(S/m)}$$

2.195   Using Problem 2.153, find the capacitance of a spherical capacitor.

▌
$$C = \frac{Q}{V_a - V_b} = \frac{4\pi\epsilon_0}{\dfrac{1}{a} - \dfrac{1}{b}} \tag{1}$$

If the space $a < r < b$ is filled with a dielectric, replace ϵ_0 in (1) by ϵ.

2.196   In the old CGS electrostatic system, capacitance had the dimension of length; in particular, the capacitance *of an isolated sphere* equaled the radius of the sphere. How did this come about?

▌ The CGS charge unit was chosen to make $E = Q/r^2$ for a point charge Q; i.e., $4\pi\epsilon_0 = 1$. An isolated sphere may be thought of as one plate of a spherical capacitor, the other plate being at infinity. As $b \to \infty$, (1) of Problem 2.195 gives $C \to 4\pi\epsilon_0 a$; or $C \to a$, in CGS units.

2.197   Starting with the fundamental energy formula

$$W = \frac{1}{2} \iiint\limits_{\text{volume}} \rho V \, d\tau$$

show that the power dissipated in a conductor under steady-state conditions is given by

$$P = \iiint \mathbf{E} \cdot \mathbf{J} \, d\tau \tag{1}$$

▌ The power dissipated is the rate of decrease of the stored energy: $P = -dW/dt$. Thus, since the voltage field does not vary with time,

$$P = \frac{1}{2} \iiint \frac{\partial \rho}{\partial t} V \, d\tau = \frac{1}{2} \iiint (\text{div } \mathbf{J}) V \, d\tau \tag{2}$$

The identity of Problem 1.139 and the divergence theorem allow (2) to be transformed to

$$P = \frac{1}{2} \iint V \mathbf{J} \cdot d\mathbf{S} + \frac{1}{2} \iiint \mathbf{E} \cdot \mathbf{J} \, d\tau \tag{3}$$

Now, since a charge element dq carries off energy $V \, dq$ as it crosses the boundary of the conductor, the surface integral in (3) represents the total rate of energy loss, which is just P. So,

$$P = \frac{1}{2} P + \frac{1}{2} \iiint \mathbf{E} \cdot \mathbf{J} \, d\tau \quad \text{or} \quad P = \iiint \mathbf{E} \cdot \mathbf{J} \, d\tau$$

2.198   For a conductive circuit element (Fig. 2-57), obtain the familiar formula $P = VI$.

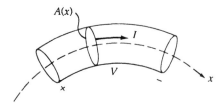

Fig. 2-57

▌ Apply (1) of Problem 2.197, assuming that $\mathbf{E}$ and $\mathbf{J}$ are everywhere in the x direction and are uniform over each cross section:

$$P = \int [E(x)J(x)]A(x) \, dx = \int E(x)I \, dx = \left(\int E(x) \, dx \right) I = VI$$

2.199   Find the power loss per unit length for the coaxial cable of Fig. 2-55, for a steady voltage difference V across it.

▌ By Problems 2.198 and 2.194,

$$p = Vi = gV^2 = \frac{2\pi\sigma V^2}{\ln(b/a)} \quad \text{(W/m)}$$

2.200 Determine the capacitance of the configuration shown in Fig. 2-58.

Fig. 2-58

▮ Capacitances in series obey

$$\frac{1}{C} = \Sigma \frac{1}{C_i}$$

Hence, by Problem 2.189,

$$\frac{1}{C} = \frac{1}{\epsilon_0 A} + \frac{1}{\epsilon A} + \frac{1}{\epsilon_0 A} \quad \text{or} \quad C = \frac{\epsilon_0 \epsilon A}{\epsilon_0 + 2\epsilon}$$

2.201 The cable shown in Fig. 2-59 is 8 km long. Calculate its capacitance.

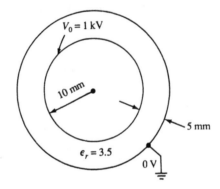

Fig. 2-59

▮ By (1) of Problem 2.192,

$$C = \frac{2\pi\epsilon L}{\ln(b/a)} = \frac{2\pi(3.5)(10^{-9}/36\pi)(8 \times 10^3)}{\ln(15/10)} = 3.8 \ \mu\text{F}$$

2.202 Determine the capacitance per unit length of the cable shown in Fig. 2-60.

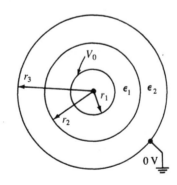

Fig. 2-60

▮ Using (1) of Problem 2.192,

$$c_1 = \frac{2\pi\epsilon_1}{\ln(r_2/r_1)} \quad \text{and} \quad c_2 = \frac{2\pi\epsilon_2}{\ln(r_3/r_2)}$$

Since the capacitances are in series,

$$c = \frac{c_1 c_2}{c_1 + c_2} = \frac{2\pi\epsilon_1\epsilon_2}{\epsilon_2 \ln(r_2/r_1) + \epsilon_1 \ln(r_3/r_2)}$$

2.203 For the coaxial cable shown in Fig. 2-60, with the conductor at potential V_0 and the outer shield grounded, determine the maximum electric intensity in each dielectric from the following data: $V_0 = 1.2 \ \text{kV}$, $\epsilon_{r1} = 4.5$, $\epsilon_{r2} = 3$, and $r_3 = 2r_2 = 4r_1 = 40 \ \text{mm}$.

▌ From Problem 2.202,

$$c_1 = \frac{2\pi \times 4.5 \times 10^{-9}}{36\pi \ln 2} = 0.36 \text{ nF/m} \qquad c_2 = \frac{2\pi \times 3 \times 10^{-9}}{36\pi \ln 2} = 0.24 \text{ nF/m}$$

Then

$$\frac{V_2}{V_1} = \frac{c_1}{c_2} = \frac{0.36}{0.24} = 1.5 \qquad \text{and} \qquad V_1 + V_2 = 1200 \text{ V}$$

from which $V_1 = 480 \text{ V}$ and $q = c_1 V_1 = 172.8 \text{ nC/m}$.

We know that $E_{r\,\text{max}}$ occurs at the inner surface of either dielectric and is given by

$$E_{r\,\text{max}} = \frac{\rho_s}{\epsilon} = \frac{q}{2\pi r \epsilon}$$

Hence, at $r = r_1$,

$$E_\text{max} = \frac{172.8 \times 10^{-9}}{2\pi(0.010)[(4.5)(10^{-9}/36\pi)]} = 69.1 \text{ kV/m}$$

and, at $r = r_2$, $E_\text{max} = (3/4)(69.1) = 51.8 \text{ kV/m}$.

2.204 A pair of 200-mm-long concentric cylindrical conductors of radii 50 and 100 mm is filled with a dielectric, $\epsilon = 10\epsilon_0$. A voltage is applied between the conductors to establish an electric field $E = (10^6/r)\mathbf{a}_r$ (V/m) between the cylinders. Calculate the energy stored.

▌ Stored energy is given by

$$W_E = \frac{1}{2}\int_\text{vol} \epsilon E^2 \, d\tau = \frac{\epsilon}{2}\int_{z=0}^{0.200}\int_{\phi=0}^{2\pi}\int_{r=0.050}^{0.100} \frac{(10^6)^2}{r^2} r \, dr \, d\phi \, dz = \frac{10 \times 10^{-9}}{36\pi \times 2} \times 0.200 \times 2\pi \times 10^{12} \ln 2 = 38.5 \text{ J}$$

2.205 Determine the capacitance and the applied voltage between the cylinders of Problem 2.204.

▌
$$C = \frac{2\pi\epsilon L}{\ln 2} = \frac{2\pi[10(10^{-9}/36\pi)](0.200)}{0.693} = 0.16 \text{ nF}$$

Since $W_E = CV^2/2$,

$$V = \left(\frac{2W_E}{C}\right)^{1/2} = \left(\frac{2 \times 38.5}{0.16 \times 10^{-9}}\right)^{1/2} = 0.694 \text{ MV}$$

2.206 Determine the capacitance of a parallel-plate capacitor filled with two dielectrics, as shown in Fig. 2-61.

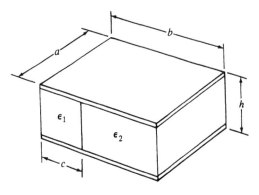

Fig. 2-61

▌
$$C_1 = \epsilon_1 \frac{ac}{h} \qquad \text{and} \qquad C_2 = \epsilon_2 \frac{a(b-c)}{h}$$

Because the capacitances are in parallel,

$$C = C_1 + C_2 = \frac{a}{h}[c(\epsilon_1 - \epsilon_2) + b\epsilon_2]$$

2.207 A parallel-plate capacitor has two layers of dielectrics, as shown in Fig. 2-62. How is the potential difference V applied between the plates divided across the dielectrics?

▌
$$\frac{V_1}{V_2} = \frac{C_2}{C_1} = \frac{\epsilon_2}{\epsilon_1}\frac{b}{a} \qquad \text{and} \qquad V_1 + V_2 = V$$

Thus,

$$V_1 = \frac{\epsilon_2 b}{\epsilon_1 a + \epsilon_2 b} V \qquad V_2 = \frac{\epsilon_1 a}{\epsilon_1 a + \epsilon_2 b} V$$

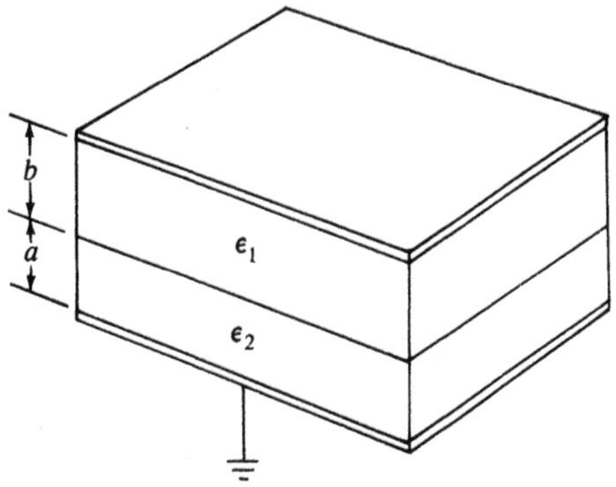

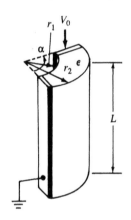

Fig. 2-62

2.208 A capacitor is formed from a segment of two coaxial cylinders, as shown in Fig. 2-63. The vertical planes form the plates of the capacitor. For the given dimensions, find the capacitance. Neglect fringing.

Fig. 2-63

$$V_0 = -\int \mathbf{E} \cdot d\mathbf{s} = -\int_0^\alpha \frac{1}{\epsilon} D_\phi \mathbf{a}_\phi \cdot r\, d\phi\, \mathbf{a}_\phi = -D_\phi \frac{r\alpha}{\epsilon}$$

But $\rho_s = -D_\phi = \epsilon V_0/r\alpha$, so that

$$Q = \int \rho_s\, dS = \frac{\epsilon V_0}{\alpha} \int_{z=0}^L \int_{r=r_1}^{r_2} \frac{dr\, dz}{r} = \frac{\epsilon L V_0}{\alpha} \ln \frac{r_2}{r_1}$$

Hence

$$C = \frac{Q}{V_0} = \frac{\epsilon L}{\alpha} \ln \frac{r_2}{r_1}$$

2.209 A region of a dielectric of uniform permittivity ϵ has a (variable) volume charge density ρ_v. Obtain a differential equation relating the potential V at any point to the charge density at that point.

$$\rho_v = \text{div } \mathbf{D} = \epsilon \text{ div } \mathbf{E} = -\epsilon \text{ div } (\text{grad } V) = -\epsilon \nabla^2 V$$

or $\nabla^2 V = -\rho_v/\epsilon$, which is known as *Poisson's equation*. *Laplace's equation* results when $\rho_v \equiv 0$.

2.210 The spherically symmetric potential distribution within a hydrogen atom is given by

$$V(r) = \frac{Ae^{-\alpha r}}{r} \left(1 + \frac{\alpha r}{2}\right)$$

where α and A are positive constants. Determine the spherical charge distribution required to produce this potential. Exclude the origin from the calculations.

In spherical coordinates, Poisson's equation yields

$$\frac{1}{r^2} \frac{\partial}{\partial r} \left(r^2 \frac{\partial V}{\partial r}\right) = -\frac{\rho}{\epsilon_0} \qquad (r > 0) \tag{1}$$

For the given V,

$$\frac{1}{r^2} \frac{\partial}{\partial r} \left(r^2 \frac{\partial V}{\partial r}\right) = \frac{A}{r^2} \frac{\partial}{\partial r} \left[r^2 \left(-\frac{e^{-\alpha r}}{r^2} + \frac{\alpha e^{-\alpha r}}{r} - \frac{\alpha^2 e^{-\alpha r}}{2}\right)\right]$$

$$= \frac{A}{r^2} \left(\alpha e^{-\alpha r} + \alpha^2 r e^{-\alpha r} - \alpha e^{-\alpha r} + \frac{\alpha^3}{2} r^2 e^{-\alpha r} - \alpha^2 r e^{-\alpha r}\right) = \frac{A}{2} \alpha^3 e^{-\alpha r} \tag{2}$$

From (1) and (2) we obtain $\rho = -\frac{1}{2} \epsilon_0 A \alpha^3 e^{-\alpha r}$.

2.211 Determine the conductance G of a spherical capacitor, of inner sphere radius a and outer sphere radius b, that is filled with a lossy dielectric having parameters σ and ϵ.

▮ By the general relation (*1*) of Problem 2.193, $G = (\sigma/\epsilon)C$. By Problem 2.195,

$$C = \frac{4\pi\epsilon ab}{b - a}$$

Hence, $G = 4\pi\sigma ab/(b - a)$.

2.212 The electrostatic energy stored in a region is changed by certain mechanical displacements. For instance, the energy stored in the capacitor shown in Fig. 2-64 will change if the movable upper plate is raised or lowered, or if the dielectric is pulled out of the capacitor. For a constant voltage across the plates obtain a general expression for the force developed in the system, assuming it to be lossless.

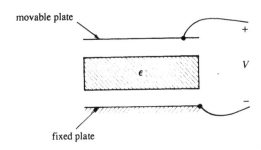

Fig. 2-64

▮ Since the system is conservative, we have

$$\begin{pmatrix} \text{input} \\ \text{electric} \\ \text{energy} \end{pmatrix} = \begin{pmatrix} \text{increase} \\ \text{in stored} \\ \text{energy} \end{pmatrix} + \begin{pmatrix} \text{mechanical} \\ \text{work done} \\ \text{by the system} \end{pmatrix} \tag{1}$$

or

$$VI\, dt = dW_e + F_e\, dx \tag{2}$$

where F_e is the mechanical force exerted by the system on the surroundings ($-F_e$ is the force exerted *on the* system *by* the surroundings) and where x describes the physical configuration. Now, $I = dQ/dt$, so that (*2*) becomes

$$dW_e = -F_e\, dx + V\, dQ \tag{3}$$

The state of the system may be specified through the independent variables x and V; thus (*3*) can be written

$$dW_e = -F_e\, dx + V\left(\frac{\partial Q}{\partial x}\, dx + \frac{\partial Q}{\partial V}\, dV\right) = \left(-F_e + V\frac{\partial Q}{\partial x}\right) dx + V\frac{\partial Q}{\partial V}\, dV \tag{4}$$

From (*4*) it follows at once that

$$\frac{\partial W_e}{\partial x} = -F_e + V\frac{\partial Q}{\partial x} \qquad \text{or} \qquad F_e = -\frac{\partial W_e}{\partial x} + V\frac{\partial Q}{\partial x} \tag{5}$$

2.213 Apply (*5*) of Problem 2.212 to determine the force between the plates of a parallel-plate capacitor.

▮ For plate separation x, the capacitance is $C(x) = \epsilon A/x$. Assume that the voltage between the plates is fixed, i.e., V is independent of x. Then

$$V\frac{\partial Q}{\partial x} = V\frac{\partial}{\partial x}(CV) = V^2\frac{\partial C}{\partial x}$$

and

$$\frac{\partial W_e}{\partial x} = \frac{\partial}{\partial x}\left(\frac{1}{2}CV^2\right) = \frac{1}{2}V^2\frac{\partial C}{\partial x}$$

Thus (*5*) for this specific case becomes

$$F_e = \frac{1}{2}V^2\frac{\partial C}{\partial x} = -\frac{V^2\epsilon A}{2x^2}$$

2.214 In Problem 2.213, F_e turns out to be negative. Why?

▮ The two plates carry opposite charges, so that they attract each other. Thus, for the setup of Fig. 2-64, the upper plate would tend to move downward ($dx < 0$). As a spontaneous motion, this corresponds to positive work *on* the surroundings ($F_e\, dx > 0$). Hence $F_e < 0$.

2.215 Show that for a capacitative system of fixed charge,

$$F_e = \frac{1}{2} V^2 \frac{dC}{dx} \qquad (1)$$

(cf. Problem 2.213).

▌ With $dQ = 0$, (3) of Problem 2.213 yields

$$F_e = -\frac{dW_e}{dx} = -\frac{d}{dx}\left(\frac{Q^2}{2C}\right) = -\frac{Q^2}{2}\frac{d}{dx}\left(\frac{1}{C}\right) = \frac{Q^2}{2C^2}\frac{dC}{dx} = \frac{1}{2}V^2\frac{dC}{dx}$$

2.216 A slab of dielectric ($\epsilon_r > 1$) is being inserted between the plates of a parallel-plate capacitor, with the charge on the capacitor held fixed. Does this require external work?

▌ No: By (1) of Problem 2.215,

$$F_e\, dx = \text{work done by system} = \tfrac{1}{2}V^2\, dC$$

But $dC > 0$, since empty space is being replaced by a medium of higher permittivity. Thus the system does positive work (e.g., on the hand guiding the slab); that is to say, the capacitor *attracts* the slab.

2.217 Outline the *separation of variables* for the three-dimensional Laplace's equation in rectangular coordinates:

$$\frac{\partial^2 V}{\partial x^2} + \frac{\partial^2 V}{\partial y^2} + \frac{\partial^2 V}{\partial z^2} = 0 \qquad (1)$$

▌ Assume a solution of the form $V = X(x)Y(y)Z(z)$. Substitute this *product solution* in (1) and divide through by V, to obtain

$$\frac{X''}{X} + \frac{Y''}{Y} + \frac{Z''}{Z} = 0 \qquad (2)$$

Each term on the left of (2) must be a constant; otherwise (2) would imply that a function of one of the independent variables was equal to a function of the other two—an impossibility. Thus, writing the constants as squares, we obtain

$$X'' - k_x^2 X = 0 \qquad Y'' - k_y^2 Y = 0 \qquad Z'' - k_z^2 Z = 0 \qquad (3)$$

where the constants are constrained to satisfy

$$k_x^2 + k_y^2 + k_z^2 = 0 \qquad (4)$$

2.218 Solve the system (3) and (4) of Problem 2.217 for k_x, k_y, and k_z real.

▌ The only real solution of (4) is $k_x = k_y = k_z = 0$. Then X, Y, and Z are linear functions, and the product solution for V is

$$V = (A_x x + B_x)(A_y y + B_y)(A_z z + B_z) \qquad (1)$$

2.219 Solve the system (3) and (4) of Problem 2.217 when two of the k's are real and one is pure imaginary.

▌ Let $k_x = j\kappa_x$, $k_y = \kappa_y$, $k_z = \kappa_z$, where $j = \sqrt{-1}$ and the k's are nonnegative real numbers that satisfy $\kappa_y^2 + \kappa_z^2 = \kappa_x^2$ (the Pythagorean relation). Equations (3),

$$X'' + \kappa_x^2 X = 0 \qquad Y'' - k_y^2 Y = 0 \qquad Z'' - \kappa_z^2 Z = 0$$

have the respective basic solutions: $\cos \kappa_x x$ and $\sin \kappa_x x$; $\cosh \kappa_y y$ and $\sinh \kappa_y y$; $\cosh \kappa_z z$ and $\sinh \kappa_z z$. These give the product solution

$$V = (a_x \cos \kappa_x x + b_x \sin \kappa_x x)(a_y \cosh \kappa_y y + b_y \sinh \kappa_y y)(a_z \cosh \kappa_z z + b_z \sinh \kappa_z z) \qquad (1)$$

2.220 Solve by separation of variables Laplace's equation in two dimensions:

$$\frac{\partial^2 V}{\partial x^2} + \frac{\partial^2 V}{\partial y^2} = 0$$

▌ Any solution of (1) of Problem 2.217 that is independent of z is a solution of the two-dimensional equation, and conversely. Thus, (1) of Problem 2.218—with $A_z = 0$, $B_z = 1$—and (1) of Problem 2.219—for $\kappa_z = 0$, $a_z = 1$—are two-dimensional product solutions.

2.221 Determine the potential within a rectangular electrode if the boundary conditions are as indicated in Fig. 2-65(a).

▌ Within the electrode the potential satisfies

$$\frac{\partial^2 V}{\partial x^2} + \frac{\partial^2 V}{\partial y^2} = 0 \qquad (1)$$

The boundary conditions suggest one of the product solutions found in Problem 2.220; namely,

$$V = V_0 \sin \kappa x \sin \kappa y \qquad (2)$$

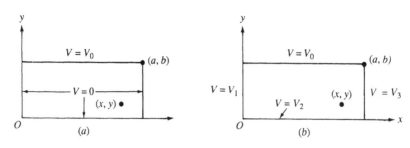

Fig. 2-65

(The conditions $\kappa_z = 0$ and $\kappa_y^2 + \kappa_z^2 = \kappa_x^2$ force $\kappa_x = \kappa_y \equiv \kappa$.) Solution (2) automatically satisfies the homogeneous boundary conditions on $x = 0$ and $y = 0$, whatever the value of κ. For V to vanish on $x = a$, we must have $\sin \kappa a = 0$, or $\kappa = n\pi/a$ ($n = 1, 2, 3, \dots$). Finally, the boundary condition on $y = b$ yields

$$V_0 = V_0 \sin \frac{n\pi x}{a} \sinh \frac{n\pi b}{a} \quad \text{or} \quad 1 = \sinh \frac{n\pi b}{a} \sin \frac{n\pi x}{a}. \tag{3}$$

It is obvious that no choice of n will make (3) valid for all $0 < x < a$. The way out of this difficulty is through superposition of product solutions, which is possible because Laplace's equation is linear. Thus we write

$$V = V_0 \sum_{n=1}^{\infty} C_n \sinh \frac{n\pi y}{a} \sin \frac{n\pi x}{a} \tag{4}$$

By construction, (4) satisfies the three homogeneous boundary conditions; the evaluation of the constants C_n from the nonhomogeneous boundary condition is left for Problem 2.231.

2.222 Solve the more general boundary-value problem shown in Fig. 2-65(b).

❚ Suppose the constants C_n in (4) of Problem 2.221 have been evaluated; then (4) can be written as

$$V = V_0 F(x, y; a, b)$$

Now the problem of Fig. 2-65(b) may be considered the superposition of four subproblems of the type of Fig. 2-65(a); in each subproblem, one of the sides is at its proper potential and the other three sides are at zero potential. Therefore, the solution to the problem of Fig. 2-65(b) is

$$V = V_0 F(x, y; a, b) + V_1 F(y, a - x; b, a) + V_2 F(a - x, b - y; a, b) + V_3 F(b - y, a - x; b, a)$$

2.223 Find the potential between the plates of a large parallel-plate capacitor, assuming that one plate, at $x = 0$, is at V_0 and the other plate, at $x = d$, is at V_d.

❚ Make an educated guess:

$$V = \frac{V_d - V_0}{d} x + V_0$$

This function obeys Laplace's equation (see Problem 2.218) and satisfies the boundary conditions. Therefore (Problem 2.226) it is *the* solution.

2.224 The cross section of a coaxial cable is shown in Fig. 2-66. For the potentials of the electrodes as shown, determine the **E**-field within the dielectric.

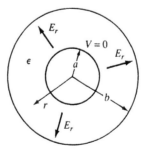

Fig. 2-66

❚ Because of symmetry, $V = V(r)$ and Laplace's equation becomes

$$\frac{1}{r} \frac{d}{dr} \left(r \frac{dV}{dr} \right) = 0 \tag{1}$$

Integrating (1) in two steps yields $V = C_1 \ln r + C_2$. If we let the conductors be at potentials V_0 and 0 as shown, we find

$$V = \frac{V_0}{\ln(b/a)} \ln \frac{r}{a} \qquad (2)$$

Then $\mathbf{E} = E_r \mathbf{a}_r$, where

$$E_r = -\frac{dV}{dr} = \frac{-V_0}{\ln(b/a)} \frac{1}{r} \quad (a < r < b) \qquad (3)$$

(The direction in Fig. 2-66 corresponds to $V_0 < 0$.)

2.225 Retrieve the solution of Problem 2.145 from Poisson's equation.

▮ By symmetry $V = V(r)$, so that Poisson's equation becomes

$$\frac{1}{r} \frac{d}{dr}\left(r \frac{dV}{dr}\right) = -\frac{\rho}{\epsilon_0} \qquad (1)$$

Note that a particular solution of (1) is $(-\rho/4\epsilon_0)r^2$; and the general solution of the homogeneous equation [viz., (1) of Problem 2.224] is $C_1 \ln r + C_2$. Therefore,

$$V(r) = -\frac{\rho}{4\epsilon_0} r^2 + C_1 \ln r + C_2$$

To keep $V(0)$ finite, we must have $C_1 = 0$. Then the condition $V(a) = V_0$ yields $C_2 = V_0 + (\rho a^2/4\epsilon_0)$, so that

$$V = V_0 + \frac{\rho}{4\epsilon_0}(a^2 - r^2)$$

as before.

2.226 Show that within a closed region, there is at most one function V that satisfies $\nabla^2 V = 0$ and assumes specified values over the boundary of the region. (This statement is known as the *uniqueness theorem* for Laplace's equation.)

▮ We prove the uniqueness theorem by contradiction; that is, first we assume that there are two possible solutions, V_1 and V_2, to the boundary value problem. In (1) of Problem 1.139, let $g \equiv V_1 - V_2$ and $\mathbf{F} \equiv \nabla g$:

$$\text{div}(g\,\nabla g) = g\,\nabla^2 g + |\nabla g|^2 = |\nabla g|^2 \qquad (1)$$

since, by linearity, $\nabla^2 g = 0$. Integrate (1) throughout the volume of the region and apply the divergence theorem:

$$\iiint_{\text{volume}} |\nabla g|^2\, d\tau = \iint_{\text{surface}} g\,\nabla g \cdot d\mathbf{S} = 0 \qquad (2)$$

since, by hypothesis, $g = 0$ at each point of the surface. By the usual continuity argument, (2) establishes that $\nabla g \equiv \mathbf{0}$ in the region; that is, $g \equiv$ const. in the region. By the homogeneous boundary condition, the constant can be only zero. Thus, $g \equiv 0$, or $V_1 \equiv V_2$.

2.227 Let $\mathscr{S}$ be the unit sphere about the origin. The boundary-value problem
$$\nabla^2 V = 0 \quad \text{inside } \mathscr{S}$$

$$V = 1 \quad \text{on } \mathscr{S}$$
seems to have two solutions, $V_1 = 1/r$ and $V_2 = 1$, in violation of the uniqueness theorem. Explain.

▮ $\nabla^2 V_1 = 0$, *except at the origin*; V_2 is the unique solution of the given problem.

2.228 Show that at any point P of a charge-free region the value of the potential is the average of its values over the surface of any (suitably small) sphere centered on P. (This is a statement of the *mean-value theorem*.)

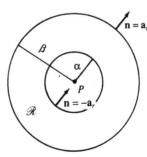

Fig. 2-67

▮ In Fig. 2-67, the point P and the concentric spheres of radii α and β lie completely within the charge-free region. In the subregion $\mathscr{R}$ between the spheres, we shall apply an identity directly derived from (1) of Problem 1.139:

$$\text{div} (g \, \nabla h) - \text{div} (h \, \nabla g) = g \, \nabla^2 h - h^2 \, \nabla^2 g \tag{1}$$

Choose $g = V$ (the potential) and $h = 1/r$:

$$\text{div} \left[V \left(-\frac{1}{r^2} \mathbf{a}_r \right) \right] + \text{div} \left(\frac{1}{r} \mathbf{E} \right) = 0 \tag{2}$$

since, in $\mathscr{R}$, both V and $1/r$ are solutions of Laplace's equation (cf. Problem 2.227). Thus, integrating (2) throughout R and using the divergence theorem, we obtain

$$-\frac{1}{\beta^2} \iint_{r=\beta} V \, dS + \frac{1}{\alpha^2} \iint_{r=\alpha} V \, dS + \frac{1}{\beta} \iint_{r=\beta} E_r \, dS - \frac{1}{\alpha} \iint_{r=\alpha} E_r \, dS = 0 \tag{3}$$

By Gauss' law, both integrals of $E_r = D_r/\epsilon$ vanish (neither sphere has any charge in its interior). Thus, after multiplication by $1/4\pi$, we can rewrite (3) as $[S(r) \equiv 4\pi r^2]$:

$$\frac{1}{S(\alpha)} \int_{r=\alpha} V \, dS = \frac{1}{S(\beta)} \iint_{r=\beta} V \, dS \tag{4}$$

Finally, let $\alpha \rightarrow 0$ in (4), obtaining in the limit

$$V(P) = \frac{1}{S(\beta)} \iint_{r=\beta} V \, dS = \frac{1}{4\pi} \int_{\phi=0}^{2\pi} \int_{\theta=0}^{\pi} V(\theta, \phi) \sin \theta \, d\theta \, d\phi$$

which is the mean-value theorem.

2.229 The mean-value theorem holds in two dimensions, with circles replacing spheres. Show that it also holds in one dimension.

❚ In one dimension,

$$\frac{d^2 V}{dX^2} = 0 \qquad \text{or} \qquad V(x) = Ax + B$$

Then, as a linear function,

$$V \left(\frac{x_1 + x_2}{2} \right) = \frac{V(x_1) + V(x_2)}{2}$$

which is the one-dimensional mean-value theorem.

2.230 For the spherical conducting shells filled with two dielectrics shown in Fig. 2-68, determine the potential distribution in the region $c < r < a$.

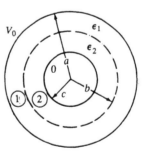

Fig. 2-68

❚ The spherically symmetric Laplace's equation,

$$\frac{d}{dr} \left(r^2 \frac{dV}{dr} \right) = 0$$

has fundamental solutions $1/r$ (for $r \neq 0$) and 1. Thus, in regions 1 and 2, respectively, we have

$$V_1 = -\frac{C_1}{r} + C_2 \qquad \text{for } b < r < a \tag{1}$$

$$V_2 = -\frac{C_3}{r} + C_4 \qquad \text{for } c < r < b \tag{2}$$

where the constants C_1, C_2, C_3, and C_4 are to be evaluated from the boundary conditions. At $r = c$, $V_2 = 0$ gives

$$C_4 = \frac{C_3}{c} \tag{3}$$

At $r = a$, $V_1 = V_0$ yields

$$V_0 = -\frac{C_1}{a} + C_2 \tag{4}$$

At $r = b$, we have $V_1 = V_2$, so that

$$-\frac{C_1}{b} + C_2 = -\frac{C_3}{b} + C_4 \tag{5}$$

Finally, at $r = b$, Problem 2.170 gives

$$\epsilon_1 E_{r1} = \epsilon_2 E_{r2} \quad \text{or} \quad \epsilon_1 \frac{dV_1}{dr}\bigg|_{r=b} = \epsilon \frac{dV_2}{dr}\bigg|_{r=b} \quad \text{or} \quad \epsilon_1 C_1 = \epsilon_2 C_3 \tag{6}$$

Solving (3)-(4)-(5)-(6) for the constants and substituting in (1) and (2) yields

$$V_1 = \left[\frac{1/a - 1/r}{(\epsilon_2/\epsilon_1)(1/c - 1/b) + (1/b - 1/a)} + 1\right]V_0 \quad (b < r < a)$$

$$V_2 = \frac{1/c - 1/r}{(1/c - 1/b) + (\epsilon_1/\epsilon_2)(1 + b - 1/a)} V_0 \quad (c < r < b)$$

2.231 Complete the solution of Problem 2.221 by evaluating the constants C_n in (4).

▮ The boundary condition at $y = b$ yields

$$1 = \sum_{n=1}^{\infty} a_n \sin \frac{n\pi x}{a}$$

where $a_n \equiv C_n \sinh(n\pi b/a)$ are seen to be the coefficients in the Fourier sine series for the function 1 over the interval $0 < x < a$. Standard formulas give

$$a_n = \begin{cases} 0 & n \text{ even} \\ 4/n\pi & n \text{ odd} \end{cases}$$

Thus

$$C_n = \begin{cases} 0 & n \text{ even} \\ 4/[n\pi \sinh(n\pi b/a)] & n \text{ odd} \end{cases}$$

2.232 Apply the method of separation of variables to Laplace's equation in cylindrical coordinates, assuming that there is no variation of the field or potential with ϕ.

▮ Since there is no variation with respect to ϕ, Laplace's equation in cylindrical coordinates becomes

$$\frac{\partial^2 V}{\partial r^2} + \frac{1}{r}\frac{\partial V}{\partial r} + \frac{\partial^2 V}{\partial z^2} = 0 \tag{1}$$

Proceeding as with rectangular coordinates, we assume a solution of the form $V(r, z) = R(r)Z(z)$, substitute in (1), and divide through the resulting equation by RZ:

$$\frac{R''}{R} + \frac{1}{r}\frac{R'}{R} = -\frac{Z''}{Z} \tag{2}$$

By the usual argument, both sides of (2) must be equal to a constant, say, $-\lambda^2$. Hence, we obtain the separated equations

$$Z'' - \lambda^2 Z = 0 \tag{3}$$

and

$$R'' + \frac{1}{r}R' + \lambda^2 R = 0 \tag{4}$$

Equation (3) obviously is satisfied by $e^{\pm\lambda z}$ (or $\cosh \lambda z$ and $\sinh \lambda z$); (4) will be treated in Problems 2.238 and 2.239.

2.233 Show that the *rectangular components* of the **E**-field satisfy Laplace's equation in a charge-free region.

▮ The potential V satisfies

$$\frac{\partial^2 V}{\partial x^2} + \frac{\partial^2 V}{\partial y^2} + \frac{\partial^2 V}{\partial z^2} = 0$$

Hence

$$\begin{aligned} 0 &= -\frac{\partial}{\partial x}\left[\frac{\partial^2 V}{\partial x^2} + \frac{\partial^2 V}{\partial y^2} + \frac{\partial^2 V}{\partial z^2}\right] \\ &= \frac{\partial^2}{\partial x^2}\left(-\frac{\partial V}{\partial x}\right) + \frac{\partial^2}{\partial y^2}\left(-\frac{\partial V}{\partial x}\right) + \frac{\partial^2}{\partial z^2}\left(-\frac{\partial V}{\partial x}\right) \\ &= \frac{\partial^2 E_x}{\partial x^2} + \frac{\partial^2 E_x}{\partial y^2} + \frac{\partial^2 E_x}{\partial z^2} \end{aligned}$$

Similarly for E_y and E_z.

2.234 Verify that the function $f(r, \phi) = (\cos \phi)/r$ satisfies Laplace's equation in cylindrical coordinates.

$\blacksquare$
$$\frac{1}{r}\frac{\partial}{\partial r}\left(r\frac{\partial f}{\partial r}\right) + \frac{1}{r^2}\frac{\partial^2 f}{\partial \phi^2} = \frac{1}{r}\frac{\partial}{\partial r}\left(-\frac{\cos \phi}{r}\right) + \frac{1}{r^2}\left(-\frac{\cos \phi}{r}\right) = \frac{\cos \phi}{r^3} - \frac{\cos \phi}{r^3} = 0$$

2.235 Interpret Problem 2.234 in the light of Problem 2.233.

$\blacksquare$ For the **E**-field of an infinite line charge along the z axis we have $\mathbf{E} \propto (1/r)\mathbf{a}_r$. Hence $E_x \propto (1/r) \cos \phi = f(r, \phi)$ must satisfy Laplace's equation.

2.236 From the mean-value theorem (Problem 2.228) obtain the *maximum-value theorem* for solutions of Laplace's equation.

$\blacksquare$ If the potential had a local maximum at a point P *inside* a charge-free region, then P could be surrounded by a small sphere at every point of the surface of which $V < V(P)$. This would contradict the mean-value theorem.
  Note that, through Problem 2.233, the maximum-value theorem also applies to the rectangular field components.

2.237 Find the potential within the semiinfinite strip shown in Fig. 2-69.

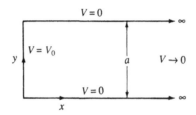

Fig. 2-69

$\blacksquare$ In the product solution (1) of Problem 2.219, interchange x and y and replace the hyperbolic functions by exponentials (the one set being a linear combination of the other):
$$V = (a_x e^{\kappa_x x} + b_x e^{-\kappa_x x})(a_y \cos \kappa_y y + b_y \sin \kappa_y y)$$
The three homogeneous boundary conditions force $a_x = 0$, $a_y = 0$, $\kappa_y = \kappa_x = n\pi/a$ $(n = 1, 2, 3, \ldots)$. The complete solution by superposition is therefore
$$V(x, y) = \sum_{n=1}^{\infty} B_n e^{-n\pi x/a} \sin \frac{n\pi y}{a}$$
The constants B_n are evaluated from the nonhomogeneous boundary condition—exactly as in Problem 2.231. The final result is
$$V(x, y) = \sum_{n \text{ odd}} \frac{4V_0}{n\pi} e^{-n\pi x/a} \sin \frac{n\pi y}{a}$$

2.238 Discuss the solutions of (4) of Problem 2.232 for $\lambda \geq 0$.

$\blacksquare$ **Case 1:** $\lambda = 0$. By Problem 2.224, a general solution is
$$R = a_0 \ln r + b_0$$
Case 2: $\lambda > 0$. Introduce the new independent variable $\rho = \lambda r$; (4) becomes
$$\frac{d^2 R}{d\rho^2} + \frac{1}{\rho}\frac{dR}{d\rho} + R = 0 \tag{1}$$
Equation (1) is recognized as *Bessel's equation of order zero*. One basic solution, regular at $\rho = 0$, is always chosen:
$$J_0(\rho) = \sum_{k=0}^{\infty} \frac{(-1)^k \rho^{2k}}{2^{2k}(k!)^2} = 1 - \frac{\rho}{4} + \frac{\rho^2}{64} - \cdots \tag{2}$$
This solution behaves like a damped sine wave as $\rho \to \infty$. For the second basic solution, which has a logarithmic singularity at the origin, a widely used choice is
$$Y_0(\rho) = \frac{2}{\pi}\left\{\left(\ln \frac{\rho}{2}\gamma\right)J_0(\rho) + \left[\frac{\rho^2}{2^2}(1) - \frac{\rho^4}{2^2 4^2}\left(1 + \frac{1}{2}\right) + \cdots\right]\right\} \tag{3}$$
where γ denotes Euler's constant.
  Reinstating the variable r, we obtain superposition solutions of (1) of Problem 2.232 of the type
$$\sum_{n=1}^{\infty} a_n J_0(\lambda_n r) \cosh \lambda_n z \qquad \text{or} \qquad \int_0^{\infty} a(\lambda) J_0(\lambda r) \cosh \lambda z \, d\lambda$$
depending on whether the boundary conditions dictate a discrete or a continuous spectrum of eigenvalues.

2.239 Discuss the solutions of (4) of Problem 2.232 for $\lambda^2 < 0$.

▮ Simply substitute $\lambda = j\alpha$ ($\alpha > 0$) in Problem 2.238, giving the *modified Bessel functions* $J_0(j\alpha r) \equiv I_0(\alpha r)$ and $K_0(\alpha r)$ [which is essentially the real part of $Y_0(j\alpha r)$], and superposition solutions such as

$$\sum_{n=1}^{\infty} a_n I_0(\alpha_n r) \cos \alpha_n z \quad \text{or} \quad \int_0^{\infty} a(\alpha) I_0(\alpha r) \cos \alpha z \, d\alpha$$

2.240 For the two-dimensional electrostatic field problem shown in Fig. 2-70(a) (cylindrical coordinates) write an elementary product solution that satisfies the homogeneous boundary conditions.

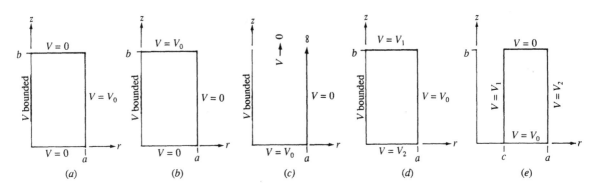

Fig. 2-70

▮ Using Problem 2.239: $V_n(r, z) = I_0(n\pi r/b) \sin(n\pi z/b)$, for $n = 1, 2, 3, \ldots$.

2.241 Repeat Problem 2.240 for the boundary conditions shown in Fig. 2-70(b).

▮ Using Problem 2.238: $V_n(r, z) = J_0(\lambda_n r) \sinh \lambda_n z$, where $\lambda_1 a, \lambda_2 a, \lambda_3 a, \ldots$ are the positive roots of $J_0(x) = 0$.

2.242 Repeat Problem 2.240 for the boundary conditions shown in Fig. 2-70(c).

▮ From Problem 2.238, $V_n(r, z) = J_0(\lambda_n r)e^{-\lambda_n z}$, with the λ_n as in Problem 2.241.

2.243 Complete the solution to Problem 2.240.

▮ Write $V(r, z) = \sum_{n=1}^{\infty} C_n V_n(r, z)$. The boundary condition at $r = a$ gives

$$V_0 = \sum_{n=1}^{\infty} C_n V_n(a, z) = \sum_{n=1}^{\infty} C_n I_0\left(\frac{n\pi a}{b}\right) \sin \frac{n\pi z}{b} \tag{1}$$

so that the C_n can be obtained from the coefficients in a Fourier sine series. Proceeding as in Problem 2.231, one finds

$$C_n = \begin{cases} 0 & n \text{ even} \\ 4V_0/n\pi I_0(n\pi a/b) & n \text{ odd} \end{cases} \tag{2}$$

[The C_n are well defined because $I_0(x) > 0$ for all $x > 0$.]

2.244 Complete the solution to Problem 2.241.

▮ In $V(r, z) = \sum_{n=1}^{\infty} C_n V_n(r, z)$, the superposition constants are now determined by

$$V_0 = \sum_{n=1}^{\infty} D_n J_0(\lambda_n r) \quad \text{where} \quad D_n \equiv C_n \sinh \lambda_n b \tag{1}$$

This is a Fourier–Bessel expansion of the constant function V_0 over the interval $0 < r < a$, with J_0 playing the role of sine. The relevant orthogonality integral is

$$\int_0^a J_0(\lambda_p r) J_0(\lambda_q r) \, r \, dr = \begin{cases} 0 & p \neq q \\ [aJ_1(\lambda_p a)]^2/2 & p = q \end{cases} \tag{2}$$

Thus, multiplying (1) through by $rJ_0(\lambda_m r)$, integrating from $r = 0$ to $r = a$, and using $\int xJ_0(x) \, dx = xJ_1(x)$, one finds

$$\frac{V_0 a}{\lambda_m} J_1(\lambda_m a) = D_m \frac{[aJ_1(\lambda_m a)]^2}{2}$$

which leads to the final result

$$C_n = \frac{2V_0}{\lambda_n a J_1(\lambda_n a) \sinh \lambda_n b}$$

[Since $J_1(x)$ has no zeros in common with $J_0(x)$, the C_n are well defined.]

2.245 Complete the solution to Problem 2.242.

▮ Imposing the boundary condition at $z = 0$ on the solution $V(r, z) = \sum_{n=1}^{\infty} C_n V_n(r, z)$, we obtain

$$V_0 = \sum_{n=1}^{\infty} C_n J_0(\lambda_n r)$$

But this is just (1) of Problem 2.244 with the factor $\sinh \lambda_n b$ replaced by 1. Therefore

$$C_n = \frac{2V_0}{\lambda_n a J_1(\lambda_n a)}$$

2.246 Outline the solution of the boundary-value problem indicated in Fig. 2-70(d).

▮ Use superposition to reduce the problem to the already solved subproblems of Fig. 2-70(a) and (b). Explicitly, let $V_0\Phi(r, z)$ and $V_0\Psi(r, z)$ respectively denote the solutions of the subproblems. Then the given problem has the solution

$$V(r, z) = V_0\Phi(r, z) + V_1\Psi(r, z) + V_2\Psi(r, b - z)$$

2.247 Solve the boundary-value problem shown in Fig. 2-71. The cylinder is bisected by the xy plane, and

$$V(a, z) = \begin{cases} V_0 & 0 < z < h/2 \\ -V_0 & -h/2 < z < 0 \end{cases}$$

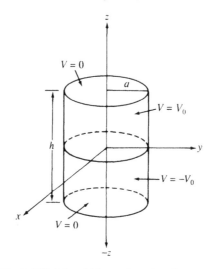

Fig. 2-71

▮ The solution to the problem of Fig. 2-70(a) is odd in z; therefore it automatically provides the solution to the present problem. Replacing b by $h/2$, we have, from Problems 2.240 and 2.243,

$$V(r, z) = \sum_{n \text{ odd}} \frac{4V_0}{n\pi} \frac{I_0(2n\pi r/h)}{I_0(2n\pi a/h)} \sin(2n\pi z/h)$$

$$= \sum_{m=2, 6, 10, \ldots} \frac{8V_0}{m\pi} \frac{I_0(m\pi r/h)}{I_0(m\pi a/h)} \sin(m\pi z/h)$$

for $0 < r < a$, $-h/2 < z < h/2$.

2.248 Set up the superposition solution for the boundary-value problem of Fig. 2-70(e).

▮ Because of the two-point boundary condition in r, both solutions of Bessel's equation—in this case I_0 and K_0—will be required. Observe that the singularity of K_0 (at $r = 0$) lies *outside* the region of interest. Taking the eigenvalues from Problem 2.240, one has

$$V(r, z) = \sum_{n=1}^{\infty} \left[A_n I_0\left(\frac{n\pi r}{b}\right) + B_n K_0\left(\frac{n\pi r}{b}\right) \right] \sin\frac{n\pi z}{b}$$

2.249 In a certain class of problems we have symmetry of fields about the z axis in spherical coordinates. Separate Laplace's equation for this case.

■ With $\partial V / \partial \phi = 0$, Laplace's equation becomes

$$\frac{\partial}{\partial r}\left(r^2 \frac{\partial V}{\partial r}\right) + \frac{1}{\sin \theta} \frac{\partial}{\partial \theta}\left(\sin \theta \frac{\partial V}{\partial \theta}\right) = 0$$

or

$$r^2 \frac{\partial^2 V}{\partial r^2} + 2r \frac{\partial V}{\partial r} + \frac{\partial^2 V}{\partial \theta^2} + \frac{1}{\tan \theta} \frac{\partial V}{\partial \theta} = 0 \qquad (1)$$

Setting $V(r, \theta) = R(r)\,\Theta(\theta)$, (1) separates into

$$r^2 \frac{R''}{R} + 2r \frac{R'}{R} = -\frac{\Theta''}{\Theta} - \frac{1}{\tan \theta} \frac{\Theta'}{\Theta} = \lambda$$

or

$$r^2 \frac{d^2 R}{dr^2} + 2r \frac{dR}{dr} - \lambda R = 0 \qquad (2)$$

and

$$\frac{d^2 \theta}{d\theta^2} + \frac{1}{\tan \theta} \frac{d\theta}{d\theta} + \lambda \Theta = 0 \qquad (3)$$

2.250 In (3) of Problem 2.249, introduce the new independent variable $\zeta = \cos \theta$. Discuss the resulting *Legendre equation*.

■ Since, operationally,

$$\frac{d}{d\theta} = (-\sin \theta) \frac{d}{d\zeta} = -\sqrt{1 - \zeta^2} \frac{d}{d\zeta}$$

(3) is transformed into a much simpler equation:

$$(1 - \zeta^2) \frac{d^2 \Theta}{d\zeta^2} - 2\zeta \frac{d\Theta}{d\zeta} + \lambda \Theta = 0 \qquad (1)$$

Because the coefficients in (1) are *polynomials*, it seems likely that—for the right choice of λ—the equation has a *polynomial solution* (very desirable for applications). The "right choice" turns out to be $\lambda = m(m + 1)$, where m is a nonnegative integer; the resulting solution is the *Legendre polynomial of degree m*:

$$P_m(\zeta) = \frac{1}{2^m m!} \frac{d^m}{d\zeta^m} (\zeta^2 - 1)^m \qquad (2)$$

[The normalizing factor makes $P_m(1) = 1$.] The first few Legendre polynomials are

$$P_0(\zeta) = 1 \qquad P_1(\zeta) = \zeta \qquad P_2(\zeta) = \tfrac{1}{2}(3\zeta^2 - 1) \qquad P_3(\zeta) = \tfrac{1}{2}(5\zeta^3 - 3\zeta)$$

2.251 Show that for $\lambda = m(m + 1)$, (2) of Problem 2.249 has the general solution

$$R = C_1 r^m + C_2 r^{-(m+1)}$$

■ $R = r^t$ will be a solution if

$$t(t - 1)r^t + 2tr^t - m(m + 1)r^t = 0$$
$$t^2 + t - m(m + 1) = 0$$
$$(t - m)[t + (m + 1)] = 0$$
$$t = m,\ -(m + 1)$$

Since the two roots t give linearly independent functions, the general solution of (2) is as indicated.

2.252 Refer to Problem 2.250. What about a second basic solution of Legendre's equation,

$$(1 - \zeta^2) \frac{d^2 \Theta}{d\zeta^2} - 2\zeta \frac{d\Theta}{d\zeta} + m(m + 1)\Theta = 0$$

as required for a general solution?

■ A general solution is given by $A_m P_m(\zeta) + B_m Q_m(\zeta)$, where $Q_m(\zeta)$ is a (nonpolynomial) *Legendre function* that has logarithmic singularities at $\zeta = \pm 1$; i.e., along the entire z axis.

2.253 Give a "general" superposition solution for (1) of Problem 2.249.

■ Previous problems—see especially Problem 2.248—have shown that cylindrical, not spherical, coordinates are called for when dealing with line charges along the z axis. Thus, a solution of (1) that omits the functions Q_m,

$$V(r, \theta) = \sum_{m=0}^{\infty} P_m(\cos \theta)(C_{1m} r^m + C_{2m} r^{-m-1}) \qquad (1)$$

will cover most physical problems.

2.254 A solid sphere of radius a and permittivity ϵ is immersed in a uniform electrostatic field $\mathbf{E} = E_0 \mathbf{a}_z$ existing in free space. Write the forms of solution for V (a) inside the sphere, (b) outside the sphere.

▌ (a) In (1) of Problem 2.253 we must take all $C_{2m} = 0$, to keep the potential finite at $r = 0$. Furthermore, C_{10}, an additive constant, can always be assumed zero. Hence,

$$V_{\text{in}}(r, \theta) = \sum_{m=1}^{\infty} C_m r^m P_m(\cos \theta) \qquad (r < a) \qquad (1)$$

(b) The potential of the applied field is

$$-E_0 z = -E_0 r \cos \theta = -E_0 r P_1(\cos \theta)$$

and V_{out} must approach this value as $r \to \infty$. Therefore, in (1) of Problem 2.253, choose $C_{11} = -E_0$ and all other $C_{1m} = 0$:

$$V_{\text{out}}(r, \theta) = D_0 r^{-1} + (D_1 r^{-2} - E_0 r)P_1(\cos \theta) + \sum_{m=2}^{\infty} D_m r^{-m-1} P_m(\cos \theta) \qquad (r > a) \qquad (2)$$

2.255 Using continuity conditions at $r = a$, determine the constants C_m and D_m in Problem 2.254.

▌ *The potential is continuous*: $V_{\text{out}}(a, \theta) = V_{\text{in}}(a, \theta)$, or

$$D_0 a^{-1} + (D_1 a^{-2} - E_0 a - C_1 a)P_1(\cos \theta) + \sum_{m=2}^{\infty} (D_m a^{-m-1} - C_m a^m)P_m(\cos \theta) = 0 \qquad (1)$$

Also, *the normal component of* **D** *is continuous*:

$$-\epsilon_0 \frac{\partial V_{\text{out}}}{\partial r}\bigg|_{r=a} = -\epsilon \frac{\partial V_{\text{in}}}{\partial r}\bigg|_{r=a}$$

or

$$\epsilon_0 D_0 a^{-2} + (\epsilon_0 2 D_1 a^{-3} + \epsilon_0 E_0 + \epsilon C_1)P_1(\cos \theta) + \sum_{m=2}^{\infty} [\epsilon_0(m+1)a^{-m-2}D_m + \epsilon m a^{m-1}C_m]P_m(\cos \theta) = 0 \qquad (2)$$

Since the Legendre polynomials are linearly independent, each coefficient in (1) and (2) must vanish: this gives $D_0 = 0$, $D_m = C_m = 0$ $(m \geq 2)$, and

$$D_1 = \frac{(\epsilon - \epsilon_0)E_0 a^3}{2\epsilon_0 + \epsilon} \qquad C_1 = \frac{-3\epsilon_0 E_0}{2\epsilon_0 + \epsilon}$$

Consequently, the solution to Problem 2.254 is

$$V_{\text{in}}(r, \theta) = \frac{-3\epsilon_0}{2\epsilon_0 + \epsilon}(E_0 r \cos \theta) \qquad V_{\text{out}}(r, \theta) = \left[\frac{\epsilon - \epsilon_0}{2\epsilon_0 + \epsilon}\left(\frac{a}{r}\right)^3 - 1\right](E_0 r \cos \theta)$$

2.256 Find the polarization field within the sphere of Problems 2.254 and 2.255.

▌ By Problem 2.161, and the fact that $V_{\text{in}} = (\text{const.})z$,

$$\mathbf{P} = (\epsilon - \epsilon_0)\mathbf{E} = +(\epsilon - \epsilon_0)\nabla V_{\text{in}} = \frac{3\epsilon_0(\epsilon - \epsilon_0)}{2\epsilon_0 + \epsilon} E_0 \mathbf{a}_z$$

2.257 By Problem 2.255, $V_{\text{in}}(0, \theta) = 0$. Does this accord with the mean-value theorem?

▌ Yes. The mean value of V_{in} over the surface of a sphere of radius $b \leq a$ is proportional to

$$\int_0^\pi \cos \theta \sin \theta \, d\theta = \tfrac{1}{2} \sin^2 \theta \bigg]_0^\pi = 0$$

2.258 Consider two regions, A and B. We wish to find the scalar potential $V_A(x, y, z)$ in region A that attains a specified value V_b on the boundary between the two regions. The *method of images* consists in the replacement of region B with an equivalent (or image) charge distribution such that the image charge and the original charge in region A together give rise to the boundary value V_b.

Use the method of images to find the **E**-field between an infinite line charge $+\rho_\ell$ and a parallel, grounded, conducting plane.

▌ Reflect the line charge $+\rho_\ell$ in the ground plane as an image line charge $-\rho_\ell$. By Problem 2.152, the two line charges force $V = 0$ over the ground plane, which can therefore be dispensed with. Thus, we have the situation of Problem 2.152, for which the **E**-field was calculated in Problems 2.183 and 2.184. As shown in Fig. 2-53, the field lines in the region of interest $(x < 0)$ are arcs of circles.

2.259 Use the method of images (Problem 2.258) to find the capacitance per unit length (out of the page) of the cylinder-and-plane system of Fig. 2-72.

▌ The conducting plane is replaced by an image cylinder carrying uniform surface charge of density $-\rho_s$. Let r denote the distance from the axis of the image cylinder. With the fields as given in Problem 2.129, we find for

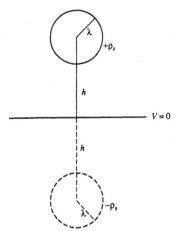

Fig. 2-72

the potential difference between the cylinders

$$V = \int_{\lambda}^{2h-\lambda} \left[\frac{\rho_s \lambda}{\epsilon_0 r} + \frac{\rho_s \lambda}{\epsilon_0 (2h-r)} \right] dr = \frac{2\rho_s \lambda}{\epsilon_0} \ln \frac{2h-\lambda}{\lambda}$$

The potential difference between the positive cylinder and the conducting plane is $V/2$; therefore, for the actual system,

$$c = \frac{q}{V/2} = \frac{2\pi\lambda\rho_s}{(\lambda\rho_s/\epsilon_0) \ln[(2h+\lambda)/\lambda]} = \frac{2\pi\epsilon_0}{\ln[(2h-\lambda)/\lambda]}$$

2.260 Find the image charges for a point charge $+Q$ midway between two infinite, grounded, conducting plates (Fig. 2-73).

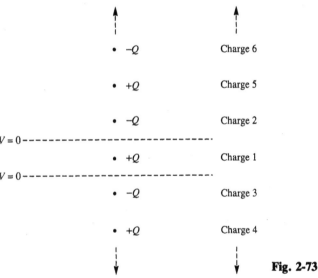

Fig. 2-73

▮ Imaging charge 1 across the upper plate produces charge 2 and zero potential along the position of the upper plate. However, the potential along the position of the lower plate is no longer zero, so we image charge 1 and charge 2 across the lower plate to produce charges 3 and 4. Now the potential along the position of the lower plate is zero, but the potential along the position of the upper plate is no longer zero, so we image across this plate to produce charges 5 and 6, and so on. This leads to an infinite set of image charges having alternating signs and separations equal to the plate separation.

2.261 Approximate Laplace's equation $\nabla^2 V = 0$ in two dimensions by a difference equation.

▮ Consider the region of potential V shown in Fig. 2-74. This region is covered by a square mesh of side h. If the unknown V's at the indicated nodes are as shown and if h is sufficiently small, then for point a midway between the nodes 0 and 1 we have

$$\frac{\partial V}{\partial x}\bigg|_a \approx \frac{1}{h}(V_1 - V_0) \tag{1}$$

Similarly, for point c we may write

$$\frac{\partial V}{\partial x}\bigg|_c \approx \frac{1}{h}(V_0 - V_3) \tag{2}$$

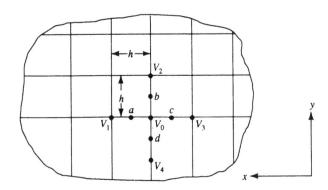

Fig. 2-74

so that

$$\left.\frac{\partial^2 V}{\partial x^2}\right|_0 \approx \frac{1}{h}\left(\left.\frac{\partial V}{\partial x}\right|_a - \left.\frac{\partial V}{\partial x}\right|_c\right) \approx \frac{1}{h^2}[(V_1 - V_0) - (V_0 - V_3)] \tag{3}$$

Similarly, if we consider the potentials along the y axis, we have

$$\left.\frac{\partial^2 V}{\partial y^2}\right|_0 \approx \frac{1}{h^2}[(V_2 - V_0) + (V_0 - V_4)] \tag{4}$$

Substituting (3) and (4) in Laplace's equation,

$$\frac{\partial^2 V}{\partial x^2} + \frac{\partial^2 V}{\partial y^2} = 0$$

yields the desired difference equation:

$$V_0 \approx \tfrac{1}{4}(V_1 + V_2 + V_3 + V_4) \tag{5}$$

2.262 Compute the potentials at nodes a through f for the region of Fig. 2-75, assuming that the boundary values are as shown.

▌ The potential function V satisfies Laplace's equation. First we inscribe the square mesh, as shown, and assume that all the specified nodes are at zero potential except node f. Then we apply (5) of Problem 2.261 to each node in succession. For instance, for the first cycle of computation, $V_f = \tfrac{1}{4}(30 + 0 + 0 + 0) = 7.5$, $V_a = \tfrac{1}{4}(10 + 20 + 0 + 0) = 7.5$, and so on. For the next cycle of computation, we use the previously calculated potentials to compute new potentials at the various nodes. Thus, for example, at the end of the second cycle we have $V_f = \tfrac{1}{4}(5.47 + 12.19 + 30 + 0) = 11.92$, and so on. This process is continued until none of the potentials deviates by more than a predetermined amount (say, 0.5 percent) from its value computed for the preceding cycle (the last value crossed out).

Clearly, such problems may be solved much more conveniently on a digital computer.

2.263 Formulate the solution to Problem 2.262 as a set of six simultaneous linear equations involving the potentials at the six nodes. Express the result in matrix notation and solve for the six potentials by matrix inversion.

▌ Labeling the nodes at the boundaries as 1 through 14, and applying (5) of Problem 2.261 at nodes a through f, we obtain

$$
\begin{aligned}
-4V_a + V_b + V_c &= -V_2 - V_{14} \\
V_a - 4V_b + V_d &= -V_3 - V_5 \\
V_a - 4V_c + V_d + V_e &= -V_{13} \\
V_b + V_c - 4V_d + V_f &= -V_6 \\
V_c - 4V_e + V_f &= -V_{10} - V_{12} \\
V_d + V_e - 4V_f &= -V_7 - V_9
\end{aligned}
$$

which may be written in matrix notation as

$$
\begin{bmatrix}
-4 & 1 & 1 & 0 & 0 & 0 \\
1 & -4 & 0 & 1 & 0 & 0 \\
1 & 0 & -4 & 1 & 1 & 0 \\
0 & 1 & 1 & -4 & 0 & 1 \\
0 & 0 & 1 & 0 & -4 & 1 \\
0 & 0 & 0 & 1 & 1 & -4
\end{bmatrix}
\begin{bmatrix}
V_a \\ V_b \\ V_c \\ V_d \\ V_e \\ V_f
\end{bmatrix}
=
\begin{bmatrix}
-V_2 - V_{14} \\
-V_3 - V_5 \\
-V_{13} \\
-V_6 \\
-V_{10} - V_{12} \\
-V_7 - V_9
\end{bmatrix}
\qquad \text{or} \qquad \mathbf{AV} = \mathbf{B}
$$

Then from $\mathbf{V} = \mathbf{A}^{-1}\mathbf{B}$ we find

$$V_a = 16.44 \qquad V_b = 21.66 \qquad V_c = 14.10 \qquad V_d = 20.19 \qquad V_e = 9.77 \qquad V_f = 14.99$$

Compare these with the results of Problem 2.262.

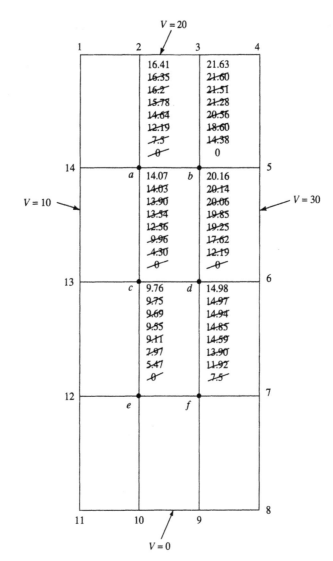

Fig. 2-75

2.264 A two-dimensional boundary-value problem is illustrated in Fig. 2-76. Obtain the potentials at the specified nodes a, b, c, d, e, f, g, and h by an iterative procedure.

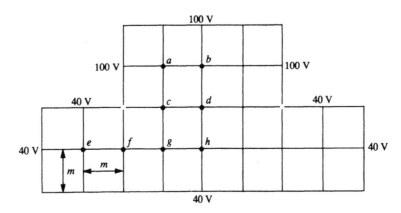

Fig. 2-76

▌ Solving the difference equations on a computer, we obtain the following potentials:

$V_a = 94\,\text{V}$ $V_b = 92\,\text{V}$ $V_c = 84\,\text{V}$ $V_d = 80\,\text{V}$ $V_e = 45.4\,\text{V}$ $V_f = 61.8\,\text{V}$ $V_g = 61.7\,\text{V}$ $V_h = 61\,\text{V}$

Check: $V_g = \frac{1}{4}(V_f + V_c + V_h + 40) = \frac{1}{4}(61.8 + 84 + 61 + 40) = 61.7\,\text{V}$.

2.265 Find the potentials at nodes *a* through *i* in the region shown in Fig. 2-77.

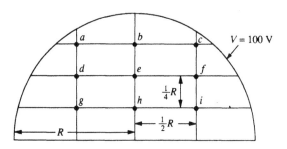

Fig. 2-77

❚ First bisect the horizontal intervals to obtain a square mesh. Then proceed as in Problem 2.264 to find

$V_a = V_c = 92.36$ V $V_b = 81.4$ V $V_d = V_f = 70.4$ V $V_e = 58.76$ V $V_g = V_i = 39.7$ V $V_h = 31.1$ V

3.1 According to *Ampère's force law,* the force $d\mathbf{F}_1$ on a current element $I_1\,d\mathbf{s}_1$, interacting with another current element $I_2\,d\mathbf{s}_2$, shown in Fig. 3-1, is given by

$$d\mathbf{F}_1 = \frac{\mu_0 I_1\,d\mathbf{s}_1 \times (I_2\,d\mathbf{s}_2 \times \mathbf{a}_R)}{4\pi R^2} \quad (\text{N}) \qquad (1)$$

Write an expression for the total force $\mathbf{F}_1$ experienced by loop 1.

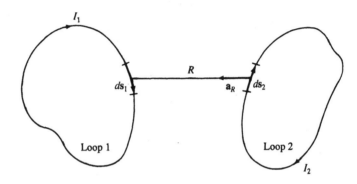

Fig. 3-1

▌ Integration about the loops gives

$$\mathbf{F}_1 = \frac{\mu_0 I_1 I_2}{4\pi} \oint_{\text{loop 1}} \oint_{\text{loop 2}} \frac{d\mathbf{s}_1 \times (d\mathbf{s}_2 \times \mathbf{a}_R)}{R^2} \quad (\text{N}) \qquad (2)$$

In (1) and (2) μ_0 is called the *permeability* of free space and has the value $4\pi \times 10^{-7}$ henries per meter (H/m).

3.2. Two infinitely long straight conductors, which are parallel to each other, are separated by a distance b. The conductors carry currents I_1 and I_2 in the directions shown in Fig. 3-2. Determine the mutual force per unit length of the conductors.

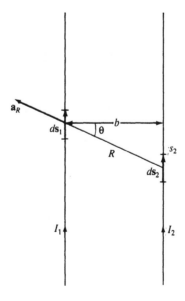

Fig. 3-2

▌ Integrate (1) of Problem 3.1 over the whole of conductor 2 and over 1 m of conductor 1. Two applications of the right-hand rule show that each $d\mathbf{F}_1$ is directed perpendicularly toward wire 2, and, at angle θ, has magnitude

$$\frac{(\mu_0 I_1\,ds_1)(I_2\,ds_2 \cos\theta)}{4\pi R^2} = \frac{\mu_0 I_1 I_2}{4\pi b}\,ds_1 \cos\theta\,d\theta$$

where we have used $s_2 = b \tan \theta$ and $R = b/(\cos \theta)$. Hence,

$$F_1 = \frac{\mu_0 I_1 I_2}{4\pi b} \int_0^1 ds_1 \int_{-\pi/2}^{\pi/2} \cos \theta \, d\theta = \frac{\mu_0 I_1 I_2}{2\pi b} \quad (\text{N/m})$$

3.3 Refer to Fig. 3-3. According to the *Biot–Savart law*, the *magnetic flux density* or *magnetic induction* $d\mathbf{B}$ at point P due to a current element $I \, d\mathbf{s}$ at Q is

$$d\mathbf{B} = \frac{\mu_0 I \, d\mathbf{s} \times \mathbf{a}_R}{4\pi R^2} \quad (\text{T}) \qquad\qquad (1)$$

where the *tesla* is defined by $1 \, \text{T} = 1 \, \text{N/A} \cdot \text{m}$. Rewrite (1) and (2) of Problem 3.1 for a current element and a current loop in an external field $\mathbf{B}_2$.

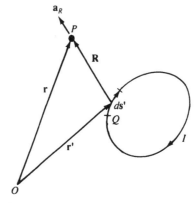

Fig. 3-3

▌ $$d\mathbf{F}_1 = I_1 \, d\mathbf{s}_1 \times d\mathbf{B}_2 \qquad \text{and} \qquad \mathbf{F}_1 = I_1 \oint_{\text{loop 1}} d\mathbf{s}_1 \times \mathbf{B}_2 \qquad\qquad (2)$$

3.4 Express the $\mathbf{B}$-field of a current loop if the origin of coordinates is chosen as in Fig. 3-4.

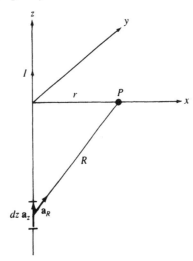

Fig. 3-4

▌ Integrate (1) of Problem 3.3 around the loop, using

$$\frac{\mathbf{a}_R}{R^2} = \frac{\mathbf{r} - \mathbf{r}'}{|\mathbf{r} - \mathbf{r}'|^3}$$

Thus

$$\mathbf{B}(P) = \oint \frac{\mu_0 I \, d\mathbf{s}' \times (\mathbf{r} - \mathbf{r}')}{4\pi \, |\mathbf{r} - \mathbf{r}'|^3}$$

where $d\mathbf{s}' = d\mathbf{r}'$.

3.5 Determine the magnetic flux density due to a very long wire carrying a current I. The observation point P is at a distance r away from the conductor (Fig. 3-5).

Fig. 3-5

▌ The necessary integration has already been carried out in Problem 3.2; the result in the present notation is

$$\mathbf{B} = \frac{\mu_0 I}{2\pi r} \mathbf{a}_\phi \qquad (1)$$

where $\mathbf{a}_\phi = \mathbf{a}_y$ for the configuration of Fig. 3-5.

3.6   Determine the flux density at a point P due to a straight conductor of length ℓ and carrying a current I, as shown in Fig. 3-6.

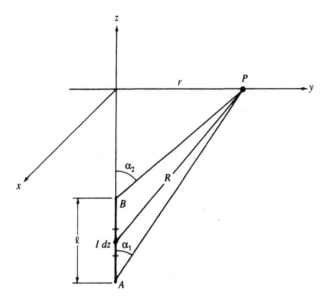

Fig. 3-6

▌ This is the same as Problem 3.2, with $\theta \equiv (\pi/2) - \alpha$ running from $(\pi/2) - \alpha_2$ to $(\pi/2) - \alpha_1$:

$$B = \frac{\mu_0 I}{4\pi r} \int_{(\pi/2)-\alpha_2}^{(\pi/2)-\alpha_1} \cos\theta \, d\theta = \frac{\mu_0 I}{4\pi r} (\cos\alpha_1 - \cos\alpha_2) \qquad (1)$$

in the direction $-\mathbf{a}_x = \mathbf{a}_\phi$.

3.7   Show that (1) of Problem 3.5 is a special case of (1) of Problem 3.6.

▌ In Problem 3.5, we have an infinitely long wire, for which, in (1) of Problem 3.6, $\alpha_1 = 0$ and $\alpha_2 = \pi$. Hence (1) of Problem 3.6 gives

$$\mathbf{B} = \frac{\mu_0 I}{4\pi r} (\cos 0 - \cos \pi)\mathbf{a}_\phi = \frac{\mu_0 I}{2\pi r} \mathbf{a}_\phi$$

3.8   Determine the magnetic flux density at the center of a square loop of side L carrying a current I, as shown in Fig. 3-7.

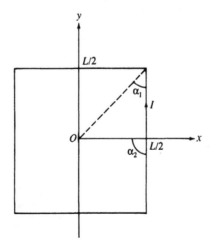

Fig. 3-7

▮ In this problem we make use of geometrical symmetry; thus, we need only determine the flux density due to each half-side. By Problem 3.6,

$$|\mathbf{B}| = 8\left[\frac{\mu_0 I}{4\pi(L/2)}\left(\frac{\sqrt{2}}{2} - 0\right)\right] = \frac{2\sqrt{2}\,\mu_0 I}{\pi L}$$

and the right-hand rule gives the direction of **B** as $+\mathbf{a}_z$.

3.9 A square loop measuring 1.5 by 1.5 m carries a 7.5-A steady current. Choose the coordinates to put the loop in the xz plane, the origin coinciding with a corner of the square. Calculate the **B**-field at a point on the y axis 0.35 m from the origin.

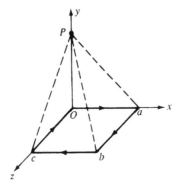

Fig. 3-8

▮ From the geometry shown in Fig. 3-8, $\overline{OP} = 0.35$ m; $\overline{aP} = \overline{cP} = 1.54$ m; and $\overline{bP} = 2.15$ m. Now using (1) of Problem 3.6, we obtain the flux densities at P due to the four sides:

$$B_{oa} = B_{co} = \frac{10^{-7} \times 7.5}{0.35}\left(\frac{1.5}{1.54} - 0\right) = 2.087\ \mu\text{T}$$

$$B_{ab} = B_{bc} = \frac{10^{-7} \times 7.5}{1.54}\left(\frac{1.5}{2.15} - 0\right) = 0.34\ \mu\text{T}$$

In terms of vector components, these become

$$\mathbf{B}_{oa} = 2.087(\mathbf{a}_x 0 + \mathbf{a}_y 0 + \mathbf{a}_z 1)\ (\mu\text{T})$$

$$\mathbf{B}_{ab} = 0.34\left(-\mathbf{a}_x\frac{0.35}{1.54} - \mathbf{a}_y\frac{1.5}{1.54} + \mathbf{a}_z 0\right)\ (\mu\text{T})$$

$$\mathbf{B}_{bc} = 0.34\left(\mathbf{a}_x 0 - \mathbf{a}_y\frac{1.5}{1.54} - \mathbf{a}_z\frac{0.35}{1.54}\right)\ (\mu\text{T})$$

$$\mathbf{B}_{ca} = 2.087(\mathbf{a}_x 1 + \mathbf{a}_y 0 + \mathbf{a}_z 0)\ (\mu\text{T})$$

giving the resultant $\mathbf{B} = 2.92(0.69\mathbf{a}_x - 0.226\mathbf{a}_y + 0.69\mathbf{a}_z)\ (\mu\text{T})$.

3.10 We define the *magnetic field intensity* **H** by

$$d\mathbf{H} = \frac{I\,d\mathbf{s} \times \mathbf{a}_R}{4\pi R^2}\quad(\text{A/m}) \tag{1}$$

Express **B**, the magnetic flux density, in terms of **H**.

▮ $\mathbf{B} = \mu_0\mathbf{H}$. Of the two vectors, **B** (the force field) is analogous to **E**, while **H** is determined by currents (Problem 3.14) just as **D** is determined by charges.

3.11 Using the Biot–Savart law (Problem 3.3) in **H**, find the magnetic field intensity at a point on the axis of a circular loop of radius a carrying a current I. The point is at a distance h (on the z axis) from the center of the loop.

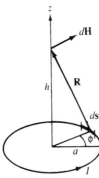

Fig. 3-9

▮ From Fig. 3-9,

$$R^2 = a^2 + h^2 \qquad \mathbf{a}_R = \frac{-a\mathbf{a}_r + h\mathbf{a}_z}{\sqrt{a^2 + h^2}} \qquad d\mathbf{s} = a\,d\phi\,\mathbf{a}_\phi$$

whence

$$d\mathbf{H} = \frac{I\,d\mathbf{s} \times \mathbf{a}_R}{4\pi R^2} = \frac{Ia\,d\phi}{4\pi(h^2 + a^2)^{3/2}}(a\mathbf{a}_z + h\mathbf{a}_r)$$

From symmetry, the radial component integrates to zero, and so

$$\mathbf{H} = H_z\mathbf{a}_z = \left[\frac{a^2 I}{4\pi(h^2 + a^2)^{3/2}} \int_0^{2\pi} d\phi\right]\mathbf{a}_z = \frac{Ia^2}{2(h^2 + a^2)^{3/2}}\mathbf{a}_z \qquad (1)$$

3.12 A semicircular loop of radius a carries a current I. The loop is situated in air. Find the magnetic field intensity at the center of the loop.

▮ Let $h \to 0$ in (1) of Problem 3.11, obtaining for a whole circle $H = I/2a$. For a semicircle, H will be half of this, or $I/4a$.

3.13 A Plexiglas disk of radius R is charged with a uniform surface charge density ρ_s. The disk is rotated at a constant speed of N rev/min. Thus we have electric charges in circular motion, which may be considered to be circular currents. Determine the magnetic field intensity at the center of the loop.

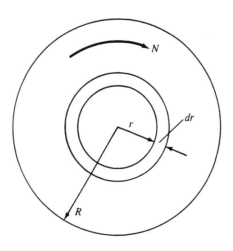

Fig. 3-10

▮ Charge on the elemental loop (shown in Fig. 3-10) is $dQ = \rho_s 2\pi r\,dr$; time for one revolution, $T = 60/N$.

Thus

$$dI = \frac{dQ}{T} = \frac{N\pi\rho_s r}{30}\,dr$$

and, from Problem 3.12,

$$dH = \frac{dI}{2r} = \frac{N\pi\rho_s}{60}\,dr \qquad H = \frac{N\pi\rho_s R}{60} \quad \text{(A/m)}$$

For the direction of rotation shown, **H** points into the paper.

3.14 *Ampère's circuital law* states that the line integral of **H** around a closed path equals the total current enclosed by the path:

$$\oint \mathbf{H} \cdot d\mathbf{s} = I_{\text{encl}} = \int \mathbf{J} \cdot d\mathbf{S} \qquad (1)$$

[The second equality in (1) identifies I_{encl} with the net current threading any open surface bounded by the given closed path.] Apply Ampère's law to verify Problem 3.5.

▮ Because of symmetry, only the ϕ component of the **H**-field exists, and it is of constant magnitude around a closed circular path of radius r about the wire. Thus

$$H_\phi(2\pi r) = I \qquad H_\phi = \frac{I}{2\pi r} \qquad \mathbf{B} = \frac{\mu_0 I}{2\pi r}\mathbf{a}_\phi$$

since $\mathbf{B} = \mu_0\mathbf{H}$.

3.15 A straight conductor of cylindrical cross section and of radius a carries a current I, which is uniformly distributed over the conductor cross section. Using Ampère's circuital law, find the magnetic flux density within and outside of the conductor.

▌ Since $\mathbf{B} = \mu_0\mathbf{H}$, we may write (1) of Problem 3.14 as $\oint \mathbf{B} \cdot d\mathbf{s} = \mu_0 I_{\text{encl}}$. By the cylindrical symmetry, $\mathbf{B} = B_\phi(r)\mathbf{a}_\phi$. So,

$$2\pi r B_\phi = \begin{cases} \mu_0(r^2/a^2)I & r < a \\ \mu_0 I & r > a \end{cases} \quad \text{or} \quad B_\phi = \begin{cases} \mu_0 Ir/2\pi a^2 & r < a \\ \mu_0 I/2\pi r & r > a \end{cases}$$

3.16 Obtain the point form of Ampère's circuital law, (1) of Problem 3.14.

▌ By the definition of the curl (Problem 1.140), we have, at point P, in any fixed direction $\mathbf{a}_n$,

$$(\text{curl } \mathbf{H}) \cdot \mathbf{a}_n = \lim_{S \to 0} \frac{\oint \mathbf{H} \cdot d\mathbf{s}}{S} = \lim_{S \to 0} \frac{1}{S} \int \mathbf{J} \cdot \mathbf{a}_n \, dS = \mathbf{J}(P) \cdot \mathbf{a}_n$$

which implies that curl $\mathbf{H} \equiv \mathbf{J}$.

3.17 Find a current distribution that produces a magnetic field of the form $\mathbf{H} = K \sin x \, \mathbf{a}_y$, where K is a constant.

▌
$$\mathbf{J} = \text{curl } \mathbf{H} = \begin{vmatrix} \mathbf{a}_x & \mathbf{a}_y & \mathbf{a}_z \\ \dfrac{\partial}{\partial x} & \dfrac{\partial}{\partial y} & \dfrac{\partial}{\partial z} \\ 0 & K\sin x & 0 \end{vmatrix} = K \cos x \, \mathbf{a}_z$$

3.18 Within a cylindrical conductor of radius a, the current density exponentially decreases with the radius, according to $\mathbf{J} = Ae^{-kr}\mathbf{a}_z$, where A and k are positive constants. Determine the resulting magnetic field intensity everywhere.

▌ From Ampère's law, (1) of Problem 3.14, we have, for $r > a$,

$$H_\phi(2\pi r) = \int_{r=0}^{a} \int_{\phi=0}^{2\pi} Ae^{-kr} r \, dr \, d\phi \quad \text{or} \quad H_\phi = \frac{A}{k^2 r}[1 - (1+ka)e^{-ka}]$$

Similarly, for $r < a$,

$$H_\phi(2\pi r) = \int_{\rho=0}^{r} \int_{\phi=0}^{2\pi} Ae^{-kr}\rho \, d\rho \, d\phi \quad \text{or} \quad H_\phi = \frac{A}{k^2 r}[1 - (1+kr)e^{-kr}]$$

3.19 Check Problem 3.15 by use of curl $\mathbf{H} = \mathbf{J}$.

▌ Here,

$$\mathbf{H} = H_\phi \mathbf{a}_\phi = \begin{cases} (Ir/2\pi a^2)\mathbf{a}_\phi & r < a \\ (I/2\pi r)\mathbf{a}_\phi & r > a \end{cases} \quad \text{and} \quad \mathbf{J} = \begin{cases} (I/\pi a^2)\mathbf{a}_z & r < a \\ 0\mathbf{a}_z & r > a \end{cases}$$

Checking:

$$\text{curl } \mathbf{H} = \frac{1}{r}\frac{d}{dr}(rH_\phi)\,\mathbf{a}_z = \begin{cases} (I/\pi a^2)\mathbf{a}_z \\ 0\mathbf{a}_z \end{cases} = \mathbf{J}$$

3.20 The current density in an electron beam is given by

$$\mathbf{J} = J_0\left(1 - \frac{r^2}{b^2}\right)\mathbf{a}_z \quad (r < b)$$

where J_0 is a constant and b is the beam radius. At $r = b/3$, the H-field is given by $\mathbf{H} = kbJ_0\mathbf{a}_\phi$; evaluate k.

▌ By Ampère's law, for $r = b/3$,

$$(kbJ_0)\left(2\pi \frac{b}{3}\right) = \int_0^{b/3} J_0\left(1 - \frac{r^2}{b^2}\right)2\pi r \, dr = 2\pi J_0\left(\frac{b^2}{18} - \frac{b^2}{324}\right) \quad \text{or} \quad k = 17/108$$

3.21 The plane current sheet $y = 0$ has a uniform linear current density $\mathbf{k} = k\mathbf{a}_z$ (A/m); see Fig. 3-11. Determine the H-field for $y > 0$ and for $y < 0$.

▌ From symmetry,

$$\mathbf{H} = \begin{cases} H_0(-\mathbf{a}_x) & y > 0 \\ H_0\mathbf{a}_x & y < 0 \end{cases}$$

Ampère's circuital law gives, for the contour shown,

$$H_0 \Delta x + 0 + H_0 \Delta x + 0 = k \Delta x \quad \text{or} \quad H_0 = \frac{k}{2}$$

3.22 In the cylindrical region $0 < r < 0.5$ m, $\mathbf{J} = 4.5e^{-2r}\mathbf{a}_z$ (A/m^2). Determine $\mathbf{H} = H_\phi \mathbf{a}_\phi$ everywhere.

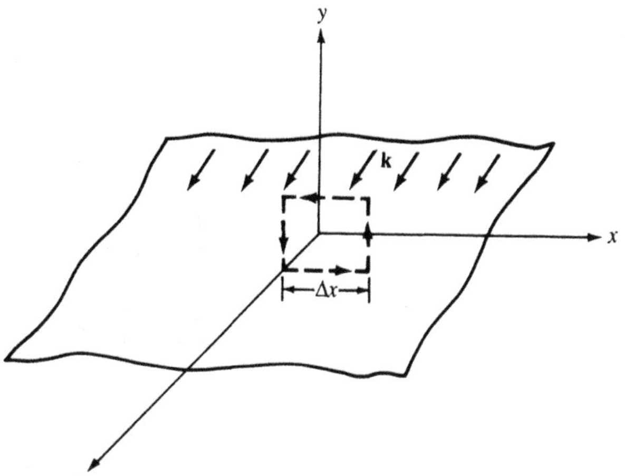

Fig. 3-11

▌ For $r > 0.5$ m,

$$H_\phi(2\pi r) = \int_0^{2\pi} \int_0^{0.5} 4.5e^{-2r} r \, dr \, d\phi \quad \text{or} \quad H_\phi = \frac{0.297}{r} \quad \text{(A/m)}$$

For $r < 0.5$ m,

$$H_\phi(2\pi r) = \int_0^{2\pi} \int_0^{r} 4.5e^{-2\rho} \, d\rho \, d\phi \quad \text{or} \quad H_\phi = \frac{1.125}{r}(1 - e^{-2r} - 2re^{-2r}) \quad \text{(A/m)}$$

3.23 An **H** due to a current source is given by $\mathbf{H} = (y \cos ax)\mathbf{a}_x + (y + e^x)\mathbf{a}_z$. Describe the current density over the yz plane.

▌ $$\mathbf{J} = \text{curl } \mathbf{H} = \mathbf{a}_x \frac{\partial}{\partial y}(y + e^x) - \mathbf{a}_y \frac{\partial}{\partial x}(y + e^x) - \mathbf{a}_z \frac{\partial}{\partial y}(y \cos ax) = \mathbf{a}_x - e^x \mathbf{a}_y - (\cos ax)\mathbf{a}_z$$

For $x = 0$, this becomes $\mathbf{J} = \mathbf{a}_x - \mathbf{a}_y - \mathbf{a}_z$, a constant vector of magnitude $\sqrt{3}$ A/m^2.

3.24 The **H**-field within a circular conductor of 1-cm radius is

$$\mathbf{H} = \frac{10^4}{r}\left(\frac{1}{a^2}\sin ar - \frac{r}{a}\cos ar\right)\mathbf{a}_\phi \quad \text{(A/m)}$$

where r is in meters and $a = 50\pi$ m^{-1}.

▌ By Ampère's law,

$$I = 2\pi(10^{-2})H_\phi(10^{-2}) = 2\pi(10^{-2})\left[\frac{10^4}{10^{-2}}\left(\frac{1}{2500\pi^2}\sin\frac{50\pi}{100} - \frac{10^{-2}}{50\pi}\cos\frac{50\pi}{100}\right)\right] = \frac{8}{\pi} = 2.546 \text{ A}$$

3.25 A cylindrical conductor carries a current that produces an $\mathbf{H} = 3r\mathbf{a}_\phi$ (A/m). Determine the current density within the conductor.

▌ By Ampère's law,

$$\mathbf{J} = \text{curl } \mathbf{H} = +\frac{1}{r}\left[\frac{d}{dr}(3r^2)\right]\mathbf{a}_z = 6\mathbf{a}_z \text{ A/m}^2$$

3.26 In an electron beam of radius b,

$$\mathbf{J} = J_0\left(1 - \frac{r}{b}\right)\mathbf{a}_z \quad (r < b)$$

Find the **H**-field at the surface of the beam.

▌ Since **H** must be circumferential, Ampère's circuital law, at the surface of the beam, gives

$$H_\phi(2\pi b) = \int_0^b J_0\left(1 - \frac{r}{b}\right)2\pi r \, dr = \frac{\pi J_0 b^2}{3}$$

or $\mathbf{H} = \frac{1}{6}J_0 b\mathbf{a}_\phi$.

3.27 Determine the total magnetic flux (of **B**) through the rectangular loop shown in Fig. 3-12. The source of the magnetic field is the long straight conductor carrying a current I.

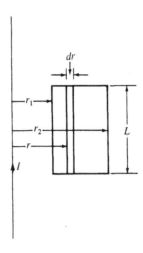

Fig. 3-12

▌ Using Problem 3.5,

$$\psi_m = \int_{r_1}^{r_2} \frac{\mu_0 I}{2\pi r} L \, dr = \frac{\mu_0 IL}{2\pi} \ln \frac{r_2}{r_1} \qquad (1)$$

The unit of ψ_m is the *weber*, where $1\,\text{Wb} = 1\,\text{T} \cdot \text{m}^2$.

3.28 A 50-cm square loop is located such that one of its sides is parallel to a straight conductor 30 cm away from the center of the square and lying in its plane. How much current must flow through the conductor for the flux through the loop to be 50 μWb?

▌ By (1) of Problem 3.27,

$$50 \times 10^{-6} = \frac{4\pi \times 10^{-7}}{2\pi} I(0.50) \ln \frac{55}{5} \qquad \text{or} \qquad I = 208.516\,\text{A}$$

3.29 Determine the core flux for the air-core toroid shown in Fig. 3-13, where $r_2/r_1 = b$.

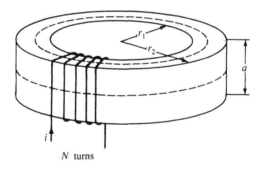

N turns

Fig. 3-13

▌ We may use (1) of Problem 3.27, with $I = Ni$ and $L = a$. Hence,

$$\psi_m = \frac{\mu_0 aNi}{2\pi} \ln b$$

3.30 For the toroid of Problem 3.29, assume the core flux density to be uniform and equal to its value at the arithmetic mean radius. What percent error would be made in the computation of the core flux by this approximation, as compared with the answer to Problem 3.29 for $b = 2$?

▌ For the approximation

$$\psi_m \approx \frac{\mu_0 aNi}{2\pi} \frac{r_2 - r_1}{(r_2 + r_1)/2} = \frac{\mu_0 aNi}{2\pi} \frac{b - 1}{(b + 1)/2}$$

one has

$$\text{percent error} = \left[\frac{b - 1}{\frac{1}{2}(b + 1) \ln b} - 1 \right] (100\%) = -3.8\% \quad \text{(for } b = 2\text{)}$$

3.31 Rework Problem 3.30, if the geometric mean radius is used.

| In this case we have

$$\psi_m \approx \frac{\mu_0 a N i}{2\pi} \frac{r_2 - r_1}{\sqrt{r_2 r_1}} = \frac{\mu_0 a N i}{2\pi} \frac{b-1}{\sqrt{b}}$$

$$\text{percent error} = \left[\frac{b-1}{\sqrt{b}\ln b} - 1\right](100\%) = +2\% \quad (\text{for } b = 2)$$

3.32 Current flows through a coaxial cable (Fig. 3-14), the inner conductor of which is solid and is of radius a. The current-density distribution within the conductor is nonuniform and may be approximated by $J(r) = J_0 e^{-(a-r)}$, where J_0 is the current density at the surface of the conductor. The return current is through the outer conductor, an infinitely thin cylindrical shell of radius b. Determine the magnetic field intensity everywhere.

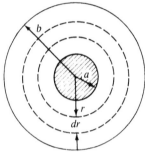

Fig. 3-14

| For $r < a$, Ampère's law gives

$$I = \int_0^r J_0 e^{-(a-u)} 2\pi u \, du = 2\pi J_0 e^{-a}[(r-1)e^r + 1] = H_\phi(2\pi r)$$

or

$$H_\phi = \frac{J_0 e^{-a}}{r}[1 + (r-1)e^r]$$

For $a < r < b$, $I = 2\pi J_0[(a-1) + e^{-a}]$, so that

$$H_\phi = \frac{J_0}{r}(a - 1 + e^{-a})$$

| For $r > b$, $I = 0$ and so $H_\phi = 0$.

3.33 The operation of a magnetic compass depends on the presence of the earth's magnetic field. At a certain location the tangential component of the earth's magnetic field is 20 μT, and this site is under a 345-kV transmission line carrying a 2000-A current. Assuming that the height of the line is 10 m, calculate the net field affecting the reading of the compass. Consider the extreme case in which the field of the transmission line is parallel to the tangential component of the earth's magnetic field.

| By Ampère's law,

$$(B_\phi)_{\text{line}} = \frac{\mu_0 I}{2\pi r} = \frac{4\pi \times 10^{-7} \times 2000}{2\pi \times 10} = 40 \ \mu\text{T}$$

Thus, the net field is $40 + 20 = 60 \ \mu$T.

3.34 A **B**-field exists in a given region. Show that the divergence of the field is zero at each point of the region.

| Because the lines of **B** are closed curves surrounding currents—unlike the lines of **E**, which terminate on charges—the flux of **B** through any closed surface must be zero. But then div **B** $\equiv 0$, by Problem 1.107.

3.35 Determine $\partial B_y/\partial y$ at point P in Fig. 3-15.

| Along the z axis, Problem 3.11 gives

$$\mathbf{B}(P) = \frac{\mu_0 I a^2}{2(z^2 + a^2)^{3/2}} \mathbf{a}_z$$

By Problem 3.34,

$$(\text{div } \mathbf{B})_P = \left(\frac{\partial B_x}{\partial x}\right)_P + \left(\frac{\partial B_y}{\partial y}\right)_P + \left(\frac{\partial B_z}{\partial z}\right)_P = 0$$

From symmetry, $\left(\dfrac{\partial B_x}{\partial x}\right)_P = \left(\dfrac{\partial B_y}{\partial y}\right)_P$; thus

$$\left(\frac{\partial B_y}{\partial y}\right)_P = -\frac{1}{2}\left(\frac{\partial B_z}{\partial z}\right)_P = -\frac{1}{2}\frac{d}{dz}\left[\frac{\mu_0 I a^2}{2(z^2 + a^2)^{3/2}}\right] = \frac{3}{4}\mu_0 I a^2 z (z^2 + a^2)^{-5/2}$$

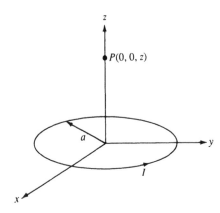

Fig. 3-15

3.36   An infinitely long straight wire carries 200 A of current, and in its vicinity a circular loop of 50 mm diameter is located, with the center of the loop 0.5 m away from the straight conductor; the wire and the loop are coplanar. The currents in the loop and the wire are such that they produce fields opposing each other (see Fig. 3-16). For what current in the loop will the **B**-field at its center be zero?

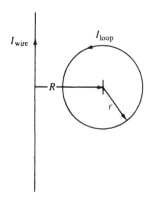

Fig. 3-16

❚ $$\frac{\mu_0 I_{\text{wire}}}{2\pi R} = \frac{\mu_0 I_{\text{loop}}}{2r} \quad \text{or} \quad I_{\text{loop}} = \frac{1}{\pi}\frac{r}{R} I_{\text{wire}} = \frac{1}{\pi}\frac{50}{500}(200) = (20/\pi)\text{ A}$$

3.37   Three long wires, parallel to each other, are located as shown in Fig. 3-17. The currents carried by these wires are also shown. Assuming that the top wire is ℓ meters long, determine the force experienced by it due to the fields of the two bottom conductors.

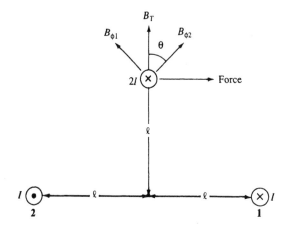

Fig. 3-17

❚   Field due to left bottom wire: $B_{\phi 1} = \dfrac{\mu_0 I}{2\pi(\sqrt{2}\,\ell)}$

Field due to right bottom wire: $B_{\phi 2} = \dfrac{\mu_0 I}{2\pi(\sqrt{2}\,\ell)}$

Because of symmetry, the net field is vertical, of magnitude

$$B_T = (B_{\phi 1} + B_{\phi 2}) \cos\theta = \frac{2\mu_0 I}{2\pi(\sqrt{2}\,\ell)}\left(\frac{1}{\sqrt{2}}\right) = \frac{\mu_0 I}{2\pi\ell}$$

By (2) of Problem 3.3, the force on the top wire is directed to the right and has magnitude

$$F = B_T \ell(2I) = \frac{\mu I^2}{\pi} \quad (\text{N})$$

3.38 Two infinitely long wires, separated by a distance 5ℓ, carry currents I in opposite directions, as shown in Fig. 3-18. Obtain an expression for the magnetic field intensity at the point P shown.

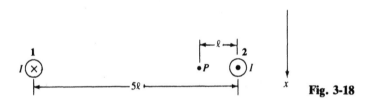

Fig. 3-18

▌ According to Ampère's law,

$$\mathbf{H}_1 = \frac{I}{2\pi(4\ell)}\,\mathbf{a}_x \qquad \mathbf{H}_2 = \frac{I}{2\pi\ell}\,\mathbf{a}_x \qquad \mathbf{H}_T = \mathbf{H}_1 + \mathbf{H}_2 = \frac{5I}{8\pi\ell}\,\mathbf{a}_x$$

3.39 Two parallel, coaxial current loops are a distance $2h$ apart (Fig. 3-19). For the data shown, obtain an expression for the **B**-field at $z = 2h$ (the center of loop 2).

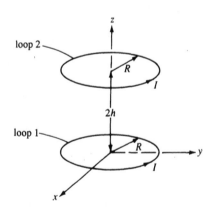

Fig. 3-19

▌ From the result of Problem 3.11,

$$\mathbf{B} = (B_1 + B_2)\mathbf{a}_z = \left[\frac{\mu_0 I}{2R} + \frac{\mu_0 I R^2}{2(R^2 + 4h^2)^{3/2}}\right]\mathbf{a}_z = \frac{\mu_0 I}{2}\left[\frac{1}{R} + \frac{R^2}{(R^2 + 4h^2)^{3/2}}\right]\mathbf{a}_z$$

3.40 For the loops of Fig. 3-19, we have $R = 0.1$ m and $I = 100$ A, and the separation between the loops is 0.2 m. Calculate the flux density at a point midway between the loops.

▌ At the midpoint, by symmetry,

$$\mathbf{B} = \frac{\mu_0 I R^2}{(R^2 + h^2)^{3/2}}\,\mathbf{a}_z = \frac{4\pi \times 10^{-7} \times 100 \times (0.1)}{(0.1^2 + 0.1^2)^{3/2}}\,\mathbf{a}_z = 0.444\mathbf{a}_z \text{ mT}$$

3.41 By (2) of Problem 3.3, the force on a straight conductor of directed length **L** which carries current I in an external **B**-field is given by

$$\mathbf{F} = I\mathbf{L} \times \mathbf{B} \tag{1}$$

Use this expression to find the force on a 2-m-long conductor located along the z axis, carrying a 20-A current in a **B**-field given by $\mathbf{B} = 0.3\mathbf{a}_x + 0.4\mathbf{a}_y$ T.

▌ $$\mathbf{F} = (20)(2)\mathbf{a}_z \times (0.3\mathbf{a}_x + 0.4\mathbf{a}_y) = -16\mathbf{a}_x + 12\mathbf{a}_y \quad \text{N}$$

3.42 A hollow cylindrical conductor carries a surface current of density $\mathbf{J}_s = J_s\mathbf{a}_z$ (A/m) as shown in Fig. 3-20. The conductor is located in a **B**-field given by $\mathbf{B} = (a/r)\mathbf{a}_r + b\mathbf{a}_\phi$ (T). Determine the magnitude and direction of the torque experienced by the conductor.

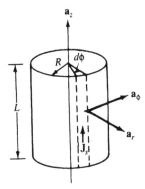

Fig. 3-20

▮ Consider the surface strip shown in Fig. 3-20; it carries current $J_s R\,d\phi$, and therefore the force on it is

$$d\mathbf{F} = (J_s R\,d\phi)L\mathbf{a}_z \times \left(\frac{a}{R}\mathbf{a}_r + b\mathbf{a}_\phi\right)$$
$$= (J_s aL\,d\phi)\mathbf{a}_\phi - (J_s RbL\,d\phi)\mathbf{a}_r$$

giving the differential torque

$$d\mathbf{T} = R\mathbf{a}_r \times d\mathbf{F} = (J_s aRL\,d\phi)\mathbf{a}_z \tag{1}$$

Since $\mathbf{a}_z$ is fixed, (1) integrates to

$$\mathbf{T} = 2\pi J_s aRL\mathbf{a}_z = (ILB_r)(R)\mathbf{a}_z \tag{2}$$

According to (2), the magnitude of the torque is equal to the force on the total current ($I = 2\pi RJ_s$) times the moment arm R; the direction $\mathbf{a}_z$ means that the torque acts to rotate the cylinder about its axis.

3.43 Show that the magnetic field intensity **H** is the gradient of a scalar function in a source-free region.

▮ $\mathbf{J} = \mathbf{0}$ implies curl $\mathbf{H} = \mathbf{J} = \mathbf{0}$, which in turn implies (Problem 1.142) the existence of a *magnetic scalar potential* F such that $\mathbf{H} = \mathbf{\nabla}(-F) = -\mathbf{\nabla}F$. (The minus sign is introduced for formal analogy with $\mathbf{E} = -\mathbf{\nabla}V$.)

3.44 Express the magnetic scalar potential in an integral form and identify its units.

▮ By analogy with (2) of Problem 2.127,

$$F_b - F_a = -\int_a^b \mathbf{H} \cdot d\mathbf{s} \tag{1}$$

From (1), F has the unit $(\text{A/m}) \cdot \text{m} = \text{A}$.

3.45 Prove that F satisfies Laplace's equation in a source-free region.

▮ It is instructive to contrast the proof with that for V:

1.	$\text{div } \mathbf{H} = \dfrac{1}{\mu_0} \text{div } \mathbf{B} = 0$ (*any region*)	$\text{div } \mathbf{E} = \dfrac{1}{\epsilon_0} \text{div } \mathbf{D} = 0$ (*source-free region*)
2.	$\mathbf{H} = -\text{grad } F$ (*source-free region*)	$\mathbf{E} = -\text{grad } V$ (*any region*)
3.	$-\text{div grad } F = -\nabla^2 F = 0$ (*source-free region*)	$-\text{div grad } V = -\nabla^2 V = 0$ (*source-free region*)

3.46 Show that, whether or not sources are present, a *magnetic vector potential* **A** exists such that $\mathbf{B} = \text{curl } \mathbf{A}$.

▮ A general proof that $\mathbf{B} = \text{curl } \mathbf{A}$ is *necessary* for div $\mathbf{B} = 0$—it is obviously sufficient—is difficult; but an important special case shows the main ideas.

In cylindrical coordinates let the **B**-field be of the form $\mathbf{B} = B_\phi\mathbf{a}_\phi$, where, because div $\mathbf{B} = 0$, $B_\phi = B_\phi(r, z)$.

We therefore suppose

$$\mathbf{A} = A_r(r, z)\mathbf{a}_r + 0\mathbf{a}_\phi + A_z(r, z)\mathbf{a}_z \qquad (1)$$

and the equation curl $\mathbf{A} = \mathbf{B}$ reduces to

$$\frac{\partial A_r}{\partial z} - \frac{\partial A_z}{\partial r} = B_\phi(r, z) \qquad (2)$$

It is evident that (2) always has a solution; in fact, an infinity of solutions (cf. Problems 1.151 and 1.152).

3.47 Show how to calculate the magnetic flux ψ_m through a given open surface in terms of the vector potential $\mathbf{A}$.

▌ Let S denote an open surface bounded by a closed contour C. Then, by Stokes' theorem,

$$\oint_C \mathbf{A} \cdot d\mathbf{s} = \iint_S (\text{curl } \mathbf{A}) \cdot d\mathbf{S} = \iint_S \mathbf{B} \cdot d\mathbf{S} = \psi_m \qquad (1)$$

(For the relation between the direction of flux and the sense of description of the contour, see Problem 1.158.)

3.48 Show that in a region with current density $\mathbf{J}$, the magnetic vector potential $\mathbf{A}$ satisfies, in rectangular coordinates, the equation

$$\nabla^2 \mathbf{A} = -\mu_0 \mathbf{J} \qquad (1)$$

▌ First, we use the vector identity

$$\nabla \times (\nabla \times \mathbf{A}) = \nabla(\nabla \cdot \mathbf{A}) - \nabla^2 \mathbf{A} \qquad (2)$$

Now, $\nabla \times \mathbf{A} = \mathbf{B}$, whence

$$\nabla \times (\nabla \times \mathbf{A}) = \nabla \times \mathbf{B} = \mu_0 \nabla \times \mathbf{H} = \mu_0 \mathbf{J}$$

Furthermore, among the infinite family of $\mathbf{A}$'s (Problem 3.46) we can always find one that is divergence-free. Then (2) becomes $\mu_0 \mathbf{J} = -\nabla^2 \mathbf{A}$.

The significance of this result is that the apparatus of potential theory carries over from the $\mathbf{E}$-field to the $\mathbf{B}$-field.

3.49 Express the magnetic vector potential directly in terms of the source currents.

▌ By the analogy between (each component of) $\nabla^2 \mathbf{A} = -\mu_0 \mathbf{J}$ and $\nabla^2 V = -(1/\epsilon_0)\rho$, we have at once, from Problem 2.126,

$$\mathbf{A}(P) = \frac{\mu_0}{4\pi} \int_V \frac{\mathbf{J}\, dV}{R} \quad (R \equiv \text{distance from } \mathbf{J}\, dV \text{ to } P) \qquad (1)$$

3.50 In the annular cylindrical space of Fig. 3-21 the magnetic vector potential is $\mathbf{A} = -k \ln r\, \mathbf{a}_z$, where k is a constant. Determine the total magnetic flux in the annular space.

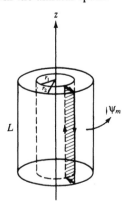

Fig. 3-21

▌ Apply (1) of Problem 3.47 to a cross section:

$$\psi_m = \int_0^L (-k \ln r_1)\, dz + 0 + \int_L^0 (-k \ln r_2)\, dz + 0 = kL \ln \frac{r_2}{r_1}$$

3.51 As shown in Fig. 3-22, the magnetic field intensity under the pole (in the air gap) of an electric motor is given by $\mathbf{H} = (10^5/r)\mathbf{a}_r$ (A/m). The pole is $L = 10$ cm long and subtends $90°$ at the axis. Calculate the flux per pole.

▌ Because H_r, and therefore B_r, is constant over the quarter-cylinder $r = a$,

$$\psi_m = B_r \times \text{area} = \mu_0 \frac{10^5}{a} \times a \frac{\pi}{2} L = 2\pi^2 \text{ mWb}$$

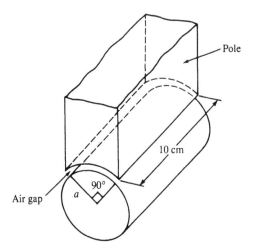

Pole

10 cm

Air gap a 90°

Fig. 3-22

3.52 Find the magnetic field, at a remote point P, produced by a current element of length L and carrying a current I, as shown in Fig. 3-23.

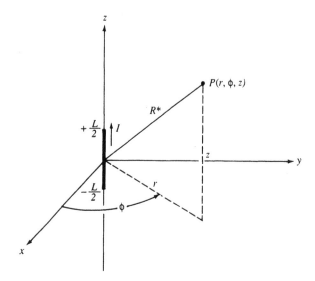

Fig. 3-23

▮ From (1) of Problem 3.49,

$$\mathbf{A} = \frac{\mu_0 I}{4\pi} \int_{-L/2}^{L/2} \frac{ds}{R} \mathbf{a}_z \qquad (1)$$

For $R^* \gg L$, (1) becomes

$$\mathbf{A} = \frac{\mu_0 I L}{4\pi R^*} \mathbf{a}_z = \frac{\mu_0 I L}{4\pi \sqrt{r^2 + z^2}} \mathbf{a}_z \qquad (2)$$

whence

$$\mathbf{B} = \operatorname{curl} \mathbf{A} = -\frac{\partial A_z}{\partial r} \mathbf{a}_\phi = \frac{\mu_0 I L r}{4\pi (r^2 + z^2)^{3/2}} \mathbf{a}_\phi \qquad (3)$$

3.53 Find the farfield magnetic vector potential for a circular current loop.

▮ Pick a rectangular coordinate system such that the observation point P lies in the plane $y = 0$ (Fig. 3-24). By (1) of Problem 3.49,

$$\mathbf{A}(P) = \frac{\mu_0 I a}{4\pi} \int_{-\pi}^{\pi} \frac{d\phi}{R} \mathbf{a}_\phi = \frac{\mu_0 I a}{4\pi} \left[-\int_{-\pi}^{\pi} \frac{\sin \phi \, d\phi}{R} \mathbf{a}_x + \int_{-\pi}^{\pi} \frac{\cos \phi \, d\phi}{R} \mathbf{a}_y \right] \qquad (1)$$

Now,

$$\mathbf{R} = \mathbf{r} - \mathbf{u} = (x\mathbf{a}_x + z\mathbf{a}_z) - (a \cos \phi \, \mathbf{a}_x + a \sin \phi \, \mathbf{a}_y)$$
$$= (x - a \cos \phi)\mathbf{a}_x - a \sin \phi \, \mathbf{a}_y + z \, \mathbf{a}_z$$

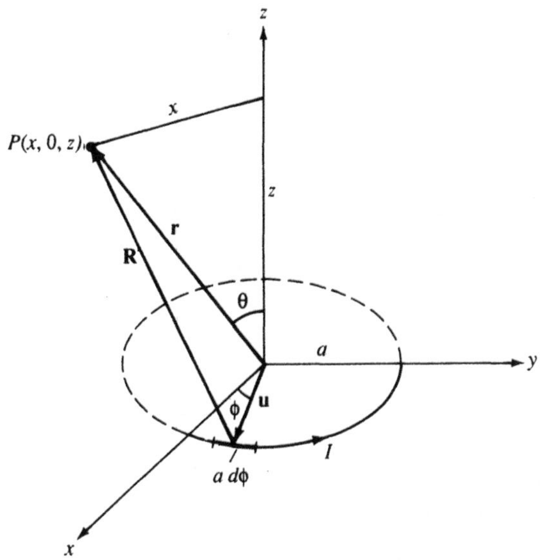

Fig. 3-24

so that, given $r^2 = x^2 + z^2 \gg a^2$,

$$R^2 = x^2 + z^2 + a^2 - 2ax\cos\phi \approx (x^2 + z^2)\left(1 - \frac{2ax\cos\phi}{x^2 + z^2}\right)$$

$$R^{-1} = (R^2)^{-1/2} \approx (x^2 + z^2)^{-1/2}\left(1 + \frac{ax\cos\phi}{x^2 + z^2}\right) \tag{2}$$

When (2) is substituted in (1), the first integral vanishes (since the integrand is odd in ϕ). Hence, $\mathbf{A}(P) = A_y\mathbf{a}_y$, with

$$A_y = \frac{\mu_0 I a}{2\pi(x^2 + z^2)^{1/2}}\int_0^\pi \left(\cos\phi + \frac{ax\cos^2\phi}{x^2 + z^2}\right)d\phi = \frac{\mu_0 I a^2 x}{4(x^2 + z^2)^{3/2}} \tag{3}$$

3.54 Transform the result of Problem 3.53 into spherical coordinates and thence into cross-product form.

$\blacksquare$ Writing $x = r\sin\theta$ and $z = r\cos\theta$, and introducing the *magnetic dipole moment* of the loop, $\mathbf{m} \equiv I\mathbf{A} = I\pi a^2\mathbf{a}_z$, we find $(\mathbf{a}_y \to \mathbf{a}_\phi)$:

$$\mathbf{A} = \frac{\mu_0 m\sin\theta}{4\pi r^2}\mathbf{a}_\phi = \frac{\mu_0\mathbf{m}\times\mathbf{r}}{4\pi r^3} \tag{1}$$

3.55 Within a cylindrical conductor the magnetic vector potential, which is attributable to a current of density J_0, is given by $\mathbf{A} = -(\mu_0 J_0/4)(x^2 + y^2)\mathbf{a}_z$. Obtain the magnetic intensity within the conductor.

$\blacksquare$
$$\mathbf{H} = \frac{1}{\mu_0}\nabla\times\mathbf{A} = -\frac{1}{2}J_0(y\mathbf{a}_x + x\mathbf{a}_y) = \frac{1}{2}J_0 r\mathbf{a}_\phi \tag{1}$$

3.56 Verify that the result (1) of Problem 3.55 is consistent with Ampère's circuital law.

$\blacksquare$ By Ampère's law, $J_0\pi r^2 = H_\phi 2\pi r$, or $\mathbf{H} = \frac{1}{2}J_0 r\mathbf{a}_\phi$.

3.57 Show that the magnetic vector potential $\mathbf{A}$ given in Problem 3.55 is a valid function.

$\blacksquare$ The validity may be established by showing that the given function satisfies Poisson's equation $\nabla^2 A_z = -\mu_0 J_z$ (Problem 3.48).

$$\frac{\partial^2 A_z}{\partial x^2} = -\frac{1}{2}\mu_0 J_0 \quad\text{and}\quad \frac{\partial^2 A_z}{\partial y^2} = -\frac{1}{2}\mu_0 J_0$$

Thus
$$\nabla^2 A_z = \frac{\partial^2 A_z}{\partial x^2} + \frac{\partial^2 A_z}{\partial y^2} = -\mu_0 J_0$$

3.58 Find the magnetic vector potential in the hole of an infinitely long wire shown in Fig. 3-25(a); assume a uniform current density $J_0\mathbf{a}_z$ in the conductor.

$\blacksquare$ Replace the hole by two equal and opposite currents; then the desired vector potential is the superposition

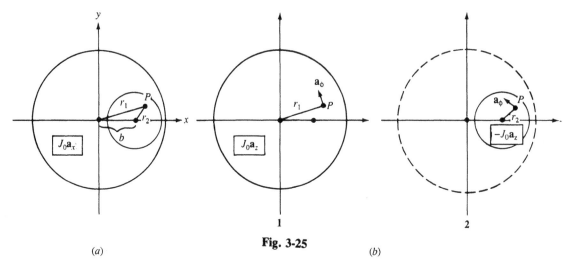

Fig. 3-25

(a)   (b)

of the vector potentials in the two conductors shown in Fig. 3-25(b). Problem 3.55 gives at once

$$\mathbf{A}_1(P) = \frac{-\mu_0 J_0 r_1^2}{4}\mathbf{a}_z \qquad \mathbf{A}_2(P) = \frac{+\mu_0 J_0 r_2^2}{4}\mathbf{a}_z$$

whence
$$\mathbf{A}(P) = \mathbf{A}_1(P) + \mathbf{A}_2(P) = \frac{\mu_0 J_0 (r_2^2 - r_1^2)}{4}\mathbf{a}_z$$

3.59 Evaluate $\mathbf{A}$ of Problem 3.58 in rectangular coordinates.

▮ From Fig. 3-25,
$$r_2^2 - r_1^2 = [(x - b)^2 + y^2] - (x^2 + y^2) = b(b - 2x)$$
so that $\mathbf{A} = \frac{1}{4}\mu_0 J_0 b(b - 2x)\mathbf{a}_z$.

3.60 Show that the flux density in the hole of Fig. 3-25(a) is uniform and is entirely y-directed.

▮
$$\mathbf{B} = \text{curl } \mathbf{A} = \tfrac{1}{2}\mu_0 J_0 b \mathbf{a}_y$$

3.61 The magnetic vector potential in the xy plane is given by $\mathbf{A} = (e^y \cos x)\mathbf{a}_x + (1 + \sin^2 x)\mathbf{a}_z$. Calculate the magnetic flux density at the origin.

▮
$$\mathbf{B}(0, 0) = (\mathbf{\nabla} \times \mathbf{A})_{x=y=0} = (-\mathbf{a}_y 2\sin x \cos x - \mathbf{a}_z e^y \cos x)_{x=y=0} = -\mathbf{a}_z$$

3.62 Two infinitely permeable semi-infinite blocks are separated from each other by a distance g (m) (Fig. 3-26). With reference to the dimensionless coordinates shown, the magnetic vector potential within the air gap is given by $\mathbf{A} = \mathbf{a}_z(k_1 \cosh y + k_2 \sinh y)$ (A). Evaluate the constants k_1 and k_2, assuming $B_x = 0.8$ T at $y = 0$.

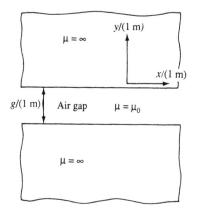

Fig. 3-26

▮ Since $\mathbf{B} = \mathbf{\nabla} \times \mathbf{A}$,

$$B_x = \frac{\partial A_z}{\partial y} = k_1 \sinh y + k_2 \cosh y \quad \text{(T)}$$

The two boundary conditions, $B_x(x, -g) = 0$ (since $\mu = \infty$) and $B_x(x, 0) = 0.8$ T, yield
$$k_1 = (0.8 \coth g)\text{ T} \qquad \text{and} \qquad k_2 = 0.8\text{ T}$$

3.63   Determine the far **B**-field for a circular current loop.

▮ Using the vector potential found in Problem 3.54 (see Fig. 3-24), we have

$$\mathbf{B} = \frac{\mu_0 m}{4\pi} \operatorname{curl} \left(\frac{\sin \theta}{r^2} \mathbf{a}_\phi \right)$$

$$= \frac{\mu_0 m}{4\pi} \left[\frac{1}{r \sin \theta} \frac{\partial}{\partial \theta} \left(\frac{\sin^2 \theta}{r^2} \right) \mathbf{a}_r - \frac{1}{r} \frac{\partial}{\partial r} \left(\frac{\sin \theta}{r} \right) \mathbf{a}_\theta \right]$$

$$= \frac{\mu_0 m}{4\pi r^3} (2 \cos \theta \, \mathbf{a}_r + \sin \theta \, \mathbf{a}_\theta)$$

3.64   The torque exerted by a uniform **B**-field on a planar loop of arbitrary shape, having directed area $\mathbf{A} = A\mathbf{a}_n$ and carrying current I, is given by

$$\mathbf{T} = \mathbf{m} \times \mathbf{B} \qquad \text{where} \qquad \mathbf{m} = \text{magnetic dipole moment} = I\mathbf{A}$$

Find the torque on loop 2 of Fig. 3-27, given that each loop has area 5 cm^2 and carries 5 A of current in 20 turns. (The axis of loop 2 is in the negative θ direction relative to the axis of loop 1.)

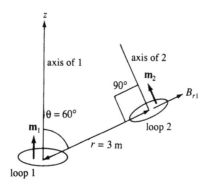

Fig. 3-27

▮ Since the loops are small and widely separated, we may assume that the far-field of loop 1 has at every point of loop 2 the value it assumes at the center of loop 2. Then, using Problem 3.63,

$$\mathbf{T} = \mathbf{m}_2 \times \mathbf{B}_1 = m_2 I_2 A_2 (-\mathbf{a}_\theta) \times \frac{\mu_0 m_1 I_1 A_1}{4\pi r^3} (2 \cos \theta \, \mathbf{a}_r + \sin \theta \, \mathbf{a}_\theta)$$

$$= \frac{\mu_0 (mIA)^2 \cos \theta}{2\pi r^3} \mathbf{a}_\phi = (9.26 \times 10^{-12} \, \text{N} \cdot \text{m}) \mathbf{a}_\phi$$

Noting that the angle between $\mathbf{m}_2$ and $\mathbf{m}_1$ is $90° - \theta$, we can generalize the above result as

$$\mathbf{T} = \frac{\mu_0}{2\pi r^3} (\mathbf{m}_2 \times \mathbf{m}_1) \tag{1}$$

which is valid regardless of whether or not $m_1 = m_2$ and whether or not the loops are circles.

3.65   Determine the magnitude of the torque between two small, widely separated, circular current loops whose planes are parallel (Fig. 3-28).

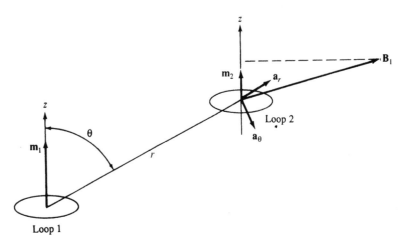

Fig. 3-28

▮ Resolve $\mathbf{B}_1$ into components parallel to and perpendicular to $\mathbf{m}_2$. Using $\mathbf{a}_z = \mathbf{a}_r \cos\theta - \mathbf{a}_\theta \sin\theta$ and the expression for $\mathbf{B}_1$ obtained in Problem 3.63,

$$B_{1z} = \mathbf{B}_1 \cdot \mathbf{a}_z = \frac{\mu_0 m_1}{4\pi r^3}(2\cos^2\theta - \sin^2\theta)$$

And, by the Pythagorean theorem,

$$B_{1\perp}^2 = B_1^2 - B_{1z}^2 = \left(\frac{\mu_0 m_1}{4\pi r^3}\right)^2 [(2\cos\theta)^2 + (\sin\theta)^2 - (2\cos^2\theta - \sin^2\theta)^2]$$

$$= \left(\frac{\mu_0 m_1}{4\pi r^3}\right)^2 (3\sin\theta\cos\theta)^2$$

or

$$B_{1\perp} = \frac{3\mu_0 m_1 \sin 2\theta}{8\pi r^3}$$

Then

$$|\mathbf{T}| = m_2 B_{1\perp} = \frac{3\mu_0 m_1 m_2 \sin 2\theta}{8\pi r^3} \tag{1}$$

Observe that (1) implies zero torque between coplanar loops $(\theta = \pi/2)$ or coaxial loops $(\theta = 0)$.

3.66 A permanent-magnet moving-coil ammeter has a uniform air-gap flux density of 0.6 T. The torque constant of the spring is $\lambda = 3\,\mu\text{N}\cdot\text{m/degree}$. The moving coil has 60 turns and is mounted on a soft-iron cylinder of diameter 15 mm and length 20 mm (Fig. 3-29). Calculate the current, assuming the meter deflection is $\alpha = 60°$.

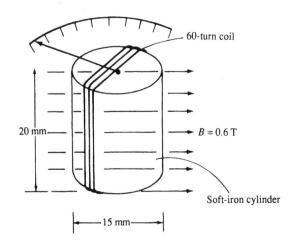

Fig. 3-29

▮
$$\text{deflecting torque} = \text{restoring torque}$$
$$NIAB = \lambda\alpha$$
$$I = \frac{\lambda\alpha}{NAB} = \frac{(3\times 10^{-6})(60)}{60(0.015)(0.020)(0.6)} = 16.67\,\text{mA}$$

3.67 For the single-turn loop of current shown in Fig. 3-30, determine the **B**-field at the center of the semicircular portion (at P).

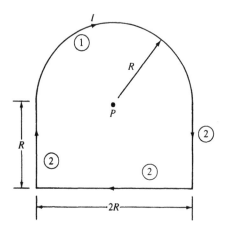

Fig. 3-30

⬛ We divide the loop into two segments as shown. From Problem 3.12, due to portion 1, we have

$$B_1 = \frac{\mu_0 I}{4R}$$

The field of portion 2 will be half that of a square loop of side $2R$; from Problem 3.8,

$$B_2 = \frac{\sqrt{2}\,\mu_0 I}{2\pi R}$$

Since both subfields are into the paper,

$$B = B_1 + B_2 = \frac{\mu_0 I}{4R}\left(1 + \frac{2\sqrt{2}}{\pi}\right)$$

3.68 A square loop, carrying a current I, is bisected by a very long straight conductor which carries a current I' (Fig. 3-31). Find the net force on the loop.

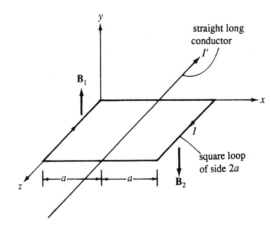

Fig. 3-31

⬛ Fields $\mathbf{B}_1$ and $\mathbf{B}_2$ due to the straight conductor are given by

$$B_1 = B_2 = \frac{\mu_0 I'}{2\pi a}$$

with their directions as shown. Hence the force on the side $x = 0$ is

$$\mathbf{F}_1 = I(2a)B_1 \mathbf{a}_x = \frac{\mu_0 I I'}{\pi}\mathbf{a}_x$$

and the force on $x = 2a$ is the same (because both the current and the field are reversed). By symmetry, there is zero force on either of the other two sides. Hence

$$\mathbf{F} = 2\mathbf{F}_1 = \frac{2\mu_0 I I'}{\pi}\mathbf{a}_x$$

3.69 The **B**-field over the xy plane is given by

$$\mathbf{B} = \sin\frac{\pi x}{2}\sin\frac{\pi y}{2}\,\mathbf{a}_z \quad (\text{T})$$

Calculate the flux passing through the square $0 < x, y < 2$ m.

⬛ $$\psi_m = \int \mathbf{B} \cdot d\mathbf{S} = \int_{x=0}^{2}\int_{y=0}^{2}\left(\sin\frac{\pi x}{2}\sin\frac{\pi y}{2}\right)\mathbf{a}_z \cdot dx\,dy\,\mathbf{a}_z = \left[-\frac{2}{\pi}\cos\frac{\pi x}{2}\right]_0^2\left[-\frac{2}{\pi}\cos\frac{\pi y}{2}\right]_0^2 = \frac{16}{\pi^2} = 1.62 \text{ Wb}$$

3.70 Two loops of radii R_1 and R_2, carrying currents I_1 and I_2, are oriented as shown in Fig. 3-32. Obtain an expression for the torque on loop 2 due to the field of loop 1, in terms of their magnetic dipole moments. Assume $R_1, R_2 \ll a$.

⬛ This is the special case $r = a$, $\theta = 0$, of (1) of Problem 3.64; hence

$$\mathbf{T} = \frac{\mu_0 m_1 m_2}{2\pi a^3}\mathbf{a}_y$$

3.71 A tightly wound flat coil, of inner and outer radii r_i and r_o, has N turns and carries a current I, as shown in Fig. 3-33. Find the flux density at the center of the coil.

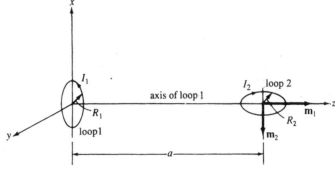

Fig. 3-32

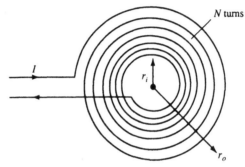

Fig. 3-33

▌ From Problem 3.12, at the center of a single-turn loop of radius r,

$$dB = \frac{\mu_0\, dI}{2r}$$

The coil may be approximated as a flat conducting annular ring having a constant linear current density

$$k = \frac{NI}{r_o - r_i}$$

Then, by integration,

$$B = \frac{\mu_0 NI}{2(r_o - r_i)} \int_{r_i}^{r_o} \frac{dr}{r} = \frac{\mu_0 NI}{2(r_o - r_i)} \ln \frac{r_o}{r_i}$$

with **B** directed normally into the paper.

3.72   The current sheet $\mathbf{k} = k\mathbf{a}_x$ occupies the xy plane. A straight conductor carrying a current I in the x direction is located at a height a above the current sheet, as shown in Fig. 3-34. Determine I such that $\mathbf{B} = \mathbf{0}$ at P.

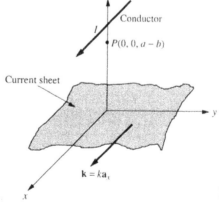

Fig. 3-34

▌ By Problem 3.21 (with relabeled axes), the field at P due to the current sheet is

$$\mathbf{B}_k = -\frac{\mu_0 k}{2}\mathbf{a}_y$$

while the field due to the straight conductor is

$$\mathbf{B}_I = \frac{\mu_0 I}{2\pi b}\mathbf{a}_y$$

For **B** = **0**, we must have

$$\frac{\mu_0 k}{2} = \frac{\mu_0 I}{2\pi b} \qquad \text{or} \qquad I = \pi b k$$

3.73   Determine the **B**-field corresponding to the magnetic vector potential $\mathbf{A} = y \cos x \, \mathbf{a}_x + (y + e^{-x})\mathbf{a}_z$.

▮ Since $\mathbf{B} = \nabla \times \mathbf{A}$,

$$\mathbf{B} = \mathbf{a}_x \frac{\partial}{\partial y}(y + e^{-x}) - \mathbf{a}_y \frac{\partial}{\partial x}(y + e^{-x}) - \mathbf{a}_z \frac{\partial}{\partial y}(y \cos x) = \mathbf{a}_x - ye^{-x}\mathbf{a}_y - \cos x \, \mathbf{a}_z$$

3.74   Given the magnetic vector potential $\mathbf{A} = k \sin \theta \, \mathbf{a}_\theta$ (spherical coordinates), evaluate the **B**-field at the point $(k, \pi/2, 0)$.

$$\mathbf{B} = \text{curl } \mathbf{A} = \frac{1}{r}\frac{\partial}{\partial r}(kr \sin \theta)\, \mathbf{a}_\phi = \frac{k}{r}\sin \theta \, \mathbf{a}_\phi$$

Thus, $\mathbf{B} = \mathbf{a}_x$ at the given point.

3.75   The flux density in a region is given by $\mathbf{B} = (k/r) \cos \phi \, \mathbf{a}_r$ (T). How much flux will pass through the portion of a cylindrical surface defined by $-\pi/4 \le \phi \le \pi/4$, $0 \le z \le L$? Show that the result is independent of the radius R of the cylinder.

▮
$$\psi_m = \int \mathbf{B} \cdot d\mathbf{S} = \int_0^L \int_{-\pi/4}^{\pi/4} \left(\frac{k}{R}\cos \phi\right) R \, d\phi \, dz = kL\sqrt{2} \ (\text{Wb})$$

which is independent of R.

3.76   Find the vector potential for the current sheet of Problem 3.21.

▮ By (*1*) of Problem 3.49, when **J** or **k** has a fixed direction, **A** has that same direction; thus, $\mathbf{A} = A\mathbf{a}_z$. By Problem 3.21,

$$\text{curl } \mathbf{A} = \frac{\partial A}{\partial y}\mathbf{a}_x - \frac{\partial A}{\partial x}\mathbf{a}_y = \mathbf{B} = \begin{cases} -(\mu_0 k/2)\mathbf{a}_x & y > 0 \\ +(\mu_0 k/2)\mathbf{a}_x & y < 0 \end{cases}$$

Thus $A = A(y)$, and integration gives, assuming $A(0) = 0$,

$$A = \begin{cases} -(\mu_0 k/2)y & y > 0 \\ +(\mu_0 k/2)y & y < 0 \end{cases}$$

or, more simply, $A = -(\mu_0 k/2)\,|y|$.

3.77   Show that the potential energy of a planar current loop in a uniform magnetic field can be defined as
$$U = -\mathbf{m} \cdot \mathbf{B} \tag{1}$$

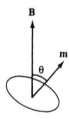

Fig. 3-35

▮ In the orientation shown in Fig. 3-35, the loop experiences a torque of magnitude $T(\theta) = mB \sin \theta$ that tends to rotate **m** into alignment with **B**. The work done in the rotation must equal the decrease in the loop's potential energy; thus

$$U(\theta) - U(0) = \int_0^\theta T(\beta)\, d\beta = mB \int_0^\theta \sin \beta \, d\beta$$
$$= mB - mB \cos \theta = mB - \mathbf{m} \cdot \mathbf{B}$$

and we are free to set $U(0) = -mB$.

3.78   A 0.5-m-long straight conductor carries 10 A of current in the positive z direction and is located 50 mm away from the z axis. The field in the region is given by $\mathbf{B} = -0.1 \sin(\phi/2)\,\mathbf{a}_r$ (T). Find the force on the conductor.

▮
$$\mathbf{F} = I\mathbf{L} \times \mathbf{B} = IL_z \mathbf{a}_z \times B_r \mathbf{a}_r = IL_z B_r \mathbf{a}_\phi$$
$$= 10 \times 0.5[0.1 \sin(\phi/2)]\,\bar{\mathbf{a}}_\phi = 0.5 \sin(\phi/2)\,\bar{\mathbf{a}}_\phi \text{ N}$$

3.79 The conductor of Problem 3.78 is rotated once about the z axis (maintaining the 50-mm radius), at constant speed in the positive ϕ direction. How much work is required?

∎ The required force is $\mathbf{F}_a = -\mathbf{F} = 0.5 \sin (\phi/2) \, \mathbf{a}_\phi$; hence the work is

$$W = \int_0^{2\pi} \mathbf{F}_a \cdot r \, d\phi \, \mathbf{a}_\phi = (0.5)(0.050) \int_0^{2\pi} \sin (\phi/2) \, d\phi = 0.1 \text{ J}$$

3.80 Determine the power required to rotate the conductor of Problems 3.78 and 3.79 at a speed of 1800 rev/min.

∎ Since power is work done per second, and 1800 rev/min corresponds to 1800/60 = 30 rev/s:

$$P = \frac{0.1 \text{ J/rev}}{\frac{1}{30} \text{ s/rev}} = 3 \text{ W}$$

3.81 A rectangular N-turn coil in the $-yz$ plane (see Fig. 3-36), carrying a current I, is immersed in the uniform field $\mathbf{B} = B_0(\mathbf{a}_x + \mathbf{a}_y)$. Determine the torque about the z axis.

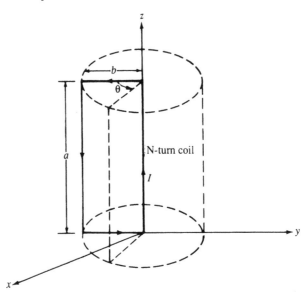

Fig. 3-36

∎
$$\mathbf{T} = \mathbf{m} \times \mathbf{B} = NIab\mathbf{a}_x \times B_0(\mathbf{a}_x + \mathbf{a}_y) = B_0 NIab\mathbf{a}_z$$

3.82 Show that if the coil of Problem 3.81 were released, it would rotate through one-eighth of a turn.

∎ In Fig. 3-36, when $\theta = \pi/4$, $\mathbf{m}$ has the direction of $\mathbf{a}_x + \mathbf{a}_y$ and is thus parallel to $\mathbf{B}$. This is the equilibrium orientation, in which the torque on the coil is zero.

3.83 For what value(s) of θ will the coil of Problems 3.81 and 3.82 experience the maximum torque? Evaluate this maximum torque.

∎ $\mathbf{T} = \mathbf{m} \times \mathbf{B}$, or $T = mB \sin \alpha$, shows that the maximum torque,
$$T_{\max} = mB = NIab\sqrt{B_0^2 + B_0^2} = NIabB_0\sqrt{2}$$
occurs when $\mathbf{m}$ and $\mathbf{B}$ are at right angles; i.e., when $\theta = -\pi/4$ or $+3\pi/4$.

3.84 Obtain an expression for the force per unit length on the conductor of Fig. 3-34, assuming that the current sheet is of finite width w ($-w/2 < y < w/2$).

∎ The force on the conductor (per unit length) due to an elemental strip of width dy of the current sheet is given by Problem 3.2 as

$$d\mathbf{F} = \frac{\mu_0 I(k \, dy)}{2\pi r} (-\mathbf{a}_r)$$

where $r = \sqrt{a^2 + y^2}$ and $-\mathbf{a}_r$ is the unit vector in the direction of decreasing r (attractive force). A similar force will occur as a result of a strip located at $-y$. Because of symmetry, the y components of the two differential forces will cancel to give

$$d\mathbf{F}_{\text{res}} = \frac{\mu_0 I(k \, dy)}{2\pi r} \left(2 \frac{a}{r}\right)(-\mathbf{a}_z) = -\frac{\mu_0 Ika}{\pi} \frac{dy}{a^2 + y^2} \mathbf{a}_z$$

The total force per unit length then becomes

$$\mathbf{F}_{\text{res}} = -\frac{\mu_0 Ika}{\pi} \int_0^{w/2} \frac{dy}{a^2 + y^2} \mathbf{a}_z = -\left(\frac{\mu_0 Ik}{\pi} \arctan \frac{w}{2a}\right)\mathbf{a}_z$$

3.85 An interface of two magnetic materials (Fig. 3-37) has a linear current density k (A/m). Obtain a relation between the tangential components of the magnetic field intensities in the two regions.

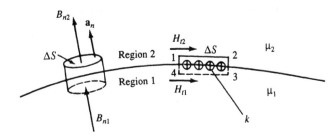

Fig. 3-37

▌ Applying Ampère's law around the infinitesimal circuit indicated,
$$H_{t2} \, \Delta s - H_{t1} \, \Delta s = k \, \Delta s \qquad \text{or} \qquad H_{t2} - H_{t1} = k$$
or, in vector form,
$$(\mathbf{H}_1 - \mathbf{H}_2) \times \mathbf{a}_n = \mathbf{k} \tag{1}$$

3.86 What is the relationship between the normal components of the **B**-field at the interface of Fig. 3-37?

▌ Applying Problem 3.34 to the infinitesimal pillbox,
$$B_{n1} \, \Delta S - B_{n2} \, \Delta S = 0 \qquad \text{or} \qquad B_{n1} = B_{n2}$$
or, in vector form,
$$(\mathbf{B}_1 - \mathbf{B}_2) \cdot \mathbf{a}_n = 0 \tag{1}$$

3.87 Relate the angles θ_1 and θ_2 in Fig. 3-38. There is no current sheet at the interface.

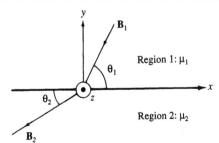

Fig. 3-38

▌ By Problem 3.85, since $\mathbf{B} = \mu \mathbf{H}$,
$$H_{t1} = H_{t2} \qquad \text{or} \qquad \frac{B_1}{\mu_1} \cos \theta_1 = \frac{B_2}{\mu_2} \cos \theta_2 \tag{1}$$
Similarly, from Problem 3.86 we obtain
$$B_{n1} = B_{n2} \qquad \text{or} \qquad B_1 \sin \theta_1 = B_2 \sin \theta_2 \tag{2}$$
Divide (2) by (1): $\mu_1 \tan \theta_1 = \mu_2 \tan \theta_2$. This is the magnetic analog to Snell's law. Just as with light, the "incident ray" $\mathbf{B}_1$, the local normal to the interface, and the "refracted ray" $\mathbf{B}_2$ all lie in the same plane; we choose it to be the xy plane.

3.88 Obtain a formula for the angular deflection in magnetic refraction (Problem 3.87).

▌ At the interface **B** is turned through an angle $\alpha = \theta_1 - \theta_2$. Let $\mu_1 \tan \theta_1 = \mu_2 \tan \theta_2 \equiv g$. Then, by trigonometry,
$$\tan \alpha = \frac{\tan \theta_1 - \tan \theta_2}{1 + (\tan \theta_1)(\tan \theta_2)} = \frac{\mu_2 \mu_1 \tan \theta_1 - \mu_1 \mu_2 \tan \theta_2}{\mu_1 \mu_2 + (\mu_1 \tan \theta_1)(\mu_2 \tan \theta_2)} = \frac{(\mu_2 - \mu_1)g}{\mu_1 \mu_2 + g^2} \tag{1}$$
Note that $\alpha > 0$ (deflection *away from* the surface normal) if and only if $\mu_2 > \mu_1$.

3.89 Refer to Fig. 3-38. Given: $\mathbf{B}_2 = 1.2\mathbf{a}_x + 0.8\mathbf{a}_y$, $\mu_1 = \mu_0$ and $\mu_2 = 15\mu_0$. Determine $\mathbf{H}_1$ and $\mathbf{B}_1$.

▌ By continuity of H_t, $H_{t1} = H_{t2} = B_{t2}/\mu_2 = 1.2/15\mu_0$; hence
$$\mathbf{H}_1 = \frac{1.2}{15\mu_0} \mathbf{a}_x + H_{n1}\mathbf{a}_y \tag{1}$$

By continuity of B_n, $B_{n1} = B_{n2} = 0.8$; hence

$$\mathbf{B}_1 = B_{t1}\mathbf{a}_x + 0.8\mathbf{a}_y \qquad (2)$$

The missing components are inferred from $\mathbf{B}_1 = \mu_1\mathbf{H}_1$:

$$B_{t1} = \frac{1.2}{15} \qquad H_{n1} = \frac{0.8}{\mu_0} \qquad (3)$$

3.90   Verify the relation $\mu_1 \tan \theta_1 = \mu_2 \tan \theta_2$ for the **B**-fields of Problem 3.89.

⬛   In this case B_2 is the incident field; but that does not matter, since the formula is symmetric. We have

$$\mu_1 \tan \theta_1 = \mu_0 \frac{0.8}{1.2/15} \qquad \mu_2 \tan \theta_2 = 15\mu_0 \frac{0.8}{1.2}$$

and these are clearly equal.

3.91   Find the angular deflection in Problem 3.89.

⬛   In the notation of Problem 3.88, $g = 10\mu_0$, and (1) gives

$$\tan \alpha = \frac{(14\mu_0)(10\mu_0)}{15\mu_0^2 + 100\mu_0^2} = 1.2174 \qquad \text{or} \qquad \alpha = 50.6°$$

3.92   Assuming, for definitness, that $\mu_2 > \mu_1$, find the largest angular deflection possible at the interface of Fig. 3-38.

⬛   Because the tangent function is strictly increasing, we can calculate $d(\tan \alpha)/dg$ from (1) of Problem 3.88 and equate it to zero. The unique root of the equation is $g_c = \sqrt{\mu_1\mu_2}$, so that

$$\tan \alpha_{\max} = \frac{\mu_2 - \mu_1}{2\sqrt{\mu_1\mu_2}} \qquad \text{or} \qquad \alpha_{\max} = \arctan \frac{\mu_2 - \mu_1}{2\sqrt{\mu_1\mu_2}}$$

3.93   Ampère's force law, (1) or (2) of Problem 3.1, endows $\mathbf{F}_1$ with the SI unit $H \cdot A^2/m$. Verify that this unit coincides with the newton.

⬛   From electric circuit theory we know that the henry is the unit of inductance; the formula $W = \frac{1}{2}LI^2$ for the energy stored in an inductor (Problem 3.122) implies that $1\,J = 1\,N \cdot m = 1\,H \cdot A^2$. Thus

$$1\,H \cdot A^2/m = 1\,N \cdot m/m = 1\,N$$

3.94   At the interface between two regions (at $x = 0$), there exists a current sheet $\mathbf{k} = 10\mathbf{a}_z$ A/m. Given $\mathbf{H}_1 = 12\mathbf{a}_y$ A/m at $x = 0^-$, determine $\mathbf{H}_2$ at $x = 0^+$.

⬛   Since $\mathbf{H}_1$ is purely tangential, $H_{n2} = H_{n1} = 0$ and we need no information regarding the permeabilities of the two regions. Now, from (1) of Problem 3.85,

$$(\mathbf{H}_1 - \mathbf{H}_2) \times \mathbf{a}_n = \mathbf{k} \qquad \text{or} \qquad (12\mathbf{a}_y - H_{y2}\mathbf{a}_y) \times \mathbf{a}_x = 10\mathbf{a}_z \qquad \text{or} \qquad (12 - H_{y2})(-\mathbf{a}_z) = 10\mathbf{a}_z$$

Hence $H_{y2} = 22$ A/m and $\mathbf{H}_2 = 22\mathbf{a}_y$ A/m.

3.95   For the regions shown in Fig. 3-39, we have $\mu_1 = 4\mu_0$, $\mu_2 = 6\mu_0$, and $\mathbf{B}_1 = 2\mathbf{a}_x + \mathbf{a}_y$ T. Find $\mathbf{B}_2$.

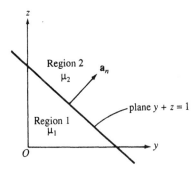

Fig. 3-39

⬛   The unit normal out of region 1 and into region 2 is

$$\mathbf{a}_n = \frac{1}{\sqrt{2}}\mathbf{a}_y + \frac{1}{\sqrt{2}}\mathbf{a}_z$$

so that

$$B_{n2} = B_{n1} = \mathbf{B}_1 \cdot \mathbf{a}_n = \frac{1}{\sqrt{2}}\,T \qquad \text{or} \qquad \mathbf{B}_{n2} = \mathbf{B}_{n1} = \frac{1}{\sqrt{2}}\mathbf{a}_n = 0.5\mathbf{a}_y + 0.5\mathbf{a}_z \quad T$$

Then

$$\mathbf{B}_{t1} = \mathbf{B}_1 - \mathbf{B}_{n1} = 2\mathbf{a}_x + 0.5\mathbf{a}_y - 0.5\mathbf{a}_z \quad \text{T}$$

$$\mathbf{H}_{t2} = \mathbf{H}_{t1} = \frac{1}{\mu_0}(0.5\mathbf{a}_x + 0.125\mathbf{a}_y - 0.125\mathbf{a}_z) \quad \text{A/m}$$

$$\mathbf{B}_{t2} = \mu_2\mathbf{H}_{t2} = 3.0\mathbf{a}_x + 0.75\mathbf{a}_y - 0.75\mathbf{a}_z \quad \text{T}$$

Finally

$$\mathbf{B}_2 = \mathbf{B}_{n2} + \mathbf{B}_{t2} = 3.0\mathbf{a}_x + 1.25\mathbf{a}_y - 0.25\mathbf{a}_z \quad \text{T}$$

3.96 Magnetic flux impinges at angle θ_1 on the interface between regions 1 and 2 of a three-layer medium, characterized by permeabilities μ_1, μ_2, and μ_3, as shown in Fig. 3-40. Show that the angle of emergence θ_4 in region 3 is independent of the permeability of region 2.

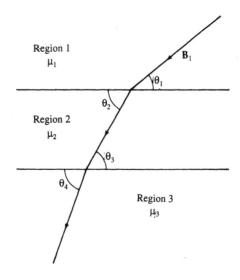

Fig. 3-40

▮ By Problem 3.87,

$$\mu_1 \tan \theta_1 = \mu_2 \tan \theta_2 \quad \text{and} \quad \mu_2 \tan \theta_3 = \mu_3 \tan \theta_4$$

But $\theta_2 = \theta_3$; hence, $\mu_1 \tan \theta_1 = \mu_3 \tan \theta_4$, independent of μ_2.

3.97 At the interface shown in Fig. 3-38, we have a current sheet $\mathbf{k} = -8\mathbf{a}_z$. Given $\mathbf{H}_1 = 12\mathbf{a}_x + 9\mathbf{a}_y$, $\mu_1 = 3\mu_0$, and $\mu_2 = 9\mu_0$, determine $\mathbf{H}_2$.

▮ From (1) of Problem 3.85,

$$(\mathbf{H}_1 - \mathbf{H}_2) \times (-\mathbf{a}_y) = \mathbf{k} \quad \text{or} \quad (12 - H_{x2})(-\mathbf{a}_z) = -8\mathbf{a}_z \quad \text{or} \quad H_{x2} = 4$$

From (1) of Problem 3.86,

$$(\mu_1\mathbf{H}_1 - \mu_2\mathbf{H}_2) \cdot (-\mathbf{a}_y) = 0 \quad \text{or} \quad (27\mu_0 - 9\mu_0 H_{y2})(-1) = 0 \quad \text{or} \quad H_{y2} = 3$$

Consequently, $\mathbf{H}_2 = 4\mathbf{a}_x + 3\mathbf{a}_y = \frac{1}{3}\mathbf{H}_1$. (For the particular data of this problem, there is zero deflection of the field.)

3.98 Solve Problem 3.97 assuming that the current sheet is removed.

▮ $\mathbf{H}_2 = 12\mathbf{a}_x + 3\mathbf{a}_y$.

3.99 The effects of a permanent magnet can be attributed to *bound currents*, of density $\mathbf{J}_b$, that give rise to a magnetic dipole moment per unit volume, or *magnetization*, through curl $\mathbf{M} = \mathbf{J}_b$. Relate $\mathbf{B}$ to $\mathbf{H}$ in free space where a magnet has been replaced by $\mathbf{M}$.

▮ Let $\mathbf{J}_f$ be the free current density. If the total current density is $\mathbf{J} = \mathbf{J}_b + \mathbf{J}_f$ and $\mathbf{B}$ is produced in free space by $\mathbf{J}$, then Ampère's law gives

$$\text{curl}\,\frac{\mathbf{B}}{\mu_0} = \mathbf{J} = \mathbf{J}_b + \mathbf{J}_f = (\text{curl}\,\mathbf{M}) + \mathbf{J}_f$$

or

$$\text{curl}\left(\frac{\mathbf{B}}{\mu_0} - \mathbf{M}\right) = \mathbf{J}_f$$

But curl $\mathbf{H} = \mathbf{J}_f$, and so we may define $\mathbf{H}$ such that

$$\mathbf{H} \equiv \frac{\mathbf{B}}{\mu_0} - \mathbf{M} \quad \text{or} \quad \mathbf{B} = \mu_0(\mathbf{M} + \mathbf{H}) \tag{1}$$

3.100 A cylindrical conductor of radius a and permeability μ carries a z-directed, uniformly distributed current I. Evaluate the magnetization $\mathbf{M}$, within the conductor.

▮ By Ampère's law, within the conductor, $0 < r < a$, we have

$$H_\phi(2\pi r) = \frac{\pi r^2}{\pi a^2} I \quad \text{or} \quad B_\phi = \mu H_\phi = \frac{\mu I r}{2\pi a^2} \quad (\text{T})$$

Thus, from (1) of Problem 3.99,

$$\mathbf{M} = M_\phi \mathbf{a}_\phi = \left(\frac{B_\phi}{\mu_0} - H_\phi\right)\mathbf{a}_\phi = \frac{I}{2\pi a^2}\left(\frac{\mu}{\mu_0} - 1\right) r\mathbf{a}_\phi \quad (\text{A/m}) \tag{1}$$

3.101 Determine the bound current density for the $\mathbf{M}$ evaluated in Problem 3.100.

▮
$$\mathbf{J}_b = \text{curl } \mathbf{M} = \frac{1}{r}\frac{\partial}{\partial r}(rM_\phi)\mathbf{a}_z = \frac{I}{\pi a^2}\left(\frac{\mu}{\mu_0} - 1\right)\mathbf{a}_z \quad (\text{A/m}^2)$$

3.102 Analogous to Problem 2.179, the energy density, w_m, at a point in a magnetic field is given by
$$w_m = \tfrac{1}{2}\mathbf{B} \cdot \mathbf{H} \quad (\text{J/m}^3) \tag{1}$$
Express this result in terms of the $\mathbf{B}$-field only; the $\mathbf{H}$-field only.

▮
$$w_m = \frac{1}{2\mu}|\mathbf{B}|^2 = \frac{\mu}{2}|\mathbf{H}|^2$$

3.103 A $\mathbf{B}$-field given by $\mathbf{B} = 0.5\mathbf{a}_x + 0.5\mathbf{a}_y$ T exists in free space. Calculate the energy density in the region.

▮
$$|\mathbf{B}|^2 = 0.5^2 + 0.5^2 = \frac{1}{2}\text{T}^2 \quad \text{and} \quad w_m = \frac{|\mathbf{B}|^2}{2\mu_0} = \frac{1/2}{2(4\pi \times 10^{-7})} = \frac{5}{8\pi}\text{MJ/m}^3$$

3.104 Obtain an expression for the *Lorentz force* on an electron moving with velocity $\mathbf{u}$ in a magnetic field $\mathbf{B}$.

▮ The moving electron is equivalent to a current element

$$I\,d\mathbf{s} = (I\,dt)\frac{d\mathbf{s}}{dt} = (-e)\mathbf{u}$$

where $-e = -1.6 \times 10^{-19}$ C is the electronic charge. Then, by (2) of Problem 3.3, $\mathbf{F} = -e\mathbf{u} \times \mathbf{B}$. In general, for a point charge q,

$$\mathbf{F} = q\mathbf{u} \times \mathbf{B} \tag{1}$$

3.105 A particle of charge q and mass m, moving at velocity $\mathbf{u}$, enters a uniform magnetic field $\mathbf{B}$ at right angles to the field. Describe the motion.

▮ The particle experiences a central force of constant magnitude, $F = quB$. This results in uniform motion around a circle of radius R. Since F is the centripetal force,

$$quB = \frac{mu^2}{R} \quad \text{or} \quad R = \frac{mu}{qB}$$

3.106 In Problem 3.105 the particle enters the field at an arbitrary angle. What then?

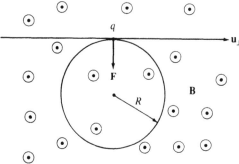

Fig. 3-41

▮ Decompose the entrance velocity $\mathbf{u}$ into a component $\mathbf{u}_\parallel$ along $\mathbf{B}$ and a component $\mathbf{u}_\perp$ perpendicular to $\mathbf{B}$ (Fig. 3-41). Only $\mathbf{u}_\perp$ interacts with the field, resulting in (Problem 3.105) uniform motion in a circle of radius $R = mu_\perp/qB$. Superimposing this motion on the uniform rectilinear motion $\mathbf{u}_\parallel$, we get uniform motion along a helix.

3.107   An electron ($e/m = 0.176 \times 10^{12}$ C/kg) enters a 0.2-T **B**-field. Find the frequency f (Hz) of its rotational motion.

▋ In the circular motion $u_\perp = 2\pi f R$; Problem 3.106 gives
$$f = \frac{u_\perp}{2\pi R} = \frac{1}{2\pi}(e/m)B = \frac{1}{2\pi}(0.176 \times 10^{12})(0.2) = 5600 \text{ MHz}$$
Note that this *cyclotron frequency* is independent of **u**.

3.108   A solenoid of length ℓ and radius R carries N turns of tightly would thin wire, as shown in Fig. 3-42. Assuming that the coil current is I, determine the **B**-field at the center of the solenoid.

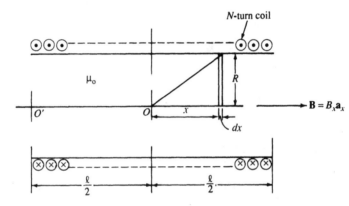

Fig. 3-42

▋ First, we replace the coil by a current sheet of linear density $k = NI/\ell$. Then, from (1) of Problem 3.11, an elemental loop of width dx (Fig. 3-42) contributes at point O
$$dB_x = \frac{\mu_0 k R^2 \, dx}{2(R^2 + x^2)^{3/2}} = \frac{\mu_0 N I R^2}{2\ell}\left[\frac{dx}{(R^2 + x^2)^{3/2}}\right] \qquad (1)$$
and so
$$B_x = \frac{\mu_0 N I R^2}{2\ell}\int_{-\ell/2}^{\ell/2}\frac{dx}{(R^2 + x^2)^{3/2}} = \frac{\mu_0 N I}{\sqrt{4R^2 + \ell^2}} \qquad (2)$$

3.109   Modify the result of Problem 3.108 for a very long solenoid.

▋ For $\ell \gg R$, (2) of Problem 3.108 becomes $B_x = \mu_0 N I / \ell = \mu_0 k$.

3.110   Find the flux density at O' in Fig. 3-42.

▋ Move the origin of x to O'; (2) of Problem 3.108 is replaced by
$$B_x = \frac{\mu_0 N I R^2}{2\ell}\int_{0}^{\ell}\frac{dx}{(R^2 + x^2)^{3/2}} = \frac{\mu_0 N I}{2\sqrt{R^2 + \ell^2}}$$

3.111   A magnetic circuit and a dc resistive circuit are compared in Fig. 3-43. The relationship between the magnetic flux ψ_m and the magnetomotive force (mmf) $\mathscr{F}$ is given by
$$\mathscr{F} = \mathscr{R}\psi_m \qquad (1)$$
Obtain an expression for the *reluctance* $\mathscr{R}$.

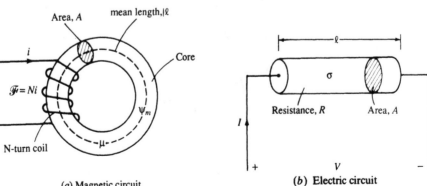

(a) Magnetic circuit   (b) Electric circuit   **Fig. 3-43**

▎ For the electric circuit we have $V = RI$; comparing this with (1), we observe that $\mathcal{R}$ is analogous to R. Since $R = \ell/\sigma A$, the analogy gives

$$\mathcal{R} = \frac{\ell}{\mu A} \tag{2}$$

By (2), the unit of reluctance is the H^{-1}.

3.112   Show that (1) of Problem 3.111 is consistent with the equation $\mathbf{B} = \mu\mathbf{H}$.

▎ From Ampère's law,

$$\oint \mathbf{H} \cdot d\mathbf{s} = H\ell = \mathcal{F}$$

and, by the definitions of ψ_m and $\mathcal{R}$,

$$\psi_m = \int \mathbf{B} \cdot d\mathbf{S} = BA = \frac{B\ell}{\mu\mathcal{R}} \quad \text{or} \quad \mathcal{R}\psi_m = \frac{B}{\mu}\ell$$

Hence, $\mathcal{F} = \mathcal{R}\psi_m$ if and only if $H = B/\mu$.

3.113   The (*self-*)*inductance* L of an N-turn coil is defined as the magnetic flux linkage per unit current; that is,

$$L \equiv \frac{\lambda}{i} = \frac{N\psi_m}{i} \tag{1}$$

where $\lambda = N\psi_m$ is the flux linkage due to i. Express the inductance of the coil of Fig. 3-43(*a*) in terms of its geometrical and magnetic properties.

▎ Substituting (1) and (2) of Problem 3.111 in (1) above yields

$$L = \frac{N}{i}\left(\frac{\mathcal{F}}{\mathcal{R}}\right) = \frac{N}{i}\left(\frac{Ni}{\mathcal{R}}\right) = \frac{N^2}{\mathcal{R}} = \frac{N^2}{\ell/\mu A} = \frac{\mu A N^2}{\ell} \tag{2}$$

3.114   The coil of a magnetic circuit has 100 turns. Calculate the current through the coil to produce 0.2 T of flux density in the core of length 30 cm, which has a relative permeability of 300.

▎ $B = \mu H = \mu(Ni/\ell)$, or

$$i = \frac{B\ell}{\mu N} = \frac{(0.2)(0.30)}{(300 \times 4\pi \times 10^{-7})(100)} = 1.6 \text{ A}$$

3.115   What must be the core cross section of the magnetic circuit of Problem 3.114 so that the coil may have a 50 μH inductance?

▎ From (2) of Problem 3.113,

$$A = \frac{L\ell}{\mu N^2} = \frac{(50 \times 10^{-6})(0.30)}{(4\pi \times 10^{-7})(100)^2} = 11.94 \text{ cm}^2$$

3.116   The core of the magnetic circuit of Fig. 3-43(*a*) is made of steel whose *B-H* characteristic is shown in Fig. 3-44. The core flux is 1.55 mWb. If the core mean length is 300 mm and the core has a circular cross section of radius 19.5 mm, determine the coil current. The coil has 100 turns.

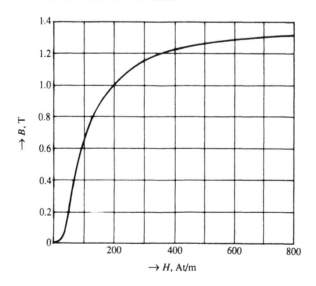

Fig. 3-44

▌ The core flux density is

$$B = \frac{\psi_m}{A} = \frac{\psi_m}{\pi r^2} = \frac{1.55 \times 10^{-3}}{\pi (0.0195)^2}$$

Then, from Fig. 3-44, $H \approx 700$ At/m and

$$i = \frac{H\ell}{N} \approx \frac{(700)(0.300)}{100} = 2.1 \text{ A}$$

3.117 What is the relative permeability of the core material at the operating point obtained in Problem 3.116?

▌
$$\mu_r = \frac{\mu}{\mu_0} = \frac{1}{\mu_0} \frac{B}{H} = \frac{1}{4\pi \times 10^{-7}} \frac{1.3}{700} = 1478$$

3.118 For the data of Problem 3.116, calculate the core flux if the coil carries 0.6 A of current.

▌
$$H = \frac{Ni}{\ell} = \frac{(100)(0.6)}{0.300} = 200 \text{ At/m} \qquad \text{and} \qquad B = 1.0 \text{ T} \quad \text{(from Fig. 3-44)}$$

Thus $\psi_m = (1.0)\pi(0.0195)^2 = 1.19 \text{ mWb}$.

3.119 A magnetic circuit with an air gap is shown in Fig. 3-45. How many turns should the exciting coil have to establish a 1.0 Wb/m² flux density in the air gap? The maximum allowable current through the coils is 10 A. The core has the *B-H* curve of Fig. 3-44. Neglect any fringing of the magnetic field at the air gap.

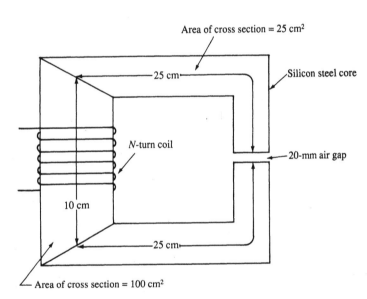

Area of cross section = 25 cm²
25 cm
Silicon steel core
N-turn coil
20-mm air gap
10 cm
25 cm
Area of cross section = 100 cm²

Fig. 3-45

▌ At $B = 1.0$ Wb/m² the *B-H* relationship is nonlinear; therefore, it is unwise to begin reluctance calculations with the iron. The total mmf can be expressed as $\mathscr{F}_{\text{tot}} = \mathscr{F}_{\text{steel}} + \mathscr{F}_{\text{air}}$, and the reluctance of the air gap is

$$\mathscr{R}_{\text{air}} = \frac{0.020}{(4\pi \times 10^{-7})(25 \times 10^{-4})} = 6.36 \times 10^5 \text{ H}^{-1}$$

The total flux is $\psi_m = (1.0)(25 \times 10^{-4}) = 2.5 \times 10^{-3}$ Wb, which should be the same for the entire magnetic circuit because no flux leaks out. Therefore,

$$\mathscr{F}_{\text{air}} = \mathscr{R}_{\text{air}}\psi_m = 1590 \text{ At}$$

From Fig. 3-44: for $B = 1.0$ Wb/m², $H = 200$ At/m. Thus, for the length $(25 + 25)$ cm, $\mathscr{F} = (200)(0.50) = 100$ At. The total flux in the limb 10 cm long is still 25×10^{-3} Wb; this gives $B = (2.5 \times 10^{-3})/(100 \times 10^{-4}) = 0.25$ Wb/m². From Fig. 3-44 again: for $B = 0.25$ Wb/m², $H = 70$ At/m; so that, for 10 cm, $\mathscr{F} = 7.0$ At. Thus

$$\mathscr{F}_{\text{steel}} = 100 + 7.0 = 107 \text{ At} \qquad \text{and} \qquad \mathscr{F}_{\text{tot}} = NI = 107 + 1590 = 1697 \text{ At}$$

The maximum allowable current is $I = 10$ A, and thus the required number of turns is $N = 169.7 \approx 170$.

3.120 From the data of Problem 3.119, calculate the flux linking the coil and its inductance.

▌ Flux linkage: $\quad \lambda = N\psi_m = (170)(2.5 \times 10^{-3}) = 0.425 \text{ Wb}$
Inductance: $\quad L = \lambda/i = 0.425/10 = 42.5 \text{ mH}$

3.121 Determine the energies stored in the various sections of the magnetic circuit of Problem 3.119.

  ▌ Energy stored in the air gap: $W_g = \dfrac{1}{2\mu_0}(B_g^2)(\text{vol})_g = \dfrac{(1.0)^2}{2(4\pi \times 10^{-7})}(0.020)(25 \times 10^{-4}) = 1.9895\ \text{J}$

  Energy stored in either 25-cm limb: $W_{25} = \dfrac{1}{2}(B_{25}H_{25})(\text{vol})_{25} = \dfrac{1.0}{2}(200)(0.25)(25 \times 10^{-4}) = 0.0625\ \text{J}$

  Energy stored in 10-cm limb: $W_{10} = \tfrac{1}{2}(B_{10}H_{10})(\text{vol})_{10} = \tfrac{1}{2}(0.25)(70)(0.10)(100 \times 10^{-4}) = 0.00875\ \text{J}$

3.122 Express the self-inductance of a coil in terms of the operating current and the associated **H**-field.

  ▌ The energy stored in an inductor L carrying a current i is given by $W_L = \tfrac{1}{2}Li^2$. This must be the same as the field energy,

$$W_m = \frac{1}{2}\iiint \mathbf{B} \cdot \mathbf{H}\, d\tau$$

from Problem 3.102. Therefore

$$L = \frac{\iiint \mathbf{B} \cdot \mathbf{H}\, d\tau}{i^2} = \frac{\mu}{i^2}\iiint H^2\, d\tau$$

3.123 Draw an electrical analog for the magnetic circuit shown in Fig. 3-46(a).

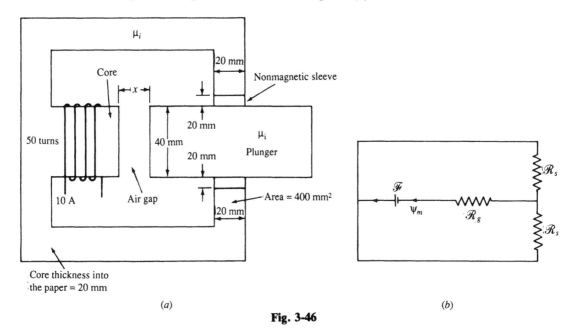

(a)   (b)

Fig. 3-46

  ▌ With mmf in the magnetic circuit corresponding to voltage in the electric circuit, the analog is as shown in Fig. 3-46(b).

3.124 From Problem 3.119 it is evident that in a magnetic circuit $\mathscr{F}_{\text{steel}} \ll \mathscr{F}_{\text{air}}$. Assuming an ideal core ($\mu_i \to \infty$), calculate the flux density and the magnetic energy stored in the air gap of the magnetic circuit of Fig. 3-46(a).

  ▌ From Fig. 3-46(b) and from (2) of Problem 3.111, we obtain

Air-gap reluctance: $\mathscr{R}_g = \dfrac{5 \times 10^{-3}}{\mu_0(20 \times 40 \times 10^{-6})} = \dfrac{50}{8\mu_0}$

Sleeve reluctance: $\mathscr{R}_s = \dfrac{2 \times 10^{-3}}{\mu_0(20 \times 20 \times 10^{-6})} = \dfrac{20}{4\mu_0}$

Total reluctance: $\mathscr{R}_t = \mathscr{R}_g + \dfrac{1}{2}\mathscr{R}_s = \dfrac{70}{8\mu_0}$

Air-gap flux: $\psi_m = \dfrac{\mathscr{F}}{\mathscr{R}_t} = \dfrac{(50)(10)}{70/8\mu_0} = \dfrac{400\mu_0}{7}$

Air-gap flux density: $B_g = \dfrac{\psi_m}{A_g} = \dfrac{400\mu_0}{7 \times (20 \times 40 \times 10^{-6})}$

Substituting $\mu_0 = 4\pi \times 10^{-7}$ H/m, we obtain $B_g = 0.0898$ T and

$$W_m = \frac{1}{2\mu_0} B_g^2 \times \text{vol}_g = 0.0128 \text{ J}$$

3.125 A composite magnetic circuit of varying cross section is shown in Fig. 3-47; the iron portion has the *B-H* characteristic of Fig. 3-44. Given: $N = 100$ turns; $\ell_1 = 4\ell_2 = 40$ cm; $A_1 = 2A_2 = 10$ cm²; $\ell_g = 2$ mm; leakage flux, $\psi'_m = 0.01$ mWb. Calculate I required establish an air-gap flux density of 0.6 T.

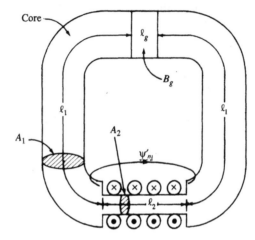

Fig. 3-47

▮ Corresponding to $B_g = 0.6$ T,

$$H_g = \frac{0.6}{\mu_0} = 4.78 \times 10^5 \text{ At/m} \qquad \mathscr{F}_g = (4.78 \times 10^5)(2 \times 10^{-3}) = 956 \text{ At} \qquad B_{\ell 1} = B_g = 0.6 \text{ T}$$

From Fig. 3-44, at $B = 0.6$ T, $H = 100$ At/m. Thus, for the two lengths ℓ_1,

$$\mathscr{F}_{\ell 1} = (100)(0.40 + 0.40) = 80 \text{ At}$$

The flux in the air gap, ψ_{mg}, is given by

$$\psi_{mg} = B_g A_1 = (0.6)(10 \times 10^{-4}) = 0.6 \text{ mWb}$$

The total flux produced by the coil is the sum of the air-gap flux and the leakage flux: $\psi_{mc} = 0.61$ mWb. The flux density in the portion ℓ_2 is, therefore,

$$B_2 = \frac{\psi_{mc}}{A_2} = \frac{0.61 \times 10^{-3}}{5 \times 10^{-4}} = 1.22 \text{ T}$$

For this flux density, from Fig. 3-44, $H = 410$ At/m and $\mathscr{F}_{\ell 2} = (410)(0.10) = 41$ At. The total required mmf $\mathscr{F}$ is thus

$$\mathscr{F} = \mathscr{F}_g + \mathscr{F}_{\ell 1} + \mathscr{F}_{\ell 2} = 956 + 80 + 41 = 1077 \text{ At}$$

For $N = 100$ turns, the desired current is, finally, $I = 1077/100 = 10.77$ A.

3.126 Draw an electrical analog for the magnetic circuit shown in Fig. 3-47.

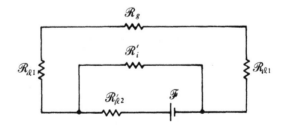

Fig. 3-48

▮ See Fig. 3-48.

3.127 Calculate the (total) self-inductance and the leakage inductance of the coil of Fig. 3-47.

▮ From Problem 3-125, the total flux and the coil current are $\psi_{mc} = 0.61$ mWb and $I = 10.77$ A. Hence

$$L = \frac{N\psi_{mc}}{I} = \frac{(100)(0.61 \times 10^{-3})}{10.77} = 5.66 \text{ mH}$$

and

$$L' = \frac{N\psi'_m}{I} = \frac{(100)(0.01 \times 10^{-3})}{10.77} = 0.093 \text{ mH}$$

3.128 Determine the magnetic energies stored in the iron and in the air gap of the magnetic circuit of Fig. 3-47.

▌ From (1) of Problem 3.102,

$$W_{air} = \frac{1}{2\mu_0} B_g^2 \times vol_g = \frac{(0.6)^2}{2\mu_0} [(10 \times 10^{-4})(2 \times 10^{-3})] = 0.286 \text{ J}$$

Then, by Problems 3.122 and 3.127,

$$W_{iron} = \tfrac{1}{2}LI^2 - W_{air} = \tfrac{1}{2}N\psi_{mc}I - W_{air} = 0.328 - 0.286 = 0.042 \text{ J}$$

3.129 A toroid of rectangular cross section is shown in Fig. 3-49. The mean radius is large compared to the core thickness in the radial direction (i.e., $3r_1 \gg r_2$), so that the core flux density is uniform. Derive an expression for the inductance of the toroid.

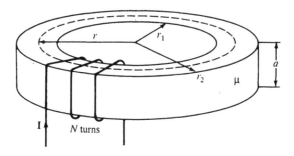

Fig. 3-49

▌ flux linkage: $\lambda = N\psi_m$

flux: $\psi_m = \dfrac{Ni}{\mathcal{R}} = \dfrac{\mu A N i}{2\pi r} = \dfrac{\mu a(r_2 - r_1)Ni}{\pi(r_2 + r_1)}$

since $A = a(r_2 - r_1)$ and $r = (r_2 + r_1)/2$. Hence

$$L = \frac{\lambda}{i} = \frac{\mu a(r_2 - r_1)N^2}{\pi(r_2 + r_1)} \tag{1}$$

3.130 For the toroid shown in Fig. 3-49, derive an expression for the core flux.

▌ Since $H(r) = NI/2\pi r$ (Ampère's law),

$$\psi_m = \int_{r_1}^{r_2} \mu \frac{NI}{2\pi r} a \, dr = \mu \frac{aNI}{2\pi} \ln \frac{r_2}{r_1}$$

3.131 For the toroid of Fig. 3-49, (a) if the core flux density is assumed to be uniform, and equal to its value at the (arithmetic) mean radius, what percent error would be made in the computation of the core flux as compared to the result of Problem 3.130? (b) If the geometric mean radius is used instead, what is the percent error?

▌ (a) Using the arithmetic mean, $B \approx \mu NI/\pi(r_2 + r_1)$, and so

$$\psi_m \approx \mu \frac{aNI}{2\pi} \frac{2(r_2 - r_1)}{r_2 + r_1}$$

Let $r_2/r_1 \equiv b$; then

$$\text{percent error} = 100\left[\frac{2(b-1)}{(b+1)\ln b} - 1\right]$$

(b) Using the geometric mean, $B \approx \mu NI/2\pi\sqrt{r_2 r_1}$, and so

$$\psi_m \approx \mu \frac{aNI}{2\pi} \frac{r_2 - r_1}{\sqrt{r_2 r_1}} \quad \text{and} \quad \text{percent error} = 100\left[\frac{b-1}{\sqrt{b}\ln b} - 1\right]$$

3.132 The *B-H* curve of a certain core material can be expressed by the *Froelich equation,*

$$B = \frac{aH}{b + H} \tag{1}$$

where $a = 1.5$ T and $b = 100$ A/m. A magnetic circuit using this material consists of two parts, of lengths ℓ_1 and ℓ_2 and cross-sectional areas A_1 and A_2. Assuming that $A_1 = 25 \text{ cm}^2 = 2A_2$ and $\ell_1 = 25 \text{ cm} = \tfrac{1}{2}\ell_2$, and if the core carries an mmf of 1000 At, calculate the core flux.

▌ From (1),

$$B = \frac{1.5H}{100 + H} \quad (\text{T})$$

For the magnetic circuit,

$$\mathscr{F} = H_1\ell_1 + H_2\ell_2 = 0.25H_1 + 0.50H_2 = 1000 \text{ At} \tag{1}$$

$$\psi_m = B_1A_1 = B_2A_2 \quad \text{or} \quad 2B_1 = B_2 \quad \text{or} \quad \frac{3.0H_1}{100 + H_1} = \frac{1.5H_2}{100 + H_2} \tag{2}$$

Eliminating H_1 between (1) and (2) yields $H_2 = 3900 \text{ At/m}$; thus

$$B_2 = \frac{(1.5)(3900)}{100 + 3900} = 1.4625 \text{ T} \quad \text{and} \quad \psi_m = (1.4625)\left(\frac{25}{2} \times 10^{-4}\right) = 1.828 \text{ mWb}$$

3.133   Using the result of Problem 3.130, obtain an approximate expression for the inductance of a very thin ring for which $r_2 - r_1 \ll r_1$.

■
$$L = \frac{N\psi_m}{I} = \frac{\mu a N^2}{2\pi} \ln\frac{r_2}{r_1} = \frac{\mu a N^2}{2\pi} \ln\left(1 + \frac{r_2 - r_1}{r_1}\right) \approx \frac{\mu a N^2}{2\pi}\left(\frac{r_2 - r_1}{r_1}\right)$$

3.134   For the ring of Problem 3.133 we have $\mu = 500\mu_0$, $r_1 = 92$ mm, $r_2 = 100$ mm, $a = 20$ mm, and $N = 100$ turns. Calculate its inductance using the exact and approximate formulas, and determine the percent error.

■
$$L_{\text{exact}} = \frac{(500\mu_0)(0.020)(100)^2}{2\pi} \ln\frac{100}{92} = 1.67 \text{ mH}$$
$$L_{\text{approx}} = \frac{(500\mu_0)(0.020)(100)^2}{2\pi}\left(\frac{100 - 92}{92}\right) = 1.74 \text{ mH}$$
$$\text{percent error} = \frac{1.74 - 1.67}{1.67} \times 100 = 4.19\%$$

3.135   Using the result of Problem 3.122, find the inductance per unit length of the coaxial cable shown in cross section in Fig. 3-50.

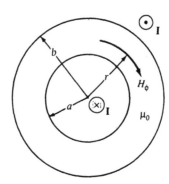

Fig. 3-50

■ It follows from Ampère's law that, for $a < r < b$, $H_\phi = I/2\pi r$. Hence
$$\frac{L}{1 \text{ m}} = \frac{\mu_0}{I^2}\int_a^b \left(\frac{I}{2\pi r}\right)^2 2\pi r \, dr = \frac{\mu_0}{2\pi}\ln\frac{b}{a} \quad \text{(H/m)}$$

3.136   Rework Problem 3.135 by the flux linkage method.

■ The flux through a $(b - a) \times (1 \text{ m})$ longitudinal section is
$$\psi_m = \int_a^b B_\phi(1 \text{ m}) \, dr = \frac{\mu_0 I(1 \text{ m})}{2\pi}\int_a^b \frac{dr}{r} = \frac{\mu_0 I(1 \text{ m})}{2\pi}\ln\frac{b}{a}$$
and $L/(1 \text{ m}) = [\psi_m/(1 \text{ m})]/I$.

3.137   For a magnetic toroid wound with n distinct coils (Fig. 3-51), n^2 inductances may be defined:
$$L_{pq} \equiv \frac{\text{flux linking the } p\text{th coil due to the current in the } q\text{th coil}}{\text{current in the } q\text{th coil}} \tag{1}$$
For $p = q$, L_{pp} is the self-inductance of the pth coil; for $p \neq q$, $L_{pq} = L_{qp}$ (see Problem 3.138) is the *mutual inductance* between coil p and coil q. Find the mutual inductance between coils p and q in terms of N_p, N_q, the permeance $\mathscr{P}$ of the core ($\mathscr{P} = 1/\mathscr{R}$), and the *coupling coefficient* $\kappa_{pq} = \kappa_{qp}$ (defined as the fraction of the flux due to coil q that links coil p).

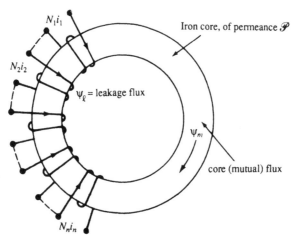

$N_1 i_1$

$N_2 i_2$

ψ_ℓ = leakage flux

Iron core, of permeance $\mathscr{P}$

ψ_{nl}

core (mutual) flux

$N_n i_n$

Fig. 3-51

▮ Let coil q be excited, producing flux $\psi_q = N_q i_q / \mathscr{R} = N_q i_q \mathscr{P}$. Then (1) gives

$$L_{pq} = \frac{N_p(\kappa_{pq}\psi_q)}{i_q} = \kappa_{pq} N_p N_q \mathscr{P} \tag{2}$$

Formula (2) may be termed the "circuit" expression for mutual inductance.

3.138   Mutual inductance can also be found from "field" considerations. Carry out the argument for the two loops of Fig. 3-1.

▮ The flux loop 1 as a result of the current in loop 2 is

$$\psi_{12} = \iint_{S_1} \mathbf{B}_2 \cdot d\mathbf{S}_1 = \oint_{\text{loop }1} \mathbf{A}_2 \cdot d\mathbf{s}_1 \tag{1}$$

(see Problem 3.47). Now, by Problem 3.49, the vector potential of loop 2 has at $d\mathbf{s}_1$ the value

$$\mathbf{A}_2 = \frac{\mu_0 I_2}{4\pi} \oint_{\text{loop }2} \frac{d\mathbf{s}_2}{R} \tag{2}$$

provided I_2 is uniform around loop 2. Substitution of (2) in (1) yields

$$\psi_{12} = \frac{\mu_0 I_2}{4\pi} \oint_{\text{loop }1} \oint_{\text{loop }2} \frac{d\mathbf{s}_2 \cdot d\mathbf{s}_1}{R} \tag{3}$$

Finally, the definition $L_{12} = \psi_{12}/I_2$ gives us *Neumann's formula*:

$$L_{12} = \frac{\mu_0}{4\pi} \oint_{\text{loop }1} \oint_{\text{loop }2} \frac{d\mathbf{s}_2 \cdot d\mathbf{s}_1}{R}$$

Neumann's formula makes it obvious that $L_{12} = L_{21}$. It also demonstrates that mutual inductance is an essentially geometric property, analogous to capacitance (see Problem 2.190).

3.139   Determine the mutual inductance between the two circular loops shown in Fig. 3-52. Assume that the spacing between the loops, D, is much larger than the loop radii. Use Neumann's formula (Problem 3.138).

▮ By the law of cosines, $\Delta^2 = r_1^2 + r_2^2 - 2r_1 r_2 \cos \alpha$, and so

$$R^2 = D^2 + \Delta^2 = D^2\left(1 + \frac{\Delta^2}{D^2}\right)$$

or

$$\frac{1}{R} = \frac{1}{D}\left(1 + \frac{\Delta^2}{D^2}\right)^{-1/2} \approx \frac{1}{D}\left(1 - \frac{\Delta^2}{2D^2}\right)$$

Moreover, since the angle between $d\mathbf{s}_2$ and $d\mathbf{s}_1$ is also α,

$$d\mathbf{s}_2 \cdot d\mathbf{s}_1 = (r_2\, d\phi_2)(r_1\, d\alpha) \cos \alpha$$

Neumann's formula then gives (note that terms in $\cos \alpha$ will integrate to zero):

$$L_{12} \approx \frac{\mu_0}{4\pi} \int_{\alpha=0}^{2\pi} \int_{\phi_2=0}^{2\pi} \frac{1}{D}\left(1 - \frac{r_1^2 + r_2^2 - 2r_1 r_2 \cos \alpha}{2D^2}\right) r_1 r_2 \cos \alpha\, d\phi_2\, d\alpha = \frac{\pi \mu_0 r_1^2 r_2^2}{2D^3} \tag{1}$$

3.140   Generalize the result (1) of Problem 3.139 to apply to two widely separated, but not necessarily coaxial or circular, loops.

▮ In terms of the areas A_1 and A_2 of the circular loops, (1) takes the form

$$L_{12} \approx \frac{\mu_0 A_1 A_2}{2\pi D^3} = \frac{\mu_0 D}{2\pi}\left(\frac{A_1}{D^2}\right)\left(\frac{A_2}{D^2}\right)$$

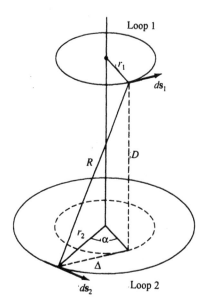

Fig. 3-52

But our basic approximation can be stated as $D^2 \gg A$, which makes

$$\frac{A_1}{D^2} \approx \Omega_1 \qquad \frac{A_2}{D^2} \approx \Omega_2$$

where Ω_1 (Ω_2) is the (small) solid angle subtended by loop 1 (loop 2) at the location of loop 2 (loop 1). Hence

$$L_{12} \approx \frac{\mu_0 D \Omega_1 \Omega_2}{\pi} \tag{1}$$

a formula which clearly extends to the larger class of configurations.

A more revealing form of (1) is obtained when one passes a sphere of diameter D through the two small loops:

$$L_{12} \approx \mu_0 \pi D f_1 f_2 \tag{2}$$

where f is the fraction of the spherical surface area within the loop.

3.141 An inductor, made of a highly permeable material, has N turns; the dimensions of the core and the coil are as shown in Fig. 3-53. Calculate the input power to the coil needed to establish a given flux density B in the air gap. The winding space factor of the coil is k_s and its conductivity is σ.

■ From Fig. 3-53(b), the mean length per turn is

$$\ell = 2a + 2d + 4\left(\frac{1}{4}\right)\left(2\pi\frac{b}{2}\right) = 2\left(a + d + \pi\frac{b}{2}\right)$$

and the total length of wire making the coil is ℓN. Let A_c denote the cross-sectional area of the wire; then its resistance is $R = \ell N/\sigma A_c$, giving an input power $P_i = I^2 R = I^2 \ell N / \sigma A_c$. But

$$\mathcal{F} = NI = \psi_m \mathcal{R} = BA\frac{g}{\mu_0 A} = \frac{Bg}{\mu_0} \qquad \text{or} \qquad I = \frac{Bg}{\mu_0 N}$$

Substituting above yields

$$P_i = \frac{B^2 g^2 \ell}{\mu_0^2 \sigma N A_c} = \frac{B^2 g^2 \ell}{\mu_0^2 \sigma b c k_s}$$

[By definition, the total volume of the wire is $(bc\ell)k_s = \ell N A_c$.]

3.142 The inductor of Problem 3.141 is made of magnet wire. Assuming that the dimensions in Fig. 3-53 are $a = b = c = d = 25$ mm and $g = 2$ mm, and the core flux density is 0.8 T, calculate the input power and the number of turns.

■ Assume $k_s = 0.8$, $\sigma = 57.8$ μS/m $(1$ S $= 1$ $\Omega^{-1})$, and a coil current of 1 A. Then Problem 3.141 gives

$$\ell = 2\left(1 + 1 + \frac{\pi}{2}\right)(25 \times 10^{-3}) = 0.1785 \text{ m}$$

$$P_i = \frac{(0.8)^2(2 \times 10^{-3})^2(0.1785)}{(4\pi \times 10^{-7})^2(57.8 \times 10^6)(25 \times 10^{-3})^2(0.8)} = 10 \text{ W}$$

$$N = \frac{Bg}{\mu_0 I} = \frac{(0.8)(2 \times 10^{-3})}{(4\pi \times 10^{-7})(1)} = 1273 \text{ turns}$$

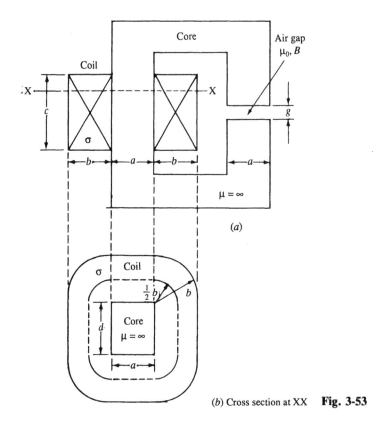

(a)

(b) Cross section at XX   **Fig. 3-53**

3.143   The core of the magnetic circuit of Fig. 3-54 has the *B-H* characteristic given in Fig. 3-44. Calculate the inductance of the coil if the air-gap flux density is 1.0 T.

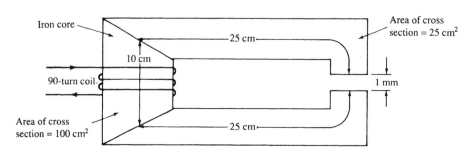

Fig. 3-54

▮   Proceeding as in Problem 3.119, we have, at $B_g = 1.0\,\text{T}$, $H_g = B_g/\mu_0 = (10^7/4\pi)$ At/m; thus,

$$\mathscr{F}_g = \frac{10^7 \times 1 \times 10^{-3}}{4\pi} = 796\,\text{At}$$

From Fig. 3-44, at $B = 1.0\,\text{T}$, $H = 200$ At/m. So, for the two 25-cm limbs, we have

$$2\mathscr{F}_{25} = 2 \times 200 \times 0.25 = 100\,\text{At}$$

The flux density in the 10-cm limb is $B = (1 \times 25)/100 = 0.25$ T, for which $H = 70$ At/m. Thus

$$\mathscr{F}_{10} = 70 \times 0.1 = 7.0\,\text{At}$$

The total mmf required to produce 1.0 T air-gap flux density is thus

$$\mathscr{F} = NI = 796 + 100 + 7.0 = 903\,\text{At}\qquad \text{whence}\qquad I = \frac{903}{90} = 10.03\,\text{A}$$

Now, $\psi_m = 0.25 \times 90 \times 10^{-4}$ Wb, so that

$$L = \frac{\psi_m}{I} = \frac{0.25 \times 90 \times 10^{-4}}{10.03} = 22.43\,\text{mH}$$

3.144   Recalculate *L* for Problem 3.143, assuming that all the magnetic energy is stored in the air gap and that the iron is infinitely permeable.

▮ Energy stored in the air gap at 1.0 T flux density is

$$W_m = \frac{B_g^2}{2\mu_0}(\text{vol}_g) = \frac{(1.0)^2 \times 1 \times 10^{-3} \times 25 \times 10^{-4}}{2 \times 4\pi \times 10^{-7}} = 0.995 \text{ J}$$

From Problem 3.143, since $\mathscr{F}_g = 796 \text{ At}$, $I = 796/90 = 8.84 \text{ A}$ and

$$L = \frac{2W_m}{I^2} = \frac{2(0.995)}{(8.84)^2} = 25.43 \text{ mH}$$

3.145 A system of three coils on an ideal core is shown in Fig. 3-55, where $N_1 = N_3 = 2N_2 = 500$ turns, $g_1 = 2g_2 = 2g_3 = 4$ mm, and $A = 1000 \text{ mm}^2$. Calculate the self-inductance of coil N_1.

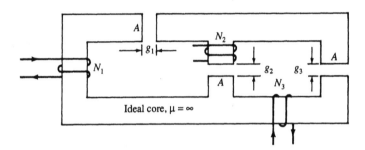

Fig. 3-55

▮ The reluctance seen by coil N_1 is

$$\mathscr{R} = \mathscr{R}_{g1} + (\mathscr{R}_{g2} \| \mathscr{R}_{g3}) = \frac{4 \times 10^{-3}}{\mu_0 A} + \frac{1}{2}\left(\frac{2 \times 10^{-3}}{\mu_0 A}\right) = \frac{5 \times 10^{-3}}{\mu_0 A} \quad (\text{H}^{-1})$$

Thus

$$L_{11} = \frac{N_1^2}{\mathscr{R}} = \frac{(500)^2 \times 4\pi \times 10^{-7} \times 1000 \times 10^{-6}}{5 \times 10^{-3}} = 62.8 \text{ mH}$$

3.146 If the gap g_1 in Fig. 3-55 were closed, what would be the mutual inductance between (a) N_1 and N_2; and (b) N_2 and N_3?

▮ (a) With g_1 closed, the entire flux produced by $N_2 i_2$ will link with N_1; hence

$$N_1 \psi_m = \frac{N_1 N_2 i_2}{\mathscr{R}_2} = \frac{N_1 N_2 i_2}{g_2/\mu_0 A}$$

and

$$L_{12} = \frac{N_1 \psi_m}{i_2} = \frac{\mu_0 A N_1 N_2}{g_2} = \frac{4\pi \times 10^{-7} \times 1000 \times 10^{-6} \times 500 \times 250}{2 \times 10^{-3}} = 78.54 \text{ mH}$$

(b) Because all the flux produced by either $N_2 i_2$ or $N_3 i_3$ will go through the short-circuited gap g_1, coils N_2 and N_3 are decoupled; hence $L_{23} = 0$.

3.147 A toroid has a core of square cross section, 2500 mm^2 in area, and a mean diameter of 250 mm. The core material is of relative permeability 1000. Calculate the number of turns to be wound on the core to obtain an inductance of 1 H.

▮ Relucance of the flux path is given by

$$\mathscr{R} = \frac{\ell}{\mu A} = \frac{\pi(250)10^{-3}}{(4\pi \times 10^{-7})(2500)10^{-6}} \quad \text{H}^{-1}$$

Then $L = 1 \text{ H} = N^2/\mathscr{R}$ yields $N = 15\,811$ turns.

3.148 Magnetic flux exists within a cylindrical conductor as it carries a current, resulting in an *internal inductance*. Show that the internal inductance per unit length of a straight cylindrical conductor is indepndent of the radius, provided the current density is uniform over a cross section.

▮ Denoting the conductor radius R, we have, by Ampère's law, $H = H_\phi = Ir/2\pi R^2$. The magnetic energy in a length ℓ of the conductor,

$$W_m = \frac{\mu}{2}\int_{\text{volume}} H^2 \, d\tau = \frac{\mu}{2}\frac{I^2}{4\pi^2 R^4}\int_0^R r^2 \, 2\pi r \ell \, dr = \frac{\mu I^2 \ell}{16\pi}$$

is independent of R; so also is

$$L_{\text{int}} = \frac{2W_m}{I^2} = \frac{\mu\ell}{8\pi} \qquad \text{or} \qquad \frac{L_{\text{int}}}{\ell} = \frac{\mu}{8\pi} \qquad\qquad (1)$$

3.149 Refer to Problem 3.148. The flux through an $R \times \ell$ longitudinal section (Fig. 3-56) is

$$\psi_m = \int_0^R B_\phi \, \ell \, dr = \frac{\mu I \ell}{4\pi} \qquad (1)$$

which seems to imply that $L_{\text{int}}/\ell = \mu/4\pi$. What's wrong here?

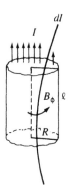

Fig. 3-56

▮ What has been forgotten is that inductance has to do with current *loops*. Thus the total current I must be decomposed into current filaments dI, each one a circular loop of infinite radius. It is easy to see that, on an average, a field line B_ϕ threads *half of* these loops. Hence the true flux linkage is $\lambda = \frac{1}{2}\psi_m$, which leads to the correct value of L_{int}/ℓ.

3.150 The current through the inner conductor of the coaxial cable of Fig. 3-50 is uniformly distributed. Determine its total inductance per unit length.

▮ This problem requires simply a superposition of the results of Problems 3.135 and 3.148:

$$L = \frac{\mu_0}{2\pi} \ln \frac{b}{a} + \frac{\mu_0}{8\pi} = \frac{\mu_0}{8\pi}\left(1 + 4 \ln \frac{b}{a}\right)$$

3.151 Determine the radius of the inner conductor of a coaxial cable, having an outer radius of 13.5 mm, for which the internal and external inductances are equal.

▮ For the internal and external inductances to be equal, we must have, from Problem 3.150,

$$4 \ln \frac{b}{a} = 1 \quad \text{or} \quad \frac{b}{a} = 1.284 \quad \text{or} \quad a = \frac{13.5}{1.284} = 10.5 \text{ mm}$$

3.152 A radio engineer asserts that the "inductance of free space" is 50 nH/m. What might be meant by this?

▮ In Problem 3.148 let $\mu = \mu_0$ and let $R \to \infty$. Then the interior of the cylinder becomes all of space, for which

$$\frac{L_{\text{int}}}{\ell} = \frac{\mu_0}{8\pi} = \frac{4\pi \times 10^{-7}}{8\pi} \text{ H/m} = 50 \text{ nH/m}$$

3.153 The **H**-field within a cylindrical conductor is given by $\mathbf{H} = 4r^3\mathbf{a}_\phi$ (A/m). Determine the corresponding current density.

▮
$$\mathbf{J} = \text{curl } \mathbf{H} = \frac{1}{r}\frac{\partial}{\partial r}(rH_\phi)\mathbf{a}_z = 16r^2\mathbf{a}_z \text{ (A/m}^2)$$

3.154 The magnetic vector potential corresponding to a current source is given by $\mathbf{A} = r^2\mathbf{a}_r + (r/2)\mathbf{a}_\phi$. Show that the resulting flux density is uniform everywhere.

▮
$$\mathbf{B} = \text{curl } \mathbf{A} = \mathbf{a}_z\left[\frac{1}{r}\frac{\partial}{\partial r}(rA_\phi) - \frac{1}{r}\frac{\partial A_r}{\partial \phi}\right] = \mathbf{a}_z\left[\frac{1}{r}\frac{\partial}{\partial r}\left(\frac{r^2}{2}\right) - \frac{1}{r}\frac{\partial}{\partial \phi}(r^2)\right] = 1\mathbf{a}_z$$

3.155 The **H**-field within a conductor has rectangular components that depend only on z. Show that there is no z-directed current present.

▮
$$J_z = (\nabla \times \mathbf{H})_z = \frac{\partial H_y}{\partial x} - \frac{\partial H_x}{\partial y} = 0 - 0 = 0$$

3.156 It is often desirable to operate a permanent magnet so that the energy density of the field is maximized. Find the maximal H for a magnet having the characteristic (**1**) of Fig. 3-57.

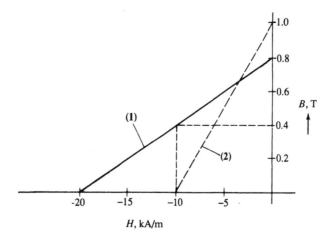

Fig. 3-57

▌ For (**1**), $B = (4 \times 10^{-5})H + 0.8$ and w_m is proportional to
$$B(-H) = -(4 \times 10^{-5})H^2 - 0.8H = 4000 - (4 \times 10^{-5})(H + 10^4)^2$$
Thus, maximal $H = -10^4 \text{ A/m} = -10 \text{ kA/m}$.

3.157 The magnetization characteristic of a sample iron is given by $B = aH + bH^2$. Determine the energy stored in the material per unit volume if H is increased from 0 to H_f.

▌
$$w_m = \int H\, dB = \int_0^{H_f} H(a + 2bH)\, dH = \frac{a}{2}H_f^2 + \frac{2b}{3}H_f^3$$

3.158 A 200-turn coil is wound on an iron ring having mean diameter $D = 0.5$ m. Given that the relative permeability of iron is 500, calculate the coil current to establish a 0.8 T average flux density in the ring.

▌
$$H = \frac{B}{\mu_0\mu_r} = \frac{NI}{\ell} = \frac{NI}{\pi D} \quad \text{or} \quad I = \frac{\pi DB}{\mu_0\mu_r N} = \frac{\pi \times 0.5 \times 0.8}{4\pi \times 10^{-7} \times 500 \times 200} = 10.0 \text{ A}$$

3.159 Rework Problem 3.158, assuming that a 2-mm air gap is cut in the ring.

▌ From Problem 3.158, for the iron, $\mathscr{F}_i = 10 \times 200 = 2000$ At. For the air gap, $B_g = \mu_0 H_g$,
$$H_g = \frac{B_g}{\mu_0} = \frac{0.8}{4\pi \times 10^{-7}} \text{ A/m} \quad \text{and} \quad \mathscr{F}_g = H_g\ell_g = \frac{0.8 \times 2 \times 10^{-3}}{4\pi \times 10^{-7}} = 1273 \text{ At}$$
Consequently, $(NI)_{\text{total}} = 2000 + 1273 = 3273$ At and $I = 3273/200 = 16.37$ A.

3.160 Draw an electrical analog, consisting of resistors and a battery, for the magnetic cricuit shown in Fig. 3-58(*a*).

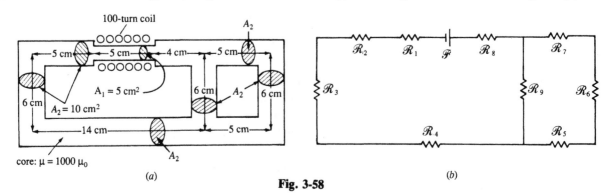

Fig. 3-58

▌ See Fig. 3-58(*b*).

3.161 Reduce the circuit of Fig. 3-58(*b*) to a single resistor and a battery, and hence obtain the equivalent reluctance of the magnetic circuit as seen by the 100-turn coil.

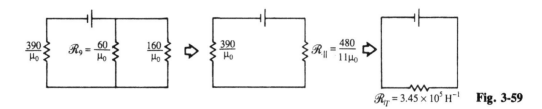

Fig. 3-59

▮ The various reluctances are

$$\mathcal{R}_1 = \frac{5 \times 10^{-2}}{\mu_0(5 \times 10^{-4})} = \frac{100}{\mu_0} \qquad \mathcal{R}_2 = \mathcal{R}_5 = \mathcal{R}_7 = \frac{5 \times 10^{-2}}{\mu_0(10 \times 10^{-4})} = \frac{50}{\mu_0} \qquad \mathcal{R}_3 = \mathcal{R}_6 = \mathcal{R}_9 = \frac{6 \times 10^{-2}}{\mu_0(10 \times 10^{-4})} = \frac{60}{\mu_0}$$

$$\mathcal{R}_4 = \frac{14 \times 10^{-2}}{\mu_0(10 \times 10^{-4})} = \frac{140}{\mu_0} \qquad \mathcal{R}_8 = \frac{4 \times 10^{-2}}{\mu_0(10 \times 10^{-4})} = \frac{40}{\mu_0}$$

from which

$$\mathcal{R}_1 + \mathcal{R}_2 + \mathcal{R}_3 + \mathcal{R}_4 + \mathcal{R}_8 = \frac{390}{\mu_0} \qquad \mathcal{R}_5 + \mathcal{R}_6 + \mathcal{R}_7 = \frac{160}{\mu_0}$$

Circuit reduction is as shown in Fig. 3-59:

$$\mathcal{R}_{11} = \frac{(160/\mu_0)\mathcal{R}_9}{\dfrac{160}{\mu_0} + \mathcal{R}_9} = \frac{(160/\mu_0)(60/\mu_0)}{(160 + 60)/\mu_0} = \frac{480}{11\mu_0} \qquad \mathcal{R}_T = \frac{390}{\mu_0} + \frac{480}{11\mu_0} = \frac{4770}{11\mu_0} = \frac{4770}{11 \times 1000(4\pi \times 10^{-7})} = 3.45 \times 10^5 \, \text{H}^{-1}$$

3.162 Find the inductance of the coil of Fig. 3-58(a).

▮ By the result of Problem 3.161,

$$L = \frac{N^2}{\mathcal{R}} = \frac{(100)^2}{3.45 \times 10^5} = 28.98 \, \text{mH}$$

3.163 Production of mechanical force by shortening of flux lines (Fig. 3-60) is a magnetic field effect. Derive an expression for F_m for a conservative (lossless) system, assuming that the coil current i and the displacement x are taken as the independent variables.

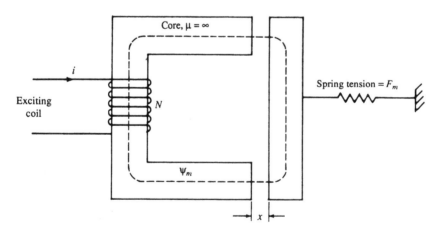

Fig. 3-60

▮ For a lossless system, the first law of thermodynamics gives

$$\begin{pmatrix} \text{input} \\ \text{electric} \\ \text{energy} \end{pmatrix} = \begin{pmatrix} \text{mechanical} \\ \text{work done} \\ \text{by system} \end{pmatrix} + \begin{pmatrix} \text{increase} \\ \text{in stored} \\ \text{energy} \end{pmatrix} \qquad (1)$$

Since the input electric energy is $dW_e = vi \, dt = (d\lambda/dt)i \, dt = i \, d\lambda$, (1) may be rewritten as

$$dW_m = -F_e \, dx + i \, d\lambda = -F_e \, dx + i \left(\frac{\partial \lambda}{\partial x} dx + \frac{\partial \lambda}{\partial i} di \right)$$

$$= \left(-F_e + i \frac{\partial \lambda}{\partial x} \right) dx + i \frac{\partial \lambda}{\partial i} di \qquad (2)$$

and (2) implies

$$-F_e + i \frac{\partial \lambda}{\partial x} = \frac{\partial W_m}{\partial x} \qquad \text{or} \qquad F_e = -\frac{\partial W_m}{\partial x} + i \frac{\partial \lambda}{\partial x} \qquad (3)$$

3.164 Assuming a linear magnetic circuit, replace the dependent variables W_m and λ of (3) of Problem 3.163 by the single dependent variable L.

◼ Since $W_m = \frac{1}{2}Li^2$ and $\lambda = Li$,

$$\frac{\partial W_m}{\partial x} = \frac{1}{2}i^2\frac{\partial L}{\partial x} \quad \text{and} \quad \frac{\partial \lambda}{\partial x} = i\frac{\partial L}{\partial x}$$

So

$$F_e = -\frac{1}{2}i^2\frac{\partial L}{\partial x} + i^2\frac{\partial L}{\partial x} = \frac{1}{2}i^2\frac{\partial L}{\partial x} \tag{1}$$

3.165 Calculate the force on the solenoid plunger shown in Fig. 3-46(a) at $x = 5$ mm.

◼ First, let the plunger be a distance x (mm) away from the core. Then, as in Problem 3.124,

$$\mathcal{R}_g = \frac{x \times 10^{-3}}{\mu_0 \times 2 \times 4 \times 10^{-4}} = \frac{10x}{8\mu_0} \quad \mathcal{R}_s = \frac{2 \times 10^{-3}}{\mu_0 \times 2 \times 2 \times 10^{-4}} = \frac{20}{4\mu_0} \quad \mathcal{R}_t = \mathcal{R}_g + \frac{1}{2}\mathcal{R}_s = \frac{5x + 10}{4\mu_0}$$

and

$$L(x) = \frac{N^2}{\mathcal{R}_t} = \frac{4\mu_0 \times 50^2}{5x + 10} = \frac{4\pi \times 10^{-3}}{5x + 10}$$

Then, (1) of Problem 3.164 gives

$$F_e = \left(\frac{1}{2} \times 10^2\right)\frac{d}{dx}\left(\frac{4\pi \times 10^{-3}}{5x + 10}\right) = (2\pi \times 10^{-1})\left[\frac{-5}{(5x + 10)^2}\right]$$

which, for $x = 5$ mm, yields $F_e = -(\pi/1225)$ N. The negative sign indicates that the force tends to decrease x.

3.166 In (3) of Problem 3.163 replace the independent variable i by the independent variable λ.

◼ Equation (2), $dW_m = -F_e\,dx + i\,d\lambda$, gives immediately $F_e = -\partial W_m/\partial x$.

3.167 Assuming that the i-λ relationship for the electromagnet shown in Fig. 3-60 is $i = a\lambda^2 + \lambda(x - b)^2$, where a and b are constants, evaluate the force on the iron mass.

◼ Use the result of Problem 3.166.

$$W_m(\lambda, x) = \frac{1}{2}\lambda i(\lambda, x) \quad \text{or} \quad F_e = -\frac{1}{2}\lambda\frac{\partial i}{\partial x} = -\frac{1}{2}\lambda[2\lambda(x - b)] = -\lambda^2(x - b)$$

3.168 An elementary reluctance motor is shown in Fig. 3-61; the exciting coil inductance is given by $L(\theta) = k_0 + k_1 \cos 2\theta$, where k_0 and k_1 are constants, and the coil current is $i = I_m \sin \omega t$. What is the instantaneous torque produced by the motor?

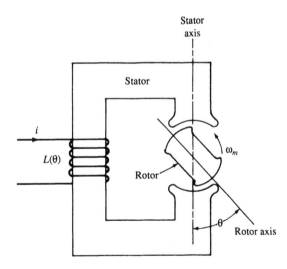

Fig. 3-61

◼ Apply the rotational form of (1) of Problem 3.164:

$$T_e = \frac{1}{2}i^2\frac{\partial L}{\partial \theta} = -k_1 I_m^2 \sin 2\theta \sin^2 \omega t \tag{1}$$

Eliminate θ from (1) via $\theta = \omega_m t - \delta$:

$$T_e = -k_1 I_m^2 \sin 2(\omega_m t - \delta) \sin^2 \omega t \tag{2}$$

3.169 A solenoid of cylindrical geometry is shown in Fig. 3-62. Given that the N-turn coil carries a current i, derive an expression for the force on the plunger.

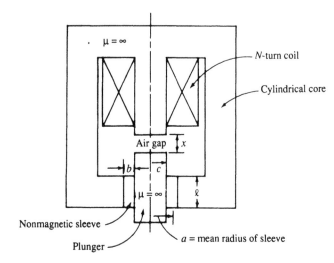

Fig. 3-62

▌ Use (*1*) of Problem 3.164. With $c = a - (b/2)$,

$$\mathscr{R} = \frac{x}{\mu_0 \pi c^2} + \frac{b}{\mu_0 2\pi a \ell} \quad \text{and} \quad L = \frac{N^2}{\mathscr{R}} = \frac{2\pi\mu_0 a \ell c^2 N^2}{2a\ell x + bc^2} \equiv \frac{k_1}{k_2 x + k_3}$$

Hence
$$F_e = -\frac{1}{2} i^2 \frac{\partial L}{\partial x} = \frac{k_1 k_2 i^2}{2(k_2 x + k_3)^2}$$

3.170 Express the result of Problem 3.166 in terms of the magnetic flux λ and the reluctance $\mathscr{R}$.

▌ As in Problem 3.167,

$$W_m(\lambda, x) = \frac{1}{2} \lambda i(\lambda, x) = \frac{1}{2} \lambda \frac{\lambda \mathscr{R}(\lambda, x)}{N^2} = \frac{\lambda^2}{2N^2} \mathscr{R}(\lambda, x)$$

so that

$$F_e = -\frac{\partial W_m}{\partial x} = -\frac{\lambda^2}{2N^2} \frac{\partial \mathscr{R}}{\partial x}$$

4.1 According to *Faraday's law* of electromagnetic induction, a time-varying magnetic flux ψ_m penetrating a loop induces an electromotive force (emf)—commonly notated ε or v—in the loop, such that

$$\varepsilon = -\frac{d\psi_m}{dt} \tag{1}$$

Express (1) in integral form, in terms of **E**- and **B**-fields.

▌ We have, for an open surface S bounded by a closed contour C,

$$\varepsilon = \oint_C \mathbf{E} \cdot d\mathbf{s} \qquad \psi_m = \int_S \mathbf{B} \cdot d\mathbf{S}$$

giving

$$\oint_C \mathbf{E} \cdot d\mathbf{s} = -\frac{d}{dt}\int_S \mathbf{B} \cdot d\mathbf{S} \tag{2}$$

4.2 Express Faraday's law in a point form.

▌ Let C be a very small square in the xy plane, with center P and side δ. Equation (2) of Problem 4.1 becomes—after application of Stokes' theorem on the left

$$(\text{curl } \mathbf{E})_z \delta^2 = -\frac{\partial B_z}{\partial t}\delta^2 \qquad \text{or} \qquad (\text{curl } \mathbf{E})_z = -\left(\frac{\partial \mathbf{B}}{\partial t}\right)_z$$

both functions being evaluated at P. Repeating the argument in the other two coordinate planes, we obtain

$$(\text{curl } \mathbf{E})_x = -\left(\frac{\partial \mathbf{B}}{\partial t}\right)_x \qquad \text{and} \qquad (\text{curl } \mathbf{E})_y = -\left(\frac{\partial \mathbf{B}}{\partial t}\right)_y$$

Hence, at P,

$$\text{curl } \mathbf{E} = -\frac{\partial \mathbf{B}}{\partial t} \tag{1}$$

It is apparent from either form of Faraday's law that, unlike the electrostatic **E**-field, the electromagnetic **E**-field is nonconservative (Problem 1.157). Thus, if we consider (1) as a vector differential equation for the **E** induced by the time change in a known **B**, we look for the "simplest" (nonconservative) solution, **E***. (The general solution would then be $\mathbf{E}^* - \nabla V$, where V is an arbitrary scalar potential.)

The derivation of (1) above implicitly assumed a *fixed* surface through which the flux changes. A correction term on the right of (1) is necessary if the surface moves or changes shape. This term is obtained in Problem 4.29.

4.3 The induced emf in the closed loop C of Problem 4.1 will cause a current in the loop. According to *Lenz's law*, the direction of the current is such that the resulting magnetic field opposes the *change* in the original magnetic field. Apply Lenz's law to determine polarities of terminals a and b in the loop of Fig. 4-1.

▌ The induced **B**-field will tend to increase the external field, and hence it will have the direction shown in Fig. 4-1(a). This will require a current i in the indicated direction. Now, if we open the loop at ab and connect these terminals to an external circuit, the polarities of a and b must be as marked in Fig. 4-1(b). (*Positive* charge must flow from $+$ to $-$ in the external circuit.)

4.4 Repeat Problem 4.3 assuming that the external field is increasing.

▌ Now a is $+$ and b is $-$.

4.5 In cylindrical coordinates, find the **E**-field induced by the flux density

$$\mathbf{B} = \begin{cases} B_0 \sin \omega t\, \mathbf{a}_z & r < r_0 \\ 0 & r > r_0 \end{cases}$$

▌ In (2) of Problem 4.1 let C be the circle $r = a$. Then,

$$\int_0^{2\pi} E_\phi a\, d\phi = \begin{cases} -\dfrac{d}{dt}[(B_0 \sin \omega t)(\pi a^2)] & a < r_0 \\[2mm] -\dfrac{d}{dt}[B_0 \sin \omega t)(\pi r_0^2)] & a > r_0 \end{cases} = (-\pi B_0 \omega \cos \omega t)\min\{a^2, r_0^2\} \tag{1}$$

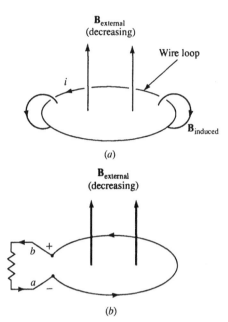

(a)

(b)   **Fig. 4-1**

Following the remarks in Problem 4.2, choose as the solution of (1) the field $\mathbf{E}^* = E_\phi \mathbf{a}_\phi$, where (replacing a by r)

$$E_\phi = -\frac{B_0 \omega \cos \omega t}{2} \min \{r, r_0^2/r\}$$

4.6   Show that another point form of Faraday's law is

$$\mathbf{E} = -\frac{\partial \mathbf{A}}{\partial t} \qquad (\text{where } \mathbf{A} = \text{magnetic vector potential}) \tag{1}$$

▮ Substitute $\mathbf{B} = \text{curl } \mathbf{A}$ in (1) of Problem 4.2:

$$\text{curl } \mathbf{E} = -\frac{\partial}{\partial t}(\text{curl } \mathbf{A}) = \text{curl}\left(-\frac{\partial \mathbf{A}}{\partial t}\right)$$

or (as we can always suppose) $\mathbf{E} = -\partial \mathbf{A}/\partial t$.

4.7   A circular loop of 10-cm radius is located in the xy plane in a $\mathbf{B}$-field given by $\mathbf{B} = (0.5 \cos 377t)(3\mathbf{a}_y + 4\mathbf{a}_z)$ (T). Determine the voltage induced in the loop.

▮ From (1) of Problem 4.1,

$$v = -\frac{d}{dt}\int \mathbf{B} \cdot d\mathbf{S} = -\frac{d}{dt}[(0.5 \cos 377t)(3\mathbf{a}_y + 4\mathbf{a}_z) \cdot \pi(0.10)^2 \mathbf{a}_z] = 0.5 \times 377 \times 4\pi \times (0.10)^2 \sin 377t$$

$$= 23.69 \sin 377t \quad (\text{V})$$

4.8   The rectangular contour in Fig. 4-2 is expanding in the field $\mathbf{B} = B_0 \mathbf{a}_z$, as a result of the uniform motion of the conducting bar. Determine the voltage available at the terminals ab.

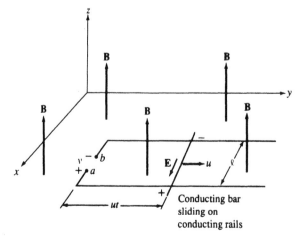

Fig. 4-2

▮ Here, $\psi_m(t) = B_z A = (B_0)(\ell)(ut)$, so that

$$v = -\frac{d\psi_m}{dt} = -B_0 \ell u \tag{1}$$

Because $v/\ell = E$, we can rewrite (1) in vector form:

$$\mathbf{E} = \mathbf{u} \times \mathbf{B} \tag{2}$$

It is readily verified that (2) gives $\mathbf{E}$ the direction required by Lenz's law.

4.9 A square loop, of side 20 cm, is located in free space adjacent to a straight conductor that carries a sinusoidal current of 0.5 A (rms) at 5 kHz (see Fig. 4-3). If a small gap is introduced into the loop, what is the induced voltage across the gap?

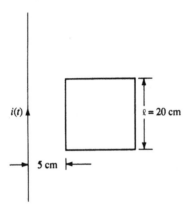

Fig. 4-3

▮ By Problem 3.27, the instantaneous magnetic flux through the loop is

$$\psi_m(t) = \frac{\mu_0 \ell \ln (25/5)}{2\pi} i(t)$$

whence

$$v(t) = -\psi_m'(t) = -\frac{\mu_0 \ell \ln 5}{2\pi} i'(t) \tag{1}$$

Writing $i(t) = I_m \cos \omega t = 0.5\sqrt{2} \cos 10^4 \pi t$, we find from (1) that $V = 1.01$ mV (rms) at 5 kHz.

4.10 A circular loop of 10-cm radius replaces the square loop in Problem 4.9. Find the voltage induced across a gap in this loop, assuming that the center of the loop is 15 cm away from the straight conductor.

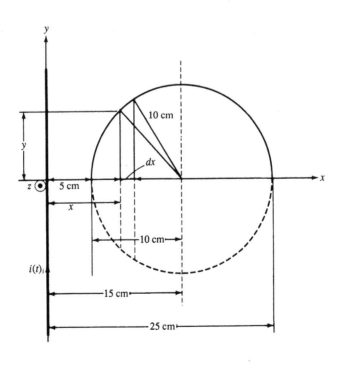

Fig. 4-4

▮ In terms of the distances shown in Fig. 4-4, $\mathbf{B} = (\mu_0 i/2\pi x)(-\mathbf{a}_z)$, so that, by symmetry,

$$\psi_m = \frac{\mu_0 i}{\pi} \int_{x=5\text{ cm}}^{25\text{ cm}} \frac{y}{x}\, dx \tag{1}$$

where the integral is over the upper semicircle.

Evaluation of $\int (y/x)\, dx$: The equation $(x-15)^2 + y^2 = 10^2$ gives $y = \sqrt{-125 + 30x - x^2} \equiv \sqrt{a + bx + cx^2} \equiv \sqrt{X}$. From integral tables,

$$\int \frac{\sqrt{X}}{x}\, dx = \sqrt{X} + \frac{b}{2}\int \frac{dx}{\sqrt{X}} + a\int \frac{dx}{x\sqrt{X}}$$
$$= \sqrt{X} + \frac{b}{2}\left(-\frac{1}{\sqrt{-c}}\sin^{-1}\frac{2cx+b}{\sqrt{-q}}\right) + a\left[\frac{1}{\sqrt{-a}}\sin^{-1}\left(\frac{bx+2a}{x\sqrt{-q}}\right)\right] \tag{2}$$

in which $q = 4ac - b^2$.

Substitution of numerical values in (2) leads to

$$\int_{5\text{ cm}}^{25\text{ cm}} \frac{y}{x}\, dx = (15 - \sqrt{125})\pi \text{ cm} = (0.15 - \sqrt{0.0125})\pi \text{ m} \approx \frac{4\pi}{100}\text{ m}$$

With the current taken from Problem 4.9, (1) now gives

$$\psi_m = \frac{4\pi \times 10^{-7}}{50}\sqrt{2}\cos 10^4 \pi t \quad \text{(Wb)}$$
$$v = -\frac{d\psi_m}{dt} = \frac{4\pi^2 \times 10^{-3}}{50}\sqrt{2}\sin 10^4 \pi t \quad \text{(V)}$$

This is an rms voltage of approximately 0.8 mV.

4.11 An inductor is formed by winding 10 turns of a thin wire around a wooden dowel rod of radius 2 cm. A uniform, sinusoidal magnetic field, of magnitude 10 mWb/m² and frequency 10 kHz, is directed along the axis of the rod; find the induced voltage in the winding.

▮ The flux through the winding is
$$\psi_m = BA = (0.010 \cos 2\pi 10^4 t)\pi(0.02)^2 = (4\pi \times 10^{-6})\cos 2\pi 10^4 t \quad \text{(Wb)}$$
This flux links the winding $N = 10$ times. Thus,
$$v = -10\frac{d\psi_m}{dt} = 7.986 \sin 2\pi 10^4 t \quad \text{(V)}$$

or $V = 5.584$ V (rms).

4.12 A uniform and constant magnetic field of 10 mWb/m² is directed along the z axis of a rectangular coordinate system. A circular contour in the xy plane centered at the origin has a radius that is decreasing at 100 m/s. Given the initial radius of 100 mm, determine the induced emf in the path as a function of time.

▮ The flux through the contour is $\psi_m = B_z(\pi r^2)$, whence
$$v = -\frac{d\psi_m}{dt} = -2\pi B_z r\frac{dr}{dt} = -2\pi(0.010)(0.100)(-100) = 0.628\text{ V}$$

4.13 Rework Problem 4.8 if $\mathbf{B} = 0.1 \cos 377t\, \mathbf{a}_z$, other data remaining unchanged.

▮ $\qquad \psi_m = B_z A = (0.1 \cos 377t)(\ell u t) \qquad v = -\frac{d\psi_m}{dt} = 37.7\ell u t \sin 377t - 0.1\ell u \cos 377t$

4.14 Repeat Problem 4.8 for $\mathbf{B} = 0.1\mathbf{a}_z$ and $\mathbf{u} = 100 \cos 10t\, \mathbf{a}_y$.

▮ $$\psi_m(t) = B_z A(t) = 0.1\ell \int_0^t u_y\, dt = \ell \sin 10t$$

and
$$v = -\frac{d\psi_m}{dt} = -10\ell \cos 10t$$

4.15 In Fig. 4-5 a rectangular wire loop of resistance 20 mΩ rotates at $\omega = 2$ rad/s in the uniform magnetic field $\mathbf{B} = 10\mathbf{a}_y$ mWb/m². Calculate the induced current.

▮ At time t the projected area normal to the field is $LW \cos\theta$; hence, $\psi_m = B_y LW \cos\theta$ and
$$v = -\frac{d\psi_m}{dt} = -\frac{d\psi_m}{d\theta}\frac{d\theta}{dt} = B_y LW\omega \sin\theta$$
$$= (10^{-2})(0.010)(0.020)(2)\sin\theta = 4\sin\theta \quad (\mu\text{V})$$

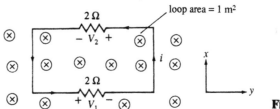

Fig. 4-5

and
$$i = \frac{v}{R} = \frac{4 \times 10^{-6}}{0.020} \sin \theta = 0.2 \sin \theta \quad (mA)$$

4.16 Calculate the voltages V_1 and V_2 across the 1-Ω and 2-Ω resistors in Fig. 4-6. The loop is located in the xy plane and $\mathbf{B} = (0.1t)\mathbf{a}_z$ (T).

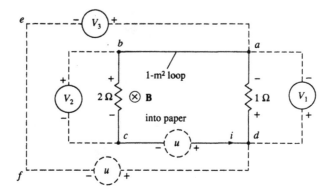

Fig. 4-6

▮ Since $\psi_m = 0.1t$ (Wb),

$$|v| = \left|\frac{d\psi_m}{dt}\right| = 0.1 \text{ V} = 100 \text{ mV}$$

where, by Lenz's law, the emf drives current in the direction indicated in Fig. 4-6. The emf is divided in the ratio of the resistances:

$$V_1 = \tfrac{1}{3}|v| = 33.3 \text{ mV} \qquad V_2 = \tfrac{2}{3}|v| = 66.7 \text{ mV}$$

4.17 A **B**-field of $0.1t$ (T) threads only the loop *abcd* of Fig. 4-7. Determine the voltmeter reading V_3.

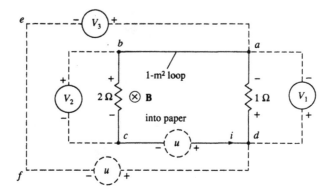

Fig. 4-7

▮ As in Problem 4.16,
$$v = 100 \text{ mV} \qquad V_1 = 33.3 \text{ mV} \qquad V_2 = 66.7 \text{ mV}$$
Hence, around loop *aefd*, $V_3 - v + V_1 = 0$, or
$$V_3 = v - V_1 = 66.7 \text{ mV} = V_2$$

4.18 With reference to Problem 4.17, how is it that voltmeters V_1 and V_3 are both across *ad* and yet give different readings?

▮ The leads of V_3 enclose a time-varying flux; the leads of V_1 do not.

4.19 Evaluate the current i in Problems 4.16 and 4.17.

❚ $$i = \frac{|v|}{R_1 + R_2} = \frac{100\,\text{mV}}{3\,\Omega} = 33.3\,\text{mA}$$

4.20 Find an expression for the emf around an arbitrary closed loop that moves in an arbitrary way through a time-invariant magnetic field.

❚ At the instant t, let s be the arc-length parameter for the loop, and let $\mathbf{v}(s)$ and $\mathbf{B}(s)$ be the velocity field and the magnetic field along the loop, respectively; the $\mathbf{v}$-field arises from all motions of the loop, including rigid motions and changes in shape. At t, equation (2) of Problem 4.8 holds pointwise on the loop: $\mathbf{E}(s) = \mathbf{v}(s) \times \mathbf{B}(s)$. Hence, the total emf is

$$v = \oint \mathbf{E}(s) \cdot d\mathbf{s} = \oint [\mathbf{u}(s) \times \mathbf{B}(s)] \cdot d\mathbf{s} \qquad (1)$$

4.21 Rework Problem 4.20 if the **B**-field changes with time.

❚ For a moving loop in a changing magnetic field, the flux changes because (i) the field is changing relative to the momentarily fixed area; (ii) the area is changing relative to the momentarily fixed field. Effect (i) gives rise to

$$v_{(i)} = -\int \frac{\partial \mathbf{B}}{\partial t} \cdot d\mathbf{S} \quad \text{(by Faraday's law)} \qquad (1)$$

and effect (ii) gives rise to

$$v_{(ii)} = \oint (\mathbf{u} \times \mathbf{B}) \cdot d\mathbf{s} \quad \text{(by Problem 4.20)} \qquad (2)$$

The total emf is $v = v_{(i)} + v_{(ii)}$.

4.22 Faraday's law may be written formally as

$$v = -\frac{d}{dt} \int \mathbf{B} \cdot d\mathbf{S} = -\int \frac{\partial \mathbf{B}}{\partial t} \cdot d\mathbf{S} - \int \mathbf{B} \cdot \frac{\partial}{\partial t}(d\mathbf{S})$$

The first integral on the right is just $v_{(i)}$ of Problem 4.21. Is the second integral equal to $v_{(ii)}$?

❚ Yes; but to prove it requires great skill in vector calculus. It is clear that Stokes' theorem must be the principal weapon.

4.23 A conducting disk rotates at a constant angular velocity ω_m in a uniform magnetic field **B** directed axially, as shown in Fig. 4-8. The inner and outer radii are r_i and r_o, respectively. Determine the induced voltage available at the terminals a and a'. (The device is known as the *Faraday disk generator*.)

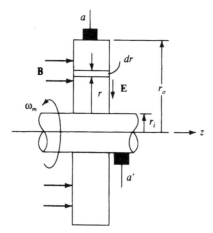

Fig. 4-8

❚ By (2) of Problem 4.8, the induced field at radius r is $\mathbf{E} = r\omega_m B(-\mathbf{a}_r)$. Hence, the voltage of a' with respect to a is

$$v = \int_{r_o}^{r_i} r\omega_m B(-\mathbf{a}_r) \cdot (dr\,\mathbf{a}_r) = \tfrac{1}{2}\omega_m B(r_o^2 - r_i^2)$$

4.24 A dc generator having a liquid-metal armature is proposed in Fig. 4-9. Determine the number of turns N_a such that the terminal voltage is independent of the load current (under steady-state conditions). The channel is c (m) long and the fluid flows at u (m/s) out of the paper.

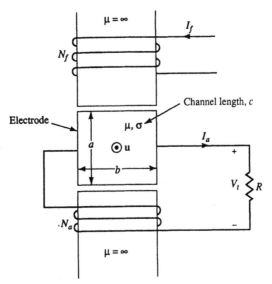

Fig. 4-9

❚ The induced voltage across the electrodes is $V_0 = uBb$. But $H = (N_f I_f + N_a I_a)/a$ and $B = \mu H$; thus

$$V_0 = \frac{\mu}{a}(N_f I_f + N_a I_a)ub$$

Since the armature resistance is $R_a = b/\sigma ac$, the terminal voltage is

$$V_t = V_0 - I_a R_a = \frac{1}{a}\mu ubN_f I_f + I_a\left(\frac{1}{a}\mu ubN_a - \frac{b}{\sigma ac}\right)$$

For V_t to be independent of I_a, we must have $N_a = 1/\mu\sigma uc$.

4.25 In a laboratory experiment, a small conducting loop is released from rest within a vertical evacuated cylinder. What voltage is induced in the falling loop, if locally the Earth's magnetic field is 10^{-5} T at a constant angle of $5°$ below the horizontal?

❚ Zero: because the free fall is purely translational, the flux through the loop remains constant.

4.26 A rectangular loop is placed in the magnetic field of a very long straight conductor carrying a current I, as shown in Fig. 4-10. Assuming that the loop moves purely in the x direction, determine the instantaneous induced voltage in the loop.

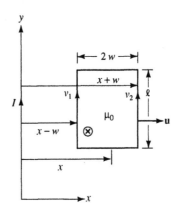

Fig. 4-10

❚ First notice that emf's will be induced only in the two sides that are parallel to the y axis. Thus, if the x speed is $v(t)$, the net induced voltage at time t is

$$v = v_1 - v_2 = B_{\phi 1}\ell u - B_{\phi 2}\ell u = \frac{\mu_0 I}{2\pi(x-w)}\ell u - \frac{\mu_0 I}{2\pi(x+w)}\ell u = \frac{\mu_0 I \ell w u}{\pi(x^2 - w^2)}$$

4.27 Rework Problem 4.26 by consideration of flux.

$\blacksquare$

$$\psi_m = \int_{x-w}^{x+w} \frac{\mu_0 I}{2\pi r} \ell \, dr = \frac{\mu_0 I \ell}{2\pi} (\ln (x + w) - \ln (x - w)]$$

$$v = -\frac{d\psi_m}{dt} = -u \frac{d\psi_m}{dx} = \frac{\mu_0 I \ell u}{2\pi} \left[\frac{1}{x - w} - \frac{1}{x + w} \right]$$

in agreement with Problem 4.26.

4.28   A thin metal sheet, of conductivity σ, is fixed in the xz plane, in a time-varying magnetic field
$$\mathbf{B} = (B_m \cos \beta x \cosh \beta y \cos \omega t) \mathbf{a}_y$$
Determine the induced electric field at a point (x, z) of the sheet.

$\blacksquare$ Assuming that $\mathbf{E} = E_z \mathbf{a}_z$ (review Problem 4.2), we have from Faraday's (point) law:
$$\text{curl } \mathbf{E} = -\frac{\partial \mathbf{B}}{\partial t} \quad \text{or} \quad \mathbf{a}_x \frac{\partial E_z}{\partial y} + \mathbf{a}_y \left(-\frac{\partial E_z}{\partial x} \right) = \mathbf{a}_x (0) + \mathbf{a}_y \left(-\frac{\partial B_y}{\partial x} \right)$$
where all derivatives are evaluated at $y = 0$. The x-equation is satisfied by choosing $E_z = E_z(x)$. Then the y-equation gives
$$\frac{dE_z}{dx} = \left(\frac{\partial B_y}{\partial t} \right)_{y=0} = -\omega B_m \cos \beta x \sin \omega t \quad \text{or} \quad E_z = -\frac{\omega B_m}{\beta} \sin \beta x \sin \omega t$$

4.29   Generalize (*1*) of Problem 4.2 to cover moving loops in time-varying magnetic fields.

$\blacksquare$ Following Problem 4.21, write $\mathbf{E} = \mathbf{E}_{(i)} + \mathbf{E}_{(ii)}$. By Problem 4.2, curl $\mathbf{E}_{(i)} = -\partial \mathbf{B}/\partial t$; by Problem 4.20, $\mathbf{E}_{(ii)} = \mathbf{u} \times \mathbf{B}$. Therefore,
$$\text{curl } \mathbf{E} = \text{curl } \mathbf{E}_{(i)} + \text{curl } \mathbf{E}_{(ii)} = -\frac{\partial \mathbf{B}}{\partial t} + \text{curl } (\mathbf{u} \times \mathbf{B}) \tag{1}$$

4.30   Assuming that the sheet of Problem 4.28 is allowed to move with a velocity $\mathbf{u} = u_x \mathbf{a}_x$, determine the induced electric field at a point of the sheet with instantaneous coordinates (x, z).

$\blacksquare$ By Problem 4.29, the total induced field is obtained by adding to the field found in Problem 4.28 the "motional" field
$$\mathbf{u} \times \mathbf{B}\big|_{y=0} = u_x \mathbf{a}_x \times B_m \cos \beta x \cos \omega t \, \mathbf{a}_y = u_x B_m \cos \beta x \cos \omega t \, \mathbf{a}_z$$

4.31   Give the "solved form" of (*1*) of Problem 4.29.

$\blacksquare$ Parallel to Problem 4.6,
$$\mathbf{E} = -\frac{\partial \mathbf{A}}{.\partial t} + \mathbf{u} \times (\text{curl } \mathbf{A}) \tag{1}$$

4.32   Generalizing the result of Problem 4.15, give an expression for the emf developed in a rigid planar loop of arbitrary shape that rotates in a uniform magnetic field which is normal to the fixed axis of rotation. (The loop may or may not intersect the axis.)

$\blacksquare$
$$v(t) = BA \sin \theta \frac{d\theta}{dt} \tag{1}$$
where A is the area of the loop and $\theta(t)$ is the angle of rotation. For rotation at constant angular velocity ω_m, $\theta = \omega_m t + \theta_0$ and (*1*) becomes
$$v(t) = BA\omega_m \sin (\omega_m t + \theta_0) \tag{2}$$

4.33   Extend (*2*) of Problem 4.32 to the case where the normal field is itself sinusoidal, with the same angular frequency ω_m; that is,
$$B(t) = B_m \sin (\omega_m t + \delta)$$

$\blacksquare$ For simplicity, suppose that $B(t)$ and $\theta(t)$ vanish at the same instant, and write
$$\theta = \omega_m t \qquad B = B_m \sin \omega_m t$$
Then, as in Problem 4.15,
$$\psi_m = BA \cos \theta = B_m A \sin \omega_m t \cos \omega_m t = \tfrac{1}{2} B_m A \sin 2\omega_m t$$
$$v = -\frac{d\psi_m}{dt} = -B_m A \omega_m \cos 2\omega_m t = B_m A \omega_m \cos (2\omega_m t - \pi) \tag{1}$$

4.34   A 10-turn rectangular loop of length 20 cm and width 16 cm rotates at 3600 rev/min in a magnetic field $B = 0.5 \sin 377t$ (T) directed normal to the axis of rotation. Calculate the rms value of the induced voltage.

∎ Problem 4.33 applies here, because the rotational frequency,

$$\left(3600 \frac{\text{rev}}{\text{min}}\right)\left(\frac{1 \text{ min}}{60 \text{ s}}\right)\left(2\pi \frac{\text{rad}}{\text{rev}}\right) = 377 \text{ rad/s}$$

equals the field frequency. Thus, since the effective area is 10 times the loop area,

$$v_{\text{rms}} = \frac{10B_m A \omega_m}{\sqrt{2}} = \frac{10(0.5)(0.20 \times 0.16)(377)}{1.414} = 42.66 \text{ V}$$

4.35 A magnetic field, $\mathbf{B} = 0.4 \sin(\pi x/2) \cos(\pi y/2) \sin 377t \, \mathbf{a}_z$ (T), passes through the stationary loop $0 < x, y < 1$ in the xy plane. Determine the voltage induced in the loop.

∎
$$\psi_m = \left[\int_0^1 \int_0^1 0.4 \sin \frac{\pi x}{2} \cos \frac{\pi y}{2} \, dx \, dy\right] \sin 377t = \frac{1.6}{\pi^2} \sin 377t \quad \text{(Wb)}$$
$$v = -\frac{d\psi_m}{dt} = -\frac{(1.6)(377)}{\pi^2} \cos 377t = -61.12 \cos 377t \quad \text{(V)}$$

4.36 Given a straight conductor Oa (Fig. 4-11) of length ℓ in the xy plane, rotating about O at angular velocity ω_m in a magnetic field $\mathbf{B} = B_0 \mathbf{a}_z$, find the induced voltage in the conductor and identify the positive terminal.

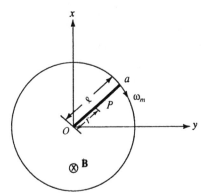

Fig. 4-11

∎ By Problem 4.20, the E-field at a point P at a radius r is $\mathbf{E}(r) = \mathbf{u}(r) \times \mathbf{B} = r\omega_m B_0 \mathbf{a}_r$, whence

$$v = \int_0^\ell r\omega_m B_0 \, dr = \frac{1}{2} \omega_m B_0 \ell^2$$

with a as the positive terminal.

4.37 The conductor of Problem 4.36 is fed with a current I at terminal a. Obtain an expression for the torque about O.

∎ The force on an elementary length dr at P (Fig. 4-11) is $d\mathbf{F} = -I \, dr \, \mathbf{a}_r \times B_0 \mathbf{a}_z = B_0 I \, dr \, \mathbf{a}_\phi$, and so

$$\mathbf{T} = \int_{r=0}^\ell r\mathbf{a}_r \times d\mathbf{F} = \int_0^\ell B_0 I r \, dr \, (\mathbf{a}_z) = \frac{1}{2} B_0 I \ell^2 \mathbf{a}_z$$

The direction $\mathbf{a}_z$ corresponds to clockwise rotation.

4.38 Let the loop of Fig. 4-10 be stationary ($u = 0$) and suppose a current $i = I \sin \omega t$ in the straight conductor. Express the mutual inductance M between the conductor and the loop in terms of the amplitudes of the current and the induced emf around the loop.

∎ The flux through the loop must have the same time dependence as the current: $\psi_m = \Psi_m \sin \omega t$. Now, by definition, $M = \psi_m/i$, or $Mi = \psi_m$, or

$$M\frac{di}{dt} = \frac{d\psi_m}{dt} \quad \text{or} \quad M\omega I \cos \omega t = \omega \Psi_m \sin \omega t = V \cos \omega t$$

Thus $M = V/\omega I$.

4.39 Show that Ampère's law for static fields, curl $\mathbf{H} = \mathbf{J}$, is inadequate for time-varying fields.

∎ The divergence of a curl is zero; but charge conservation requires that the divergence of $\mathbf{J}$ equal $-\partial\rho/\partial t \neq 0$.

4.40 Modify curl $\mathbf{H} = \mathbf{J}$ so that it becomes valid for time-varying fields (see Problem 4.39).

∎ Add the vector $\partial\mathbf{D}/\partial t$ (which has the same units, A/m², as $\mathbf{J}$, and which vanishes for a static field) to $\mathbf{J}$. Then, because

$$\operatorname{div}\left(\mathbf{J}+\frac{\partial\mathbf{D}}{\partial t}\right)=\operatorname{div}\mathbf{J}+\frac{\partial}{\partial t}\operatorname{div}\mathbf{D}=-\frac{\partial\rho}{\partial t}+\frac{\partial\rho}{\partial t}=0$$

the contradiction of Problem 4.39 is avoided. The term $\partial\mathbf{D}/\partial t$ is known as the *displacement current density*, in contrast to $\mathbf{J}$, the *conduction current density*.

4.41 Give in point form *Maxwell's equations* for time-varying electric and magnetic fields.

∎
$$\operatorname{curl}\mathbf{E}=-\frac{\partial\mathbf{B}}{\partial t} \qquad \text{(Faraday's law; fixed surface)} \tag{1}$$

$$\operatorname{curl}\mathbf{H}=\mathbf{J}+\frac{\partial\mathbf{D}}{\partial t} \qquad \text{(Modified Ampère's law)} \tag{2}$$

$$\operatorname{div}\mathbf{D}=\rho \qquad \text{(Gauss' law for }\mathbf{D}) \tag{3}$$

$$\operatorname{div}\mathbf{B}=0 \qquad \text{(Gauss' law for }\mathbf{B}) \tag{4}$$

4.42 Express Maxwell's equations, (1)–(4) of Problem 4.41, in integral form.

∎
$$\oint_{C}\mathbf{E}\cdot d\mathbf{s}=-\frac{d}{dt}\int_{S}\mathbf{B}\cdot d\mathbf{S} \tag{1}$$

$$\oint_{C}\mathbf{H}\cdot d\mathbf{s}=\int_{S}\mathbf{J}\cdot d\mathbf{S}+\frac{d}{dt}\int_{S}\mathbf{D}\cdot d\mathbf{S} \tag{2}$$

$$\int_{S}\mathbf{D}\cdot d\mathbf{S}=\int_{V}\rho\,dV \tag{3}$$

$$\int_{S}\mathbf{B}\cdot d\mathbf{S}=0 \tag{4}$$

4.43 A portion of a circuit containing a capacitor is shown in Fig. 4-12. Show that the displacement current i_d between the capacitor plates is precisely equal to the conduction current i outside the plates.

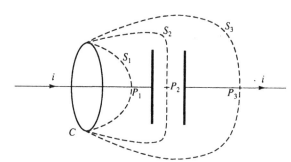

Fig. 4-12

∎ In Fig. 4-12, the three open surfaces S_1, S_2, S_3 all pass through the closed contour C. Noting that $\mathbf{J}=\mathbf{0}$ between the plates, apply (2) of Problem 4.42 to C and S_2:

$$\int_{C}\mathbf{H}\cdot d\mathbf{s}=0+i_d$$

However, for C and S_1 (over which $\partial\mathbf{D}/\partial t=\mathbf{0}$),

$$\int_{C}\mathbf{H}\cdot d\mathbf{s}=i+0$$

Therefore, $i_d=i$.

4.44 For a material medium characterized by conductivity σ and permittivity ϵ exposed to a sinusoidally varying $\mathbf{E}$-field of frequency ω, obtain the ratio of the conduction current density, $|\mathbf{J}_c|$, to the displacement current density, $|\mathbf{J}_d|$.

∎ Let the field (in phasor form) be $\mathbf{E}=E_0e^{j\omega t}$. Then

$$\mathbf{J}_c=\sigma\mathbf{E}=\sigma E_0e^{j\omega t} \qquad \text{and} \qquad -\mathbf{J}_d=\frac{\partial\mathbf{D}}{\partial t}=j\omega\mathbf{D}=j\omega\epsilon E=j\omega\epsilon E_0e^{j\omega t}$$

so that, by division,

$$\frac{|\mathbf{J}_c|}{|\mathbf{J}_d|}=\frac{\sigma}{\omega\epsilon} \tag{1}$$

4.45 Using the result of Problem 4.14, obtain a criterion for a material medium to act as a conductor or an insulator in the presence of an alternating electric field.

▌ For a conductor, $|\mathbf{J}_c| \gg |\mathbf{J}_d|$, or $\omega \ll \sigma/\epsilon$; for an insulator, $|\mathbf{J}_c| \ll |\mathbf{J}_d|$, or $\omega \gg \sigma/\epsilon$.

4.46 For copper $\sigma = 58\,\text{MS/m}$ and $\epsilon = \epsilon_0$; for Teflon, $\sigma = 30\,\text{nS/m}$ and $\epsilon = 2.1\epsilon_0$. Verify that at 1 MHz, copper is a good conductor and Teflon a good insulator.

▌ Use the test of Problem 4.45: for copper,

$$\frac{\sigma}{\omega\epsilon} = \frac{58 \times 10^6}{2\pi \times 10^6 \times \dfrac{1}{36\pi} \times 10^{-9}} = 10^{12} \gg 1$$

and for Teflon,

$$\frac{\sigma}{\omega\epsilon} = \frac{30 \times 10^{-9}}{2\pi \times 10^6 \times 2.1 \times \dfrac{1}{36\pi} \times 10^{-9}} = 2.57 \times 10^{-4} \ll 1$$

4.47 A material has $\sigma = 10^{-2}\,\text{S/m}$ and $\epsilon = 3\epsilon_0$. At what frequency (Hz) will the conduction current equal the displacement current?

▌ For the two currents to be equal, (1) of Problem 4.44 gives

$$\sigma = \omega\epsilon = 2\pi f\epsilon \quad \text{or} \quad f = \frac{\sigma}{2\pi\epsilon} = \frac{10^{-2}}{2\pi \times 3 \times (10^{-9}/36\pi)} = 60\,\text{MHz}$$

4.48 Calculate the displacement current through a parallel-plate air-filled capacitor having plates of area $10\,\text{cm}^2$ separated by a distance of 2 mm and connected to a 360-V, 1-MHz source.

▌ Since $J_d = \omega\epsilon_0 E$ (see Problem 4.44) and $E = V/d$, we have

$$J_d = \omega\epsilon_0 \frac{V}{d} \quad \text{and} \quad I_d = \omega\epsilon_0 \frac{VA}{d} = 2\pi \times 10^6 \times \frac{1}{36\pi} \times 10^{-9} \times 360 \times \frac{10 \times 10^{-4}}{2 \times 10^{-3}} = 10\,\text{mA}$$

4.49 Write the displacement current through a capacitor in an ac circuit, in terms of its capacitance C.

▌ For the capacitor, $q = Cv$, so that the charging (conductive) current is

$$i_c = \frac{dq}{dt} = C\frac{dv}{dt} \quad \text{or} \quad I_c = \omega CV$$

But $I_d = I_c$, by Problem 4.43.

4.50 Given, in free space,

$$\mathbf{E} = E_0 \cos(\omega t - \beta z)\mathbf{a}_x \quad \text{and} \quad \mathbf{H} = \frac{E_0}{\eta}\cos(\omega t - \beta z)\mathbf{a}_y$$

Determine η in terms of ω, μ_0, and ϵ_0 so that the fields satisfy all of Maxwell's equations (Problem 4.41).

▌ Equation (1) of Problem 4.41, requires

$$\frac{\partial E_x}{\partial z}\mathbf{a}_y = -\mu_0\frac{\partial H_y}{\partial t}\mathbf{a}_y \quad \text{or} \quad \beta E_0 \sin(\omega t - \beta z) = \omega\mu_0\frac{E_0}{\eta}\sin(\omega t - \beta z)$$

Thus

$$\beta = \frac{\omega\mu_0}{\eta} \tag{1}$$

Similarly, (2) of Problem 4.41 gives, for $\mathbf{J} = \mathbf{0}$,

$$-\frac{\partial H_y}{\partial z}\mathbf{a}_x = \epsilon_0\frac{\partial E_x}{\partial t}\mathbf{a}_x \quad \text{or} \quad -\frac{\beta E_0}{\eta}\sin(\omega t - \beta z) = -\omega\epsilon_0 E_0 \sin(\omega t - \beta z)$$

Hence

$$\beta = \omega\epsilon_0\eta \tag{2}$$

Combining (1) and (2) gives $\eta = \sqrt{\mu_0/\epsilon_0}$ (independent of ω); moreover, $\omega/\beta = 1/\sqrt{\epsilon_0\mu_0}$.

4.51 Do the fields

$$\mathbf{E} = E_m \sin x \sin t\, \mathbf{a}_y \quad \text{and} \quad \mathbf{H} = \frac{E_m}{\mu_0}\cos x \cos t\, \mathbf{a}_z$$

satisfy Maxwell's equations?

▌ For (2) of Problem 4.41 to hold,

$$-\frac{\partial H_z}{\partial x}\mathbf{a}_y = \epsilon_0\frac{\partial E_y}{\partial t}\mathbf{a}_y \quad \text{or} \quad \frac{E_m}{\mu_0}\sin x \cos t = \epsilon_0 E_m \sin x \cos t \quad \text{or} \quad \epsilon_0\mu_0 = 1$$

which last is clearly false.

4.52 Find the magnetic field corresponding to the free-space electric field $\mathbf{E} = E_m \sin \alpha x \cos(\omega t - \beta z)\, \mathbf{a}_y$.

▮ Faraday's law, (1) of Problem 4.41, gives

$$-\beta E_m \sin \alpha x \sin(\omega t - \beta z)\, \mathbf{a}_x + \alpha E_m \cos \alpha x \cos(\omega t - \beta z)\, \mathbf{a}_z = -\mu_0 \frac{\partial \mathbf{H}}{\partial t}$$

Thus, to within a nonessential static field,

$$\mathbf{H} = -\frac{\beta}{\mu_0 \omega} E_m \sin \alpha x \cos(\omega t - \beta z)\, \mathbf{a}_x - \frac{\alpha}{\mu_0 \omega} E_m \cos \alpha x \sin(\omega t - \beta z)\, \mathbf{a}_z$$

4.53 Verify that the fields of Problem 4.52 satisfy (3) and (4) of Problem 4.41.

▮
$$\operatorname{div} \mathbf{D} = \epsilon_0 \operatorname{div} \mathbf{E} = \epsilon_0 \frac{\partial E_y}{\partial y} = 0$$

so that (3) holds (free space). Also,

$$\operatorname{div} \mathbf{B} = \mu_0 \operatorname{div} \mathbf{H} = -\frac{\alpha \beta}{\omega} E_m \cos \alpha x \cos(\omega t - \beta z) + \frac{\alpha \beta}{\omega} E_m \cos \alpha x \cos(\omega t - \beta z) = 0$$

Thus (4) is satisfied.

4.54 Under what condition do the fields of Problem 4.52 satisfy Ampère's law, (2) of Problem 4.41?

▮ The second Maxwell equation yields

$$\left[-\frac{\beta^2}{\omega \mu_0} E_m \sin \alpha x \sin(\omega t - \beta z) - \frac{\alpha^2}{\mu_0 \omega} E_m \sin \alpha x \sin(\omega t - \beta z) \right] \mathbf{a}_y = \epsilon_0 [-\omega E_m \sin \alpha x \sin(\omega t - \beta z)] \mathbf{a}_y$$

and the required condition is $\alpha^2 + \beta^2 = \mu_0 \epsilon_0 \omega^2$.

4.55 Given an electric field $\mathbf{E} = E_0 \sin \beta z \cos \omega t\, \mathbf{a}_z$ in free space, determine the corresponding charge density.

▮
$$\rho = \operatorname{div} \mathbf{D} = \epsilon_0 \operatorname{div} \mathbf{E} = \epsilon_0 \frac{\partial E_z}{\partial z} = \epsilon_0 \beta E_0 \cos \beta z \cos \omega t$$

4.56 Find the magnetic field associated with $\mathbf{E} = E_0 \cos \beta x \cos \omega t\, \mathbf{a}_z$.

▮ Refer to Problem 4.41. By (1),

$$-\frac{\partial E_z}{\partial x} \mathbf{a}_y = -\frac{\partial B_y}{\partial t} \mathbf{a}_y \qquad \text{or} \qquad \beta E_0 \sin \beta x \cos \omega t = -\frac{\partial B_y}{\partial t}$$

or, by integration,

$$H_y = -\frac{\beta}{\omega \mu_0} E_0 \sin \beta x \sin \omega t$$

By (2),

$$\frac{\partial H_y}{\partial x} \mathbf{a}_z = -\omega \epsilon_0 E_0 \cos \beta x \sin \omega t\, \mathbf{a}_z \qquad \text{or} \qquad H_y = -\frac{\omega}{\beta} \epsilon_0 E_0 \sin \beta x \sin \omega t$$

Under the consistency condition $\omega/\beta = 1/\sqrt{\epsilon_0 \mu_0}$, our solution becomes

$$\mathbf{H} = -\sqrt{\frac{\epsilon_0}{\mu_0}} E_0 \sin \beta x \sin \omega t\, \mathbf{a}_y$$

Observe that the amplitude ratio, E/H, is the same as in Problem 4.50.

4.57 For sinusoidally varying fields it is convenient to carry out the analysis in terms of complex exponentials (or *phasors*). Represent $\mathbf{E} = E_0 \cos(\omega t - \beta x)\, \mathbf{a}_z$ (V/m) as a phasor.

▮
$$\mathbf{E} = \mathbf{a}_z E_0 \operatorname{Re}[e^{j(\omega t - \beta x)}] \rightarrow \mathbf{a}_z E_0 e^{-j\beta x} \quad \text{(V/m)}$$

with $e^{j\omega t}$ suppressed, and Re (or *real part of*) implied. To avoid confusion, we shall often write the phasor representation of a sinusoidal vector $\mathbf{A}$ as $\hat{\mathbf{A}}$.

4.58 Express in the time domain the phasor $\hat{\mathbf{E}} = -j30\, \mathbf{a}_x + j10\, \mathbf{a}_y$, at 10 MHz.

▮
$$\mathbf{E} = \operatorname{Re}[\hat{\mathbf{E}} e^{j\omega t}] = \operatorname{Re}[-j30 e^{j\omega t} \mathbf{a}_x + j10 e^{j\omega t} \mathbf{a}_y] = \operatorname{Re}[30 e^{j(\omega t - \pi/2)} \mathbf{a}_x + 10 e^{j(\omega t + \pi/2)} \mathbf{a}_y]$$
$$= 30 \sin \omega t\, \mathbf{a}_x - 10 \sin \omega t\, \mathbf{a}_y$$

in which $\omega = 2\pi \times 10^7$ rad/s.

4.59 Given: $\hat{\mathbf{A}} = j3 e^{-j2x} \mathbf{a}_y + 2 e^{-3x} \mathbf{a}_z$ and $\hat{\mathbf{B}} = -j e^{-jx} \mathbf{a}_x - (1 + j) e^{-jx} \mathbf{a}_z$, where the phasors represent fields varying at 10 MHz. Determine $\mathbf{A} \times \mathbf{B}^*$ in the time domain, where * stands for *complex conjugate*.

❚ First, we calculate the cross product in the phasor domain and then transform the result into the time domain.

$$\hat{\mathbf{A}} \times \hat{\mathbf{B}}^* = \begin{vmatrix} \mathbf{a}_x & \mathbf{a}_y & \mathbf{a}_z \\ 0 & j3e^{-j2x} & 2e^{-3x} \\ +je^{+jx} & 0 & -(1-j)e^{+jx} \end{vmatrix}$$
$$= -\mathbf{a}_x\, e^{-jx}(3+j3) + \mathbf{a}_y\, 2je^{(-3+j)x} + \mathbf{a}_z\, 3e^{-jx}$$

$$\mathrm{Re}\,[(\hat{\mathbf{A}} \times \hat{\mathbf{B}}^*)e^{j\omega t}] = \mathbf{a}_x[-3\cos(2\pi \times 10^7 t - x) + 3\sin(2\pi \times 10^7 t - x)] + \mathbf{a}_y[-2e^{-3x}\sin(2\pi \times 10^7 t + x)]$$
$$+ \mathbf{a}_z[3\cos(2\pi \times 10^7 t - x)]$$

4.60 Given: $\hat{\mathbf{B}} = j2e^{-j4\pi z/3}e^{-20z}\mathbf{a}_x$ at 10 MHz. Express **B** in the time domain.

❚ The required expression is (with $\omega = 2\pi \times 10^7$):
$$\mathbf{B} = \mathrm{Re}\,[j2e^{-j4\pi z/3}e^{-20z}e^{j\omega t}]\,\mathbf{a}_x = \mathrm{Re}\,[2e^{-20z}e^{j(\omega t - 4\pi z/3 + \pi/2)}]\,\mathbf{a}_x$$
$$= -2e^{-20z}\sin(2\pi \times 10^7 t - 4\pi z/3)\,\mathbf{a}_x$$

4.61 Express the fields of Problem 4.51 as phasors.

❚
$$\hat{\mathbf{E}} = e^{-j\pi/2}E_m \sin x\, \mathbf{a}_y \qquad \text{and} \qquad \hat{\mathbf{H}} = \frac{E_m}{\mu_0}\cos x\, \mathbf{a}_z$$

4.62 Let **V** be a vector with sinusoidal time dependence, and let $\hat{\mathbf{V}}$ be the phasor representation of **V**. Show that the rms magnitude of **V** is given by
$$V_{\mathrm{rms}}^2 = \tfrac{1}{2}\hat{\mathbf{V}} \cdot \hat{\mathbf{V}}^* \qquad \text{where} \qquad * \equiv \text{complex conjugate} \tag{1}$$

❚ Write $\mathbf{V} = \mathbf{a}_x A \cos(\omega t + \alpha) + \mathbf{a}_y B \cos(\omega t + \beta) + \mathbf{a}_z C \cos(\omega t + \gamma)$, and denote by $\langle\ \rangle$ the average value over one period, $T = 2\pi/\omega$. Then,
$$V_{\mathrm{rms}}^2 \equiv \langle \mathbf{V} \cdot \mathbf{V} \rangle = A^2 \langle \cos^2(\omega t + \alpha) \rangle + B^2 \langle \cos^2(\omega t + \beta) \rangle + C^2 \langle \cos^2(\omega t + \gamma) \rangle = \tfrac{1}{2}(A^2 + B^2 + C^2)$$
But $\hat{\mathbf{V}} = \mathbf{a}_x Ae^{j\alpha} + \mathbf{a}_y Be^{j\beta} + \mathbf{a}_z Ce^{j\gamma}$, so that
$$\hat{\mathbf{V}} \cdot \hat{\mathbf{V}}^* = A^2 + B^2 + C^2$$

and the proof of (1) is complete.

4.63 A lossy dielectric has $\mu_r = 1$, $\epsilon_r = 10$, and $\sigma = 20$ nS/m. An electric field $\mathbf{E} = 200 \sin \omega t\, \mathbf{a}_z$ (V/m) exists in the dielectric. At what frequency will the conduction current density and the displacement current density have equal magnitudes?

❚ From (1) of Problem 4.44, we must have $\omega = \sigma/\epsilon$, or
$$f = \frac{1}{2\pi}\frac{\sigma}{\epsilon} = \frac{1}{2\pi}\frac{20 \times 10^{-9}}{10 \times (10^{-9}/36\pi)} = 36\ \text{Hz}$$

4.64 At the frequency obtained in Problem 4.63, determine the instantaneous displacement current density in the dielectric.

❚ From Maxwell's equation,
$$\mathbf{J}_d = \frac{\partial \mathbf{D}}{\partial t} \qquad \text{or} \qquad J_d \mathbf{a}_z = \epsilon \frac{\partial E}{\partial t}\mathbf{a}_z = \epsilon\omega\, 200 \cos \omega t\, \mathbf{a}_z$$
But $\omega = 2\pi \times 36 = 72\pi$ rad/s, from Problem 4.63. Thus
$$J_d \mathbf{a}_z = \frac{10^{-8}}{36\pi} \times 72\pi \times 200 \cos 72\pi t\, \mathbf{a}_z = 4 \cos 72\pi t\, \mathbf{a}_z \quad (\mu\text{A/m}^2)$$

4.65 Show that for a sinusoidally varying field the conduction current and the displacement current are always displaced from each other by 90° in time.

❚ Let $\mathbf{E} \sim \sin \omega t$; then $\mathbf{J}_c = \sigma\mathbf{E} \sim \sin \omega t$, but
$$\mathbf{J}_d = \frac{\partial \mathbf{D}}{\partial t} = \epsilon\frac{\partial \mathbf{E}}{\partial t} \sim \cos \omega t = \sin(\omega t + 90°)$$

4.66 Graph the voltage available at the terminals 1 and 2 of the loop shown in Fig. 4-13(a). Notice that the **B**-field is confined within the square of side $a + b$.

❚ In the position shown, $0 < x < b$ and only conductor 2 will have an induced emf ($v_2 = Bcu$, with terminal 1 positive). This voltage remains constant until conductor 1 enters the field. The voltages induced in 1 and 2 cancel out, and we have zero voltage at the terminals until conductor 2 emerges from the field. Subsequently we

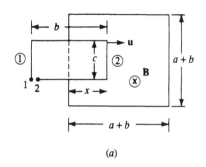

(a)

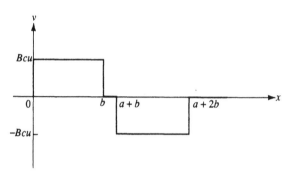

(b)

Fig. 4-13

have the voltage at 12 due to the emf induced in conductor 1, until it too emerges from the square. Hence,

$$v = \begin{cases} Bcu & 0 < x < b \\ 0 & b < x < a + b \\ -Bcu & a + b < x < a + 2b \\ 0 & a + 2b < x \end{cases}$$

The sketch is shown in Fig. 4-13(*b*).

4.67 A magnetic field $\mathbf{H} = H_x \cos 2x \cos (\omega t - \beta y)\, \mathbf{a}_x$ exists within a dielectric of permittivity ϵ. Determine the corresponding displacement current density.

❚
$$\mathbf{J}_d = \frac{\partial \mathbf{D}}{\partial t} = \operatorname{curl} \mathbf{H} = \mathbf{a}_y \frac{\partial H_x}{\partial z} - \mathbf{a}_z \frac{\partial H_x}{\partial y} = -\beta H_x \cos 2x \sin (\omega t - \beta y)\, \mathbf{a}_z$$

4.68 Obtain the charge density and the electric field corresponding to the **H**-field of Problem 4.67.

❚ Integrate $\partial \mathbf{D} / \partial t = \mathbf{J}_d$ to obtain

$$\mathbf{D} = \frac{\beta}{\omega} H_x \cos 2x \cos (\omega t - \beta y)\, \mathbf{a}_z \quad \text{or} \quad \mathbf{E} = \frac{\beta}{\omega \epsilon} H_x \cos 2x \cos (\omega t - \beta y)\, \mathbf{a}_z$$

Then, by Gauss' law, $\rho = \nabla \cdot \mathbf{D} = 0 \text{ C/m}^3$.

4.69 Does Faraday's law hold for the fields of Problems 4.67 and 4.68?

❚
$$\operatorname{curl} \mathbf{E} = \mathbf{a}_x \frac{\partial E_z}{\partial y} - \mathbf{a}_y \frac{\partial E_z}{\partial x} = \mathbf{a}_x \frac{\beta^2}{\omega \epsilon} H_x \cos 2x \sin (\omega t - \beta y) + \mathbf{a}_y \frac{2\beta}{\omega \epsilon} H_x \sin 2x \cos (\omega t - \beta y)$$

$$-\frac{\partial \mathbf{B}}{\partial t} = -\mu_0 \frac{\partial \mathbf{H}}{\partial t} = \mathbf{a}_x \, \omega \mu_0 H_x \cos 2x \sin (\omega t - \beta y)$$

The fields obviously *do not* satisfy Faraday's law; this situation arises from the *x*-dependence of H_x.

4.70 At the interface of two regions shown in Fig. 4-14, we have (in region 1): $\mathbf{B}_1 = 0.6\mathbf{a}_x + 1.1\mathbf{a}_y$ (T). Determine $\mathbf{B}_2$ (in region 2).

Fig. 4-14

■ Since $\mathbf{B} = \mu\mathbf{H}$, we have

$$\mathbf{H}_1 = \frac{0.6}{\mu_1}\mathbf{a}_x + \frac{1.1}{\mu_1}\mathbf{a}_y \quad (\text{A/m})$$

The interface conditions $B_{n1} = B_{n2}$ and $H_{t1} = H_{t2}$ require

$$B_{y1} = B_{y2} = 1.1\,\text{T} \quad \text{and} \quad H_{x1} = H_{x2} = 0.6/\mu_1 \text{ A/m}$$

But $H_{x2} = B_{x2}/\mu_2$, therefore, $\mathbf{B}_2 = (0.6\mu_2/\mu_1)\mathbf{a}_x + 1.1\mathbf{a}_y$ (T).

4.71 Suppose three-dimensional space divided into region 1 ($x < 0$) and region 2 ($x > 0$). Given $\sigma_1 = \sigma_2 = 0$ and $\mathbf{E}_1 = \alpha\mathbf{a}_x + \beta\mathbf{a}_y + \delta\mathbf{a}_z$, determine $\mathbf{E}_2$.

■ From $D_{n1} = D_{n2}$, we obtain $D_{x1} = D_{x2}$, or $\epsilon_1 E_{x1} = \epsilon_2 E_{x2}$, whence

$$E_{x2} = \frac{\epsilon_1}{\epsilon_2} E_{x1} = \alpha\frac{\epsilon_1}{\epsilon_2}$$

Similarly, $E_{t1} = E_{t2}$ requires that $E_{y1} = E_{y2} = \beta$ and $E_{z1} = E_{z2} = \delta$. Consequently,

$$\mathbf{E}_2 = \alpha\frac{\epsilon_1}{\epsilon_2}\mathbf{a}_x + \beta\mathbf{a}_y + \delta\mathbf{a}_z$$

4.72 Given: $\mathbf{B}_1 = \alpha\mathbf{a}_x + \beta\mathbf{a}_y + \delta\mathbf{a}_z$ for region 1 of Problem 4.71. Find $\mathbf{B}_2$.

■ At $x = 0$, $H_{t1} = H_{t2}$ requires that

$$\frac{B_{y1}}{\mu_1} = \frac{B_{y2}}{\mu_2} \qquad \frac{B_{z1}}{\mu_1} = \frac{B_{z2}}{\mu_2}$$

which give $B_{y2} = \beta(\mu_2/\mu_1)$ and $B_{z2} = \delta(\mu_2/\mu_1)$. Similarly, $B_{n1} = B_{n2}$ implies that $B_{x2} = \alpha$. Thus,

$$\mathbf{B}_2 = \alpha\mathbf{a}_x + \beta\frac{\mu_2}{\mu_1}\mathbf{a}_y + \delta\frac{\mu_2}{\mu_1}\mathbf{a}_z$$

4.73 Regions 1 and 2 of Problem 4.71 are perfect dielectrics. At $x = 0^-$, $\mathbf{D}_1 = \alpha\mathbf{a}_x + \beta\mathbf{a}_y + \delta\mathbf{a}_z$. Determine $\mathbf{D}_2$ at $x = 0^+$.

■ Since $D_{n1} = D_{n2}$, we have $D_{x1} = D_{x2} = \alpha$. The condition $E_{t1} = E_{t2}$ requires that $E_{y1} = E_{y2}$ and $E_{z1} = E_{z2}$, whence

$$D_{y2} = \frac{\epsilon_2}{\epsilon_1}\beta \qquad \text{and} \qquad D_{z2} = \frac{\epsilon_2}{\epsilon_1}\delta$$

Hence $\mathbf{D}_2 = \alpha\mathbf{a}_x + (\epsilon_2/\epsilon_1)(\beta\mathbf{a}_y + \delta\mathbf{a}_z)$.

4.74 At $x = 0^-$ in Problem 4.71, $\mathbf{H}_1 = \alpha\mathbf{a}_x + \beta\mathbf{a}_y + \delta\mathbf{a}_z$. Determine $\mathbf{H}_2$ at $x = 0^+$.

■ Since $B_{n1} = B_{n2}$ and $\mathbf{B} = \mu\mathbf{H}$,

$$\mu_1 H_{x1} = \mu_2 H_{x2} \qquad \text{or} \qquad H_{x2} = \frac{\mu_1}{\mu_2}\alpha$$

Similarly, $H_{t1} = H_{t2}$ leads to $H_{y2} = \beta$ and $H_{z2} = \delta$. Hence $\mathbf{H}_2 = \alpha(\mu_1/\mu_2)\mathbf{a}_x + \beta\mathbf{a}_y + \delta\mathbf{a}_z$.

4.75 Obtain a set of boundary conditions for time-varying electromagnetic fields at the surface of a perfect conductor ($\sigma = \infty$, finite μ and ϵ).

■ Let the interior of the conductor be region 1. Ohm's law, $\mathbf{J}_1 = \sigma_1\mathbf{E}_1$, implies $\mathbf{E}_1 = \mathbf{0}$ (otherwise, J_1 would be infinite); hence also, $\mathbf{D}_1 = \epsilon_1\mathbf{E}_1 = \mathbf{0}$. Now, curl $\mathbf{E}_1 = -\partial\mathbf{B}_1/\partial t$ implies that $\partial\mathbf{B}_1/\partial t = 0$. In other words, there can be no time-varying magnetic field within a perfect conductor ($\mathbf{B}_1 = \mathbf{H}_1 = \mathbf{0}$). Combination of these results with Gauss' law for $\mathbf{D}$ and Ampère's circuital law for $\mathbf{H}$ yields the desired conditions just outside the conductor:

$$E_{t2} = 0 \qquad H_{t2} = k \qquad D_{n2} = \rho_s \qquad B_{n2} = 0 \qquad\qquad (1)$$

Here, k and ρ_s are, respectively, the current density (A/m) and static charge density (C/m^2) on the outer surface of the conductor. Either of these point functions may, of course, vanish identically.

4.76 Region 1 of Problem 4.71 is free space and region 2 is a perfect conductor having a surface charge density ρ_s and surface current density $\mathbf{k}$. Given

$$\mathbf{E}_1 = \alpha_e\mathbf{a}_x + \beta_e\mathbf{a}_y + \gamma_e\mathbf{a}_z \qquad \text{and} \qquad \mathbf{H}_1 = \alpha_h\mathbf{a}_x + \beta_h\mathbf{a}_y + \gamma_h\mathbf{a}_z$$

determine the α's, β's, and γ's.

■ Note that the labeling of the two regions is opposite to that used in Problem 4.75. The four conditions (1) of that problem give, respectively,

$$\beta_e = \gamma_e = 0 \qquad \beta_h = k_y, \ \gamma_h = k_z \qquad \alpha_e = -\rho_s/\epsilon_0 \qquad \alpha_h = 0$$

(The minus sign in the third equation arises from the fact that the unit normal out of the conductor is $-\mathbf{a}_x$.)

4.77 In region 1 of Fig. 4-15, we have (at the interface) $\mathbf{E}_1 = 2\mathbf{a}_y + 3\mathbf{a}_z$ V/m. Obtain $\mathbf{E}_2$ (in region 2) at the interface.

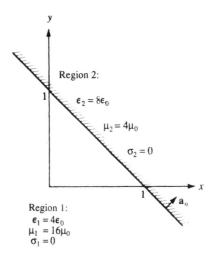

Region 2:
$\epsilon_2 = 8\epsilon_0$
$\mu_2 = 4\mu_0$
$\sigma_2 = 0$

Region 1:
$\epsilon_1 = 4\epsilon_0$
$\mu_1 = 16\mu_0$
$\sigma_1 = 0$

Fig. 4-15

▮ The unit normal to the surface (pointing into region 2) is $\mathbf{a}_n = (1/\sqrt{2})\mathbf{a}_x + (1/\sqrt{2})\mathbf{a}_y$. Thus, the continuity of tangential E yields

$$\mathbf{E}_2 - (\mathbf{E}_2 \cdot \mathbf{a}_n)\mathbf{a}_n = \mathbf{E}_1 - (\mathbf{E}_1 \cdot \mathbf{a}_n)\mathbf{a}_n$$

$$E_{x2}\mathbf{a}_x + E_{y2}\mathbf{a}_y + E_{z2}\mathbf{a}_z - \left(\frac{E_{x2}}{\sqrt{2}} + \frac{E_{y2}}{\sqrt{2}}\right)\left(\frac{1}{\sqrt{2}}\mathbf{a}_x + \frac{1}{\sqrt{2}}\mathbf{a}_y\right) = 2\mathbf{a}_y + 3\mathbf{a}_z - (\sqrt{2})\left(\frac{1}{\sqrt{2}}\mathbf{a}_x + \frac{1}{\sqrt{2}}\mathbf{a}_y\right)$$

Equating coefficients of the unit vectors gives $E_{y2} - E_{x2} = 2$, $E_{z2} = 3$. In addition, continuity of normal $\mathbf{D} = \epsilon\mathbf{E}$ yields

$$\epsilon_2\mathbf{E}_2 \cdot \mathbf{a}_n = \epsilon_1\mathbf{E}_1 \cdot \mathbf{a}_n$$

$$8\epsilon_0\left(\frac{E_{x2}}{\sqrt{2}} + \frac{E_{y2}}{\sqrt{2}}\right) = 4\epsilon_0(\sqrt{2})$$

$$E_{y2} + E_{x2} = 1$$

Hence, $E_{x2} = -1/2$, $E_{y2} = 3/2$, and so

$$\mathbf{E}_2 = -\tfrac{1}{2}\mathbf{a}_x + \tfrac{3}{2}\mathbf{a}_y + 3\mathbf{a}_z \quad \text{V/m}$$

4.78 Given, in Problem 4.77, $\mathbf{H}_1 = 0.1\mathbf{a}_x + 0.2\mathbf{a}_z$ A/m, evaluate $\mathbf{B}_2$ at the interface.

▮ This is just like Problem 4.77, only with $\mathbf{H}$ in place of $\mathbf{E}$ and $\mathbf{B}$ in place of $\mathbf{D}$. The final result is
$$\mathbf{B}_2 = \mu_0(\mathbf{a}_x + 0.6\mathbf{a}_y + 0.8\mathbf{a}_z) \quad \text{T}$$

4.79 In Problem 4.77, $\sigma_2 = 0$ is changed to $\sigma_2 = \infty$ (making region 2 a perfect conductor); all other data remain the same. Determine the charge density on the surface of the conductor.

▮
$$\rho_s = -D_{n1} = -\epsilon_1\mathbf{E}_1 \cdot \mathbf{a}_n = -(4\epsilon_0)(\sqrt{2}) \quad (\text{C/m}^2)$$

4.80 For the data of Problem 4.79 and with $\mathbf{H}_1$ as in Problem 4.78, determine the current density vector on the surface of the conductor.

▮ By Problem 4.75,

$$\mathbf{k} = \mathbf{H}_{t1} = \mathbf{H}_1 - [\mathbf{H}_1 \cdot (-\mathbf{a}_n)](-\mathbf{a}_n) = 0.1\mathbf{a}_x + 0.2\mathbf{a}_z - \left(-\frac{0.1}{\sqrt{2}}\right)\left(-\frac{1}{\sqrt{2}}\mathbf{a}_x - \frac{1}{\sqrt{2}}\mathbf{a}_y\right) = 0.05\mathbf{a}_x - 0.05\mathbf{a}_y + 0.2\mathbf{a}_z \quad \text{A/m}$$

4.81 A bar magnet is moved axially at a velocity u (m/s) through the center of a one-turn circular loop of radius b (see Fig. 10-7). Determine the emf induced in the loop when one of the poles of the magnet is at the center of the loop. Assume that the field of the magnet pole is radial, is spherically symmetric, and is given by $B = c/r^2$, where c is a constant.

▮ From Problem 10.29 we have

$$\psi_m = 2\pi c\left(1 - \frac{x}{\sqrt{x^2 + b^2}}\right) \quad \text{(Wb)} \qquad \text{and} \qquad v = -\frac{d\psi_m}{dt} = 2\pi c\left[\frac{1}{\sqrt{x^2 + b^2}}\frac{dx}{dt} - \frac{x^2}{(x^2 + b^2)^{3/2}}\frac{dx}{dt}\right] \quad \text{(V)}$$

Substituting $dx/dt = u$ and $x = 0$ yields $v = (2\pi cu/b)$ (V).

4.82 A two-dimensional **E**-field is given by $\mathbf{E} = x^2\mathbf{a}_x + x\mathbf{a}_y$. Show that this field cannot arise from a static distribution of charge.

▌

$$\nabla \times \mathbf{E} = \begin{vmatrix} \mathbf{a}_x & \mathbf{a}_y & \mathbf{a}_z \\ \dfrac{\partial}{\partial x} & \dfrac{\partial}{\partial y} & \dfrac{\partial}{\partial z} \\ x^2 & x & 0 \end{vmatrix} = 1\mathbf{a}_z \neq \mathbf{0}$$

4.83 Determine the induced voltage in the Faraday disk of Problem 4.23 (Fig. 4-8) assuming that the steady, dc **B**-field is replaced by $\mathbf{B} = B_m \sin \omega t\, \mathbf{a}_z$.

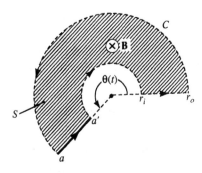

Fig. 4-16

▌ While it is possible to use Problem 4.21 here, the procedure is tricky. Much better is to analyze the Faraday disk as a pivoted "bar," of length $r_o - r_i$, as shown in Fig. 4-16 (looking in the z direction). The instantaneous position of the bar defines a planar surface S, through which the magnetic flux is

$$\psi_m = -B_z \times \text{area} = (-B_m \sin \omega t)\frac{\theta}{2\pi}\pi(r_o^2 - r_i^2)$$

Hence, the emf around C, in the direction indicated, is (using $d\theta/dt = \omega_m$):

$$v = -\frac{d\psi_m}{dt} = \frac{B_m(r_0^2 - r_i^2)}{2}(\omega \cos \omega t\, \theta + \omega_m \sin \omega t)$$

where $\theta = \omega_m t + \theta_0$. Because aa' is the only conducting segment of the contour C, the entire emf appears across it.

4.84 In a rectangular duct (Fig. 4-17) known as a *waveguide*, we have

$$E_x = E_z = 0 \qquad E_y = C\frac{\omega \mu_0 a}{\pi}\sin\frac{\pi x}{a}\sin(\omega t - \beta z)$$

$$H_x = -C\frac{\beta a}{\pi}\sin\frac{\pi x}{a}\sin(\omega t - \beta z) \qquad H_y = 0 \qquad H_z = C\cos\frac{\pi x}{a}\cos(\omega t - \beta z)$$

Under what condition will these fields satisfy Maxwell's equations within the waveguide?

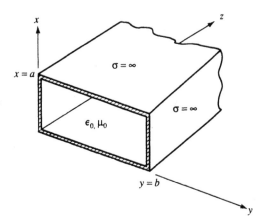

Fig. 4-17

▌ First we check Faraday's law:

$$\nabla \times \mathbf{E} = -\mu_0\frac{\partial \mathbf{H}}{\partial t} \qquad \text{or} \qquad -\frac{\partial E_y}{\partial z}\mathbf{a}_x + \frac{\partial E_y}{\partial x}\mathbf{a}_z = -\mu_0\frac{\partial H_x}{\partial t}\mathbf{a}_x - \mu_0\frac{\partial H_z}{\partial t}\mathbf{a}_z$$

Evaluating the partial derivatives, we see that Faraday's law is satisfied. Checking Ampère's law ($\mathbf{J}=\mathbf{0}$),

$$\nabla \times \mathbf{H} = \epsilon_0 \frac{\partial \mathbf{E}}{\partial t} \quad \text{or} \quad \frac{\partial H_z}{\partial y}\mathbf{a}_x + \left(\frac{\partial H_x}{\partial z} - \frac{\partial H_z}{\partial x}\right)\mathbf{a}_y - \frac{\partial H_x}{\partial y}\mathbf{a}_z = \epsilon_0 \frac{\partial E_y}{\partial t}\mathbf{a}_y$$

we obtain the component equations $0 = 0$,

$$\beta C \frac{\beta a}{\pi} \sin (\pi x/a) \cos (\omega t - \beta z) + \frac{\pi}{a} C \sin (\pi x/a) \cos (\omega t - \beta z) = \epsilon_0 \omega C \frac{\omega \mu_0 a}{\pi} \sin (\pi x/a) \cos (\omega t - \beta z) \quad (1)$$

and $0 = 0$. From (1), the required condition is:

$$\beta^2 \frac{a}{\pi} + \frac{\pi}{a} = \omega^2 \epsilon_0 \mu_0 \frac{a}{\pi} \quad \text{or} \quad \beta = \sqrt{\omega^2 \mu_0 \epsilon_0 - (\pi/a)^2} \quad (2)$$

4.85 Notice that the walls of the waveguide of Problem 4.84 are perfectly conducting. Verify that the fields of Problem 4.84 satisfy the boundary conditions.

▮ The only enforceable boundary conditions are that tangential $\mathbf{E}$ and normal $\mathbf{B}$ equal zero along the (perfectly conducting) walls of the waveguide (Problem 4.75). It is easy to see that these two conditions are satisfied.

4.86 Evaluate the surface charge densities and the linear current densities on the walls of the waveguide of Problem 4.84.

▮ By Problem 3.85, $\mathbf{k} = \mathbf{H} \times \mathbf{a}_n$. Thus, on the left wall ($y = 0$),

$$\mathbf{k} = -(H_x)_{y=0}\mathbf{a}_z + (H_z)_{y=0}\mathbf{a}_x = C\frac{\beta a}{\pi} \sin \left(\frac{\pi x}{a}\right) \sin (\omega t - \beta z) \, \mathbf{a}_z + C \cos \left(\frac{\pi x}{a}\right) \cos (\omega t - \beta z) \, \mathbf{a}_x$$

On the right wall ($y = b$) we have a similar result since H_x and H_z are independent of y:

$$\mathbf{k} = (H_x)_{y=b}\mathbf{a}_z - (H_z)_{y=b}\mathbf{a}_x = -C\frac{\beta a}{\pi} \sin \left(\frac{\pi x}{a}\right) \sin (\omega t - \beta z) \, \mathbf{a}_z - C \cos \left(\frac{\pi x}{a}\right) \cos (\omega t - \beta z) \, \mathbf{a}_x$$

Along the top wall ($x = a$),

$$\mathbf{k} = (H_z)_{x=a}\mathbf{a}_y = -C \cos (\omega t - \beta z) \, \mathbf{a}_y$$

Along the bottom wall ($x = 0$),

$$\mathbf{k} = -(H_z)_{x=0}\mathbf{a}_y = -C \cos (\omega t - \beta z) \, \mathbf{a}_y$$

The surface charge density on a wall is $\rho_s = -\epsilon_0 E_n$. Hence, on the left wall ($y = 0$),

$$\rho_s = \epsilon_0 (E_y)_{y=0} = C\epsilon_0 \mu_0 \frac{\omega a}{\pi} \sin \left(\frac{\pi x}{a}\right) \sin (\omega t - \beta z)$$

Along the right wall ($y = b$),

$$\rho_s = -\epsilon_0 (E_y)_{y=b} = -C\epsilon_0 \mu_0 \frac{\omega a}{\pi} \sin \left(\frac{\pi x}{a}\right) \sin (\omega t - \beta z)$$

Along the top and bottom walls, $\rho_s = 0$ (because $E_x = 0$).

4.87 Show that the last two of Maxwell's point equations are contained in the first two equations; that is, (4) of Problem 4.41 is implicit in (1), and (3) is implicit in (2).

▮ Take the divergence of (1):

$$0 = \text{div} \left(-\frac{\partial \mathbf{B}}{\partial t}\right) = -\frac{\partial}{\partial t}(\text{div } \mathbf{B})$$

Thus, within a static field, div $\mathbf{B} = 0$. Similarly, take the divergence of (2):

$$0 = \text{div } \mathbf{J} + \text{div} \frac{\partial \mathbf{D}}{\partial t} = -\frac{\partial \rho}{\partial t} + \frac{\partial}{\partial t}(\text{div } \mathbf{D}) = \frac{\partial}{\partial t}(\text{div } \mathbf{D} - \rho)$$

or div $\mathbf{D} = \rho$.

4.88 Express the fields given in Problem 4.84 as phasors.

▮ Componentwise, the phasors are

$$\hat{E}_x = \hat{E}_z = 0 \qquad \hat{E}_y = -jC\omega \frac{\mu_0 a}{\pi} \sin \frac{\pi x}{a} e^{-j\beta z}$$

$$\hat{H}_x = jC \frac{\beta a}{\pi} \sin \frac{\pi x}{a} e^{-j\beta z} \qquad \hat{H}_y = 0 \qquad \hat{H}_z = C \cos \frac{\pi x}{a} e^{-j\beta z}$$

4.89 Verify the Maxwell equation div $\mathbf{H} = 0$ for the phasor field of Problem 4.88.

▮

$$\frac{\partial H_x}{\partial x} + \frac{\partial H_y}{\partial y} + \frac{\partial H_z}{\partial z} = jC\beta \cos \frac{\pi x}{a} e^{-j\beta z} + 0 - j\beta C \cos \frac{\pi x}{a} e^{-j\beta z} = 0$$

4.90 Show that $\mathbf{E} \cdot \dfrac{\partial \mathbf{D}}{\partial t}$ and $\mathbf{H} \cdot \dfrac{\partial \mathbf{B}}{\partial t}$ respectively denote the rate of change of energy densities stored in electric and magnetic fields.

$\blacksquare$ By Problems 2.180 and 3.102, the instantaneous densities are

$$w_e = \frac{\epsilon}{2} \mathbf{E} \cdot \mathbf{E} \qquad \text{and} \qquad w_m = \frac{\mu}{2} \mathbf{H} \cdot \mathbf{H}$$

Hence

$$\frac{\partial w_e}{\partial t} = \epsilon \mathbf{E} \cdot \frac{\partial \mathbf{E}}{\partial t} = \mathbf{E} \cdot \frac{\partial \mathbf{D}}{\partial t} \qquad \frac{\partial w_m}{\partial t} = \mu \mathbf{H} \cdot \frac{\partial \mathbf{H}}{\partial t} = \mathbf{H} \cdot \frac{\partial \mathbf{B}}{\partial t} \tag{1}$$

4.91 Define the *Poynting vector* $\mathbf{S} \equiv \mathbf{E} \times \mathbf{H}$ (W/m^2). Show that the net efflux of electromagnetic energy through a closed surface Σ is given by $\int_\Sigma \mathbf{S} \cdot d\Sigma$.

$\blacksquare$ Apply the vector identity div $(\mathbf{A} \times \mathbf{B}) = \mathbf{B} \cdot \text{curl } \mathbf{A} - \mathbf{A} \cdot \text{curl } \mathbf{B}$ to the Poynting vector, and use (1) and (2) of Problem 4.41, to obtain

$$\text{div } \mathbf{S} = -\mathbf{H} \cdot \frac{\partial \mathbf{B}}{\partial t} - \mathbf{E} \cdot \mathbf{J} - \mathbf{E} \cdot \frac{\partial \mathbf{D}}{\partial t} \tag{1}$$

Integrate (1) through the volume V enclosed by Σ, and apply the divergence theorem:

$$\int_\Sigma \mathbf{S} \cdot d\Sigma = -\int_V \left(\mathbf{E} \cdot \frac{\partial \mathbf{D}}{\partial t} + \mathbf{H} \cdot \frac{\partial \mathbf{B}}{\partial t} \right) d\tau - \int_V \mathbf{E} \cdot \mathbf{J} \, d\tau$$

or

$$\int_\Sigma \mathbf{S} \cdot d\Sigma + \int_V \mathbf{E} \cdot \mathbf{J} \, d\tau = -\frac{d}{dt}[W_e + W_m] \tag{2}$$

We interpret (2) in the light of conservation of energy. The right-hand side is clearly the rate of decrease of stored electromagnetic energy within Σ; therefore, the left-hand side must represent the total flux of energy out of V. Now, there are just two mechanisms for power loss from V: (i) internal dissipation (as heat), which is accounted for by

$$\int_V \mathbf{E} \cdot \mathbf{J} \, d\tau = \sigma \int_V E^2 \, d\tau$$

and (ii) radiation of energy through Σ, which, by elimination, must equal $\int_\Sigma \mathbf{S} \cdot d\Sigma$.

4.92 An antenna in free space is centered at the origin of a spherical coordinate system, as shown in Fig. 4-18. The fields produced by the antenna at a radial distance r are given by

$$\mathbf{E} = \frac{E_0}{r} \sin\theta \, \sin\omega(t - r\sqrt{\epsilon_0 \mu_0}) \, \mathbf{a}_\theta \qquad \mathbf{H} = \frac{E_0}{r\sqrt{\mu_0/\epsilon_0}} \sin\theta \, \sin\omega(t - r\sqrt{\epsilon_0 \mu_0}) \, \mathbf{a}_\phi$$

Calculate the time-average power radiated by the antenna.

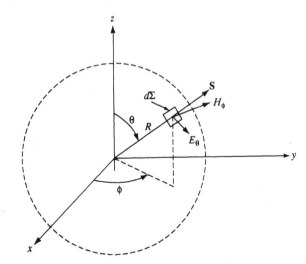

Fig. 4-18

$\blacksquare$ The electric field is in the θ-direction and the magnetic field is in the ϕ-direction; thus, the Poynting vector is in the radial direction:

$$\mathbf{S} = \mathbf{E} \times \mathbf{H} = \frac{E_0^2}{r^2 \sqrt{\mu_0/\epsilon_0}} \sin^2\theta \, \sin^2\omega(t - r\sqrt{\epsilon_0 \mu_0}) \, \mathbf{a}_r \tag{1}$$

By Problem 4.91, the power leaving the antenna can be obtained by integrating S_r over the surface of a sphere of radius R:

$$P_{\text{rad}} = \int_{\phi=0}^{2\pi} \int_{\theta=0}^{\pi} S_r\big|_{r=R} R^2 \sin\theta \, d\theta \, d\phi = \frac{2\pi E_0^2}{\sqrt{\mu_0/\epsilon_0}} \sin^2 \omega(t - R\sqrt{\epsilon_0\mu_0}) \int_0^{\pi} \sin^3\theta \, d\theta = \frac{8\pi E_0^2}{3\sqrt{\mu_0/\epsilon_0}} \sin^2 \omega(t - R\sqrt{\epsilon_0\mu_0})$$

The time average of the sine-squared over one period is 1/2, yielding

$$\langle P_{\text{rad}} \rangle = \frac{4\pi E_0^2}{3\sqrt{\mu_0/\epsilon_0}} \tag{2}$$

4.93   Verify (2) of Problem 4.91 for a wire of conductivity σ and radius r_w carrying a dc current i, as shown in Fig. 4-19(a).

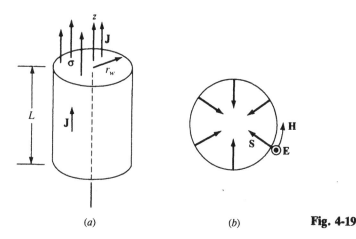

(a)   (b)   **Fig. 4-19**

▮ For steady dc, $W_e + W_m = \text{const.}$, so that the relation to be checked is

$$\int_{\Sigma} \mathbf{S} \cdot d\Sigma + \int_V \mathbf{E} \cdot \mathbf{J} \, d\tau = 0 \tag{1}$$

where Σ is the cylindrical surface $r = r_w$ plus the end caps $z = \pm L/2$, and where $d\Sigma$ has the direction of the *outer* normal. Writing $A \equiv \pi r_w^2$ as the cross-sectional area, we have

$$\mathbf{E} = \frac{1}{\sigma}\mathbf{J} = \frac{I}{\sigma A}\mathbf{a}_z \qquad \mathbf{H} = \frac{I}{2\pi r_w}\mathbf{a}_\phi = \frac{I r_w}{2A}\mathbf{a}_\phi$$

whence

$$\mathbf{S} = \mathbf{E} \times \mathbf{H} = \frac{I^2 r_w}{2\sigma A^2}(-\mathbf{a}_r)$$

and

$$\int_{\Sigma} \mathbf{S} \cdot d\Sigma = \int_{\phi=0}^{2\pi} \frac{I^2 r_w}{2\sigma A^2}(-\mathbf{a}_r) \cdot (L r_w \, d\phi \, \mathbf{a}_r) + 0 + 0 = -\frac{I^2 L}{\sigma A}$$

On the other hand,

$$\int_V \mathbf{E} \cdot \mathbf{J} \, d\tau = (\sigma E^2)(AL) = \frac{I^2 L}{\sigma A}$$

and (1) is verified.

4.94   From Problem 4.93 obtain the usual formula for the power dissipated in a linear resistive element.

▮ Since $R \equiv L/\sigma A$, $\int_V \mathbf{E} \cdot \mathbf{J} \, d\tau = I^2 R$.

4.95   Show that for fields varying sinusoidally in time, the time-average Poynting vector is given by

$$\langle \mathbf{S} \rangle = \tfrac{1}{2} \text{Re}[\hat{\mathbf{S}}] \tag{1}$$

where $\hat{\mathbf{S}} \equiv \hat{\mathbf{E}} \times \hat{\mathbf{H}}^*$ is the *phasor Poynting vector*.

▮ Because

$$\mathbf{E} = \text{Re}[\hat{\mathbf{E}}e^{j\omega t}] = \tfrac{1}{2}(\hat{\mathbf{E}}e^{j\omega t} + \hat{\mathbf{E}}^* e^{-j\omega t}) \qquad \text{and} \qquad \mathbf{H} = \tfrac{1}{2}(\hat{\mathbf{H}}e^{j\omega t} + \hat{\mathbf{H}}^* e^{-j\omega t})$$

the distributive law for the cross product gives

$$\mathbf{S} = \mathbf{E} \times \mathbf{H} = \tfrac{1}{4}(\hat{\mathbf{E}} \times \hat{\mathbf{H}}^* + \hat{\mathbf{E}}^* \times \hat{\mathbf{H}}) + \tfrac{1}{4}(\hat{\mathbf{E}} \times \hat{\mathbf{H}}e^{j2\omega t} + \hat{\mathbf{E}}^* \times \hat{\mathbf{H}}^* e^{-j2\omega t}) \tag{2}$$

Since $(\hat{\mathbf{E}} \times \hat{\mathbf{H}}^*)^* = \hat{\mathbf{E}}^* \times \hat{\mathbf{H}}$ and $\hat{\mathbf{E}}^* \times \hat{\mathbf{H}}^* = (\hat{\mathbf{E}} \times \hat{\mathbf{H}})^*$, (2) may be transformed into

$$\mathbf{S} = \tfrac{1}{2}\text{Re}[\hat{\mathbf{E}} \times \hat{\mathbf{H}}^*] + \tfrac{1}{2}\text{Re}[\hat{\mathbf{E}} \times \hat{\mathbf{H}}e^{j2\omega t}] \tag{3}$$

The first term on the right-hand side of (3) is independent of time, and the second term has average value zero; this establishes (1).

4.96   If the field vectors of a wave in free space are given by

$$\mathbf{E} = 100 \cos\left(\omega t + \frac{4\pi}{3}x\right)\mathbf{a}_z \quad (\text{V/m}) \qquad \mathbf{H} = \frac{100}{120\pi}\cos\left(\omega t + \frac{4\pi}{3}x\right)\mathbf{a}_y \quad (\text{A/m})$$

where $f = \omega/2\pi = 200$ MHz, determine the phasor Poynting vector.

▌ $$\hat{\mathbf{E}} = 100 e^{(j4\pi/3)x}\mathbf{a}_z \qquad \hat{\mathbf{H}} = \frac{100}{120\pi}e^{(j4\pi/3)x}\mathbf{a}_y \qquad \hat{\mathbf{S}} = \hat{\mathbf{E}} \times \hat{\mathbf{H}}^* = -\frac{(100)^2}{120\pi}\mathbf{a}_x \quad \text{W/m}^2$$

4.97   For the wave of Problem 4.96, compute the average power crossing a 4-m^2 patch of the yz plane.

▌   In this case $\hat{\mathbf{S}}$ is real and is constant in magnitude and direction; therefore,

$$\langle P \rangle = \langle \mathbf{S} \rangle \cdot \mathbf{A} = \frac{1}{2}\hat{\mathbf{S}} \cdot \mathbf{A} = \frac{1}{2}\left(-\frac{100^2}{120\pi}\mathbf{a}_x\right)\cdot(4\mathbf{a}_x) = -53 \text{ W}$$

The minus sign indicates flux in the negative x-direction.

4.98   Verify (1) of Problem 4.95 for the antenna of Problem 4.92.

▌   From (1) of Problem 4.92,

$$\langle \mathbf{S} \rangle = \left(\frac{E_0^2}{r^2\sqrt{\mu_0/\epsilon_0}}\sin^2\theta\right)\left(\frac{1}{2}\right)\mathbf{a}_r$$

On the other hand,

$$\hat{\mathbf{E}} = -j\frac{E_0}{r}\sin\theta\, e^{-j\omega r\sqrt{\epsilon_0\mu_0}}\mathbf{a}_\theta \qquad \text{and} \qquad \hat{\mathbf{H}} = -j\frac{E_0}{r\sqrt{\mu_0/\epsilon_0}}\sin\theta\, e^{-j\omega r\sqrt{\epsilon_0\mu_0}}\mathbf{a}_\phi$$

so that

$$\hat{\mathbf{S}} = \mathbf{E} \times \hat{\mathbf{H}}^* = +\frac{E_0^2}{r^2\sqrt{\mu_0/\epsilon_0}}\sin^2\theta\, \mathbf{a}_r$$

and (1) of Problem 4.95 clearly holds.

4.99   Refer to Problems 4.93 and 4.94. Prove, without appeal to Ohm's law, that the power consumption of the element of Fig. 4-19 is given by $I\,\Delta V$, where ΔV is the potential drop across the element.

▌   We have $\mathbf{E} = (\Delta V/L)\mathbf{a}_z$; also, Ampère's law yields, as in Problem 4.93, $\mathbf{H} = (Ir_w/2A)\mathbf{a}_\phi$. Hence

$$\mathbf{S} = \mathbf{E} \times \mathbf{H} = \frac{Ir_w}{2LA}\frac{\Delta V}{}(-\mathbf{a}_r)$$

and this gives

$$\int_\Sigma \mathbf{S} \cdot d\Sigma = -I\,\Delta V$$

4.100   For the waveguide of Problem 4.84, find the time-averaged z-component of the Poynting vector.

▌   By (1) of Problem 4.95,

$$\langle S_z \rangle = \tfrac{1}{2}\,\text{Re}\,[(\hat{\mathbf{E}} \times \hat{\mathbf{H}}^*)_z] = \tfrac{1}{2}\,\text{Re}\,[\hat{E}_x\hat{H}_y^* - \hat{E}_y\hat{H}_x^*]$$

Substitution of the phasor components from Problem 4.88 yields

$$\langle S_z \rangle = \frac{C^2\omega\mu_0\beta a^2}{2\pi^2}\sin^2\frac{\pi x}{a}$$

4.101   From the result of Problem 4.100, determine the total average power over the cross section of the waveguide.

▌ $$\langle P \rangle = \iint \langle S_z \rangle\, dA = \frac{C^2\omega\mu_0\beta a^2}{2\pi^2}\int_{x=0}^a \int_{y=0}^b \sin^2\left(\frac{\pi x}{a}\right)dy\, dx = \frac{C^2\omega\mu_0\beta a^3 b}{4\pi^2} \quad (\text{W}) \tag{1}$$

Note that (2) of Problem 4.84 can be used to eliminate either β or ω from the power expression.

4.102   The waveguide of Problem 4.84 has the dimensions $a = 22.9$ mm and $b = 10.2$ mm. The guide operates at 7 GHz, with the electric field having an amplitude of 1000 V/m. Calculate the total average power over the guide cross section.

▌   From the given amplitude of the E-field, $C\omega\mu_0 a/\pi = 1000$ V/m, and from Problem 4.84,

$$\beta = \sqrt{\omega^2\mu_0\epsilon_0 - (\pi/a)^2}$$

Thus, from (1) of Problem 4.101,

$$\langle P \rangle = \left(\frac{C\omega\mu_0 a}{\pi}\right)^2\left(\frac{ab}{4}\right)\left(\frac{\beta}{\omega\mu_0}\right) = (1000)^2\frac{ab}{4}\sqrt{\frac{\epsilon_0}{\mu_0} - \left(\frac{\pi}{\omega\mu_0 a}\right)^2}$$

Substituting numerical values, we obtain $\langle P \rangle = 54.6$ mW.

4.103 The electric and magnetic fields in free space in a spherical coordinate system are given by

$$\mathbf{E} = \frac{10}{r} \sin \theta \cos \left(\omega t - \frac{4\pi}{3} \right) \mathbf{a}_\theta \quad \text{(V/m)} \qquad \mathbf{H} = \frac{10}{120\pi r} \sin \theta \cos \left(\omega t - \frac{4\pi}{3} \right) \mathbf{a}_\phi \quad \text{(A/m)}$$

Determine the instantaneous power flow.

▮
$$\mathbf{S} = \mathbf{E} \times \mathbf{H} = \frac{100}{120\pi r^2} \sin^2 \theta \cos^2 \left(\omega t - \frac{4\pi}{3} r \right) \mathbf{a}_r \quad \text{(W/m}^2\text{)}$$

4.104 For the fields given in Problem 4.103, evaluate the time-average power leaving a sphere of radius $R = 100$ m, the center of which is at the origin.

▮ The total power through the sphere is given by

$$P = \iint\limits_{r=R} \mathbf{S} \cdot d\Sigma = \frac{100}{120\pi R^2} \int_{\theta=0}^{\pi} \int_{\phi=0}^{2\pi} \sin^2 \theta \cos^2 \left(\omega t - \frac{4\pi}{3} R \right) R^2 \sin \theta \, d\theta \, d\phi = \frac{800}{360} \cos^2 \left(\omega t - \frac{4\pi}{3} R \right) \quad \text{(W)}$$

As is required by conservation of energy, only the *phase* of the total power depends on R. Integrating over one cycle, we obtain

$$\langle P \rangle = \frac{1}{2} \left(\frac{800}{360} \right) = \frac{10}{9} \text{ W}$$

4.105 Develop the phasor formula

$$\langle P \rangle = \frac{1}{2} \text{Re} \left[\iint\limits_{\Sigma} \hat{\mathbf{S}} \cdot d\Sigma \right] \qquad (1)$$

for time-averaged, integrated power (cf. Problem 4.104).

▮ By (*1*) of Problem 4.95,

$$\langle P \rangle = \left\langle \iint\limits_{\Sigma} \mathbf{S} \cdot d\Sigma \right\rangle = \iint\limits_{\Sigma} \langle \mathbf{S} \rangle \cdot d\Sigma = \iint\limits_{\Sigma} \left(\frac{1}{2} \text{Re} \left[\hat{\mathbf{S}} \right] \right) \cdot d\Sigma = \frac{1}{2} \text{Re} \left[\iint\limits_{\Sigma} \hat{\mathbf{S}} \cdot d\Sigma \right]$$

4.106 The following phasor fields exist in free space:
$$\hat{\mathbf{E}} = 10 e^{-200x} e^{-j200x} \mathbf{a}_z \quad \text{(V/m)} \qquad \hat{\mathbf{H}} = -\tfrac{1}{2} e^{-200x} e^{-j(200x + \pi/4)} \mathbf{a}_y \quad \text{(A/m)}$$
Calculate the time-average energy flux (power) through the surface of the cube shown in Fig. 4-20.

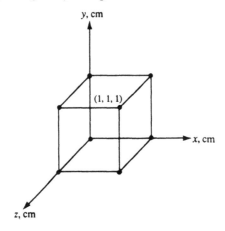

Fig. 4-20

▮
$$\hat{\mathbf{S}} = \hat{\mathbf{E}} \times \hat{\mathbf{H}}^* = 5 e^{-400x} e^{j\pi/4} \mathbf{a}_x$$

and (*1*) of Problem 4.105 gives

$$\langle P \rangle = - \left(\frac{5\sqrt{2}}{4} e^{-0} \right) (10^{-4}) + \left(\frac{5\sqrt{2}}{4} e^{-4} \right) (10^{-4}) = -0.173 \text{ mW}$$

The minus sign betokens a flow of energy *into* the cube. Thus, the stored electromagnetic energy within the cube must be increasing at the average rate of 0.173 mW. (This corresponds to the case of $\sigma = 0$ of Problem 4.91.)

4.107 In a certain region of free space, the **B**-field is given by $\mathbf{B} = B_0 \cos (\omega t - \beta x) \mathbf{a}_y$, and the corresponding **E**-field, as given by Maxwell's equations, is z-directed. Show that equal densities of electric and magnetic energy are associated with the wave.

▌ By Problem 4.50 the amplitude ratio is given by

$$\frac{E}{B} = \frac{1}{\mu_0}\frac{E}{H} = \frac{1}{\mu_0}\sqrt{\frac{\mu_0}{\epsilon_0}} = \frac{1}{\sqrt{\epsilon_0\mu_0}}$$

and so

$$\frac{w_e}{w_m} = \frac{\epsilon_0 E^2/2}{B^2/2\mu_0} = \epsilon_0\mu_0\left(\frac{E}{B}\right)^2 = 1$$

4.108 A traveling **B**-wave, $\mathbf{B} = B_m e^{j(\omega t - \beta x)}\mathbf{a}_y$, where $\omega/\beta = 1/\sqrt{\epsilon\mu}$, is impressed on the upper surface of the infinite plate shown in Fig. 4-21; the lower surface is in contact with a material of infinite permeability. Find a differential equation for the electric field within the plate, supposing it to be z-directed.

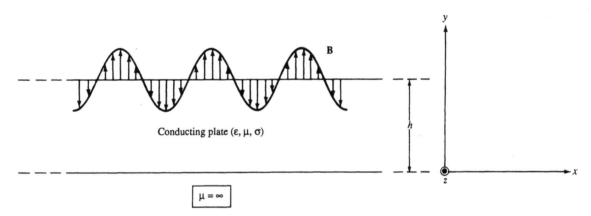

Fig. 4-21

▌ Assume: (1) all fields vary as $e^{j\omega t}$ and are independent of z; (2) $H_z \equiv 0$. Then Maxwell's equations within the plate are: curl $\mathbf{E} = -j\omega\mathbf{B}$, or

$$\frac{\partial E_z}{\partial y} = -j\omega\mu H_x \qquad \frac{\partial E_z}{\partial x} = j\omega\mu H_y \qquad (1)$$

and curl $\mathbf{H} = (\sigma + j\omega\epsilon)\mathbf{E}$, or

$$\frac{\partial H_y}{\partial x} - \frac{\partial H_x}{\partial y} = (\sigma + j\omega\epsilon)E_z \qquad (2)$$

Calculate $\partial^2 E_z/\partial y^2$ and $\partial^2 E_z/\partial x^2$ from (1), and then use (2), to find

$$\frac{\partial^2 E_z}{\partial x^2} + \frac{\partial^2 E_z}{\partial y^2} = (-\omega^2\epsilon\mu + j\omega\mu\sigma)E_z \equiv \gamma^2 E_z \qquad (3)$$

which is the required equation.

4.109 Explicitly determine E_z within the plate of Problem 4.108.

▌ It is clear that E_z must show the same x-variation as does the surface field B_y:

$$E_z(x, y) = e^{-j\beta x}Y(y)$$

Substitution in (3) of Problem 4.108 yields the ordinary differential equation

$$Y'' = j\omega\mu\sigma Y \qquad (1)$$

of which the general solution is $Y = C_1 e^{\alpha y} + C_2 e^{-\alpha y}$, where

$$\alpha \equiv \sqrt{j\omega\mu\sigma} = \sqrt{\frac{\omega\mu\sigma}{2}}(1 + j) \equiv \frac{1 + j}{\delta} \qquad (2)$$

For the present problem, the boundary condition at $y = 0$ (continuity of H_x) requires

$$Y'(0) = 0 \qquad \text{or} \qquad C_1 = C_2 = C/2 \qquad \text{or} \qquad Y = C\cosh\alpha y$$

The boundary condition at $y = h$ (continuity of B_y) then demands

$$\left.\frac{\partial E_z}{\partial x}\right|_{y=h^-} = j\omega B_y|_{y=h} \qquad \text{or} \qquad -j\beta C\cosh\alpha h = j\omega B_m$$

Consequently,

$$C = -\frac{\omega}{\beta}\frac{B_m}{\cosh\alpha h} \qquad \text{and} \qquad E_z(x, y) = -\frac{\omega}{\beta}B_m e^{-j\beta x}\frac{\cosh\alpha y}{\cosh\alpha h}$$

4.110 Show that if an electromagnetic field exists in a current-free region, the conservation of energy may be expressed by

$$\text{div } \mathbf{S} + \frac{\partial w}{\partial t} = 0 \tag{1}$$

where $\mathbf{S}$ is the Poynting vector (power density) and w is the electromagnetic energy density.

▮ Follows at once from (1) of Problem 4.91 (with $\mathbf{J} = \mathbf{0}$) and (1) of Problem 4.90; in fact, it is just the "point form" of (2) of Problem 4.91. A similar equation holds for any flow field, in the absence of sources or sinks. For example, as applied to the streamline flow of water, (1) is precisely Bernoulli's law.

4.111 In a source-free, nonconducting, magnetizable medium characterized by μ, ϵ, with a prescribed magnetization $\mathbf{M}(x, y, z, t)$, define a magnetic vector potential by $\mathbf{A} = \text{curl } \mathbf{C}$. Show that the electromagnetic field

$$\mathbf{B} = \text{curl } \mathbf{A} = \text{curl curl } \mathbf{C} \qquad \mathbf{E} = \text{curl}\left(-\frac{\partial \mathbf{C}}{\partial t}\right)$$

automatically satisfies Maxwell's equations (1), (3), and (4) of Problem 4.41.

▮ Equation (1) becomes an identity, and (3) and (4) hold because the divergence of a curl is zero ($\rho = 0$).

4.112 Obtain conditions on the vector field $\mathbf{C}$ of Problem 4.111 such that (2) of Problem 4.41 (with $\mathbf{J} = \mathbf{0}$) is also satisfied.

▮ First impose the condition div $\mathbf{C} = 0$, which is no trouble. Then, in rectangular coordinates,
$$\mathbf{B} = \mu \mathbf{H} + \mu \mathbf{M} = \text{curl curl } \mathbf{C} = -\nabla^2 \mathbf{C}$$

whence
$$\text{curl } \mathbf{H} = -\text{curl}\left(\frac{1}{\mu}\nabla^2 \mathbf{C} + \mathbf{M}\right)$$

Now,
$$\frac{\partial \mathbf{D}}{\partial t} = \epsilon \frac{\partial \mathbf{E}}{\partial t} = -\text{curl}\left(\epsilon \frac{\partial^2 \mathbf{C}}{\partial t^2}\right)$$

so that Maxwell's equation (2) will certainly be satisfied if

$$\frac{1}{\mu}\nabla^2 \mathbf{C} + \mathbf{M} = \epsilon \frac{\partial^2 \mathbf{C}}{\partial t^2} \qquad \text{or} \qquad \nabla^2 \mathbf{C} - \mu\epsilon \frac{\partial^2 \mathbf{C}}{\partial t^2} = -\mu \mathbf{M} \tag{1}$$

This is an inhomogeneous wave equation for $\mathbf{C}$.

4.113 Show that the power density corresponding to the field
$$\mathbf{E} = \mathbf{a}_x \cos(\beta z - \omega t) + \mathbf{a}_y \sin(\beta z - \omega t)$$
is constant everywhere.

▮ By Faraday's law,

$$-\frac{\partial \mathbf{B}}{\partial t} = \nabla \times \mathbf{E} = -\mathbf{a}_x \frac{\partial E_y}{\partial z} + \mathbf{a}_y \frac{\partial E_x}{\partial z} = -\mathbf{a}_x \beta \cos(\beta z - \omega t) - \mathbf{a}_y \beta \sin(\beta z - \omega t)$$

or
$$\mathbf{B} = -\frac{\beta}{\omega}[\mathbf{a}_x \sin(\beta z - \omega t) - \mathbf{a}_y \cos(\beta z - \omega t)]$$

Then
$$\mathbf{S} = \mathbf{E} \times \mathbf{H} = \frac{1}{\mu}\mathbf{E} \times \mathbf{B} = \frac{\beta}{\mu\omega}\mathbf{a}_z = \sqrt{\frac{\epsilon}{\mu}}\mathbf{a}_z \approx \frac{1}{120\pi}\mathbf{a}_z$$

where the final approximation applies to free space.

4.114 Show that, in free space, the magnetic vector potential $\mathbf{A}$ satisfies the wave equation

$$\nabla^2 \mathbf{A} = \mu_0\epsilon_0 \frac{\partial^2 \mathbf{A}}{\partial t^2}$$

▮ Refer to Problem 4.112. In free space, $\mathbf{C}$ satisfies the wave equation; hence, the space derivatives of C_x, C_y, C_z satisfy it; hence, the curl of $\mathbf{C}$, or $\mathbf{A}$, satisfies it.

4.115 Obtain an equation governing the $\mathbf{H}$-field in an unmagnetized medium characterized by (μ, σ, ϵ).

▮ From $\nabla \times \mathbf{H} = \mathbf{J} + \partial \mathbf{D}/\partial t$, $\mathbf{J} = \sigma \mathbf{E}$, and $\mathbf{D} = \epsilon \mathbf{E}$, we have

$$\nabla \times (\nabla \times \mathbf{H}) \equiv \nabla(\nabla \cdot \mathbf{H}) - \nabla^2 \mathbf{H} = \sigma \nabla \times \mathbf{E} + \epsilon \frac{\partial}{\partial t}(\nabla \times \mathbf{E}) \tag{1}$$

Substituting $\nabla \times \mathbf{E} = -\partial \mathbf{B}/\partial t$, $\nabla \cdot \mathbf{H} = 0$, and $\mathbf{B} = \mu \mathbf{H}$ in (1) yields

$$-\nabla^2 \mathbf{H} = -\mu\sigma \frac{\partial \mathbf{H}}{\partial t} - \mu\epsilon \frac{\partial^2 \mathbf{H}}{\partial t^2} \qquad \text{or} \qquad \nabla^2 \mathbf{H} - \mu\sigma \frac{\partial \mathbf{H}}{\partial t} - \mu\epsilon \frac{\partial^2 \mathbf{H}}{\partial t^2} = 0 \tag{2}$$

4.116 Rework Problem 4.115 for the **E**-field, in a source-free region.

▌ Equation (2) of Problem 4.115 also governs **E**.

4.117 Show that $\psi = F(x - ut) + G(x + ut)$ is a general solution of the one-dimensional wave equation

$$\frac{\partial^2 \psi}{\partial x^2} = \frac{1}{u^2}\frac{\partial^2 \psi}{\partial t^2}$$

▌ $\dfrac{\partial^2 \psi}{\partial x^2} = F''(x - ut) + G''(x + ut)$ and $\dfrac{\partial^2 \psi}{\partial t^2} = u^2 F''(x - ut) + u^2 G''(x + ut)$

4.118 A voltage v is applied across the plates of the circular-plate capacitor shown in Fig. 4-22(a). If the resulting current is i, show that the input power is vi (cf. Problem 4.99).

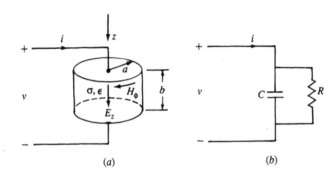

(a)   (b)   **Fig. 4-22**

▌ At radius a, from Ampère's law, $\mathbf{H} = (i/2\pi a)\mathbf{a}_\phi$, and at any radius (very nearly) $\mathbf{E} = (v/b)\mathbf{a}_z$. Thus

$$\mathbf{S} = \mathbf{E} \times \mathbf{H} = \frac{iv}{2\pi ab}(-\mathbf{a}_r)$$

and the net power *into* the capacitor is

$$P = |\mathbf{S}| \times (\text{lateral area}) = iv$$

4.119 The dielectric in the capacitor of Problem 4.118 is lossy and has a conductivity σ; thus, the capacitor may be represented by the equivalent circuit of Fig. 4-22(b). Write the v-i relationship for this circuit, and show that the result is consistent with that obtained from Maxwell's equation.

▌ From Fig. 4-22(b),

$$i = \frac{v}{R} + C\frac{dv}{dt} \tag{1}$$

Since $R = b/\pi a^2 \sigma$ and $C = \epsilon \pi a^2/b$, (1) may be written as

$$\frac{i}{\pi a^2} = \sigma\frac{v}{b} + \epsilon\frac{d}{dt}\left(\frac{v}{b}\right)$$

or $$J_{\text{total}} = \sigma E_z + \epsilon\frac{\partial E_z}{\partial t} = \sigma E_z + \frac{\partial D_z}{\partial t} = \text{conduction current} + \text{displacement current}$$

4.120 A decaying current $i = I_0 e^{-\alpha t}$ $(\alpha > 0)$ is carried by an infinitely long wire running in the y-direction. A triangular area lies in the xy plane in free space near the wire, as shown in Fig. 4-23. Find the emf induced between points A and B.

▌ By Ampère's law, $B_z = \mu_0 i/2\pi x$ at a distance x from the wire in the xy plane. Thus

$$\psi_m = \int \mathbf{B} \cdot d\mathbf{S} = \int_L^{2L}\left(\frac{\mu_0 i}{2\pi x}\right)^2 (x - L)\, dx = \frac{\mu_0 iL}{\pi}(1 - \ln 2)$$

and $$\text{emf} = -\frac{d\psi_m}{dt} = \alpha\psi_m$$

4.121 State the polarities of A and B Problem 4.120.

▌ For the current direction shown, ψ_m is out of the page but decreasing. Thus, according to Lenz's law, induced current would flow counterclockwise in the loop; i.e., A is positive and B is negative.

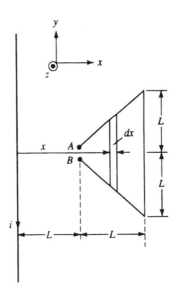

Fig. 4-23

4.122 Transform the phasor $\hat{\mathbf{B}} = (3 + 2j)e^{-j\beta x}\mathbf{a}_y$ to the time domain.

❚ Since $3 + 2j = \sqrt{13}\,e^{j\delta}$, where $\delta = \tan^{-1}(2/3)$,
$$\mathbf{B} = \text{Re}\,[\hat{\mathbf{B}}e^{j\omega t}] = \text{Re}\,[\sqrt{13}\,e^{j(\omega t - \beta x + \delta)}\,\mathbf{a}_y] = \sqrt{13}\cos(\omega t - \beta x + \delta)\,\mathbf{a}_y$$

4.123 Obtain the average magnitude of the power density associated with the wave of Problem 4.122.

❚ For a plane wave,
$$S_{\text{avg}} = \frac{1}{2}E_{\max}B_{\max} = \frac{1}{2}\left(\frac{\omega}{\beta}\right)B_{\max}^2 = \frac{13\omega}{2\beta}$$

4.124 An **E**-field in free space is given by $\mathbf{E} = 5\cos(\omega t - \beta y)\,\mathbf{a}_x$. Determine the corresponding **H**-field in complex form.

❚ For $\hat{\mathbf{E}} = 5e^{-j\beta y}\mathbf{a}_x$, Maxwell's equation curl $\hat{\mathbf{E}} = -j\omega\mu_0\hat{\mathbf{H}}$ yields
$$\hat{\mathbf{H}} = -\frac{5\beta}{\omega\mu_0}\,e^{-j\beta y}\mathbf{a}_z$$

4.125 For the field of Problem 4.124, determine the time-average Poynting vector.

❚ By Problems 4.95 and 4.124,
$$\langle \mathbf{S} \rangle = \frac{1}{2}\text{Re}\,[\mathbf{E} \times \mathbf{H}^*] = \frac{25\beta}{2\omega\mu_0}\,\mathbf{a}_y$$

4.126 The field $\mathbf{B} = B_0(1 - \alpha t)\mathbf{a}_z$ (T) links the circular loop of Fig. 4-24. Determine the voltage induced in the loop. Also find the polarity of the voltage across a small gap AB.

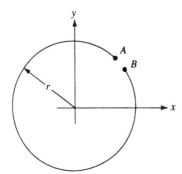

Fig. 4-24

❚
$$\psi_m = \int \mathbf{B} \cdot d\mathbf{S} = B_0\pi r^2(1 - \alpha t) \quad \text{(Wb)} \qquad \text{emf} = -\frac{d\psi_m}{dt} = \pi B_0\alpha r^2 \quad \text{(V)}$$

On the assumption that B_0 and α are positive, Lenz's law requires counterflux in the $+z$ direction. Thus, terminal B is positive.

4.127   A moving conductor of length L (Fig. 4-25) is oriented in the x-direction in a magnetic field

$$\mathbf{B} = B_1\left(2 - \frac{z^2}{2L^2}\right)\mathbf{a}_y$$

Assuming that the conductor moves with velocity $\mathbf{u} = u\mathbf{a}_z$ on conducting rails as shown, evaluate the induced emf around the loop at the instant when $z = L/2$.

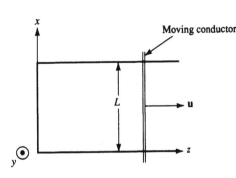

Fig. 4-25

▌ At distance z, the total flux through the loop is

$$\psi_m = B_1 L \int_0^z \left(2 - \frac{\zeta^2}{2L^2}\right) d\zeta = B_1 L\left(2z - \frac{z^3}{6L^2}\right)$$

Thus

$$\text{emf} = -\frac{d\psi_m}{dt} = -\frac{d\psi_m}{dz} u = -B_1 Lu\left(2 - \frac{z^2}{2L^2}\right)$$

or $-15B_1 Lu/8$, for $z = L/2$.

4.128   In a lossy dielectric material, of relative permittivity 12, the displacement current is 25 times greater than the conduction current at 100 MHz. Calculate the conductivity of the dielectric.

▌ By (1) of Problem 4.44,

$$\sigma = \omega\epsilon\frac{|\mathbf{J}_c|}{|\mathbf{J}_d|} = (2\pi \times 10^8)(12 \times 10^{-9}/36\pi)(1/25) = 0.00267 \text{ S/m}$$

4.129   Find the magnetic field corresponding to the electric field $\mathbf{E} = E_m \sin(\omega t - \beta z + \beta y)\mathbf{a}_x$ (V/m) in free space.

▌ From Maxwell's equation,

$$-\mu_0\frac{\partial\mathbf{H}}{\partial t} = \nabla \times \mathbf{E} = +\mathbf{a}_y\frac{\partial E_x}{\partial z} - \mathbf{a}_z\frac{\partial E_x}{\partial y} = -\beta E_m \cos(\omega t - \beta z + \beta y)(\mathbf{a}_y + \mathbf{a}_z)$$

Hence

$$\mathbf{H} = \frac{\beta E_m}{\mu_0\omega} \sin(\omega t - \beta z + \beta y)(\mathbf{a}_y + \mathbf{a}_z)   \text{(A/m)}$$

4.130   Determine the direction of power flow for the electromagnetic field of Problem 4.129.

▌ The Poynting vector $\mathbf{S} = \mathbf{E} \times \mathbf{H}$ has the direction $\mathbf{a}_x \times (\mathbf{a}_y + \mathbf{a}_z) = \mathbf{a}_z - \mathbf{a}_y$; i.e., the flow is at 45° to the $+z$- and the $-y$-directions.

4.131   In a time-varying field the current-density distribution within a conductor is nonuniform, and the current is said to *diffuse* into the conductor. Obtain a differential equation for $\mathbf{J}$ within a conducting medium characterized by (μ, σ).

▌ From Faraday's law (with $\partial\mathbf{D}/\partial t = \mathbf{0}$) and Ohm's law we have

$$\nabla \times \mathbf{E} = -\frac{\partial\mathbf{B}}{\partial t} = -\mu\frac{\partial\mathbf{H}}{\partial t}   \text{or}   \nabla \times \mathbf{J} = -\mu\sigma\frac{\partial\mathbf{H}}{\partial t}$$

Taking the curl of both sides of the second equation yields

$$\nabla \times (\nabla \times \mathbf{J}) \equiv \nabla(\nabla \cdot \mathbf{J}) - \nabla^2\mathbf{J} = -\mu\sigma\frac{\partial}{\partial t}(\nabla \times \mathbf{H})$$

Since $\nabla \cdot \mathbf{J} = 0$ and $\nabla \times \mathbf{H} = \mathbf{J}$, we finally obtain the required *diffusion equation*

$$\nabla^2\mathbf{J} = \mu\sigma\frac{\partial\mathbf{J}}{\partial t} \tag{1}$$

Note that (1) corresponds to the case $\epsilon \to 0$ in (2) of Problem 4.115.

4.132 Solve the one-dimensional diffusion equation for a harmonic current density. Assume that the current flows only in the z direction and varies only with y.

❚ For $J = J_z(y)e^{j\omega t}\mathbf{a}_z$, (1) of Problem 4.131 gives $J_z'' = j\omega\mu\sigma J_z$, which is precisely the equation studied in Problem 4.109. The general solution may be written in the equivalent forms:
$$J_z = C_1 e^{\alpha y} + C_2 e^{-\alpha y} \qquad \text{and} \qquad J_z = D_1 \cosh \alpha y + D_2 \sinh \alpha y \qquad (1)$$

4.133 An infinite plate, $-b < y < +b$, is immersed in a magnetic field $\mathbf{H} = H_z(y)e^{j\omega t}\mathbf{a}_z$ (Fig. 4-26). Determine $\mathbf{H}$ within the plate.

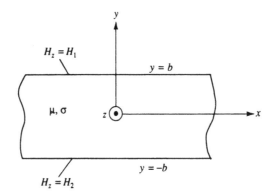

Fig. 4-26

❚ Since $\mathbf{H}$ satisfies the diffusion equation in y and t, Problem 4.132 gives
$$H_z = C_1 e^{\alpha y} + C_2 e^{-\alpha y}$$
The two boundary conditions yield
$$H_1 = C_1 e^{\alpha b} + C_2 e^{-\alpha b} \qquad H_2 = C_1 e^{-\alpha b} + C_2 e^{\alpha b}$$
whence
$$C_1 = \frac{H_1 e^{\alpha b} - H_2 e^{-\alpha b}}{e^{2\alpha b} - e^{-2\alpha b}} \qquad C_2 = \frac{H_2 e^{\alpha b} - H_1 e^{-\alpha b}}{e^{2\alpha b} - e^{-2\alpha b}}$$

4.134 Solve Problem 4.133 in the special case $H_1 = H_2 = H_0$.

❚ For symmetry about $y = 0$, use the cosh solution:
$$H_z = H_0 \frac{\cosh \alpha y}{\cosh \alpha b}$$

4.135 Determine the $\mathbf{E}$-field within the conducting plate of Problem 4.134.

❚ By Maxwell's equation,
$$\mathbf{E} = \frac{\mathbf{J}}{\sigma} = \frac{1}{\sigma}\operatorname{curl}\mathbf{H} = -\frac{1}{\sigma}\frac{dH_z}{dy}\mathbf{a}_x = \frac{\alpha H_0}{\sigma}\frac{\sinh \alpha y}{\cosh \alpha b}\mathbf{a}_x$$

4.136 The quantity δ defined in (2) of Problem 4.109 is known as the *skin depth*. Verify that δ has the dimension of length.

❚ Let $[X]$ denote the physical dimensions of X. Then, since $\delta^{-2} = \omega\mu\sigma/2$,
$$[\delta]^{-2} = [\omega][\mu][\sigma] = \mathrm{T}^{-1} \times \mathrm{M \cdot L \cdot C^{-2}} \times \mathrm{M^{-1} \cdot L^{-3} \cdot T \cdot C^2} = \mathrm{L^{-2}}$$
or $[\delta] = \mathrm{L}$.

4.137 For a good conductor, show that the skin depth, δ, is much smaller than the wavelength, λ, of the impinging electromagnetic wave.

❚ This follows at once from the criterion for a good conductor (Problem 4.45) and the definition of wavelength $[\lambda(\omega/2\pi) = u = 1/\sqrt{\mu\epsilon}]$.

4.138 Suppose the half-space $x > 0$ occupied by a metal (properties μ, σ). An $\mathbf{E}$-wave, $\mathbf{E} = E_0 e^{j(\omega t - \beta x)}\mathbf{a}_z$, hits the surface $x = 0$ and diffuses into the metal (see Fig. 4-27). Determine $\mathbf{E}$ within the metal.

❚ Except for the boundary conditions, this is the same as Problem 4.133 (with the coordinates x, y, z renamed z, $-x$, $-y$). Thus, $E_z = C_1 e^{-\alpha x} + C_2 e^{\alpha x}$. Because $\operatorname{Re}[\alpha] = 1/\delta > 0$, the boundary condition $E_z(+\infty) = 0$ requires

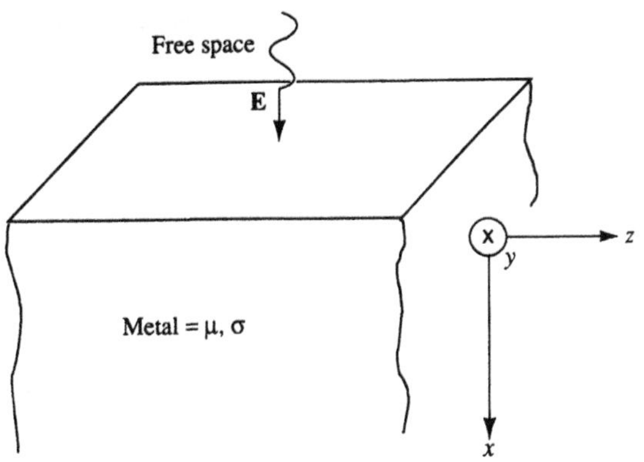

Fig. 4-27

$C_2 = 0$. Now, Maxwell's equation curl $\mathbf{E} = -j\omega\mathbf{B}$ reduces to

$$\frac{\partial E_z}{\partial x} = -\alpha C_1 e^{-\alpha x} = j\omega\mu H_y \qquad (x>0) \tag{1}$$

$$\frac{\partial E_z}{\partial x} = -j\beta E_0 e^{-j\beta x} = j\omega\mu_0 H_y \qquad (x<0) \tag{2}*$$

Continuity of H_y at $x = 0$ then gives as the second boundary condition:

$$C_1 = j\frac{\beta}{\alpha}\frac{\mu}{\mu_0}E_0 = \frac{\beta\delta\mu_r E_0}{\sqrt{2}}e^{j\pi/4} \equiv E_m e^{j\pi/4} \tag{3}$$

Thus, in the time domain,

$$\mathbf{E} = E_m e^{-x/\delta}e^{j(\omega t - x/\delta + \pi/4)}\mathbf{a}_z \tag{4}$$

4.139 From the form of (4) of Problem 4.138, infer the physical significance of the parameter δ.

❚ It is seen that at depth $x = \delta$ in the metal, $|\mathbf{E}|$ has been reduced to $e^{-1} \approx 37$ percent of its surface value—hence the name *skin depth*. Moreover, the propagation speed is given by

$$u = \omega\delta = \sqrt{\frac{2\omega}{\mu\sigma}}$$

which shows that (unlike free space) a conductor is a dispersive medium, with each frequency traveling at its own velocity.

4.140 Determine the $\mathbf{H}$-field within the metal of Problem 4.138.

❚ Maxwell's equation curl $\mathbf{E} = -\partial\mathbf{B}/\partial t$ gives the magnetic field as

$$H_y = \frac{\alpha}{\omega\mu}E_z = \frac{1+j}{\delta\omega\mu}E_z = \frac{\sqrt{2}\,e^{j\pi/4}}{\delta\omega\mu}E_z$$

or

$$\mathbf{H} = \frac{\sqrt{2}\,E_m}{\delta\omega\mu}e^{-x/\delta}e^{j(\omega t - x/\delta + \pi/2)}\mathbf{a}_y$$

4.141 Define the *internal impedance* Z per unit transverse length (y-direction) for the metal block of Problem 4.138 by $Z = E_m/I_z$, where E_m is the amplitude of the $\mathbf{E}$-field at the surface of the block ($x = 0$) and I_z is the total current per y-unit. Obtain an expression for the internal impedance in terms of σ and δ, and show that the internal reactance and the resistance of the block are equal at any frequency.

❚ Let J_m be the value of the current density at $x = 0$. Then, as governed by the diffusion equation,
$$J_z(x) = J_m e^{-(1+j)x/\delta} = \sigma E_m e^{-(1+j)x/\delta}$$

Then the total current per y-unit is

$$I_z = \int_0^\infty (J_z)(1)\,dx = \sigma E_m \int_0^\infty e^{-(1+j)x/\delta}\,dx = \frac{\sigma E_m \delta}{1+j} \tag{1}$$

so that

$$Z = \frac{E_m}{I_z} = \frac{1}{\sigma\delta}(1+j) \equiv R_s + j\omega L_i \tag{2}$$

where $R_s = 1/\sigma\delta = \omega L_i$.

* Actually, there will also be a reflected wave in $x < 0$, so that (2) is not quite right. But it's close enough.

4.142 Obtain an expression for the time-average power loss per unit surface area of metal in Problem 4.138, in terms of the current amplitude $|I_z|$ and the resistance R_s defined in Problem 4.141.

▮ At $x = 0$, we have from Problems 4.138 and 4.140,

$$\hat{\mathbf{E}} = E_m e^{j\pi/4}\, \mathbf{a}_z \qquad \hat{\mathbf{H}} = \frac{\sqrt{2}\, E_m}{\delta\omega\mu} e^{j\pi/2}\mathbf{a}_y$$

so that

$$S_{av} = \left|\frac{1}{2}\operatorname{Re}[\hat{\mathbf{E}} \times \hat{\mathbf{H}}^*]\right| = \frac{1}{2}\frac{\sqrt{2}\, E_m^2}{\delta\omega\mu}\operatorname{Re}[e^{-j\pi/4}] = \frac{E_m^2}{2\delta\omega\mu}$$

However, from Problems 4.141 and 4.109,

$$E_m^2 = \frac{2}{\sigma^2\delta^2}|I_z|^2 \qquad \text{and} \qquad \frac{1}{\omega\mu} = \frac{\sigma\delta^2}{2}$$

which when substituted yield

$$S_{av} = \frac{1}{2\sigma\delta}|I_z|^2 = \frac{1}{2}R_s\,|I_z|^2 \tag{1}$$

Note that R_s is in Ω and I_z in A/m, giving S_{avg} the proper unit W/m^2.

4.143 Current flows at angular frequency ω in a wire of circular cross section, in the axial (z-) direction. If the wire radius is a, obtain an expression for the current density Δ_z within the wire, given $\Delta_z = \Delta_0$ at $r = a$. (The reason for this queer notation will become evident below.)

▮ Since there is no variation with z or ϕ, the diffusion equation for the current density may be written in cylindrical coordinates as

$$\frac{d^2\Delta_z}{dr^2} + \frac{1}{r}\frac{d\Delta_z}{dr} + \lambda^2\Delta_z = 0 \tag{1}$$

with

$$\lambda^2 \equiv -j\omega\mu\sigma \equiv \frac{2}{j\delta^2} \qquad \text{or} \qquad \lambda = j^{-1/2}\sqrt{2}/\delta \tag{2}$$

From Problems 2.232 and 2.238 (with λ now complex, instead of pure real or pure imaginary), the solution of (1) that is regular at $r = 0$ and that satisfies the boundary condition at $r = a$ is

$$\Delta_z = \Delta_0 \frac{J_0(\lambda r)}{J_0(\lambda a)} \tag{3}$$

Here, $J_0(\ )$ is the Bessel function of order zero.

4.144 With reference to Problem 4.143, express $|\Delta_z|$ in terms of the auxiliary Bessel functions

$$\operatorname{ber}(v) \equiv \operatorname{Re}[J_0(j^{-1/2}v)] \qquad \operatorname{bei}(v) \equiv \operatorname{Im}[J_0(j^{-1/2}v)]$$

▮

$$|\Delta_z| = |\Delta_0| \frac{|J_0(j^{-1/2}r\sqrt{2}/\delta)|}{|J_0(j^{-1/2}a\sqrt{2}/\delta)|} = |\Delta_0|\left[\frac{\operatorname{ber}^2(r\sqrt{2}/\delta) + \operatorname{bei}^2(r\sqrt{2}/\delta)}{\operatorname{ber}^2(a\sqrt{2}/\delta) + \operatorname{bei}^2(a\sqrt{2}/\delta)}\right]^{1/2}$$

4.145 Use the recurrence formula

$$\frac{d}{dw}[wJ_1(w)] = wJ_0(w)$$

to calculate the total current I_z in Problem 4.143.

▮

$$I_z = \int_0^a \Delta_z 2\pi r\, dr = \frac{2\pi\Delta_0}{\lambda^2 J_0(\lambda a)}\int_0^{\lambda a} J_0(w)w\, dw = \frac{2\pi a\Delta_0}{\lambda}\frac{J_1(\lambda a)}{J_0(\lambda a)}$$

4.146 Calculate the internal impedance per unit length,

$$Z = \frac{E_z|_{r=a}}{I_z}$$

(cf. Problem 4.141) for the wire of Problems 4.143 and 4.145.

▮ Since $E_z = \Delta_z/\sigma$,

$$Z_1 = \frac{\Delta_0/\sigma}{I_z} = \frac{\lambda}{2\pi a\sigma}\frac{J_0(\lambda a)}{J_1(\lambda a)}$$

5.1 Obtain a relationship between the space and time variations of an **H**-field in an unmagnetized medium characterized by (μ, σ, ϵ).

▮ See Problem 4.115.

5.2 Repeat Problem 5.1 for the **E**-field, assuming the absence of free charge.

▮ See Problem 4.116. The common equation,

$$\nabla^2 \mathbf{F} - \mu\sigma \frac{\partial \mathbf{F}}{\partial t} - \mu\epsilon \frac{\partial^2 \mathbf{F}}{\partial t^2} = 0 \tag{1}$$

is known as the *Helmholtz equation* or *dissipative wave equation*.

5.3 Write (*1*) of Problem 5.2 for the special case when the **F**-field varies sinusoidally in time at an angular frequency ω.

▮ In this case, $\partial/\partial t = j\omega$, giving

$$\nabla^2 \mathbf{F} = \mu\sigma(j\omega\mathbf{F}) + \mu\epsilon(j\omega)^2 \mathbf{F} = j\omega\mu(\sigma + j\omega\epsilon)\mathbf{F} \equiv \gamma^2 \mathbf{F} \tag{1}$$

5.4 In (*1*) of Problem 5.3, the complex number γ is known as the *propagation constant*. Evaluate γ for a lossless medium.

▮ For a lossless medium $\sigma = 0$, so that $\gamma = \sqrt{j\omega\mu(j\omega\epsilon)} = j\omega\sqrt{\mu\epsilon}$, which is pure imaginary.

5.5 Power flows in the z-direction in a region characterized by (μ, σ, ϵ). The corresponding **E**- and **H**-fields may be represented by vectors which always lie in an xy plane and are constant over each such plane. The paired fields constitute a *plane wave*. Given $\mathbf{E} = E_x(z)e^{j\omega t}\mathbf{a}_x$, find the corresponding **H**-field. Show that **E** and **H** are mutually orthogonal and that power flows in the z-direction.

▮ With

$$\frac{\partial E_x}{\partial x} = \frac{\partial E_x}{\partial y} = 0$$

Faraday's law becomes

$$\nabla \times \mathbf{E} = E_x'(z)\mathbf{a}_y = -j\omega\mu\mathbf{H}, \qquad \text{or} \qquad \mathbf{H} = \frac{j}{\omega\mu} E_x'(z)\mathbf{a}_y$$

which shows that **H** depends only on z and is orthogonal to **E**. Also, the direction of $\mathbf{S} = \mathbf{E} \times \mathbf{H}$ is $\mathbf{a}_z$.

5.6 A plane wave exists in a lossless medium. For an **E**-field of the form $\mathbf{E} = E_m e^{j(\omega t - \beta z)}\mathbf{a}_x$, relate β, ω, μ, and ϵ.

▮ Set $\sigma = 0$ and substitute the expression for E_x in (*1*) of Problem 5.2:

$$-\beta^2 E_x - \mu\epsilon(-\omega^2)E_x = 0 \qquad \text{or} \qquad \beta^2 = \mu\epsilon\omega^2$$

which is called the *dispersion relation*.

5.7 The *wavelength* λ of the **E**-field of Problem 5.6 is defined as the minimal distance, at any instant, between points of equal **E**. Obtain an expression for λ.

▮ If $e^{j(\omega t_0 - \beta z_1)} = e^{j(\omega t_0 - \beta z_2)}$ and $|z_1 - z_2| = \lambda$ is minimal, then $\beta |z_1 - z_2| = \beta\lambda = 2\pi$, or $\lambda = 2\pi/\beta$.

5.8 Find the (phase) velocity of the **E**-wave of Problem 5.6.

▮ A surface of constant phase is given by

$$\omega t - \beta z = \text{const.} \quad \text{or} \quad \frac{d}{dt}(\omega t - \beta z) = 0 \quad \text{or} \quad \omega - \beta \frac{dz}{dt} = 0 \quad \text{or} \quad \frac{dz}{dt} \equiv u = \frac{\omega}{\beta}$$

5.9 Determine the phase velocity of an electromagnetic wave in free space.

▮ From Problems 5.8 and 5.6,

$$u = \frac{\omega}{\beta} = \frac{1}{\sqrt{\mu_0\epsilon_0}} = \frac{1}{\sqrt{4\pi \times 10^{-7} \times (10^{-9}/36\pi)}} = 3 \times 10^8 \text{ m/s}$$

5.10 Evaluate the ratio $|\mathbf{E}|/|\mathbf{H}|$ for a plane wave in free space.

▌ By Problem 5.5, assuming the z-variation $e^{-j\beta z}$,

$$\frac{|\mathbf{E}|}{|\mathbf{H}|} = \frac{|E_x|}{(1/\omega\mu_0)(\beta\,|E_x|)} = \omega\mu_0/\beta \equiv \eta_0$$

But, by Problem 5.6, $\omega/\beta = 1/\sqrt{\mu_0\epsilon_0}$. Hence,

$$\eta_0 = \frac{|\mathbf{E}|}{|\mathbf{H}|} = \sqrt{\frac{\mu_0}{\epsilon_0}} = \sqrt{\frac{4\pi \times 10^{-7}}{10^{-9}/36\pi}} = 120\pi\ \Omega$$

5.11 Display the general forms of $E_x(z)$ and $H_y(z)$ in Problem 5.5.

▌ The fields must satisfy (1) of Problem 5.3; hence,

$$E_x = E_m^+ e^{-\gamma z} + E_m^- e^{\gamma z} = E_m^+ e^{-\alpha z} e^{-j\beta z} + E_m^- e^{\alpha z} e^{j\beta z} \tag{1}$$

$$H_y = H_m^+ e^{-\gamma z} + H_m^- e^{\gamma z} = H_m^+ e^{-\alpha z} e^{-j\beta z} + H_m^- e^{\alpha z} e^{j\beta z} \tag{2}$$

for undetermined constants E_m^+, E_m^-, H_m^+, H_m^-, and where $\alpha > 0$ and $\beta > 0$ are respectively the real and imaginary parts of γ.

5.12 Relate the integration constants of Problem 5.11 to each other.

▌ From Problem 5.5,

$$\frac{dE_x}{dz} = -j\omega\mu H_y \tag{1}$$

Substituting (1) and (2) of Problem 5.11 in (1) yields

$$-\gamma E_m^+ e^{-\gamma z} + \gamma E_m^- e^{\gamma z} = -j\omega\mu(H_m^+ e^{-\gamma z} + H_m^- e^{\gamma z})$$

whence $\qquad\qquad \gamma E_m^+ = j\omega\mu H_m^+ \qquad$ and $\qquad \gamma E_m^- = -j\omega\mu H_m^- \tag{2}$

5.13 Let a phasor plane wave $\hat{\mathbf{E}}^+$, $\hat{\mathbf{H}}^+$ traverse a lossy medium in the $+z$-direction. Evaluate the *intrinsic impedance*, $\eta \equiv |\hat{\mathbf{E}}^+|/|\hat{\mathbf{H}}^+|$, of the medium in terms of its physical properties.

▌ By (2) of Problem 5.12,

$$\eta = \frac{E_m^+ e^{-\gamma z}}{H_m^+ e^{-\gamma z}} = \frac{j\omega\mu}{\gamma} = \sqrt{\frac{j\omega\mu}{\sigma + j\omega\epsilon}} \equiv |\eta|\, e^{j\theta_\eta} \tag{1}$$

For a highly conductive medium, $\theta_\eta = \pi/4$ (see Problem 5.21).

5.14 Express the forward-traveling fields of Problem 5.13 in the time domain.

▌ Writing $E_m^+ = |E_m^+|\, e^{j\theta^+}$,

$$\mathbf{E}^+ = \mathrm{Re}\,[|E_m^+|\, e^{j\theta^+} e^{-\alpha z} e^{-j\beta z} \mathbf{a}_x e^{j\omega t}] = |E_m^+|\, e^{-\alpha z} \cos{(\omega t - \beta z + \theta^+)}\, \mathbf{a}_x \tag{1}$$

and, since $H_m^+ = E_m^+/\eta$,

$$\mathbf{H}^+ = \frac{|\mathbf{E}_m^+|}{|\eta|}\, e^{-\alpha z} \cos{(\omega t - \beta z + \theta^+ - \theta_\eta)}\, \mathbf{a}_y \tag{2}$$

It is clear that—by proper choice of rectangular coordinates—any planar solution of the Helmholtz equation can be reduced to the canonical form (1) and (2).

5.15 Graph the fields (1) and (2) of Problem 5.14 at $t = 0$; assume free space.

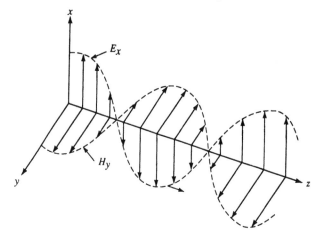

Fig. 5-1

▌ For free space $\theta_\eta = 0$ and the two waves are in phase; see Fig. 5-1.

5.16 Determine the wavelength of an electromagnetic wave traveling in free space at 30 GHz.

▌ In free space $\mu_0 = 4\pi \times 10^{-7}$ H/m and $\epsilon_0 \approx 10^{-9}/36\pi$ F/m, so that
$$u_0 = \frac{1}{\sqrt{\epsilon_0/\mu_0}} \approx 3 \times 10^8 \text{ m/s}$$
and
$$\lambda_0 = \frac{u_0}{f} = \frac{3 \times 10^8 \text{ m/s}}{30 \times 10^9 \text{ s}^{-1}} = 0.01 \text{ m}$$

5.17 Express the phase velocity u, the intrinsic impedance η, the wavelength λ, and the *wave number* or *phase constant* $\beta \equiv 2\pi/\lambda$ for a wave traveling in a lossless material medium characterized by (μ_r, ϵ_r), in terms of the free-space values.

▌ $$u = \frac{u_0}{\sqrt{\mu_r\epsilon_r}} \qquad \eta = \eta_0\sqrt{\frac{\mu_r}{\epsilon_r}} \qquad \lambda = \frac{\lambda_0}{\sqrt{\mu_r\epsilon_r}} \qquad \beta = \beta_0\sqrt{\mu_r\epsilon_r}$$

5.18 A plane wave of frequency 1 GHz is traveling in a large block of Teflon ($\epsilon_r \approx 2.1$, $\mu_r \approx 1$, and $\sigma \approx 0$). Determine u, η, λ, and β.

▌ From the results of Problem 5.17 and the values $u_0 = 3 \times 10^8$ m/s, $\eta_0 = 120\pi = 377$ Ω, $\lambda_0 = 0.3$ m, we find:
$$u = 2.07 \times 10^8 \text{ m/s} \qquad \eta = 260 \ \Omega \qquad \lambda = 0.207 \text{ m} \qquad \beta = 2\pi/\lambda = 30.4 \text{ rad/m}$$

5.19 Give expressions for the backward-traveling counterparts of the fields of Problem 5.14.

▌ In (1) and (2) of Problem 5.14, replace z by $-z$; E_m^+ by E_m^-; and (to reverse the Poynting vector) $\mathbf{a}_x$, $\mathbf{a}_y$ by $\mathbf{a}_x$, $-\mathbf{a}_y$.
$$\mathbf{E}^- = |E_m^-| e^{\alpha z} \cos(\omega t + \beta z + \theta^-) \mathbf{a}_x \tag{1}$$
$$\mathbf{H}^- = -\frac{|E_m^-|}{|\eta|} e^{\alpha z} \cos(\omega t + \beta z + \theta^- - \theta_\eta) \mathbf{a}_y \tag{2}$$
Note that replacing θ^- by $\theta^- + \pi$ takes care of the other possibility for reversing $\mathbf{S}$: $(\mathbf{a}_x, \mathbf{a}_y) \rightarrow (-\mathbf{a}_x, \mathbf{a}_y)$.

5.20 Compute the phase velocity, attenuation constant, phase constant, and intrinsic impedance for a 1-MHz plane wave in a large block of copper ($\sigma = 5.8 \times 10^7$, $\epsilon_r \approx 1$, and $\mu_r \approx 1$).

▌ The propagation constant is
$$\gamma = \sqrt{j\omega\mu(\sigma + j\omega\epsilon)} = \sqrt{j(2\pi \times 10^6)(4\pi \times 10^{-7})[5.8 \times 10^7 + j(2\pi \times 10^6)(10^{-9}/36\pi)]}$$
$$= (2.14 \times 10^4)\underline{/45°} = (1.513 \times 10^4) + j(1.513 \times 10^4) \text{ m}^{-1}$$
from which the attenuation and phase constants are $\alpha = 1.513 \times 10^4$ Np/m and $\beta = 1.513 \times 10^4$ rad/m. The wavelength and wave speed in copper at this frequency become
$$\lambda = \frac{2\pi}{\beta} = 415.3 \ \mu\text{m} \qquad u = \frac{\omega}{\beta} = 415.3 \text{ m/s} \quad (= \lambda f)$$
The intrinsic impedance becomes
$$\eta = \sqrt{\frac{j\omega\mu}{\sigma + j\omega\epsilon}} = \sqrt{\frac{j(2\pi \times 10^6)(4\pi \times 10^{-7})}{5.8 \times 10^7 + j(2\pi \times 10^6)(10^{-9}/36\pi)}} = 3.689 \times 10^{-4}\underline{/45°} \ \Omega$$

5.21 Prove that $\theta_\eta = \pi/4$ for a good conductor.

▌ For such a medium, $\omega\epsilon \ll \sigma$; hence
$$\eta = \sqrt{\frac{j\omega\mu/\sigma}{1 + j\omega\epsilon/\sigma}} \approx \sqrt{j\omega\mu/\sigma} = \sqrt{\omega\mu/\sigma} \, e^{j\pi/4}$$

5.22 Verify that the pair of field vectors
$$\mathbf{E}_1 = E_{10}\cos(\omega t - \beta z) \mathbf{a}_x \qquad (\beta = \omega\sqrt{\mu\epsilon})$$
$$\mathbf{H}_1 = \frac{E_{10}}{\eta}\cos(\omega t - \beta z) \mathbf{a}_y \qquad (\eta = \sqrt{\mu/\epsilon})$$
satisfies the time-domain Maxwell's equations for a lossless, linear, homogeneous, isotropic, charge-free medium characterized by μ, ϵ.

▌ In the paradigm, (1) and (2) of Problem 5.14, choose E_m^+ real and let $\sigma \rightarrow 0$ [which makes $\alpha \rightarrow 0$, $\beta \rightarrow \omega\sqrt{\mu\epsilon}$, and $\eta \rightarrow \sqrt{\mu/\epsilon}$ (a real number)].

5.23 The amplitude and phase of a plane wave are both constant over any plane that is normal to the direction of propagation. In a nonplanar wave surfaces of constant amplitude and surfaces of constant phase are not the same. Show that the following nonplanar wave satisfies the phasor form of Maxwell's equations (current-free region):

$$\hat{\mathbf{E}} = E_0 e^{-\alpha y} e^{-j\beta z} \mathbf{a}_x$$

$$\hat{\mathbf{H}} = \frac{\beta}{\omega\mu} E_0 e^{-\alpha y} e^{-j\beta z} \mathbf{a}_y + j\frac{\alpha}{\omega\mu} E_0 e^{-\alpha y} e^{-j\beta z} \mathbf{a}_z$$

where $\alpha > 0$ (otherwise, a planar wave) and $\alpha^2 - \beta^2 = -\omega^2\mu\epsilon$.

▌ By Problem 4.87, only the two curl equations need be considered. For the given fields, curl $\hat{\mathbf{E}} = -j\omega\mu\hat{\mathbf{H}}$ becomes

$$-j\beta E_0 e^{-\alpha y} e^{-j\beta z}\mathbf{a}_y + \alpha E_0 e^{-\alpha y} e^{-j\beta z}\mathbf{a}_z = -j\omega\mu\left(\frac{\beta}{\mu\omega}\mathbf{a}_y + j\frac{\alpha}{\mu\omega}\mathbf{a}_z\right)E_0 e^{-\alpha y} e^{-j\beta z}$$

which is an identity. Also, curl $\hat{\mathbf{H}} = j\omega\epsilon\hat{\mathbf{E}}$ becomes

$$j\frac{\beta^2}{\mu\omega}E_0 e^{-\alpha y} e^{-j\beta z}\mathbf{a}_x - j\frac{\alpha^2}{\mu\omega}E_0 e^{-\alpha y} e^{-j\beta z}\mathbf{a}_x = j\omega\epsilon E_0 e^{-\alpha y} e^{-j\beta z}\mathbf{a}_x$$

which is valid if $\alpha^2 - \beta^2 = -\omega^2\mu\epsilon$.

5.24 Refer to Problems 5.14 and 5.19. Determine the time-average Poynting vector for the complete wave $\mathbf{E} = \mathbf{E}^+ + \mathbf{E}^-$, $\mathbf{H} = \mathbf{H}^+ + \mathbf{H}^-$.

▌ Use (1) of Problem 4.95, with

$$\hat{\mathbf{E}} = \hat{\mathbf{E}}^+ + \hat{\mathbf{E}}^- = \{|E_m^+|\, e^{-\gamma z + j\theta^+} + |E_m^-|\, e^{\gamma z + j\theta^-}\}\mathbf{a}_x$$

$$\hat{\mathbf{H}} = \hat{\mathbf{H}}^+ + \hat{\mathbf{H}}^- = \left\{\frac{|E_m^+|}{|\eta|}e^{-\gamma z + j(\theta^+ - \theta_\eta)} - \frac{|E_m^-|}{|\eta|}e^{\gamma z + j(\theta^- - \theta_\eta)}\right\}\mathbf{a}_y$$

Thus,

$$\hat{\mathbf{S}} = \hat{\mathbf{E}}\times\hat{\mathbf{H}}^* = \left\{\frac{|E_m^+|^2}{|\eta|}e^{-2\alpha z}e^{j\theta_\eta} - \frac{|E_m^+ E_m^-|}{|\eta|}e^{-j(2\beta z - \theta^+ + \theta^- - \theta_\eta)} + \frac{|E_m^+ E_m^-|}{|\eta|}e^{j(2\beta z - \theta^+ + \theta^- + \theta_\eta)} - \frac{|E_m^-|^2}{|\eta|}e^{2\alpha z}e^{j\theta_\eta}\right\}\mathbf{a}_z$$

and

$$\langle\mathbf{S}\rangle = \frac{1}{2}\operatorname{Re}[\hat{\mathbf{S}}]$$

$$= \frac{|E_m^+|^2}{2|\eta|}e^{-2\alpha z}\cos\theta_\eta\,\mathbf{a}_z - \frac{|E_m^+|\,|E_m^-|}{|\eta|}\sin(2\beta z - \theta^+ + \theta^-)\sin\theta_\eta\,\mathbf{a}_z - \frac{|E_m^-|^2}{2|\eta|}e^{2\alpha z}\cos\theta_\eta\,\mathbf{a}_z$$

$$\equiv \langle\mathbf{S}^+\rangle + \langle\mathbf{S}^\pm\rangle + \langle\mathbf{S}^-\rangle$$

with an obvious notation for the Poynting vectors of the forward wave by itself, the cross-coupled forward and backward waves, and the backward wave by itself.

5.25 Specialize Problem 5.24 to a lossless medium. Give a physical interpretation of your result.

▌ When $\sigma = \theta_\eta = \alpha = 0$,

$$\langle\mathbf{S}\rangle = \langle\mathbf{S}^+\rangle + \langle\mathbf{S}^-\rangle = \frac{|E_m^+|^2 - |E_m^-|^2}{2|\eta|}\mathbf{a}_z$$

Note that $\langle\mathbf{S}\rangle$ is independent of z: the net average power traveling in the $+z$ direction is conserved for lossless media, as it must be.

5.26 Define a *complex permittivity* $\hat{\epsilon}$ such that Ampère's law in a lossy dielectric becomes curl $\hat{\mathbf{H}} = j\omega\hat{\epsilon}\hat{\mathbf{E}}$ (formally the same as for a lossless dielectric).

▌ Because curl $\hat{\mathbf{H}} = (\sigma + j\omega\epsilon)\hat{\mathbf{E}}$, we must have

$$j\omega\hat{\epsilon} = \sigma + j\omega \qquad \text{or} \qquad \hat{\epsilon} = \epsilon\left(1 - j\frac{\sigma}{\omega\epsilon}\right)$$

5.27 A 100-V/m plane wave of frequency 300 MHz travels in the $+z$-direction in an infinite, lossless medium having $\epsilon_r = 9$, $\mu_r = 1$, and $\sigma = 0$. Write complete time-domain expressions for the field vectors.

▌ By Problem 5.17,

$$\eta = (120\pi\ \Omega)\sqrt{1/9} = 40\pi\ \Omega$$

$$\beta = \left(\frac{2\pi\times 300\times 10^6}{3\times 10^8}\ \text{rad/m}\right)\sqrt{(1)(9)} = 6\pi\ \text{rad/m}$$

Then, by Problem 5.14, with $\theta^+ = \theta_\eta = \alpha = 0$,

$$\mathbf{E} = 100\cos(600\pi\times 10^6 t - 6\pi z)\,\mathbf{a}_x \quad\text{(V/m)} \qquad \mathbf{H} = \frac{5}{2\pi}\cos(600\pi\times 10^6 t - 6\pi z)\,\mathbf{a}_y \quad\text{(A/m)}$$

5.28   Calculate the average power density in the wave of Problem 5.27.

▮ By Problem 5.25,

$$\langle \mathbf{S}^+ \rangle = \frac{(100)^2}{2(40\pi)} = \frac{125}{\pi} \; \text{W/m}^2$$

5.29   Assuming that the material of Problem 5.27 has $\sigma = 10$ S/m, calculate its intrinsic impedance and propagation constant.

▮

$$\eta = \sqrt{\frac{j\omega\mu}{\sigma + j\omega\epsilon}} = \sqrt{236.87 \, \underline{/89.14^\circ}} = 15.4 \, \underline{/44.57^\circ} \quad \Omega$$
$$\gamma = \sqrt{j\omega\mu(\sigma + j\omega\epsilon)} = \sqrt{23\,689.72 \, \underline{/90.86^\circ}} = 153.91 \, \underline{/45.43^\circ} = 108.01 + j109.65 \quad \text{m}^{-1}$$

5.30   Determine the average power density for the wave of Problems 5.27 and 5.29.

▮ From Problems 5.24 and 5.29:

$$\langle \mathbf{S}^+ \rangle = \frac{(100)^2}{2(15.4)} e^{-2(108.01)z} \cos 44.57^\circ \, \mathbf{a}_z = 231.3 e^{-216z} \mathbf{a}_z \quad (\text{W/m}^2)$$

5.31   A plane wave is traveling in the x-direction in a lossless medium, with a 100-V/m electric field in the z-direction. Given that the wavelength is 25 cm and the velocity of propagation is 2×10^8 m/s, determine the frequency of the wave and the relative permittivity of the medium. Assume that $\mu_r = 1$.

▮

$$f = \frac{u}{\lambda} = \frac{2 \times 10^8}{25 \times 10^{-2}} = 800 \, \text{MHz}$$
$$u = \frac{1}{\sqrt{\mu\epsilon}} = \frac{3 \times 10^8}{\sqrt{\epsilon_r}} \quad \text{or} \quad \epsilon_r = \left(\frac{3 \times 10^8}{u}\right)^2 = \left(\frac{3}{2}\right)^2 = \frac{9}{4}$$

5.32   Express the electric and magnetic field vectors of Problem 5.31 in the time domain.

▮ The intrinsic impedance and the wave number are (Problem 5.17):

$$\eta = (120\pi)\sqrt{4/9} = 80\pi \; \Omega \qquad \beta = \frac{2\pi}{\lambda} = 8\pi \; \text{m}^{-1}$$

Consequently, the field vectors are

$$\mathbf{E} = 100 \cos (1.6\pi \times 10^9 t - 8\pi x) \, \mathbf{a}_z \quad (\text{V/m}) \qquad \mathbf{H} = \frac{-100}{80\pi} \cos (1.6\pi \times 10^9 t - 8\pi x) \, \mathbf{a}_y \quad (\text{A/m})$$

5.33   Write a time-domain expression for the electric field of a plane wave if the magnetic field is given by
$$\mathbf{H} = 0.1 e^{-200y} \cos (2\pi \times 10^{10} t - 300y) \, \mathbf{a}_x$$
and the medium is characterized by free-space permeability.

▮ The electric field must have the form
$$\mathbf{E} = |\eta| \, 0.1 \, e^{-200y} \cos (2\pi \times 10^{10} t - 300y + \theta_\eta) \, \mathbf{a}_z$$
so all that is needed is η. But, given $\mu = \mu_0 = 4\pi \times 10^{-7}$, we have from Problem 5.13:
$$\eta = \frac{j\omega\mu_0}{\alpha + j\beta} = \frac{j(2\pi \times 10^{10})(4\pi \times 10^{-7})}{200 + j300} = \frac{80\pi^2}{\sqrt{13}} \left(\frac{\pi}{2} - \tan^{-1}(3/2)\right)$$

5.34   For a harmonic wave in a lossy medium, one defines the *loss tangent* (a dimensionless number) as $\tau \equiv \sigma/\omega\epsilon$. Evaluate this number in terms of the attenuation and phase constants of the wave.

▮ By definition of the propagation constant,
$$\gamma^2 = (\alpha + j\beta)^2 = j\omega\mu(\sigma + j\omega\epsilon)$$
which is equivalent to the two equations
$$\beta^2 - \alpha^2 = \omega^2\mu\epsilon \quad \text{and} \quad 2\alpha\beta = \omega\mu\sigma \tag{1}$$
Divide the second equation (*1*) by the first, to obtain
$$\tau = \frac{2\alpha\beta}{\beta^2 - \alpha^2} \tag{2}$$

5.35   A 5-GHz plane wave is propagating in a material characterized by $\epsilon_r = 2.53$, $\mu_r = 1$, and $\sigma = 0$. Assuming that the electric field is given by $\mathbf{E} = 10 \cos (10\pi \times 10^9 t - \beta z) \, \mathbf{a}_x$ (V/m), determine u, λ, and β.

$$u = \frac{1}{\sqrt{\mu\epsilon}} = \frac{3 \times 10^8}{\sqrt{\epsilon_r}} = \frac{3 \times 10^8}{\sqrt{2.53}} = 1.89 \times 10^8 \text{ m/s}$$

$$\lambda = \frac{u}{f} = \frac{1.89 \times 10^8}{5 \times 10^9} = 3.78 \text{ cm}$$

$$\beta = \frac{2\pi}{\lambda} = \frac{2\pi}{3.78 \times 10^{-2}} = 166.6 \text{ rad/m}$$

5.36 For the wave of Problem 5.35, write the time-domain expression for the magnetic field.

❚ From Problem 5.17, the intrinsic impedance is
$$\eta = \frac{120\pi}{\sqrt{\epsilon_r}} = 236.96 \ \Omega$$
The amplitude of **H**-field is $|\mathbf{H}| = |\mathbf{E}|/\eta = 0.0422$ A/m, giving
$$\mathbf{H} = 0.0422 \cos(10\pi \times 10^9 t - 166.6z) \, \mathbf{a}_y \quad (\text{A/m})$$

5.37 Reverse Problem 5.34 to obtain exact expressions for α and β in terms of τ.

❚ Eliminate α between the two equations (*1*) of Problem 5.34 to obtain
$$(\beta^2)^2 - \omega^2\mu\epsilon\beta^2 - \omega^2\mu^2\epsilon^2/4 = 0 \tag{1}$$
Using $\tau = \sigma/\omega\epsilon$, we find for the positive root of (*1*)
$$\beta^2 = (\omega^2\mu\epsilon)\frac{1 + (1 + \tau^2)^{1/2}}{2}$$
Hence,
$$\beta = \omega\sqrt{\mu\epsilon}\left[\frac{1 + (1 + \tau^2)^{1/2}}{2}\right]^{1/2}$$
$$\alpha = \frac{\omega\mu\sigma}{2\beta} = \omega\sqrt{\mu\epsilon}\,\frac{\tau}{2}\left[\frac{1 + (1 + \tau^2)^{1/2}}{2}\right]^{-1/2} \tag{2}$$

5.38 Using the results of Problem 5.37, express the intrinsic impedance (magnitude and polar angle) in terms of τ.

❚ From the basic relation $\eta = j\omega\mu/\gamma$,
$$|\eta|^2 = \frac{\omega^2\mu^2}{\alpha^2 + \beta^2} \tag{1}$$
Substitution of (*2*) of Problem 5.37 in (*1*) yields, after some algebra,
$$|\eta|^2 = \frac{\mu}{\epsilon}(1 + \tau^2)^{-1/2} \quad \text{or} \quad |\eta| = \sqrt{\frac{\mu}{\epsilon}}(1 + \tau^2)^{-1/4} \tag{2}$$
Also,
$$\tan\theta_\eta = \frac{\alpha}{\beta} = \frac{\tau}{1 + (1 + \tau^2)^{1/2}} \tag{3}$$

5.39 Justify the name "loss tangent" for $\tau \equiv \sigma/\omega\epsilon$.

❚ Define an angle ϕ by $\tau = \tan\phi$. Then, by (*3*) of Problem 5.38,
$$\tan\theta_\eta = \frac{\tan\phi}{1 + \sec\phi} = \frac{\sin\phi}{1 + \cos\phi} \equiv \tan\frac{\phi}{2}$$
or $\phi = 2\theta_\eta$. It is therefore entirely appropriate to call ϕ the *loss angle*, because it is zero (along with θ_η) for a lossless medium; τ is then the tangent of the loss angle, or the *loss tangent*.

5.40 Rewrite the results of Problem 5.38 in terms of the loss angle ϕ.

❚
$$|\eta| = \sqrt{\frac{\mu}{\epsilon}}\cos\phi \qquad \theta_\eta = \frac{\phi}{2} \tag{1}$$

5.41 A plane wave travels at velocity u in the $+z$-direction through a lossy medium. Show that the time-average power density is given by
$$\langle S_z \rangle = \frac{E_m^2}{2\mu u} e^{-2\alpha z} \tag{1}$$

❚ By Problem 5.24,
$$\langle S_z \rangle = \frac{E_m^2}{2|\eta|} e^{-2\alpha z} \cos\theta_\eta \tag{2}$$

Now, by (1) and (3) of Problem 5.38,

$$|\eta|^2 = \frac{\omega^2 \mu^2}{\alpha^2 + \beta^2} = \frac{(\omega/\beta)^2 \mu^2}{(\alpha/\beta)^2 + 1} = \frac{u^2 \mu^2}{\tan^2 \theta_\eta + 1} = u^2 \mu^2 \cos^2 \theta$$

or
$$|\eta| = \mu u \cos \theta_\eta = \mu u \cos \frac{\phi}{2} \qquad (3)$$

Substitution of (3) in (2) yields (1).

It is easy to show that (3) is equivalent to (1) of Problem 5.40 or (2) of Problem 5.38.

5.42 Approximate to the first order (in τ or ϕ) the propagation constant for a good dielectric ($\tau \ll 1$).

❚ By (2) of Problem 5.37,

$$\gamma = \alpha + j\beta \approx \omega\sqrt{\mu\epsilon}\,\frac{\tau}{2} + j\omega\sqrt{\mu\epsilon} = \frac{\sigma}{2}\sqrt{\frac{\mu}{\epsilon}} + j\omega\sqrt{\mu\epsilon}$$

5.43 Approximate the propagation constant for a good conductor ($\tau \gg 1$).

❚ By (2) of Problem 5.37,

$$\gamma = \alpha + j\beta \approx \omega\sqrt{\mu\epsilon}\,\frac{\tau}{2}\left(\frac{\tau}{2}\right)^{-1/2} + j\omega\sqrt{\mu\epsilon}\left(\frac{\tau}{2}\right)^{1/2} = \sqrt{\frac{\omega\mu\sigma}{2}}(1+j)$$

5.44 Repeat Problem 5.42 for the intrinsic impedance.

❚ By (2) and (3) of Problem 5.38,

$$|\eta| \approx \sqrt{\frac{\mu}{\epsilon}} \qquad \text{and} \qquad \tan \theta_\eta \approx \sin \theta_\eta \approx \frac{\tau}{2} = \frac{\sigma}{2\omega\epsilon}$$

Thus
$$\eta = |\eta|\,(\cos \theta_\eta + j \sin \theta_\eta) \approx \sqrt{\frac{\mu}{\epsilon}}\left(1 + j\frac{\sigma}{2\omega\epsilon}\right)$$

5.45 Repeat Problem 5.43 for the intrinsic impedance.

❚ By (2) and (3) of Problem 5.38,

$$|\eta| \approx \sqrt{\frac{\mu}{\epsilon}}\,\tau^{-1/2} = \sqrt{\frac{\omega\mu}{\sigma}} \qquad \text{and} \qquad \tan \theta_\eta \approx 1$$

Thus
$$\eta = |\eta|\,(\cos \theta_\eta + j \sin \theta_\eta) \approx \sqrt{\frac{\omega\mu}{2\sigma}}(1+j)$$

which is in agreement with Problem 5.21.

5.46 Calculate α and $|\eta|$ for a 2-GHz plane wave propagating in a medium with $\epsilon_r = 2.25$ and $\mu_r = 1$, assuming a loss tangent $\tau = 0.01$.

❚ Under these conditions, the medium is a good dielectric. By Problem 5.42,

$$\alpha \approx \omega\sqrt{\mu\epsilon}\,\frac{\tau}{2} = \omega\frac{\sqrt{\epsilon_r}}{u_0}\frac{\tau}{2} = (4\pi \times 10^9)\left(\frac{1.5}{3 \times 10^8}\right)(0.005) = \frac{\pi}{10}\,\text{Np/m}$$

and by Problem 5.44,

$$|\eta| \approx \sqrt{\frac{\mu}{\epsilon}} = \frac{\eta_0}{\sqrt{\epsilon_r}} = \frac{120\pi}{1.5} = 80\pi\,\Omega \qquad .$$

5.47 Repeat Problem 5.46 for $\tau = 200$.

❚ Now the medium is a good conductor. By Problem 5.43,

$$\alpha \approx \omega\frac{\sqrt{\epsilon_r}}{u_0}\left(\frac{\tau}{2}\right)^{1/2} = (4\pi \times 10^9)\left(\frac{1.5}{3 \times 10^8}\right)(10) = 200\pi\,\text{Np/m}$$

and by Problem 5.45,

$$|\eta| \approx \frac{\eta_0}{\sqrt{\epsilon_r}\,\sqrt{\tau}} = \frac{120\pi}{(1.5)(10\sqrt{2})} = (4\sqrt{2})\pi\,\Omega$$

5.48 A plane wave is propagating in a medium having the properties $\epsilon_r = 36$, $\mu_r = 4$, and $\sigma = 1\,\text{S/m}$. The electric field is given by

$$\mathbf{E} = 100e^{-\alpha x}\cos(10^9 \pi t - \beta x)\,\mathbf{a}_z \quad \text{(V/m)}$$

Determine the associated magnetic field.

❚ By Problem 5.14, all that is needed is to compute the complex impedance. With

$$\tau = \frac{\sigma}{\omega\epsilon} = \frac{1}{(10^9\pi)(36 \times 10^{-9}/36\pi)} = 1 \quad \text{or} \quad \phi = 45°$$

Problem 5.40 yields

$$|\eta| = \sqrt{\frac{\mu}{\epsilon}} \sqrt{\cos \phi} = \sqrt{\frac{4}{36}} (120\pi) \frac{1}{2^{1/4}} = 105.67 \; \Omega$$

and $\theta_\eta = 22.5°$. Then,

$$\mathbf{H} = \frac{100}{105.67} e^{-\alpha x} \cos (10^9\pi t - \beta x - 22.5°) \, (-\mathbf{a}_y) \quad (\text{A/m})$$

5.49 For what frequency range may seawater ($\sigma \approx 4 \, \text{S/m}$, $\mu_r \approx 1$, and $\epsilon_r \approx 81$) be considered a good conductor?

❚ Using the criterion $\sigma/\omega\epsilon \geq 100$, we have

$$\frac{4}{2\pi f \times \frac{81}{36\pi} \times 10^{-9}} \geq 100 \quad \text{or} \quad f \leq \frac{4 \times 18 \times 10^7}{81} = 8.89 \, \text{MHz}$$

5.50 Wet, marshy soil is characterized by $\sigma = 10^{-2} \, \text{S/m}$, $\epsilon_r = 15$, and $\mu_r = 1$. At the frequencies 60 Hz, 1 MHz, 100 MHz, and 10 GHz, indicate whether soil may be considered a conductor, a dielectric, or neither.

❚ Since

$$\frac{\sigma}{\omega\epsilon} = \frac{10^{-2}}{2\pi f \left(\frac{15}{36\pi} \times 10^{-9}\right)} = \frac{1.2 \times 10^7}{f}$$

we may prepare the following table:

f	$\sigma/\omega\epsilon$	comments
60 Hz	2×10^5	good conductor
1 MHz	12	quasi-conductor
100 MHz	0.12	quasi-dielectric
10 GHz	1.2×10^{-3}	good dielectric

5.51 For the soil of Problem 5.50, approximate the propagation constant, the skin depth, and the intrinsic impedance at 60 Hz.

❚ At 60 Hz the soil is a good conductor, for which (Problem 5.43)

$$\alpha \approx \beta \approx \sqrt{\mu\sigma\omega/2} = \sqrt{4\pi \times 10^{-7} \times 10^{-2} \times 2\pi \times 60/2} = 1.539 \times 10^{-3} \quad \text{or} \quad \gamma \approx 1.539 \times 10^{-3}(1+j1) \, \text{m}^{-1}$$

Skin depth:

$$\delta \equiv \frac{1}{\alpha} \approx \frac{1}{1.539 \times 10^{-3}} = 650 \, \text{m}$$

Intrinsic impedance: $\eta \approx \sqrt{\mu\omega/\sigma} \underline{/45°} = \sqrt{4\pi \times 10^{-7} \times 2\pi \times 60/10^{-2}} \underline{/45°} = 0.2176 \underline{/45°} \; \Omega$
(see Problem 5.45).

5.52 How far must the 60-Hz wave of Problem 5.51 proceed for its amplitude to be reduced by 20 dB?

❚ The reduction in amplitude is governed by the attenuation factor α. By definition, 20 dB of attenuation corresponds to a factor 1/10; therefore,

$$e^{-\alpha d} = \frac{1}{10} \quad \text{or} \quad d = \frac{\ln 10}{\alpha} = \frac{2.303}{1.539 \times 10^{-3}} = 1496 \, \text{m}$$

5.53 Repeat Problem 5.51 for a 100-MHz wave.

❚ At 100 MHz, Problems 5.42 and 5.44 yield

$$\alpha \approx \frac{\sigma}{2} \sqrt{\frac{\mu}{\epsilon}} = \frac{10^{-2}}{2} \sqrt{\frac{4\pi \times 10^{-7}}{(10^{-9}/36\pi) \times 15}} = 0.4867 \, \text{Np/m}$$

$$\beta \approx \omega\sqrt{\mu\epsilon} = 2\pi \times 100 \times 10^6 \sqrt{4\pi \times 10^{-7} \times (15 \times 10^{-9}/36\pi)} = 8.112 \, \text{rad/m}$$

$$\delta \equiv \frac{1}{\alpha} \approx \frac{1}{0.4867} = 2.055 \, \text{m}$$

$$\eta \approx \sqrt{\frac{\mu}{\epsilon}} \underline{/\sigma/2\omega\epsilon} = \frac{120\pi}{\sqrt{15}} \underline{/3.43°} \; \Omega$$

5.54 Repeat Problem 5.52 for a 100-MHz wave.

▮
$$e^{-\alpha d} = e^{-0.4867d} = \frac{1}{10} \quad \text{or} \quad d = \frac{\ln 10}{0.4867} = 4.732 \text{ m}$$

5.55 A plane wave has a wavelength of 2 cm in free space and 1 cm in a perfect dielectric ($\sigma = 0$, $\mu_r \approx 1$). Determine the relative permittivity of the dielectric.

▮ Since $\lambda = 2\pi/\beta = 2\pi/\omega\sqrt{\mu\epsilon}$,
$$\frac{1}{2} = \frac{\lambda}{\lambda_0} = \frac{\sqrt{\epsilon_0}}{\sqrt{\epsilon}} = \frac{1}{\sqrt{\epsilon_r}} \quad \text{or} \quad \epsilon_r = 4$$

5.56 Refer to Problem 5.49. Assuming that $\tau \geq 100$, obtain a relationship between the frequency, f, of a plane wave and the distance, d, traveled in seawater for an amplitude reduction of 80 dB.

▮ For a good conduction,
$$\alpha \approx \sqrt{\mu\sigma\omega/2} = \sqrt{4\pi \times 10^{-7} \times 4 \times 2\pi f/2} = 3.974 \times 10^{-3}\sqrt{f}$$
Now, 80 dB = 4(20 dB) corresponds to the factor $(1/10)^4$; hence $e^{-\alpha d} = 10^{-4}$, or
$$d = \frac{9.21}{\alpha} = \frac{2317}{\sqrt{f}} \quad (f \leq 8.89 \text{ MHz}) \tag{1}$$
where d is in meters for f in hertz.

5.57 For seawater (Problems 5.49 and 5.56), calculate the maximum frequency which may be used in submarine communication, if the signal is to attenuate by at most 20 dB over a 30-m depth.

▮ From Problem 5.56, $\alpha = 3.974 \times 10^{-3}\sqrt{f}$. For 20-dB attenuation over $d = 30$ m,
$$e^{-\alpha d} = e^{-30\alpha} = \frac{1}{10} \quad \text{or} \quad \alpha = \frac{\ln 10}{30} = 7.67 \times 10^{-2}$$
Equating the values for α:
$$3.974 \times 10^{-3}\sqrt{f} = 7.67 \times 10^{-2} \quad \text{or} \quad f \approx 400 \text{ Hz}$$

5.58 A large copper conductor ($\sigma = 58$ MS/m, $\epsilon_r \approx 1$, $\mu_r \approx 1$) supports a 60-Hz plane wave. Compute the attenuation constant, propagation constant, intrinsic impedance, wavelength, and phase velocity of propagation.

▮
$$\tau = \frac{\sigma}{\omega\epsilon} = \frac{5.8 \times 10^7}{2\pi \times 60 \times 10^{-9}/36\pi} = 1.74 \times 10^{16} \gg 1$$
and Problems 5.43 and 5.45 apply.

$$\alpha \approx \beta \approx \sqrt{\omega\mu\sigma/2} = 117.21 \text{ m}^{-1} \qquad \eta \approx \sqrt{\frac{\omega\mu}{\sigma}}\underline{/45°} = 2.86 \times 10^{-6}\underline{/45°} \text{ } \Omega$$

$$\lambda \equiv \frac{2\pi}{\beta} \approx 5.36 \times 10^{-2} \text{ m} \qquad u = \lambda f \approx 3.216 \text{ m/s}$$

5.59 Repeat Problem 5.58 for a frequency of 10 GHz.

▮ Now $\tau = 1.044 \times 10^8$, so that copper remains a very good conductor. As α, β, $|\eta|$, $1/\lambda$, and u all vary as $\sqrt{f}$, merely multiply the values found in Problem 5.58 by $10^5/\sqrt{60} \approx 13\,000$.

5.60 A 100-V/m, 7-GHz plane **E**-wave travels in seawater (Problem 5.49). Compute the average power dissipated in a volume of seawater presenting an area (normal to the Poynting vector) $A = 1000$ mm^2 and extending five skin depths (parallel to the Poynting vector).

▮ For $\tau = \sigma/\omega\epsilon = 0.127$ (dielectric), the phase velocity may be approximated as
$$u \approx \frac{1}{\sqrt{\mu\epsilon}} = \frac{1}{3} \times 10^8 \text{ m/s}$$
Then, by (1) of Problem 5.41,
$$\langle P_{\text{diss}}\rangle = [\langle S_{\text{in}}\rangle - \langle S_{\text{out}}\rangle]A = \frac{E_m^2}{2\mu u}[1 - e^{-(2\alpha)(5/\alpha)}]A$$
$$\approx \frac{(100)^2(3 \times 10^{-8})}{2(4\pi \times 10^{-7})}[1](10^{-3}) = \frac{3}{8\pi} \text{ W}$$

5.61 Repeat Problem 5.60 for a frequency of 10 kHz.

∎ Now $\tau = 8.89 \times 10^4$ (conductor), and

$$u \approx \sqrt{\frac{2\omega}{\mu\sigma}} = 1.58 \times 10^5 \text{ m/s}$$

Hence

$$\langle P_{\text{diss}} \rangle = \frac{\frac{1}{3} \times 10^8}{1.58 \times 10^5} \left(\frac{3}{8\pi} \right) = 25.2 \text{ W}$$

5.62 A 100-MHz plane wave traveling in a lossy dielectric ($\mu_r \approx 1$) has the magnetic field phasor

$$\hat{\mathbf{H}} = (1\mathbf{a}_y + j2\mathbf{a}_z)e^{-0.2x}e^{-j2x} \equiv e^{-\gamma x}\mathbf{a}_y + 2je^{-\gamma x}\mathbf{a}_z$$

Write complete time-domain expressions for the electric and magnetic field vectors.

∎ We have at once

$$\mathbf{H} = \text{Re}\,[\hat{\mathbf{H}}e^{j\omega t}] = e^{-0.2x}\cos(\omega t - 2x)\,\mathbf{a}_y - 2e^{-0.2x}\sin(\omega t - 2x)\,\mathbf{a}_z$$

The safest way to find **E** (with the correct direction) is to appeal to Maxwell's equation curl $\hat{\mathbf{H}} = (\sigma + j\omega\epsilon)\hat{\mathbf{E}}$, which gives

$$\hat{\mathbf{E}} = \eta(2je^{-\gamma x}\mathbf{a}_y - e^{-\gamma x}\mathbf{a}_z)$$

in which, as usual,

$$\eta = \frac{j\omega\mu}{\gamma} = \frac{j(2\pi \times 10^8)(4\pi \times 10^{-7})}{0.2 + j2} = 393\,\underline{/5.7^\circ}$$

We then obtain

$$\mathbf{E} = \text{Re}\,[\hat{\mathbf{E}}e^{j\omega t}] = -786e^{-0.2x}\sin(\omega t - 2x + 5.7^\circ)\,\mathbf{a}_y - 393e^{-0.2x}\cos(\omega t - 2x + 5.7^\circ)\,\mathbf{a}_z$$

5.63 Formalize the direction rule implicit in Problem 5.62.

∎ Relative to the basis $\{\mathbf{a}_x, \mathbf{a}_y\}$, $\{\mathbf{a}_y, \mathbf{a}_z\}$, or $\{\mathbf{a}_z, \mathbf{a}_x\}$, let the electric field in a plane wave be given as

$$\hat{\mathbf{E}} = (E_1, E_2)$$

Then the magnetic field is

$$\hat{\mathbf{H}} = (-E_2/\eta, E_1/\eta)$$

5.64 When is a plane wave said to be *plane-polarized*?

∎ When, at any instant, the **E**-vectors along a given ray are all parallel and thus define a plane containing the ray. For example, the wave of Fig. 5-1 is plane-polarized, with xz as the plane of polarization. For a plane-polarized wave it is also true that, at any fixed location, the **E**-vector oscillates in a fixed line; hence the wave is also termed *linearly polarized*.

5.65 Why isn't the **H**-field also involved in the concept of polarization?

∎ In a plane wave **E** and **H** are perpendicular to each other and to the direction of propagation, so that the directions of both are fixed when the direction of one is fixed. It is **E** that is always chosen as the "one."

5.66 Show that a plane wave with

$$\mathbf{E} = E_{m1}\cos(\omega t - \beta z)\,\mathbf{a}_x + E_{m2}\cos(\omega t - \beta z + \theta)\,\mathbf{a}_y \tag{1}$$

is plane-polarized if and only if $\theta = 0$.

∎ The angle ϕ between **E** and the x-direction is given by

$$\tan\phi = \frac{E_y}{E_x} = \frac{E_{m2}}{E_{m1}} \frac{\cos(\omega t - \beta z + \theta)}{\cos(\omega t - \beta z)} \tag{2}$$

It is clear that ϕ will be independent of z when and only when $\theta = 0$.

5.67 Consider the wave of Problem 5.66, with $E_{m1} = E_{m2} = E_m$ and $\theta = -90^\circ$. Show that at any fixed point z_0, the tip of the **E**-vector traces a circle. Sketch the circle.

∎ From (1) of Problem 5.66,

$$\mathbf{E}(z_0, t) = E_m[\cos(\omega t - \beta z_0)\,\mathbf{a}_x + \sin(\omega t - \beta z_0)\,\mathbf{a}_y]$$

The length of this vector is E_m at all times; moreover, (2) of Problem 5.66 gives

$$\tan\phi_0 = \tan(\omega t - \beta z_0) \quad \text{or} \quad \phi_0 = \omega t - \beta z_0$$

Thus, the tip of the vector, rotating at constant angular velocity ω, traces out a circle. This is referred to as *circular polarization*; in fact, *right-circular polarization*, because if one places the fingers of the right hand in the direction of rotation, the thumb will point in the direction of propagation (here, the $+z$-direction). See Fig. 5-2(a).

5.68 Choose $\theta = +90^\circ$ (and $E_{m1} = E_{m2}$) in Problem 5.67 and rework the problem.

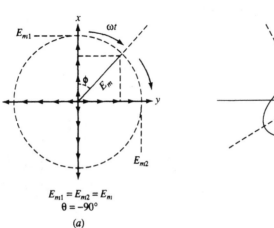

$$E_{m1} = E_{m2} = E_m$$
$$\theta = -90°$$
(a)

$$E_{m1} \neq E_{m2}$$
$$\theta \text{ arbitrary}$$
(b)

Fig. 5-2

▌ The sign of E_y is reversed and *left-circular polarization* results.

5.69   Show that the wave of Problem 5.66, with $E_{m1} \neq E_{m2}$, is *elliptically polarized*.

▌ For convenience, let $v \equiv \omega t - \beta z_0$; then, at location z_0, the locus of the tip of **E** is given by
$$x = E_{m1} \cos v \qquad y = E_{m2} \cos (v + \theta)$$
Eliminate the parameter v as follows:
$$\frac{y}{E_{m2}} = \cos v \cos \theta - \sqrt{1 - \cos^2 v} \sin \theta = \frac{x}{E_{m1}} \cos \theta - \sqrt{1 - \frac{x^2}{E_{m1}^2}} \sin \theta$$
or
$$\frac{x^2}{E_{m1}^2} + \frac{y^2}{E_{m2}^2} - \frac{2 \cos \theta}{E_{m1} E_{m2}} xy = \sin^2 \theta$$
This is the equation of an ellipse such as shown in Fig. 5-2(b). Elliptical polarization, like circular, may be right- or left-; the governing parameter is θ.

5.70   Show that a linearly polarized wave can be obtained as the superposition of two circularly polarized waves rotating in opposite directions but at the same angular rate.

▌ Consider the two circularly polarized waves
$$\mathbf{E}_{\text{left}} = E \cos (\omega t - \beta z) \mathbf{a}_x + E \cos \left(\omega t - \beta z + \frac{\pi}{2} \right) \mathbf{a}_y = E \cos (\omega t - \beta z) \mathbf{a}_x - E \sin (\omega t - \beta z) \mathbf{a}_y$$

$$\mathbf{E}_{\text{right}} = E \cos (\omega t - \beta z) \mathbf{a}_x + E \cos \left(\omega t - \beta z - \frac{\pi}{2} \right) \mathbf{a}_y = E \cos (\omega t - \beta z) \mathbf{a}_x + E \sin (\omega t - \beta z) \mathbf{a}_y$$
Then $\mathbf{E}_{\text{left}} + \mathbf{E}_{\text{right}} = 2E \cos (\omega t - \beta z) \mathbf{a}_x$, a linearly polarized wave.

5.71   Repeat Problem 5.70 with "a linearly" replaced by "an elliptically."

▌ Choose the axes of the ellipse as the x and y axes, putting the wave in the canonical form
$$\mathbf{E} = E_1 \cos (\omega t - \beta z) \mathbf{a}_x - E_2 \sin (\omega t - \beta z) \mathbf{a}_y$$
But then **E** is the vector sum of the two circularly polarized waves
$$\mathbf{E}_{\text{left}} = \frac{E_1 + E_2}{2} \cos (\omega t - \beta z) \mathbf{a}_x - \frac{E_1 + E_2}{2} \sin (\omega t - \beta z) \mathbf{a}_y$$
$$\mathbf{E}_{\text{right}} = \frac{E_1 - E_2}{2} \cos (\omega t - \beta z) \mathbf{a}_x + \frac{E_1 - E_2}{2} \sin (\omega t - \beta z) \mathbf{a}_y$$

5.72   Determine the polarizations of the following plane waves:
$$(a)\ \mathbf{E} = 1 \cos (\omega t + \beta z) \mathbf{a}_x + 1 \sin (\omega t + \beta z) \mathbf{a}_y$$
$$(b)\ \mathbf{E} = 1 \cos (\omega t + \beta z) \mathbf{a}_x - 1 \sin (\omega t + \beta z) \mathbf{a}_y$$
$$(c)\ \mathbf{E} = 1 \cos (\omega t + \beta z) \mathbf{a}_x - 2 \sin (\omega t + \beta z - 45°) \mathbf{a}_y$$

▌ Subtract 90° from the phase of a sine to convert it to a cosine, and note that the wave propagates in the $-z$-direction.

(a) left-circular   (b) right-circular   (c) right-elliptical

5.73 Consider the interference between two plane waves
$$E_y' = E_m \cos\left[(\omega + \Delta\omega)t - (\beta + \Delta\beta)z\right] \quad \text{and} \quad E_y'' = E_m \cos\left[(\omega - \Delta\omega)t - (\beta - \Delta\beta)z\right]$$
which have equal amplitudes and very nearly equal frequencies and wave numbers. Express $E_y = E_y' + E_y''$ as a product of two cosine terms and sketch $E_y(z, 0)$.

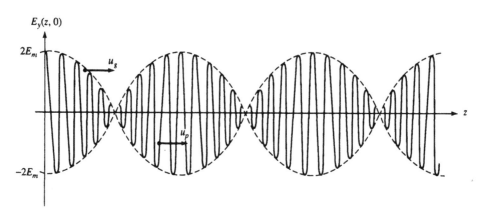

Fig. 5-3

$\blacksquare$
$$E_y = E_m[\cos(\omega + \Delta\omega)t \cos(\beta + \Delta\beta)z + \sin(\omega + \Delta\omega)t \sin(\beta + \Delta\beta)z$$
$$+ \cos(\omega - \Delta\omega)t \cos(\beta - \Delta\beta)z + \sin(\omega - \Delta\omega)t \sin(\beta - \Delta\beta)z]$$
$$= 2E_m \cos(\omega t - \beta z) \cos[(\Delta\omega)t - (\Delta\beta)z]$$
The required sketch is given in Fig. 5-3.

5.74 Notice from Problem 5.73 that the composite wave consists of the carrier, traveling at the phase velocity u_p, and a modulation envelope, traveling at the *group velocity* u_g. Obtain expressions for u_p and u_g.

$\blacksquare$ The condition of constant carrier phase is expressed by
$$\frac{d}{dt}(\omega t - \beta z) = 0 \quad \text{or} \quad \omega - \beta \frac{dz}{dt} = 0 \quad \text{or} \quad \frac{dz}{dt} \equiv u_p = \frac{\omega}{\beta}$$
Similarly, $u_g = \Delta\omega/\Delta\beta$.

5.75 Depict the results of Problem 5.74 on an ω-β diagram.

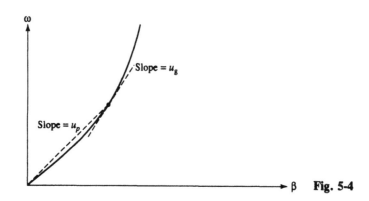

Fig. 5-4

$\blacksquare$ See Fig. 5-4.

5.76 In the language of telecommunications, the wave of Problem 5.73 exhibits two sideband frequencies, $\omega + \Delta\omega$ and $\omega - \Delta\omega$, due to the modulation of the carrier frequency ω by a frequency $\Delta\omega$. For a particular frequency, given the bandwidth $\Delta\omega \to 0$, obtain a relationship between the phase and group velocities.

$\blacksquare$ As $\Delta\omega \to 0$,
$$\frac{1}{u_g} = \frac{\Delta\beta}{\Delta\omega} \to \frac{d\beta}{d\omega} \quad \text{or} \quad u_g \to \frac{d\omega}{d\beta} \tag{1}$$
Since $\omega = \beta u_p$, we may write (1) as
$$u_g = \frac{d}{d\beta}(\beta u_p) = u_p + \beta \frac{du_p}{d\beta} \tag{2}$$

Now, since $\beta = 2\pi/\lambda$, (2) may also be expressed as

$$u_g = u_p - \lambda \frac{du_p}{d\lambda} \tag{3}$$

Normally, $du_p/d\lambda > 0$ and therefore $u_g < u_p$.

5.77 Refer to Problem 4.84, which presented a plane-wave solution of Maxwell's equations within a rectangular waveguide. Assuming that this guide is carrying a narrowband information signal centered around a carrier frequency of 7 GHz, and the guide dimension is $a = 22.9$ mm, determine the phase and group velocities of this signal.

■ At 7 GHz the dispersion relation (2) of Problem 4.84 yields $\beta = 51.7$ rad/m and, by differentiation,

$$\frac{1}{u_g} = \frac{d\beta}{d\omega} = \frac{\omega\mu_0\epsilon_0}{\beta} = \frac{u_p}{u_0^2} \quad \text{or} \quad u_p u_g = u_0^2$$

Thus $u_p = (14\pi \times 10^9)/51.7 = 8.51 \times 10^8$ m/s and $u_g = (3 \times 10^8)^2/(8.51 \times 10^8) = 1.058 \times 10^8$ m/s. Note the simplicity of the u_p-u_g relation here.

5.78 Show that in an unbounded, highly conductive medium $u_g \approx 2u_p$. (Dispersion with $u_g > u_p$ is called *anomalous*.)

■ By Problem 5.43, the dispersion relation for a good conductor is (very nearly) $\beta = \sqrt{\omega\mu\sigma/2}$, or

$$\omega = \frac{2\beta^2}{\mu\sigma}$$

Consequently

$$u_g = \frac{d\omega}{d\beta} = \frac{4\beta}{\mu\sigma} = 2\frac{\omega}{\beta} = 2u_p$$

5.79 Verify that a perfect dielectric is nondispersive.

■ $u_p (= 1/\sqrt{\mu\epsilon})$ is independent of ω.

5.80 Characterize the dispersion exhibited by a medium in which phase velocity is proportional to the square root of wavelength.

■ If $u_p = a\sqrt{\lambda}$, (3) of Problem 5.76 gives

$$u_g = a\sqrt{\lambda} - \lambda\left(\frac{a}{2\sqrt{\lambda}}\right) = \frac{a}{2}\sqrt{\lambda} = \frac{1}{2}u_p$$

Because $u_g < u_p$, this is normal dispersion.

5.81 Show that the group velocity of a plane wave of frequency f and wavelength λ may be expressed as

$$u_g = -\lambda^2 \frac{df}{d\lambda} \tag{1}$$

■

$$u_g \equiv \frac{d\omega}{d\beta} = 2\pi\frac{df}{d\beta} = 2\pi\frac{df}{d\lambda}\bigg/\frac{d\beta}{d\lambda} \tag{2}$$

But $\beta \equiv 2\pi/\lambda$, so that $d\beta/d\lambda = -2\pi/\lambda^2$. When this is substituted in (2), (1) results.

5.82 Establish the general relation

$$u_g = 2u_p - \mu\epsilon u_p^3 \tag{1}$$

for a plane wave propagating in an unbounded medium characterized by (σ, μ, ϵ).

■ The dispersion relation is given implicitly by (1) of Problem 5.37:

$$\beta^4 - \mu\epsilon\omega^2\beta^2 - \frac{\mu^2\sigma^2}{4}\omega^2 = 0 \tag{2}$$

or

$$1 - \mu\epsilon u_p^2 - \frac{\mu^2\sigma^2}{4\beta^2}u_p^2 = 0 \tag{3}$$

Differentiate (2) with respect to β:

$$4\beta^3 - \mu\epsilon\left(2\omega\frac{d\omega}{d\beta}\beta^2 + \omega^2 2\beta\right) - \frac{\mu^2\sigma^2}{2}\omega\frac{d\omega}{d\beta} = 0$$

or

$$2 - \mu\epsilon(u_p u_g + u_p^2) - \frac{\mu^2\sigma^2}{4\beta^2}u_p u_g = 0 \tag{4}$$

Multiply (3) by u_g and (4) by $-u_p$, and add:

$$u_g - 2u_p + \mu\epsilon u_p^3 = 0$$

which is just (1).

5.83 Show that in Problem 5.82 dispersion is always anomalous (!)

❚ Rewrite (1) of Problem 5.82 as

$$u_g - u_p = u_p(1 - \mu \epsilon u_p^2) = \frac{\mu^2 u_p^3}{4\beta^2} \sigma^2$$

where the last step follows from (3) of Problem 5.82. Thus, when $\sigma = 0$, $u_g = u_p$, and there is no dispersion (cf. Problem 5.79); whereas, when $\sigma > 0$, $u_g > u_p$, and there is anomalous dispersion.

5.84 Figure 5-5 shows an interface ($z = 0$) between two media. A normally incident wave ($\mathbf{E}_i$, $\mathbf{H}_i$) in medium 1 is partly reflected at the interface as ($\mathbf{E}_r$, $\mathbf{H}_r$), and partly transmitted into medium 2 as ($\mathbf{E}_t$, $\mathbf{H}_t$). In the phasor domain, evaluate the ratios $\hat{E}_{rx}(0)/\hat{E}_{ix}(0)$ and $\hat{E}_{tx}(0)/\hat{E}_{ix}(0)$ in terms of the intrinsic impedances of the two media.

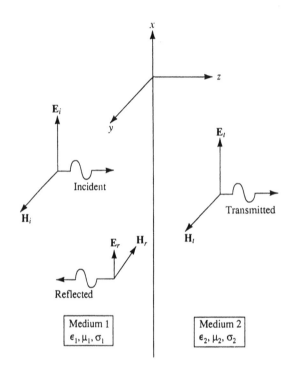

Fig. 5-5

❚ At $z = 0$ the tangential components of the total electric and magnetic phasors must be continuous:
$$\hat{E}_{ix}(0) + \hat{E}_{rx}(0) = \hat{E}_{tx}(0) \qquad \hat{H}_{iy}(0) + \hat{H}_{ry}(0) = \hat{H}_{ty}(0)$$
Divide the first equation by $\hat{E}_{ix}(0)$; the second, by $\hat{H}_{iy}(0)$; and use (2) of Problem 5.12, to find
$$1 + \frac{\hat{E}_{rx}(0)}{\hat{E}_{ix}(0)} = \frac{\hat{E}_{tx}(0)}{\hat{E}_{ix}(0)} \qquad 1 - \frac{\hat{E}_{rx}(0)}{\hat{E}_{ix}(0)} = \frac{\eta_1}{\eta_2}\frac{\hat{E}_{tx}(0)}{\hat{E}_{ix}(0)}$$
From these,
$$\text{reflection coefficient} \equiv \frac{\hat{E}_{rx}(0)}{\hat{E}_{ix}(0)} = \frac{\eta_2 - \eta_1}{\eta_2 + \eta_1} \equiv \Gamma \qquad \frac{\hat{E}_{tx}(0)}{\hat{E}_{ix}(0)} = \frac{2\eta_2}{\eta_2 + \eta_1} \equiv \text{T} \equiv \text{transmission coefficient}$$

5.85 State the relationship between Γ and T.

❚ $1 + \Gamma = \text{T}$.

5.86 The incident wave in Problem 5.84 may always be put in the form $\hat{\mathbf{E}}_i = E_m e^{-\gamma_1 z}\mathbf{a}_x$, with E_m real and positive. Give the implied forms of the remaining field phasors.

❚
$$\hat{\mathbf{E}}_i = E_m e^{-\gamma_1 z}\mathbf{a}_x \qquad \hat{\mathbf{E}}_r = \Gamma E_m e^{\gamma_1 z}\mathbf{a}_x \qquad \hat{\mathbf{E}}_t = \text{T} E_m e^{-\gamma_2 z}\mathbf{a}_x$$
$$\hat{\mathbf{H}}_i = \frac{E_m}{\eta_1} e^{-\gamma_1 z}\mathbf{a}_y \qquad \hat{\mathbf{H}}_r = -\frac{\Gamma E_m}{\eta_1} e^{\gamma_1 z}\mathbf{a}_y \qquad \hat{\mathbf{H}}_t = \frac{\text{T} E_m}{\eta_2} e^{-\gamma_2 z}\mathbf{a}_y$$

5.87 Express the field vectors of Problem 5.86 in the time domain.

❚ $\mathbf{E}_i = E_m e^{-\alpha_1 z} \cos(\omega t - \beta_1 z)\mathbf{a}_x$

$$\mathbf{H}_i = \frac{E_m}{|\eta_1|} e^{-\alpha_1 z} \cos{(\omega t - \beta_1 z - \theta_{\eta 1})} \, \mathbf{a}_y$$

$$\mathbf{E}_r = |\Gamma| \, E_m e^{\alpha_1 z} \cos{(\omega t + \beta_1 z + \theta_\Gamma)} \, \mathbf{a}_x$$

$$\mathbf{H}_r = -\frac{|\Gamma| \, E_m}{|\eta_1|} e^{\alpha_1 z} \cos{(\omega t + \beta_1 z + \theta_\Gamma - \theta_{\eta 1})} \, \mathbf{a}_y$$

$$\mathbf{E}_t = |\mathbf{T}| \, E_m e^{-\alpha_2 z} \cos{(\omega t - \beta_2 z + \theta_{\mathrm{T}})} \, \mathbf{a}_x$$

$$\mathbf{H}_t = \frac{|\mathbf{T}| \, E_m}{|\eta_2|} e^{-\alpha_2 z} \cos{(\omega t - \beta_2 z + \theta_{\mathrm{T}} - \theta_{\eta 2})} \, \mathbf{a}_y$$

In the event that η_1 and η_2 are real, with $\eta_1 > \eta_2$, the preceding expressions show that, at the interface, $\mathbf{E}_r$ is 180° out of phase with $\mathbf{E}_i$ (while $\mathbf{H}_r$ remains in phase with $\mathbf{H}_i$).

5.88 Determine the time-average power density of the transmitted wave of Problem 5.86.

$\blacksquare$

$$\langle \mathbf{S} \rangle = \frac{1}{2} \operatorname{Re} [\hat{\mathbf{E}}_t \times \hat{\mathbf{H}}_t^*] = \frac{1}{2} \operatorname{Re} \left[\mathbf{T} E_m e^{-\gamma_2 z} \frac{\mathbf{T}^* E_m e^{-\gamma_2^* z}}{\eta_2^*} \right] \mathbf{a}_z$$

$$= \frac{|\mathbf{T}|^2 \, E_m^2 e^{-2\alpha_2 z} \cos{\theta_{\eta 2}}}{2 \, |\eta_2|} \, \mathbf{a}_z = \frac{E_t^2 \cos{\theta_{\eta 2}}}{2 \, |\eta_2|} \, \mathbf{a}_z$$

where E_t is the amplitude of $\mathbf{E}_t$.

5.89 Refer to Fig. 5-5. Per unit of xy area and averaged over time, how much power is lost by the transmitted wave in penetrating medium 2 to a depth $z = \ell$?

$\blacksquare$ From Problem 5.88,

$$P(\text{lost}) = P(\text{in at } z = 0) - P(\text{out at } z = \ell)$$

$$= \frac{|\mathbf{T}|^2 \, E_m^2 \cos{\theta_{\eta 2}}}{2 \, |\eta_2|} (1 - e^{-2\alpha_2 \ell})$$

5.90 A plane wave whose electric field is given by $\mathbf{E}_i = 100 \cos{(\omega t - 6\pi x)} \, \mathbf{a}_z$ passes normally from a material having $\epsilon_r = 4$, $\mu_r = 1$, and $\sigma = 0$ to a material having $\epsilon_r = 9$, $\mu_r = 4$, and $\sigma = 0$. Write complete time-domain expressions for incident, reflected, and transmitted electric and magnetic fields.

$\blacksquare$ First, we calculate the material parameters:

$$\eta_1 = \sqrt{\frac{\mu_1}{\epsilon_1}} = \frac{120\pi}{\sqrt{4}} = 60\pi \qquad \eta_2 = 120\pi \sqrt{\frac{4}{9}} = 80\pi$$

$$\mathbf{T} = \frac{2\eta_2}{\eta_1 + \eta_2} = \frac{8}{7} \qquad \Gamma = \mathbf{T} - 1 = \frac{1}{7}$$

$$\beta_1 = \omega\sqrt{\mu_1 \epsilon_1} = 6\pi \qquad \beta_2 = \omega\sqrt{\mu_2 \epsilon_2} = \omega\sqrt{\mu_1 \epsilon_1} \frac{\sqrt{\mu_2 \epsilon_2}}{\sqrt{\mu_1 \epsilon_1}} = 6\pi \frac{\sqrt{9 \cdot 4}}{\sqrt{4 \cdot 1}} = 18\pi$$

Therefore, from Problem 5.87, with $(x, y, z) \rightarrow (z, -y, x)$,

$$\mathbf{H}_i = \frac{100}{60\pi} \cos{(\omega t - 6\pi x)} \, (-\mathbf{a}_y)$$

$$\mathbf{E}_r = \frac{100}{7} \cos{(\omega t + 6\pi x)} \, \mathbf{a}_z$$

$$\mathbf{H}_r = -\frac{100}{420\pi} \cos{(\omega t + 6\pi x)} \, (-\mathbf{a}_y) = \frac{100}{420\pi} \cos{(\omega t + 6\pi x)} \, \mathbf{a}_y$$

$$\mathbf{E}_t = \frac{800}{7} \cos{(\omega t - 18\pi x)} \, \mathbf{a}_z$$

$$\mathbf{H}_t = \frac{10}{7\pi} \cos{(\omega t - 18\pi x)} \, (-\mathbf{a}_y)$$

5.91 Determine the time-average power transmitted through a 2 m^2 area of material 2 of Problem 5.90.

$\blacksquare$ From Problem 5.88,

$$\langle P \rangle = \langle S \rangle A = \frac{(800/7)^2 \cos{0}}{2(80\pi)} (2) = \frac{8000}{49\pi}$$

5.92 A 10-V/m plane wave, of frequency 3 MHz, is incident from free space (medium 1) normal to the surface of a material (medium 2) having $\epsilon_r = 4$, $\mu_r = 1$, and $\sigma = 10^3$ S/m. Determine the average power dissipated per unit of surface area in a 1-mm penetration of medium 2.

▮ The intrinsic impedance for medium 1 is $\eta_1 = 120\pi \ \Omega$; since

$$\tau_2 = \frac{\sigma_2}{\omega\epsilon_2} = \frac{10^3}{2\pi \times 3 \times 10^6 \times 4 \times 10^{-9}/36\pi} = \frac{3}{2} \times 10^6 \gg 1$$

the constants for medium 2 are given approximately by

$$\alpha_2 = \beta_2 = \sqrt{\tfrac{1}{2}\omega\mu_2\sigma_2} = \sqrt{\tfrac{1}{2} \times 2\pi \times 3 \times 10^6 \times 4\pi \times 10^{-7} \times 10^3} = 108.83 \ \mathrm{m}^{-1}$$

$$\eta_2 = \sqrt{\frac{\omega\mu_2}{\sigma_2}} \underline{/45°} = \sqrt{2\pi \times 3 \times 10^6 \times 4\pi \times 10^{-7}/10^3} = 0.154 \ \underline{/45°} \ \Omega$$

$$T = \frac{2\eta_2}{\eta_1 + \eta_2} = 8.16 \times 10^{-4} \underline{/44.98°}$$

Problem 5.89 then yields

$$P(\text{lost}) = \frac{(8.16 \times 10^{-4})^2 (10)^2 (1/\sqrt{2})}{2(0.154)} [1 - e^{-2(108.83)(0.001)}] = 29.9 \ \mu\mathrm{W}$$

5.93 A plane wave of 200 MHz, traveling in free space, impinges normally on a large block of a material having $\epsilon_r = 4$, $\mu_r = 9$, and $\sigma = 0$. Determine η_1, η_2, β_1, β_2, Γ, and T.

▮

$$\eta_1 = 120\pi \ \Omega \qquad\qquad \eta_2 = 120\pi\sqrt{\mu_{r2}/\epsilon_{r2}} = 180\pi \ \Omega$$

$$\beta_1 = \frac{\omega}{u_1} = \frac{4\pi}{3} \ \mathrm{rad/m} \qquad\qquad \beta_2 = \sqrt{\mu_{r2}\epsilon_{r2}} \ \beta_1 = 8\pi \ \mathrm{rad/m}$$

$$\Gamma = \frac{\eta_2 - \eta_1}{\eta_2 + \eta_1} = \frac{\frac{3}{2} - 1}{\frac{3}{2} + 1} = \frac{1}{5} \qquad\qquad T = \frac{2\eta_2}{\eta_2 + \eta_1} = \frac{3}{5/2} = \frac{6}{5}$$

5.94 At the interface of the media of Problem 5.93, the incident **H**-vector is given by $\mathbf{H}_i = 1\cos(\omega t - \beta_1 y)\,\mathbf{a}_z$. Write complete time-domain expressions for the incident, reflected, and transmitted field vectors.

▮ By Problem 5.87, with $(x, y, z) \to (-x, z, y)$,

$$\mathbf{H}_i = 1\cos\left(\omega t - \frac{4\pi}{3}y\right)\mathbf{a}_z$$

$$\mathbf{E}_i = 120\pi\cos\left(\omega t - \frac{4\pi}{3}y\right)(-\mathbf{a}_x)$$

$$\mathbf{E}_r = \frac{120\pi}{5}\cos\left(\omega t + \frac{4\pi}{3}y\right)(-\mathbf{a}_x)$$

$$\mathbf{H}_r = -\frac{1}{5}\cos\left(\omega t + \frac{4\pi}{3}y\right)\mathbf{a}_z$$

$$\mathbf{E}_t = \left(\frac{6}{5}\right)(120\pi)\cos(\omega t - 8\pi y)(-\mathbf{a}_x)$$

$$\mathbf{H}_t = \frac{(6/5)(120\pi)}{180\pi}\cos(\omega t - 8\pi y)\mathbf{a}_z$$

5.95 Calculate the average power transmitted through a 5 m² area of the interface of Problems 5.93 and 5.94.

▮ By Problem 5.88,

$$\langle P \rangle = \langle S \rangle A = \frac{E_t^2 A}{2\eta_2} = \frac{(144\pi)^2(5)}{2(180\pi)} = 288\pi$$

5.96 A 300-MHz plane wave traveling in free space strikes a large block of copper ($\mu_r \approx 1$, $\epsilon_r \approx 1$, and $\sigma = 58\ \mathrm{MS/m}$), normal to the surface. Assuming that the surface of the copper lies in the yz plane and the wave is propagating in the x direction, write complete time-domain expressions for the incident, reflected, and transmitted field vectors. Assume that the magnitude of the incident electric field is 1 V/m and that the field is in the z-direction.

▮ First, we calculate the constants of the two media, recognizing that copper (medium 2) is an excellent conductor:

$$\alpha_1 = 0 \ \mathrm{Np/m} \qquad \beta_1 = \omega\sqrt{\mu_0\epsilon_0} = \frac{2\pi \times 3 \times 10^8}{3 \times 10^8} = 2\pi \ \mathrm{rad/m} \qquad \eta_1 = 120\pi \ \Omega$$

$$\alpha_2 = \beta_2 = \sqrt{\tfrac{1}{2}\omega\mu\sigma} = 2.621 \times 10^5 \ \mathrm{Np/m} \qquad \eta_2 = \sqrt{\frac{\omega\mu}{\sigma}} \underline{/45°} = 6.39 \times 10^{-3} \ \underline{/45°} \ \Omega$$

$$T = \frac{2\eta_2}{\eta_2 + \eta_1} \approx \frac{2\eta_2}{\eta_1} = 3.39 \times 10^{-5} \ \underline{/45°} \qquad \Gamma = T - 1 \approx -1$$

Then, from Problem 5.87, with $(x, y, z) \rightarrow (z, -y, x)$,

$$\mathbf{E}_i = 1 \cos{(\omega t - 2\pi x)} \, \mathbf{a}_z \ \text{V/m}$$

$$\mathbf{H}_i = \frac{1}{120\pi} \cos{(\omega t - 2\pi x)} \, (-\mathbf{a}_y) \ \text{A/m}$$

$$\mathbf{E}_r = -1 \cos{(\omega t + 2\pi x)} \, \mathbf{a}_z \ \text{V/m}$$

$$\mathbf{H}_r = +\frac{1}{120\pi} \cos{(\omega t + 2\pi x)} \, (-\mathbf{a}_y) \ \text{A/m}$$

$$\mathbf{E}_t = 3.39 \times 10^{-5} e^{-262100x} \cos{(\omega t - 262100x + 45°)} \, \mathbf{a}_z \ \text{V/m}$$

$$\mathbf{H}_t = 5.31 \times 10^{-3} e^{-262100x} \cos{(\omega t - 262100x)} \, (-\mathbf{a}_y) \ \text{A/m}$$

5.97 Calculate the average power dissipated in a rectangular volume of the copper of Problem 5.96 that lies between the surface and a parallel plane three skin depths in the interior, and has a cross-sectional area of $2 \ \text{m}^2$.

$\blacksquare$ Set $\ell = 3/\alpha_2$ in Problem 5.89:

$$\text{power} = \text{flux} \times \text{area}$$
$$= \frac{(3.39 \times 10^{-5})^2 (1)^2 (1/\sqrt{2})}{2(6.39 \times 10^{-3})} (1 - e^{-6}) \times 2 = 0.1267 \ \mu\text{W}$$

5.98 A plane wave is incident normally on the surface of a semi-infinite conducting medium; let δ denote the skin depth. The current density induced in the conducting medium by the transmitted electric field, $\mathbf{J}_t = \sigma \mathbf{E}_t$, decays exponentially with depth. Show that the total current per unit width in the plane of the surface is the same as would be obtained if $\mathbf{J}_t$ were constant at its surface value to a depth of $\delta/\sqrt{2}$, and zero beyond that point.

$\blacksquare$ By (1) of Problem 4.141,

$$I_z = \frac{J_m \delta}{1 + j}$$

where J_m is the surface amplitude. Then

$$|I_z| = \frac{J_m \delta}{\sqrt{2}} = J_m \frac{\delta}{\sqrt{2}}$$

5.99 A plane wave traveling in free space is normally incident on the surface of a perfect conductor. If the total electric field is zero 1 m away from the surface of the perfect conductor, what is the lowest possible frequency of the incident wave?

$\blacksquare$ For a perfect conductor, E_t vanishes at distances $0, \lambda/2, \lambda, \ldots$, from the surface; thus

$$n\frac{\lambda}{2} = 1 \ \text{m} \quad \text{or} \quad \frac{nu}{2f} = 1 \quad \text{or} \quad f_{\min} = 1\frac{u}{2} = 1.5 \times 10^8 \ \text{Hz} = 150 \ \text{MHz}$$

5.100 An airplane flies over the surface of the ocean. The airplane transmits a signal in the form of a 1-MHz plane wave having an electric field intensity of 1000 V/m and propagating vertically downward. If a submarine requires a minimum signal level of $10 \ \mu\text{V/m}$, how deep can it be submerged and still be reached by the airplane?

$\blacksquare$ At 1 MHz seawater (medium 2) is a good conductor (Problem 5.49); thus

$$\eta_2 \approx \sqrt{\frac{\omega\mu}{\sigma}} \underline{/45°} = \sqrt{\frac{2\pi \times 10^6 \times 4\pi \times 10^{-7}}{4}} \underline{/45°} = 1.4 \underline{/45°} \ \Omega$$

$$\alpha_2 \approx \beta_2 \approx \sqrt{\frac{\omega\mu\sigma}{2}} = \sqrt{\frac{2\pi \times 10^{-6} \times 4\pi \times 10^{-7} \times 4}{2}} = 3.97 \ \text{m}^{-1}$$

The corresponding terms for free space (medium 1) are

$$\eta_1 = 120\pi \ \Omega \qquad \alpha_1 = 0 \qquad \beta_1 = \frac{\omega}{u} = 2.09 \times 10^{-2} \ \text{m}^{-1}$$

The transmission and reflection coefficients become

$$T = \frac{2\eta_2}{\eta_2 + \eta_1} \approx 7.43 \times 10^{-3} \underline{/44.80°} \quad \text{and} \quad \Gamma = T - 1 \approx 0.995 \underline{/180°}$$

Consequently, the electric field transmitted into the ocean will be $1000 \times |T| = 7.43 \ \text{V/m}$. This will be reduced in magnitude as it propagates a distance d into the ocean by the factor $e^{-\alpha_2 d}$. Thus, we require d such that

$$10 \times 10^{-6} = 7.43 e^{-3.97d} \quad \text{or} \quad d = 3.41 \ \text{m}$$

5.101 Relative to the coordinate system shown in Fig. 5-6, write complete time-domain expressions for all the field vectors in Problem 5.100.

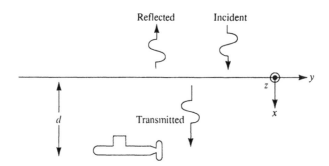

Fig. 5-6

▮ Assuming the incident electric field to be polarized in the z-direction, we obtain from Problem 5.87, with $(x, y, z) \to (z, -y, x)$:

$\mathbf{E}_i = E_m \cos(\omega t - \beta_1 x)\, \mathbf{a}_z = 1000 \cos(2\pi \times 10^6 t - 2.09 \times 10^{-2} x)\, \mathbf{a}_z$ (V/m)

$\mathbf{E}_r = |\Gamma|\, E_m \cos(\omega t + \beta_1 x + \theta_\Gamma)\, \mathbf{a}_z = 995 \cos(2\pi \times 10^6 t + 2.09 \times 10^{-2} x + 180°)\, \mathbf{a}_z$ (V/m)

$\mathbf{E}_t = |\mathrm{T}|\, E_m e^{-\alpha_2 x} \cos(\omega t - \beta_2 x + \theta_\mathrm{T})\, \mathbf{a}_z = 7.43\, e^{-3.97 x} \cos(2\pi \times 10^6 t - 3.97 x + 44.8°)\, \mathbf{a}_z$ (V/m)

$\mathbf{H}_i = \dfrac{E_m}{|\eta_1|} \cos(\omega t - \beta_1 x)\,(-\mathbf{a}_y) = -2.65 \cos(2\pi \times 10^6 t - 2.09 \times 10^{-2} x)\, \mathbf{a}_y$ (A/m)

$\mathbf{H}_r = -\dfrac{|\Gamma|\, E_m}{|\eta_1|} \cos(\omega t + \beta_1 x + \theta_\Gamma)\,(-\mathbf{a}_y) = 2.64 \cos(2\pi \times 10^6 t + 2.09 \times 10^{-2} x + 180°)\, \mathbf{a}_y$ (A/m)

$\mathbf{H}_t = \dfrac{|\mathrm{T}|\, E_m}{|\eta_2|} e^{-\alpha_2 x} \cos(\omega t - \beta_2 x + \theta_\mathrm{T} - \theta_{\eta 2})\,(-\mathbf{a}_y) = -5.31 e^{-3.97 x} \cos(2\pi \times 10^6 t - 3.97 x + 44.8° - 45°)\, \mathbf{a}_y$ (A/m)

5.102 A harmonic plane wave falls normally on the planar interface between two different media. Develop expressions for the time-average power densities into and out of the interface.

▮ Let the field phasors be specified as in Problem 5.86. In Problem 5.24 make the identifications $\hat{\mathbf{E}}^+ \equiv \hat{\mathbf{E}}_i$ and $\hat{\mathbf{E}}^- \equiv \hat{\mathbf{E}}_r$, so that $|E_m^+| \equiv E_m$, $\theta^+ \equiv 0$; $|E_m^-| \equiv |\Gamma|\, E_m$, $\theta^- \equiv \theta_\Gamma$. Then we have, for the power influx,

$$P_{\text{in}} = \langle S_{\text{inc}+\text{refl}} \rangle = \frac{E_m^2}{2 |\eta_1|} [(1 - |\Gamma|^2) \cos \theta_{\eta 1} - 2 |\Gamma| \sin \theta_\Gamma \sin \theta_{\eta 1}]$$

$$= \frac{E_m^2}{2 |\eta_1|^2} [(1 - |\Gamma|^2) \operatorname{Re}(\eta_1) - 2 \operatorname{Im}(\Gamma) \operatorname{Im}(\eta_1)] \tag{1}$$

and for the power efflux

$$P_{\text{out}} = \langle S_{\text{trans}} \rangle = \frac{|\mathrm{T}|^2 E_m^2}{2 |\eta_2|} \cos \theta_{\eta 2} = \frac{|\mathrm{T}|^2 E_m^2}{2 |\eta_2|^2} \operatorname{Re}(\eta_2) \tag{2}$$

Note the presence of an interference term in (1) when η_1 is complex.

5.103 Refer to Problem 5.102. Prove that $P_{\text{in}} = P_{\text{out}}$, as demanded by conservation of energy.

▮ The simplest method here is brute force. Starting with

$$\Gamma = \frac{\eta_2 - \eta_1}{\eta_2 + \eta_1} \qquad \mathrm{T} = \frac{2\eta_2}{\eta_2 + \eta_1}$$

calculate

$$1 - |\Gamma|^2 = 1 - \frac{\eta_2 - \eta_1}{\eta_2 + \eta_1} \frac{\eta_2^* - \eta_1^*}{\eta_2^* + \eta_1^*} = 1 - \frac{|\eta_2|^2 - 2\operatorname{Re}(\eta_1 \eta_2^*) + |\eta_1|^2}{|\eta_2|^2 + 2\operatorname{Re}(\eta_1 \eta_2^*) + |\eta_1|^2} = \frac{4\operatorname{Re}(\eta_1 \eta_2^*)}{|\eta_2|^2 + 2\operatorname{Re}(\eta_1 \eta_2^*) + |\eta_1|^2}$$

$$\operatorname{Im}(\Gamma) = \frac{1}{2j}\left(\frac{\eta_2 - \eta_1}{\eta_2 + \eta_1} - \frac{\eta_2^* - \eta_1^*}{\eta_2^* + \eta_1^*}\right) = \frac{-2\operatorname{Im}(\eta_1 \eta_2^*)}{|\eta_2|^2 + 2\operatorname{Re}(\eta_1 \eta_2^*) + |\eta_1|^2}$$

$$|\mathrm{T}|^2 = \frac{4|\eta_2|^2}{|\eta_2|^2 + 2\operatorname{Re}(\eta_1 \eta_2^*) + |\eta_1|^2}$$

Substitute these expressions in (1) and (2) of Problem 5.102, and divide the resulting equations to obtain

$$\frac{P_{\text{in}}}{P_{\text{out}}} = \frac{\operatorname{Re}(\eta_1 \eta_2^*) \operatorname{Re}(\eta_1) + \operatorname{Im}(\eta_1 \eta_2^*) \operatorname{Im}(\eta_1)}{|\eta_1|^2 \operatorname{Re}(\eta_2)} = \frac{\operatorname{Re}(\eta_1 \eta_2^* \cdot \eta_1^*)}{|\eta_1|^2 \operatorname{Re}(\eta_2)} = \frac{|\eta_1|^2 \operatorname{Re}(\eta_2^*)}{|\eta_1|^2 \operatorname{Re}(\eta_2)} = 1$$

5.104 Specialize the results of Problem 5.102 to the common case where the medium of the incident wave is lossless (e.g., free space).

▮ When $\sigma_1 = 0$, $\eta_1 = \sqrt{\mu_1/\epsilon_1}$ is real, and (1) simplifies to

$$P_{\text{in}} = \frac{E_m^2(1 - |\Gamma|^2)}{2\eta_1} = P_{\text{inc}} - P_{\text{refl}} \tag{1}$$

In this case the incident and reflected waves are uncoupled.

5.105 Rewrite (*1*) of Problem 5.104 as a relation between P_{trans} and P_{inc}.

▐ By energy conservation (Problem 5.103),
$$P_{\text{trans}} = (1 - |\Gamma|^2)P_{\text{inc}} \tag{1}$$
(irrespective of the value of σ_2).

5.106 Use (*1*) of Problem 5.105 to check Problems 5.90 and 5.91.

▐ The incident power in Problem 5.90 is
$$P_{\text{inc}} = \frac{E_i^2}{2\eta_1} = \frac{(100)^2}{2(60\pi)} = \frac{250}{3\pi}$$
The reflection coefficient is $\Gamma = 1/7$, and the transmitted power in Problem 5.91 is $P_{\text{trans}} = 4000/49\pi$. These values satisfy (*1*) of Problem 5.105.

5.107 Use (*1*) of Problem 5.105 to check Problems 5.93, 5.94, and 5.95.

▐ The incident power in Problem 5.93 is
$$P_{\text{inc}} = \frac{\eta_1 H_i^2}{2} = \frac{(120\pi)(1)^2}{2} = 60\pi$$
The reflection coefficient is $\Gamma = 1/5$, and the transmitted power in Problem 5.95 is $P_{\text{trans}} = 288\pi/5$. These values satisfy (*1*) of Problem 5.105.

5.108 Assuming that both media in Problems 5.84 and 5.86 are lossless, give phasor expressions for the total electric and magnetic fields in medium 1.

▐ With $\gamma_1 = j\beta_1$, and with η_1 positive and Γ real, we have
$$\hat{\mathbf{E}}_1 = \hat{\mathbf{E}}_i + \hat{\mathbf{E}}_r = [E_m e^{-j\beta_1 z} + E_m \Gamma e^{j\beta_1 z}]\,\mathbf{a}_x = E_m e^{-j\beta_1 z}[1 + \Gamma e^{j2\beta_1 z}]\,\mathbf{a}_x \tag{1}$$
$$\hat{\mathbf{H}}_1 = \hat{\mathbf{H}}_i + \hat{\mathbf{H}}_r = \frac{E_m}{\eta_1} e^{-j\beta_1 z}[1 - \Gamma e^{j2\beta_1 z}]\,\mathbf{a}_y \tag{2}$$

5.109 Represent the electric field vector $\mathbf{E}_1$ given by (*1*) of Problem 5.108 graphically.

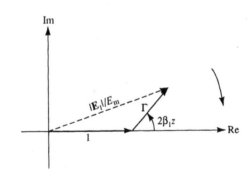

Fig. 5-7

▐ From (*1*), the amplitude of $\mathbf{E}_1$ varies with z according to
$$|\mathbf{E}_1| = E_m |1 + \Gamma e^{j2\beta_1 z}| = E_m \begin{cases} |1\,\underline{/0} + |\Gamma|\,\underline{/2\beta_1 z}| & \Gamma > 0 \\ |1\,\underline{/0} + |\Gamma|\,\underline{/2\beta_1 z + \pi}| & \Gamma < 0 \end{cases} \tag{1}$$

Figure 5-7 is the complex graph of $|\mathbf{E}_1|/E_m$ for $\Gamma > 0$; this is known as the *crank diagram*. For distances away from the boundary, z is increasingly negative, so that the vector $\Gamma e^{j2\beta_1 z}$ rotates in the clockwise direction about the tip of the $1\,\underline{/0°}$ vector.

5.110 From Fig. 5-7 obtain the maxima and minima of $|\mathbf{E}_1|$, for $\Gamma > 0$.

▐ The maximum of $|\mathbf{E}_1|$, which equals $E_m(1 + \Gamma)$, will occur at $2\beta_1 z = -2n\pi$, for $n = 0, 1, 2, \ldots$. Since $\beta_1 = 2\pi/\lambda_1$, the maximum occurs at $z = -n\lambda_1/2$, or at multiples of one-half wavelength away from the boundary.
   The minimum of $|\mathbf{E}_1|$, which equals $E_m(1 - \Gamma)$, will occur when $2\beta_1 z = -m\pi$, for $m = 1, 3, 5, \ldots$, or $z = -m\lambda_1/4$; that is, at odd multiples of a quarter-wavelength away from the boundary.

5.111 Show that for $\Gamma = \pm 1$ the **E**-wave of Problem 5.108 is a *standing wave*.

▮ Transforming (1) of Problem 5.108 into the time domain,
$$E_{1x} = E_m \text{ Re } [e^{j(\omega t - \beta_1 z)} + \Gamma e^{j(\omega t + \beta_1 z)}]$$
$$= E_m \cos(\omega t - \beta_1 z) + E_m \Gamma \cos(\omega t + \beta_1 z)$$
$$= [(1 + \Gamma) E_m \cos \beta_1 z] \cos \omega t + [(1 - \Gamma) E_m \sin \beta_1 z] \sin \omega t \qquad (1)$$
When $\Gamma = \pm 1$, and only then, the z- and t-variations of E are uncoupled, resulting in a standing wave. When $-1 < \Gamma < +1$, the disturbance, though progressive, can nevertheless be considered the sum of *two* standing waves.

5.112 Let $\Gamma = 1$ in Problems 5.108 and 5.111. Plot E_{1x} over the first wavelength away from the boundary, at times $\omega t = 0, \pi/4, \pi/2, 3\pi/4, \pi$. Identify the *nodes* of the standing wave and describe the time-variation of E_{1x} over an internodal segment.

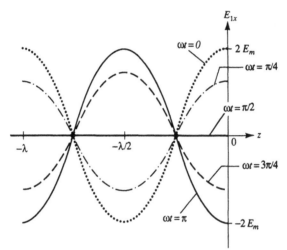

Fig. 5-8

▮ See Fig. 5-8. The nodes, or points of *permanently* zero field, are $z = -\lambda/4, -3\lambda/4, \ldots$ (cf. Problem 5.110). Between successive nodes all the **E**-vectors achieve their maximum magnitudes simultaneously and achieve their minimum magnitudes (zero) simultaneously. This is in contrast to a traveling wave, which has no nodes and in which the oscillations are exactly in phase only at points separated by an integral number of wavelengths.

5.113 Verify that (1) of Problem 5.111 and the crank diagram Fig. 5-7 are in agreement.

▮ By standard trigonometry we can always determine an angle δ such that
$$A \cos \theta + B \sin \theta = \sqrt{A^2 + B^2} \cos(\theta + \delta)$$
Thus, (1) of Problem 5.111 implies
$$(|\mathbf{E}_1|/E_m)^2 = (1 + \Gamma)^2 \cos^2 \beta_1 z + (1 - \Gamma)^2 \sin^2 \beta_1 z$$
$$= (1 + \Gamma^2)(\cos^2 \beta_1 z + \sin^2 \beta_1 z) + 2\Gamma(\cos^2 \beta_1 z - \sin^2 \beta_1 z)$$
$$= 1 + \Gamma^2 + 2\Gamma \cos 2\beta_1 z \qquad (1)$$
But this is precisely the statement of the law of cosines for the triangle of Fig. 5-7.

5.114 When $-1 < \Gamma < 1$, the two standing waves composing the **E**-wave (1) of Problem 5.111 have different amplitudes. Evaluate the *standing wave ratio*, defined as the ratio of the larger amplitude to the smaller.

▮ For $0 < \Gamma < 1$, $S = (1 + \Gamma)/(1 - \Gamma)$; for $-1 < \Gamma < 0$, $S = (1 - \Gamma)/(1 + \Gamma)$. Both formulas are subsumed in
$$S = \frac{1 + |\Gamma|}{1 - |\Gamma|} \qquad (1)$$

5.115 Show that the standing wave ratio S may be equivalently defined as the ratio $|\mathbf{E}_1|_{max}/|\mathbf{E}_1|_{min}$; i.e., the ratio of the maximum magnitude of the field (over all space and all time) to the minimum magnitude.

▮ This follows at once from Problems 5.113 and 5.110.

5.116 Consider a planar interface ($z = 0$) between free space and glass ($\sigma \approx 0$, $\epsilon_r \approx 4$, and $\mu_r = 1$). Given a plane wave with $\mathbf{E} = 1 \cos(\omega t - \beta_1 z) \mathbf{a}_x$ (V/m) incident from free space (medium 1) normal to the glass, determine the transmission and reflection coefficients.

$$\eta_1 = \sqrt{\frac{\mu_0}{\epsilon_0}} = 120\pi \ \Omega \qquad \eta_2 = \frac{\eta_1}{\sqrt{\epsilon_r}} = 60\pi \ \Omega$$
$$\Gamma = \frac{60\pi - 120\pi}{60\pi + 120\pi} = -\frac{1}{3} \qquad T = 1 + \Gamma = \frac{2}{3}$$

5.117 Refer to Problem 5.116. Determine the time-average power transmitted through a 1-m² patch of the interface.

❙ By Problem 5.104,

$$P_{\text{in}} = P_{\text{out}} = \frac{E_m^2 (1 - |\Gamma|^2)}{2\eta_1} = \frac{1^2 (1 - \frac{1}{9})}{2(120\pi)} = 1.18 \ \text{mW}$$

5.118 Find the standing wave ratio for medium 1 (free space) of Problem 5.116.

❙ By (1) of Problem 5.114,

$$S = \frac{1 + |\Gamma|}{1 - |\Gamma|} = \frac{1 + 1/3}{1 - 1/3} = 2$$

5.119 Show that the standing wave ratio has the same value on either side of an interface between media of properties $(\sigma_1, \mu_1, \epsilon_1)$ and $(\sigma_2, \mu_2, \epsilon_2)$.

❙ The standing wave ratio S depends on $|\Gamma|$. But

$$|\Gamma_1| = \frac{|\eta_2 - \eta_1|}{|\eta_2 + \eta_1|} = \frac{|\eta_1 - \eta_2|}{|\eta_1 + \eta_2|} = |\Gamma_2|$$

5.120 Airplanes use radar altimeters to determine their low-level altitude accurately. Given that an airplane is flying over the ocean (medium 2) and the radar frequency is 7 GHz, determine the percent reduction in received power due to the ocean.

❙ For 7 GHz seawater is a poor conductor (Problem 5.49). Thus, from Problem 5.44,

$$\eta_2 = \sqrt{\frac{\mu_2}{\epsilon_2}} \left(1 + j\frac{\sigma_2}{2\omega\epsilon_2}\right) = \frac{120\pi}{9}(1 + j0.064) = 41.72 \ \underline{/3.618°}$$

while $\eta_1 = 120\pi = 377 \ \underline{/0°}$. Hence,

$$\Gamma = \frac{\eta_2 - \eta_1}{\eta_2 + \eta_1} \approx -0.8$$

The lost power is that transmitted into the ocean; by Problem 5.105,

$$P_{\text{trans}}/P_{\text{inc}} = 1 - |\Gamma|^2 \approx 36\%$$

5.121 In Problem 5.84, assume that medium 2 is a perfect conductor. Calculate the reflection and transmission coefficients for medium 1.

❙ By Problem 5.38, as $\tau_2 \to \infty$, $\eta_2 \to 0$. Therefore,

$$\Gamma = \frac{\eta_2 - \eta_1}{\eta_2 + \eta_1} = \frac{0 - \eta_1}{0 + \eta_1} = -1 \qquad T = \frac{2\eta_2}{\eta_2 + \eta_1} = \frac{0}{0 + \eta_1} = 0$$

5.122 Obtain expressions for the total fields in medium 1 of Problem 5.121.

❙ Problem 5.86 gives:

$$\hat{\mathbf{E}}_1 = \hat{\mathbf{E}}_i + \hat{\mathbf{E}}_r = E_m(e^{-\gamma_1 z} - e^{\gamma_1 z})\mathbf{a}_x$$
$$\hat{\mathbf{H}}_1 = \hat{\mathbf{H}}_i + \hat{\mathbf{H}}_r = \frac{E_m}{\eta_1}(e^{-\gamma_1 z} + e^{\gamma_1 z})\mathbf{a}_y$$

5.123 Rewrite the results of Problem 5.122 under the additional assumption that $\sigma_1 = 0$.

❙ Setting $\gamma_1 = j\beta_1$:

$$\hat{\mathbf{E}}_1 = -2jE_m \sin \beta_1 z \ \mathbf{a}_x \qquad \hat{\mathbf{H}}_1 = \frac{2E_m}{\eta_1} \cos \beta_1 z \ \mathbf{a}_y \quad (\eta_1 > 0)$$

5.124 Write the fields of Problem 5.123 in the time domain.

❙
$$\mathbf{E}_1 = \text{Re}\,(\hat{\mathbf{E}}_1 e^{j\omega t}) = 2E_m \sin \beta_1 z \sin \omega t \ \mathbf{a}_x \qquad \mathbf{H}_1 = \frac{2E_m}{\eta_1} \cos \beta_1 z \cos \omega t \ \mathbf{a}_y$$

It is seen that both fields in medium 1 are standing waves.

5.125 Reconcile the results of Problems 5.121–5.124 and 5.111.

▌ Looking again at the solution of Problem 5.111, one sees that the assumption $\sigma_2 = 0$ (made in the parent Problem 5.108) was never actually used. Thus, provided only that $\alpha_1 = 0$ and $\Gamma = \pm 1$, there will be a standing wave in medium 1; this agrees with Problems 5.121–5.124. One might say that the incident wave "doesn't care" where the reflected wave is coming from, so long as the two have equal amplitudes (and a phase difference of 0° or 180° at the interface).

5.126 Determine the surface current density at the interface of the material media of Problem 5.123.

▌ Just outside the perfect conductor (at $z = 0^-$) the tangential magnetic field is, from Problem 5.124,

$$\mathbf{H}_1 = \frac{2E_m}{\eta_1} \cos \omega t \, \mathbf{a}_y$$

whereas, just inside, $\mathbf{H}_2 = \mathbf{0}$. The linear current density at the surface is then given by (1) of Problem 3.85 as

$$\mathbf{k} = \frac{2E_m}{\eta_1} \cos \omega t \, \mathbf{a}_x$$

5.127 A coating of lossless dielectric, of (small) thickness τ, is applied to the surface of a perfect conductor. Determine the effective reflection coefficient at the air interface of this dielectric for normal incidence.

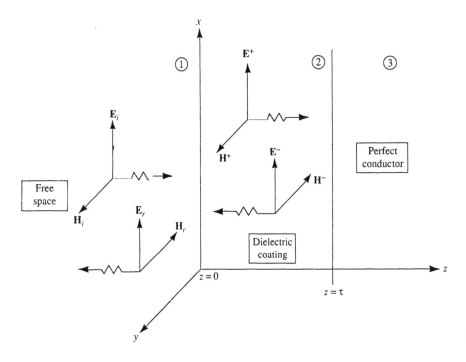

Fig. 5-9

▌ The interfaces between the various material media are shown in Fig. 5-9. The incident and reflected field vectors in region 1 are

$$\mathbf{E}_i = E_i e^{-j\beta_1 z}\mathbf{a}_x \qquad \mathbf{E}_r = E_r e^{j\beta_1 z}\mathbf{a}_x$$

$$\mathbf{H}_i = \frac{E_i}{\eta_1} e^{-j\beta_1 z}\mathbf{a}_y \qquad \mathbf{H}_r = -\frac{E_r}{\eta_1} e^{j\beta_1 z}\mathbf{a}_y$$

where $\eta_1 = 120\pi \ \Omega$. Similarly, within the dielectric coating (region 2),

$$\mathbf{E}^+ = E^+ e^{-j\beta_2 z}\mathbf{a}_x \qquad \mathbf{E}^- = E^- e^{j\beta_2 z}\mathbf{a}_x$$

$$\mathbf{H}^+ = \frac{E^+}{\eta_2} e^{-j\beta_2 z}\mathbf{a}_y \qquad \mathbf{H}^- = \frac{-E^-}{\eta_2} e^{j\beta_2 z}\mathbf{a}_y$$

where $\eta_2 = (\sqrt{\mu_r/\epsilon_r})\eta_1$. Continuity of tangential $\mathbf{E}$ and tangential $\mathbf{H}$ at $z = 0$ requires

$$E_i + E_r = E^+ + E^- \qquad \text{and} \qquad \frac{E_i}{\eta_1} - \frac{E_r}{\eta_1} = \frac{E^+}{\eta_2} - \frac{E^-}{\eta_2} \tag{1}$$

and continuity of tangential $\mathbf{E}$ at $z = \tau$ requires

$$E^+ e^{-j\beta_2 \tau} + E^- e^{j\beta_2 \tau} = 0 \tag{2}$$

Elimination of E^+ and E^- among the three equations (1) and (2) leads to

$$\Gamma_{\text{eff}} \equiv \frac{E_r}{E_i} = \frac{\eta_2 \sin \beta_2 \tau - j\eta_1 \cos \beta_2 \tau}{\eta_2 \sin \beta_2 \tau + j\eta_1 \cos \beta_2 \tau} \tag{3}$$

According to (3), $|\Gamma_{\text{eff}}| = 1$, for any value of τ.

5.128 Show that the choice $\tau = \lambda_2/4$ in Problem 5.127 renders the dielectric "invisible" (and produces a standing wave in medium 1).

▎ For $\beta_2 \tau = (2\pi/\lambda_2)(\lambda_2/4) = \pi/2$, $\Gamma_{\text{eff}} = \eta_2/\eta_2 = +1$. On the other hand,
$$\lim_{\tau \to 0} \Gamma_{\text{eff}} = -1 \quad \text{(cf. Problem 5.121)}$$

Thus, the sole effect of a quarter-wavelength coating is to change the sign of E_r. (A half-wavelength coating would not even do this, and so would be *completely* invisible.)

5.129 Effective shielding of electronic equipment from the effects of external electromagnetic fields can be obtained with conductive barriers. The shielding effectiveness (SE) is often expressed as the ratio of the magnitudes of the transmitted and incident electric fields, as shown in Fig. 5-10:

$$\text{SE} \equiv \frac{|\mathbf{E}_t|_{z=w}}{|\mathbf{E}_i|_{z=0}}$$

Derive an expression for the shielding effectiveness, assuming normal incidence.

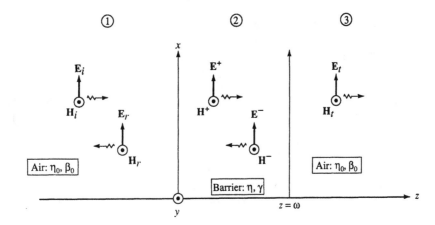

Fig. 5-10

▎ Proceeding as in Problem 5.127, we obtain:

Region 1

$$\mathbf{E}_i = E_i e^{-j\beta_0 z}\mathbf{a}_x \qquad \mathbf{E}_r = E_r e^{j\beta_0 z}\mathbf{a}_x$$
$$\mathbf{H}_i = \frac{E_i}{\eta_0} e^{-j\beta_0 z}\mathbf{a}_y \qquad \mathbf{H}_r = -\frac{E_r}{\eta_0} e^{j\beta_0 z}\mathbf{a}_y$$

Region 2

$$\mathbf{E}^+ = E^+ e^{-\gamma z}\mathbf{a}_x \qquad \mathbf{E}^- = E^- e^{\gamma z}\mathbf{a}_x$$
$$\mathbf{H}^+ = \frac{E^+}{\eta} e^{-\gamma z}\mathbf{a}_y \qquad \mathbf{H}^- = -\frac{E^-}{\eta} e^{\gamma z}\mathbf{a}_y$$

Region 3

$$\mathbf{E}_t = E_t e^{-j\beta_0(z-w)}\mathbf{a}_x \qquad \mathbf{H}_t = \frac{E_t}{\eta_0} e^{-j\beta_0(z-w)}\mathbf{a}_y$$

Boundary conditions at $z = 0$ yield

$$E_i + E_r = E^+ + E^- \tag{1}$$
$$\frac{E_i}{\eta_0} - \frac{E_r}{\eta_0} = \frac{E^+}{\eta} - \frac{E^-}{\eta} \tag{2}$$

Similarly, at $z = w$, we have

$$E^+ e^{-\gamma w} + E^- e^{\gamma w} = E_t \tag{3}$$
$$\frac{E^+}{\eta} e^{-\gamma w} - \frac{E^-}{\eta} e^{\gamma w} = \frac{E_t}{\eta_0} \tag{4}$$

Eliminate E^+, E^-, and E_r from among (1)–(4), as follows. Multiply (3) by $1/\eta$ and add to (4):

$$\frac{2E^+}{\eta}e^{-\gamma w} = \left(\frac{1}{\eta}+\frac{1}{\eta_0}\right)E_t \quad \text{or} \quad E_t = (1+\Gamma_2)e^{-\gamma w}E^+ \tag{5}$$

where, as usual, $\Gamma_2 = (\eta_0 - \eta)/(\eta_0 + \eta)$. Next, multiply (3) by $1/\eta_0$ and subtract (4), to find

$$E^- = \Gamma_2 e^{-2\gamma w}E^+ \tag{6}$$

Similarly, multiply (1) by $1/\eta_0$ and add to (2), obtaining

$$E_i = \frac{1}{2\eta}(\eta_0 + \eta)E^+ - \frac{1}{2\eta}(\eta_0 - \eta)E^- = \frac{1}{1-\Gamma_2}[E^+ - \Gamma_2 E^-] \tag{7}$$

Finally, divide (5) by (7), and use (6) to evaluate E^-/E^+:

$$\frac{E_t}{E_i} = \frac{(1-\Gamma_2^2)e^{-\gamma w}}{1-\Gamma_2^2 e^{-2\gamma w}} \quad \text{or} \quad \text{SE} = \frac{|1-\Gamma_2^2|\,e^{-\alpha w}}{|1-\Gamma_2^2 e^{-2\gamma w}|} \tag{8}$$

5.130 Approximate the expression for SE found in Problem 5.129 under the assumptions that the barrier is a good conductor (at the incident frequency) and that w is large compared to the skin depth $\delta = 1/\alpha$.

▮ By Problem 5.45,

$$\left|\frac{\eta}{\eta_0}\right| \approx \sqrt{\omega\mu_r\epsilon_0/\sigma} \ll 1$$

Hence

$$\Gamma_2 = \frac{1-(\eta/\eta_0)}{1+(\eta/\eta_0)} \approx \left(1-\frac{\eta}{\eta_0}\right)^2$$

$$\Gamma_2^2 \approx \left(1-\frac{\eta}{\eta_0}\right)^4 \approx 1-4\frac{\eta}{\eta_0}$$

Thus, for the numerator in (8) of Problem 5.129, we have

$$|1-\Gamma_2^2|\,e^{-w/\delta} \approx 4\sqrt{\omega\mu_r\epsilon_0/\sigma}\,e^{-w/\delta}$$

To the same order of approximation, the absolute value of the denominator in (8) is unity. Consequently,

$$\text{SE} \approx 4\sqrt{\frac{\omega\mu_r\epsilon_0}{\sigma}}\,e^{-w/\delta} = \frac{4\sqrt{2}}{\eta_0\sigma\delta}e^{-w/\delta} \tag{1}$$

The SE is an inverse measure: the smaller the SE, the smaller is E_t and the better is the shielding. As would be expected, (1) shows that SE decreases as σ and/or w increases (ω staying fixed).

5.131 Figure 5-11 depicts a case of oblique incidence on an interface between two *lossless* media. The incident wave is polarized perpendicular to the *plane of incidence* (the plane determined by the incident Poynting vector and the normal to the interface). The plane of incidence is here referred to two cartesian coordinate systems, xy and $x'y'$. Give the coordinate transformations and the unit-vector transformations between the two systems.

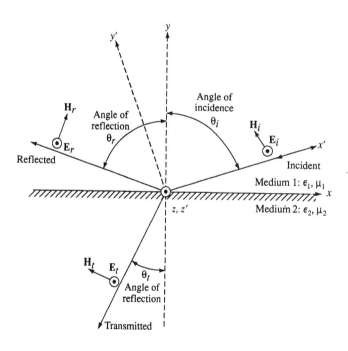

Fig. 5-11

▌ Starting with the unit vectors, one sees from Fig. 5-11 that
$$\mathbf{a}_{x'} = \quad \sin \theta_i \, \mathbf{a}_x + \cos \theta_i \, \mathbf{a}_y$$
$$\mathbf{a}_{y'} = -\cos \theta_i \, \mathbf{a}_x + \sin \theta_i \, \mathbf{a}_y \tag{1}$$
Now, in the vector equality $x'\mathbf{a}_{x'} + y'\mathbf{a}_{y'} = x\mathbf{a}_x + y\mathbf{a}_y$ substitute for $\mathbf{a}_{x'}$ and $\mathbf{a}_{y'}$ the expressions (1), and equate coefficients of $\mathbf{a}_x$ and coefficients of $\mathbf{a}_y$:
$$x = \sin \theta_i \, x' - \cos \theta_i \, y'$$
$$y = \cos \theta_i \, x' + \sin \theta_i \, y' \tag{2}$$
or, inversely,
$$x' = \quad \sin \theta_i \, x + \cos \theta_i \, y$$
$$y' = -\cos \theta_i \, x + \sin \theta_i \, y \tag{3}$$
Note that (1) and (3) are identical in form.

5.132   Refer to Problem 5.131. Assuming that the incident **E**-field is given by $\mathbf{E}_i = E_0 e^{j\beta_1 x'} \mathbf{a}_{z'}$, express $\mathbf{E}_i$ and $\mathbf{H}_i$ in the (x, y, z) coordinate system.

▌ Using (1) and (3) of Problem 5.131:
$$\mathbf{E}_i = E_0 e^{j\beta_1 x'} \mathbf{a}_{z'} = E_0 e^{j\beta_1(x \sin \theta_i + y \cos \theta_i)} \mathbf{a}_z \tag{1}$$
$$\mathbf{H}_i = \frac{E_0}{\eta_1} e^{j\beta_1 x'} \mathbf{a}_{y'} = \frac{E_0}{\eta_1} e^{j\beta_1(x \sin \theta_i + y \cos \theta_i)}(-\cos \theta_i \, \mathbf{a}_x + \sin \theta_i \, \mathbf{a}_y) \tag{2}$$

5.133   For the incident plane wave of Problem 5.132, write expressions in the (x, y, z) system for the reflected and transmitted **E**- and **H**-field vectors, on the (true) assumption that the reflected and transmitted waves are also perpendicularly polarized.

▌ Define, in the usual fashion, reflection and transmission coefficients for perpendicular polarization:
$$\Gamma_\perp \equiv \frac{|\mathbf{E}_r|}{|\mathbf{E}_i|} \qquad T_\perp \equiv \frac{|\mathbf{E}_t|}{|\mathbf{E}_i|}$$
Then the reflected and transmitted amplitudes will be
$$|\mathbf{E}_r| = \Gamma_\perp E_0 \qquad |\mathbf{E}_t| = T_\perp E_0$$
$$|\mathbf{H}_r| = \frac{\Gamma_\perp E_0}{\eta_1} \qquad |\mathbf{H}_t| = \frac{T_\perp E_0}{\eta_2}$$
Now, the phase of the *incoming* incident fields is given by
$$\exp\left[j\beta_1(x \sin \theta_i + y \cos \theta_i)\right]$$
This must give the phase of the reflected fields under the replacements $j \to -j$ (the reflected wave is *outgoing*) and $\theta_i \to -\theta_r$ (the clockwise angle from the positive y-axis to the reflected direction is $-\theta_r$). Hence
$$\mathbf{E}_r = \Gamma_\perp E_0 e^{j\beta_1(x \sin \theta_r - y \cos \theta_r)} \mathbf{a}_z$$
$$\mathbf{H}_r = -\frac{\Gamma_\perp E_0}{\eta_1} e^{j\beta_1(x \sin \theta_r - y \cos \theta_r)}(-\cos \theta_r \, \mathbf{a}_x - \sin \theta_r \, \mathbf{a}_y) \tag{1}$$
$$= \frac{\Gamma_\perp E_0}{\eta_1} e^{j\beta_1(x \sin \theta_r - y \cos \theta_r)}(\cos \theta_r \, \mathbf{a}_x + \sin \theta_r \, \mathbf{a}_y)$$
Similarly, the replacements $j \to -j$, $\theta_i \to \theta_t - \pi$, and $\beta_1 \to \beta_2$ yield
$$\mathbf{E}_t = T_\perp E_0 e^{j\beta_2(x \sin \theta_t + y \cos \theta_t)} \mathbf{a}_z$$
$$\mathbf{H}_t = \frac{T_\perp E_0}{\eta_2} e^{j\beta_2(x \sin \theta_t + y \cos \theta_t)}(-\cos \theta_t \, \mathbf{a}_x + \sin \theta_t \, \mathbf{a}_y) \tag{2}$$

5.134   Write boundary conditions for the fields of Problem 5.133.

▌ Two boundary conditions consist in the continuity of the tangential components of **E** and **H** at the origin. Thus, from (1) and (2) of Problems 5.132 and 5.133,
$$1 + \Gamma_\perp = T_\perp \tag{1}$$
$$\frac{1}{\eta_1}(-\cos \theta_i) + \frac{\Gamma_\perp}{\eta_1}(\cos \theta_r) = \frac{T_\perp}{\eta_2}(-\cos \theta_t) \tag{2}$$
On the assumption that the interface is current-free, a third boundary condition expresses the continuity of normal **H**:
$$\frac{1}{\eta_1}(\sin \theta_i) + \frac{\Gamma_\perp}{\eta_1}(\sin \theta_r) = \frac{T_\perp}{\eta_2}(\sin \theta_t) \tag{3}$$

5.135 From Fig. 5-11 obtain the *law of reflection*, $\theta_i = \theta_r$.

❚ The argument is very simple. Suppose, on the contrary, that $\theta_r < \theta_i$ $(\theta_r > \theta_i)$; i.e., the reflection process rotates the ray toward (away from) the surface normal. Now consider the reversed reflected wave being reflected as the reversed incident wave (light paths are reversible). In this case the rotation is *away from* (*toward*) the normal—an impossibility for isotropic media 1 and 2.

Observe that the law of reflection, as a matter of symmetry, is quite independent of the boundary conditions listed in Problem 5.134.

5.136 Using Problems 5.134 and 5.135, derive the *law of refraction* (*Snell's law*).

❚ In (3) of Problem 5.134 substitute θ_i for θ_r, and $1 + \Gamma_\perp$ for $T_\perp$, to find

$$\frac{1}{\eta_1} \sin \theta_i = \frac{1}{\eta_2} \sin \theta_t \qquad (1)$$

This is one form of Snell's law.

5.137 From Problem 5.134 deduce the *Fresnel equations* for $\Gamma_\perp$ and $T_\perp$.

❚ Eliminate $T_\perp$ between (2) and (3) of Problem 5.134, obtaining

$$\Gamma_\perp = -\frac{\sin (\theta_i - \theta_t)}{\sin (\theta_i + \theta_t)} \qquad (1)$$

from which
$$T_\perp = 1 + \Gamma_\perp = \frac{2 \cos \theta_i \sin \theta_t}{\sin (\theta_i + \theta_t)} \qquad (2)$$

Only geometric quantities appear in this version of the Fresnel equations; however, the material constants are implicitly involved.

5.138 (*a*) Define the *index of refraction, n*, of a perfect dielectric. (*b*) What is the relation between n and the characteristic impedance η?

❚ (*a*)
$$n \equiv \frac{u_0}{u} = \frac{\text{wave velocity in free space}}{\text{wave velocity in the dielectric}} \qquad (1)$$
Because $u_0 = 1/\sqrt{\mu_0 \epsilon_0}$ and $u = 1/\sqrt{\mu \epsilon}$, $n = \sqrt{\mu_r \epsilon_r}$, and is thus a material property.
(*b*) By (3) of Problem 5.41, $\eta = \mu u = u_0 \mu / n$, or

$$\frac{1}{\eta} = \left(\frac{1}{u_0 \mu}\right) n \qquad (2)$$

5.139 For most common dielectrics $\mu \approx \mu_0$, and (2) of Problem 5.138 becomes an absolute proportionality between $1/\eta$ and n. Give the usual forms of Snell's law (Problem 5.136) and the Fresnel equations (Problem 5.137), in which the explicit physical parameter is $n = \sqrt{\epsilon_r}$.

❚ Equation (1) of Problem 5.136 becomes
$$n_1 \sin \theta_i = n_2 \sin \theta_t \qquad (1)$$
Equation (1) of Problem 5.137 becomes
$$\begin{aligned}
\Gamma_\perp &= -\frac{\sin \theta_i \cos \theta_t - \cos \theta_i \sin \theta_t}{\sin \theta_i \cos \theta_t + \cos \theta_i \sin \theta_t} \\
&= \frac{\cos \theta_i [(n_1/n_2) \sin \theta_i] - \sin \theta_i \cos \theta_t}{\cos \theta_i [(n_1/n_2) \sin \theta_i] + \sin \theta_i \cos \theta_t} \\
&= \frac{n_1 \cos \theta_i - n_2 \cos \theta_t}{n_1 \cos \theta_i + n_2 \cos \theta_t} \qquad (2)
\end{aligned}$$
and so
$$T_\perp = 1 + \Gamma_\perp = \frac{2n_1 \cos \theta_i}{n_1 \cos \theta_i + n_2 \cos \theta_t} \qquad (3)$$

5.140 Define the *critical angle* of incidence for a plane wave in medium 1 that impinges on a medium 2 with $n_2 < n_1$.

❚ The critical angle, θ_c, is the smallest value of θ_i for which the incident wave is totally reflected back into medium 1. By Snell's law,

$$n_1 \sin \theta_c = n_2 \sin 90° \qquad \text{or} \qquad \theta_c = \sin^{-1} (n_2/n_1)$$

Clearly, total reflection is impossible if $n_2 > n_1$.

5.141 A submerged light source is used to illuminate the surface of a swimming pool. Assuming that the light is placed at a depth of $d = 1$ m, determine the area of brightness, as viewed from above the surface. The relative permittivity of water at optical frequencies is $\epsilon_r = 1.77$.

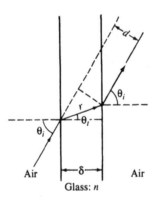

Water: $\epsilon_r = 1.77$

Light source

Fig. 5-12

∎ The problem is illustrated in Fig. 5-12. Only light within a cone of half-angle θ_c (Problem 5.140) will be transmitted into the air. Thus there will be a circle of brightness, of area

$$\pi(d \tan \theta_c)^2 = \pi d^2 \frac{\sin^2 \theta_c}{1 - \sin^2 \theta_c} = \pi d^2 \frac{1/\epsilon_r}{1 - (1/\epsilon_r)} = 4.08 \text{ m}^2$$

5.142 A light ray traveling in air is incident at an angle θ_i on a sheet of glass, of thickness δ and index of refraction n, as shown in Fig. 5-13. Determine the displacement d.

Air   Air

Glass: n **Fig. 5-13**

∎ By reversibility, the entering and exiting rays must be parallel. The upper triangle in Fig. 5-13 yields

$$d = r \sin (\theta_i - \theta_t) \tag{1}$$

and the lower triangle gives

$$r \cos \theta_t = \delta \tag{2}$$

From (1) and (2), and $\sin \theta_t = (1/n) \sin \theta_i$,

$$d = r \sin \theta_i \cos \theta_t - r \cos \theta_i \sin \theta_t$$
$$= \delta \sin \theta_i - \cos \theta_i \left(\frac{\delta \sin \theta_i}{n \cos \theta_t} \right)$$
$$= \delta \sin \theta_i \left[1 - \frac{\cos \theta_i}{\sqrt{n^2 - \sin^2 \theta_i}} \right]$$

5.143 A glass isosceles prism is used to change the path of a light ray, as shown in Fig. 5-14. Assuming that the index of refraction of glass is 1.5, determine the ratio of transmitted and incident power densities.

∎ The angle of incidence on the back face ($\sin \theta_i = 1/\sqrt{2}$) exceeds the critical angle ($\sin \theta_c = 1/1.5$). Thus, all light is reflected at the back face. At the front face, (3) of Problem 5.139 gives

$$T_{1\perp} = \frac{2(1)}{1 + 1.5} = \frac{4}{5}$$

At the bottom face

$$T_{2\perp} = \frac{2(1.5)}{1.5 + 1} = \frac{6}{5}$$

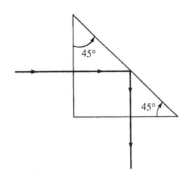

Fig. 5-14

Hence $\qquad \dfrac{\text{transmitted power}}{\text{incident power}} = T_{1\perp}^2 T_{2\perp}^2 = \dfrac{576}{625} = 0.9216$

5.144 A man standing in a boat observes a fish feeding on the bottom of a shallow lake. The man's height is 6 ft, and the lake depth at this location is 10 ft. The fish appears to be 20 ft from the boat. Determine the true horizontal distance of the fish from the boat.

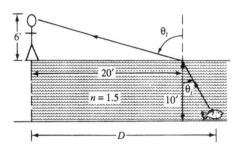

Fig. 5-15

▌ By Snell's law and from the geometry of Fig. 5-15,

$$D = 20 + 10 \tan \theta_i = 20 + 10 \frac{\sin \theta_i}{\sqrt{1 - \sin^2 \theta_i}}$$

$$= 20 + 10 \frac{\sin \theta_t}{\sqrt{n^2 - \sin^2 \theta_t}}$$

Substituting $n = 1.5$ and $\sin \theta_t = 20/\sqrt{6^2 + 20^2}$, we obtain $D = 28.3$ ft.

5.145 A fiber-optic "transmission line" is shown in Fig. 5-16. Determine the minimum ϵ_r of the fiber such that for any angle of incidence of light entering one end of the fiber, the light will be totally contained within the fiber until it exits the other end.

Fig. 5-16

▌ We essentially require that the angle $\theta = 90° - \theta_t$ shown in Fig. 5-16 be greater than or equal to the critical angle for any incident angle θ_i; thus

$$\cos \theta_t = \sin \theta \geq \sin \theta_c = \frac{1}{\sqrt{\epsilon_r}} \qquad \text{or} \qquad \cos^2 \theta_t \geq \frac{1}{\epsilon_r} \tag{1}$$

However, by Snell's law, $\cos^2 \theta_t = 1 - \sin^2 \theta_t = 1 - (1/\epsilon_r) \sin^2 \theta_i$; hence

$$1 - \frac{1}{\epsilon_r} \sin^2 \theta_i \geq \frac{1}{\epsilon_r} \qquad \text{or} \qquad \epsilon_r \geq 1 + \sin^2 \theta_i \tag{2}$$

which will hold for all θ_i if and only if $\epsilon_r \geq 2$.

5.146 Figure 5-11 shows a perpendicularly polarized plane wave incident in the negative x'-direction. Redraw the sketch if the wave travels in the positive z'-direction, and express the incident, reflected, and transmitted field phasors. Again, both media are lossless.

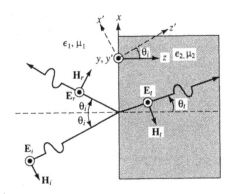

Fig. 5-17

▮ See Fig. 5-17. By analogy with Problem 5.131, the transformation between the primed and unprimed systems is

$$x' = x \cos \theta_i - z \sin \theta_i$$
$$z' = x \sin \theta_i + z \cos \theta_i$$

Therefore

$$\hat{\mathbf{E}}_i = E_i e^{-j\beta_1 z'} \mathbf{a}_y = E_i e^{-j\beta_1(x \sin \theta_i + z \cos \theta_i)} \mathbf{a}_y \tag{1}$$

$$\hat{\mathbf{H}}_i = \frac{E_i}{\eta_1} e^{-j\beta_1(x \sin \theta_i + z \cos \theta_i)}(-\cos \theta_i \, \mathbf{a}_x + \sin \theta_i \, \mathbf{a}_z) \tag{2}$$

$$\hat{\mathbf{E}}_r = E_r e^{j\beta_1(-x \sin \theta_i + z \cos \theta_i)} \mathbf{a}_y \tag{3}$$

$$\hat{\mathbf{H}}_r = \frac{E_r}{\eta_1} e^{j\beta_1(-x \sin \theta_i + z \cos \theta_i)}(\cos \theta_i \, \mathbf{a}_x + \sin \theta_i \, \mathbf{a}_z) \tag{4}$$

$$\hat{\mathbf{E}}_t = E_t e^{-j\beta_2(x \sin \theta_t + z \cos \theta_t)} \mathbf{a}_y \tag{5}$$

$$\hat{\mathbf{H}}_t = \frac{E_t}{\eta_2} e^{-j\beta_2(x \sin \theta_t + z \cos \theta_t)}(-\cos \theta_t \, \mathbf{a}_x + \sin \theta_t \, \mathbf{a}_z) \tag{6}$$

5.147 Repeat Problem 5.146 for *parallel polarization* (Fig. 5-18).

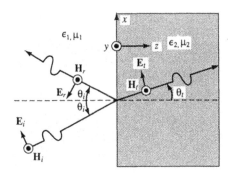

Fig. 5-18

▮ Comparing Figs. 5-17 and 5-18, we have the correspondences $\mathbf{E}_\perp \leftrightarrow \mathbf{H}_\parallel$ and $\mathbf{H}_\perp \leftrightarrow -\mathbf{E}_\parallel$. Hence, introducing the reflection and transmission coefficients $\Gamma_\parallel$ and $T_\parallel$,

$$\hat{\mathbf{E}}_i = E_i e^{-j\beta_1(x \sin \theta_i + z \cos \theta_i)}(\cos \theta_i \, \mathbf{a}_x - \sin \theta_i \, \mathbf{a}_z) \tag{1}$$

$$\hat{\mathbf{H}}_i = \frac{E_i}{\eta_1} e^{-j\beta_1(x \sin \theta_i + z \cos \theta_i)} \, \mathbf{a}_y \tag{2}$$

$$\hat{\mathbf{E}}_r = \Gamma_\parallel E_i e^{j\beta_1(-x \sin \theta_i + z \cos \theta_i)}(-\cos \theta_i \, \mathbf{a}_x - \sin \theta_i \, \mathbf{a}_z) \tag{3}$$

$$\hat{\mathbf{H}}_r = \frac{\Gamma_\parallel E_i}{\eta_1} e^{j\beta_1(-x \sin \theta_i + z \cos \theta_i)} \mathbf{a}_y \tag{4}$$

$$\hat{\mathbf{E}}_t = T_\parallel E_i e^{-j\beta_2(x \sin \theta_t + z \cos \theta_t)}(\cos \theta_t \, \mathbf{a}_x - \sin \theta_t \, \mathbf{a}_z) \tag{5}$$

$$\hat{\mathbf{H}}_t = \frac{T_\parallel E_i}{\eta_2} e^{-j\beta_2(x \sin \theta_t + z \cos \theta_t)} \mathbf{a}_y \tag{6}$$

5.148 Repeat Problem 5.134 for parallel polarization.

❚ *tangential* **E:**

$$(1 - \Gamma_{\parallel}) \cos \theta_i = T_{\parallel} \cos \theta_t \tag{1}$$

tangential **H:**

$$n_1(1 + \Gamma_{\parallel}) = n_2 T_{\parallel} \tag{2}$$

In (2), $1/\eta$ has been replaced by n; see Problems 5.138 and 5.139. If, in addition, it is assumed that the interface is charge-free, then we must have continuity of

normal D:

$$n_1^2(1 + \Gamma_{\parallel}) \sin \theta_i = n_2^2 T_{\parallel} \sin \theta_t \tag{3}$$

5.149 Verify that Snell's law also holds for parallel polarization.

❚ Divide (3) of Problem 5.148 by (2).

5.150 Repeat Problem 5.139 for parallel polarization.

❚ Solve (1) and (2) of Problem 5.148 simultaneously, to find

$$\Gamma_{\parallel} = \frac{n_2 \cos \theta_i - n_1 \cos \theta_t}{n_2 \cos \theta_i + n_1 \cos \theta_t} \tag{1}$$

$$T_{\parallel} = \frac{2n_1 \cos \theta_i}{n_2 \cos \theta_i + n_1 \cos \theta_t} \tag{2}$$

Formulas (1) and (2) were derived under the assumption $\mu_1 = \mu_2$ (whereby $1/\eta \propto n$). They are unconditionally valid provided n is replaced by $1/\eta$ throughout. The same is true of expressions (2) and (3) of Problem 5.139.

5.151 Relate $\Gamma_{\parallel}$ to $\Gamma_{\perp}$ and $T_{\parallel}$ to $T_{\perp}$ [for general μ_1 and μ_2].

❚ $$\Gamma_{\parallel}(\eta_1, \eta_2) = \Gamma_{\perp}(\eta_2, \eta_1) \qquad T_{\parallel}(\eta_1, \eta_2) = \frac{\eta_2}{\eta_1} T_{\perp}(\eta_2, \eta_1)$$

5.152 Show that a perpendicularly polarized incident wave can never be totally transmitted into another medium.

❚ For $\Gamma_{\perp} = 0$ (the criterion for total transmission), (2) and (3) of Problem 5.134 become

$$\frac{1}{\eta_1} \cos \theta_i = \frac{1}{\eta_2} \cos \theta_t \qquad \text{and} \qquad \frac{1}{\eta_1} \sin \theta_i = \frac{1}{\eta_2} \sin \theta_t$$

Square these two equations and add, to get

$$\frac{1}{\eta_1^2} = \frac{1}{\eta_2^2} \qquad \text{or} \qquad \eta_1 = \eta_2$$

which is false.

5.153 For a particular angle of incidence—called the *Brewster angle* θ_B—a *parallel-polarized* incident wave can be totally transmitted into another medium. Determine θ_B.

❚ Setting $\Gamma_{\parallel} = 0$ and $\theta_i = \theta_B$, the three equations of Problem 5.148, in their general form, become

$$\cos \theta_B = T_{\parallel} \cos \theta_t$$
$$\frac{1}{\eta_1} = \frac{1}{\eta_2} T_{\parallel}$$
$$\epsilon_1 \sin \theta_B = \epsilon_2 T_{\parallel} \sin \theta_t$$

Eliminating θ_t and $T_{\parallel}$, we find

$$\sin^2 \theta_B = \frac{1 - (\eta_2/\eta_1)^2}{1 - (\epsilon_1/\epsilon_2)^2} = \frac{(\epsilon_2/\epsilon_1) - (\mu_2/\mu_1)}{(\epsilon_2/\epsilon_1) - (\epsilon_1/\epsilon_2)} \tag{1}$$

5.154 Prove that in the common case $\mu_1 = \mu_2$ the Brewster angle for parallel polarization is given by

$$\theta_B = \tan^{-1}(n_2/n_1) \tag{1}$$

❚ From (1) of Problem 5.153,

$$\sin^2 \theta_B = \frac{(\epsilon_2/\epsilon_1) - 1}{(\epsilon_2/\epsilon_1) - (\epsilon_1/\epsilon_2)} = \frac{\epsilon_2(\epsilon_2 - \epsilon_1)}{\epsilon_2^2 - \epsilon_1^2} = \frac{\epsilon_2}{\epsilon_2 + \epsilon_1} = \frac{n_2^2}{n_2^2 + n_1^2}$$

where the last step follows from the fact that $n = \sqrt{\mu_r \epsilon_r} \propto \sqrt{\epsilon}$, for fixed μ_r. Thus

$$\sin \theta_B = \frac{n_2}{\sqrt{n_2^2 + n_1^2}} \qquad \text{or} \qquad \tan \theta_B = \frac{n_2}{n_1}$$

5.155 Calculate the Brewster angle for a parallel-polarized wave passing from air into a glass of refractive index 2.

▌ By (1) of Problem 5.154, $\theta_B = \tan^{-1} 2 = 63.4°$.

5.156 In Fig. 5-17 replace medium 2 by a perfect conductor. Find the total fields in medium 1.

▌ Relations (1) through (6) of Problem 5.146 hold for this case, with $E_t = 0$ and $E_r = -E_i$. Hence,

$$\hat{\mathbf{E}}_1 = \hat{\mathbf{E}}_i + \hat{\mathbf{E}}_r = -2jE_i e^{-j\beta_1 x \sin \theta_i} \sin (\beta_1 z \cos \theta_i) \, \mathbf{a}_y \tag{1}$$

$$\hat{\mathbf{H}}_1 = \hat{\mathbf{H}}_i + \hat{\mathbf{H}}_r = -\frac{2E_i}{\eta_1} e^{-j\beta_1 x \sin \theta_i} [\cos \theta_i \cos (\beta_1 z \cos \theta_i) \, \mathbf{a}_x + j \sin \theta_i \sin (\beta_1 z \cos \theta_i) \, \mathbf{a}_z] \tag{2}$$

5.157 Express the fields of Problem 5.156 in the time domain.

▌ The time-domain results are obtained by multiplying (1) and (2) of Problem 5.156 by $e^{j\omega t}$ and taking the real part of the result:

$$\mathbf{E}_1 = 2E_i \sin (\beta_1 z \cos \theta_i) \sin (\omega t - \beta_1 x \sin \theta_i) \, \mathbf{a}_y \tag{1}$$

$$\mathbf{H}_1 = -2\frac{E_i}{\eta_1} [\cos \theta_i \cos (\beta_1 z \cos \theta_i) \cos (\omega t - \beta_1 x \sin \theta_i) \, \mathbf{a}_x - \sin \theta_i \sin (\beta_1 z \cos \theta_i) \sin (\omega t - \beta_1 x \sin \theta_i) \, \mathbf{a}_z] \tag{2}$$

5.158 Determine the time-average power density of the wave of Problem 5.156.

▌ $\langle \mathbf{S} \rangle = \frac{1}{2} \text{Re} \, |\hat{\mathbf{E}}_1 \times \hat{\mathbf{H}}_1^*] = \frac{1}{2} \text{Re} \, [E_y \mathbf{a}_y \times (H_x^* \mathbf{a}_x + H_z^* \mathbf{a}_z)] = -\frac{1}{2} \text{Re} \, [E_y H_x^*] \, \mathbf{a}_z + \frac{1}{2} \text{Re} \, [E_y H_z^*] \, \mathbf{a}_x$

Substituting the phasor components from (1) and (2) of Problem 5.156 yields

$$\langle \mathbf{S} \rangle = 2\frac{E_i^2}{\eta_1} \sin \theta_i \sin^2 (\beta_1 z \cos \theta_i) \, \mathbf{a}_x \tag{1}$$

5.159 Do the fields of Problem 5.157 constitute a *plane* wave in (lossless) medium 1? What is the phase velocity?

▌ No. The propagation direction is $+\mathbf{a}_x$—as evidenced by the phase, $\omega t - kx$ ($k \equiv \beta_1 \sin \theta_i$), and by the result of Problem 5.158—but over yz planes the instantaneous magnitudes are not constant, but vary with z.

The velocity of the (nonplanar) wave is

$$u_x = \frac{\omega}{k} = \frac{\omega}{\beta_1 \sin \theta_i} = \frac{u_1}{\sin \theta_i}$$

5.160 Could the results of Problem 5.159 have been anticipated?

▌ Yes. Because an external electromagnetic field cannot penetrate a perfect conductor (medium 2), one expects the resultant Poynting vector in medium 1 to be parallel to the interface; i.e., in the x-direction. Moreover, the oblique incidence puts the planes of constant $|\mathbf{E}_i|$ at an angle (of $\pi - 2\theta_i$) to the planes of constant $|\mathbf{E}_r|$; under such circumstances, constructive interference to produce a new plane wave is impossible.

5.161 Refer to Problems 5.156 and 5.157. At what distances from the perfectly conducting slab does the resultant E-field vanish?

▌ From (1) of Problem 5.157, $|\mathbf{E}_1| = 0$ at

$$\beta_1 \cos \theta_i = -n\pi \qquad \text{or} \qquad z = -\frac{n\lambda_1}{2 \cos \theta_i}$$

for $n = 0, 1, 2, \ldots$.

5.162 Repeat Problem 5.156 for the case of parallel polarization (Fig. 5-18).

▌ The field phasors are given by Problem 5.147 if we set $\Gamma_\parallel = +1$ and $T_\parallel = 0$ ($\sigma_2 = \infty$). Consequently,

$$\hat{\mathbf{E}}_1 = \hat{\mathbf{E}}_i + \hat{\mathbf{E}}_r = -2E_i e^{-j\beta_1 x \sin \theta_i} [j \cos \theta_i \sin (\beta_1 z \cos \theta_i) \, \mathbf{a}_x + \sin \theta_i \cos (\beta_1 z \cos \theta_i) \, \mathbf{a}_z] \tag{1}$$

$$\hat{\mathbf{H}}_1 = \hat{\mathbf{H}}_i + \hat{\mathbf{H}}_r = \frac{2E_i}{\eta_1} e^{-j\beta_1 x \sin \theta_i} \cos (\beta_1 z \cos \theta_i) \, \mathbf{a}_y \tag{2}$$

5.163 Express the fields of Problem 5.162 in the time domain.

▌ The time-domain expressions are obtained by multiplying the phasors by $e^{j\omega t}$ and taking the real part of the result:

$$\mathbf{E}_1 = 2E_i \cos \theta_i \sin (\beta_1 z \cos \theta_i) \sin (\omega t - \beta_1 x \sin \theta_i) \, \mathbf{a}_x - 2E_i \sin \theta_i \cos (\beta_1 z \cos \theta_i) \cos (\omega t - \beta_1 x \sin \theta_i) \, \mathbf{a}_z$$

$$\mathbf{H}_1 = \frac{2E_i}{\eta_1} \cos (\beta_1 z \cos \theta_i) \cos (\omega t - \beta_1 x \sin \theta_i) \, \mathbf{a}_y$$

5.164 Determine the average power in the wave of Problem 5.162.

$$\langle \mathbf{S} \rangle = \frac{1}{2} \text{Re} \left[\hat{\mathbf{E}}_1 \times \hat{\mathbf{H}}_1^* \right] = 2 \frac{E_i^2}{\eta_1} \sin \theta_i \cos^2 (\beta_1 z \cos \theta_i) \, \mathbf{a}_x$$

5.165 What is the phase velocity of the parallel-polarized wave of Problem 5.163?

$$u_x = \frac{u_1}{\sin \theta_i}$$

as for perpendicular polarization (Problem 5.159).

5.166 Examine the perpendicularly polarized **E**-wave of Problem 5.157 in the limit as $\theta_i \to 0$.

▌ Compare Problem 5.111. Here, $\mathbf{E}_1 \to 2E_i \sin \beta_1 z \sin \omega t \, \mathbf{a}_y$; this is the standing wave of Fig. 5-8, but shifted by a quarter-wavelength to put a node at the conductor surface. Correspondingly, by (1) of Problem 5.158, $\langle \mathbf{S} \rangle \to \mathbf{0}$.

5.167 Determine the amplitude (A) and frequency (f) of the free-space electric wave
$$\mathbf{E}_i = 1.732 \cos (\omega t - 0.5\pi y - 0.866\pi x) \, \mathbf{a}_y - \cos (\omega t - 0.5\pi y - 0.866\pi x) \, \mathbf{a}_x \quad \text{(V/m)} \qquad (1)$$

▌ Choose rotated coordinates,
$$x' = x \cos \phi + y \sin \phi$$
$$y' = -x \sin \phi + y \cos \phi$$
in the plane of polarization, such that the wave travels in the $+x'$-direction; then
$$\hat{\mathbf{E}}_i = A e^{-j\beta_1 x'} \mathbf{a}_{y'} = A e^{-j\beta_1 (x \cos \phi + y \sin \phi)} (-\sin \phi \, \mathbf{a}_x + \cos \phi \, \mathbf{a}_y)$$
or
$$\mathbf{E}_i = A \cos (\omega t - \beta_1 x \cos \phi - \beta_1 y \sin \phi)(-\sin \phi \, \mathbf{a}_x + \cos \phi \, \mathbf{a}_y) \qquad (2)$$
Comparison of (2) with (1) yields
$$\beta_1 \cos \phi = 0.866\pi \qquad A \sin \phi = 1$$
$$\beta_1 \sin \phi = 0.5\pi \qquad A \cos \phi = 1.732$$
These four equations are consistent and yield $\phi = 30°$; thus
$$A = 1 \csc 30° = 2 \text{ m} \qquad \beta_1 = 0.5\pi \csc 30° = \pi \text{ rad/m}$$
and
$$f = \frac{\beta_1 u_0}{2\pi} = 150 \text{ MHz}$$

5.168 The parallel-polarized wave of Problem 5.167 is incident on the perfect dielectric half-space $x > 0$, as indicated in Fig. 5-19. Express the incident magnetic field $\mathbf{H}_i$ in the time domain.

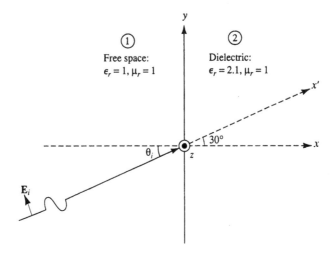

Fig. 5-19

▌ The results of Problem 5.167, together with $|\mathbf{E}_i| / |\mathbf{H}_i| = \eta_1 = 377 \ \Omega$, yield
$$\mathbf{H}_i = \frac{2}{377} \cos (\omega t - 0.5\pi y - 0.866\pi x) \, \mathbf{a}_z \quad \text{(A/m)}$$
in which $\omega = 3 \times 10^8 \pi$ rad/s.

5.169 Determine the reflection and transmission coefficients for the wave of Problem 5.168.

    ❙ Referring to Fig. 5-19, we have $\theta_i = 30°$; $n_1 = 1$; $n_2 = \sqrt{2.1}$; and, by Snell's law,

$$\sin \theta_t = \frac{1}{2\sqrt{2.1}} = 0.345 \qquad \cos \theta_t = 0.939$$

Substitution in (1) and (2) of Problem 5.150 yields $\Gamma_{\parallel} = 0.144$ and $T_{\parallel} = 0.789$.

5.170 Write the reflected and transmitted phasors for the wave of Problems 5.167 and 5.168.

    ❙ Figure 5-18 and (1) through (6) of Problem 5.147 apply to the present problem after the relabeling $(z, x, y) \rightarrow (x, y, z)$. Using the constants calculated in Problems 5.167–5.169, and additionally

$$\beta_2 = \omega\sqrt{\mu_2 \epsilon_2} = \sqrt{2.1}\,\beta_1 = 1.45\pi \text{ rad/m} \qquad \eta_2 = \sqrt{\frac{\mu_2}{\epsilon_2}} = \frac{1}{\sqrt{2.1}}\,\eta_1 = 260\ \Omega$$

we have

$$\hat{\mathbf{E}}_r = -(0.249\mathbf{a}_y + 0.144\mathbf{a}_x)e^{j(-0.5\pi y + 0.866\pi x)} \quad \text{(V/m)}$$

$$\hat{\mathbf{H}}_r = \frac{0.288}{377}\,\mathbf{a}_z e^{j(-0.5\pi y + 0.866\pi x)} \quad \text{(A/m)}$$

$$\hat{\mathbf{E}}_t = (1.483\mathbf{a}_y - 0.544\mathbf{a}_x)e^{-j(0.725\pi y + 1.25\pi x)} \quad \text{(V/m)}$$

$$\hat{\mathbf{H}}_t = \frac{1.578}{260}\,\mathbf{a}_z e^{-j(0.725\pi y + 1.25\pi x)} \quad \text{(A/m)}$$

5.171 Obtain time-domain expressions for the reflected and transmitted field vectors found in Problem 5.170.

    ❙

$$\mathbf{E}_r = -0.249 \cos(\omega t - 0.5\pi y + 0.866\pi x)\,\mathbf{a}_y - 0.144 \cos(\omega t - 0.5\pi y + 0.866\pi x)\,\mathbf{a}_x \quad \text{(V/m)}$$

$$\mathbf{H}_r = \frac{0.288}{377} \cos(\omega t - 0.5\pi y + 0.866\pi x)\,\mathbf{a}_z \quad \text{(A/m)}$$

$$\mathbf{E}_t = 1.483 \cos(\omega t - 0.725\pi y - 1.25\pi x)\,\mathbf{a}_y - 0.544 \cos(\omega t - 0.725\pi y - 1.25\pi x)\,\mathbf{a}_x \quad \text{(V/m)}$$

$$\mathbf{H}_t = \frac{1.578}{260} \cos(\omega t - 0.725\pi y - 1.25\pi x)\,\mathbf{a}_z \quad \text{(A/m)}$$

5.172 A plane wave in free space, with

$$\mathbf{E}_i = (2\mathbf{a}_x - 4\mathbf{a}_y + 3\mathbf{a}_z) \cos(\omega t - 1.5\pi z - 1.12\pi y)$$

impinges on the perfectly conducting half-space $y > 0$ (Fig. 5-20). Determine the angle of incidence and the amplitudes of the perpendicular- and parallel-polarized components of the wave.

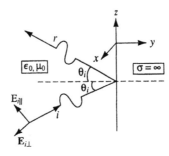

Fig. 5-20

    ❙ We have $\mathbf{E}_{i\perp} = 2\mathbf{a}_x \cos(\omega t - 1.5\pi x - 1.12\pi y)$, of amplitude 2. Comparing the given and the canonical representations

$$\mathbf{E}_{i\parallel} = (-4\mathbf{a}_y + 3\mathbf{a}_z) \cos(\omega t - 1.12\pi y - 1.5\pi z) \quad \text{and} \quad \mathbf{E}_{i\parallel} = A(-\sin \theta_i\,\mathbf{a}_y + \cos \theta_i\,\mathbf{a}_z) \cos(\omega t - 1.12\pi y - 1.5\pi z)$$

we see that $A = \sqrt{4^2 + 3^2} = 5$; $\sin \theta_i = 4/5$, $\cos \theta_i = 3/5$, or $\theta_i = 53.5°$.

5.173 For Problem 5.172, express $\mathbf{H}_i$.

    ❙ In any plane wave, $\mathbf{E}$, $\mathbf{H}$, and $\mathbf{S} = \mathbf{E} \times \mathbf{H}$ form a right-handed orthogonal system; hence

$$\mathbf{a}_H = \mathbf{a}_S \times \mathbf{a}_E \qquad (1)$$

Now, for either component of the present wave,

$$\mathbf{a}_S = \cos \theta_i\,\mathbf{a}_y + \sin \theta_i\,\mathbf{a}_z = \tfrac{3}{5}\mathbf{a}_y + \tfrac{4}{5}\mathbf{a}_z$$

Therefore, for $\mathbf{H}_{i\perp}$,

$$\mathbf{a}_H = (\tfrac{3}{5}\mathbf{a}_y + \tfrac{4}{5}\mathbf{a}_z) \times \mathbf{a}_x = \tfrac{4}{5}\mathbf{a}_y - \tfrac{3}{5}\mathbf{a}_z$$

and

$$\mathbf{H}_{i\perp} = \tfrac{2}{377}(\tfrac{4}{5}\mathbf{a}_y - \tfrac{3}{5}\mathbf{a}_z) \cos(\omega t - 1.12\pi y - 1.5\pi z) \qquad (1)$$

Likewise, for $\mathbf{H}_{i\perp}$,
$$\mathbf{a}_H = (\tfrac{3}{5}\mathbf{a}_y + \tfrac{4}{5}\mathbf{a}_z) \times (-\tfrac{4}{5}\mathbf{a}_y + \tfrac{3}{5}\mathbf{a}_z) = \mathbf{a}_x$$
and
$$\mathbf{H}_{i\parallel} = \tfrac{5}{377}\mathbf{a}_x \cos(\omega t - 1.12\pi y - 1.5\pi z) \qquad (2)$$
The vector sum of (1) and (2) is $\mathbf{H}_i$.

5.174   Obtain the reflected fields $\mathbf{E}_r$ and $\mathbf{H}_r$ for Problem 5.172.

▌ Once again the two polarizations are treated separately. Since $\Gamma_\perp = -1 = 1\,\underline{/\pi}$, we have
$$\mathbf{E}_{r\perp} = 2\mathbf{a}_x \cos(\omega t - 1.5\pi z - 1.12\pi y + \pi)$$
$$\mathbf{a}_H = (-\tfrac{3}{5}\mathbf{a}_y + \tfrac{4}{5}\mathbf{a}_z) \times \mathbf{a}_x = \tfrac{4}{5}\mathbf{a}_y + \tfrac{3}{5}\mathbf{a}_z$$
$$\mathbf{H}_{r\perp} = \tfrac{2}{377}(\tfrac{4}{5}\mathbf{a}_y + \tfrac{3}{5}\mathbf{a}_z) \cos(\omega t - 1.5\pi z - 1.12\pi y + \pi)$$
Since $\Gamma_\parallel = +1$, we have (cf. Problem 5.147)
$$\mathbf{E}_{r\parallel} = (-4\mathbf{a}_y - 3\mathbf{a}_z) \cos(\omega t - 1.12\pi y - 1.5\pi z)$$
$$\mathbf{a}_H = (-\tfrac{3}{5}\mathbf{a}_y + \tfrac{4}{5}\mathbf{a}_z) \times (-\tfrac{4}{5}\mathbf{a}_y - \tfrac{3}{5}\mathbf{a}_z) = \mathbf{a}_x$$
$$\mathbf{H}_{r\parallel} = \tfrac{5}{377}\mathbf{a}_x \cos(\omega t - 1.12\pi y - 1.5\pi z)$$
The total fields are $\mathbf{E}_r = \mathbf{E}_{r\perp} + \mathbf{E}_{r\parallel}$ and $\mathbf{H}_r = \mathbf{H}_{r\perp} + \mathbf{H}_{r\parallel}$.

5.175   A plane wave having $|\mathbf{E}| = 2$ V/m impinges normally on a copper block at a frequency of 2 GHz. For copper $\mu = \mu_0$ and $\sigma = 58$ MS/m. Calculate the time-average power density absorbed by the copper block.

▌ By Problem 5.45,
$$|\eta| = \sqrt{\frac{\mu\omega}{\sigma}} = \sqrt{\frac{4\pi \times 10^{-7} \times 2\pi \times 2 \times 10^9}{58 \times 10^6}} = 1.65 \times 10^{-2}\ \Omega$$
and so
$$\langle \mathbf{S} \rangle = \frac{|\mathbf{E}|^2}{2|\eta|} = 121.2\ \text{W/m}^2$$

5.176   A 10-MHz plane wave, $\mathbf{E} = E_0 \cos(\omega t + \beta x)\,\mathbf{a}_z$, travels in a nonmagnetic medium ($x > 0$) of high permittivity ($\epsilon_r = 9$). Calculate the wavelength λ in the dielectric.

▌ $n = \sqrt{\epsilon_r} = 3$, so that $u = u_0/n = 10^8$ m/s and
$$\lambda = \frac{u}{f} = \frac{10^8}{10 \times 10^6} = 10\ \text{m}$$

5.177   Assuming that the space $x < 0$ is air, calculate the reflection and transmission coefficients for the wave of Problem 5.176.

▌ Here, $n_1 = 3$, $n_2 = 1$; Problem 5.84 gives, since $\eta \propto 1/n$,
$$\Gamma = \frac{n_1 - n_2}{n_1 + n_2} = \frac{3-1}{3+1} = \frac{1}{2} \qquad T = 1 + \Gamma = \frac{3}{2}$$

5.178   The magnetic field for a linearly polarized wave in a lossless magnetic material having the permittivity of free space is given by
$$\hat{\mathbf{H}} = 4.5e^{j(\omega t - \beta y)}\mathbf{a}_x \quad \text{(A/m)}$$
For this material the intrinsic impedance is 1500 Ω. Express the $\mathbf{E}$-field in the time domain.

▌ Because $|\mathbf{E}| = |\mathbf{H}|\,\eta = 4.5 \times 1500 = 6750$ V/m and $\mathbf{a}_E = \mathbf{a}_H \times \mathbf{a}_S = \mathbf{a}_x \times \mathbf{a}_y = \mathbf{a}_z$,
$$\mathbf{E} = 6750 \cos(\omega t - \beta y)\,\mathbf{a}_z \quad \text{(V/m)}$$

5.179   Calculate the relative permeability of the magnetic material of Problem 5.178.

▌
$$\eta = \sqrt{\mu_r}\,\eta_0 \qquad \text{or} \qquad \mu_r = \left(\frac{\eta}{\eta_0}\right)^2 = \left(\frac{1500}{377}\right)^2 = 15.9$$

5.180   Determine the time-average Poynting vector for the wave of Problems 5.178 and 5.179.

▌
$$\langle \mathbf{S} \rangle = \frac{|\mathbf{E}|^2}{2\eta}\,\mathbf{a}_y = 15.2\mathbf{a}_y\ \text{kW/m}^2$$

5.181   Refer to Problems 5.129 and 5.130 and Fig. 5-10. Obtain an approximate expression for the ratio of the input power to the power transmitted through the conductive barrier.

■ Since power $\propto |\mathbf{E}|^2/\eta$ and since $\eta = \eta_0$ for regions 1 and 3 (air),

$$\text{power ratio} = (\text{SE})^2 \approx \frac{32}{\eta_0^2 \sigma^2 \delta^2} e^{-2\omega/\delta} \tag{1}$$

where the approximation comes from (1) of Problem 5.130.

5.182 What barrier thickness in Problem 5.181, measured in skin depths, allows 1 percent transmission of the incident power? The conductivity of the material is 10^4 S/m, $\mu_r \approx 1$, and the wavelength of the incident radiation is 300 nm.

■ Substituting

$$\eta_0 = \sqrt{\frac{\mu_0}{\epsilon_0}} = 120\pi \qquad \delta = \sqrt{\frac{2}{\omega\mu\sigma}} \qquad \omega = \frac{2\pi}{\lambda\sqrt{\mu_0\epsilon_0}}$$

in (1) of Problem 5.181, we obtain

$$\text{power ratio} = \frac{4\mu_r}{15\sigma\lambda} e^{-2(w/\delta)}$$

Thus, we need only solve the equation

$$0.01 = \frac{4(1)}{15(10^4)(300 \times 10^{-9})} e^{-2(w/\delta)}$$

to find $w/\delta \approx 4.5$.

5.183 The phase velocity u_p of a plane wave is given as a function of frequency, $u_p = u_p(\omega)$; derive an equation for the group velocity u_g.

■ By definition,

$$u_g = \frac{d\omega}{d\beta} = \frac{d}{d\beta}(\beta u_p) = u_p + \beta \frac{du_p}{d\beta}$$

$$= u_p + \beta\left(\frac{du_p}{d\omega}\frac{d\omega}{d\beta}\right) = u_p + \beta u_g \frac{du_p}{d\omega} = u_p + \frac{\omega}{u_p}u_g \frac{du_p}{d\omega} \tag{1}$$

Solution of (1) for u_g yields

$$u_g = \frac{u_p}{1 - \dfrac{\omega}{u_p}\dfrac{du_p}{d\omega}} \tag{2}$$

5.184 Derive a general expression for the time-average Poynting vector of an elliptically polarized plane wave traveling in the z-direction through a lossless medium.

■ By Problems 5.66 and 5.69 the **E**-phasor may be written as

$$\hat{\mathbf{E}} = E_1 e^{-j\beta z}\mathbf{a}_x + E_2 e^{-j(\beta z - \theta)}\mathbf{a}_y$$

where $E_1 \neq E_2$ and $\theta \neq 0$. Correspondingly (Problem 5.63),

$$\hat{\mathbf{H}} = -\frac{E_2}{\eta}e^{-j(\beta z - \theta)}\mathbf{a}_x + \frac{E_1}{\eta}e^{-j\beta z}\mathbf{a}_y$$

with η real and positive.

$$\langle \mathbf{S} \rangle = \frac{1}{2}\operatorname{Re}(\hat{\mathbf{E}} \times \hat{\mathbf{H}}^*) = \frac{E_1^2 + E_2^2}{2\eta}\mathbf{a}_z \tag{1}$$

5.185 Comment on the simple result (1) of Problem 5.184.

■ When η is real (lossless medium), Problem 5.24 shows that interference between plane waves is impossible. Hence, the average power of a sum of plane waves (e.g., an elliptically polarized wave) must equal the sum of the average powers.

5.186 The average power density carried (in the z-direction) by an elliptically polarized wave in free space is 0.240 W/m^2. Obtain an expression for the **E**-field, assuming that its x-amplitude is twice its y-amplitude.

■ By (1) of Problem 5.184,

$$0.240 = \frac{(2E_2)^2 + E_2^2}{2(377)}$$

giving $E_2 = 6$ V/m, $E_1 = 12$ V/m. Hence,

$$\mathbf{E} = 12\cos(\omega t - \beta z)\mathbf{a}_x + 6\cos(\omega t - \beta z + \theta)\mathbf{a}_y \quad (\text{V/m})$$

5.187 Refer to Fig. 5-11, where medium 1 is air and medium 2 is liquid with $\epsilon_r = 100$, $\mu_r = 1$. A plane wave is incident from the liquid onto the interface at 45°. If the magnitude of the electric field in the liquid is 2 V/m, what is its magnitude in the air?

▮ Relabeling the liquid as medium 1, we have $n_1 = \sqrt{100} = 10$ and $n_2 = 1$. By Problem 5.140, the critical angle in this case is $\theta_c = \sin^{-1}(1/10) = 5.74°$. Since $45° > 5.74°$, there is no transmitted wave; i.e., $E_{\text{air}} = 0$.

5.188 Obtain an expression for the reflection coefficient in terms of θ_i and permittivities for a perpendicularly polarized plane wave traveling from one lossless nonmagnetic dielectric to another.

▮ In (2) of Problem 5.139 substitute $n_1 \propto \sqrt{\epsilon_1}$, $n_2 \propto \sqrt{\epsilon_2}$, and
$$\cos \theta_t = \sqrt{1 - \sin^2 \theta_t} = \sqrt{1 - \frac{n_1^2}{n_2^2}\sin^2 \theta_i} = \sqrt{1 - \frac{\epsilon_1}{\epsilon_2}\sin^2 \theta_i}$$
to obtain:
$$\Gamma_\perp = \frac{\cos \theta_i - \sqrt{(\epsilon_2/\epsilon_2) - \sin^2 \theta_i}}{\cos \theta_i + \sqrt{(\epsilon_2/\epsilon_1) - \sin^2 \theta_i}} \tag{1}$$

5.189 Repeat Problem 5.188 for parallel polarization.

▮ By Problems 5.188 and 5.151,
$$\Gamma_\parallel = \frac{\cos \theta_i - \sqrt{(\epsilon_1/\epsilon_2) - \sin^2 \theta_i}}{\cos \theta_i + \sqrt{(\epsilon_1/\epsilon_2) - \sin^2 \theta_i}} \tag{1}$$

5.190 Assuming that $\epsilon_2 > \epsilon_1$, plot the reflection coefficient (1) of Problem 5.188 for $0° \le \theta_i \le 90°$.

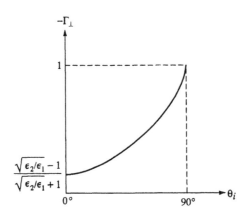

▮ See Fig. 5-21.

Fig. 5-21

5.191 Problems 5.108 through 5.115 investigate the superposition of an incident and a reflected wave in a lossless medium. Summarize the results by graphing the *standing wave envelope,* the (stationary) locus of the tips of the longest **E**-vectors all along the negative z axis.

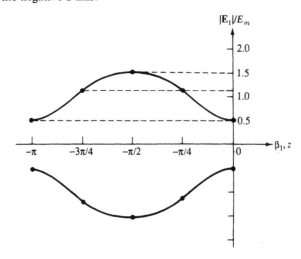

Fig. 5-22

▮ What is asked for is just a real form of the crank diagram (Problem 5.109). By Problem 5.113 the desired function is

$$|\mathbf{E}_1|/E_m = \pm(1 + \Gamma^2 + 2\Gamma\cos 2\beta_1 z)^{1/2} \tag{1}$$

This is graphed in Fig. 5-22, where, for definiteness, it is assumed that $\Gamma = -\frac{1}{2}$. Note that the envelope, as a function of $\beta_1 z$, is periodic of period π.

5.192 Verify that Fig. 5-21 bears out Problems 5.114 and 5.115.

▮ For $\Gamma = -\frac{1}{2}$, $S = (1 + \frac{1}{2})/(1 - \frac{1}{2}) = 3$. From Fig. 5-21, $|\mathbf{E}_1|_{max}/|\mathbf{E}_1|_{min} = 1.5/0.5 = 3$.

5.193 Redefine the standing wave ratio in terms of the incident and reflected amplitudes.

▮ By definition $|\Gamma| = |\mathbf{E}_r|/|\mathbf{E}_i|$, so that

$$S = \frac{1 + |\Gamma|}{1 - |\Gamma|} = \frac{|\mathbf{E}_i| + |\mathbf{E}_r|}{|\mathbf{E}_i| - |\mathbf{E}_r|}$$

5.194 Refer to Problem 5.144 (Fig. 5-15). At what location would the fish's true distance be twice its apparent distance?

▮ The true distance must satisfy

$$D = \frac{D}{2} + 10\frac{\sin\theta_t}{\sqrt{1.5^2 - \sin^2\theta_t}} = \frac{D}{2} + 10\frac{(D/2)/\sqrt{D^2/4 + 6^2}}{\sqrt{1.5^2 - \dfrac{D^2/4}{D^2/4 + 6^2}}}$$

or

$$\frac{D}{2} = \frac{5D}{\sqrt{\dfrac{5D^2}{16} + 81}} \qquad \text{or} \qquad \frac{5D^2}{16} + 81 = 100$$

or

$$D^2 = \frac{19.16}{5} \qquad \text{or} \qquad D = 4\sqrt{\frac{19}{5}} \approx 8 \text{ ft}$$

The above neglects the trivial solution $D = 0$.

5.195 A plane wave in lossless medium 1 impinges normally on lossy medium 2. Express the ratio of transmitted to incident power in terms of the standing wave ratio.

▮ By Problem 5.105,

$$P_{trans}/P_{inc} = 1 - |\Gamma|^2$$

But

$$S = \frac{1 + |\Gamma|}{1 - |\Gamma|} \qquad \text{or} \qquad |\Gamma| = \frac{S - 1}{S + 1}$$

Hence

$$P_{trans}/P_{inc} = 1 - \left(\frac{S - 1}{S + 1}\right)^2 = \frac{4S}{(S + 1)^2}$$

5.196 For the situation of Problem 5.195, what is the ratio of reflected to incident power?

▮ Because $P_{inc} = P_{refl} + P_{trans}$,

$$\frac{P_{refl}}{P_{inc}} = 1 - \frac{P_{trans}}{P_{inc}} = \left(\frac{S - 1}{S + 1}\right)^2$$

5.197 Consider a lossy dielectric, characterized by (μ, σ, ϵ), located in an electromagnetic field. Show that $\mathbf{E}$ and $\mathbf{H}$ within the dielectric may be related to each other via a complex permittivity.

▮ From Maxwell's equation (Ampère's law),

$$\text{curl } \mathbf{H} = \mathbf{J} + \frac{\partial \mathbf{D}}{\partial t} = (\sigma + j\omega\epsilon)\mathbf{E} \equiv j\omega\epsilon_c\mathbf{E}$$

where ϵ_c is the complex permittivity, given by $\epsilon_c = \epsilon - j(\sigma/\omega)$.

5.198 A plane wave traveling in free space has an average power density of 30 W/m². Determine the amplitudes of the $\mathbf{E}$- and $\mathbf{B}$-fields.

▮

$$\frac{|\mathbf{E}|^2}{2(120\pi)} = 30 \qquad \text{or} \qquad |\mathbf{E}| = 150.4 \text{ V/m}$$

Then $|\mathbf{B}| = \mu_0|\mathbf{H}| = \mu_0|\mathbf{E}|/\eta_0 = |\mathbf{E}|/u_0 = 0.5\,\mu\text{T}$.

5.199 The **E**-field of a plane wave in a lossless dielectric is given by $\mathbf{E} = 15 \cos(10^9 t - 20z)\, \mathbf{a}_y$ (V/m). Specify (*a*) the amplitude, (*b*) the frequency, and (*c*) the direction in which the wave is traveling.

▮ (*a*) 15 V/m; (*b*) $10^9/2\pi = 159.15$ MHz; (*c*) the positive z-direction.

5.200 Determine the wavelength and the phase velocity of the wave of Problem 5.199.

▮ $$\lambda = \frac{2\pi}{20} = 0.31416 \text{ m} \qquad u = \lambda f = (0.31416)(159.15 \times 10^6) = 5 \times 10^7 \text{ m/s}$$

5.201 Calculate the relative permittivity of the dielectric of Problem 5.200.

▮ In $u_0/u = \sqrt{\mu_r \epsilon_r}$, substitute $\mu_r = 1$ and $u = 5 \times 10^7$ m/s, to find $\epsilon_r = 36$.

5.202 Express the **H**-field in Problem 5.199.

▮ $$\mathbf{H} = -\frac{1}{\eta} E_y \mathbf{a}_x = -\frac{\sqrt{\epsilon_r}}{\eta_0} E_y \mathbf{a}_x = -\frac{3}{4\pi} \cos(10^9 t - 20z)\, \mathbf{a}_x \quad \text{(A/m)}$$

5.203 A 0.5-MHz plane wave is normally incident on wet earth characterized by $\sigma = 10^{-3}$ S/m, $\mu_r = 1$, and $\epsilon_r = 10$. Calculate the attenuation of the wave, in Np/km.

▮ First we evaluate

$$\tau = \frac{\sigma}{\omega \epsilon} = \frac{10^{-3}}{2\pi(0.5 \times 10^6)\left(\dfrac{10^{-9}}{36\pi} \times 10\right)} = 3.6$$

Since $\tau \approx 1$, we must use the exact formula, (2) of Problem 5.37, to calculate α; it gives
$$\alpha = 0.0509 \text{ Np/m} = 50.9 \text{ Np/km}$$

5.204 A 50-MHz plane wave traveling in a lossless medium, with $\mu_r = 3$ and $\epsilon_r = 3$, has a time-average power density of 5 W/m². Determine the velocity of the wave and its wavelength and the intrinsic impedance of the medium.

▮ The velocity of the wave is given by

$$u = \frac{u_0}{\sqrt{\mu_r \epsilon_r}} = \frac{3 \times 10^8}{\sqrt{3 \times 3}} = 10^8 \text{ m/s}$$

and the wavelength is obtained as

$$\lambda = \frac{u}{f} = \frac{10^8}{50 \times 10^6} = 2.0 \text{ m}$$

The intrinsic impedance is

$$\eta = \left(\sqrt{\frac{\mu_r}{\epsilon_r}}\right)\eta_0 = \left(\sqrt{\frac{3}{3}}\right)(377) = 377 \ \Omega$$

5.205 Determine the rms values of the electric and magnetic fields of the wave of Problem 5.204.

▮ The average power and the rms **E**-field are related by

$$\langle S \rangle = \frac{|\mathbf{E}|^2}{2\eta} \equiv \frac{E_{\text{rms}}^2}{\eta}$$

whence
$$E_{\text{rms}} = \sqrt{\eta \langle S \rangle} = \sqrt{377 \times 5} = 43.4 \text{ V/m}$$
$$H_{\text{rms}} = \frac{E_{\text{rms}}}{\eta} = \frac{43.4}{377} = 0.115 \text{ A/m}$$

6.1 A transverse electromagnetic (*TEM*) field structure is one in which the electric and magnetic field vectors at each point in space have no components in the direction of propagation. Verify that a coaxial cable (Fig. 6-1) supports the TEM mode.

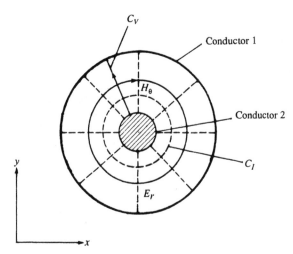

Fig. 6-1

▌ The axis of the cable (having two conductors 1 and 2) is the z axis of a rectangular coordinate system, and the cable cross section lies in the xy plane. If the two conductors of the cable are infinite in length (so that we may neglect fringing of the fields at the conductor endpoints) and we apply a dc (0-Hz) voltage between the two conductors, the electric field lines will lie in the xy plane, transverse to the line (z) axis, as shown in Fig. 6-1. A dc current along these conductors in the z direction will also produce magnetic field lines transverse to the cable axis. As the frequency of excitation is increased from zero, we would expect the field lines to maintain (to a reasonable approximation) this TEM structure.

6.2 Apply Maxwell's equations in integral form to the TEM mode, using a contour C_{xy} and a surface S_{xy} in an xy plane of Fig. 6-1. Show that these Maxwell's equations reduce to those for static fields.

▌ Maxwell's equations

$$\oint \mathbf{E} \cdot d\mathbf{s} = -\mu \frac{d}{dt} \int \mathbf{H} \cdot d\mathbf{S} \qquad \text{and} \qquad \oint \mathbf{H} \cdot d\mathbf{s} = \int \mathbf{J} \cdot d\mathbf{S} + \epsilon \frac{d}{dt} \int \mathbf{E} \cdot d\mathbf{S}$$

when applied to C_{xy} and S_{xy} yield

$$\oint_{C_{xy}} (E_x \, dx + E_y \, dy) = -\mu \frac{d}{dt} \int_{S_{xy}} H_z \, dx \, dy \qquad \text{and} \qquad \oint_{C_{xy}} (H_x \, dx + H_y \, dy) = \int_{S_{xy}} J_z \, dx \, dy + \epsilon \frac{d}{dt} \int_{S_{xy}} E_z \, dx \, dy$$

But, for the TEM mode, $E_z = H_z = 0$. Thus, we obtain

$$\oint_{C_{xy}} (E_x \, dx + E_y \, dy) = 0 \qquad \text{and} \qquad \oint_{C_{xy}} (H_x \, dx + H_y \, dy) = \int_{S_{xy}} J_z \, dx \, dy$$

which are precisely the equations that obtain in the static case.

6.3 What conclusions may be drawn from Problem 6.2?

▌ (1) The electric field of the TEM mode *in any transverse* (xy) *plane* is conservative and satisfies an electrostatic field distribution.

(2) The magnetic field of the TEM mode satisfies a magnetostatic field distribution *in any transverse* (xy) *plane*.

6.4 Relate the voltage $V(z, t)$ between the two conductors of the cable of Fig. 6-1 and the current $I(z, t)$ along the cable to the **E**- and **H**-fields, and show that the voltage and the current are uniquely defined.

▌ From Problems 6.2 and 6.3 we conclude that it is possible to uniquely define a voltage between the conductors at each point along the cable, as follows:

$$V(z, t) = -\int_{C_V} \mathbf{E}_T \cdot d\mathbf{s} \tag{1}$$

where contour C_V is shown in Fig. 6-1 and $\mathbf{E}_T = E_x\mathbf{a}_x + E_y\mathbf{a}_y$ is the transverse electric field. Similarly, we may uniquely define the current on either conductor surface as the line integral of $\mathbf{H}$ around any closed contour in the transverse xy plane which encircles that conductor; that is,

$$I(z, t) = \oint_{C_I} \mathbf{H}_T \cdot d\mathbf{s} \tag{2}$$

where contour C_I is shown in Fig. 6-1 and $\mathbf{H}_T = H_x\mathbf{a}_x + H_y\mathbf{a}_y$ is the transverse magnetic field.

6.5 A portion of a two-conductor, uniform transmission line is shown in Fig. 6-2(a), which also shows the **H**- and **E**-fields. Characterize this section of the line via a lumped inductance and a lumped capacitance.

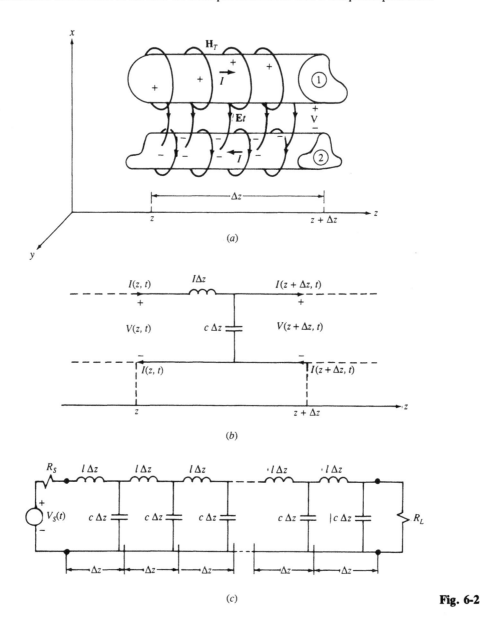

Fig. 6-2

▌ Over a small length Δz, the current of one conductor and the returning current on the other conductor produce a transverse magnetostatic field $\mathbf{H}_T$ (Problem 6.2). If this Δz-section of the line is considered to be a loop, the magnetic flux passing between the conductors links the current of the loop, which constitutes an inductance L. The per-unit-length inductance of this section, $l = L/\Delta z$, will characterize all other sections of the line, since the line is uniform. Similarly, the transverse electrostatic field $\mathbf{E}_T$ results from the separation of charge on the conductor surfaces; i.e., the Δz-section acts as a capacitance C. We may therefore define a per-unit-length capacitance by $c = C/\Delta z$. Again, all other sections of the line will be characterized by this per-unit-length capacitance, since the line is uniform.

Equivalent circuits for a Δz-section and for the complete line are shown in Fig. 6-2(b) and (c).

6.6 Obtain a pair of equations giving the voltage-current relationship for the section Δz of the line diagramed in Fig. 6-2(b).

❚ From Fig. 6-2(b) we have

$$V(z + \Delta z, t) - V(z, t) = -l\,\Delta z\,\frac{\partial}{\partial t}I(z, t) \tag{1}$$

$$I(z + \Delta z, t) - I(z, t) = -c\,\Delta z\,\frac{\partial}{\partial t}V(z + \Delta z, t) = -c\,\Delta z\,\frac{\partial V(z, t)}{\partial t} + lc(\Delta z)^2\frac{\partial^2 I(z, t)}{\partial t^2} \tag{2}$$

where we have substituted $V(z + \Delta z, t)$ from (1) into (2). Dividing both sides of (1) and (2) by Δz and taking the limit as $\Delta z \to 0$, we obtain the (*lossless*) *transmission-line equations*

$$\frac{\partial V(z, t)}{\partial z} = -l\frac{\partial I(z, t)}{\partial t} \tag{3}$$

$$\frac{\partial I(z, t)}{\partial z} = -c\frac{\partial V(z, t)}{\partial t} \tag{4}$$

6.7 Show that $V(z, t)$ and $I(z, t)$ (Problem 6.6) separately satisfy the wave equation.

❚ Differentiating (3) and (4) of Problem 6.6 with respect to z and t, respectively, we obtain

$$\frac{\partial^2 V(z, t)}{\partial z^2} = -l\frac{\partial^2 I(z, t)}{\partial z\,\partial t} \quad \text{and} \quad \frac{\partial^2 I(z, t)}{\partial z\,\partial t} = -c\frac{\partial^2 V(z, t)}{\partial t^2}$$

Elimination of the mixed derivative yields

$$\frac{\partial^2 V(z, t)}{\partial z^2} = lc\frac{\partial^2 V(z, t)}{\partial t^2}$$

which is the one-dimensional wave equation, with propagation velocity $1/\sqrt{lc}$. Similarly,

$$\frac{\partial^2 I(z, t)}{\partial z^2} = lc\frac{\partial^2 I(z, t)}{\partial t^2}$$

6.8 Refer to Problem 6.7. Verify that $1/\sqrt{lc}$ has the units of velocity.

❚ Let $[X]$ represent the units of the physical quantity X. Then $[l] = \text{H/m} = \text{V} \cdot \text{s/A} \cdot \text{m}$ and $[c] = \text{F/m} = \text{A} \cdot \text{s/V} \cdot \text{m}$, so that

$$[lc] = [l][c] = \text{s}^2/\text{m}^2$$

and

$$\left[\frac{1}{\sqrt{lc}}\right] = \frac{1}{\sqrt{[lc]}} = \frac{\text{m}}{\text{s}}$$

6.9 Letting $u \equiv 1/\sqrt{lc}$, indicate general solutions for $V(z, t)$ and $I(z, t)$ in Problem 6.7.

❚ $$V(z, t) = V^+(t - z/u) + V^-(t + z/u) \qquad I(z, t) = I^+(t - z/u) + I^-(t + z/u)$$

6.10 With reference to Problem 6.9, verify that $F(t \mp z/u)$ really is a solution of the one-dimensional wave equation.

❚ $$\frac{\partial^2 F}{\partial z^2} = F''(t \mp z/u)\left(\mp\frac{1}{u}\right)^2 \quad \text{and} \quad \frac{\partial^2 F}{\partial t^2} = F''(t \mp z/u)$$

6.11 The voltage and current solutions of Problem 6.9 may not be chosen independently. Infer the necessary connections from Problem 6.6.

❚ By linearity, the forward-moving and backward-moving waves must separately satisfy the transmission-line equations; thus

$$-\frac{1}{u}(V^+)' = -l(I^+)' \tag{1}$$

$$-\frac{1}{u}(I^+)' = -c(V^+)' \tag{2}$$

with a prime denoting differentiation with respect to the argument $t - z/u$. Since $u = 1/\sqrt{lc}$, (1) and (2) are consistent; either one gives, after neglect of a constant of integration,

$$I^+(t - z/u) = \frac{1}{\sqrt{l/c}} V^+(t - z/u) \equiv \frac{1}{R_C} V^+(t - z/u) \tag{3}$$

Similarly, one obtains

$$I^-(t + z/u) = -\frac{1}{R_C} V^-(t + z/u) \tag{4}$$

6.12 The "Ohm's laws" (3) and (4) of Problem 6.11 are formally identical to the results of Problems 5.14 and 5.19. Comment.

▮ The TEM mode can be analyzed as plane waves traveling in a hypothetical lossless medium of intrinsic impedance $\eta = R_C$. (We call R_C the *characteristic resistance* of the line.)

6.13 Are the two ways of notating the phase—$\omega t - \beta z$ in Chapter 5, $t - z/u$ in Chapter 6—equivalent?

▮ Yes; in fact

$$f(\omega t - \beta z) = f\left(\omega\left(t - \frac{z}{\omega/\beta}\right)\right) = g(t - z/c)$$

However, one would always use $t - z/c$ for the argument of a nonperiodic wave function, because the notions of frequency and wavelength would be empty.

6.14 Determine the velocity of wave propagation and the characteristic resistance of a lossless cable, if $l = 0.25\ \mu\text{H/m}$ and $c = 100\ \text{pF/m}$.

▮ $$u = \frac{1}{\sqrt{lc}} = 2 \times 10^8\ \text{m/s} \qquad R_C = \sqrt{l/c} = 50\ \Omega$$

6.15 Repeat Problem 6.14, assuming that the cable has $c = 10\ \text{pF/m}$, $\epsilon_r = 2.1$, and $\mu_r = 1$.

▮ $$u = \frac{u_0}{\sqrt{\mu_r \epsilon_r}} = \frac{3 \times 10^8}{\sqrt{2.1}} = 2.07 \times 10^8\ \text{m/s}$$
$$R_C = \frac{1}{cu} = \frac{1}{10 \times 10^{-12} \times 2.07 \times 10^8} = 483.09\ \Omega$$

6.16 A transmission line of total length ζ that is terminated in a resistance R_L is driven by a pulse voltage source having an open-circuit waveform $V_S(t)$ and an internal resistance R_S (Fig. 6-3). Show that in general both forward- and backward-traveling voltage waves must exist in the line.

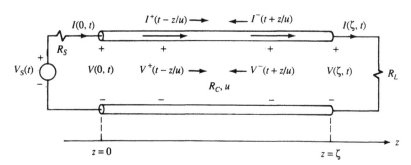

Fig. 6-3

▮ The end condition $V(\zeta, t) = R_L I(\zeta, t)$ cannot be met by forward-traveling waves alone ([see (3) of Problem 6.11] or backward-traveling waves alone [see (4) of Problem 6.11].

6.17 State the sole exception to the general result of Problem 6.16.

▮ When the load is *matched* to the line—i.e., when $R_L = R_C$—the end condition requires purely forward-traveling waves.

6.18 For the line of Problem 6.16, define a *voltage reflection coefficient* as the ratio of the backward- to the forward-traveling voltage wave at $z = \zeta$:

$$\Gamma_L \equiv \frac{V^-(t + \zeta/u)}{V^+(t - \zeta/u)} \tag{1}$$

Express the voltage and current at the load in terms of the forward voltage wave and the voltage reflection coefficient.

▌ By Problems 6.9 and 6.11,

$$V(\zeta, t) = V^+(t - \zeta/u) + V^-(t + \zeta/u) = V^+(t - \zeta/u)(1 + \Gamma_L) \tag{2}$$

$$I(\zeta, t) = I^+(t - \zeta/u) + I^-(t + \zeta/u) = \frac{V^+(t - \zeta/u)}{R_C} - \frac{V^-(t + \zeta/u)}{R_C} = \frac{V^+(t - \zeta/u)}{R_C}(1 - \Gamma_L) \tag{3}$$

6.19 In terms of the voltage reflection coefficient Γ_L, obtain an expression for the current reflection coefficient $I^-(t + \zeta/u)/I^+(t - \zeta/u)$.

▌

$$\frac{I^-(t + \zeta/u)}{I^+(t - \zeta/u)} = -\frac{V^-(t + \zeta/u)/R_C}{V^+(t - \zeta/u)/R_C} = -\Gamma_L$$

6.20 Evaluate the voltage reflection coefficient Γ_L in terms of the load resistance R_L and the characteristic resistance R_C.

▌ Substituting (2) and (3) of Problem 6.18 into $V(\zeta, t) = R_L I(\zeta, t)$ yields

$$V^+(t - \zeta/u)(1 + \Gamma_L) = \frac{R_L}{R_C}[V^+(t - \zeta/u)](1 - \Gamma_L)$$

Solving for Γ_L yields

$$\Gamma_L = \frac{R_L - R_C}{R_L + R_C} \tag{1}$$

6.21 A transmission line with a characteristic resistance of 50 Ω is connected to a 100-Ω resistive load and to a 30-V dc source with zero internal resistance. Calculate the voltage reflection coefficients at the load and at the source.

▌ By (1) of Problem 6.20,

$$\Gamma_L = \frac{R_L - R_C}{R_L + R_C} = \frac{100 - 50}{100 + 50} = \frac{1}{3} \qquad \Gamma_S = \frac{R_S - R_C}{R_S + R_C} = \frac{0 - 50}{0 + 50} = -1$$

6.22 A transmission line, terminated by its characteristic resistance, is connected at $t = 0$ to a dc source of voltage V_S with internal resistance R_S. Plot the voltage on the line at a fixed distance z_1 against time.

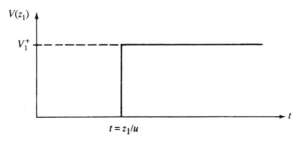

Fig. 6-4

▌ Since $R_L = R_C$, the line is matched and there will be no reflected wave. The height of the voltage pulse is

$$V_1^+ = \frac{R_C V_S}{R_C + R_S}$$

The wave travels in the $+z$-direction at velocity $u = 1/\sqrt{lc}$, as shown in Fig. 6-4. As $t \to \infty$ (steady state), the voltage becomes V_1^+ everywhere.

6.23 Generalize Problem 6.22 to the case $R_S \neq R_C \neq R_L$. Specifically, obtain expressions for **(a)** V_1^-, the height of the first reflected voltage pulse and **(b)** V_L, the steady-state voltage across the load.

▌ The original forward pulse is given in Problem 6.22, and the reflection coefficients at the two ends of the line are as in Problem 6.21.
(a) $V_1^- = \Gamma_L V_1^+$.
(b) As $t \to \infty$, the voltage at any point is given by the superposition of the first forward, first reflected, second

forward, . . . , waves:

$$V_L = V_1^+ + V_1^- + V_2^+ + V_2^- + \cdots$$
$$= V_1^+(1 + \Gamma_L + \Gamma_S\Gamma_L + \Gamma_S\Gamma_L^2 + \Gamma_S^2\Gamma_L^2 + \Gamma_S^2\Gamma_L^3 + \cdots)$$
$$= V_1^+[(1 + \Gamma_S\Gamma_L + \Gamma_S^2\Gamma_L^2 + \cdots) + \Gamma_L(1 + \Gamma_S\Gamma_L + \Gamma_S^2\Gamma_L^2 + \cdots)]$$
$$= V_1^+\left[\left(\frac{1}{1 - \Gamma_S\Gamma_L}\right) + \left(\frac{\Gamma_L}{1 - \Gamma_S\Gamma_L}\right)\right] = V_1^+\left(\frac{1 + \Gamma_L}{1 - \Gamma_S\Gamma_L}\right)$$

6.24 Obtain an expression for the steady-state load current for the line of Problem 6.23.

❚ Following Problem 6.23,

$$I_L = I_1^+ + I_1^- + I_2^+ + I_2^- + \cdots \tag{1}$$

Now, since $I_1^+ = V_1^+/R_C$ and $I_1^- = -V_1^-/R_C$ and so on, (1) may be written as

$$I_L = \frac{V_1^+}{R_C} - \Gamma_L\frac{V_1^+}{R_C} + \frac{\Gamma_S\Gamma_L V_1^+}{R_C} - \frac{\Gamma_S\Gamma_L^2 V_1^+}{R_C} + \cdots$$
$$= \frac{V_1^+}{R_C}(1 - \Gamma_L + \Gamma_S\Gamma_L - \Gamma_S\Gamma_L^2 + \Gamma_S^2\Gamma_L^2 - \Gamma_S^2\Gamma_L^3 + \cdots)$$
$$= \frac{V_1^+}{R_C}[(1 + \Gamma_S\Gamma_L + \Gamma_S^2\Gamma_L^2 + \cdots) - \Gamma_L(1 + \Gamma_S\Gamma_L + \Gamma_S^2\Gamma_L^2 + \cdots)]$$
$$= \frac{V_1^+}{R_C}\left(\frac{1 - \Gamma_L}{1 - \Gamma_S\Gamma_L}\right) \tag{1}$$

Alternatively, $I_L = V_L/R_L$ [check equivalance to (1)].

6.25 A transmission line of 30 Ω characteristic resistance, having a load of 90 Ω, is connected to a 100-V dc source at $t = 0$. The internal resistance of the source is 60 Ω. Determine the steady-state voltage across the load.

❚

$$V_1^+ = \frac{R_C V_S}{R_C + R_S} = \frac{(30)(100)}{30 + 60} = 33.33 \text{ V}$$
$$\Gamma_S = \frac{R_S - R_C}{R_S + R_C} = \frac{60 - 30}{60 + 30} = 0.33$$
$$\Gamma_L = \frac{R_L - R_C}{R_L + R_C} = \frac{90 - 30}{90 + 30} = 0.5$$

Then, by Problem 6.23(b),

$$V_L = (33.33)\left[\frac{1 + 0.5}{1 - (0.33)(0.5)}\right] = 60 \text{ V}$$

6.26 Calculate the steady-state load current for the transmission line of Problem 6.25.

❚ $I_L = V_L/R_L = (60 \text{ V})/(90 \text{ Ω}) = 0.667 \text{ A}$.

6.27 For the transmission line of Problem 6.25, sketch the voltage along the line at three instants: $0 < t_1 < T$; $T < t_2 < 2T$; $2T < t_3 < 3T$. Here, $T \equiv \zeta/u$, the transit time in either direction.

❚ See Fig. 6-5.

6.28 Repeat Problem 6.27 for the current wave.

❚ See Fig. 6-6. Note that Problem 6.24, and the fact that here Γ_L and Γ_S are positive, implies that I_1^- (and all other backward waves) is negative.

6.29 At $t = 0$, a 30-V battery with zero source resistance is connected to a transmission line, as shown in Fig. 6-7. The line has a 2 μs transit time. Sketch the distribution of voltage along the line at (a) $t = 1$ μs; (b) $t = 2.5$ μs; (c) $t = 4.5$ μs; (d) $t = 6.5$ μs.

❚ The load and source reflection coefficients are

$$\Gamma_L = \frac{R_L - R_C}{R_L + R_C} = \frac{100 - 50}{100 + 50} = \frac{1}{3} \qquad \Gamma_S = \frac{R_S - R_C}{R_S + R_c} = \frac{0 - 50}{0 + 50} = -1$$

(a) The wavefront, of height 30 V, is at the midpoint of the line [Fig. 6-8(a)].
(b) The wavefront, of height $(1 + \frac{1}{3})(30) = 40$ V, is at the three-quarter point [Fig. 6-8(b)].
(c) The wavefront, of height $(1 + \frac{1}{3} - \frac{1}{3})(30) = 30$ V, is at the one-quarter point [Fig. 6-8(c)].
(d) The wavefront, of height $(1 + \frac{1}{3} - \frac{1}{3} - \frac{1}{9})(30) = 26.67$ V, is at the three-quarter point [Fig. 6-8(d)].

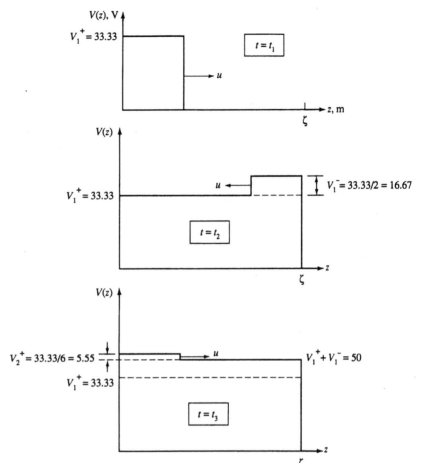

Fig. 6-5

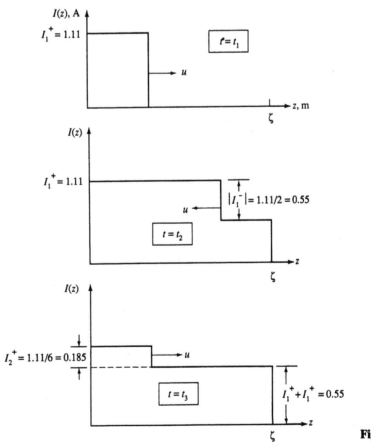

Fig. 6-6

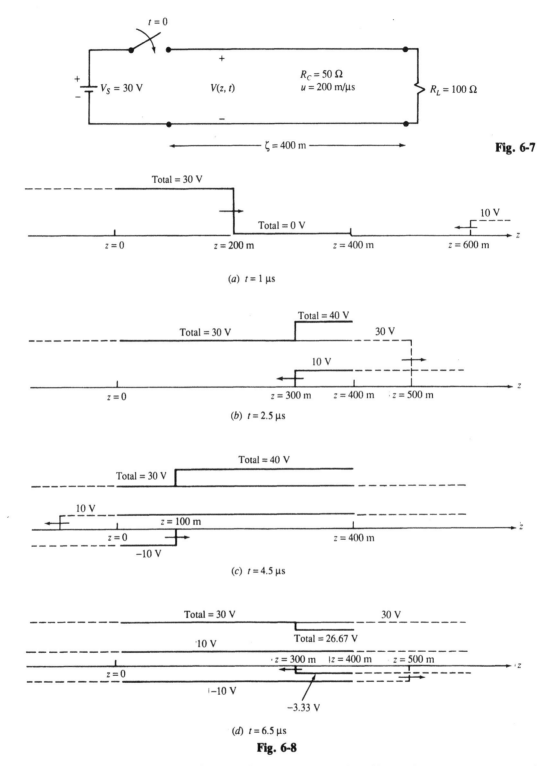

Fig. 6-7

(a) $t = 1 \, \mu\text{s}$

(b) $t = 2.5 \, \mu\text{s}$

(c) $t = 4.5 \, \mu\text{s}$

(d) $t = 6.5 \, \mu\text{s}$

Fig. 6-8

6.30   In Problem 6.29, replace the constant 30-V source by a rectangular pulse of height 30 V and duration 1 μs (Fig. 6-9). Sketch the voltage distribution at (**a**) $t = 1 \, \mu\text{s}$; (**b**) $t = 2.75 \, \mu\text{s}$; (**c**) $t = 4.75 \, \mu\text{s}$; (**d**) $t = 6.75 \, \mu\text{s}$.

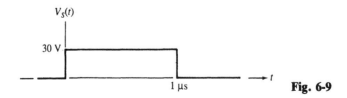

Fig. 6-9

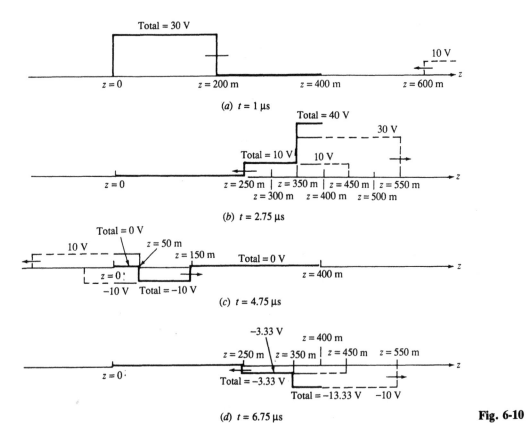

Total = 30 V

$z = 0$ $z = 200$ m $z = 400$ m $z = 600$ m

10 V

(a) $t = 1$ μs

Total = 40 V

30 V

Total = 10 V   10 V

$z = 0$ $z = 250$ m | $z = 350$ m | $z = 450$ m | $z = 550$ m

$z = 300$ m $z = 400$ m $z = 500$ m

(b) $t = 2.75$ μs

Total = 0 V

10 V $z = 50$ m

$z = 150$ m

$z = 0$ Total = 0 V

-10 V   Total $= -10$ V $z = 400$ m

(c) $t = 4.75$ μs

-3.33 V $z = 400$ m

$z = 250$ m \ $z = 350$ m | $z = 450$ m $z = 550$ m

$z = 0$

Total $= -3.33$ V

Total $= -13.33$ V -10 V

(d) $t = 6.75$ μs

Fig. 6-10

▌ See Fig. 6-10. Note that since the voltage source does not remain constant for $t > 0$ but returns to zero at $t = 1$ μs, the traveling waves are zero along portions of the line. For instance, in the 0.75 μs after the leading edge of the original pulse has reached $z = \zeta = 400$ m, three-quarters of the pulse has "passed through" $z = \zeta$, leaving the trailing edge at $z = 350$ m; during the same time the first reflected pulse has reached $z = 250$ m. The superposition of these two pulses is the two-step distribution shown in Fig. 6-10(b).

6.31 A transmission line [Fig. 6-11(a)] has $R_S = 300$ Ω, $R_L = 60$ Ω, $R_C = 100$ Ω, $u = 400$ m/μs, $\zeta = 400$ m, and $V_S(t) = 400u(t)$ V, where $u(t)$ is the unit step function. Sketch $V(0, t)$ for $0 < t \leqslant 10$ μs.

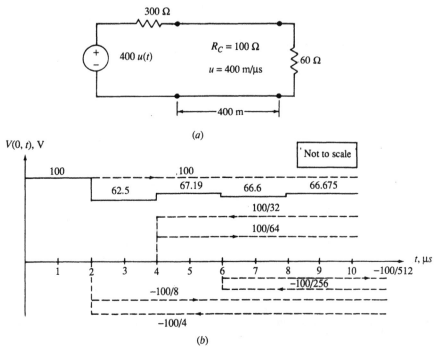

300 Ω

$400\,u(t)$

$R_C = 100$ Ω

$u = 400$ m/μs

60 Ω

400 m

(a)

$V(0, t)$, V

Not to scale

100   .100

62.5   67.19   66.6   66.675

100/32

100/64

1 2 3 4 5 6 7 8 9 10 t, μs $-100/512$

$-100/8$ $-100/256$

$-100/4$

(b)

Fig. 6-11

■ The voltage reflection coefficients are

$$\Gamma_s = \frac{300 - 100}{300 + 100} = \frac{1}{2} \qquad \Gamma_L = \frac{60 - 100}{60 + 100} = -\frac{1}{4}$$

The initial voltage is

$$V_1^+ = \frac{R_C V_s(0^+)}{R_C + R_s} = \frac{(100)(400)}{100 + 300} = 100 \text{ V}$$

$V(0, t)$ is sketched in Fig. 6-11(b); the steady-state voltage is

$$V(0, \infty) = \frac{60}{300 + 60} \times 400 = 66.67 \text{ V}$$

6.32 For the line of Problem 6.31, sketch $I(0, t)$ for $0 < t \le 10 \ \mu s$.

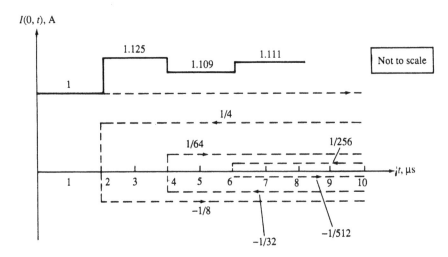

Fig. 6-12

■ The current reflection coefficients are the negatives of the voltage reflection coefficients (Problem 6.19), and the initial current is

$$I_1^+ = \frac{400}{300 + 100} = 1 \text{ A}$$

The required $I(0, t)$ is sketched in Fig. 6-12, where the steady-state current is

$$I(0, \infty) = \frac{400}{300 + 60} = 1.11 \text{ A}$$

6.33 For the line of Problem 6.31, sketch $V(\zeta, t) = V(400, t)$ over the interval $0 < t \le 10 \ \mu s$.

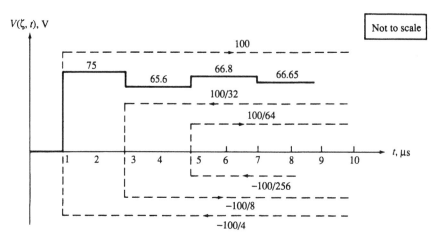

Fig. 6-13

■ With the voltage reflection coefficients from Problem 6.31, using the procedure of Problem 6.29 we obtain the plot shown in Fig. 6-13.

6.34   Repeat Problem 6.33 for $I(\zeta, t)$.

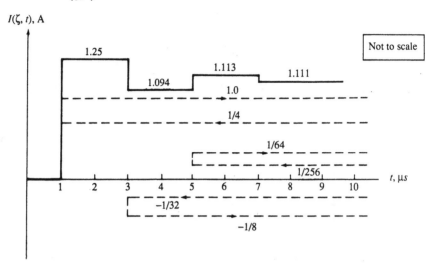

Fig. 6-14

▮ The current reflection coefficients are $\Gamma_s = -1/2$ and $\Gamma_L = 1/4$. Proceeding as in Problem 6.29, we obtain the sketch shown in Fig. 6-14.

6.35   Repeat Problem 6.31 for a short-circuited line.

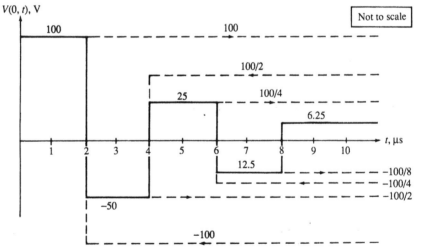

Fig. 6-15

▮ In this case, $R_L = 0$; thus, $\Gamma_s = 1/2$ and $\Gamma_L = -1$. With $V_1^+ = 100$ V, as in Problem 6.31, we obtain the sketch of Fig. 6-15.

6.36   Sketch $I(0, t)$ for $0 < t \le 10$ μs, assuming that the line of Problem 6.31 is short-circuited.

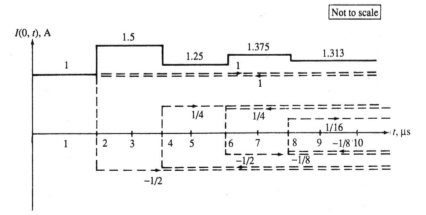

Fig. 6-16

■ As in Problem 6.35, with $R_L = 0$, the current reflection coefficients are $\Gamma_s = -1/2$ and $\Gamma_L = 1$. With $I_1^+ = 1.0$ A, we obtain the sketch shown in Fig. 6-16.

6.37 Sketch $V(\zeta, t)$ for $0 < t \leq 10\ \mu s$, assuming that the line of Problem 6.31 is short-circuited.

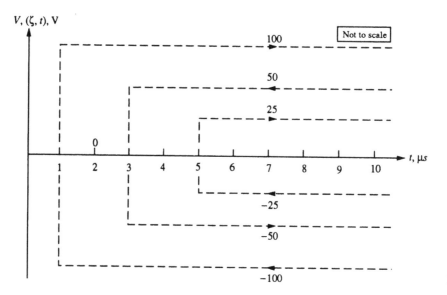

Fig. 6-17

■ With $R_L = 0$, the voltage reflection coefficients are $\Gamma_s = 1/2$ and $\Gamma_L = -1$. With $V_1^+ = 100$ V, we obtain the sketch shown in Fig. 6-17.

6.38 Sketch $I(\zeta, t)$ for $0 < t \leq 10\ \mu s$, assuming that the line of Problem 6.31 is short-circuited.

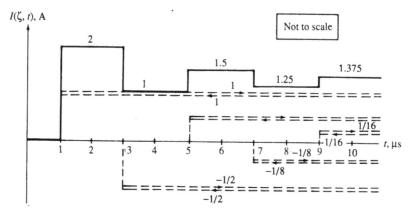

Fig. 6-18

■ With $R_L = 0$ and the current reflection coefficients $\Gamma_s = -1/2$ and $\Gamma_L = 1$, we obtain the sketch of Fig. 6-18.

6.39 Sketch $V(0, t)$ for $0 < t \leq 10\ \mu s$, assuming that the line of Problem 6.31 is open-circuited.

■ For an open-circuited line, $R_L = \infty$. Thus, the voltage reflection coefficients are $\Gamma_s = 1/2$ and $\Gamma_L = 1$. With $V_1^+ = 100$ V as the initial voltage, we obtain the sketch of Fig. 6-19.

6.40 Sketch $V(\zeta, t)$ for $0 < t \leq 10\ \mu s$, assuming that the line of Problem 6.31 is open-circuited.

■ In this case, $R_L = \infty$, $\Gamma_s = 1/2$, and $\Gamma_L = 1$. With $V_1^+ = 100$ V, we obtain the sketch shown in Fig. 6-20.

6.41 Sketch $I(0, t)$ for $0 < t \leq 10\ \mu s$, assuming that the line of Problem 6.31 is open-circuited.

■ With $R_L = \infty$, $\Gamma_s = -1/2$, $\Gamma_L = -1$, and $I_1^+ = 1.0$ A, we obtain the sketch shown in Fig. 6-21.

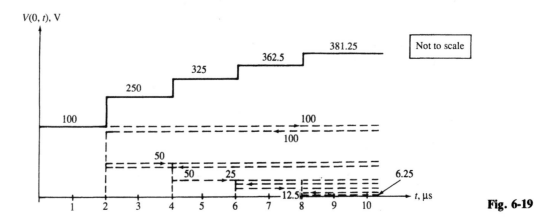

Fig. 6-19

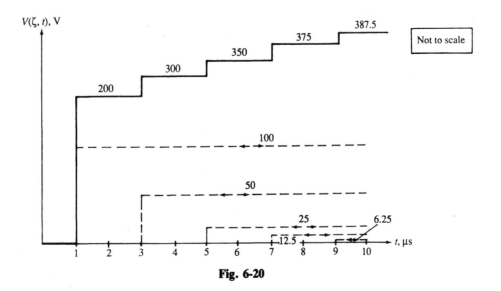

Fig. 6-20

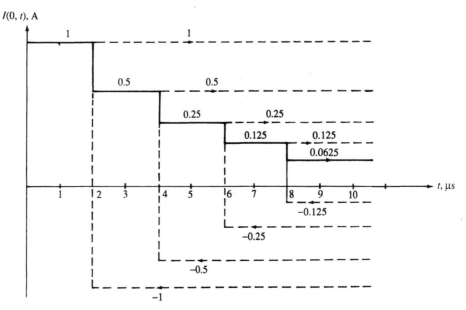

Fig. 6-21

6.42 Sketch $I(\zeta, t)$ for $0 < t \le 10 \ \mu s$, assuming that the line of Problem 6.31 is open-circuited.

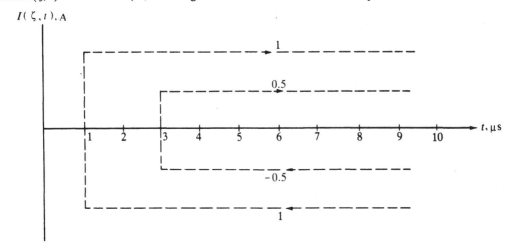

Fig. 6-22

❚ The parameters are as in Problem 6.41; the sketch is given in Fig. 6-22.

6.43 A time-domain reflectometer (TDR) launches a pulse down a transmission line and records the transit time for that pulse to be reflected at the load and to return to the line input. Suppose a TDR having a source impedance of 50 Ω is attached to a 50-Ω coaxial cable of unknown length and load resistance. The dielectric of the cable is Teflon ($\epsilon_r = 2.1$). The open-circuit voltage of the TDR is a pulse of duration 10 μs. Assuming that the recorded voltage at the input to the line is as shown in Fig. 6-23(a), determine the load resistance.

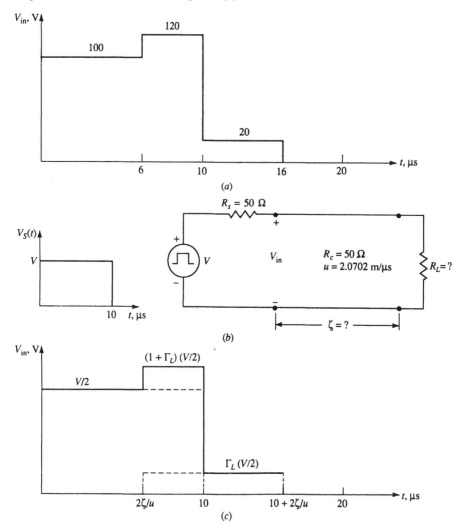

Fig. 6-23

▌ The pulse from the source and the line are shown in Fig. 6-23(b). In Fig. 6-23(c), V_{in} is graphed in terms of the source voltage and the voltage reflection coefficient; we see that $V/2 = 100\ \text{V}$, or $V = 200\ \text{V}$; and $\Gamma_L(100\ \text{V}) = 20\ \text{V}$, or $\Gamma_L = 1/5$.

But
$$\Gamma_L = \frac{R_L - R_C}{R_L + R_C} \quad \text{or} \quad \frac{1}{5} = \frac{R_L - 50}{R_L + 50} \quad \text{or} \quad R_L = 75\ \Omega$$

6.44   Calculate the length of the line of Problem 6.43.

▌ Comparing Fig. 6-23(c) and (a), we have
$$10 + 2\zeta/u = 16 \quad \text{or} \quad \zeta/u = 3\ \mu\text{s}$$
But $u = (3 \times 10^8)/\sqrt{2.1} = 2.07 \times 10^8\ \text{m/s} = 207\ \text{m}/\mu\text{s}$, so that $\zeta = (207)(3) = 621\ \text{m}$.

6.45   A 12-V battery with negligible internal resistance is connected to an unknown length of transmission line that is terminated in a resistance. Assuming that the input current to the line is as shown in Fig. 6-24, determine the line characteristic resistance.

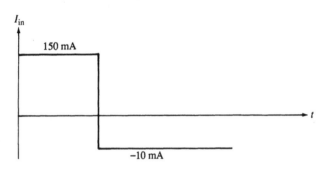

Fig. 6-24

▌
$$0.150\ \text{A} = \frac{12\ \text{V}}{R_C} \quad \text{or} \quad R_C = 80\ \Omega$$

6.46   Calculate the load resistance on the transmission line of Problem 6.45.

▌ Since $R_S = 0$,

$$\Gamma_S = \frac{R_S - R_C}{R_S + R_C} = -1 \tag{1}$$

and Problem 6.24 gives
$$-10\ \text{mA} = (150\ \text{mA})(1 - 2\Gamma_L) \quad \text{or} \quad \Gamma_L = 8/15$$

But
$$\Gamma_L = \frac{8}{15} = \frac{R_L - R_C}{R_L + R_C} = \frac{R_L - 80}{R_L + 80}$$

whence $R_L = 262.9\ \Omega$.

6.47   A transmission line has $R_S = 0$ and $R_L = \infty$. Assume that the source voltage is the ramp waveform
$$V_S(t) = \begin{cases} 0 & t < 0 \\ t/\tau_r & 0 < t < \tau_r \\ 1 & t > \tau_r \end{cases}$$
where τ_r is the *rise time*. Sketch the load voltage for a transmission line of one-way transit time $T = 10\tau_r$.

▌ The input voltage and the line are shown in Fig. 6-25(a). The voltage reflection coefficients at the source and at the load are, respectively, $\Gamma_S = -1$ and $\Gamma_L = 1$. Hence load voltage appears as in Fig. 6-25(b).

6.48   Repeat Problem 6.47 for $T = \tau_r/2$.

▌ See Fig. 6-26.

6.49   Repeat Problem 6.47 for $T = \tau_r/3$.

▌ See Fig. 6-27.

6.50   Repeat Problem 6.47 for $T = \tau_r/4$.

▌ See Fig. 6-28.

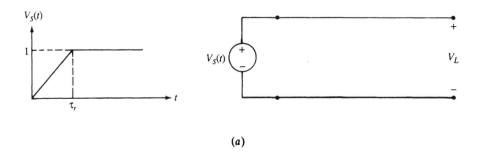

(a)

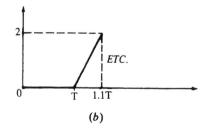

(b)

Fig. 6-25

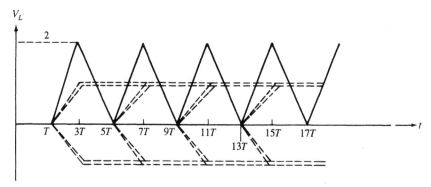

Fig. 6-26

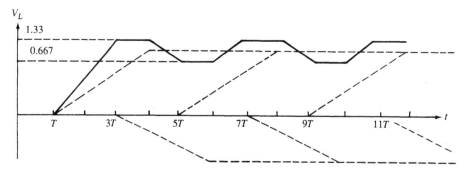

Fig. 6-27

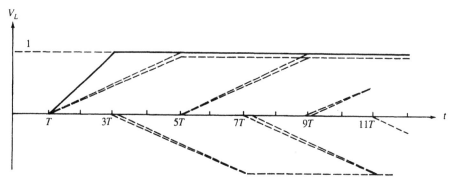

Fig. 6-28

6.51 A step voltage $V_S(t) = V_0[u(t) - u(t - T)]$ (Fig. 6-29(a)) is applied to a line having a one-way delay T and resistances $R_S = R_C = R_L/3$. Plot the input current to the line as a function of time, for $0 < t < 4T$.

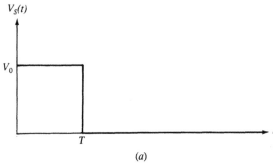

(a)

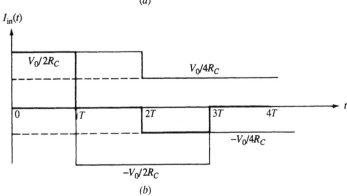

(b)   **Fig. 6-29**

▮ The current reflection coefficients are $\Gamma_S = 0$ and $\Gamma_L = -\frac{1}{2}$. The input or load current is plotted in Fig. 6-29(b).

6.52 Repeat Problem 6.51 for the source voltage $V_S(t) = V_0[2u(t) - u(t - T) - u(t - 2T)]$ [Fig. 6-30(a)].

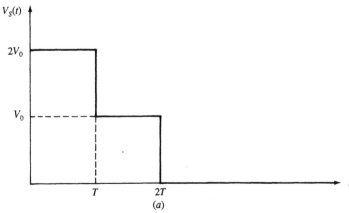

(a)

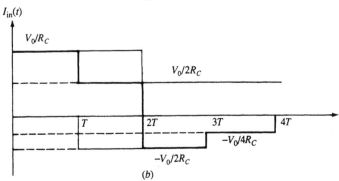

(b)   **Fig. 6-30**

▮ See Fig. 6-30(b).

6.53   Repeat Problem 6.51 for the source voltage given in Fig. 6-31(a).

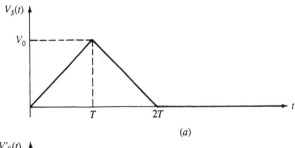

(a)

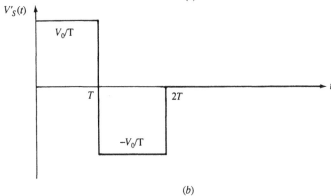

(b)

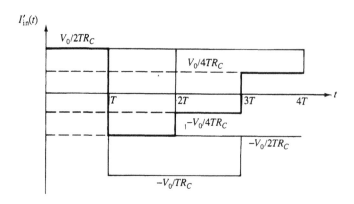

(c)

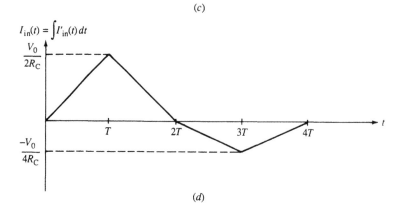

(d)

Fig. 6-31

▌ Observe that the given voltage of Fig. 6-31(a) may be synthesized from that of Fig. 6-31(b) by integration; that is, $V_S(t) = \int V_S' \, dt$, where

$$V_S'(t) = \frac{V_0}{T} u(t) - \frac{2V_0}{T} u(t-T) + \frac{V_0}{T} u(t-2T)$$

Now, $I_{in}'(t)$ due to $V_S'(t)$ is obtained as shown in Fig. 6-31(c). The required $I_{in}(t)$ is then obtained by integrating $I_{in}'(t)$, yielding the result of Fig. 6-31(d).

6.54 A step voltage source, $V_S(t) = 400u(t)$ V, having $R_s = 350\ \Omega$, is applied to a 50-Ω line that has a short-circuit load. Plot the load current as a function of the line's one-way delay T, for $0 < t < 12T$.

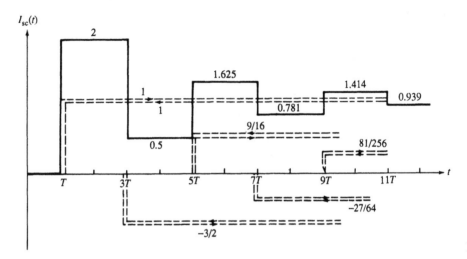

Fig. 6-32

   ❚ The current reflection coefficients are $\Gamma_s = -3/4$ and $\Gamma_L = 1$, since $R_L = 0$. The short-circuit current is plotted in Fig. 6-32.

6.55 At what time will the short-circuit current in Problem 6.54 be within 10 percent of its steady-state value?

   ❚ The steady-state current is $400/350 = 1.143$ A, so that a 10 percent spread is represented by the interval $(1.029, 1.257$ A). Calculating the current at $t = T, 3T, 5T, \ldots$, we have:

time	current
T	2
$3T$	$2(1 + \Gamma_s) = 2(1 - 3/4) = 0.5$
$5T$	$2(1 + \Gamma_s + \Gamma_s^2) = 2(1 - 3/4 + 4/16) = 1.625$
$7T$	$2(1 + \Gamma_s + \Gamma_s^2 + \Gamma_s^3) = 0.781$
$9T$	$2(1 + \cdots + \Gamma_s^4) = 1.414$
$11T$	$2(1 + \cdots + \Gamma_s^5) = 0.939$
$13T$	$2(1 + \cdots + \Gamma_s^6) = 1.295$
$15T$	$2(1 + \cdots + \Gamma_s^7) = 1.028$

Thus, by $15T$, the current converges to within 10 percent of the steady-state value.

6.56 The voltage pulses shown in Fig. 6-33(a) are applied to the ends of a transmission line, as shown in Fig. 6-33(b). Sketch the voltages at the two ends for $0 < t < 7T$, where T is the line's one-way delay.

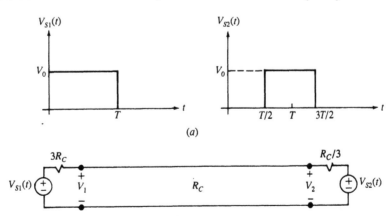

Fig. 6-33

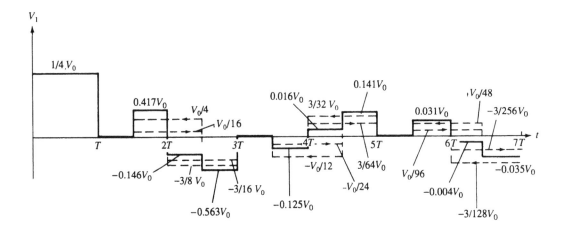

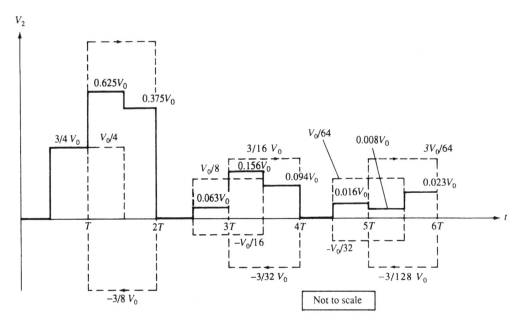

Not to scale

Fig. 6-34

▌ The voltage reflection coefficients are $\Gamma_{S1} = \Gamma_{L2} = \frac{1}{2}$, $\Gamma_{S2} = \Gamma_{L1} = -\frac{1}{2}$. The initial voltages at the two ends are $(V_1)_{\text{initial}} = \frac{1}{4}V_0$ (at $t = 0$) and $(V_2)_{\text{initial}} = \frac{3}{4}V_0$ (at $t = T/2$). Consequently we obtain the voltages V_1 and V_2 shown in Fig. 6-34.

6.57   A transmission line having $\zeta = 200$ m, $u = 200$ m/μs, $R_C = 50\ \Omega$, and $R_L = 20\ \Omega$ is driven by a source which has $R_S = 100\ \Omega$ and has a voltage that is a rectangular pulse of magnitude 6 V and duration 3 μs. Sketch the input current to the line over the first 5 μs.

▌ The current reflection coefficients are $\Gamma_S = -1/3$ and $\Gamma_L = 3/7$. With the initial current $I_1^+ = 6/150 = 40$ mA, we obtain the sketch shown in Fig. 6-35.

6.58   The coaxial cable shown in Fig. 6-36(a) is 400 m long, is terminated in a short circuit ($R_L = 0$), and is driven by a pulse source [Fig. 6-36(b)] having an internal resistance of 150 Ω. Sketch the voltage at the input to the line, $V(0, t)$, for $0 < t < 14\ \mu$s.

▌ The coaxial cable has parameters $c = 100$ pF/m, $l = 0.25\ \mu$H/m, $R_C = \sqrt{l/c} = 50\ \Omega$,

$$u = \frac{1}{\sqrt{lc}} = 200 \times 10^6 \text{ m/s} = 200 \text{ m/}\mu\text{s} \qquad \text{and} \qquad T = \frac{400 \text{ m}}{200 \text{ m/}\mu\text{s}} = 2\ \mu\text{s}$$

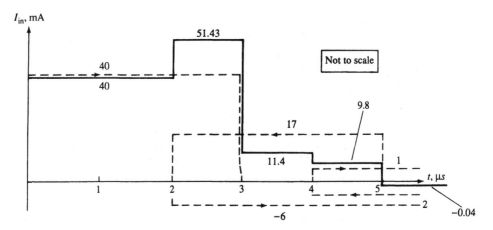

Fig. 6-35

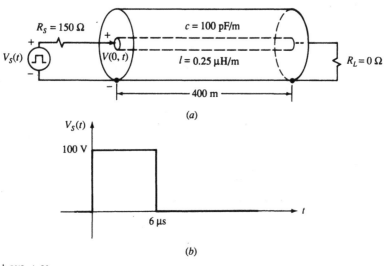

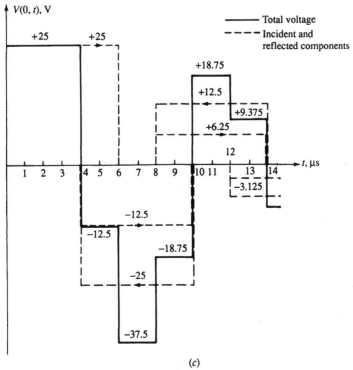

Fig. 6-36

The voltage reflection coefficients are

$$\Gamma_L = \frac{R_L - R_C}{R_L + R_C} = -1 \qquad \Gamma_S = \frac{R_S - R_C}{R_S + R_C} = \frac{1}{2}$$

The source initially "sees" a line resistance of $R_C = 50\ \Omega$; thus, the initially launched voltage wave is a pulse of $6\ \mu s$ duration and magnitude

$$V_1^+ = \frac{R_C}{R_C + R_S}\ V_S = 25\ V$$

This pulse propagates toward the load, which reflects a pulse with magnitude $\Gamma_L V_1^+ = -25\ V$; this reflected pulse is re-reflected at the source, producing a pulse traveling back toward the load of $\Gamma_S(\Gamma_L V_1^+) = -12.5\ V$; and so forth. In Fig. 6-36(c), dashed lines indicate the forward-traveling ($\rightarrow$) and backward-traveling ($\leftarrow$) components. The total voltage at the source end is plotted as a solid line; it is the instantaneous combination of all the components present at $z = 0$.

6.59 Repeat Problem 6.58 if the source voltage is of the form shown in Fig. 6-37(a).

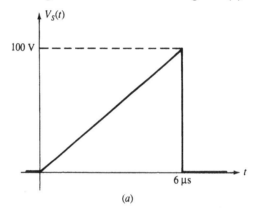

(a)

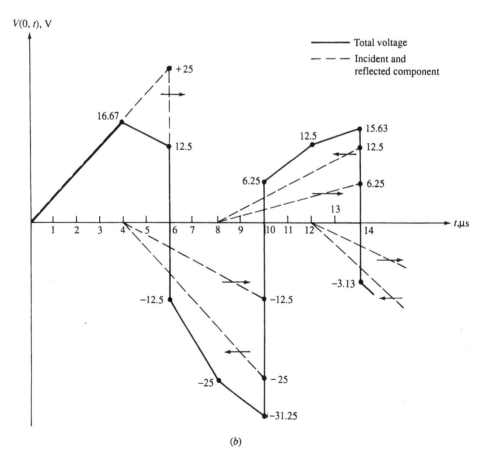

(b)

Fig. 6-37

❚ Proceeding exactly as in Problem 6.58, we obtain the desired result, Fig. 6-37(b).

6.60 For the transmission line and the source voltage shown in Fig. 6-38(a), sketch the voltage at the load as a function of time, for $0 < t < 16$ μs.

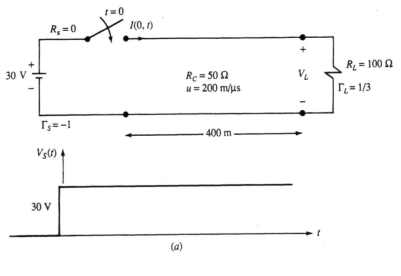

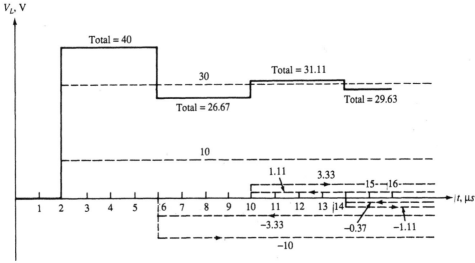

(b)

Fig. 6-38

❚ For this line $T = 2$ μs, $\Gamma_L = \frac{1}{3}$, $\Gamma_S = -1$. The solid line in Fig. 6-38(b) gives the load voltage.

6.61 Sketch the input current $I(0, t)$ for the line of Problem 6.60.

❚ Using the current reflection coefficients $\Gamma_S = 1$ and $\Gamma_L = -1/3$, and an initial current pulse of $(30 \text{ V})/R_C = 0.6$ A, we sketch the forward- and backward-traveling waves, as shown in Fig. 6-39, and combine them to obtain the total input current.

6.62 Obtain transmission-line equations, similar to those of Problem 6.6, assuming that the line operates under sinusoidal steady state.

❚ In this case express the line voltage and current in phasor form; that is,
$$V(z, t) = \text{Re}\,[\hat{V}(z)e^{j\omega t}] \qquad I(z, t) = \text{Re}\,[\hat{I}(z)e^{j\omega t}]$$
Equations (3) and (4) of Problem 6.6 become
$$\frac{d\hat{V}(z)}{dz} = -j\omega l \hat{I}(z) \tag{1}$$
$$\frac{d\hat{I}(z)}{dz} = -j\omega c \hat{V}(z) \tag{2}$$

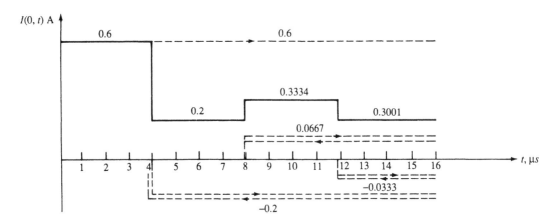

Fig. 6-39

6.63 For a transmission line operating under sinusoidal steady state, obtain a differential equation for the voltage along the line or the current along the line.

▌ Differentiating (1) and (2) of Problem 6.62 with respect to z, we obtain the single equation

$$\frac{d^2\hat{F}}{dz^2} = -\beta^2\hat{F} \qquad [\beta \equiv \omega\sqrt{lc} \equiv \omega/u] \tag{1}$$

for either $\hat{V}(z)$ or $\hat{I}(z)$.

6.64 Give general expressions for $V(z, t)$ and $I(z, t)$ in Problem 6.62.

▌ From (1) of Problem 6.63 and the relations of Problem 6.11,

$$\hat{V}(z) = V_m^+ e^{-j\beta z} + V_m^- e^{-j\beta z} \qquad \text{and} \qquad \hat{I}(z) = \frac{V_m^+}{R_C} e^{-j\beta z} - \frac{V_m^-}{R_C} e^{j\beta z}$$

where V_m^+ and V_m^- are, in general, complex numbers: $V_m^+ = |V_m^+| e^{j\theta^+}$, $V_m^- = |V_m^-| e^{j\theta^-}$. Consequently,

$$V(z, t) = |V_m^+| \cos(\omega t - \beta z + \theta^+) + |V_m^-| \cos(\omega t + \beta z + \theta^-) \tag{1}$$

$$I(z, t) = \frac{|V_m^+|}{R_C} \cos(\omega t - \beta z + \theta^+) - \frac{|V_m^-|}{R_C} \cos(\omega t + \beta z + \theta^-) \tag{2}$$

6.65 Rewrite the phasors of Problem 6.64 in terms of the *complex voltage reflection coefficient*, $\Gamma(z)$.

▌ Defining

$$\Gamma(z) \equiv \frac{V_m^- e^{j\beta z}}{V_m^+ e^{-j\beta z}} \tag{1}$$

we have

$$\hat{V}(z) = V_m^+ e^{-j\beta z}[1 + \Gamma(z)] \tag{2}$$

$$\hat{I}(z) = \frac{V_m^+}{R_C} e^{-j\beta z}[1 - \Gamma(z)] \tag{3}$$

6.66 Define the *input impedance* Z_{in} to a sinusoidally driven transmission line as the function

$$Z_{in}(z) \equiv \frac{\hat{V}(z)}{\hat{I}(z)} \tag{1}$$

Relate $Z_{in}(z)$ and $\Gamma(z)$.

▌ Divide (2) of Problem 6.65 by (3), to find

$$Z_{in}(z) = R_C \frac{1 + \Gamma(z)}{1 - \Gamma(z)} \qquad \text{or} \qquad \Gamma(z) = \frac{Z_{in}(z) - R_C}{Z_{in}(z) + R_C} \tag{2}$$

6.67 Derive an expression for the voltage reflection coefficient at the load Γ_L in terms of the load impedance Z_L and the characteristic resistance, R_C.

▌ For $z = \zeta$, (2) of Problem 6.66 gives

$$\Gamma_L = \frac{Z_L - R_C}{Z_L + R_C} \tag{1}$$

6.68 Obtain an expression for the pointwise voltage reflection coefficient, $\Gamma(z)$, in terms of the voltage reflection coefficient at the load, Γ_L.

▮ From (1) of Problem 6.65,

$$\Gamma(z) = \frac{V_m^-}{V_m^+} e^{j2\beta z} = \left(\frac{V_m^-}{V_m^+} e^{j2\beta \zeta}\right) e^{j2\beta(z-\zeta)} = \Gamma_L e^{j2\beta(z-\zeta)} \tag{1}$$

6.69 A 10-m section of lossless transmission line having $R_C = 50\ \Omega$ and $u = 200\ \text{m}/\mu\text{s}$ is driven by a 26-MHz source with open-circuit voltage $\hat{V}_S = 100\ \underline{/0°}$ V and impedance $Z_S = 50 + j0\ \Omega$. If the line is terminated in a load impedance of $Z_L = 100 + j50\ \Omega$, determine the input impedance to the line.

▮ The phase constant is $\beta = \omega/u = 8.168 \times 10^{-1}\ \text{rad/m}$, whence $2\beta\zeta = 16.34\ \text{rad} = 936°$, which is equivalent to $936° - 720° = 216°$. The load reflection coefficient is

$$\Gamma_L = \frac{Z_L - R_C}{Z_L + R_C} = \frac{50 + j50}{150 + j50} = 0.4472\ \underline{/26.57°}$$

Thus, from (1) of Problem 6.68 and (2) of Problem 6.66,

$$\Gamma(0) = \Gamma_L e^{-j2\beta\zeta} = (0.4472\ \underline{/26.57°})(1\ \underline{/-216°}) = 0.4472\ \underline{/-189.4°}$$

$$Z_{\text{in}}(0) = R_C \frac{1 + \Gamma(0)}{1 - \Gamma(0)} = 19.53\ \underline{/10.39°} = 19.21 + j3.52\ \Omega$$

6.70 Determine the input voltage in the time domain for the line of Problem 6.69.

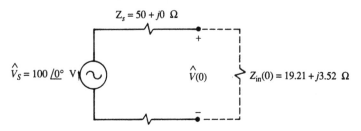

Fig. 6-40

▮ By voltage division from Fig. 6-40,

$$\hat{V}(0) = \frac{Z_{\text{in}}(0)}{Z_{\text{in}}(0) + Z_S} \hat{V}_S = 28.18\ \underline{/7.48°}\ \text{V}$$

and the time-domain voltage becomes $V(0, t) = 28.18 \cos(\omega t + 7.48°)$ (V), in which $\omega = 52\pi \times 10^6\ \text{rad/s}$.

6.71 Find the voltage (in the time domain) at the load of the line of Problems 6.69 and 6.70.

▮ From (2) of Problem 6.65,

$$\hat{V}(\zeta) = \hat{V}(0) e^{-j\beta\zeta} \frac{1 + \Gamma_L}{1 + \Gamma(0)} \tag{1}$$

Substitution of numerical values from Problems 6.69 and 6.70 yields $\hat{V}(\zeta) = 70.71\ \underline{/-99.84°}$ V, or

$$V(\zeta, t) = 70.71 \cos(\omega t - 99.84°)\quad (\text{V})$$

6.72 Show that the input impedance to a transmission line may be expressed as

$$Z_{\text{in}}(z) = R_C \frac{Z_L + jR_C \tan \beta(\zeta - z)}{R_C + jZ_L \tan \beta(\zeta - z)} \tag{1}$$

▮ Substitute (1) of Problem 6.68 and (1) of Problem 6.67 in (2) of Problem 6.66:

$$Z_{\text{in}}(z) = R_C \frac{1 + [(Z_L - R_C)/(Z_L + R_C)] e^{j2\beta(z-\zeta)}}{1 - [(Z_L - R_C)/(Z_L + R_C)] e^{j2\beta(z-\zeta)}}$$

$$= R_C \frac{Z_L + R_C[(1 - e^{j2\beta(z-\zeta)})/(1 + e^{j2\beta(z-\zeta)})]}{R_C + Z_L[(1 - e^{j2\beta(z-\zeta)})/(1 + e^{j2\beta(z-\zeta)})]}$$

$$= R_C \frac{Z_L + R_C j \tan \beta(\zeta - z)}{R_C + Z_L j \tan \beta(\zeta - z)}$$

6.73 What is the input impedance to a transmission line, of length ζ, as seen by the source?

▌ Set $z = 0$ in (1) of Problem 6.72:

$$Z_{in}(0) = R_C \frac{Z_L + jR_C \tan \beta \zeta}{R_C + jZ_L \tan \beta \zeta} \qquad (1)$$

6.74 The *electrical length* of a transmission line having physical length ζ is defined as
$$\zeta_e \equiv \zeta/\lambda = \beta\zeta/2\pi \qquad (1)$$
where λ is the wavelength in the medium surrounding the line conductors. Plot the input impedance of a short-circuited line versus its electrical length.

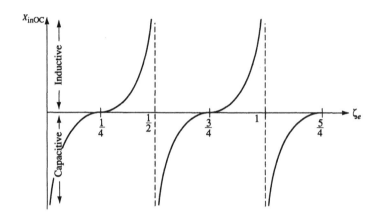

Fig. 6-41

▌ Setting $Z_L = 0$ in (1) of Problem 6.73 yields
$$Z_{inSC} = jR_C \tan 2\pi\zeta_e \equiv jX_{inSC} \qquad (1)$$
The reactance X_{inSC} is graphed in Fig. 6-41.

6.75 Repeat Problem 6.74 for an open-circuited line.

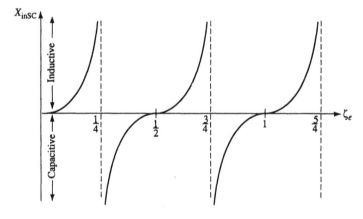

Fig. 6-42

▌ Corresponding to $Z_L = \infty$,
$$Z_{inOC} = -jR_C \cot 2\pi\zeta_e \equiv jX_{inOC} \qquad (1)$$
See Fig. 6-42.

6.76 Obtain an expression for the characteristic resistance R_C of a transmission line in terms of its short-circuit and open-circuit input impedances.

▌ Multiply (1) of Problem 6.74 by (1) of Problem 6.75, to find
$$R_C = \sqrt{Z_{inSC} Z_{inOC}} \qquad (1)$$

6.77 Investigate the short-circuit and open-circuit input impedances of a transmission line of length $\lambda/4$.

▌ For $\zeta_e = 1/4$, Figs. 6-41 and 6-42 give
$$Z_{inSC} = j\infty \quad \text{(line looks like an open circuit)}$$
$$Z_{inOC} = j0 \quad \text{(line looks like a short circuit)}$$

6.78 Verify that the results of Problem 6.77 also follow from (2) of Problem 6.66.

▌ When $2\beta\zeta = \pi$, (1) of Problem 6.68 gives $\Gamma(0) = \Gamma_L e^{-j\pi} = -\Gamma_L$. For a short-circuit load, $\Gamma_L = -1$ and so $\Gamma(0) = 1$. Thus

$$\frac{1 + \Gamma(0)}{1 - \Gamma(0)} = \infty \quad \text{and} \quad Z_{\text{inSC}} = \infty$$

Similarly, for an open-circuit load, $\Gamma_L = 1$ and $\Gamma(0) = -1$. Thus

$$\frac{1 + \Gamma(0)}{1 - \Gamma(0)} = 0 \quad \text{and} \quad Z_{\text{inOC}} = 0$$

6.79 A quarter-wavelength transmission line has a purely inductive termination. Show that the input impedance at the source appears as a capacitive reactance.

▌ From (1) of Problem 6.73, with $Z_L = j\omega L$ and $\beta\zeta = \pi/2$,

$$Z_{\text{in}}(0) = \frac{R_C^2}{Z_L} = -\frac{jR_C^2}{\omega L}$$

so that the input impedance appears as a capacitive reactance ($C = L/R_C^2$).

6.80 Verify that the input impedance is periodic of period $\lambda/2$.

▌ This follows at once from (1) of Problem 6.72, since the tangent function has period π.

6.81 For a certain lossless transmission line under sinusoidal steady state, we have:

$$\zeta = 1\,\text{m} \quad f = 262.5\,\text{MHz} \quad R_C = 50\,\Omega \quad Z_L = (30 - j200)\,\Omega \quad Z_S = (100 + j50)\,\Omega \quad u = 300\,\text{m}/\mu\text{s}$$

Determine the electrical length and calculate the voltage reflection coefficients at the load and at the input to the line.

▌

$$\zeta_e = \frac{\zeta}{\lambda} = \frac{\zeta f}{u} = \frac{(1)(262.5 \times 10^6)}{3 \times 10^8} = 0.875$$

At the load,

$$\Gamma_L = \frac{Z_L - R_C}{Z_L + R_C} = \frac{(30 - j200) - 50}{(30 - j200) + 50} = 0.933\,\underline{/-27.5°}$$

and (1) of Problem 6.68 gives

$$\Gamma(0) = \Gamma_L e^{-j4\pi\zeta_e} = (0.933\,\underline{/-27.5°})(1\,\underline{/+90°}) = 0.933\,\underline{/62.5°}$$

6.82 Repeat Problem 6.81 for the following data:

$$\zeta = 36\,\text{m} \quad f = 28\,\text{MHz} \quad R_C = 150\,\Omega \quad Z_S = (500 + j0)\,\Omega \quad Z_L = -j30\,\Omega \quad u = 300\,\text{m}/\mu\text{s}$$

▌

$$\zeta_e = \frac{(36)(28 \times 10^6)}{3 \times 10^8} = 3.36$$

$$\Gamma_L = \frac{-j30 - 150}{-j30 + 150} = 1\,\underline{/-157.38°}$$

$$\Gamma(0) = \Gamma_L e^{-j4\pi\zeta_e} = (1\,\underline{/-157.38°})(1\,\underline{/100.74°}) = 1\,\underline{/-56.64°}$$

6.83 Repeat Problem 6.81 for the following data:

$$\zeta = 2\,\text{m} \quad f = 175\,\text{MHz} \quad u = 200\,\text{m}/\mu\text{s} \quad R_C = 100\,\Omega \quad Z_S = 50\,\Omega; \quad Z_L = (200 - j30)\,\Omega$$

▌

$$\zeta_e = \frac{2(175 \times 10^6)}{2 \times 10^8} = 1.75$$

$$\Gamma_L = \frac{(200 - j30) - 100}{(200 - j30) + 100} = 0.346\,\underline{/-11.0°}$$

$$\Gamma(0) = \Gamma_L e^{-j4\pi\zeta_e} = (0.346\,\underline{/-11.0°})(1\,\underline{/180.0°}) = 0.346\,\underline{/169.0}$$

6.84 For a lossless transmission line whose length is $\lambda/4$, show that, in sinusoidal steady state,

$$\frac{\hat{V}(\lambda/4)}{Z_L} = \frac{\hat{V}(0)}{jR_C} \tag{1}$$

▌ For $\zeta = \lambda/4$, (1) of Problem 6.71 gives

$$\frac{\hat{V}(\lambda/4)}{\hat{V}(0)} = -j\frac{1 + \Gamma_L}{1 + \Gamma(0)} \tag{2}$$

But, from Problem 6.78, $\Gamma(0) = -\Gamma_L$; so that

$$Z_L = R_C \frac{1 + \Gamma_L}{1 - \Gamma_L} = R_C \frac{1 + \Gamma_L}{1 + \Gamma(0)} \tag{3}$$

Together, (2) and (3) imply (1).

6.85   Show that the magnitude of the reflection coefficient for a lossless line having a purely reactive load is exactly unity.

▮   The voltage reflection coefficient at the load is given by

$$\Gamma_L = \frac{jX - R_C}{jX + R_C} \qquad \text{whence} \qquad |\Gamma_L| = \frac{\sqrt{R_C^2 + X^2}}{\sqrt{R_C^2 + X^2}} = 1$$

6.86   A transmission line of length $\zeta = 3\lambda/4$ is formed into a pair of closed loops by connecting the appropriate input and output terminals [Fig. 6-43(a)]. Determine the impedance between two points, one on one loop and the other the same distance along the other loop.

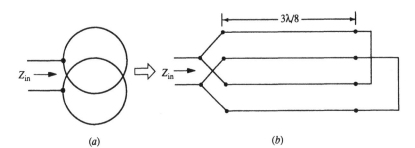

Fig. 6-43

(a)   (b)

▮   The transmission-line representation of the loops is shown in Fig. 6-43(b), where Z_{in} is the input impedance of an open-circuited line of length $3\lambda/8$. By (1) of Problem 6.75,

$$Z_{\text{inOC}} = -jR_C \cot 2\pi(\tfrac{3}{8}) = jR_C$$

6.87   A section of lossless coaxial cable having $R_C = 50\ \Omega$ and $u = 200$ m/μs is terminated in a short circuit and operated at a frequency of 10 MHz. Determine the shortest length of the lines such that, at the input terminals, the line appears to be a 100-pF capacitor.

▮   From Problem 6.74,

$$Z_{\text{inSC}} = jR_C \tan 2\pi\zeta_e = +\frac{1}{j\omega C} \qquad \text{or} \qquad \tan 2\pi\zeta_e = \frac{-1}{\omega C R_C} = -3.183$$

from which

$$(2\pi\zeta_e)_{\min} = \frac{\zeta_{\min} 2\pi f}{u} = 1.876\ \text{rad} \qquad \text{or} \qquad \zeta_{\min} = \frac{(1.876)(2 \times 10^8)}{2\pi(10 \times 10^6)} = 5.97\ \text{m}$$

6.88   Determine the shortest length of the line of Problem 6.87 such that the the line appears to be a 1-μH inductor.

▮   Paralleling Problem 6.87,

$$Z_{\text{inSC}} = jR_C \tan \frac{2\pi f \zeta_{\min}}{u} = j\omega L \qquad \text{or} \qquad \tan \frac{2\pi f \zeta_{\min}}{u} = \frac{\omega L}{R_C} = 1.257 \qquad \text{or} \qquad \frac{2\pi f \zeta_{\min}}{u} = 0.899$$

whence $\zeta_{\min} = 2.86$ m.

6.89   Repeat Problem 6.87 for an open-circuited transmission line.

▮   From Problem 6.75 we have

$$Z_{\text{inOC}} = -jR_C \cot \frac{2\pi f \zeta_{\min}}{u} = \frac{1}{j\omega C} = -j159.15$$

or

$$\tan \frac{2\pi f \zeta_{\min}}{u} = \frac{R_C}{159.15} = 0.314 \qquad \text{or} \qquad \frac{2\pi f \zeta_{\min}}{u} = 0.304$$

or $\zeta_{\min} = 0.969$ m.

6.90   Repeat Problem 6.88 for an open-circuited line.

▋ Now $Z_{\text{inOC}} = -jR_C \cot (2\pi f\zeta_{\min}/u) = j\omega L$, which leads to

$$\frac{2\pi f\zeta_{\min}}{u} = 2.47 \qquad \text{or} \qquad \zeta_{\min} = 7.86 \text{ m}$$

6.91 For the transmission line shown in Fig. 6-44, express the voltage and current phasors a distance d back from the load in terms of the voltage phasor and the reflection coefficient at the load.

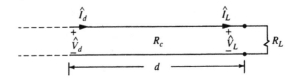

Fig. 6-44

▋ From (2) of Problem 6.65 and (1) of Problem 6.68,

$$\frac{\hat{V}_d}{\hat{V}_L} = \frac{\hat{V}(\zeta - d)}{\hat{V}(\zeta)} = e^{j\beta d} \frac{1 + \Gamma(\zeta - d)}{1 + \Gamma_L} = e^{j\beta d} \frac{1 + \Gamma_L e^{-j2\beta d}}{1 + \Gamma_L} \tag{1}$$

Similarly,

$$\frac{\hat{I}_d}{\hat{I}_L} = e^{j\beta d} \frac{1 - \Gamma_L e^{-j2\beta d}}{1 - \Gamma_L} \tag{2}$$

6.92 Obtain a "crank diagram," similar to that shown in Fig. 5-7, for the transmission line of Problem 6.91.

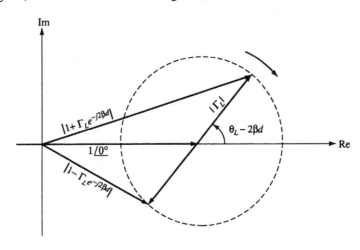

Fig. 6-45

▋ From Problem 6.91,

$$|\hat{V}_d| = \frac{|1 + \Gamma_L e^{-j2\beta d}|}{|1 + \Gamma_L|} |\hat{V}_L| \tag{1}$$

$$|\hat{I}_d| = \frac{|1 - \Gamma_L e^{-j2\beta d}|}{|1 - \Gamma_L|} |\hat{I}_L| \tag{2}$$

The variations with distance d are contained in the terms $|1 \pm \Gamma_L e^{-j2\beta d}|$; these can be viewed as the magnitudes of the sum or difference of the phasors $1\,\underline{/0°}$ and $|\Gamma_L|\,\underline{/\theta_L - 2\beta d}$, as displayed in the crank diagram, Fig. 6-45. As d increases, the latter phasor rotates (in the clockwise direction) about the tip of the former.

6.93 Using the results of Problem 6.92, sketch $|\hat{V}_d|$ and $|\hat{I}_d|$ for the transmission line if terminated with a short circuit.

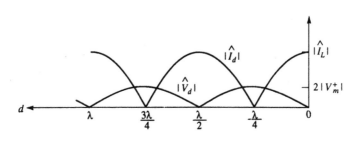

Fig. 6-46

▮ In this case we have $R_L = 0$ and $\Gamma_L = -1$, and (2) of Problem 6.92 becomes

$$|\hat{I}_d| = \frac{|1 + e^{-j2\beta d}|}{2}|\hat{I}_L| = |e^{-j\beta d}|\frac{|e^{j\beta d} + e^{-j\beta d}|}{2}|\hat{I}_L| = \left|\cos\frac{2\pi d}{\lambda}\right||\hat{I}_L|$$

since $\beta d = 2\pi d/\lambda$. This is plotted in Fig. 6-46.

Since $V_L = 0 = 1 + \Gamma_L$, (1) of Problem 6.92 becomes indeterminate. Returning to (2) of Problem 6.65 and (1) of Problem 6.68,

$$\hat{V}_d = V_m^+ e^{-j\beta(\zeta - d)}[1 + \Gamma_L e^{-j2\beta d}]$$

Thus $|\hat{V}_d| = |V_m^+| |1 + \Gamma_L e^{-j2\beta d}|$, and substituting $\Gamma_L = -1$ gives

$$|\hat{V}_d| = |V_m^+| |1 - e^{-j2\beta d}| = |V_m^+| |e^{-j\beta d}| |e^{j\beta d} - e^{-j\beta d}| = 2|V_m^+|\left|\sin\frac{2\pi d}{\lambda}\right|$$

This relation is also plotted in Fig. 6-46.

6.94 Figure 6-46 shows that the current and voltage magnitudes for a short-circuited line are periodic of period $\lambda/2$, with adjacent maxima and minima separated by $\lambda/4$. Infer from the crank diagram (Fig. 6-45) that these results are valid for arbitrary load impedance.

▮ Values will repeat after one revolution in the crank diagram:
$$-2\beta\,\Delta d = -2\pi \qquad \text{or} \qquad \Delta d = \pi/\beta = \lambda/2$$
Also it is obvious that maxima and minima are separated by a half-revolution, or $\lambda/4$.

6.95 Repeat Problem 6.93 for an open-circuited transmission line.

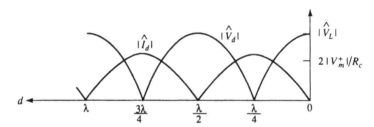

Fig. 6-47

▮ In this case, $R_L = \infty$, $\Gamma_L = +1$, and $I_L = 0$. Then (1) of Problem 6.92 becomes
$$|\hat{V}_d| = \left|\cos\frac{4\pi d}{\lambda}\right||\hat{V}_L|$$

Similarly, (3) of Problem 6.65 and (1) of Problem 6.68 yield
$$|\hat{I}_d| = 2\frac{|V_m^+|}{R_C}\left|\sin\frac{2\pi d}{\lambda}\right|$$

These results are plotted in Fig. 6-47.

6.96 Verify from Figs. 6-46 and 6-47 that a quarter-wavelength line that is terminated in an open circuit appears as a short circuit, and vice versa.

▮ The figures show that at a quarter-wavelength away from an open circuit, the voltage goes to zero, whereas the current is nonzero: a short circuit. Similarly, for a quarter-wavelength line terminated in a short circuit, the current goes to zero, whereas the voltage is nonzero: an open circuit.

6.97 Repeat Problem 6.93, assuming that the transmission line is terminated with a purely resistive load, $R_L > R_C$.

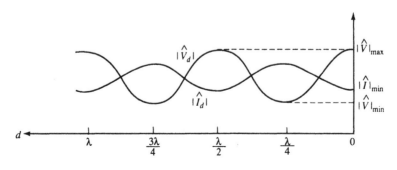

Fig. 6-48

▮ In this case, Γ_L is real and positive so $\theta_L = 0$. It is clear from the crank diagram (Fig. 6-45) that the voltage maximum and the current minimum will occur exactly at the load ($d = 0$); this is illustrated in Fig. 6-48. Problem 6.94, of course, applies to this special case.

6.98 Repeat Problem 6.93 for $R_L = R_C$.

▮ For a matched line, $\Gamma_L = 0$, which implies $\Gamma(z) \equiv 0$, which in turn implies $|\hat{V}(z)| = |V_m^+| = $ const. and $|\hat{I}(z)| = |V_m^+|/R_C = $ const. along the line.

6.99 The degree of mismatch of a transmission line is measured by the *voltage standing-wave ratio* (VSWR), defined as the ratio of the magnitude of the maximum voltage on the line to the magnitude of the minimum voltage on the line:

$$\text{VSWR} \equiv \frac{|\hat{V}|_{max}}{|\hat{V}|_{min}} \qquad (1)$$

[Compare Problem 5.115.] Determine the range of values of VSWR for various terminations on a transmission line.

▮ For a matched line ($\Gamma_L = 0$), VSWR $= |V_m^+|/|V_m^+| = 1$. For a totally mismatched line ($\Gamma_L = \pm 1$), VSWR $= |\hat{V}|_{max}/0 = \infty$. Thus, the VSWR will always lie between unity and infinity.

6.100 Express the VSWR in terms of $|\Gamma(z)| = |\Gamma_L|$ (see Problem 6.68).

▮ By (1) of Problem 6.92 or Fig. 6-45,

$$\text{VSWR} = \frac{|\hat{V}|_{max}}{|\hat{V}|_{min}} = \frac{1 + |\Gamma_L|}{1 - |\Gamma_L|} = \frac{1 + |\Gamma(z)|}{1 - |\Gamma(z)|} \qquad (1)$$

which is in agreement with Problem 5.114.

6.101 Express the magnitude of the reflection coefficient of a transmission line in terms of the VSWR.

▮ Inverting (1) of Problem 6.100,

$$|\Gamma(z)| = \frac{\text{VSWR} - 1}{\text{VSWR} + 1} \qquad (1)$$

6.102 Derive an expression for the time-average power flow along a lossless transmission line operating in the sinusoidal steady state.

▮ The average power flow in the $+z$-direction is given by

$$P_{av} = \tfrac{1}{2} \text{Re}\, [\hat{V}(z)\hat{I}^*(z)] \qquad (1)$$

Substitution for the phasors from (2) and (3) of Problem 6.65 yields (since $\Gamma - \Gamma^*$ is pure imaginary and $\Gamma\Gamma^* = |\Gamma|^2$ is pure real)

$$P_{av} = \frac{1}{2} \text{Re}\left[\frac{V_m^+ V_m^{+*}}{R_C}(1 + \Gamma)(1 - \Gamma^*)\right] = \frac{|V_m^+|^2}{2R_C} \text{Re}\,(1 + \Gamma - \Gamma^* - \Gamma\Gamma^*) = \frac{|V_m^+|^2}{2R_C}(1 - |\Gamma|^2) = \frac{|V_m^+|^2}{2R_C}(1 - |\Gamma_L|^2) \quad (2)$$

where the last step follows from Problem 6.68. As expected for a lossless line, the power flow is independent of z.

6.103 Find the average powers carried by the forward wave and by the backward wave on the line of Problem 6.102. Show that the total average power, obtained as the sum of the two powers, is the same as given by (2) of Problem 6.102.

▮ The two waves have voltage and current phasors

$$\hat{V}^+(z) = V_m^+ e^{-j\beta z} \qquad \hat{V}^-(z) = V_m^- e^{j\beta z}$$
$$\hat{I}^+(z) = \frac{V_m^+}{R_C} e^{-j\beta z} \qquad \hat{I}^-(z) = -\frac{V_m^-}{R_C} e^{j\beta z}$$

Correspondingly,

$$P_{av}^+ = \frac{|V_m^+|^2}{2R_C}$$
$$P_{av}^- = -\frac{|V_m^-|^2}{2R_C} = -\frac{|V_m^+|^2}{2R_C}|\Gamma(z)|^2 = -\frac{|V_m^+|^2}{2R_C}|\Gamma_L|^2$$

and these indeed have the required sum.

6.104 Show that when a transmission line is terminated in an open circuit or in a short circuit, no net average power flows along the line.

■ Substitute $\Gamma_L = \pm 1$ in (2) of Problem 6.102.

6.105 Find the power transmitted by a matched transmission line.

■ If the line is matched at the load, i.e., $Z_L = R_C$, then $\Gamma_L = 0$, and (2) of Problem 6.102 gives $P_{av} = |V_m^+|^2/2R_C = P_{av}^+$.

6.106 (*a*) Represent the lossless transmission line shown in Fig. 6-49(*a*) by its equivalent circuit. (*b*) Express the average power delivered to the line in terms of the input voltage to the line and its input impedance.

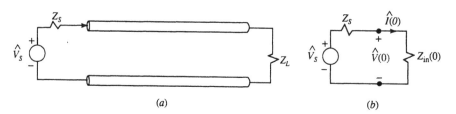

(a)   (b)   **Fig. 6-49**

■ (*a*) See Fig. 6-49(*b*). (*b*) By (1) of Problem 6.102,

$$P_{av} = \frac{1}{2} \text{Re}\,[\hat{V}(0)\hat{I}^*(0)] = \frac{1}{2} \text{Re}\left[\hat{V}(0)\frac{\hat{V}^*(0)}{Z_{in}^*(0)}\right] = \frac{|\hat{V}(0)|^2}{2\,|Z_{in}(0)|}\cos\theta_{in0} \tag{1}$$

where θ_{in0} is the polar angle of $Z_{in}(0)$ and where

$$\hat{V}(0) = \frac{Z_{in}(0)}{Z_{in}(0)+Z_s}\hat{V}_s \tag{2}$$

6.107 Determine the VSWR and the input impedance to the line of Problem 6.81.

■ From Problem 6.81, $\Gamma(0) = 0.933\,\underline{/62.5^\circ}$. By Problem 6.100,

$$\text{VSWR} = \frac{1+0.933}{1-0.933} = 28.85$$

By (2) of Problem 6.66,

$$Z_{in}(0) = R_C\frac{1+\Gamma(0)}{1-\Gamma(0)} = 50\frac{1+(0.431-j0.828)}{1-(0.431-j0.828)} = 82.3\,\underline{/85.5^\circ} = (6.42+j82.1)\;\Omega$$

6.108 Repeat Problem 6.107 for the line of Problem 6.82.

■ In this case we have $R_C = 150\;\Omega$ and $\Gamma(0) = 1\,\underline{/-56.64^\circ}$. Since $|\Gamma(z)| = 1$, VSWR $= \infty$; also,

$$Z_{in} = R_C\frac{1+\Gamma(0)}{1-\Gamma(0)} = 150\frac{1+(0.55-j0.84)}{1-(0.55-j0.84)} = 150\frac{1.761\,\underline{/-28.323^\circ}}{0.949\,\underline{/61.678^\circ}} = 278.35\,\underline{/-90^\circ}\;\Omega$$

6.109 Repeat Problem 6.107 for the line of Problem 6.83.

■ With $\Gamma(0) = 0.346\,\underline{/-191^\circ}$ and $R_C = 100\;\Omega$:

$$\text{VSWR} = \frac{1+0.346}{1-0.346} = 2.058$$

$$Z_{in} = 100\frac{1+(-0.34+j0.066)}{1-(-0.34+j0.066)} = 100\frac{0.664\,\underline{/5.704^\circ}}{1.341\,\underline{/-2.819^\circ}} = 49.477\,\underline{/8.522^\circ}\;\Omega$$

6.110 For the line of Problem 6.81, determine the time-domain voltage at the line input and at the load.

■ From the results of Problem 6.107 the input voltage is given by

$$\hat{V}(0) = \frac{Z_{in}(0)}{Z_s+Z_{in}(0)}\hat{V}_s = \frac{82.3\,\underline{/85.5^\circ}}{169.6\,\underline{/51.1^\circ}}(10\,\underline{/30^\circ}) = 4.85\,\underline{/64.4^\circ}\;\;\text{V}$$

so that $V(0,t) = 4.85\cos(\omega t + 64.4^\circ)$ (V). Now, applying (1) of Problem 6.71, with $\beta\zeta = 135^\circ$, $\Gamma_L = 0.933\,\underline{/-27.5^\circ}$, and $\Gamma(0) = 0.933\,\underline{/62.5^\circ}$

$$\hat{V}(\zeta) = 5.513\,\underline{/-113.9^\circ} \qquad \text{or} \qquad V(\zeta,t) = 5.513\cos(\omega t - 113.9^\circ)\;\;\text{(V)}$$

6.111 Repeat Problem 6.110 for the line of Problem 6.82.

■ From Problem 6.82 we have

$$\hat{V}(0) = \frac{Z_{in}(0)}{Z_s+Z_{in}(0)}\hat{V}_s = \frac{278.35\,\underline{/-90^\circ}}{572.251\,\underline{/-29.105^\circ}}(100\,\underline{/0^\circ}) = 48.64\,\underline{/-61.0^\circ}\;\;\text{V}$$

Thus, $V(0, t) = 48.64 \cos(\omega t - 61.0°)$ (V). With $\beta\zeta = 129.63°$, $\Gamma_L = 1 \underline{/-157.38°}$, and $\Gamma(0) = 1 \underline{/-56.64°}$, we have from (1) of Problem 6.71:

$$\hat{V}(\zeta) = 10.82 \underline{/241°} \quad \text{or} \quad V(\zeta, t) = 10.82 \cos(\omega t + 241°) \quad \text{(V)}$$

6.112 Repeat Problem 6.110 for the line of Problem 6.83.

❚
$$\hat{V}(0) = \frac{Z_{in}(0)}{Z_{in}(0) + Z_s} \hat{V}_s = \frac{49.477 \underline{/8.322°}}{99.202 \underline{/4.138°}} (10 \underline{/0°}) = 4.987 \underline{/4.183°} \quad \text{V}$$

or, $V(0, t) = 4.987 \cos(\omega t + 4.183°)$ (V). With $\beta\zeta = 135°$, $\Gamma_L = 0.346 \underline{/-11.0°}$, and $\Gamma(0) = 0.346 \underline{/169.0°}$,
$$\hat{V}(\zeta) = 10.07 \underline{/85.66°} \quad \text{or} \quad V(\zeta, t) = 10.07 \cos(\omega t + 85.66°) \quad \text{(V)}$$

6.113 Transform (1) of Problem 6.106 from the input to the load.

❚ For a lossless line, P_{av} is independent of distance along the line; therefore,

$$P_{av} = \frac{|\hat{V}(\zeta)|^2}{2 |Z_{in}(\zeta)|} \cos \theta_{in\zeta} \tag{1}$$

6.114 For a lossless transmission line having a purely resistive load R_L, show that

$$\text{VSWR} = \frac{\max(R_L, R_C)}{\min(R_L, R_C)} \tag{1}$$

❚
$$\text{VSWR} \equiv \frac{1 + |\Gamma_L|}{1 - |\Gamma_L|} = \frac{1 + |(R_L - R_C)/(R_L + R_C)|}{1 - |(R_L - R_C)/(R_L + R_C)|}$$

Hence, for $R_L \geq R_C$,

$$\text{VSWR} = \frac{1 + (R_L - R_C)/(R_L + R_C)}{1 - (R_L - R_C)/(R_L + R_C)} = \frac{R_L}{R_C} = \frac{\max(R_L, R_C)}{\min(R_L, R_C)}$$

Similarly, for $R_L \leq R_C$,

$$\text{VSWR} = \frac{R_C}{R_L} = \frac{\max(R_L, R_C)}{\min(R_L, R_C)}$$

6.115 An antenna having an input impedance at 100 MHz of $(72 + j40)$ Ω is connected to a 100-MHz generator via a section of 300-Ω air-filled line of length 1.75 m. Given that the generator has a source voltage of 10 V and a source impedance of 50 Ω, determine the voltage reflection coefficients at the load and the line input terminals.

❚ The reflection coefficient at the load is

$$\Gamma_L = \frac{(72 + j40) - 300}{(72 + j40) + 300} = 0.62 \underline{/163.91°}$$

Now
$$\zeta_e = \frac{\zeta f}{u} = \frac{(1.75)(100 \times 10^6)}{3 \times 10^8} = 0.5833$$

and so
$$\Gamma(0) = \Gamma_L e^{-j4\pi\zeta_e} = 0.62 \underline{/163.91° - 420°} = 0.62 \underline{/-256.09°}$$

6.116 Calculate the average power delivered to the antenna of Problem 6.115.

❚ Apply (1) and (2) of Problem 6.106:

$$Z_{in}(0) = R_C \frac{1 + \Gamma(0)}{1 - \Gamma(0)} = (300) \frac{1.04 \underline{/35.27°}}{1.30 \underline{/-27.64°}} = 241.06 \underline{/62.91°} \quad \Omega$$

$$\hat{V}(0) = \frac{Z_{in}(0)}{Z_s + Z_{in}(0)} \hat{V}_s = \frac{241.06 \underline{/62.91°}}{267.55 \underline{/53.34°}} (10 \underline{/0°}) = 9.01 \underline{/9.57°} \quad \text{V}$$

$$P_{av} = \frac{(9.01)^2}{2(241.06)} \cos 62.91° = 76.68 \text{ mW}$$

6.117 For the transmission line of Fig. 6-50, determine the voltage reflection coefficient at the input to the line.

❚
$$R_C = \sqrt{\frac{l}{c}} = \sqrt{\frac{0.5 \times 10^{-6}}{200 \times 10^{-12}}} = 50 \, \Omega$$

$$\Gamma_L = \frac{Z_L - R_C}{Z_L + R_C} = \frac{(100 + j50) - 50}{(100 + j50) + 50} = 0.447 \underline{/26.6°}$$

$$\zeta_e = \frac{\zeta f}{u} = \zeta f \sqrt{lc}$$

$$= 1(30 \times 10^6)\sqrt{(0.5 \times 10^{-6})(200 \times 10^{-12})} = 0.3$$

$$\Gamma(0) = \Gamma_L e^{-j4\pi\zeta_e} = 0.447 \underline{/-189.4°}$$

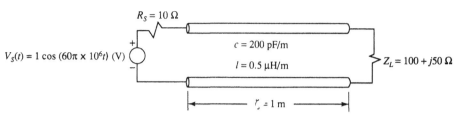

Fig. 6-50

6.118   For the line of Problem 6.117, calculate the input impedance.

$$Z_{\text{in}}(0) = R_C \frac{1 + \Gamma(0)}{1 - \Gamma(0)} = 19.54 \underline{/10.34°} \quad \Omega$$

6.119   Determine the input voltage and input current, in the time domain, to the line of Problem 6.117.

   The data of Fig. 6-50 give

$$\hat{V}(0) = \frac{Z_{\text{in}}(0)}{Z_{\text{in}}(0) + R_S} \hat{V}_s = 0.664 \underline{/3.5°} \quad \text{V}$$

and

$$\hat{I}(0) = \frac{\hat{V}(0)}{Z_{\text{in}}(0)} = 34 \times 10^{-3} \underline{/-6.8°} \quad \text{A}$$

Thus, in the time domain,
$$V(0, t) = 0.664 \cos (60\pi \times 10^6 t + 3.5°) \quad \text{(V)} \qquad I(0, t) = 0.034 \cos (60\pi \times 10^6 t - 6.8°) \quad \text{(A)}$$

6.120   Find the average input power to the lossless line of Problem 6.117.

   By (*1*) of Problem 6.106,
$$P_{\text{av}} = \frac{|\hat{V}(0)|^2}{2 |Z_{\text{in}}(0)|} \cos \theta_{\text{in}0} = \frac{(0.664)^2}{2(19.54)} \cos 10.34° = 11.1 \text{ mW}$$

6.121   Obtain the time-domain expression for the load voltage for the line of Problem 6.117.

   By (*1*) of Problem 6.71,
$$\hat{V}(\zeta) = \hat{V}(0) e^{-j2\pi \zeta_e} \frac{1 + \Gamma_L}{1 + \Gamma(0)}$$
$$= (0.664 \underline{/3.5°})(1 \underline{/-108°}) \frac{1 + 0.447 \underline{/26.6°}}{1 + 0.447 \underline{/-189.4°}}$$
$$= 1.668 \underline{/-103.8°} \quad \text{V}$$
Hence $V(\zeta, t) = 1.668 \cos (60\pi \times 10^6 t - 103.8°) \quad \text{(V)}$.

6.122   Using the load voltage and load current, determine the average power delivered to the load in Problem 6.117. Verify that the result is consistent with that of Problem 6.120.

   By (*1*) of Problem 6.113,
$$P_{\text{av}} = \frac{|\hat{V}(\zeta)|^2}{2 |Z_{\text{in}}(\zeta)|} \cos \theta_{\text{in}\zeta} = \frac{(1.668)}{2(111.8)} \cos 26.57° = 11.1 \text{ mW}$$

6.123   A lossless transmission line has a load impedance $Z_L = R_C + jX$. Can X be completely determined in terms of the measured VSWR?

   The voltage reflection coefficient at the load is
$$\Gamma_L = \frac{jX}{2R_C + jX}$$

We then have
$$|\Gamma_L| = \frac{\text{VSWR} - 1}{\text{VSWR} + 1} = \frac{X}{\sqrt{4R_C^2 + X^2}} \tag{1}$$
Squaring both sides of (*1*), we can solve uniquely for X^2. Thus, $|X|$ is uniquely determined, but the sign of X remains unknown.

6.124   A generator with a 50-Ω source resistance is being operated at 100 MHz. A load of $(50 - j100) \ \Omega$ (Fig. 6-51) is placed across the terminals of this generator. Assuming that a section of 150-Ω, lossless, air-filled line is placed in parallel with this load, determine the shortest length of line for which maximum time-average power is transferred to the load when the line has an open-circuit load.

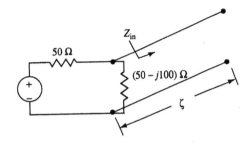

Fig. 6-51

▌ Maximum power is drawn when the total impedance or admittance is purely resistive. Since

$$Y_L = \frac{1}{50 - j100} = 0.004 + j0.008 \text{ S}$$

and $Y_{\text{tot}} = Y_L + Y_{\text{inOC}}$, we require (see Problem 6.75)

$$Y_{\text{inOC}} = j\frac{\tan(2\pi f\zeta/u)}{R_C} = -j0.008 \qquad \text{or} \qquad \tan\frac{2\pi\zeta}{3} = -1.2 \qquad\qquad (1)$$

The smallest positive solution of (1) is $\zeta_{\min} = 1.082$ m.

6.125 Repeat Problem 6.124 for a short-circuit load.

▌ Now the condition reads

$$Y_{\text{inSC}} = -j\frac{\cot(2\pi f\zeta/u)}{R_C} = -j0.008 \qquad \text{or} \qquad \cot\frac{2\pi\zeta}{3} = 1.2$$

whence $\zeta_{\min} = 0.332$ m.

6.126 A 100-Ω lossless transmission line of unknown length is terminated in an unknown impedance. The input impedance is measured as $(200 - j50)$ Ω. Assuming that the input impedance becomes $-j150$ Ω on no-load, determine the unknown impedance.

▌ On open-circuit, from Problem 6.75,

$$-j150 = -j100 \cot \beta\zeta \qquad \text{or} \qquad \tan \beta\zeta = 2/3$$

With $\tan \beta\zeta$, R_C, and $Z_{\text{in}}(0)$ known, (1) of Problem 6.73 may be solved to give $Z_L = (130 + j85)$ Ω.

6.127 Two antennas, each with impedance $100 - j30$ Ω, are connected to a transmitter with three equal lengths of identical lossless transmission line, as shown in Fig. 6-52. Determine the power delivered to either antenna.

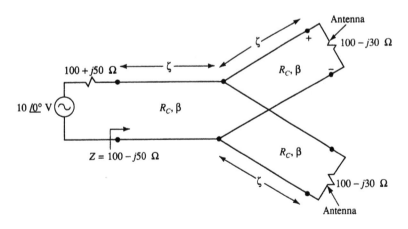

Fig. 6-52

▌ The input current to the line is

$$I_{\text{in}} = \frac{10 \,\underline{/0^\circ}}{(100 + j50) + (100 - j50)} = \frac{1}{20}\underline{/0^\circ} \text{ A}$$

Thus, average power input to the line becomes

$$(P_{\text{av}})_{\text{in}} = \tfrac{1}{2} |I_{\text{in}}|^2 \times 100 = 125 \text{ mW}$$

with 62.5 mW going to either antenna.

6.128 Calculate the magnitude of the voltage across either antenna of Problem 6.127.

▮ With $|Z_{\text{ant}}| = 104.4\ \Omega$ and $\cos\theta_{Z_{\text{ant}}} = 0.958$, the average power delivered to an antenna may also be written as

$$62.5\ \text{mW} = \frac{|\hat{V}_{\text{ant}}|^2}{2\,|Z_{\text{ant}}|}\cos\theta_{Z_{\text{ant}}} = \frac{|\hat{V}_{\text{ant}}|^2}{2(104.4)}(0.958)$$

Solving, we obtain $|\hat{V}_{\text{ant}}| = 3.69\ \text{V}$.

6.129 Transmission lines whose lengths are much less than a wavelength (i.e., $\zeta \ll \lambda$ or $\zeta_e \ll 1$) are said to be "electrically short." For such a line, the distributed-parameter effects are negligible and the line may be modeled as the lumped circuit shown in Fig. 6-53. To investigate the adequacy of the model, determine the normalized input impedance, $Z_{\text{in}}(0)/R_C$, to a line of length $\lambda/10$ having load resistance $R_L = 10R_C$, by using a standard lumped-circuit computer program, and compare with the exact result.

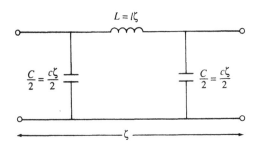

Fig. 6-53

▮ The exact normalized impedance is given by (*1*) of Problem 6.73 as $1.3671\ \underline{/-78°}$.
By computer modeling using SPICE II,

$$R_C = \sqrt{l/c} \qquad u = 1/\sqrt{lc} \qquad R_C = ul = 1/uc$$

or

$$l = \frac{R_C}{u} \qquad \lambda = \frac{u}{f} \qquad \omega l\zeta = 2\pi f\frac{R_C}{u}\zeta$$

and

$$\omega c\zeta = 2\pi f\frac{1}{R_C u}\zeta = \frac{2\pi\zeta}{\lambda}\frac{1}{R_C}$$

so use $f = \zeta/\lambda$, $L = R_C$, $C = 1/R_C$.

6.130 Repeat the exact solution of Problem 6.129 for $R_L = R_C$ (and $\zeta = \lambda/10$).

▮ 1 (independent of ζ/λ).

6.131 Repeat the exact solution of Problem 6.129 for $R_C = 10R_L$ and $\zeta = \lambda/10$.

▮ $0.7314\ \underline{/78°}$.

6.132 Repeat the exact solution of Problem 6.129 for $R_L = 10R_C$ and $\zeta = \lambda/100$.

▮ $8.46\ \underline{/-31.80°}$.

6.133 For a lossless transmission line, of electrical length ζ_e with a purely resistive load, establish the formula

$$\hat{V}(\zeta) = \frac{\hat{V}(0)}{\cos 2\pi\zeta_e + j\rho\sin 2\pi\zeta_e} \tag{1}$$

in which $\rho \equiv R_C/R_L$.

▮ In (*1*) of Problem 6.91 set $d = \zeta$, to find

$$\frac{\hat{V}(\zeta)}{\hat{V}(0)} = \frac{1 + \Gamma_L}{e^{j\beta\zeta} + \Gamma_L e^{-j\beta\zeta}} \tag{2}$$

But $\Gamma_L = (R_L - R_C)/(R_L + R_C) = (1 - \rho)/(1 + \rho)$; substituting in (*2*):

$$\frac{\hat{V}(\zeta)}{\hat{V}(0)} = \frac{2}{(1 + \rho)e^{j\beta\zeta} + (1 - \rho)e^{-j\beta\zeta}}$$

$$= \frac{1}{[(e^{j\beta\zeta} + e^{-j\beta\zeta})/2] + \rho[(e^{j\beta\zeta} - e^{-j\beta\zeta})/2]}$$

$$= \frac{1}{\cos\beta\zeta + \rho j\sin\beta\zeta}$$

which is equivalent to (*1*) above.

6.134 A lossless transmission line is driven by a 1-V ideal voltage source (zero source impedance) and is terminated in a resistive load, $R_L = 10R_C$. If the effects of the line are negligible, the load voltage should be approximately $1 \underline{/0°}$ V. In order to investigate this, compute the load voltage for line length $\lambda/10$.

▮ Substitution of $\hat{V}(0) = 1 \underline{/0°}$ V, $\zeta_e = 1/10$, and $\rho = 1/10$ in (1) of Problem 6.133 yields $\hat{V}(\zeta) = 1.2328 \underline{/-4.1555°}$ V, which is nowhere near $1.0 \underline{/0°}$ V.

6.135 Repeat Problem 6.134 for $\zeta = \lambda/100$.

▮ In this case we find $\hat{V}(\zeta) = 1.002 \underline{/-0.3605°}$ V, which is very close to $1.0 \underline{/0°}$ V.

6.136 Check (1) of Problem 6.133 in the case $\rho = 1$.

▮ In this case we obtain $\hat{V}(\zeta) = \hat{V}(0)e^{-j\beta\zeta}$, the correct matched-line relation.

6.137 Approximate (1) of Problem 6.133 for a very short (electrically) line having $\rho < 1$.

▮ For small angles x, $\cos x \approx 1$ and $\sin x \approx x$. Thus
$$\frac{\hat{V}(\zeta)}{\hat{V}(0)} \approx \frac{1}{1 + j\rho 2\pi\zeta_e} \approx 1 - j\rho 2\pi\zeta_e$$

6.138 A $\frac{3}{8}\lambda$ length of a 50-Ω line is driven by a voltage source that has $\hat{V}_S = 100 \underline{/0°}$ V and $Z_S = (50 + j0)$ Ω. Assuming that the far end of the line is terminated in a short circuit, determine the time-domain current through the short circuit.

▮ Choosing $d = \zeta$ in (2) of Problem 6.91, we have, since $\Gamma_L = -1$,
$$\hat{I}_L = \frac{\hat{I}(0)}{\cos 2\pi\zeta_e} \qquad (1)$$

But
$$\hat{I}(0) = \frac{\hat{V}_S}{Z_{\text{inSC}} + Z_S} \qquad (2)$$

and, from Problem 6.74,
$$Z_{\text{inSC}} = jR_C \tan 2\pi\zeta_e \qquad (3)$$

Combination of (1), (2), and (3) gives
$$\hat{I}_L = \frac{\hat{V}_S}{Z_S \cos 2\pi\zeta_e + jR_C \sin 2\pi\zeta_e}$$
$$= \frac{100}{50 \cos 135° + j50 \sin 135°} = 2 \underline{/-135°} \quad \text{A}$$
or $I(\zeta, t) = 2 \cos(\omega t - 135°)$ (A).

6.139 Two antennas, each of input impedance $Z_{\text{ant}} = (73 + j0)$ Ω, are fed with lossless lines from a generator, as shown in Fig. 6-54. Determine the average power delivered to either antenna.

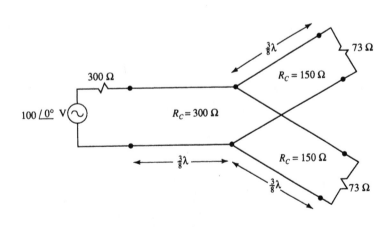

Fig. 6-54

▮ The input impedance of either line is, from Problem 6.72,
$$Z'_{\text{in}} = (150) \frac{73 + j150 \tan(2\pi \times \frac{3}{8})}{150 + j73 \tan(2\pi \times \frac{3}{8})} = (150) \frac{73 - j150}{150 - j73} = 118.04 - j92.55 \quad \Omega$$

Thus, for the generator (impedances in parallel), $Z_L = \frac{1}{2}Z'_{in} = 59.02 - j46.28 \ \Omega$ and

$$Z_{in} = (300)\frac{(59.02 - j46.28) + j300\tan\left(2\pi \times \frac{3}{8}\right)}{300 + j(59.02 - j46.28)\tan\left(2\pi \times \frac{3}{8}\right)} = 156.57 - j357.96 \quad \Omega$$

Hence

$$I_{in} = \frac{100}{300 + (156.57 - j357.96)} = 0.172\underline{/38.10°} \ \text{A}$$

and

$$P_{av} = \frac{1}{2}|\hat{I}_{in}|^2 R_{in} = \frac{1}{2}(0.172)^2(156.57) = 2.32 \ \text{W}$$

or 1.16 W per antenna.

6.140 A lossy coaxial cable has $r = 2.25\ \Omega/\text{m}$, $l = 1\ \mu\text{H/m}$, $c = 100\ \text{pF/m}$, and $g \approx 0$ at 500 MHz. Determine the attenuation constant of the line at 500 MHz; also, determine the attenuation per 100 ft of cable length, in decibels.

❚ The propagation constant is given by
$$\gamma = \alpha + j\beta = \sqrt{(r + j\omega l)(j\omega c)} = \sqrt{(2.25 + j2\pi \times 5 \times 10^8 \times 10^{-6})(j2\pi \times 5 \times 10^8 \times 10^{-10})}$$
$$= 31.42\underline{/89.98°} = 1.125 \times 10^{-2} + j31.42 \quad \text{m}^{-1}$$
Thus, $\alpha = 1.125 \times 10^{-2}\ \text{Np/m} = 3.429 \times 10^{-3}\ \text{Np/ft}$, and attenuation in dB $= 10\log_{10}e^{2\alpha 100} \approx 3\ \text{dB}$.

6.141 Compute the input impedance to a 10-m section of the cable of Problem 6.140 at 500 MHz, assuming that the load impedance is an open circuit.

❚ Under open circuit, the voltage reflection coefficient at the load is $\Gamma_L = 1$, so that
$$\Gamma(0) = \Gamma_L e^{-2\gamma\xi} \approx 0.7985\underline{/0°}$$
$$Z_C = \sqrt{\frac{r + j\omega l}{j\omega c}} = \sqrt{\frac{2.25 + j2\pi \times 5 \times 10^8 \times 10^{-6}}{j2\pi \times 5 \times 10^8 \times 10^{-10}}} \approx 100\underline{/0°} \quad \Omega$$
$$Z_{in}(0) = Z_C\frac{1 + \Gamma(0)}{1 - \Gamma(0)} \approx (100)\frac{1 + 0.7985}{1 - 0.7985} = 892.56\underline{/0°} \quad \Omega$$

6.142 Repeat Problem 6.141 for a short-circuit load on the line.

❚ In this case $\Gamma_L = -1$, so that $\Gamma(0) \approx -0.7985\underline{/0°}$, and
$$Z_{in}(0) \approx (100)\frac{1 - 0.7985}{1 + 0.7985} = 11.204\underline{/0°} \quad \Omega$$

6.143 Repeat Problem 6.141 for a resistive load of 10 Ω.

❚ Proceeding as in Problem 6.141, the voltage reflection coefficient at the load is
$$\Gamma_L = \frac{Z_L - Z_C}{Z_L + Z_C} = \frac{10 - 100}{10 + 100} = -0.818 \quad\text{and}\quad \Gamma(0) \approx (-0.818)(0.7985) = -0.653$$
Thus
$$Z_{in}(0) \approx (100)\frac{1 - 0.653}{1 + 0.653} = 20.97\ \Omega$$

6.144 A portion of a lossy transmission line is shown in Fig. 6-55. Beginning with Maxwell's equations, obtain a differential equation relating the voltage variation along the line to the current through the line.

❚ First, we apply Faraday's law,
$$\oint_C \mathbf{E} \cdot d\mathbf{s} = -j\omega \int_S \mathbf{B} \cdot d\mathbf{S}$$
to the rectangle indicated in Fig. 6-55; thus
$$\int_{x_1}^{x_2}[\hat{E}_x(x, z + \Delta z) - \hat{E}_x(x, z)]\,dx + \int_z^{z+\Delta z}[\hat{E}_z(x_1, z) - \hat{E}_z(x_2, z)]\,dz = -j\omega\int_z^{z+\Delta z}\int_{x_1}^{x_2}\hat{B}_y(x, z)\,dx\,dz \quad (1)$$
If we define the phasor voltage between the wires as
$$\hat{V}(z) = -\int_{x_1}^{x_2}\hat{E}_x(x, z)\,dx$$
then the first integral in (1) equals $-[\hat{V}(z + \Delta z) - \hat{V}(z)]$.

Suppose a phasor current $\hat{I}(z)$ flows through the upper wire in the positive z-direction and returns in the lower wire, and the losses (represented by $r\,\Delta z$, between z and $z + \Delta z$) in the wires can be lumped as an impedance through which $\hat{I}$ passes. Since the wires are not perfect conductors, the current $\hat{I}$ is not confined to the surface of the wires but penetrates into them. Thus there is internal flux-linkage, resulting in an internal inductance per unit of wire length, l_i. This being the case, the second integral in (1) can be written as
$$\int_z^{z+\Delta z}[\hat{E}_z(x_1, z) - \hat{E}_z(x_2, z)]\,dz = +(r + j\omega l_i)\,\Delta z\,\hat{I}(z)$$

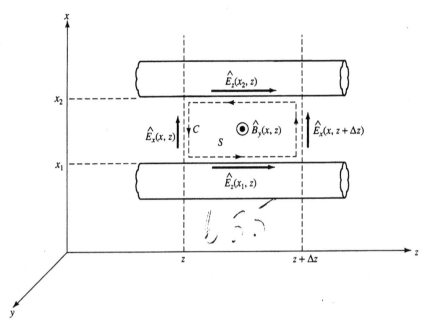

Fig. 6-55

Also, the magnetic flux external to the wires penetrates the surface S, resulting in an external inductance per unit length given by

$$l_e \, \Delta z = -\frac{\int_z^{z+\Delta z} \int_{x_1}^{x_2} \hat{B}_y(x, z) \, dx \, dz}{\hat{I}(z)}$$

Substituting these three expressions in (1) results in

$$\hat{V}(z + \Delta z) - \hat{V}(z) = -[(r + j\omega l_i) + j\omega l_e] \, \Delta z \, \hat{I}(z) \tag{2}$$

Dividing both sides of (2) by Δz and taking the limit as $\Delta z \to 0$, we obtain the *first (lossy) transmission-line equation*:

$$\frac{d\hat{V}(z)}{dz} = -(r + j\omega l)\hat{I}(z) \tag{3}$$

in which $l \equiv l_i + l_e$.

6.145 Obtain the *second (lossy) transmission-line equation,* relating the rate of change of the current $\hat{I}(z)$ through the line to the voltage $\hat{V}(z)$ across the line.

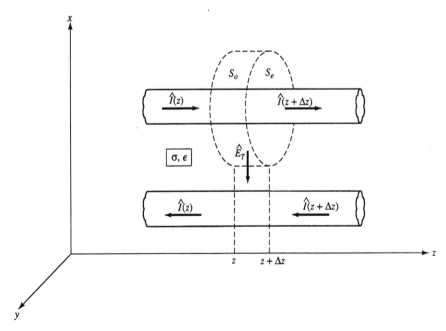

Fig. 6-56

❚ Redraw the two conductors of the line as shown in Fig. 6-56, and consider a cylinder of length Δz enclosing the upper wire. The portion of this surface just off the periphery of the cylinder is denoted by S_o, and the portions on the ends of the cylinder are denoted by S_e. The continuity equation applied to the surface S of the cylinder is

$$\int_S \hat{\mathbf{J}} \cdot d\mathbf{S} = -j\omega\hat{Q} \tag{1}$$

where $Q(t) = \mathrm{Re}\,[\hat{Q}e^{j\omega t}]$ is the charge enclosed by S. Over the ends of the cylinder,

$$\int_{S_e} \hat{\mathbf{J}} \cdot d\mathbf{S} = \hat{I}(z + \Delta z) - \hat{I}(z) \tag{2}$$

Over the sides of the cylinder, the conductivity of the medium results in a transverse conduction current through S_o (or imaginary displacement current due to polarization loss, which results in the same effect); thus

$$\int_{S_o} \hat{\mathbf{J}} \cdot d\mathbf{S} = g\,\Delta z\,\hat{V}(z) \tag{3}$$

where g is the per-unit-length conductance between the wires produced by the lossy medium. In addition, over the length Δz,

$$\hat{Q} = c\,\Delta z\,\hat{V}(z) \tag{4}$$

where c is the per-unit-length capacitance of the configuration. Inserting (4), (3), and (2) into (1) yields

$$\hat{I}(z + \Delta z) - \hat{I}(z) + g\,\Delta z\,\hat{V}(z) = -j\omega c\,\Delta z\,\hat{V}(z) \tag{5}$$

Dividing (5) by Δz, we obtain, in the limit as $\Delta z \to 0$, the required equation:

$$\frac{d\hat{I}(z)}{dz} = -(g + j\omega c)\hat{V}(z) \tag{6}$$

6.146 From the results of Problems 6.144 and 6.145, write the transmission-line equations for a lossy line, similar to those for a lossless line given as (3) and (4) of Problem 6.6.

❚
$$\frac{d\hat{V}(z)}{dz} = -Z\hat{I}(z) \tag{1}$$

$$\frac{d\hat{I}(z)}{dz} = -Y\hat{V}(z) \tag{2}$$

where $Z \equiv r + j\omega l$ and $Y \equiv g + j\omega c$ are the per-unit-length impedance and admittance, respectively.

6.147 Construct a circuit representation for the transmission line of Problem 6.144.

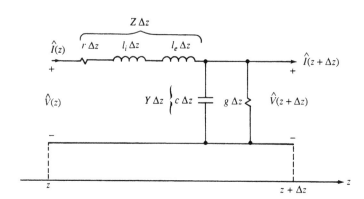

Fig. 6-57

❚ See Fig. 6-57.

6.148 Write the forms of solutions to (1) and (2) of Problem 6.146.

❚
$$\hat{V}(z) = V_m^+ e^{-\gamma z} + V_m^- e^{\gamma z} \qquad \text{and} \qquad \hat{I}(z) = \frac{V_m^+}{Z_C}e^{-\gamma z} - \frac{V_m^-}{Z_C}e^{\gamma z}$$

where the *propagation constant* and the *characteristic impedance* are defined by

$$\gamma \equiv \sqrt{ZY} \equiv \alpha + j\beta \qquad \text{and} \qquad Z_C \equiv \sqrt{Z/Y}$$

Note that if $r = g = 0$, then $\alpha = 0$ and the above solutions reduce to those of Problem 6.64.

6.149 Comparing Problems 6.64 and 6.148, comment on the effects of losses in a transmission line.

▋ (i) Forward- and backward-traveling waves are attenuated as $\exp(-\alpha|z|)$. (ii) In either direction, the voltage and current waves are no longer in phase, since Z_C is complex. (iii) For lossy lines, β will be larger than for a lossless line, so that $u = \omega/\beta$ will be smaller (at any given frequency).

6.150 Obtain an expression for the input impedance of the lossy line of Problem 6.144.

▋ See the remark concluding Problem 6.148. Make the substitutions $R_C \to Z_C$ and $j\beta \to \gamma$ in (1) of Problem 6.72, to obtain

$$Z_{\text{in}}(z) = Z_C \frac{Z_L + Z_C \tanh \gamma(\zeta - z)}{Z_C + Z_L \tanh \gamma(\zeta - z)} \qquad (1)$$

6.151 Obtain expressions for $\hat{V}(z)$ and $\hat{I}(z)$ along a lossy line in terms of V_m^+, the voltage reflection coefficient, and the characteristic impedance.

▋ Analogous to Problem 6.65,

$$\Gamma(z) \equiv \frac{V_m^- e^{\gamma z}}{V_m^+ e^{-\gamma z}} \qquad (1)$$

and

$$\hat{V}(z) = V_m^+ e^{-\gamma z}[1 + \Gamma(z)] \qquad (2)$$

$$\hat{I}(z) = \frac{V_m^+}{Z_C} e^{-\gamma z}[1 - \Gamma(z)] \qquad (3)$$

6.152 Obtain an expression for the input impedance of a lossy line in terms of the voltage reflection coefficient.

▋ Divide (2) of Problem 6.151 by (3):

$$Z_{\text{in}}(z) \equiv \frac{\hat{V}(z)}{\hat{I}(z)} = Z_C \frac{1 + \Gamma(z)}{1 - \Gamma(z)}$$

6.153 Evaluate the voltage reflection coefficient at the load in terms of impedances.

▋ Problem 6.152 gives, for $z = \zeta$,

$$Z_L = Z_C \frac{1 + \Gamma_L}{1 - \Gamma_L} \qquad \text{or} \qquad \Gamma_L = \frac{Z_L - Z_C}{Z_L + Z_C}$$

6.154 For a lossy line, evaluate $\Gamma(z)$ in terms of $\Gamma(\zeta) = \Gamma_L$. Obtain an expression for the voltage reflection coefficient at any point on a lossy line, $\Gamma(z)$, in terms of the voltage reflection coefficient at the load, Γ_L. Compare the results with that for a lossless line.

▋ Simply replace $j\beta$ in (1) of Problem 6.68 by γ, to obtain

$$\Gamma(z) = \Gamma_L e^{2\gamma(z - \zeta)} \qquad (1)$$

From (1),

$$|\Gamma(z)| = |\Gamma_L| e^{-2\alpha(\zeta - z)} \qquad (2)$$

i.e., the magnitude of the reflection coefficient is not constant along the lossy line, but grows exponentially.

6.155 A lossy transmission line has $Z_L = Z_C$; that is, the line is matched. Obtain an expression for the average power flow along the line.

▋ Since the line is matched, there will be no backward wave. Consequently, the voltage and current are

$$\hat{V}(z) = V_m^+ e^{-\gamma z} \qquad \hat{I}(z) = \frac{V_m^+}{Z_C} e^{-\gamma z}$$

and the average power flow in the $+z$-direction is

$$P_{\text{av}}(z) = \frac{1}{2} \text{Re}\left[\hat{V}(z)\hat{I}^*(z)\right] = \frac{|V_m^+|^2}{2|Z_C|} e^{-2\alpha z} \cos \theta_{Z_C} \qquad (1)$$

6.156 For a matched line, evaluate the *attenuation per longitudinal distance d*, defined as the ratio $P_{\text{av}}(z + d)/P_{\text{av}}(z)$, expressed in decibels.

▋ From (1) of Problem 6.155, $P_{\text{av}}(z + d)/P_{\text{av}}(z) = e^{-2\alpha d}$; hence

$$\text{dB} = -10 \log_{10} \frac{P_{\text{av}}(z + d)}{P_{\text{av}}(z)} = 10 \log_{10} e^{2\alpha d} = 8.686 \alpha d \qquad (1)$$

6.157 For a "low-loss line," such that $r \ll \omega l$ and $g \ll \omega c$, find an approximate expression for the propagation constant γ.

▌ Use the binomial theorem:

$$\gamma = \sqrt{(r + j\omega l)(g + j\omega c)} = j\omega\sqrt{lc}\left(1 + \frac{r}{j\omega l}\right)^{1/2}\left(1 + \frac{g}{j\omega c}\right)^{1/2}$$

$$\approx j\omega\sqrt{lc}\left(1 + \frac{r}{2j\omega l}\right)\left(1 + \frac{g}{2j\omega c}\right) \approx j\omega\sqrt{lc}\left[1 + \frac{1}{2j\omega}\left(\frac{r}{l} + \frac{g}{c}\right)\right]$$

6.158 For the line of Problem 6.157, obtain approximate expressions for the attenuation constant and the phase constant.

▌
$$\alpha = \text{Re}\,[\gamma] \approx \frac{j\omega\sqrt{lc}}{2j\omega}\left(\frac{r}{l} + \frac{g}{c}\right)$$

$$= \frac{r}{2}\sqrt{\frac{c}{l}} + \frac{g}{2}\sqrt{\frac{l}{c}} \approx \frac{r}{2R_C} + \frac{gR_C}{2}$$

since $R_C = \sqrt{l_e/c} \approx \sqrt{l/c}$, the internal inductance being negligible except at low frequencies. Similarly,

$$\beta = \text{Im}\,[\gamma] \approx \omega\sqrt{lc} \approx \omega c R_C$$

6.159 Derive an approximate expression for the characteristic impedance of the line of Problem 6.157.

▌ Using $R_C \approx \sqrt{l/c}$, we have

$$Z_C = \sqrt{\frac{r + j\omega l}{g + j\omega c}} \approx R_C\left(1 + \frac{r}{j\omega l}\right)^{1/2}\left(1 + \frac{g}{j\omega c}\right)^{-1/2}$$

$$\approx R_C\left(1 + \frac{r}{2j\omega l}\right)\left(1 - \frac{g}{2j\omega c}\right) \approx R_C\left[1 + \frac{1}{2j\omega}\left(\frac{r}{l} - \frac{g}{c}\right)\right]$$

$$\approx R_C \quad \left[\omega \gg \frac{1}{2}\left(\frac{r}{l} - \frac{g}{c}\right)\right]$$

6.160 For a low-loss line, find approximate expressions for (a) the characteristic impedance Z_C, in terms of the velocity of propagation along the line, and (b) the phase constant, in terms of the relative permittivity ϵ_r and u_0, the velocity of an electromagnetic wave in free space.

▌ (a) From Problem 6.159,
$$Z_C \approx R_C \approx \sqrt{l/c} = ul = 1/uc \tag{1}$$
(b) Since $u = 1/\sqrt{lc} \approx \sqrt{\mu/\epsilon} = u_0/\sqrt{\epsilon_r}$, we have, from Problem 6.158,
$$\beta \approx \omega\sqrt{lc} = \frac{\omega}{u} \approx \frac{\omega}{u_0}\sqrt{\epsilon_r} \tag{2}$$

6.161 A low-loss coaxial cable has a characteristic resistance of 53.5 Ω. The interior dielectric has $\epsilon_r = 2.25$. Determine the per-unit-length external inductance and capacitance.

▌ The velocity of propagation is $u = u_0/\sqrt{\epsilon_r} = 2 \times 10^8$ m/s; therefore,
$$l_e = R_C/u = 0.268\ \mu\text{H/m} \qquad c = 1/uR_C = 93.5\ \text{pF/m}$$

6.162 What is the attenuation constant at 100 MHz of the cable of Problem 6.161, if the attenuation per 100 ft at 100 MHz is 4.5 dB?

▌ From the given attenuation and the result of Problem 6.156,
$$4.5 = (8.686)\alpha(100) \qquad \text{or} \qquad \alpha = 5.18 \times 10^{-3}\ \text{Np/ft} = 0.017\ \text{Np/m}$$

6.163 Calculate the input impedance to a 1-m section of the cable of Problem 6.161 at 100 MHz, if the cable is terminated in a 300-Ω resistive load.

▌ The propagation constant at 100 MHz is, from Problems 6.161 and 6.162,
$$\gamma = \alpha + j\beta = \alpha + j\omega\sqrt{\epsilon_r}/u_0 = 0.017 + j\pi \quad \text{m}^{-1}$$
For a 300-Ω resistive load, the load reflection coefficient is
$$\Gamma_L = \frac{300 - 53.5}{300 + 53.5} = 0.7$$
At the input to the 1-m length of line, the reflection coefficient is (Problem 6.154)
$$\Gamma_{\text{in}} = \Gamma_L e^{-2\gamma(1)} = 0.7 e^{-2(0.017)} e^{-j2\pi} = 0.6766\ \underline{/0°}$$

Thus, the input impedance is

$$Z_{in} = Z_C \frac{1 + \Gamma_{in}}{1 - \Gamma_{in}} = (53.5) \frac{1 + 0.6766}{1 - 0.6766} = 277.4 \ \Omega$$

6.164 Repeat Problem 6.163 for a 1.2-m-long cable, other data remaining unchanged.

▌ From Problem 6.163, $\gamma = 0.017 + j\pi$ m^{-1}, $\Gamma_L = 0.7$ and $Z_C = 53.5 \ \Omega$; thus,

$$\Gamma_{in} = \Gamma_L e^{-2\gamma(1.2)} = 0.7 e^{-2(0.017)(1.2)} e^{-j2(3.14)(1.2)} = 0.672 \ \underline{/-71.78°}$$

and

$$Z_{in} = Z_C \frac{1 + \Gamma_{in}}{1 - \Gamma_{in}} = (53.5) \frac{1 + 0.67 \ \underline{/-71.78°}}{1 - 0.67 \ \underline{/-71.78°}} = 71.8 \ \underline{/-66.7°} \ \Omega$$

6.165 A lossy transmission line is operated at 100 MHz and has $Z_C = (75 + j0) \ \Omega$, $\alpha = 0.02$ Np/m, and $\beta = 3$ rad/m. Determine the per-unit-length resistance, inductance, capacitance, and conductance of the line.

▌ By the definitions of Problems 6.146 and 6.148,

$$r + j\omega l \equiv Z = \gamma Z_C = (0.02 + j3)(75 + j0) = 1.5 + j225 \quad \Omega/m$$

$$g + j\omega c \equiv Y = \frac{\gamma}{Z_C} = \frac{0.02 + j3}{75 + j0} = 2.667 \times 10^{-4} + j0.04 \quad \text{S/m}$$

which give

$$r = 1.5 \ \Omega/m \qquad g = 266.7 \ \mu\text{S/m}$$
$$l = 358 \ \text{nH/m} \qquad c = 63.7 \ \text{pF/m}$$

6.166 If a 7-m length of the line of Problem 6.165 is terminated in $Z_L = (150 + j0) \ \Omega$ and is driven by a source having $\hat{V}_s = 10 \ \underline{/0°}$ V and $Z_s = (75 + j0) \ \Omega$, determine the average power delivered to the line.

▌ The reflection coefficient at the load is

$$\Gamma_L = \frac{150 - 75}{150 + 75} = \frac{1}{3}$$

Thus, by (1) of Problem 6.154,

$$\Gamma_{in} = \tfrac{1}{3} e^{-2\alpha 7} e^{-j2\beta 7} = 0.25 \ \underline{/2406.42°} = 0.25 \ \underline{/246.42°}$$

and

$$Z_{in} = Z_C \frac{1 + \Gamma_{in}}{1 - \Gamma_{in}} = (75) \frac{1 + 0.25 \ \underline{/246.42°}}{1 - 0.25 \ \underline{/246.42°}} = 61.91 \ \underline{/-26.25°} = 55.53 - j27.38 \quad \Omega$$

Then

$$\hat{I}_{in} = \frac{10 \ \underline{/0°}}{75 + Z_{in}} = 75 \ \underline{/+11.85°} \ \text{mA} \qquad \text{and} \qquad \langle P_{in} \rangle = \frac{1}{2} |\hat{I}_{in}|^2 \text{Re} [Z_{in}] = 156.1 \ \text{mW}$$

6.167 What is the average power delivered to the load of Problem 6.166?

▌ Replace $j\beta$ by γ in (1) of Problem 6.71 to obtain

$$\hat{V}(\zeta) = \hat{V}(0) e^{-\gamma \zeta} \frac{1 + \Gamma_L}{1 + \Gamma_{in}} \tag{1}$$

Also, by voltage division,

$$\hat{V}(0) = \frac{Z_{in}}{Z_{in} + Z_s} \hat{V}_s \tag{2}$$

From (1) and (2), and the fact that Z_L is real,

$$\langle P_L \rangle = \frac{|\hat{V}(\zeta)|^2}{2R_L} = \frac{|Z_{in}|^2 |\hat{V}_s|^2}{2R_L |Z_{in} + Z_s|^2} e^{-2\alpha \zeta} \frac{|1 + \Gamma_L|^2}{|1 + \Gamma_{in}|^2}$$
$$= \frac{(61.91)^2 (10)^2}{2(150)[(130.53)^2 + (27.38)^2]} e^{-2(0.02)(7)} \frac{(4/3)^2}{(0.9)^2 + (0.229)^2} = 0.112 \ \text{W}$$

6.168 Repeat Problem 6.166 for a matched line; that is, $Z_L = Z_C = (75 + j0) \ \Omega$.

▌ Now $\Gamma_L = 0$, $\Gamma_{in} = 0$, $Z_{in} = (75 + j0) \ \Omega$,

$$\hat{I}_{in} = \frac{10 \ \underline{/0°}}{150} \ \text{A} \qquad \text{and} \qquad \langle P_{in} \rangle = \frac{1}{2} \left(\frac{10}{150} \right)^2 (75) = \frac{1}{6} \ \text{W}$$

6.169 Repeat Problem 6.167 for a matched line.

▌ Substitute the new parameter-values in the expression developed in Problem 6.167:

$$\langle P_L \rangle = \frac{(75)^2 (10)^2}{2(75)(150)^2} e^{-0.28} = \frac{1}{6} e^{-0.28} = 125.96 \ \text{mW}$$

6.170 Verify the consistency of the results of Problems 6.168 and 1.169.

▌ For a matched line, Problem 6.155 shows that average power decreases as $e^{-2\alpha z}$; thus $\langle P_L \rangle = \langle P_{in} \rangle e^{-2\alpha\zeta}$.

6.171 Express the characteristic impedance of a lossy line in terms of its input impedances, under short-circuit and open-circuit conditions.

▌ **Short circuit:** In (1) of Problem 6.150, set $Z_L = 0$ and $z = 0$:

$$Z_{inSC} = Z_C \tanh \gamma\zeta \tag{1}$$

Open circuit: Set $Z_L = \infty$ and $z = 0$:

$$Z_{inOC} = \frac{Z_C}{\tanh \gamma\zeta} \tag{2}$$

Together (1) and (2) imply

$$Z_{inSC} Z_{inOC} = Z_C^2 \tag{3}$$

6.172 Given a 10-m-long lossy line for which $Z_{inSC} = 235\underline{/30°}$ Ω and $Z_{inOC} = 120\underline{/-45°}$ Ω, determine Z_C, α, and β for this cable.

▌ From (3) of Problem 6.171,

$$Z_C = \sqrt{(120\underline{/-45°})(235\underline{/30°})} = 167.93\underline{/-7.5°} = 166.49 - j21.92 \quad \Omega$$

Now

$$\Gamma_{inSC} = \frac{Z_{inSC} - Z_C}{Z_{inSC} + Z_C} = \frac{235\underline{/30°} - 167.93\underline{/7.5°}}{235\underline{/30°} + 167.93\underline{/7.5°}} = 0.38\underline{/60.64°} = 0.38\underline{/0.337\pi \text{ rad}} \tag{1}$$

But, using $\Gamma_L = -1$ in (1) of Problem 6.154,

$$\Gamma_{inSC} = e^{j\pi}e^{-20(\alpha+j\beta)} = e^{-20\alpha}\underline{/\pi - 20\beta} \tag{2}$$

Comparing (1) and (2), we find that $\alpha = 0.0485$ Np/m and $\beta = 0.104$ rad/m.

6.173 For a certain cable, $Z_C = 50 + j0$ Ω, $\alpha\zeta = 2.21 \times 10^{-2}$ Np, and $\beta\zeta = 4.082$ rad. If the cable is terminated in a short circuit ($\Gamma_L = -1$) and driven by a source having $\hat{V}_S = 10\underline{/0°}$ V and $Z_S = (50 + j0)$ Ω, determine the short-circuit current in the time domain.

▌ The current along the line is

$$\hat{I}(z) = \frac{V_m^+}{Z_C} e^{-\gamma z}[1 - \Gamma(z)]$$

Thus, the input current to the line and to the load become

$$\hat{I}(0) = \frac{V_m^+}{Z_C}(1 - \Gamma_{in}) \qquad \hat{I}(\zeta) = \frac{V_m^+}{Z_C} e^{-\gamma\zeta}(1 - \Gamma_L) = \frac{2V_m^+ e^{-\gamma\zeta}}{Z_C}$$

Now,

$$\Gamma_{in} = \Gamma_L e^{-2\gamma\zeta} = (-1)e^{-2\alpha\zeta}e^{-j2\beta\zeta} = -0.96\underline{/-467.76°}$$

$$Z_{in} = Z_C \frac{1 + \Gamma_{in}}{1 - \Gamma_{in}} = 68.5\underline{/87.34°} \quad \Omega$$

$$\hat{I}(0) = \frac{\hat{V}_S}{Z_S + Z_{in}} = \frac{10\underline{/0°}}{50\underline{/0°} + 68.5\underline{/87.34°}} = 116.1\underline{/-52.14°} \quad \text{mA}$$

$$V_m^+ = \frac{Z_C\hat{I}(0)}{1 - \Gamma_{in}} = 5.02\underline{/0.14°} \quad \text{V}$$

$$\hat{I}(\zeta) = \hat{I}_L = \frac{V_m^+}{Z_C} e^{-\alpha\zeta} e^{-j\beta\zeta}(1 - \Gamma_L) = \frac{5.02\underline{/0.14°}}{50} e^{-2.21 \times 10^{-2}} e^{-j4.082} \times 2 = 0.2\underline{/234.02°} \quad \text{A}$$

Consequently, $i_L(t) = 0.2 \cos(\omega t + 234.02)$ A.

6.174 Give differential equations for the (transverse) **E**- and **H**-fields in a homogeneous lossy medium surrounding the conductors of a transmission line. Assume the time dependence $e^{j\omega t}$.

▌ By Problem 5.3, we have, in a rectangular coordinate system of which the z axis is along the transmission line,

$$\frac{d^2\hat{\mathbf{E}}}{dz^2} - \gamma^2\hat{\mathbf{E}} = 0 \qquad \frac{d^2\hat{\mathbf{H}}}{dz^2} - \gamma^2\hat{\mathbf{H}} = 0 \tag{1}$$

where $\gamma = \sqrt{j\omega\mu(\sigma + j\omega\epsilon)}$ is the propagation constant of the medium.

6.175 From (1) of Problem 6.174, retrieve the results of Problem 6.148.

▌ Equations (1) have the general integrals

$$\hat{\mathbf{E}} = \hat{\mathbf{E}}^+ e^{-\gamma z} + \hat{\mathbf{E}}^- e^{\gamma z} \qquad \hat{\mathbf{H}} = \hat{\mathbf{H}}^+ e^{-\gamma z} + \hat{\mathbf{H}}^- e^{\gamma z}$$

Then, by (*1*) and (*2*) of Problem 6.4,

$$\hat{V}(z) = \left(-\int_{C_V} \hat{\mathbf{E}}^+ \cdot d\mathbf{s}\right)e^{-\gamma z} + \left(-\int_{C_V} \hat{\mathbf{E}}^- \cdot d\mathbf{s}\right)e^{\gamma z} \equiv V_m^+ e^{-\gamma z} + V_m^- e^{\gamma z}$$

$$\hat{I}(z) = \left(\oint_{C_I} \hat{\mathbf{H}}^+ \cdot d\mathbf{s}\right)e^{-\gamma z} + \left(\oint_{C_I} \hat{\mathbf{H}}^- \cdot d\mathbf{s}\right)e^{\gamma z} \equiv I_m^+ e^{-\gamma z} + I_m^- e^{\gamma z}$$

6.176 Show that, if the line of Problem 6.17 is itself lossless ($r = 0$ and $l = l_e$), then $g/c = \sigma/\epsilon$, where g and c are per-unit-length circuit parameters and σ and ϵ are per-unit-length field parameters.

▌ Problems 6.146, 6.148, and 6.174 give
$$\gamma^2 = j\omega\mu(\sigma + j\omega\epsilon) = (j\omega l_e)(g + j\omega c)$$
and the desired result follows on equating real parts and imaginary parts.

6.177 Recall that a lossy medium can be characterized by a complex permittivity

$$\hat{\epsilon} = \epsilon\left(1 - j\frac{\sigma}{\omega\epsilon}\right) \qquad (1)$$

and that the per-unit-length capacitance of a lossless line may be written as
$$c = \epsilon K \qquad (2)$$
where K is a factor that depends only on the cross-sectional geometry of the line. On the basis of (*1*) and (*2*), re-solve Problem 6.176.

▌ The complex capacitance corresponding to $\hat{\epsilon}$ is

$$\hat{c} = \hat{\epsilon}K = \epsilon\left(1 - j\frac{\sigma}{\omega\epsilon}\right)K \qquad (3)$$

In terms of $\hat{c}$, (*2*) of Problem 6.146 becomes $\hat{I}' = -j\omega\hat{c}\hat{V}$; hence

$$g + j\omega c = j\omega\hat{c} = j\omega\epsilon\left(1 - j\frac{\sigma}{\omega\epsilon}\right)K = \sigma K + j\omega\epsilon K$$

from which $g = \sigma K$ and $c = \epsilon K$, or $g/c = \sigma/\epsilon$.

6.178 Determine l_e, the external inductance per unit length, of the cable shown in Fig. 6-58(*a*).

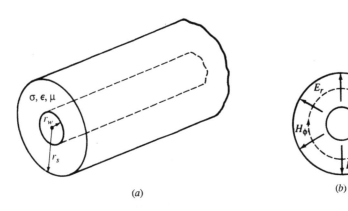

(*a*) (*b*) **Fig. 6-58**

▌ Let the inner conductor carry a current I that returns on the interior of the outer conductor. By symmetry, these currents will be uniformly distributed around each conductor periphery and the magnetic field intensity will have only a ϕ-component, given by Ampère's law as

$$H_\phi = \frac{I}{2\pi r} \qquad (r_w < r < r_s)$$

The total magnetic flux per unit length linking the current is the flux through a rectangle $r_w < r < r_s$, $a < z < a + 1$:

$$\psi_m = \int \mathbf{B} \cdot d\mathbf{S} = \int_{r_w}^{r_s} \mu H_\phi(1)\, dr = \frac{\mu I}{2\pi}\ln\frac{r_s}{r_w}$$

whence
$$l_e = \frac{\psi_m}{I} = \frac{\mu}{2\pi}\ln\frac{r_s}{r_w} \qquad (1)$$

6.179 Find c, the capacitance per unit length, of the cable of Problem 6.178.

∎ The per-unit-length capacitance can be found by assuming a charge per unit length $+q$ on the surface of the inner conductor and $-q$ on the inner surface of the outer conductor. By symmetry, the electric field will be radial; Gauss's law yields

$$E_r = \frac{q}{2\pi\epsilon r} \quad (r_w < r < r_s)$$

The voltage between the two conductors is

$$V = -\int_{r_s}^{r_w} E_r \, dr = \frac{q}{2\pi\epsilon} \ln \frac{r_s}{r_w}$$

Thus, the per-unit-length capacitance becomes

$$c = \frac{q}{V} = \frac{2\pi\epsilon}{\ln(r_s/r_w)} \tag{1}$$

6.180 Show that $l_e c = \mu\epsilon$ for the cable of Problem 6.178.

∎ The result follows directly from (1) of Problem 6.178 and (1) of Problem 6.179.

6.181 Determine g, the conductance per unit length, of the cable of Problem 6.178.

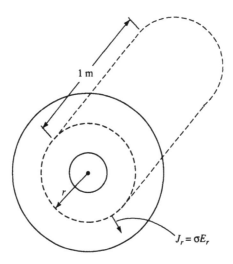

Fig. 6-59

∎ The total transverse conduction current per meter can be obtained by integrating

$$J_r = \sigma E_r = \frac{\sigma q}{2\pi\epsilon r}$$

over the cylindrical surface shown in Fig. 6-59:

$$I_T = \int_{\phi=0}^{2\pi} \frac{\sigma q}{2\pi\epsilon r}\,(1)\,r\,d\phi = \frac{\sigma q}{\epsilon}$$

Thus, the per-unit-length conductance is

$$g = \frac{I_T}{V} = \frac{\sigma q/\epsilon}{(q/2\pi\epsilon)\ln(r_s/r_w)} = \frac{2\pi\sigma}{\ln(r_s/r_w)} \tag{1}$$

6.182 Show that $g/c = \sigma/\epsilon$ for the cable of Problem 6.178 (compare Problems 6.176 and 6.177).

∎ The result follows from (1) of Problem 6.179 and (1) of Problem 6.181.

6.183 Show that the dc internal inductance per unit length of a cylindrical conductor has the value $\mu/8\pi$, independent of the conductor radius. What assumption underlies the result?

∎ Review Problems 3.148 and 3.149. It is assumed that J_z is uniform over a cross section of the conductor.

6.184 Obtain an expression for the high-frequency (surface) impedance of the round wire of Problem 6.183.

∎ At high frequencies essentially all the current will be found within one skin depth of the surface, as indicated in Fig. 6-60. Thus, a surface impedance may be defined by dividing the internal impedance Z (Problem 4.141) by the circumference $2\pi r_w$ of the conductor (consider the surface as made up of narrow strips that are

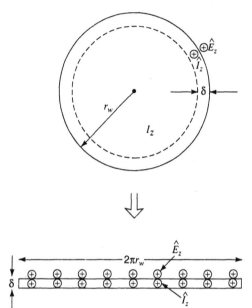

Fig. 6-60

electrically connected in parallel):

$$Z_{sHF} = \frac{Z}{2\pi r_w} = \frac{1}{2\pi r_w \sigma \delta}(1+j) \equiv r_{HF} + j\omega l_{iHF}$$

Here, as always, $\delta = \sqrt{2/\omega\mu\sigma}$.

6.185 Express r_{HF} and l_{iHF} (Problem 6.184) in terms of dc values.

▮ The dc resistance per unit length is $r_{DC} = 1/\sigma\pi r_w^2$; therefore

$$r_{HF} = \frac{r_w}{2\delta} r_{DC} \qquad (1)$$

The dc internal inductance per unit length is $l_{iDC} = \mu/8\pi$ (Problem 6.183); therefore

$$l_{iHF} = \frac{4}{r_w \sigma \delta \mu \omega} l_{iDC} = \frac{2\delta}{r_w} l_{iDC} \qquad (2)$$

6.186 By Problem 6.185, $r_{HF} \propto 1/\delta$ and $l_{iHF} \propto \delta$. Does this imply that the conductor may be treated as purely resistive at very high frequencies (very small δ-values)?

▮ No: the resistance and the inductive *reactance* are equal at all frequencies, by Problem 6.184.

6.187 For the cable of Problem 6.161, compute the per-unit-length total resistance and internal inductance at 1 kHz. The inner wire is 20-gauge ($r_w = 16$ mils, or 0.406 mm). The overall shield is constructed by weaving 36-gauge wires (radius 2.5 mils, or 0.0636 mm), giving an approximate thickness $t = 5$ mils, or 0.127 mm. The inner radius of the shield is 60 mils, or 1.52 mm. All conductors are copper ($\mu = \mu_0$, $\sigma = 58$ MS/m).

▮ At 1 kHz, the skin depth in copper is

$$\delta = \frac{1}{\sqrt{\pi f \mu \sigma}} = \frac{1}{\sqrt{\pi(10^3)(4\pi \times 10^{-7})(58 \times 10^6)}} \approx 1 \text{ mm}$$

Since $\delta > r_w > t$, this is basically a dc situation:

$$r = \frac{1}{\sigma \pi r_w^2} + \frac{1}{2\pi r_s t \sigma} = \frac{1}{5.8 \times 10^7 \pi}\left[\frac{1}{(0.406)^2 \times 10^{-6}} + \frac{1}{2 \times 1.52 \times 1.27}\right] = 47.5 \text{ m}\Omega/\text{m}$$

and, neglecting the internal inductance of the shield,

$$l_i = \frac{\mu_0}{8\pi} = \frac{4\pi \times 10^{-7}}{8\pi} = 50 \text{ nH/m}$$

6.188 Repeat Problem 6.187 for a frequency of 100 MHz.

▮ At 100 MHz, both the wire radius and the shield thickness are much larger than a skin depth ($\delta \approx 1 \ \mu$m).

Thus, the high-frequency formulas of Problem 6.184 apply.

$$r_{HF} = \frac{1}{2\pi r_w \sigma \delta} + \frac{1}{2\pi r_s \sigma \delta} = \frac{1}{2\pi \times 5.8 \times 10^7 \times 6.609 \times 10^{-6}}\left(\frac{1}{0.406} + \frac{1}{1.52}\right) = 1.296 \text{ m}\Omega/\text{m}$$

$$l_{iHF} = \frac{r_{HF}}{\omega} = 2.06 \text{ pH/m}$$

6.189 Evaluate the ratio l_i/l_e for the cable of Problems 6.161 and 6.188.

▮
$$\frac{l_i}{l_e} = \frac{2.06 \times 10^{-12}}{0.268 \times 10^{-6}} \approx 10^{-5}$$

which confirms that the internal inductance is completely negligible with respect to the external, at 100 MHz.

6.190 The characteristic resistance of coaxial cables, as a logarithmic function, typically falls within a rather narrow range of values. To illustrate this, compute R_C for a coaxial cable having an inner dielectric of polyethylene ($\epsilon_r = 2.25$), for dimensional ratios $r_s/r_w = 1.5, 2, 5, 10$.

▮ From Problems 6.178 and 6.179,

$$l_e = \frac{\mu_0}{2\pi} \ln \frac{r_s}{r_w} \qquad c = \frac{2\pi\epsilon}{\ln r_s/r_w}$$

Thus
$$R_C = \sqrt{\frac{l_e}{c}} = \frac{1}{2\pi} \sqrt{\frac{\mu_0}{\epsilon}} \ln \frac{r_s}{r_w} = (40 \, \Omega) \ln \frac{r_s}{r_w}$$

and we have the following table:

r_s/r_w	R_C, Ω
1.5	16.2
2	27.73
5	64.38
10	92.10

6.191 Repeat Problem 6.190 for an air-filled, parallel-wire line, for the following ratios of wire separation to wire radius: $D/r_w = 2.5, 5, 10, 20$.

▮ For a parallel-wire line we have

$$l_e = \frac{\mu_0}{\pi} \cosh^{-1} \frac{D}{2r_w} \qquad c = \frac{\pi\epsilon_0}{\cosh^{-1} \frac{D}{2r_w}} \qquad R_C = (120 \, \Omega) \cosh^{-1} \frac{D}{2r_w}$$

Using the identity $\cosh^{-1} x = \ln(x + \sqrt{x^2 - 1})$ $(x \geq 1)$, we find:

D/r_w	R_C, Ω
2.5	83.18
5	188.02
10	275.09
20	359.19

6.192 In many types of coaxial cables, the outer shield is a very thin aluminum foil coated with a Mylar backing instead of a woven braid. A typical cable has the following dimensions:

Shield (aluminum) thickness, t	9 μm
Shield inner radius, r_s	813 μm
Inner wire (copper) radius, r_w	127 μm
Dielectric (foamed polyethylene)	$\epsilon_r = 1.5$

Calculate r, l_e, and c for this cable.

▮ For the wire,

$$\delta_w = \frac{1}{\sqrt{\pi f \mu_0 \sigma_w}} = 8.14 \, \mu\text{m} \ll r_w$$

so that
$$r_{wire} = \frac{1}{2\pi r_w \sigma_w \delta_w} = 2.65 \, \Omega/\text{m}$$

For the shield,

$$\delta_s = \frac{1}{\sqrt{\pi f \mu_0 \sigma_s}} = 6.61 \ \mu m \approx t$$

so that, as in Problem 6.187,

$$r_{\text{shield}} = \frac{1}{2\pi r_s t \sigma_s} = 0.924 \ \Omega/m$$

Thus $r = r_{\text{wire}} + r_{\text{shield}} = 3.574 \ \Omega/m$.

The per-unit-length inductance and capacitance are

$$l_e = \frac{\mu_0}{2\pi} \ln \frac{r_s}{r_w} = 371 \ nH/m \qquad c = \frac{2\pi\epsilon}{\ln(r_s/r_w)} = 44.9 \ pF/m$$

6.193 Calculate the attenuation constant for the cable of Problem 6.192 at 100 MHz.

▮ $$\gamma = \alpha + j\beta = \sqrt{(r + j\omega l)j\omega c} = 4.01 \times 10^{-2} + j2.57 \quad m^{-1}$$
Thus, $\alpha = 4.01 \times 10^{-2} \ Np/m$.

6.194 A line is "electrically short" ($\beta\zeta \ll 1$) and has small losses ($\alpha\zeta \ll 1$). Show that the input impedance to this line is approximately $Z\zeta \equiv (r + j\omega l)\zeta$ when the load is a short circuit.

▮ By (1) of Problem 6.171,

$$Z_{\text{inSC}} = Z_C \tanh \gamma\zeta$$

Now, $|\gamma\zeta|^2 = (\alpha\zeta)^2 + (\beta\zeta)^2 \ll 1$; so that a Taylor's series gives

$$\tan \gamma\zeta = \gamma\zeta - \frac{(\gamma\zeta)^3}{3} + \cdots \approx \gamma\zeta$$

and $Z_{\text{inSC}} \approx Z_C \gamma\zeta = Z\zeta$, by Problem 6.148.

6.195 For the line of Problem 6.194, show that the input admittance is approximately $Y\zeta \equiv (g + j\omega c)\zeta$ when the load is an open circuit.

▮ Start with (2) of Problem 6.171 and proceed as in Problem 6.194; note that $Y = \gamma/Z_C$ and $Y_{\text{in}} \equiv 1/Z_{\text{in}}$.

6.196 Verify the result of Problem 6.194 for a line with $r = 0.1 \ \Omega/m$, $g = 270 \ \mu S/m$, $l = 360 \ nH/m$, $c = 64 \ pF/m$, $f = 1 \ MHz$, and $\zeta = 1 \ m$.

▮ For the data,

$$Z = 2.26 \underline{/87.5°} \ \Omega/m \qquad Y = 4.84 \times 10^{-4} \underline{/56.1°} \ S/m$$

whence

$$\gamma = \sqrt{ZY} = 1.03 \times 10^{-2} + j3.15 \times 10^{-2} \quad m^{-1}$$
$$\Gamma_{\text{in}} = \Gamma_L e^{-2\gamma\zeta} = (-1)e^{-2\gamma(1)} = -0.98 \underline{/-3.61°}$$
$$Z_C = \sqrt{Z/Y} = 68.4 \underline{/15.7°} \ \Omega$$
$$Z_{\text{in}} = Z_C \frac{1 + \Gamma_{\text{in}}}{1 - \Gamma_{\text{in}}} = 2.27 \underline{/87.6°} \ \Omega$$

This exact value is very well approximated by $Z\zeta = 2.26 \underline{/87.5°} \ \Omega$.

6.197 Verify the result of Problem 6.195 for the line of Problem 6.196.

▮ Under open circuit we have $\Gamma_L = 1$, so that $\Gamma_{\text{in}} = +0.98 \underline{/-3.61°}$; hence
$$Z_{\text{in}} = (68.4 \underline{/15.7°})(30.2 \underline{/-71.9°}) = 2066 \underline{/-56.2°} \ \Omega \qquad \text{and} \qquad Y_{\text{in}} = 4.84 \times 10^{-4} \underline{/56.2°} \ S$$
The approximate value is $Y\zeta = 4.84 \times 10^{-4} \underline{/56.1°} \ S$.

6.198 Normalize the input impedance Z_{in} to a transmission line with respect to its characteristic impedance Z_C. Show that the loci of constant normalized resistance r and constant normalized reactance x are circles.

▮ By definition,

$$z_{\text{in}}(z) \equiv \frac{Z_{\text{in}}(z)}{Z_C} = \frac{1 + \Gamma(z)}{1 - \Gamma(z)} \equiv r + jx \tag{1}$$

where r and x are the real and imaginary parts of z_{in}. Similarly, the reflection coefficient at point z may be written in terms of its real and imaginary parts, p and q, as

$$\Gamma(z) = |\Gamma(z)| \underline{/\theta_\Gamma(z)} = p + jq \tag{2}$$

Substituting (2) in (1), we obtain

$$r + jx = \frac{1 + p + jq}{1 - p - jq} \tag{3}$$

If in (3) we equate the real parts and the imaginary parts, we obtain two families of circles in the pq plane:

$$\left(p - \frac{r}{r+1}\right)^2 + q^2 = \frac{1}{(r+1)^2} \qquad \textbf{circles of constant } r \tag{4}$$

$$(p-1)^2 + \left(q - \frac{1}{x}\right)^2 = \frac{1}{x^2} \qquad \textbf{circles of constant } x \tag{5}$$

6.199 Sketch (4) of Problem 6.198 for $r = 0, 1/2, 1, 2$.

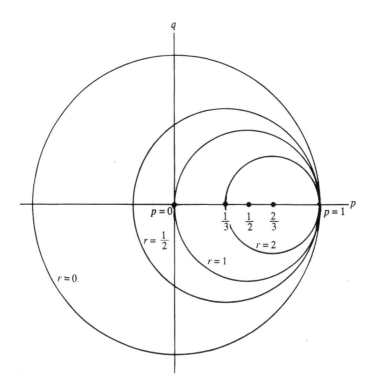

Fig. 6-61

▌ See Fig. 6-61.

6.200 Sketch (5) of Problem 6.198 for $x = \pm 2/3, \pm 1, \pm 2$.

▌ The six circles are shown in Fig. 6-62.

6.201 Superimpose Figs. 6-61 and 6-62 and hence obtain the *Smith chart.*

▌ See Fig. 6-63. Notice that on the outer rim of the chart is a scale labeled "Angle of Reflection Coefficient in Degrees" from which $\theta_\Gamma = \arctan(q/p)$ can be read.

6.202 For a given z_{in} the corresponding Γ can be read from the Smith chart [see Problem 6.203]. What is the algebraic alternative?

▌ $$\Gamma = (z_{\text{in}} - 1)/(z_{\text{in}} + 1).$$

6.203 Given a *lossless* transmission line ($Z_C = R_C$) of length ζ, having a load impedance Z_L and a phase constant β. Use the Smith chart to find $Z_{\text{in}}(0)$.

▌ See Fig. 6-64. First locate $z_{\text{in}}(\zeta) = Z_L/R_C$ on the Smith chart; the polar coordinates of this point will be $|\Gamma_L|$, θ_{Γ_L}. By Problem 6.68, $\Gamma(0) = \Gamma_L e^{-j2\beta\zeta}$. Hence the point $z_{\text{in}}(0)$ has polar coordinates $|\Gamma_L|$, $\theta_{\Gamma_L} - 2\beta\zeta$. Knowing $z_{\text{in}}(0)$, we have at once $Z_{\text{in}}(0) = R_C z_{\text{in}}(0)$.

6.204 A lossless transmission line of length 10 m is immersed in polyethylene ($\epsilon_r = 2.25$, $\mu_r = 1$) and operated at 34 MHz. Given the load impedance $Z_L = (50 + j100)\ \Omega$ and the characteristic resistance $R_C = 50\ \Omega$, determine the input impedance to the line.

▌ Proceed as in Problem 6.203, with the necessary clockwise rotation of $2\beta\zeta = 4\pi\zeta_e$ accomplished via the

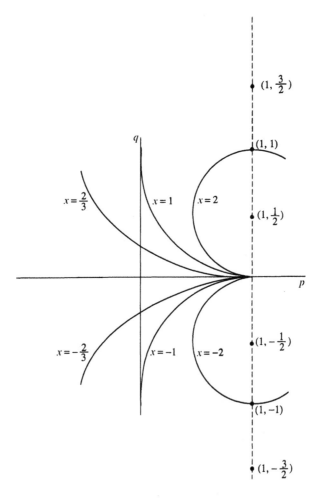

Fig. 6-62

WAVELENGTHS TOWARD GENERATOR, or T.G., scale, found on the outer periphery of the Smith chart. (The factor 4π is absorbed in the scale.) Thus

$$\zeta_e = \frac{\zeta}{\lambda} = \frac{\zeta f \sqrt{\epsilon_r}}{u_0} = \frac{(10)(34 \times 10^6)(1.5)}{3 \times 10^8} = 1.7$$

On the Smith chart (Fig. 6-65), start at $z_{in}(\zeta) = (50 + j100)/50 = 1 + j2$, with T.G.-coordinate 0.1875. Note that the T.G. scale runs from 0 to 0.5—consistent with the fact (Problem 6.94) that electrical properties repeat every $\lambda/2$ along the line. Thus, make a clockwise rotation of $1.7 - 3(0.5) = 0.2$ T.G., to find, at 0.3875 T.G., $z_{in}(0) = 0.29 - j0.82$. Then,

$$Z_{in}(0) = (50)(0.29 - j0.82) = 14.5 - j41 \quad \Omega$$

6.205 Determine $\Gamma_L = \Gamma(\zeta)$ for the line of Problem 6.204.

❚ From Fig. 6-65, $\Gamma_L = 0.71 \underline{/45°}$.

6.206 Determine $\Gamma(0)$ for the line of Problem 6.204.

❚ From Fig. 6-65, $\Gamma(0) = 0.71 \underline{/-99°}$.

6.207 The input impedance to a line is $Z_{in} = (30 - j40) \, \Omega$ and the load impedance is $Z_L = (20 + j40) \, \Omega$. If the line has characteristic resistance $R_C = 100 \, \Omega$ and velocity of propagation $u = 250 \, \text{m}/\mu s$ and is operated at 30 MHz, what is the minimum length of the line?

❚ On a Smith chart such as Fig. 6-65, the normalized impedances plot as
$$z_{in}(0) = 0.3 - j0.4 \quad (0.435 \text{ T.G.})$$
$$z_{in}(\zeta) = 0.2 + j0.4 \quad (0.062 \text{ T.G.})$$
The smallest possible electrical length is thus $(\zeta_e)_{min} = 0.435 - 0.062 = 0.373$, and so

$$\zeta_{min} = \frac{u}{f}(\zeta_e)_{min} = 3.11 \text{ m}$$

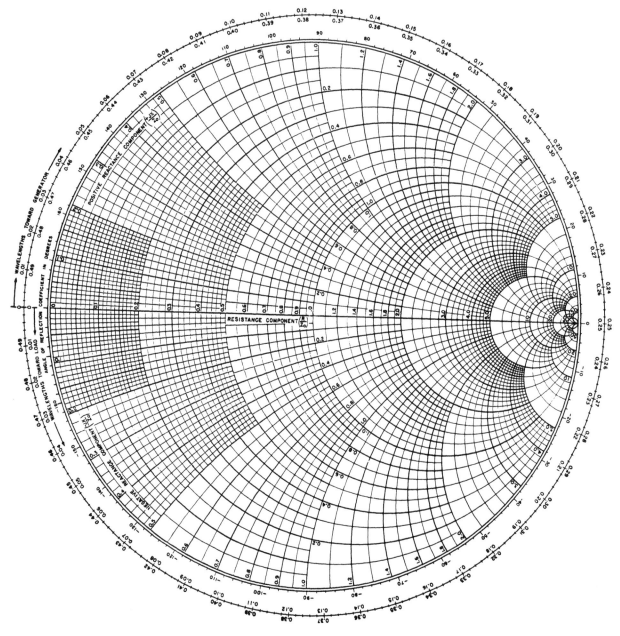

Fig. 6-63

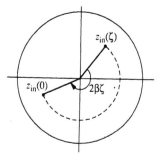

Fig. 6-64

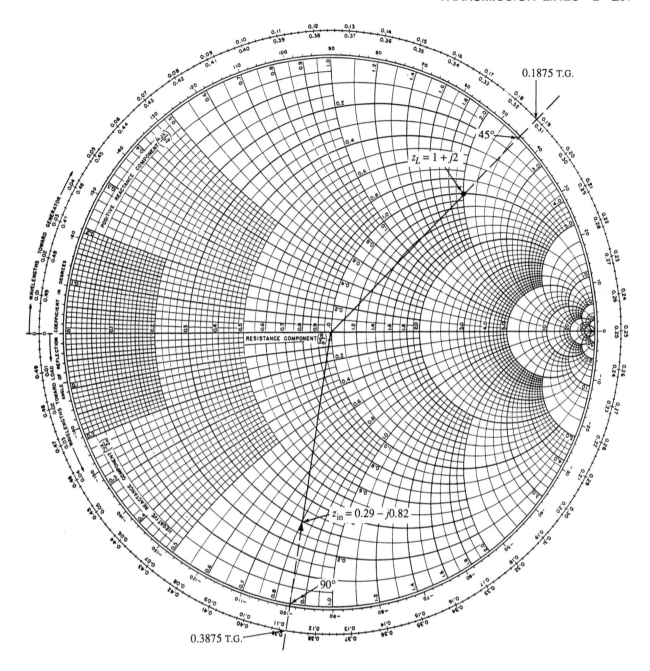

0.1875 T.G.

$45°$

$z_L = 1 + j2$

$z_{in} = 0.29 - j0.82$

$90°$

0.3875 T.G.

Fig. 6-65

6.208 Show that at a voltage maximum on a lossless transmission line, $z_{in} = $ VSWR, a positive real number; i.e., z_{in} is purely resistive.

▌ By definition, and (1) of Problem 6.68,

$$z_{in} = \frac{1 + \Gamma}{1 - \Gamma} = \frac{1 + \Gamma_L e^{-j2\beta d}}{1 - \Gamma_L e^{-j2\beta d}} \qquad (d \equiv \zeta - z)$$

But the crank diagram (Fig. 6-45) shows that voltage maxima occur at distances d such that

$$2\beta d = \theta_L + 2\pi n \qquad (n \text{ integral})$$

Thus

$$(z_{in})_{\text{volt.max.}} = \frac{1 + |\Gamma_L| e^{j\theta_L} e^{-j\theta_L}}{1 - |\Gamma_L| e^{j\theta_L} e^{-j\theta_L}} = \frac{1 + |\Gamma_L|}{1 - |\Gamma_L|} = \text{VSWR}$$

where the last step follows from (1) of Problem 6.100.

6.209   Show that at a voltage minimum on a lossless line, $z_{\text{in}} = 1/\text{VSWR}$.

▮ Proof as in Problem 6.208; this time, $2\beta d = \theta_L + (2n + 1)\pi$.

6.210   Explain how the VSWR on a lossless transmission line may be obtained from the Smith chart.

▮ First, plot the normalized load impedance $z_L = z_{\text{in}}(\zeta)$, which determines $|\Gamma_L|$. Rotating this vector until it crosses the horizontal axis of the Smith chart results in an input impedance that is purely real. Two possible values, separated by $\lambda/4$, may be located; the larger value is the VSWR, and the smaller is $1/\text{VSWR}$. The VSWR may also be found directly by using the scale at the bottom of the chart labeled "Standing-Wave Vol. Ratio." If $|\Gamma|$ is transferred from the chart to this scale, the VSWR may be read directly.

6.211   Consider a lossless transmission line terminated in $Z_L = (150 - j200)\,\Omega$. Assuming that the line has characteristic resistance $R_C = 100\,\Omega$, find the VSWR and the shortest distance from the load at which the input impedance to the line is purely resistive.

▮ Follow Problem 6.210. The normalized load impedance, $z_L = 1.5 - j2$, corresponds to 0.302 T.G. Now, on the T.G. scale the input impedance to the line appears purely resistive at 0.5. Thus, the shortest distance from the load at which this occurs is $0.5 - 0.302 = 0.198$. At this point, the normalized input impedance is

$$z_{\text{in}} = \frac{1}{\text{VSWR}} = 0.22 \quad \text{whence} \quad \text{VSWR} = 4.55$$

(See Fig. 6-66.)

6.212   Given Z, a complex number, its reciprocal Y may be obtained from the Smith chart as follows: Choose a characteristic resistance R_C such that $z \equiv Z/R_C$ may be accurately plotted on the Smith chart. Rotate z one-half revolution ($\lambda/4$) around the chart and read the value of $y = 1/z$. Then $Y = y/R_C$.
Use the same procedure to find the admittance corresponding to the impedance $Z = (100 + j150)\,\Omega$.

▮ The choice $R_C = 100\,\Omega$ seems reasonable, resulting in $z = 1 + j1.5$. The results are shown in Fig. 6-67. The normalized admittance is $y = 0.3 - j0.47$, or $Y = y/R_C = (3 - j4.7)\,\text{mS}$. The exact value is $(3.08 - j4.62)\,\text{mS}$.

6.213   An unknown impedance is to be measured by a device connected to the impedance by a cable of characteristic resistance R_C, as shown in Fig. 6-68(a). Because the connection cable is a significant portion of a wavelength at the measurement frequency, the impedance being measured is not the desired impedance; we must have a way of "calibrating out" the connection cable. A method of doing this consists in a short-circuit test, as shown in Fig. 6-68(b); it is found that $Z_{\text{inSC}} = j100\,\Omega$. Find the unknown impedance Z, assuming that the characteristic resistance of the line is $R_C = 80\,\Omega$ and the impedance measured with Z attached is $(50 - j200)\,\Omega$.

▮ The normalized input impedance is $z_{\text{inSC}} = j100/80 = j1.25$. Locating this on the Smith chart, we determine the electrical length of the line to be $0.5 - 0.3575 = 0.1425$, as shown in Fig. 6-68(c). The normalized measured impedance is $z_{\text{in}} = 0.625 - j2.5$. Rotating this counterclockwise (toward the load) through the electrical length of the line [see Fig. 6-68(d)], we arrive at $0.193 + 0.1425 = 0.3355$ T.L. and obtain $z = 0.36 + j1.62$. The unknown impedance then becomes

$$Z = zR_C = 28.8 + j129.6 \quad \Omega$$

and the effect of the connection cable has been "calibrated out."

6.214   Determine from the Smith chart the input impedance, VSWR, and voltage reflection coefficient at the load for a transmission line having $Z_L = (25 - j50)\,\Omega$; $R_C = 50\,\Omega$; $\zeta = 0.4\,\lambda$.

▮ Starting at $z_L = Z_L/R_C = 0.5 - j1$, we obtain, from Fig. 6-69(a),

$$z_{\text{in}} = 3.65 - j1.4 \qquad \text{VSWR} = 4.3$$
$$Z_{\text{in}} = 182.5 - j70 \quad \Omega \qquad \Gamma_L = 0.62\,\underline{/-83°}$$

6.215   Repeat Problem 6.214 for a line with $Z_L = (100 + j50)\,\Omega$; $R_C = 100\,\Omega$; $\zeta = 1.3\,\lambda$.

▮ Now $z_L = 1 + j0.5$, and Fig. 6-69(b) yields

$$z_{\text{in}} = 0.67 + j0.21 \qquad \text{VSWR} = 1.61$$
$$Z_{\text{in}} = 67 - j21 \quad \Omega \qquad \Gamma_L = 0.24\,\underline{/-140°}$$

6.216   Determine the load impedance, VSWR, and the load reflection coefficient for a transmission line with $Z_{\text{in}} = (30 - j100)\,\Omega$; $R_C = 50\,\Omega$; $\zeta = 0.4\,\lambda$. Use the Smith chart.

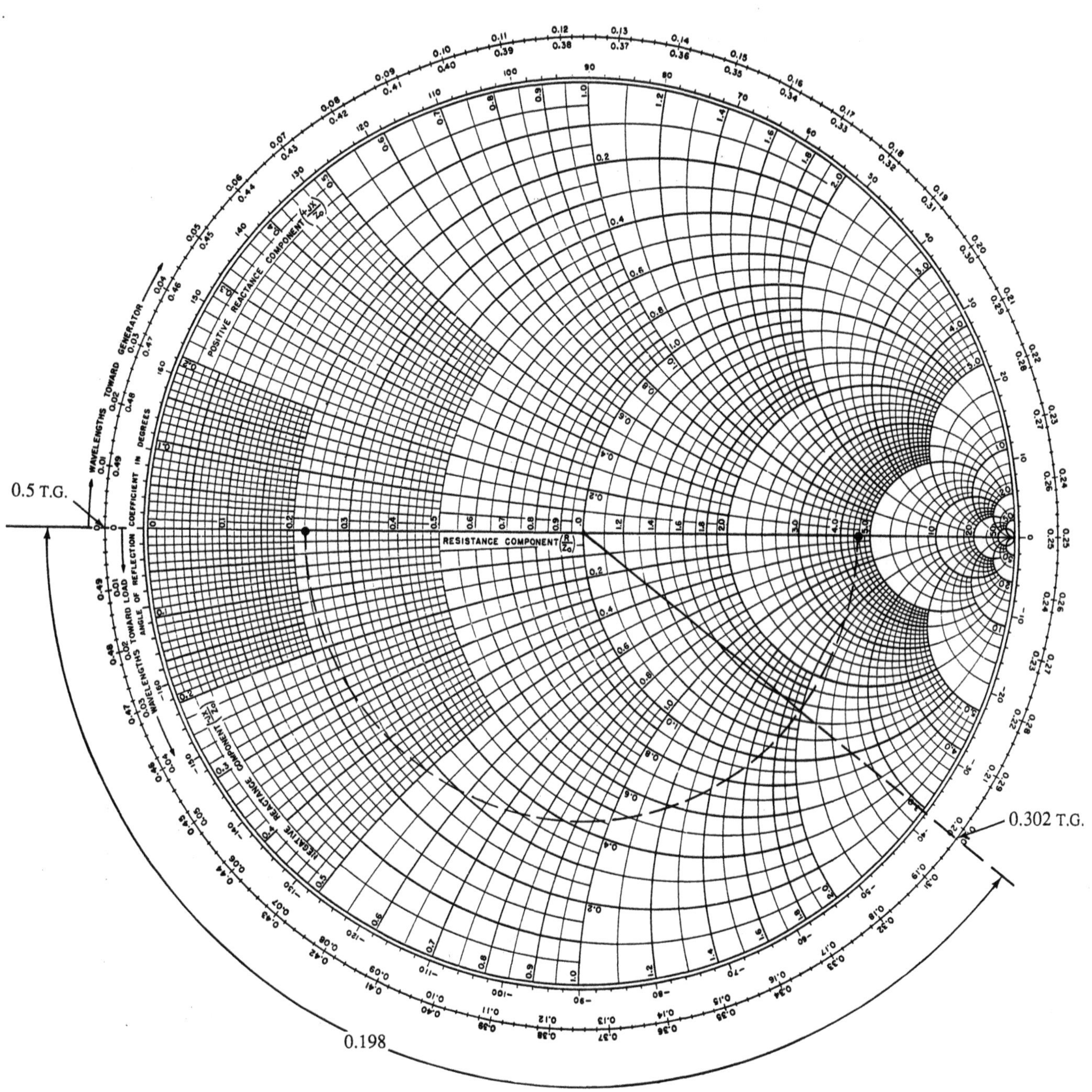

Fig. 6-66

▌ The normalized input impedance is $z_{in} = 0.6 - j2$; thus, from Fig. 6-70,

$$z_L = 0.145 - j0.545 \qquad \text{VSWR} = 9$$
$$Z_L = 7.25 - j27.25 \quad \Omega \qquad \Gamma_L = 0.8 \underline{/-122°}$$

6.217 Repeat Problem 6.216 for a line with $Z_{in} = 50 \ \Omega$; $R_C = 75 \ \Omega$; $\zeta = 1.3 \ \lambda$.

▌ Now $z_{in} = 0.667 + j0$, and Fig. 6-71 gives

$$z_L = 1.35 + j0.34 \qquad \text{VSWR} = 1.5$$
$$Z_L = 101.25 + j25.5 \quad \Omega \qquad \Gamma_L = 0.205 \underline{/36°}$$

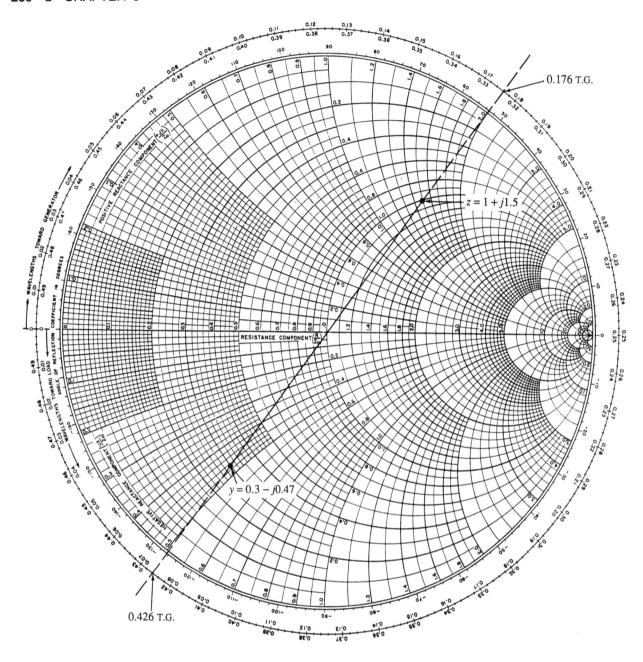

Fig. 6-67

6.218 A transmission line has $Z_{\text{in}} = j20\ \Omega$; $Z_L = j50\ \Omega$; $R_C = 100\ \Omega$. Find its shortest length, the VSWR, and Γ_L.

▌ The normalized input impedance is $z_{\text{in}} = j0.2$. Thus, from Fig. 6-72,

$$z_L = j0.5 \qquad\qquad \text{VSWR} = \infty$$
$$\zeta_e = 0.4265 - 0.031 = 0.396 \qquad\qquad \Gamma_L = 1\ \underline{/127°}$$

6.219 Repeat Problem 6.218 for a line with $Z_{\text{in}} = (50 - j200)\ \Omega$; $Z_L = (60 - j45)\ \Omega$; $R_C = 100\ \Omega$.

▌ Now $z_{\text{in}} = 0.5 - j2.0$; so that, from Fig. 6-73,

$$z_L = 0.6 - j0.45 \qquad\qquad \text{VSWR} = 2.17$$
$$\zeta_e = 0.5 - 0.226 + 0.089 = 0.363 \qquad\qquad \Gamma_L = 0.37\ \underline{/-115.5°}$$

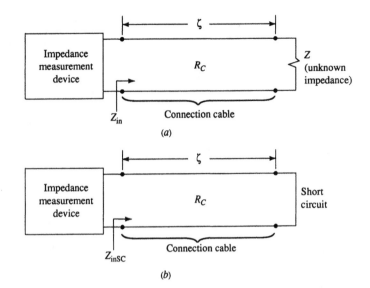

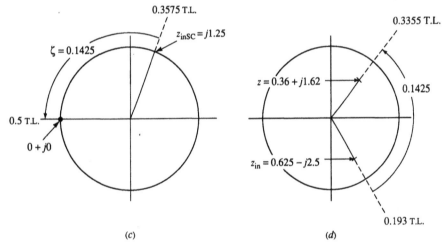

Fig. 6-68

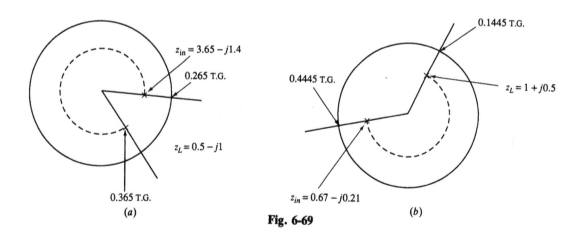

Fig. 6-69

6.220 Using the Smith chart, find the inverse of $C = 100 + j30$.

❚ Choose $R_C = 50$; then $c = 2 + j0.6$, and, from Fig. 6-74,

$$\frac{1}{c} = 0.455 - j0.14$$

$$\frac{1}{C} = \frac{1/c}{R_C} = \frac{0.455 - j0.14}{50} = (9.1 - j2.8) \times 10^{-3}$$

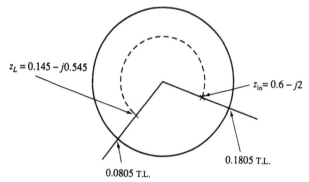

$z_L = 0.145 - j0.545$

$z_{in} = 0.6 - j2$

0.0805 T.L.

0.1805 T.L.

Fig. 6-70

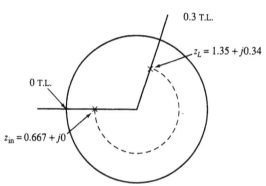

0.3 T.L.

$z_L = 1.35 + j0.34$

0 T.L.

$z_{in} = 0.667 + j0$

Fig. 6-71

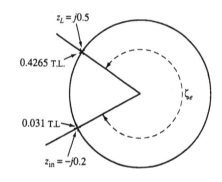

$z_L = j0.5$

0.4265 T.L.

ζ_e

0.031 T.L

$z_{in} = -j0.2$

Fig. 6-72

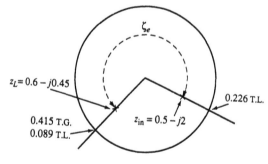

ζ_e

$z_L = 0.6 - j0.45$

0.226 T.L.

0.415 T.G.
0.089 T.L.

$z_{in} = 0.5 - j2$

Fig. 6-73

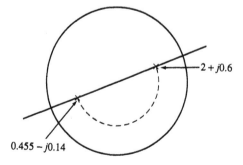

$2 + j0.6$

$0.455 - j0.14$

Fig. 6-74

6.221   Repeat Problem 6.220 for $C = (25 - j50) \times 10^{-3}$.

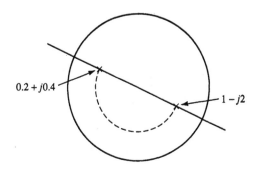

Fig. 6-75

▌ Choose $R_C = 25 \times 10^{-3}$; then $c = 1 - j2$. From Fig. 6-75,
$$\frac{1}{c} = 0.2 + j0.4 \quad \text{and} \quad \frac{1}{C} = \frac{1/c}{R_C} = 8 + j16$$

6.222   A lossless 100-Ω transmission line is connected between a 50-Ω source and a load of $Z_L = (100 + j70)$ Ω.
(a) From the Smith chart, determine a length of this line such that maximum power is delivered from the source to the load. **(b)** Can you always find such a length of line, for any load impedance?

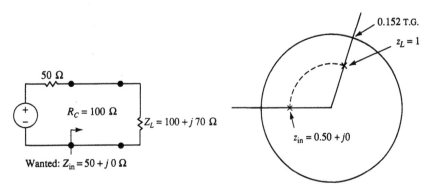

Fig. 6-76

▌ The *lossless* line and a sketch of the Smith chart are shown in Fig. 6-76.
(a) $\qquad\qquad\qquad\qquad (P_{av})_{line} = (P_{av})_{load} \qquad z_L = 1 + j0.70 \qquad \zeta = 0.152\lambda$
(b) No: not every z_L can be rotated into $0.50 + j0$.

6.223   We want to determine from the Smith chart the value of an unknown impedance Z_L, attached to a length of transmission line having a characteristic resistance of 100 Ω. Removing the load yields an input impedance of $-j80$ Ω. With the unknown impedance attached, the input impedance is $(30 + j40)$ Ω.

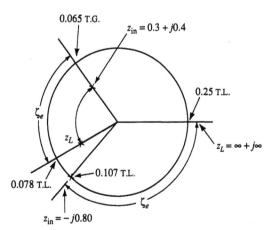

Fig. 6-77

∎ See Fig. 6-77.

$$z_{inOC} = -j0.80 \quad (0.107 \text{ T.L.}) \qquad \text{whence} \qquad \zeta_e = 0.25 - 0.107 = 0.143$$

With load,

$$z_{in} = 0.3 + j0.4 \qquad z_L = 0.32 - j0.49 \qquad Z_L = (32 - j49) \ \Omega$$

6.224 Given the unmatched $(Z_L \neq R_C)$ transmission line of Fig. 6-78(a), use the Smith chart to find the distance d from the load at which the combined input admittance of the line and a parallel-connected auxiliary network has the value $1/R_C$ (i.e., $Z_{in}(d) = R_C$ for the combination), as in Fig. 6-78(c). The line of Fig. 6-78(c) is said to be *stub-matched* (at $z = \zeta - d$).

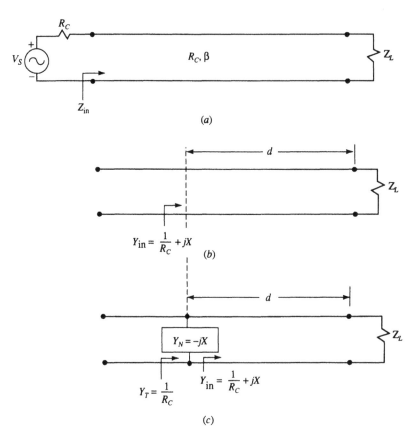

(a)

(b)

(c)

Fig. 6-78

∎ The distance d is determined by the condition $\text{Re}\,[Y_{in}] = 1/R_C$ [Fig. 6-78(b)]; thus

$$Y_{in} = \frac{1}{R_C} + jX \qquad \text{or} \qquad y_{in} = Y_{in}R_C = 1 + jXR_C \tag{1}$$

If we plot the normalized load admittance $y_L = 1/z_L$ on a Smith chart, then d is the distance, in wavelengths, that y_L must be rotated to hit the unity-real-part circle. All that remains is to attach a network having admittance $Y_N = jX$ in parallel with the line at $z = \zeta - d$, to cancel the reactive portion of Y_{in}:

$$Y_T = Y_{in} + Y_N + \left(\frac{1}{R_C} + jX\right) - jX = \frac{1}{R_C}$$

At all points to the left of this point of attachment the line appears matched.

6.225 Discuss the nature of the stub-matching network of Problem 6.224.

∎ First, the network must provide an admittance that is purely reactive. A way of doing this would be to parallel the line with either a capacitor or an inductor having the appropriate value of admittance. The difficulty with this idea is that these lumped elements behave as simple capacitors or inductors only at medium frequencies. In the higher-frequency ranges it is more desirable to use a section of transmission line called a *stub* or *stub tuner* for this matching network. A section of line having R_C, β, and length l, with a short-circuit load, will also satisfy the requirements, since the input admittance is

$$Y_{NSC} = -j\frac{1}{R_C \tan \beta l} \tag{1}$$

Similarly, the input admittance of a line having R_C, β, and length l, with an open-circuit load, is

$$Y_{NOC} = j\frac{\tan \beta l}{R_C} \qquad (2)$$

6.226 Explain how short-circuited stub-matching (Problem 6.225) is done, using a Smith chart.

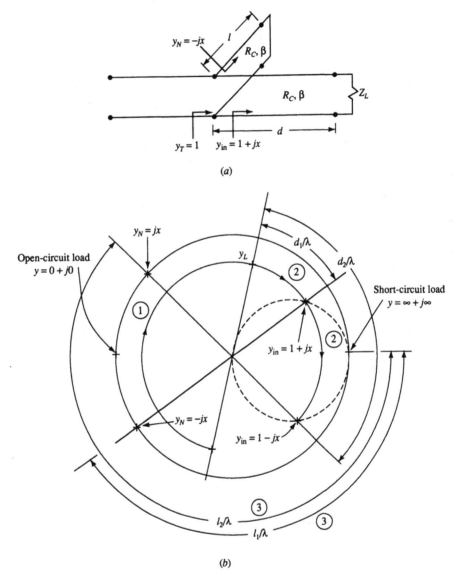

(a)

(b)   **Fig. 6-79**

▌ Referring to Fig. 6-79, we first plot the normalized load impedance $z_L = Z_L/R_C$ on the Smith chart and rotate 180° to find the normalized load admittance $y_L = Y_L R_C$, as shown in Fig. 6-79(b) (step 1). Then we move a distance d_1/λ (toward the generator) until we intersect the unity-real-part circle (step 2). At this point the normalized input impedance to the line is $y_{\text{in}} = 1 + jx$. In other words, the actual admittance is given by (1) of Problem 6.224. We could also have rotated further a total distance d_2/λ and found an input admittance of $y_{\text{in}} = 1 - jx$, which is also shown in Fig. 6-79(b).

Now we need to find the length of a short-circuited stub which will cancel the reactive part of this input admittance. If we place the stub at distance d_1 from the load, the stub must have a normalized input admittance of $y_N = -jx$; denote this length of stub as l_1. If we place the stub at distance d_2 from the load, the stub must have a normalized input admittance of $y_N = +jx$; denote this length of stub as l_2. Plot the stub load admittance (normalized), which is $\infty + j\infty$, as shown in Fig. 6-79(b). We move toward the generator along the stub until we find this reactance (on the outer circle of the chart) (step 3). For d_1 we need $y_N = -jx$, or stub length l_1. If we had chosen d_2, we would need $y_N = jx$, or a stub length l_2. Since the stub is assumed lossless, maximum power will be transferred from the source having $Z_S = R_C$ to the load Z_L.

6.227 Consider a transmission line having $R_C = 50\ \Omega$ and $Z_L = (25 - j50)\ \Omega$. Determine two locations and lengths of short-circuited stubs to match this line. Use the Smith chart.

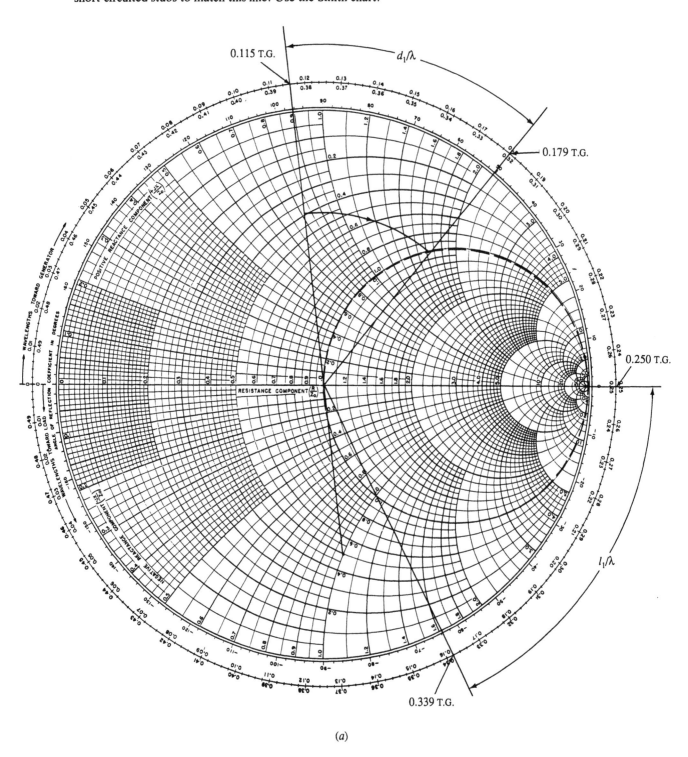

(a)

Fig. 6-80

▮ Refer to Problem 6.226. The normalized impedance, $z_L = 0.5 - j1.0$, is plotted in Fig. 6-80(a). The normalized load admittance is $y_L = 0.4 + j0.8$, and it occurs at 0.115 on the T.G. scale. Rotating to the unity-real-part circle produces $y_{in} = 1 + j1.6$, located at 0.179 on the T.G. scale, Thus, one solution is
$$d_1 = 0.179\lambda - 0.115\lambda = 0.064\lambda \qquad \text{and} \qquad l_1 = 0.339\lambda - 0.250\lambda = 0.089\lambda$$

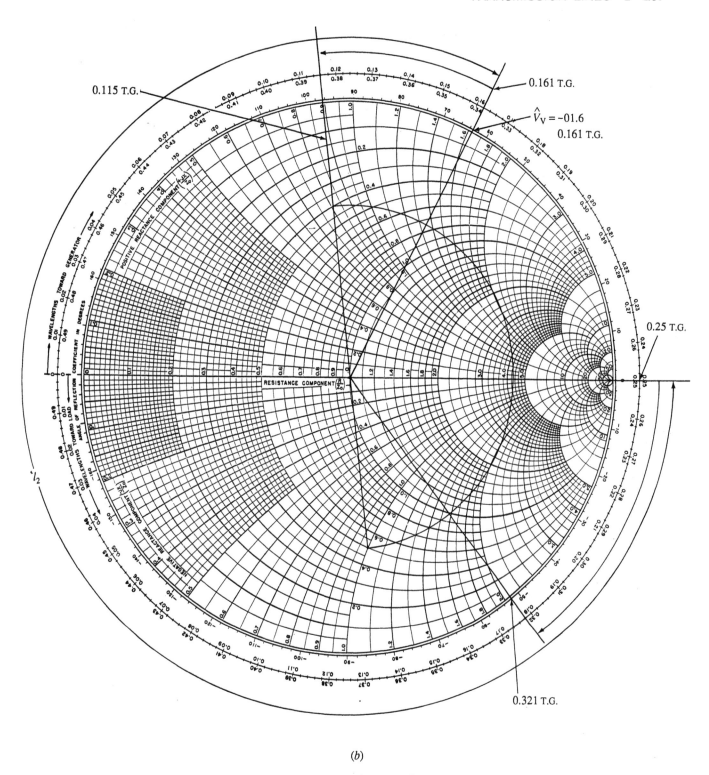

0.115 T.G.

0.161 T.G.

$\widehat{V}_V = -01.6$

0.161 T.G.

0.25 T.G.

0.321 T.G.

(b)

Fig. 6-80—(continued)

The second solution is illustrated in Fig. 6-80(b):
$$d_2 = 0.321\lambda - 0.115\lambda = 0.206\lambda \qquad \text{and} \qquad l_2 = 0.161\lambda + 0.25\lambda = 0.411\lambda$$

6.228 An air-filled, 100-Ω, parallel-wire line is driven by a 300-MHz source and terminated in a mismatched load of $(40 - j20)$ Ω. To effect maximum power transfer from the source to the load either an inductor or a capacitor is to be connected across the line at a distance d from the load. Determine the shortest distance d at which to place this element; also determine the type of element and its value. Use the Smith chart.

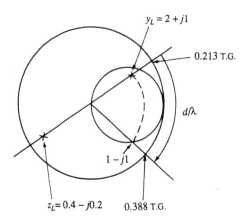

Fig. 6-81

▮ The normalized load impedance is $z_L = 0.4 - j0.2$; from Fig. 6-81,
$$d = 0.338\lambda - 0.213\lambda = 0.125\lambda = 0.125 \text{ m}$$

We require
$$y_N = j1 \quad \text{or} \quad Y_N = j10^{-2} \text{ S}$$
This admittance is provided by a capacitor such that $j\omega C = j10^{-2}$, or $C = 5.305$ pF.

6.229 Repeat Problem 6.228 assuming that the load impedance is changed to $(25 - j25)\ \Omega$ and the line characteristic resistance is changed to $50\ \Omega$.

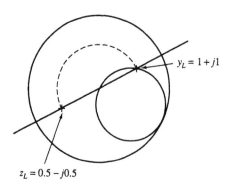

Fig. 6-82

▮ Now $z_L = 0.5 - j0.5$; from Fig. 6-82, $y_L = 1 + j1$ and $d = 0$. We require $y_N = -j1$, $Y_N = -j2 \times 10^{-2}$ S. Thus, we want an inductor, such that
$$-\frac{j}{\omega L} = -j2 \times 10^{-2} \quad \text{or} \quad L = 26.53 \text{ nH}$$

Note Where confusion is unlikely, problems of Chapters 7, 8, and 9 employ the same symbols for time-sinusoidal vectors and their phasor representations.

7.1 Figure 7-1 represents a rectangular waveguide, in the lossless interior of which the fields are governed by

$$\nabla^2 \mathbf{E} = -\omega^2 \mu \epsilon \mathbf{E} \qquad (1)$$
$$\nabla^2 \mathbf{H} = -\omega^2 \mu \epsilon \mathbf{H} \qquad (2)$$

Assuming the field components $E_x(x, y, z), \ldots, H_z(x, y, z)$ all to be of the form $F(x, y)e^{-\gamma z}$, obtain a partial differential equation for the functions F.

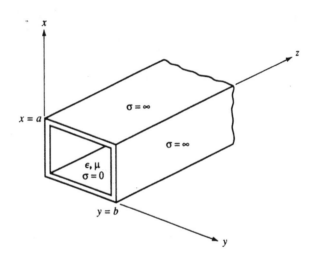

Fig. 7-1

$\blacksquare$
$$\frac{\partial^2 F}{\partial x^2} + \frac{\partial^2 F}{\partial y^2} = -(\gamma^2 + \omega^2 \mu \epsilon)F \qquad (3)$$

7.2 Defining the transverse vector operator

$$\boldsymbol{\nabla}_T \equiv \mathbf{a}_x \frac{\partial}{\partial x} + \mathbf{a}_y \frac{\partial}{\partial y}$$

replace equations (3) of Problem 7.1 by a pair of vector equations.

$\blacksquare$ With the usual notation $\nabla_T^2 \equiv \boldsymbol{\nabla}_T \cdot \boldsymbol{\nabla}_T$,

$$\nabla_T^2 \mathbf{E} = -(\gamma^2 + \omega^2 \mu \epsilon)\mathbf{E} \qquad (1)$$
$$\nabla_T^2 \mathbf{H} = -(\gamma^2 + \omega^2 \mu \epsilon)\mathbf{H} \qquad (2)$$

7.3 Apply Faraday's law and Ampère's law to the interior of the waveguide of Fig. 7-1 and obtain coupled partial differential equations governing the components of the **E**- and **H**-fields.

$\blacksquare$ Faraday's and Ampère's laws in the interior of the guide may be expressed as

$$\nabla \times \mathbf{E} = -j\omega\mu\mathbf{H} \qquad \nabla \times \mathbf{H} = j\omega\epsilon\mathbf{E}$$

Substituting the assumed form of the field components given in Problem 7.1 yields

$$\frac{\partial E_z}{\partial y} + \gamma E_y = -j\omega\mu H_x$$

$$-\gamma E_x - \frac{\partial E_z}{\partial x} = -j\omega\mu H_y \qquad (1)$$

$$\frac{\partial E_y}{\partial x} - \frac{\partial E_x}{\partial y} = -j\omega\mu H_z$$

$$\frac{\partial H_z}{\partial y} + \gamma H_y = j\omega\epsilon E_x$$

$$-\gamma H_x - \frac{\partial H_z}{\partial x} = j\omega\epsilon E_y \qquad (2)$$

$$\frac{\partial H_y}{\partial x} - \frac{\partial H_x}{\partial y} = j\omega\epsilon E_z$$

7.4   Combine the three equations (*1*) of Problem 7.3 into a single vector equation; do the same for the three equations (*2*).

❚ Multiply the first equation (*1*) by $\mathbf{a}_x$, the second by $\mathbf{a}_y$, the third by $\mathbf{a}_z$, and add:
$$\mathbf{\nabla}_T \times \mathbf{E} - \gamma\mathbf{a}_z \times \mathbf{E} = -j\omega\mu\mathbf{H} \qquad (1)$$
where the operator $\mathbf{\nabla}_T$ is as defined in Problem 7.2. Similarly,
$$\mathbf{\nabla}_T \times \mathbf{H} - \gamma\mathbf{a}_z \times \mathbf{H} = j\omega\epsilon\mathbf{E} \qquad (2)$$

7.5   Show that all field components may be derived from E_z and H_z.

❚ From the first equation (*1*) and the second equation (*2*) of Problem 7.3,
$$E_y = -j\frac{\omega\mu}{\gamma} H_x - \frac{1}{\gamma}\frac{\partial E_z}{\partial y} \qquad \text{and} \qquad H_x = -j\frac{\omega\epsilon}{\gamma} E_y - \frac{1}{\gamma}\frac{\partial H_z}{\partial x}$$
Combining these gives
$$E_y = \frac{1}{\gamma^2 + \omega^2\mu\epsilon}\left(j\omega\mu\frac{\partial H_z}{\partial x} - \gamma\frac{\partial E_z}{\partial y}\right) \qquad (1)$$
Similarly, by combining the other relations given in Problem 7.3, we obtain
$$E_x = -\frac{1}{\gamma^2 + \omega^2\mu\epsilon}\left(j\omega\mu\frac{\partial H_z}{\partial y} + \gamma\frac{\partial E_z}{\partial x}\right) \qquad (2)$$
$$H_y = -\frac{1}{\gamma^2 + \omega^2\mu\epsilon}\left(j\omega\epsilon\frac{\partial E_z}{\partial x} + \gamma\frac{\partial H_z}{\partial y}\right) \qquad (3)$$
$$H_x = \frac{1}{\gamma^2 + \omega^2\mu\epsilon}\left(j\omega\epsilon\frac{\partial E_z}{\partial y} - \gamma\frac{\partial H_z}{\partial x}\right) \qquad (4)$$

7.6   What conclusion can be drawn from Problems 7.5 and 7.1?

❚ Two solutions, E_z and H_z, of (*3*) of Problem 7.1 suffice completely to describe the rectangular waveguide.

7.7   Set up the solution of (*3*) of Problem 7.1 by separation of variables.

❚ Substituting $F(x, y) = X(x)Y(y)$ in (*3*) yields, after division by XY,
$$\frac{1}{X(x)}\frac{d^2X(x)}{dx^2} + \frac{1}{Y(y)}\frac{d^2Y(y)}{dy^2} + (\gamma^2 + \omega^2\mu\epsilon) = 0 \qquad (1)$$
Since X is a function only of x, and Y is a function only of y, each term on the left side of (*1*) must be constant; thus
$$\frac{1}{X(x)}\frac{d^2X(x)}{dx^2} = -M^2 \qquad (2)$$
$$\frac{1}{Y(y)}\frac{d^2Y(y)}{dy^2} = -N^2 \qquad (3)$$
(the form of these constants is chosen to simplify the solution), where the separation constants and γ are constrained to obey
$$M^2 + N^2 - \gamma^2 = \omega^2\mu\epsilon \qquad (4)$$

7.8   Give general solutions to (*2*) and (*3*) of Problem 7.7.

❚ $X(x) = A \sin Mx + B \cos Mx \quad (M \neq 0)$ or $X(x) = A_0 x + B_0 \quad (M = 0)$
$Y(y) = C \sin Ny + D \cos Ny \quad (N \neq 0)$ or $Y(y) = C_0 y + D_0 \quad (N = 0)$
for undetermined constants $A, B, C, D; A_0, B_0, C_0, D_0$.

7.9   Using the results of Problem 7.8, write the forms of solutions for E_z and H_z. Divide these solutions into two fundamental types, or *modes*.

❚ By (*4*) of Problem 7.7, γ depends on M and N; we write $\gamma = \gamma(M, N)$. Then there are four forms of solutions for E_z:

$$(E_z)_{0,0} = (A_0 x + B_0)(C_0 y + D_0)e^{-\gamma(0,0)z} \tag{1}$$

$$(E_z)_{0,N} = (A_0 x + B_0)(C \sin Ny + D \cos Ny)e^{-\gamma(0,N)z} \tag{2}$$

$$(E_z)_{M,0} = (A \sin Mx + B \cos Mx)(C_0 y + D_0)e^{-\gamma(M,0)z} \tag{3}$$

$$(E_z)_{M,N} = (A \sin Mx + B \cos Mx)(C \sin Ny + D \cos Ny)e^{-\gamma(M,N)z} \tag{4}$$

Exactly the same four forms are possible for H_z.

We distinguish two cases:

1. $E_z = 0$, $H_z \neq 0$. Because the electric fields for this case are transverse to the guide axis, these field structures are called *transverse electric* (TE) *modes*.
2. $H_z = 0$, $E_z \neq 0$. Similarly, these field structures are called *transverse magnetic* (TM) *modes*.

Note that setting $E_z = H_z = 0$ (*the* TEM *mode*) in Problem 7.5 appears to render all field components zero. However, for the special choice $\gamma = j\omega\sqrt{\mu\epsilon}$, (2), (3), and (4) become indeterminate. Thus, we cannot immediately conclude that the TEM mode is not a possible mode of propagation.

7.10 Show that the TEM mode cannot exist within a hollow waveguide.

▮ Suppose, on the contrary, the TEM mode existed. In this case the magnetic field must lie solely in the transverse (xy) plane. The magnetic field lines must form closed paths in this transverse plane, since $\nabla \cdot \mathbf{H} = 0$. From Ampère's law, the integral of this transverse magnetic field around these closed paths must yield the axial (z-directed) conduction or displacement current. But $E_z = 0$ for the TEM mode, so that no axial displacement current can exist. Also, since there is no center conductor, no axial conduction current can exist.

7.11 Specify the boundary conditions on E_z in TM modes of a rectangular waveguide.

▮ The tangential electric field must be zero at the walls of the guide; that is,

$$E_z(0, y, z) = E_z(a, y, z) = E_z(x, 0, z) = E_z(x, b, z) = 0 \tag{1}$$

These boundary conditions immediately disallow solution forms (1), (2), and (3) of Problem 7.9; moreover, $(E_z)_{M,N}$ will obey (1) only if

$$B = 0 \qquad Ma = m\pi \quad (m = 1, 2, 3, \ldots)$$
$$D = 0 \qquad Nb = n\pi \quad (n = 1, 2, 3, \ldots)$$

We emphasize that neither $m = 0$ nor $n = 0$ is allowed by the boundary conditions for a TM mode.

7.12 Display the superposition solution for E_z arising from Problem 7.11.

▮

$$E_z(x, y, z) = \sum_{m=1}^{\infty} \sum_{n=1}^{\infty} E_{mm} \sin\frac{m\pi x}{a} \sin\frac{n\pi y}{b} e^{-\gamma_{mn}z} \tag{1}$$

where

$$\gamma_{mn} \equiv \sqrt{\left(\frac{m\pi}{a}\right)^2 + \left(\frac{n\pi}{b}\right)^2 - \omega^2\mu\epsilon} \tag{2}$$

7.13 Given a pair of eigenvalues m, n (Problem 7.12), what is the condition (on ω) for wave propagation in the z-direction?

▮ For wave propagation γ_{mn} must be an imaginary number; or

$$\omega^2\mu\epsilon > \left(\frac{m\pi}{a}\right)^2 + \left(\frac{n\pi}{b}\right)^2 \tag{1}$$

Frequencies obeying (1) are said to excite "the TM$_{mn}$ mode."

7.14 Describe the (m, n) mode when (1) of Problem 7.13 is not obeyed.

▮ In this case $\gamma_{mn} = \alpha_{mn}$, a positive real number, and, in the time domain,

$$E_z = |E_{mn}| \sin\frac{m\pi x}{a} \sin\frac{n\pi y}{b} e^{-\alpha_{mn}z} \cos(\omega t + \theta_{mn})$$

It is seen that E_z does not propagate but decays exponentially with z (an *evanescent mode*).

7.15 Give the *cutoff frequency* (the boundary frequency between attenuation and wave propagation) for the TM$_{mn}$ mode.

▮ Set $\gamma_{mn} = 0$ to find

$$f_{c,mn} = \frac{1}{2\pi\sqrt{\mu\epsilon}} \sqrt{\left(\frac{m\pi}{a}\right)^2 + \left(\frac{n\pi}{b}\right)^2} = \frac{u}{2} \sqrt{\frac{m^2}{a^2} + \frac{n^2}{b^2}} \tag{1}$$

where $u = 1/\sqrt{\mu\epsilon}$.

7.16   Express the phase constant β_{mn} for the TM_{mn} mode in the waveguide of Problem 7.11 in terms of the cutoff frequency.

    ❚ From (2) of Problem 7.12,

$$\beta_{mn} = \sqrt{\omega^2\mu\epsilon - \left[\left(\frac{m\pi}{a}\right)^2 + \left(\frac{n\pi}{b}\right)^2\right]} = \beta\sqrt{1 - \left(\frac{f_{c.mn}}{f}\right)^2} \tag{1}$$

where $\beta = \omega\sqrt{\mu\epsilon} = \omega/u$.

7.17   A typical rectangular waveguide has dimensions 0.9 in (22.86 mm) and 0.4 in (10.16 mm). Assuming that the guide is air-filled and the frequency of operation is 25 GHz, list all possible propagating TM modes.

    ❚ From (1) of Problem 7.15,

$$f_{c.mn} = \frac{u_0}{2}\sqrt{\left(\frac{m}{a}\right)^2 + \left(\frac{n}{b}\right)^2}$$

We seek all positive integers m, n for which $f = 25\,\text{GHz} > f_c$; thus

$$\frac{4f^2}{u_0^2} > \left(\frac{m}{a}\right)^2 + \left(\frac{n}{b}\right)^2 \qquad \text{or} \qquad 1914m^2 + 9688n^2 < 27\,780 \tag{1}$$

Condition (1) is satisfied only for TM_{11}, TM_{21}, and TM_{31}.

7.18   List, in ascending order, the cutoff frequencies of TM_{11}, TM_{12}, TM_{21}, and TM_{22}, normalized with respect to that of TM_{11}, for a square waveguide ($a = b$).

    ❚

$$\frac{f_{c.mn}}{f_{c.11}} = \frac{\sqrt{m^2 + n^2}}{\sqrt{2}}$$

Hence $TM_{11} = 1$, $TM_{12} = TM_{21} = \sqrt{5/2}$, $TM_{22} = 2$.

7.19   Repeat Problem 7.18 for $a = 2b$.

    ❚

$$\frac{f_{c.mn}}{f_{c.11}} = \frac{\sqrt{m^2 + 4n^2}}{\sqrt{5}}$$

Hence $TM_{11} = 1$, $TM_{21} = \sqrt{8/5}$, $TM_{12} = \sqrt{17/5}$, $TM_{22} = 2$.

7.20   Repeat Problem 7.19 for $a = 2.25b$.

    ❚

$$\frac{f_{c.mn}}{f_{c.11}} = \frac{\sqrt{16m^2 + 81n^2}}{\sqrt{97}}$$

Hence $TM_{11} = 1$, $TM_{21} = \sqrt{145/97}$, $TM_{12} = \sqrt{340/97}$, $TM_{22} = 2$.

7.21   Obtain expressions for E_x, E_y, H_x, and H_y in the rectangular waveguide of Fig. 7-1, operating in the TM_{mn} mode.

    ❚ Substituting

$$E_z = E_{mn}\sin\frac{m\pi x}{a}\sin\frac{n\pi y}{b}e^{-j\beta_{mn}z} \qquad \text{and} \qquad H_z = 0$$

in the formulas of Problem 7.5, we obtain:

$$E_x = \frac{-j\beta_{mn}(m\pi/a)E_{mn}}{(m\pi/a)^2 + (n\pi/b)^2}\cos\frac{m\pi x}{a}\sin\frac{n\pi y}{b}e^{-j\beta_{mn}z} \tag{1}$$

$$E_y = \frac{-j\beta_{mn}(n\pi/b)E_{mn}}{(m\pi/a)^2 + (n\pi/b)^2}\sin\frac{m\pi x}{a}\cos\frac{n\pi y}{b}e^{-j\beta_{mn}z} \tag{2}$$

$$H_x = \frac{j\omega\epsilon(n\pi/b)E_{mn}}{(m\pi/a)^2 + (n\pi/b)^2}\sin\frac{m\pi x}{a}\cos\frac{n\pi y}{b}e^{-j\beta_{mn}z} \tag{3}$$

$$H_y = \frac{-j\omega\epsilon(m\pi/a)E_{mn}}{(m\pi/a)^2 + (n\pi/b)^2}\cos\frac{m\pi x}{a}\sin\frac{n\pi y}{b}e^{-j\beta_{mn}z} \tag{4}$$

7.22   The phase velocity of a wave of TM_{mn} mode is $u_{mn} = \omega/\beta_{mn}$. Express this in terms of the cutoff frequency and the characteristic velocity u of the interior medium.

    ❚ By (1) of Problem 7.16,

$$u_{mn} = \frac{\omega}{\beta\sqrt{1 - (f_{c.mn}/f)^2}} = \frac{u}{\sqrt{1 - (f_{c.mn}/f)^2}} \tag{1}$$

7.23 Show that the group velocity, $u_{g.mn}$, of narrowband signals propagating within a rectangular waveguide satisfies $u_{mn}u_{g.mn} = u^2$.

▮ We write $\beta_{mn} = \omega/u_{mn}$ as

$$\beta_{mn} = \frac{\omega}{u}\sqrt{1 - (f_{c.mn}/f)^2} = \frac{\sqrt{\omega^2 - \omega_{c.mn}^2}}{u}$$

Then

$$\frac{1}{u_{g.mn}} \equiv \frac{d\beta_{mn}}{d\omega} = \frac{1}{2u}\frac{1}{\sqrt{\omega^2 - \omega_{c.mn}^2}}2\omega = \frac{1}{u\sqrt{1-(f_{c.mn}/f)^2}}$$

or

$$u_{g.mn} = u\sqrt{1 - (f_{c.mn}/f)^2} \tag{1}$$

From (1) of Problem 7.22 and from (1) above we obtain

$$u_{mn}u_{g.mn} = u^2 \tag{2}$$

7.24 Determine the group velocity of a 12-GHz AM signal propagating in the TM_{11} mode in an air-filled square guide of side 2 cm.

▮ From (1) of Problem 7.15,

$$f_{c.11} = (1.5 \times 10^8)\sqrt{\frac{1}{0.02^2} + \frac{1}{0.02^2}} = 10.6 \text{ GHz}$$

Thus, from (1) of Problem 7.23,

$$u_{g.11} = u_0\sqrt{1 - (f_{c.11}/f)^2} = 1.406 \times 10^8 \text{ m/s}$$

7.25 Refer to Problem 7.22. Express the wavelength λ_{mn} in terms of the cutoff frequency $f_{c.mn}$ and the wavelength $\lambda = 2\pi/\beta$ of a plane wave in the interior of the guide.

▮ By (1) of Problem 7.16,

$$\lambda_{mn} = \frac{2\pi}{\beta_{mn}} = \frac{2\pi/\beta}{\sqrt{1 - (f_{c.mn}/f)^2}} = \frac{\lambda}{\sqrt{1 - (f_{c.mn}/f)^2}} \tag{1}$$

7.26 Refer to Problem 7.22. Express the wave impedance $\eta_{\text{TM}.mn}$ in TM_{mn} in terms of $f_{c.mn}$ and the intrinsic impedance η of the medium within the waveguide.

▮ From (1) and (4) of Problem 7.21,

$$\eta_{\text{TM}.mn} \equiv \frac{E_x}{H_y} = \frac{\beta_{mn}}{\omega\epsilon} = \eta\frac{\beta_{mn}}{\beta} = \eta\sqrt{1 - (f_{c.mn}/f)^2} \tag{1}$$

the last step following from (1) of Problem 7.16.

7.27 Sketch $\eta_{\text{TM}.mn}$ as a function of frequency $(f > f_{c.mn})$.

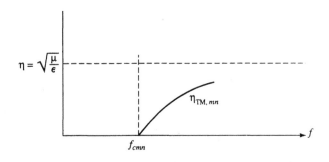

Fig. 7-2

▮ See Fig. 7-2.

7.28 Express the field components of Problem 7.21 in the time domain.

▮ For convenience, write $h_{mn}^2 \equiv (m\pi/a)^2 + (n\pi/b)^2$; for simplicity, suppose E_{mn} to be real.

$$E_z = E_{mn}\sin\frac{m\pi x}{a}\sin\frac{n\pi y}{b}\cos(\omega t - \beta_{mn}z) \tag{1}$$

$$E_x = \frac{\beta_{mn}}{h_{mn}^2}\frac{m\pi}{a}E_{mn}\cos\frac{m\pi x}{a}\sin\frac{n\pi y}{b}\sin(\omega t - \beta_{mn}z) \tag{2}$$

$$E_y = \frac{\beta_{mn}}{h_{mn}^2}\frac{n\pi}{b}E_{mn}\sin\frac{m\pi x}{a}\cos\frac{n\pi y}{b}\sin(\omega t - \beta_{mn}z) \tag{3}$$

$$H_x = -\frac{\omega\epsilon}{h_{mn}^2}\frac{n\pi}{b}E_{mn}\sin\frac{m\pi x}{a}\cos\frac{n\pi y}{b}\sin(\omega t - \beta_{mn}z) \tag{4}$$

$$H_y = \frac{\omega\epsilon}{h_{mn}^2}\frac{m\pi}{a}E_{mn}\cos\frac{m\pi x}{a}\sin\frac{n\pi y}{b}\sin(\omega t - \beta_{mn}z) \tag{5}$$

$$H_z = 0 \tag{6}$$

7.29 A rectangular waveguide is air-filled and operated at 30 GHz. Assuming that the cutoff frequency of the TM_{21} mode is 18 GHz, determine the wavelength, phase constant, phase velocity, and intrinsic impedance of this mode.

▮ Define $F_{21} \equiv \sqrt{1 - (f_{c.21}/f)^2} = \sqrt{1 - (18/30)^2} = 0.8$; then

$$u_{21} = u_0/F_{21} = \frac{3 \times 10^8}{0.8} = 3.75 \times 10^8 \text{ m/s}$$

$$\lambda_{21} = \lambda_0/F_{21} = \frac{1}{0.8} = 1.25 \text{ cm}$$

$$\beta_{21} = 2\pi/\lambda_{21} = 160\pi \text{ rad/m}$$

$$\eta_{TM.21} = \eta_0 F_{21} = (120\pi)(0.8) = 301.6 \ \Omega$$

7.30 For the waveguide of Problem 7.29 determine the distance over which the fields of this mode will be reduced in magnitude by 20 dB (a factor of 10), given an operating frequency of 15 GHz.

▮ For operation below the cutoff frequency, (2) of Problem 7.12 gives

$$\gamma_{mn} = \alpha_{mn} = \sqrt{4\pi^2\mu\epsilon(f_{c.mn}^2 - f^2)} = \beta\sqrt{(f_{c.mn}/f)^2 - 1} \tag{1}$$

Here, $\beta = 2\pi f/u_0 = 2\pi(15 \times 10^9)/(3 \times 10^8) = 100\pi$ rad/m, so that

$$\alpha_{21} = (100\pi)\sqrt{(18/15)^2 - 1} = 208.4 \text{ Np/m}$$

Then, solving $\exp(-\alpha_{21}d) = 1/10$, we obtain $d = 11.05$ mm.

7.31 Repeat Problem 7.29 assuming that the waveguide is filled with polyethylene for which $\epsilon_r = 2.25$.

▮ Correct the results of Problem 7.29 by $\sqrt{\epsilon_r} = 1.5$:

$$u_{21} = \frac{3.75 \times 10^8}{1.5} = 2.5 \times 10^8 \text{ m/s}$$

$$\lambda_{21} = \frac{1.25}{1.5} = 0.833 \text{ cm}$$

$$\beta_{21} = (160\pi)(1.5) = 240\pi \text{ rad/m}$$

$$\eta_{TM.21} = \frac{301.6}{1.5} = 201.067 \ \Omega$$

7.32 Repeat Problem 7.30 assuming that the waveguide is filled with a dielectric of $\epsilon_r = 2.25$.

▮ Reasoning as in Problem 7.31, $d = 11.04/1.5 = 7.36$ mm.

7.33 For an air-filled waveguide, having the dimensions $a = 2.29$ cm and $b = 1.02$ cm, determine the cutoff frequencies of the TM_{11}, TM_{21}, and TM_{12} modes.

▮
$$f_{c.mn} = \frac{u_0}{2}\sqrt{\left(\frac{m}{a}\right)^2 + \left(\frac{n}{b}\right)^2} = (1.5 \times 10^8)\sqrt{\left(\frac{m}{2.29 \times 10^{-2}}\right)^2 + \left(\frac{n}{1.02 \times 10^{-2}}\right)^2}$$

Thus $f_{c.11} = 16.16$ GHz, $f_{c.21} = 19.75$ GHz, $f_{c.12} = 30.25$ GHz.

7.34 For the waveguide of Problem 7.33, determine $\eta_{TM.11}$, β_{11}, u_{11}, and λ_{11}, for the TM_{11} mode at 18 GHz.

▮ Follow Problem 7.29, taking $f_{c.11}$ from Problem 7.33.

$$F_{11} = \sqrt{1 - \left(\frac{16.16}{18}\right)^2} = 0.44 \qquad \lambda_0 = \frac{u_0}{f} = \frac{3 \times 10^8}{18 \times 10^9} = \frac{1}{60} \text{ m}$$

$$u_{11} = \frac{3 \times 10^8}{0.44} = 6.81 \times 10^8 \text{ m/s}$$

$$\lambda_{11} = \frac{1}{(60)(0.44)} = 3.78 \text{ cm}$$

$$\beta_{11} = 2\pi/\lambda_{11} = 166 \text{ rad/m}$$

$$\eta_{TM.11} = (120\pi)(0.44) = 166 \ \Omega$$

(The numerical equality between β_{11} and $\eta_{TM.11}$ is exact and fortuitous.)

7.35 For an air-filled waveguide in the TM_{mn} mode, calculate the propagation constant for a frequency that is one-half the cutoff frequency.

▌ By (1) of Problem 7.30,
$$\gamma_{mn} = \alpha_{mn} = \beta_0\sqrt{(f_{c.mn}/f)^2 - 1} = (120\pi)\sqrt{3} = 652.97\,\text{Np/m}$$

7.36 Compare the cutoff frequencies for the TM_{12} mode in air-filled waveguides whose dimensions are (a) $a = 0.9$ in, $b = 0.4$ in; (b) $a = 0.4$ in, $b = 0.9$ in; (c) $a = b = 1$ cm; (d) $a = b = 10$ cm.

▌ The cutoff frequencies are given by
$$f_{c.12} = \frac{u_0}{2}\sqrt{\left(\frac{1}{a}\right)^2 + \left(\frac{2}{b}\right)^2}$$

(a) $a = 0.9$ in $= 2.286$ cm, $b = 0.4$ in $= 1.016$ cm; $f_{c.12} = 30.215$ GHz.
(b) $a = 1.016$ cm, $b = 2.286$ cm; $f_{c.12} = 19.75$ GHz.
(c) $a = b = 1$ cm; $f_{c.12} = 33.54$ GHz.
(d) $a = b = 10$ cm; $f_{c.12} = 3.354$ GHz.

7.37 For the field components in the TM_{11} mode as given in Problem 7.28, sketch the *field lines* at the instant $t = 0$ at the longitudinal location $\beta_{11}z = 3\pi/2$ (270°).

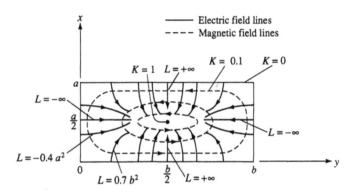

Fig. 7-3

▌ A field line of **E** (**H**) is, in general, a space curve at each point of which **E** (**H**) is tangential. For a TM mode, $H_z \equiv 0$, which means that the field lines of **H** can be sketched in the xy plane $\beta_{11}z = 3\pi/2$ (or in any other xy plane); they are given, for $m = n = 1$, by
$$\frac{dy}{dx} = \frac{H_y}{H_x} = -\frac{b\tan(\pi y/b)}{a\tan(\pi x/a)} \tag{1a}$$
which integrates to
$$\sin\frac{\pi x}{a}\sin\frac{\pi y}{b} = K \qquad (0 \le K \le 1) \tag{1b}$$

The magnetic field lines (1b) are shown as dashed lines in Fig. 7-3. Note that the line $K = 0$ coincides with the inner surface of the waveguide, and that $K = 1$ degenerates into the centerpoint $(a/2, b/2)$.

As it happens, the particular t- and z-values specified render E_z zero, so that the lines of **E** can also be sketched in the same xy plane. They are given by
$$\frac{dy}{dx} = \frac{E_y}{E_x} = \frac{a\tan(\pi x/a)}{b\tan(\pi y/b)} \tag{2a}$$
or
$$a^2\ln\left|\cos\frac{\pi x}{a}\right| - b^2\ln\left|\cos\frac{\pi y}{b}\right| = L \qquad (-\infty < L < +\infty) \tag{2b}$$

The electric field lines (2b) are shown as solid lines in Fig. 7-3. Note that the lines leave the guide surface perpendicularly (tangential **E** must vanish at a perfect conductor).

Remark 1 We know in advance from Maxwell's equations that the lines of **E** and the lines of **H** compose two orthogonal families. This fact is especially useful when the differential equations of the lines are not readily integrated; then we need only calculate one family—basically, by a numerical integration of its differential equation, aided by a knowledge of the points where dy/dx vanishes or is infinite—after which the other family is simply sketched in as the orthogonal trajectories of the first. Often the E-lines are chosen for calculation, since their direction at the boundary is known a priori.

Remark 2 The vanishing or nonvanishing of E_z in a TM mode is of no importance. The lines of the *transverse* E-fields, $\mathbf{E}_T = E_x\mathbf{a}_x + E_y\mathbf{a}_y$, are always given by an equation of the form (2a). On the sketch of these lines one may superpose symbols $\odot$ or $\oplus$ to indicate the local direction of (nonzero) $E_z\mathbf{a}_z$.

7.38 For the waveguide of Problems 7.28 and 7.37, plot the TM_{11} field lines in the centerplane $x = a/2$ at the instant $t = 0$.

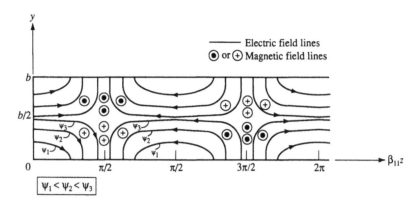

Fig. 7-4

▮ The field components are, from Problem 7.28,

$$E_z = E_{11} \sin \frac{\pi y}{b} \cos \beta_{11} z \qquad H_x = \frac{\omega \epsilon \pi E_{11}}{h_{11}^2 b} \cos \frac{\pi y}{b} \sin \beta_{11} z$$

$$E_x = 0 \qquad H_y = 0$$

$$E_y = \frac{-\beta_{11} \pi E_{11}}{h_{11}^2 b} \cos \frac{\pi y}{b} \sin \beta_{11} z \qquad H_z = 0$$

Thus, the **E**-lines are given by

$$\frac{dz}{dy} = \frac{E_z}{E_y} = -\frac{h_{11}^2 b \tan (\pi y / b)}{\beta_{11} \pi \tan \beta_{11} z} \tag{1a}$$

where $h_{11}^2 = (\pi^2/a^2) + (\pi^2/b^2)$. Integrating, we obtain

$$\left(1 + \frac{b^2}{a^2}\right) \ln \left| \cos \frac{\pi y}{b} \right| + \ln |\cos \beta_{11} z| = -\psi \qquad (0 < \psi < +\infty) \tag{1b}$$

The **E**-lines ($1b$) are graphed in Fig. 7-4. Observe that the pattern is periodic in z of period $\pi/\beta_{11} = \lambda_{11}/2$; this accords with the general behavior of transmission lines (Problem 6.94).

Since the **H**-lines are parallel to the x axis, they project as the circled points of Fig. 7-4. The general rule stated in Remark 1 of Problem 7.37 holds in the yz plane too: $\mathbf{E} \cdot \mathbf{H} = 0$ (because $\mathbf{H} = 0$).

7.39 (*a*) Repeat Problem 7.37 for the TM_{21} mode. (*b*) Predict the field pattern for TM_{mn}.

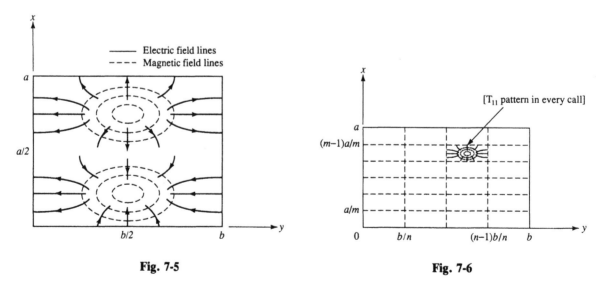

Fig. 7-5   **Fig. 7-6**

▮ (*a*) The procedure of Problem 7.37 leads to Fig. 7-5. (*b*) A comparison of Fig. 7-3 for TM_{11} and Fig. 7-5 for TM_{21} makes it apparent that the pattern for TM_{mn} may be obtained as mn repetitions of the TM_{11} pattern; see Fig. 7-6.

7.40 Specify the boundary conditions on the **E**-field for TE modes in the rectangular waveguide of Fig. 7-1.

▮ $E_x = 0$ at $y = 0$, b; $E_y = 0$ at $x = 0$, a.

7.41 (*a*) What are the boundary conditions on H_z for a TE mode? (*b*) Give the product solutions for H_z that satisfy the boundary conditions.

▮ (*a*) Problems 7.5 and 7.40, together with $E_z \equiv 0$, imply

$$\left.\frac{\partial H_z}{\partial y}\right|_{y=0,b} = 0 \qquad \left.\frac{\partial H_z}{\partial x}\right|_{x=0,a} = 0$$

(*b*) All four solution forms of Problem 7.9 can be specialized to meet the boundary conditions:

$$(H_z)_{0,0} = H_{0,0} e^{-j\beta z} \tag{1}$$
$$(H_z)_{0,N} = H_{0,N} \cos Ny\, e^{-\gamma(0,N)z} \tag{2}$$
$$(H_z)_{M,0} = H_{M,0} \cos Mx\, e^{-\gamma(M,0)z} \tag{3}$$
$$(H_z)_{M,N} = H_{M,N} \cos Mx \cos Ny\, e^{-\gamma(M,N)z} \tag{4}$$

where, as before, $M = m\pi/a$ and $N = n\pi/b$. Now, (1) is nothing but an ordinary plane wave in the hollow (the guide might as well be absent); it is therefore disregarded. Also, if we allow M or N to be zero in (4), then (4) will subsume (2) and (3). Thus we are left with the uniform solution

$$H_z = H_{mn} \cos\frac{m\pi x}{a} \cos\frac{n\pi y}{b} e^{-\gamma_{mn}z} \qquad (m, n = 0, 1, 2, \dots) \tag{1}$$

with only $m = n = 0$ prohibited. This in contrast to TM, where neither eigenvalue could be zero.

7.42 Write the forms of the field components in a propagating TE_{mn}.

▮ Substitute H_z as given by (1) of Problem 7.41, with $\gamma_{mn} = j\beta_{mn}$, in the relations of Problem 7.5, to get

$$H_z = H_{mn} \cos\frac{m\pi x}{a} \cos\frac{n\pi y}{b} e^{-j\beta_{mn}z} \tag{1}$$

$$H_x = \frac{j\beta_{mn}}{h_{mn}^2}\frac{m\pi}{a} H_{mm} \sin\frac{m\pi x}{a} \cos\frac{n\pi y}{b} e^{-j\beta_{mn}z} \tag{2}$$

$$H_y = \frac{j\beta_{mn}}{h_{mn}^2}\frac{n\pi}{b} H_{mn} \cos\frac{m\pi x}{a} \sin\frac{n\pi y}{b} e^{-j\beta_{mn}z} \tag{3}$$

$$E_x = \frac{j\omega\mu}{h_{mn}^2}\frac{n\pi}{b} H_{mn} \cos\frac{m\pi x}{a} \sin\frac{n\pi y}{b} e^{-j\beta_{mn}z} \tag{4}$$

$$E_y = \frac{-j\omega\mu}{h_{mn}^2}\frac{m\pi}{a} H_{mn} \sin\frac{m\pi x}{a} \cos\frac{n\pi y}{b} e^{-j\beta_{mn}z} \tag{5}$$

$$E_z = 0$$

where h_{mn}^2 is defined as in Problem 7.28.

7.43 Express the field components of Problem 7.42 in the time domain.

▮ On the assumption that H_{mn} is real:

$$H_z = H_{mn} \cos\frac{m\pi x}{a} \cos\frac{n\pi y}{b} \cos(\omega t - \beta_{mn}z)$$

$$H_x = -\frac{\beta_{mn}}{h_{mn}^2}\frac{m\pi}{a} H_{mn} \sin\frac{m\pi x}{a} \cos\frac{n\pi y}{b} \sin(\omega t - \beta_{mn}z)$$

$$H_y = -\frac{\beta_{mn}}{h_{mn}^2}\frac{n\pi}{b} H_{mn} \cos\frac{m\pi x}{a} \sin\frac{n\pi y}{b} \sin(\omega t - \beta_{mn}z)$$

$$E_x = -\frac{\omega\mu}{h_{mn}^2}\frac{n\pi}{b} H_{mn} \cos\frac{m\pi x}{a} \sin\frac{n\pi y}{b} \sin(\omega t - \beta_{mn}z)$$

$$E_y = \frac{\omega\mu}{h_{mn}^2}\frac{m\pi}{a} H_{mn} \sin\frac{m\pi x}{a} \cos\frac{n\pi y}{b} \sin(\omega t - \beta_{mn}z)$$

$$E_z = 0$$

7.44 Obtain an expression for the intrinsic wave impedance in the TE_{mn} mode.

▮ From (3) and (4) of Problem 7.42,

$$\eta_{\text{TE}.mn} \equiv \frac{E_x}{H_y} = \frac{\omega\mu}{\beta_{mn}} = \frac{\eta}{\sqrt{1 - (f_{c.mn}/f)^2}} \tag{1}$$

where $\eta = \sqrt{\mu/\epsilon}$ is the intrinsic impedance of the core medium, and where the cutoff frequency is given by (1) of Problem 7.16.

7.45 Define the *dominant mode* of propagation in a rectangular waveguide. Can it be a TM mode?

▮ The dominant mode is that of lowest cutoff frequency. Now, $f_{c,mn}$ is clearly minimized when the eigenvalue associated with the smaller dimension is 0 and the other eigenvalue is 1. Hence the dominant mode is TE_{10} if $a > b$, or TE_{01} if $b > a$.

7.46 Obtain the field equations (time domain) for the dominant mode in a rectangular waveguide. Assume that $a > b$.

▮ Take $m = 1$, $n = 0$ in Problem 7.43:

$$H_z = H_{10} \cos \frac{\pi x}{a} \cos (\omega t - \beta_{10} z) \qquad H_x = -\frac{\beta_{10} a}{\pi} H_{10} \sin \frac{\pi x}{a} \sin (\omega t - \beta_{10} z) \qquad H_y = 0$$

$$E_z = 0 \qquad E_y = \frac{\omega \mu a}{\pi} H_{10} \sin \frac{\pi x}{a} \sin (\omega t - \beta_{10} z) \qquad E_x = 0$$

7.47 For what range of frequencies will a rectangular guide with $a > b$ support only the dominant mode?

▮ The condition is $f_{c.10} < f < f_{c.01}$, or, by Problem 7.15,

$$\frac{u}{2a} < f < \frac{u}{2b} \tag{1}$$

7.48 Refer to Problem 7.47. Determine guide dimensions $(a > b)$ such that the operating frequency is 25 percent above the TE_{10} cutoff frequency and 25 percent below the TE_{01} cutoff frequency.

▮
$$f = 1.25 \frac{u}{2a} = 0.75 \frac{u}{2b} \qquad \text{or} \qquad \frac{a}{b} = \frac{5}{3}$$

Any pair of dimensions in this ratio will do.

7.49 What is the propagation velocity in the guide of Problem 7.48, assuming that it is air-filled?

▮ It is given that $f = (5/4) f_{c.10}$; therefore (Problem 7.22),

$$F_{10} \equiv \sqrt{1 - (f_{c.10}/f)^2} = 3/5 \qquad \text{and} \qquad u_{10} = \frac{u_0}{F_{10}} = 5 \times 10^8 \text{ m/s}$$

7.50 What is the wave impedance for the waveguide of Problems 7.48 and 7.49?

▮ By Problem 7.44,

$$\eta_{TE.10} = \frac{\eta_0}{F_{10}} = \frac{120\pi}{3/5} = 200\pi \ \Omega$$

7.51 If the operating frequency in Problem 7.49 is 15 GHz, what is the wavelength?

$$\lambda_{10} = \frac{u_{10}}{f} = \frac{5 \times 10^8}{15 \times 10^9} = 3.33 \text{ cm}$$

7.52 Determine the dimension a of a square air-filled waveguide having a cutoff frequency for the TE_{10} (TE_{01}) mode of (**a**) 30 GHz, (**b**) 3 GHz, (**c**) 300 MHz, (**d**) 30 MHz.

▮ From (1) of Problem 7.15,

$$a = \frac{1.5 \times 10^8 \text{ m/s}}{f_{c.10} \text{ (s}^{-1})}$$

Hence: (**a**) 5 mm, (**b**) 5 cm, (**c**) 0.5 m, (**d**) 5 m.

7.53 Determine the cutoff frequencies of the dominant (TE_{10}) mode for the following air-filled waveguides:

    (**a**)   6.25 in × 3.25 in (*L band*)   (**d**)   0.9 in × 0.4 in (*X band*)

    (**b**)   2.84 in × 1.34 in (*S band*)   (**e**)   0.42 in × 0.21 in (*K band*)

    (**c**)   1.872 in × 0.872 in (*C band*)   (**f**)   0.148 in × 0.074 in (*V band*)

▮ Convert the larger dimension to meters and substitute in

$$f_{c.10} = \frac{1.5 \times 10^8 \text{ m/s}}{a \text{ (m)}}$$

(**a**) 945 MHz; (**b**) 2.079 GHz; (**c**) 3.155 GHz; (**d**) 6.562 GHz; (**e**) 14.06 GHz; (**f**) 39.9 GHz.

7.54 Refer to Problem 7.46. Sketch the field lines at $t = 0$ in the two centerplanes of the guide ($x = a/2$ and $y = b/2$).

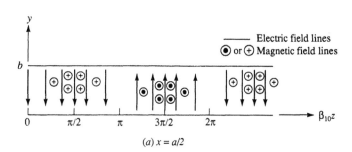

$(a)\ x = a/2$

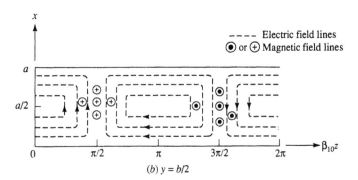

$(b)\ y = b/2$ **Fig. 7-7**

❚ The procedure described in Problem 7.37 applies equally to TE modes. The required sketches are given in Fig. 7-7.

7.55 Evaluate the time-average Poynting vector for the TE_{10} mode.

❚ The average Poynting vector (average power density) is given by $\mathbf{S}_{av} = \frac{1}{2}\,\mathrm{Re}\,(\hat{\mathbf{E}} \times \hat{\mathbf{H}}^*)$. In this case, the phasor components are given by Problem 7.42, for $m = 1$, $n = 0$. Thus,

$$\hat{\mathbf{E}} \times \hat{\mathbf{H}}^* = (E_y\mathbf{a}_y) \times (H_x^*\mathbf{a}_x + H_z^*\mathbf{a}_z) = -E_y H_x^*\mathbf{a}_z + E_y H_z^*\mathbf{a}_x$$

$$= \frac{\beta_{10}\omega\mu a^2}{\pi^2}|H_{10}|^2\sin^2\frac{\pi x}{a}\,\mathbf{a}_z - \frac{j\omega\mu a}{\pi}|H_{10}|^2\sin\frac{\pi x}{a}\cos\frac{\pi x}{a}\,\mathbf{a}_x$$

so that

$$\mathbf{S}_{av} = \frac{\beta_{10}\omega\mu a^2}{2\pi^2}|H_{10}|^2\sin^2\frac{\pi x}{a}\,\mathbf{a}_z = \frac{|E_y|^2}{2\eta_{\mathrm{TE.}10}}\,\mathbf{a}_z = \frac{\eta_{\mathrm{TE.}10}|H_x|^2}{2}\,\mathbf{a}_z$$

where the last step uses Problem 7.44.

7.56 Generalize Problem 7.55 by ascertaining that the average power flow in TE_{mn} is entirely in the z-direction.

❚ It is easy to see that this must be the case, because the x- and y-components in Problem 7.42 all contain the factor j, while $E_z = 0$ [express the cross product as a determinant]. Indeed, we find

$$\mathbf{S}_{av} = \beta_{mn}\left[\frac{|E_x|^2}{2\omega\mu} + \frac{|E_y|^2}{2\omega\mu}\right]\mathbf{a}_z = \frac{|E_x|^2 + |E_y|^2}{2\eta_{\mathrm{TE.}mn}}\,\mathbf{a}_z = \frac{\eta_{\mathrm{TE.}mn}(|H_x|^2 + |H_y|^2)}{2}\,\mathbf{a}_z \qquad (1)$$

Note that $\mathbf{S}_{av}$ is independent of z, which presumes a lossless waveguide.

7.57 Obtain the analog of (1) of Problem 7.56 for the TM_{mn} mode.

❚ Taking the phasor components from Problem 7.21 and the wave impedance from Problem 7.26, we get

$$\mathbf{S}_{av} = \frac{|E_x|^2 + |E_y|^2}{2\eta_{\mathrm{TM.}mn}}\,\mathbf{a}_z = \frac{\eta_{\mathrm{TM.}mn}(|H_x|^2 + |H_y|^2)}{2}\,\mathbf{a}_z$$

7.58 The cross-sectional dimensions of an air-filled waveguide are $a = 2.29$ cm and $b = 1.02$ cm. What is the cutoff frequency of the dominant mode? Also calculate the phase constant, phase velocity, wavelength, and intrinsic impedance of this mode at 7 GHz.

❚ Since $a > b$, the dominant mode is TE_{10}, with cutoff frequency

$$f_{c.10} = \frac{u_0}{2a} = \frac{3 \times 10^8}{2 \times 2.29 \times 10^{-2}} = 6.55\ \mathrm{GHz}$$

Then $F_{10} \equiv \sqrt{1 - (f_{c.10}/f)^2} = 0.3528$; whence

$$u_{10} = \frac{u_0}{F_{10}} = \frac{3 \times 10^8}{0.3528} = 8.503 \times 10^8 \text{ m/s}$$

$$\lambda_{10} = \frac{u_{10}}{f} = \frac{8.503 \times 10^8}{7 \times 10^9} = 12.1 \text{ cm}$$

$$\beta_{10} = \frac{2\pi}{\lambda_{10}} = \frac{6.28}{0.121} = 51.9 \text{ rad/m}$$

$$\eta_{\text{TE}.10} = \frac{\eta_0}{F_{10}} = \frac{377}{0.3528} = 1069 \text{ }\Omega$$

7.59 Determine the average power transmitted down the waveguide of Problem 7.58 assuming that the amplitude of the electric field is 1000 V/m.

❚ From Problem 7.46 we infer that for this mode,

$$\mathbf{E} = E_y \mathbf{a}_y = -j1000 \sin \frac{\pi x}{a} e^{-j\beta_{10}z} \mathbf{a}_y \text{ (V/m)}$$

Then Problems 7.55 and 7.58 give

$$\mathbf{S}_{\text{av}} = \frac{|E_y|^2}{2\eta_{\text{TE}.10}} \mathbf{a}_z = \frac{[1000 \sin (\pi x/a)]^2}{2(1082)} \mathbf{a}_z = 462.1 \sin^2 \frac{\pi x}{a} \mathbf{a}_z \text{ (W/m}^2)$$

Hence the total average power through any cross section of the waveguide is

$$P_{\text{av}} = \int_{x=0}^{a} \int_{y=0}^{b} \mathbf{S}_{\text{av}} \cdot \mathbf{a}_z \, dx \, dy = 462.1b \int_{x=0}^{a} \sin^2 \frac{\pi x}{a} \, dx$$
$$= 231.05ab = 53.67 \text{ mW}$$

7.60 A square, air-filled waveguide operates in the TE_{22} mode at twice the cutoff frequency. If both components of the electric field have amplitude 100 V/m, what is the average power transmitted down the guide?

❚ From (4) and (5) of Problem 7.42 ($a = $ side of square),

$$|E_x|^2 = (100)^2 \cos^2 \frac{2\pi x}{a} \sin^2 \frac{2\pi y}{a} \qquad |E_y|^2 = (100)^2 \sin^2 \frac{2\pi x}{a} \cos^2 \frac{2\pi y}{a}$$

From Problem 7.44,

$$\eta_{\text{TE}.22} = \frac{\eta_0}{\sqrt{1 - (f_{c.22}/f)^2}} = \frac{377}{\sqrt{1 - \frac{1}{4}}} = 435.3 \text{ }\Omega$$

Hence (Problem 7.56) $\mathbf{S}_{\text{av}} = S_{\text{av}}\mathbf{a}_z$, where

$$S_{\text{av}} = \frac{|E_x|^2}{2\eta_{\text{TE}.22}} + \frac{|E_y|^2}{2\eta_{\text{TE}.22}}$$

$$= \frac{(100)^2}{2(435.3)} \cos^2 \frac{2\pi x}{a} \sin^2 \frac{2\pi y}{a} + \frac{(100)^2}{2(435.3)} \sin^2 \frac{2\pi x}{a} \cos^2 \frac{2\pi y}{a} \text{ (W/m}^2)$$

By symmetry, the two terms composing S_{av} have the same integral over a cross section. Therefore,

$$P_{\text{av}} = \int_0^a \int_0^a S_{\text{av}} \, dx \, dy = \frac{(100)^2}{435.3} \int_0^a \int_0^a \cos^2 \frac{2\pi x}{a} \sin^2 \frac{2\pi y}{a} \, dx \, dy$$

$$= (22.97)\left(\frac{a}{2}\right)\left(\frac{a}{2}\right) = 5.743a^2 \text{ (W)}$$

7.61 Obtain an expression for the average power transmitted down a guide in the TM_{11} mode.

❚ By Problem 7.57,

$$P_{\text{av}} = \frac{\eta_{\text{TM}.11}}{2} \int_{y=0}^{b} \int_{x=0}^{a} (|H_x|^2 + |H_y|^2) \, dx \, dy$$

Now, by Problems 7.26 and 7.21,

$$\eta_{\text{TM}.11} = \eta \sqrt{1 - (f_{c.11}/f)^2}$$

$$|H_x|^2 = \frac{\omega^2 \epsilon^2 a^4 b^2 |E_{11}|^2}{\pi^2 (a^2 + b^2)^2} \sin^2 \frac{\pi x}{a} \cos^2 \frac{\pi y}{b}$$

$$|H_y|^2 = \frac{\omega^2 \epsilon^2 b^4 a^2 |E_{11}|^2}{\pi^2 (a^2 + b^2)^2} \sin^2 \frac{\pi y}{b} \cos^2 \frac{\pi x}{a}$$

Performing the integration, we obtain

$$P_{\text{av}} = \frac{\eta}{2}\sqrt{1 - (f_{c.11}/f)^2}\,\frac{\omega^2\epsilon^2 a^2 b^2\,|E_{11}|^2}{\pi^2(a^2 + b^2)}(a^2 + b^2)\left(\frac{a}{2}\right)\left(\frac{b}{2}\right)$$

$$= \frac{\eta}{8\pi^2}\sqrt{1 - (f_{c.11}/f)^2}\,(\omega^2\epsilon^2\,|E_{11}|^2)\frac{a^3 b^3}{a^2 + b^2}$$

7.62 For a square air-filled waveguide, operating well below the cutoff frequency of the dominant mode, express the attenuation in dB, in terms of the guide length (l) and width (w).

▮ The attenuation constant is given by

$$\alpha_{10} = \beta_0\sqrt{(f_{c.10}/f)^2 - 1} \approx \beta_0 f_{c.10}/f = 2\pi f_{c.10}/u_0 = \pi/w$$

where the last step follows from Problem 7.15. The attenuation is measured by $\exp(\alpha_{10}l)$. Thus

$$A_{\text{dB}} = 20\log_{10}\left[\exp(\alpha_{10}l)\right] = 20\alpha_{10}l\log_{10}e \approx (20)(\pi/w)l(0.4343) = 27.29(l/w)$$

7.63 The fields within a waveguide may be attributed to currents and charges on the (perfectly conducting) walls of the guide. The linear current density **K** is related to the magnetic field **H** by

$$\mathbf{K} = \mathbf{a}_n \times \mathbf{H} \tag{1}$$

Apply this condition to find the linear current densities on the four walls of a rectangular waveguide for the TE$_{10}$ mode.

▮ Applying the boundary condition (1) to the field equations of Problem 7.46 yields

Left wall $(y = 0)$: $\quad \mathbf{K} = -H_x|_{y=0}\,\mathbf{a}_z + H_z|_{y=0}\,\mathbf{a}_x$

$$= \frac{\beta_{10}a}{\pi}H_{10}\sin\frac{\pi x}{a}\sin(\omega t - \beta_{10}z)\,\mathbf{a}_z + H_{10}\cos\frac{\pi x}{a}\cos(\omega t - \beta_{10}z)\,\mathbf{a}_x$$

Right wall $(y = b)$: $\quad \mathbf{K} = H_x|_{y=b}\,\mathbf{a}_z - H_z|_{y=b}\,\mathbf{a}_x$

$$= -\frac{\beta_{10}a}{\pi}H_{10}\sin\frac{\pi x}{a}\sin(\omega t - \beta_{10}z)\,\mathbf{a}_z - H_{10}\cos\frac{\pi x}{a}\cos(\omega t - \beta_{10}z)\,\mathbf{a}_x$$

Top wall $(x = a)$: $\quad \mathbf{K} = H_z|_{x=a}\,\mathbf{a}_y = -H_{10}\cos(\omega t - \beta_{10}z)\,\mathbf{a}_y$
Bottom wall $(x = 0)$: $\quad \mathbf{K} = -H_z|_{x=0}\,\mathbf{a}_y = -H_{10}\cos(\omega t - \beta_{10}z)\,\mathbf{a}_y$

7.64 Sketch the wall currents at $t = 0$ in the guide of Problem 7.63.

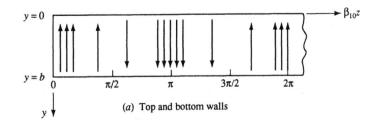

(a) Top and bottom walls

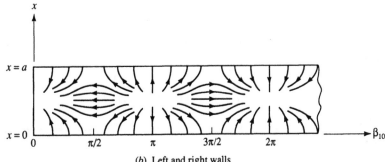

(b) Left and right walls

Fig. 7-8

▮ See Fig. 7-8.

7.65 Refer to Problem 7.63. Infer the surface charge densities from the condition

$$\rho_s = \epsilon E_{\text{normal}} \tag{1}$$

❚ Apply this condition to the field equations of Problem 7.46:

Left wall $(y = 0)$: $\quad \rho_s = \epsilon E_y\big|_{y=0} = \epsilon \dfrac{\omega\mu a}{\pi} H_{10} \sin\dfrac{\pi x}{a}\sin(\omega t - \beta_{10}z)$

Right wall $(y = b)$: $\quad \rho_s = -\epsilon E_y\big|_{y=b} = -\epsilon \dfrac{\omega\mu a}{\pi} H_{10} \sin\dfrac{\pi x}{a}\sin(\omega t - \beta_{10}z)$

Top wall $(x = a)$: $\quad \rho_s = 0$
Bottom wall $(x = 0)$: $\quad \rho_s = 0$

7.66 Sketch at $t = 0$ the surface current and surface charge densities on the top wall $(x = a)$ of a rectangular waveguide, for the TM_{11} mode.

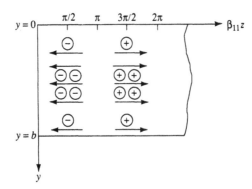

Fig. 7-9

❚ Application of (1) of Problem 7.63 and (1) of Problem 7.65 to the expressions of Problem 7.28 yields, at $t = 0$ and $x = a$,

$$\mathbf{K} = -H_y\mathbf{a}_z = -\dfrac{\omega\epsilon}{h_{11}^2}\dfrac{\pi}{a}E_{11}\sin\dfrac{\pi y}{b}\sin\beta_{11}z\,\mathbf{a}_z$$

$$\rho_s = -\epsilon E_x = \epsilon\eta_{TM.11}K_z$$

Because K_z and ρ_s are proportional, a single sketch suffices; see Fig. 7-9.

7.67 Is it possible to define a *cutoff wavelength* for a given mode in an air-filled waveguide?

❚ It would seem not; for the relation

$$\lambda_{mn} = \dfrac{u_0}{\sqrt{f^2 - f_{c.mn}^2}} \tag{1}$$

shows that λ_{mn} decreases from infinity to zero as f increases from $f_{c.mn}$ to ∞; that is, there is no cutoff for the wavelength *in the guide*. Nevertheless, one does speak of a cutoff wavelength, meaning by this the wavelength *in the outside air* corresponding to the cutoff frequency of the mode:

$$\lambda_{c.mn} \equiv u_0/f_{c.mn} = 2/\sqrt{(m/a)^2 + (n/b)^2} \tag{2}$$

In terms of (2), (1) takes the form

$$\lambda_{mn} = \dfrac{\lambda_{c.mn}}{\sqrt{(f/f_{c.mn})^2 - 1}} \tag{3}$$

7.68 What is the phase velocity in an air-filled rectangular waveguide that operates in a certain mode at 1.6 times the cutoff frequency?

❚ $$u_{mn} = \dfrac{u}{\sqrt{1 - (f_{c.mn}/f)^2}} = \dfrac{3 \times 10^8}{\sqrt{1 - (1/1.6)^2}} = 3.84 \times 10^8\,\text{m/s}$$

7.69 An air-filled rectangular waveguide is operating in TM_{mn} at a frequency twice the cutoff frequency. Determine the wave impedance.

❚ From (1) of Problem 7.26, $\eta_{TM.mn} = 377\sqrt{1 - (1/2)^2} = 326.48\,\Omega$.

7.70 Repeat Problem 7.69 for TE_{mn}.

❚ From (1) of Problem 7.44,

$$\eta_{TE.mn} = \dfrac{377}{\sqrt{1 - (1/2)^2}} = 435.33\,\Omega$$

7.71 Find an expression for the impedance in an evanescent TM mode.

▌ In Problem 7.21 replace $j\beta_{mn}$ by $\alpha_{mn} = \beta\sqrt{(f_{c.mn}/f)^2 - 1}$ [see (1) of Problem 7.30]. Then

$$\eta_{\text{TM}.mn} \equiv \frac{E_x}{H_y} = \frac{-\alpha_{mn}}{-j\omega\epsilon} = -j\frac{\beta}{\omega\epsilon}\sqrt{(f_{c.mn}/f)^2 - 1} = -j\eta\sqrt{(f_{c.mn}/f)^2 - 1}$$

It is seen that (1) of Problem 7.26 also applies to evanescent modes, provided the correct square root is chosen.

7.72 Repeat Problem 7.71 for an evanescent TE mode.

▌ Start with Problem 7.42 and proceed as in Problem 7.71, to find

$$\eta_{\text{TE}.mn} = \frac{\eta}{-j\sqrt{(f_{c.mn}/f)^2 - 1}}$$

Thus (1) of Problem 7.44 also extends to evanescent modes.

7.73 Obtain an expression for the phase velocity of TEM waves in an air-filled waveguide.

▌ For the TEM mode $E_z = H_z = 0$; so we obtain a trivial solution unless

$$\gamma_{\text{TEM}}^2 + \omega^2\mu_0\epsilon_0 = 0 \quad \text{or} \quad \gamma_{\text{TEM}} = j\beta_0 \quad \text{or} \quad u_{\text{TEM}} = \omega/\beta_0 = u_0$$

7.74 Show that the wave impedance in the TEM mode is η (the intrinsic impedance of the material of the core).

▌ See Problem 7.41(b); the TEM mode corresponds to $m = n = 0$.

7.75 Refer to Problem 7.74. With $m = n = 0$, why isn't the TEM mode dominant?

▌ By definition, the dominant mode is that of lowest cutoff frequency. But plane TEM waves of *any* frequency can propagate in the waveguide's interior; that is to say, a cutoff frequency does not exist there.

7.76 Sketch the ω-β diagram for the propagating TE and TM modes, and for the TEM mode, in a waveguide.

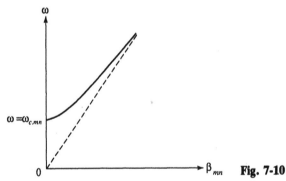

Fig. 7-10

▌ For the TEM mode $\omega/\beta = u = $ const., the dashed line in Fig. 7-10. For the propagating TE and TM modes, from (1) of Problem 7.16 we have

$$\beta_{mn} = \omega\sqrt{\mu\epsilon}\sqrt{1 - (f_{c.mn}/f)^2} \quad \text{or} \quad \omega^2 - u^2\beta_{mn}^2 = \omega_{c.mn}^2$$

which plots as a quadrant of a hyperbola in Fig. 7-10.

7.77 Plot the α-ω relationship for evanescent modes in a waveguide.

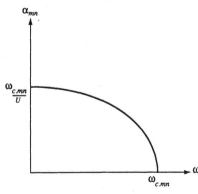

Fig. 7-11

▌ By (1) of Problem 7.30,

$$\alpha_{mn} = \frac{2\pi}{u}\sqrt{f_{c.mn}^2 - f^2} \quad \text{or} \quad u^2\alpha^2 + \omega^2 = \omega_{c.mn}^2$$

which plots as a quadrant of an ellipse (Fig. 7-11).

7.78 Obtain an expression for the attenuation constant, α_d, for TM modes in a parallel-plate waveguide filled with a lossy dielectric ($\epsilon_d = \epsilon - j\sigma/\omega$).

▌ A rectangular waveguide with $a \to \infty$ becomes a parallel-plate configuration with separation b. Thus, by (2) of Problem 7.12,

$$\gamma = j\sqrt{\omega^2\mu\epsilon_d - (n\pi/b)^2} = j\left[\omega^2\mu\epsilon\left(1 - \frac{j\sigma}{\omega\epsilon}\right) - \left(\frac{n\pi}{b}\right)^2\right]^{1/2}$$

$$= j\sqrt{\omega^2\mu\epsilon - (n\pi/b)^2}\ \{1 - j\omega\mu\sigma[\omega^2\mu\epsilon - (n\pi/b)^2]^{-1}\}^{1/2} \tag{1}$$

Provided $\omega\mu\sigma \ll \omega^2\mu\epsilon - (n\pi/b)^2$, a binomial expansion of the one-half power in (1) yields

$$\gamma \approx j\sqrt{\omega^2\mu\epsilon - (n\pi/b)^2}\left\{1 - \frac{j\omega\mu\sigma}{2}[\omega^2\mu\epsilon - (n\pi/b)^2]^{-1}\right\} \tag{2}$$

Substitution of $n\pi/b = 2\pi f_{c.n}\sqrt{\mu\epsilon}$ and $\eta = \sqrt{\mu/\epsilon}$ in (2) leads to

$$\gamma \approx \frac{\sigma\eta}{2\sqrt{1 - (f_{c.n}/f)^2}} + j\omega\sqrt{\mu\epsilon}\ \sqrt{1 - (f_{c.n}/f)^2}$$

whence

$$\alpha_d = \text{Re}\ \gamma \approx \frac{\sigma\eta}{2\sqrt{1 - (f_{c.n}/f)^2}} \tag{3}$$

7.79 Because of losses due to imperfectly conducting walls, the time-average power along a waveguide may be written as

$$P_{av}(z) = P_{av}(0)e^{-2\alpha_{mn}z}$$

Express α_{mn} in terms of the per-unit-length power loss P_{loss} and P_{av}.

▌ Since $P_{loss}(z) = -P_{av}'(z) = 2\alpha_{mn}P_{av}(z)$,

$$\alpha_{mn} = \frac{P_{loss}(z)}{2P_{av}(z)} = \frac{P_{loss}(0)}{2P_{av}(0)} \tag{1}$$

In (1), $P_{av}(0)$ may be calculated from the Poynting vector for a *lossless* guide (Problem 7.56 or 7.57).

7.80 Using the result of Problem 7.79 derive an expression for α_{10} (for the TE_{10} mode) in a lossy waveguide.

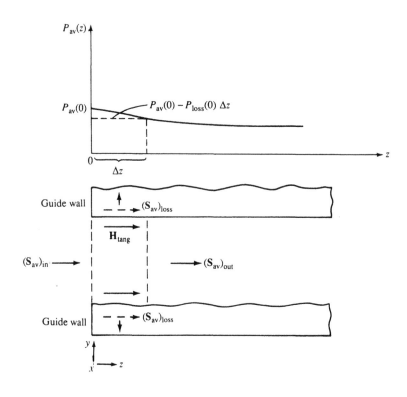

Fig. 7-12

▌ By Problems 7.55 (for S_{av}), 7.44 (for $\eta_{\text{TE.10}}$), and 7.46 (for $|H_x|$):

$$P_{\text{av}}(0) = \int_{y=0}^{b} \int_{x=0}^{a} S_{\text{av}}\, dx\, dy = \frac{\eta \beta^2 a^3 b\, |H_{10}|^2}{4\pi^2} \sqrt{1 - (f_{c.10}/f)^2} \qquad (1)$$

This gives the denominator in (1) of Problem 7.79; to find the numerator, consider Fig. 7-12. We assume the fields of the guide to be essentially unaltered in the presence of small losses. If the walls of the guide are not perfect conductors, the tangential magnetic field $\mathbf{H}_{\text{tang}}$ will be continuous across a wall. This magnetic field will induce a tangential electric field

$$\mathbf{E}_{\text{tang}} = \mathbf{a}_n \times \eta_c \mathbf{H}_{\text{tang}}$$

within the wall; here, $\mathbf{a}_n$ is the outer normal and η_c is the intrinsic impedance of the metal wall, given in Problem 5.45 as

$$\eta_c = \sqrt{\frac{\omega \mu_c}{2\sigma_c}}(1 + j1) = \frac{1}{\sigma_c \delta_c}(1 + j1) \qquad (\delta_c \equiv \text{skin depth})$$

These fields just interior to the wall result in a time-average power-density vector directed into the wall:

$$(S_{\text{av}})_{\text{loss}} = \frac{1}{2}|\mathbf{H}_{\text{tang}}|^2 \operatorname{Re} \eta_c = \frac{1}{2\sigma_c \delta_c}|\mathbf{H}_{\text{tang}}|^2 \qquad (2)$$

The average power loss through the first 1 m of a given wall is obtained as the integral of (2) over the appropriate cross-sectional dimension. Combining the losses for all four walls yields

$$P_{\text{loss}}(0) = \frac{1}{2\sigma_c \delta_c}\left[\int_{x=0}^{a}(|H_x|^2 + |H_z|^2)\big|_{y=0}\, dx + \int_{x=0}^{a}(|H_x|^2 + |H_z|^2)\big|_{y=b}\, dx \right.$$
$$\left. + \int_{y=0}^{b}(|H_y|^2 + |H_z|^2)\big|_{x=0}\, dy + \int_{y=0}^{b}(|H_y|^2 + |H_z|^2)\big|_{x=a}\, dy\right]$$

Substitution of the field components from Problem 7.46 yields

$$P_{\text{loss}}(0) = \frac{|H_{10}|^2}{2\sigma_c \delta_c}\left[2\int_0^a \left(\frac{\beta_{10}^2 a^2}{\pi^2}\sin^2\frac{\pi x}{a} + \cos^2\frac{\pi x}{a}\right) dx + 2\int_0^b (1)\, dy\right]$$
$$= \frac{|H_{10}|^2}{2\sigma_c \delta_c}\left[\frac{\beta_{10}^2 a^3}{\pi^2} + a + 2b\right] = \frac{|H_{10}|^2}{2\sigma_c \delta_c}[2b + a(f/f_{c.10})^2] \qquad (3)$$

This last result is obtained by recalling that $\beta_{10}^2 = \omega^2 \mu \epsilon - (\pi/a)^2$ and $f_{c.10} = (\pi/a)/(2\pi\sqrt{\mu\epsilon})$. Then, by (1) and (3),

$$\alpha_{10} = \frac{P_{\text{loss}}(0)}{2P_{\text{av}}(0)} = \frac{\pi^2[2b + a(f/f_{c.10})^2]}{\sigma_c \delta_c \eta \beta^2 a^3 b\sqrt{1 - (f_{c.10}/f)^2}} = \frac{\sqrt{\pi\mu_c/\sigma_c}}{\eta b}(\sqrt{f})\frac{1 + (2b/a)(f_{c.10}/f)^2}{\sqrt{1 - (f_{c.10}/f)^2}} \qquad (4)$$

7.81 The walls of an air-core waveguide of dimensions $a = 0.9$ in and $b = 0.4$ in (1.02 cm) are made of brass ($\sigma_c = 15.7\,\text{MS/m}$, $\mu_c = \mu_0$). Calculate the loss, in dB/m, in the guide walls at an operating frequency of 9 GHz for the TE_{10} mode.

▌ From Problem 7.58, $f_{c.10} = 6.56$ GHz. Thus, from (4) of Problem 7.80,

$$\alpha_{10} = \frac{\sqrt{\pi(4\pi \times 10^{-7})/(15.7 \times 10^6)}}{(120\pi)(1.02 \times 10^{-2})}(\sqrt{9 \times 10^9})\frac{1 + (8/9)(6.56/9)^2}{\sqrt{1 - (6.56/9)^2}} = 2.66 \times 10^{-2}$$

Now, from Problem 7.79,

$$\text{loss (dB/m)} = 10\log_{10}\frac{P_{\text{av}}(0)}{P_{\text{av}}(1)} = (8.69)\alpha_{10} = 0.2312\,\text{dB/m}$$

7.82 For the waveguide of Problem 7.81, sketch the wall loss versus frequency.

▌ From the results of Problems 7.80 and 7.58,

$$\alpha_{10} = \frac{4.14 \times 10^{-3}\sqrt{f}\,[1 + (8/9)(6.56/f)^2]}{\sqrt{1 - (6.56/f)^2}}$$

where f is expressed in gigahertz. The loss, in decibels per meter, is sketched in Fig. 7-13. Note that the attenuation shows a minimum at around 15 GHz. This is slightly below the cutoff frequency of the TE_{11} and TM_{11} modes (16.16 GHz) and slightly above the cutoff frequency of the TE_{01} mode (14.76 GHz).

7.83 Calculate the normalized frequency $\psi \equiv f/f_{c.10}$ for which the wall loss of the TE_{10} mode in a rectangular waveguide is a minimum.

▌ To within a multiplicative constant, (4) of Problem 7.80 gives

$$\alpha_{10} = \frac{\psi^2 + (2b/a)}{(\psi^3 - \psi)^{1/2}}$$

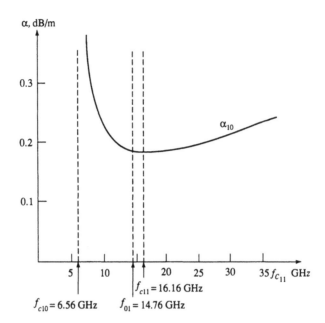

$f_{c10} = 6.56$ GHz $f_{01} = 14.76$ GHz

Fig. 7-13

Setting $d\alpha_{10}/d\psi = 0$ and solving for the minimizing ψ, we obtain

$$\psi = [k + \sqrt{k^2 - (2b/a)}]^{1/2} \qquad \text{where} \qquad k \equiv \frac{3}{2} + \frac{3b}{a} \qquad (1)$$

Note that the chosen root is real and greater than unity, corresponding to a propagating mode.

7.84 Calculate the ratio of the operating frequency to the cutoff frequency for the TE_{10} mode in a rectangular waveguide of dimensions $a = 0.9$ in and $b = 0.4$ in such that the wall losses are a minimum.

▌ By (1) of Problem 7.83,

$$k = \frac{3}{2} + \frac{3(0.4)}{0.9} = 2.83 \qquad \psi = 2.346$$

7.85 Repeat Problem 7.84 for $b = 0.04$ in.

▌
$$k = \frac{3}{2} + \frac{3(0.04)}{0.9} = 1.63 \qquad \psi = 2.052$$

7.86 Repeat Problem 7.80 for TM_{11}.

▌ This time the components of **H** come from Problem 7.21; the wave impedance, from Problem 7.26; the time-average Poynting vector, from Problem 7.57. The analog to (1) of Problem 7.80 has already been developed in Problem 7.61:

$$P_{av}(0) = \frac{\eta \omega^2 \epsilon^2 |E_{11}|^2}{8\pi^2} \frac{a^3 b^3}{a^2 + b^2} \sqrt{1 - (f_{c.11}/f)^2} \qquad (1)$$

Analogous to (3) of Problem 7.80, we calculate

$$P_{loss}(0) = \frac{1}{2\sigma_c \delta_c} \left[2 \int_0^a (|H_x|^2)_{y=0}\, dx + 2 \int_0^b (|H_y|^2)_{x=0}\, dy \right]$$

$$= \frac{\omega^2 \epsilon^2 |E_{11}|^2}{2\pi^2 \sigma_c \delta_c} \frac{a^2 b^2 (a^3 + b^3)}{(a^2 + b^2)^2} \qquad (2)$$

Then, by division and use of

$$\sigma_c \delta_c = \sqrt{\frac{\sigma_c}{\pi \mu_c f}}$$

we find

$$\alpha_{11} = \frac{P_{loss}(0)}{2P_{av}(0)} = \frac{2(a^3 + b^3)}{\eta \sigma_c \delta_c (a^3 b + b^3 a)\sqrt{1 - (f_{c.11}/f)^2}}$$

$$= \frac{2\sqrt{\pi f \mu_c}}{\eta \sqrt{\sigma_c}\, b \sqrt{1 - (f_{c.11}/f)^2}} \frac{(b/a)^3 + 1}{(b/a)^2 + 1} \qquad (3)$$

7.87 Verify that α_{11} as found in Problem 7.86 has the correct units (m^{-1}).

▮ In the symmetric expression (3), replace each quantity by its physical units:

$$(\alpha_{11}) = \frac{(m^3)}{(\Omega)(\Omega^{-1} \cdot m^{-1})(m)(m^4)} = (m^{-1})$$

7.88 An electromagnetic wave propagates in a TE mode between two extensive parallel plates having a separation of 3 cm. The region between these plates is filled with a nonmagnetic dielectric material having a relative dielectric constant of 100. Find the cutoff frequency of the lowest TE mode.

▮ The propagation constant γ is given by

$$\gamma^2 = \left(\frac{n\pi}{a}\right)^2 - \mu\epsilon\omega^2 \qquad (1)$$

where a is the separation between the plates. At cutoff $\gamma = 0$, so that

$$\omega_c^2 = \frac{1}{\mu\epsilon}\left(\frac{n\pi}{a}\right)^2$$

In the lowest TE mode, $n = 1$; thus

$$\omega_c = \frac{1}{\sqrt{\mu\epsilon}}\left(\frac{\pi}{a}\right) \qquad \text{or} \qquad f_c = \frac{u}{2a} = \frac{u_0/\sqrt{\epsilon_r}}{2a} \qquad (2)$$

Substitution of the data in (2) yields $f_c = 500 \text{ MHz}$.

7.89 A 1-GHz signal is imposed on the transmission system of Problem 7.88 and the lowest TE mode is propagated. Determine the propagation constant for this electromagnetic wave.

▮ Rewriting (1) of Problem 7.88 we have, with $n = 1$ and $f = 10^9$ Hz,

$$\gamma^2 = \left(\frac{\pi}{a}\right)^2 - \frac{4\pi^2 f^2}{u^2} = \pi^2\left[\left(\frac{1}{3}\right)^2 - \frac{4(10^9)^2}{(3\times10^9)^2}\right] \qquad \text{or} \qquad \gamma = j\frac{\pi}{\sqrt{3}} \approx j1.81 \text{ m}^{-1}$$

7.90 A rectangular waveguide ($a = 3$ cm, $b = 2$ cm in Fig. 7-1) transmits microwave power in the TE_{11} mode at a frequency of 12 GHz. The electric field is given by

$$E_y = \frac{K_x}{K_c} E_0 \sin K_x x \cos K_y y\, e^{j(\omega t - \beta z)} \qquad E_x = \frac{-K_y}{K_c} E_0 \cos K_x x \sin K_y y\, e^{j(\omega t - \beta z)}$$

where $K_c^2 = K_x^2 + K_y^2$ and $E_0 = 100$ kV/m. What is the waveguide cutoff frequency for the lowest-order mode?

▮ The cutoff frequency is given by

$$f_c = \frac{u}{2}\sqrt{\left(\frac{m}{a}\right)^2 + \left(\frac{n}{b}\right)^2} \qquad (1)$$

The lowest-order mode is the TE_{10} mode. Substituting $u = 3 \times 10^{10}$ cm/s, $m = 1$, $a = 3$ cm and $n = 0$ in (1) yields

$$f_c = \frac{3 \times 10^{10}}{2}\sqrt{\left(\frac{1}{3}\right)^2} = 5 \text{ GHz}$$

7.91 Calculate the TE_{11} mode cutoff frequency in the waveguide of Problem 7.90.

▮ Substituting $m = n = 1$, $u = 3 \times 10^{10}$ cm/s, $a = 3$ cm and $b = 2$ cm, we obtain

$$f_c = \frac{3 \times 10^{10}}{2}\sqrt{\left(\frac{1}{3}\right)^2 + \left(\frac{1}{2}\right)^2} = 9 \text{ MHz}$$

7.92 Find H_x in the waveguide of Problem 7.90.

▮ From Maxwell's equation we have

$$(\nabla \times \mathbf{E})_x = -\mu\frac{\partial H_x}{\partial t} \qquad \text{or} \qquad -\frac{\partial E_y}{\partial z} = +j\beta E_y = -j\omega\mu H_x$$

Thus

$$H_x = -\frac{\beta}{\omega\mu} E_y = -\frac{\beta}{\omega\mu}\frac{K_x}{K_c} E_0 \sin K_x x \cos K_y y\, e^{j(\omega t - \beta z)} \qquad (1)$$

7.93 Calculate β for the waveguide of Problem 7.90, for the TE_{11} mode.

▮ Taking f_c from Problem 7.91,

$$\beta = \frac{2\pi}{u}\sqrt{f^2 - f_c^2} = \frac{2\pi}{3\times10^8}\sqrt{(12\times10^9)^2 - (9\times10^9)^2} = 167 \text{ m}^{-1}$$

7.94 Show that the top and bottom of the waveguide of Problem 7.90 are always at the same potential.

▮ The voltage difference is given by

$$\Delta V = \int_0^b E_y \, dy \propto \sin K_y y \Big|_0^b = 0$$

because $K_y = \pi/b$.

7.95 In the waveguide of Problem 7.90, b is now decreased. What is the effect on the TE_{11} mode cutoff frequency?

▮ By (1) of Problem 7.90, the cutoff frequency increases.

7.96 What is the effect of decreasing b on the amplitude of H_x in the waveguide of Problem 7.90?

▮ Notice that β/K_c decreases with b. Thus, from (1) of Problem 7.92, the amplitude of H_x decreases.

7.97 What is the effect of varying b on the waveguide cutoff frequency, for the waveguide of Problem 7.90?

▮ From (1) of Problem 7.90, for the lowest mode, f_c is independent of b. Thus, varying b will have no effect on the waveguide cutoff frequency.

7.98 In the waveguide of Problem 7.90, the frequency of the microwave signal is decreased. What is the effect on: (a) the phase constant for the TE_{11} mode? (b) the amplitude of H_x? (c) the group velocity for the TE_{11} mode?

▮ (a) Since $\beta^2 = (\omega^2/u^2) - K_c^2$, β will decrease. (b) From Problem 7.96, H_x will decrease in amplitude. (c) See Problem 7.23; group velocity will decrease.

7.99 An X-band microwave signal of 10 GHz propagates down an air-filled waveguide of 1.0-by-2.5-cm cross section. Determine the cutoff frequencies of TE_{10}, TE_{01}, and TE_{20} modes.

▮ TE_{10}: $\quad f_c = \dfrac{u}{2} \sqrt{\left(\dfrac{1}{a}\right)^2} = \dfrac{3 \times 10^{10}}{2 \times 2.5} = 6 \text{ GHz}$

TE_{01}: $\quad f_c = \dfrac{u}{2} \sqrt{\left(\dfrac{1}{1}\right)^2} = \dfrac{3 \times 10^{10}}{2} = 15 \text{ GHz}$

TE_{20}: $\quad f_c = \dfrac{u}{2} \sqrt{\left(\dfrac{2}{2.5}\right)^2 + \left(\dfrac{0}{1}\right)^2} = 12 \text{ GHz}$

7.100 Which modes in the waveguide of Problem 7.99 will propagate? Determine the corresponding group velocities.

▮ Since $f = 10$ GHz, from Problem 7.99, only the TE_{10} mode will propagate. The group velocity is calculated from

$$v_g = u\sqrt{1 - (f_c/f)^2} = (3 \times 10^8)\sqrt{1 - (6/10)^2} = 2.4 \times 10^8 \text{ m/s}$$

7.101 Calculate the length of the waveguide of Problem 7.99 such that the signal propagated down the guide is delayed by 2 μs relative to that propagated in the air outside.

▮ If L is the unknown length, then, using Problem 7.100,

$$\frac{L}{2.4 \times 10^8} - \frac{L}{3 \times 10^8} = 2 \times 10^{-6} \quad \text{or} \quad L = 2.4 \text{ km}$$

7.102 A wave propagates between two parallel plates of infinite extent, which are separated by a distance w (Fig. 7-14). The electric field for a TE mode is

$$E_y = E_0 \sin \alpha_e x \, e^{j(\omega t - \beta z)} \tag{1}$$

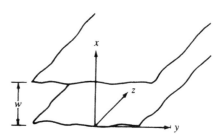

Fig. 7-14

and the magnetic field for a TM mode is

$$H_y = H_0 \cos \alpha_m x \, e^{j(\omega t - \beta z)} \qquad (2)$$

Obtain the rms electric field components for the TM mode.

▌ For the TM mode, $H_z = 0$ and H_y is given by (2). Thus, Maxwell's equation gives

$$(\nabla \times \mathbf{H})_x = -\frac{\partial H_y}{\partial z} = \epsilon \frac{\partial E_x}{\partial t} \quad \text{or} \quad E_x = \frac{\beta}{\omega \epsilon} H_y = \frac{\beta}{\omega \epsilon} H_0 \cos \alpha_m x \, e^{j(\omega t - \beta z)}$$

Therefore

$$(E_x)_{\text{rms}} = \frac{\beta H_0}{\sqrt{2} \, \omega \epsilon} \cos \alpha_m x \qquad (E_y)_{\text{rms}} = 0$$

Similarly, using $H_x \equiv 0$,

$$(\nabla \times \mathbf{H})_z = \frac{\partial H_y}{\partial x} = \epsilon \frac{\partial E_z}{\partial t} \quad \text{or} \quad E_z = \frac{\alpha_m H_0}{\omega \epsilon} \sin \alpha x \, e^{j(\omega t - \beta z + \pi/2)}$$

whence

$$(E_z)_{\text{rms}} = \frac{\alpha_m H_0}{\sqrt{2} \, \omega \epsilon} \sin \alpha_m x$$

7.103 From the data of Problem 7.102, find the rms magnetic field components for the TE mode.

▌ For the TE mode, $E_z = 0$ and E_y is given by (1) of Problem 7.102. Thus, Maxwell's equation gives

$$(\nabla \times \mathbf{E})_x = -\frac{\partial E_y}{\partial z} = -\mu \frac{\partial H_x}{\partial t} \quad \text{or} \quad H_x = -\frac{\beta}{\mu \omega} E_y = -\frac{\beta}{\mu \omega} E_0 \sin \alpha_e x \, e^{j(\omega t - \beta z)}$$

Thus

$$(H_x)_{\text{rms}} = \frac{\beta E_0}{\sqrt{2} \, \mu \omega} \sin \alpha_e x \qquad (H_y)_{\text{rms}} = 0$$

Similarly, using $E_x \equiv 0$,

$$(\nabla \times \mathbf{E})_z = -\frac{\partial E_y}{\partial x} = -\mu \frac{\partial H_z}{\partial t} \quad \text{or} \quad H_z = \frac{\alpha_e E_0}{\mu \omega} \cos \alpha_e x \, e^{j(\omega t - \beta z - \pi/2)}$$

whence

$$(H_z)_{\text{rms}} = \frac{\alpha_e E_0}{\sqrt{2} \, \mu \omega} \cos \alpha_e x$$

7.104 For the TE mode of Problem 7.102, find the time-average and instantaneous Poynting vectors.

▌ From Problems 7.102 and 7.103,

$$(E_y)_{\text{rms}} = \frac{E_0}{\sqrt{2}} \sin \alpha_e x \qquad \text{and} \qquad (H_x)_{\text{rms}} = \frac{\beta E_0}{\sqrt{2} \, \mu \omega} \sin \alpha_e x$$

Thus, the time-average Poynting vector becomes

$$(S_z)_{\text{av}} = E_y H_x = \frac{\beta E_0^2}{2 \mu \omega} \sin^2 \alpha_e x \qquad (1)$$

and the instantaneous Poynting vector is

$$S_z = \frac{\beta E_0^2}{\mu \omega} \sin^2 \alpha_e x \cos^2 (\omega t - \beta z) \qquad (2)$$

7.105 For the parallel-plate configuration of Problem 7.102, we have $w = 6$ cm, $E_0 = 1$ kV/m, and the operating frequency is $f = 3.5$ GHz. Calculate β for the lowest-order TE mode.

▌
$$\beta^2 = \frac{\omega^2}{u^2} - \left(\frac{1\pi}{w}\right)^2 = \frac{(2\pi \times 3500 \times 10^6)^2}{(3 \times 10^{10})^2} - \left(\frac{\pi}{6}\right)^2 = 0.0266\pi^2$$

or $\beta = 51.2 \text{ m}^{-1}$.

7.106 From the data of Problem 7.105, determine the cutoff frequency of the lowest mode.

▌
$$f_c = \frac{u}{2w} = \frac{3 \times 10^{10}}{2 \times 6} = 2.5 \text{ GHz}$$

7.107 How much time-averaged power is transmitted between the plates of Problem 7.105 over a 1-m width, in the lowest mode?

▌ The power per unit width is given by

$$P = \int_0^w (S_z)_{\text{av}} \, dz$$

Substituting from (1) of Problem 7.104, with $\alpha_e = \pi/w$, and integrating, we obtain

$$P = \frac{\beta w E_0^2}{4 \mu \omega} = \frac{52.1 \times 6 \times 10^{-2} \times (10^3)^2}{4 \times 4\pi \times 10^{-7} \times 2\pi \times 3.5 \times 10^9} = 28.28 \text{ W/m}$$

7.108 A waveguide having a square cross section of 9 cm on each side propagates energy in the TE$_{22}$ mode at 12 GHz: Given

$$H_z = H_0 \cos \alpha_x \, x \cos \alpha_y \, y \, e^{j(\omega t - \beta z)}$$

Determine the cutoff frequency.

▮ The cutoff frequency is obtained from

$$\omega_c = u\sqrt{\alpha_x^2 + \alpha_y^2} = u\sqrt{2(2\pi/a)^2} = 2\pi u\sqrt{2}/a$$

or

$$f_c = \frac{u\sqrt{2}}{a} = \frac{(3 \times 10^{10})\sqrt{2}}{9} = 4.71 \text{ GHz}$$

7.109 Determine the wavelength of the TE$_{22}$ wave in the guide of Problem 7.108.

▮ First we determine β from

$$\beta = \frac{1}{u}\sqrt{\omega^2 - \omega_c^2} = \frac{2\pi \times 10^9}{3 \times 10^8}\sqrt{(12)^2 - (4.71)^2} = 231.1 \text{ m}^{-1}$$

Then

$$\lambda = \frac{2\pi}{\beta} = \frac{2\pi}{231.1} = 2.72 \text{ cm}$$

7.110 Calculate the maximum amplitude of the total electric field for the TE$_{22}$ mode in the guide of Problems 7.108 and 7.109.

▮ The maximum amplitudes of the electric field components are given by

$$|E_x|_{\max} = \frac{\omega\mu}{h^2}H_0\alpha_x \qquad |E_y|_{\max} = \frac{\omega\mu}{h^2}H_0\alpha_y$$

where $h \equiv \omega_c/u$. Now, since E_x and E_y are out of phase spatially,

$$|E|_{\max} = |E_x|_{\max} = |E_y|_{\max}$$
$$= \frac{\omega u^2}{\omega_c^2}\mu H_0\left(\frac{2\pi}{a}\right) = \left(\frac{f}{f_c^2}\right)\frac{u^2\mu H_0}{a}$$

Substituting numerical values yields

$$|E|_{\max} = \frac{12}{(4.71)^2}\frac{(3 \times 10^8)^2 4\pi \times 10^{-7} \times 15}{9 \times 10^{-2}} = 10.2 \text{ kV/m}$$

7.111 A wave propagates in a waveguide in the z-direction in the lowest TE mode at a frequency 1.5 times cutoff frequency. The guide dimensions are $a = 5$ cm and $b = 1$ cm, as shown in Fig. 7-1. Calculate β for the given conditions.

▮ For TE$_{10}$,

$$\beta = \frac{\pi}{a}\sqrt{(f/f_c)^2 - 1} = \frac{\pi}{0.05}\sqrt{1.25} = 70.25 \text{ m}^{-1}$$

7.112 In the waveguide of Problem 7.111, the maximum amplitude of the electric field for the given mode is $|E_y|_{\max} = 1$ kV/m. Calculate the maximum amplitude of the magnetic field, assuming the guide filled with air.

▮ From Maxwell's equations,

$$H_x = -\frac{\beta}{\omega\mu}E_y$$

Substituting for β from Problem 7.111 and using $\omega = 2\pi(1.5u/2a) = 9\pi \times 10^9$ rad/s,

$$|H_x|_{\max} = \frac{70.25 \times 10^3}{9\pi \times 10^9 \times 4\pi \times 10^{-7}} = 1.98 \text{ A/m}$$

7.113 Calculate the cutoff frequency of the next-lowest TE mode in the waveguide of Problem 7.111.

▮ For TE$_{20}$,

$$\omega_c = u\sqrt{\left(\frac{2\pi}{a}\right)^2 + \left(\frac{0\pi}{b}\right)^2} = \frac{u\pi}{a}(2) = 2\pi f_c$$

or

$$f_c = \frac{u}{a} = \frac{3 \times 10^{10}}{5} = 6 \text{ GHz}$$

7.114 What is the cutoff frequency of the next-lowest TM mode in the waveguide of Problem 7.111?

■ For TM$_{21}$,

$$2\pi f_c = (3 \times 10^{10})\sqrt{\left(\frac{2\pi}{5}\right)^2 + \left(\frac{1\pi}{1}\right)^2} \quad \text{or} \quad f_c = 16.16 \text{ GHz}$$

7.115 If the signal frequency of the waveguide of Problem 7.111 is 10 GHz, list the modes (TE and TM) that may propagate.

■ TE$_{10}$ ($f_c = 3$ GHz); TE$_{20}$ ($f_c = 6$ GHz); TE$_{30}$ ($f_c = 9$ GHz).

7.116 The phase velocity of a wave in a certain material medium is

$$v_p = u_0(\lambda_0/\lambda)^2 \qquad (1)$$

where $u_0 = 3 \times 10^8$ m/s. Derive an expression for the group velocity.

■ From (1) and $v_p = \omega/\beta = \omega\lambda/2\pi$, one has

$$\omega = \frac{u_0\lambda_0^2}{4\pi^2}\beta^3 \qquad (2)$$

Thus

$$v_g = \frac{d\omega}{d\beta} = 3\left(\frac{u_0\lambda_0^2}{4\pi^2}\right)\beta^2 = 3u_0\left(\frac{\lambda_0}{\lambda}\right)^2 = 3v_p$$

7.117 A TE wave propagates between two parallel plates of infinite extent which are $a = 10$ cm apart (Fig. 7-15). The electric field of the wave is

$$E_z = E_0 \sin \alpha_x x \cos(\beta y - \omega t) \qquad (1)$$

If $\alpha_x = 30\pi$ m^{-1}, what mode is propagating?

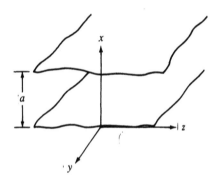

Fig. 7-15

■ The mode number is given by

$$\alpha_x = 30\pi = \frac{m\pi}{a} \quad \text{or} \quad m = 30a = 30(0.10) = 3$$

7.118 For the structure of Problem 7.117, find the cutoff frequency for the propagating mode.

■ The cutoff frequency for the mth mode is given by $f_c = m(u/2a)$. Hence, for $m = 3$,

$$f_c = 3\left(\frac{3 \times 10^{10}}{20}\right) = 4.5 \text{ GHz}$$

7.119 For the structure of Problem 7.117, write the magnetic field components in a TE mode.

■ Maxwell's equations

$$\frac{\partial E_z}{\partial y} = -\mu\frac{\partial H_x}{\partial t} \quad \text{and} \quad -\frac{\partial E_z}{\partial x} = -\mu\frac{\partial H_y}{\partial t}$$

give

$$H_x = \frac{\beta E_0}{\omega\mu}\sin \alpha_x x \cos(\beta y - \omega t) \quad \text{and} \quad H_y = -\frac{\alpha E_0}{\omega\mu}\cos \alpha_x x \sin(\beta y - \omega t)$$

and $E_x = E_y = 0$ implies $H_z = 0$.

7.120 The structure of Problem 7.117 is excited at a frequency equal to one-half the lowest cutoff frequency. Find the attenuation constant α for the resulting disturbance.

■ Because $\omega_c \equiv \alpha_x u$,

$$\alpha = \sqrt{\alpha_x^2 - (\omega^2/u^2)} = \alpha_x\sqrt{1 - (\omega/\omega_c)^2} \qquad (1)$$

For $m = 1$, $\alpha_x = \pi/a$; hence

$$\alpha = \frac{\pi}{0.10} \sqrt{\frac{3}{4}} = 27.2 \text{ m}^{-1}$$

7.121 What is the wavelength in Problem 7.120?

▮ Undefined; see Problem 7.14.

7.122 For the structure of Problem 7.117, assume that a frequency of 1.5 times the lowest cutoff frequency is applied. What is the value of β for the lowest mode?

▮ Analogous to (1) of Problem 7.120,

$$\beta = \alpha_x \sqrt{(\omega/\omega_c)^2 - 1} = \frac{\pi}{0.10} \sqrt{1.25} = 35.12 \text{ m}^{-1}$$

7.123 For the wave of Problem 7.122, what is the phase shift over a y-distance of $\lambda/4$, where λ is the free-space wavelength at the applied frequency?

▮ The phase shift is given by $\theta = \beta(\lambda/4)$, where β is as evaluated in Problem 7.122. Now, since $f_c = u/2a$ for the lowest mode,

$$\lambda = \frac{u}{f} = \frac{u/f_c}{f/f_c} = \frac{2a}{f/f_c} = \frac{0.20}{1.5} = \frac{2}{15} \text{ m}$$

Hence, $\theta = (35.12)(1/30) = 1.17 \text{ rad} = 67.07°$.

Observe that in problems of the type 7.120–7.123 it is never necessary to know f and f_c separately.

7.124 For the structure of Problem 7.117, write the field equations for the TEM mode.

▮ For the TEM mode, $E = E_x = E_0 \cos(\beta y - \omega t)$ and, from Chapter 5,

$$H_z = -\frac{E_x}{\eta_0} = -\frac{E_0}{\eta_0} \cos(\beta y - \omega t)$$

where $\eta_0 = \sqrt{\mu_0/\epsilon_0}$.

7.125 A waveguide of square cross section measures 6 cm on each side. A TE wave is identified by the following electric field:

$$E_y = E_0 \sin \alpha_1 x \cos \alpha_2 y \, e^{j(\omega t - \beta z)}$$

Determine the cutoff frequency of the waveguide in the lowest TE mode.

▮ The cutoff frequency of an (m, n) mode is given by

$$f_c = \frac{u}{2} \sqrt{\left(\frac{m}{a}\right)^2 + \left(\frac{n}{b}\right)^2} \qquad (1)$$

For the lowest TE mode, $m = 1$ and $n = 0$. With $u = 3 \times 10^{10}$ cm/s we have

$$f_c = \frac{3 \times 10^{10}}{2} \sqrt{\left(\frac{1}{6}\right)^2} = 2.5 \text{ GHz}$$

7.126 What is the lowest cutoff frequency for the TM modes in the waveguide of Problem 7.125?

▮ For the lowest TM mode, $m = 1$ and $n = 1$. Thus, (1) of Problem 7.125 yields

$$f_c = \frac{3 \times 10^{10}}{2} \sqrt{\left(\frac{1}{6}\right)^2 + \left(\frac{1}{6}\right)^2} = 3.536 \text{ GHz}$$

7.127 Calculate the maximum amplitude of H_x for the lowest TE mode, at a frequency twice the cutoff frequency, in the waveguide of Problem 7.125, if $E_0 = 50$ kV/m.

▮ Maxwell's equations imply

$$H_x = -\frac{\beta E_y}{\omega \mu}$$

However, by Problem 7.44,

$$\frac{\omega \mu}{\beta} = \frac{\eta_0}{\sqrt{1 - (f_c/f)^2}}$$

Thus

$$|H_x|_{\max} = E_0 \frac{\sqrt{1 - (f_c/f)^2}}{\eta_0} = (50 \times 10^3) \frac{\sqrt{3/4}}{120\pi} = 114.86 \text{ A/m}$$

7.128   Determine β for the waveguide of Problem 7.125, at a frequency of 3 GHz, for the lowest TE mode.

❚ $$\beta = \sqrt{\left(\frac{\pi}{a}\right)^2 - \frac{\omega^2}{u^2}} = \pi\sqrt{\left(\frac{1}{0.06}\right)^2 - \left(\frac{2\times 3\times 10^9}{3\times 10^8}\right)^2} = 34.73 \text{ m}^{-1}$$

7.129   Which modes could be propagated in the waveguide of Problem 7.125 at 5.2 GHz?

❚   By Problems 7.125 and 7.126, TE_{10}, TE_{01}, TE_{11}, and TM_{11} will propagate. Now try TE_{21} (etc.):
$$f_c = \frac{3\times 10^{10}}{2}\sqrt{\left(\frac{2}{6}\right)^2 + \left(\frac{1}{6}\right)^2} = 5.6 \text{ GHz} > 5.2 \text{ GHz}$$

Thus, TE_{21} (etc.) will not propagate. Next, try TE_{20} and TE_{02}:
$$f_c = \frac{3\times 10^{10}}{2}\sqrt{\left(\frac{2}{6}\right)^2 + 0} = 5 \text{ GHz}$$

and these modes will propagate. Thus, the allowed modes are TE_{10}, TE_{01}, TE_{20}, TE_{02}, TE_{11}, and TM_{11}. The TEM mode is not allowed at all (Problem 7.10).

7.130   An electromagnetic wave propagates in the lowest mode of an air-filled waveguide whose dimensions are 3×4 cm. The frequency is 1.2 times the lowest cutoff frequency, f_c. Find f_c.

❚   For TE_{10},
$$f_c = \frac{u}{2a} = \frac{3\times 10^{10}}{8} = 3.75 \text{ GHz}$$

7.131   Over what frequency band does only the lowest mode propagate in the waveguide of Problem 7.130?

❚   From Problem 7.130, for the lowest mode, $f_c = 3.75$ GHz. For the next-lowest mode, we have from (1) of Problem 7.125:
$$f_c' = \frac{u}{2b} = \frac{3\times 10^{10}}{2\times 3} = 5 \text{ GHz}$$

Hence, $\Delta f = f_c' - f_c = 1.25$ GHz.

7.132   If the larger waveguide dimension in Problem 7.130 *alone* is changed—from 4 to 5 cm—what will be the effect on the Poynting vector?

❚   The amplitude of the Poynting vector is given by
$$|\mathbf{S}| = \frac{\beta E_0^2}{2\omega\mu}$$
But
$$\frac{\beta}{\omega} = \frac{(\sqrt{\omega^2 - \omega_c^2})/u}{\omega} = \frac{1}{u}\sqrt{1 - (\omega_c/\omega)^2} = \frac{1}{u}\sqrt{1 - (1/1.2)^2}$$
so that $|\mathbf{S}|$ is unchanged. (The *total* power will, of course, increase by 25 percent.)

7.133   The waveguide of Fig. 7-1 is filled with a dielectric having $\epsilon_r = 4$. In a certain mode of operation,
$$H_z = 0 \qquad \text{and} \qquad H_x = 3\sin\frac{\pi x}{2b}\cos\frac{3\pi y}{b}\sin(\pi 10^{11}t - \beta z) \quad \text{(A/m)}$$
Calculate the cutoff frequency of this mode, given $b = 7.5$ mm.

❚   The mode is a TM ($H_z = 0$), and the form of H_x specifies it as TM_{13}. Therefore
$$f_{c.13} = \frac{u_0/\sqrt{\epsilon_r}}{2}\sqrt{\left(\frac{1}{2b}\right)^2 + \left(\frac{3}{b}\right)^2} = \frac{(3\times 10^8)/2}{2(7.5\times 10^{-3})}\sqrt{\frac{1}{4} + 9} = 22.81 \text{ GHz}$$

7.134   Evaluate β in Problem 7.133.

❚ $$\beta = \beta_{13} = \frac{\omega\sqrt{\epsilon_r}}{u_0}\sqrt{1 - (f_{c.13}/f)^2} = 1863.7 \text{ rad/m}$$

7.135   Show that in a waveguide operating below the cutoff frequency there is no net average power flow down the waveguide.

❚   Propagation of energy requires a propagating mode (cf. Problem 7.121).

7.136   An air-filled waveguide of square cross section, 10 cm on a side, transports energy in the lowest TE mode at the rate of 1 kW/s. If the operating frequency is 20 GHz, what is the peak value of the electric field in the guide?

❙ The electric and magnetic fields may be expressed as

$$E_x = E_0 \sin k_y y \cos(\omega t - \beta z) \qquad H_y = \frac{\beta E_x}{\omega \mu_0}$$

(This corresponds to TE_{01} for the waveguide of Fig. 7-1.) Calculating the time-average power density $\langle E_x H_y \rangle$ and integrating over the cross section yield

$$\langle P \rangle = 1000 \text{ W} = \frac{\beta E_0^2}{4\omega \mu_0} \times \text{area} \qquad (1)$$

In (1) substitute $\omega = 40\pi \times 10^9$ rad/s, $\mu_0 = 4\pi \times 10^{-7}$ H/m, area $= 0.01$ m^2, $u_0 = 3 \times 10^8$ m/s, $k_y = 10\pi$ m^{-1}, and

$$\beta = \sqrt{(\omega/u_0)^2 - k_y^2} = 277 \text{ rad/m}$$

to obtain $E_0 = 15.1$ kV/m.

7.137 Show that for a TE_{10} wave in a waveguide operating just above cutoff frequency, $\beta \approx (\pi/a)\sqrt{2(\Delta\omega/\omega_c)}$, where $\Delta\omega \equiv \omega - \omega_c$ and $a \equiv$ larger dimension of guide.

❙ As in Problem 7.122,

$$\beta = \frac{\pi}{a}\sqrt{\left(\frac{\omega}{\omega_c}\right)^2 - 1} = \frac{\pi}{a}\sqrt{\left(1 + \frac{\Delta\omega}{\omega_c}\right)^2 - 1} \approx \frac{\pi}{a}\sqrt{2\frac{\Delta\omega}{\omega_c}}$$

7.138 X-band microwave equipments operate between 8.2 GHz and 12.4 GHz. What are the dimensions of air-filled X-band waveguides to meet this frequency range?

❙ For $a > b$, operation in the dominant (TE_{10}) mode over the given frequency range is possible provided

$$f_{c.10} = \frac{u}{2a} = 8.2 \text{ GHz} \qquad \text{or} \qquad a = 1.83 \text{ cm}$$

and

$$f_{c.01} = \frac{u}{2b} = 12.4 \text{ GHz} \qquad \text{or} \qquad b = 1.21 \text{ cm}$$

7.139 For the air-filled waveguide of Fig. 7-1, $a = 1$ cm, $b = 3$ cm. Over what frequency band can only one mode be propagated?

❙ For the dominant (TE_{01}) mode:

$$f_{c.01} = \frac{u}{2b} = \frac{3 \times 10^{10}}{2 \times 3} = 5 \text{ GHz}$$

Similarly,

$$f_{c.02} = \frac{u}{b} = 10 \text{ GHz} \qquad \text{and} \qquad f_{c.10} = \frac{u}{2a} = \frac{3 \times 10^{10}}{2} = 15 \text{ GHz}$$

Thus, for only one (TE_{01}) mode, the required frequency spread is $10 - 5 = 5$ GHz.

7.140 An X-band microwave signal of frequency 10 GHz propagates down a waveguide of (1 cm) $\times$ (2.5 cm) cross section. How long should the waveguide be so that the signal down the guide emerges 1 μs after the one propagated in the air outside the guide?

❙
$$f_{c.10} = \frac{u}{2a} = \frac{3 \times 10^{10}}{2 \times 2.5} = 6 \text{ GHz}$$

$$f_{c.01} = \frac{u}{2b} = \frac{3 \times 10^{10}}{2 \times 1} = 15 \text{ GHz}$$

$$f_{c.20} = \frac{u}{a} = \frac{3 \times 10^{10}}{2.5} = 12 \text{ GHz}$$

Since the signal frequency is 10 GHz, only the TE_{10} mode will propagate. The group velocity of this mode is

$$v_g = u\sqrt{1 - (f_{c.10}/f)^2} = (3 \times 10^8)\sqrt{1 - (6/10)^2} = 2.4 \times 10^8 \text{ m/s}$$

For a guide of length L the delay is given by

$$\Delta = \frac{L}{v_g} - \frac{L}{u} \qquad \text{whence} \qquad L = 1.2 \text{ km}$$

7.141 A rectangular waveguide has dimensions $a = 6$ cm, $b = 2$ cm. What is the wavelength in the guide at 6 GHz in the TE_{10} mode?

❙ We have

$$f_{c.10} = \frac{u}{2a} = \frac{3 \times 10^{10}}{2 \times 6} = 2.5 \text{ GHz}$$

and
$$\lambda_{10} = \frac{u}{\sqrt{f^2 - f_{c.10}^2}} = \frac{30}{\sqrt{(6)^2 - (2.5)^2}} = 5.5 \text{ cm}$$

7.142 At what frequency for the TE_{10} mode in the waveguide of Problem 7.141, is the group velocity 90 percent of the phase velocity?

▮ The group velocity and the phase velocity are respectively given by
$$v_g = u\sqrt{1 - (f_c/f)^2} \qquad v_p = \frac{u}{\sqrt{1 - (f_c/f)^2}}$$

Thus
$$u\sqrt{1 - (f_c/f)^2} = \frac{0.9u}{\sqrt{1 - (f_c/f)^2}} \qquad \text{or} \qquad \frac{f_c}{f} = 0.316$$

whence $f = 2.5/0.316 = 7.9 \text{ GHz}$.

7.143 A cylindrical waveguide, referred to cylindrical coordinates, is shown in Fig. 7-16. Write the scalar Maxwell's equations within the guide, assuming that each field component is of the form $F(r, \phi, z) = F_0(r, \phi)e^{-\gamma z}$.

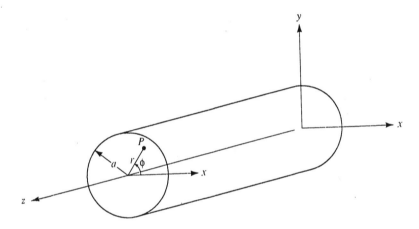

Fig. 7-16

▮
$$\frac{1}{r}\frac{\partial E_{z0}}{\partial \phi} + \gamma E_{\phi 0} + j\omega\mu H_{r0} = 0 \tag{1}$$

$$-\gamma E_{r0} - \frac{\partial E_{z0}}{\partial r} + j\omega\mu H_{\phi 0} = 0 \tag{2}$$

$$\frac{\partial E_{\phi 0}}{\partial r} + \frac{1}{r}E_{\phi 0} - \frac{1}{r}\frac{\partial E_{r0}}{\partial \phi} + j\omega\mu H_{z0} = 0 \tag{3}$$

$$\frac{1}{r}\frac{\partial H_{z0}}{\partial \phi} + \gamma H_{\phi 0} - j\omega\epsilon E_{r0} = 0 \tag{4}$$

$$-\gamma H_{r0} - \frac{\partial H_{z0}}{\partial r} - j\omega\epsilon E_{\phi 0} = 0 \tag{5}$$

$$\frac{\partial H_{\phi 0}}{\partial r} + \frac{1}{r}H_{\phi 0} - \frac{1}{r}\frac{\partial H_{r0}}{\partial \phi} - j\omega\epsilon E_{z0} = 0 \tag{6}$$

$$\frac{\partial E_{r0}}{\partial r} + \frac{E_{r0}}{r} + \frac{1}{r}\frac{\partial E_{\phi 0}}{\partial \phi} - \gamma E_{z0} = 0 \tag{7}$$

$$\frac{\partial H_{r0}}{\partial r} + \frac{H_{r0}}{r} + \frac{1}{r}\frac{\partial H_{\phi 0}}{\partial \phi} - \gamma H_{z0} = 0 \tag{8}$$

7.144 Write the governing equation for E_{z0} within the waveguide of Problem 7.143.

▮ In cylindrical coordinates the z-component of the **E**-field (or **H**-field)—and only that component—coincides with a rectangular component; hence it satisfies a scalar Helmholtz equation:
$$(\nabla^2 + \omega^2\mu\epsilon)(E_{z0}e^{-\gamma z}) = 0 \tag{1}$$
Transforming the Laplacian in (1) to cylindrical coordinates produces
$$\frac{1}{r}\frac{\partial}{\partial r}\left(r\frac{\partial E_{z0}}{\partial r}\right) + \frac{1}{r^2}\frac{\partial^2 E_{z0}}{\partial \phi^2} + h^2 E_{z0} = 0 \qquad (h^2 \equiv \gamma^2 + \omega^2\mu\epsilon) \tag{2}$$
The trivial solution to (2), $E_{z0} = 0$, corresponds to TE modes; nontrivial solutions correspond to TM modes.

7.145 Solve (2) of Problem 7.144 by separation of variables.

▮ Proceed as in Problem 2.232 and apply the constraints
   (i) $E_{z0}(r, \phi + 2\pi) = E_{z0}(r, \phi)$.
   (ii) $E_{z0}(0, \phi)$ finite.
   (iii) $E_{z0}(a, \phi) = 0$.
to find, for the TM_{np} mode,

$$E_{z0} = E_{np}J_n(h_{np}r)\cos n\phi \tag{1}$$

where $n = 0, 1, 2, \ldots$ and $h_{np}a$ is the pth positive root ($p = 1, 2, \ldots$) of the equation $J_n(v) = 0$.

7.146 Express E_{r0} and $E_{\phi 0}$ for TM_{np} in the waveguide of Problem 7.143.

▮ Since $H_z = 0$, (2) and (4) of Problem 7.143 become

$$\gamma_{np}E_{r0} - j\omega\mu H_{\phi 0} = -\frac{\partial E_{z0}}{\partial r} = -E_{np}h_{np}J'_n(h_{np}r)\cos n\phi$$

$$\gamma_{np}H_{\phi 0} - j\omega\epsilon E_{r0} = 0$$

where $\gamma_{np}^2 = h_{np}^2 - \omega^2\mu\epsilon$. Elimination of $H_{\phi 0}$ gives

$$E_{r0} = -\frac{\gamma_{np}E_{np}}{h_{np}}J'_n(h_{np}r)\cos n\phi \tag{1}$$

in which, for a propagating mode, $\gamma_{np} = j\beta_{np}$ (a pure imaginary).
   In the same way, elimination of H_{r0} between (1) and (5) of Problem 7.143 leads to

$$E_{\phi 0} = \frac{n\gamma_{np}E_{np}}{h_{np}^2 r}J_n(h_{np}r)\sin n\phi \qquad (\gamma_{np} = j\beta_{np}) \tag{2}$$

7.147 Calculate the magnetic field components in the waveguide of Problem 7.143, for the TM_{np} mode.

▮ Refer to Problem 7.146. By (5),

$$H_{r0} = -\frac{j\omega\epsilon}{\gamma_{np}}E_{\phi 0} = -\frac{j\omega\epsilon n E_{np}}{h_{np}^2 r}J_n(h_{np}r)\sin n\phi \tag{1}$$

and by (4),

$$H_{\phi 0} = \frac{j\omega\epsilon}{\gamma_{np}}E_{r0} = -\frac{j\omega\epsilon E_{np}}{h_{np}}J'_n(h_{np}r)\cos n\phi \tag{2}$$

and, of course, $H_{z0} = 0$. In vector form, $\mathbf{E}_T \cdot \mathbf{H}_T = 0$.

7.148 Evaluate h_{01} for a grid of 1-cm inner radius.

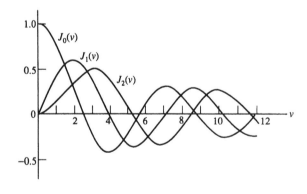

Fig. 7-17

▮ By Problem 7.145, $h_{01}a$ is the first positive zero of $J_0(v)$; i.e., 2.405 (Fig. 7-17). Hence

$$h_{01} = \frac{2.405}{0.01} = 240.5 \text{ m}^{-1}$$

7.149 If the waveguide of Problem 7.143 is air-filled and has an inner radius of 1.2 cm, what is the cutoff frequency for the TM_{01} mode?

▮ See Problem 7.146; the cutoff frequency for TM_{np} is given by $\gamma_{np} = 0$, or

$$f_{c.np} = h_{np}\frac{u}{2\pi} \tag{1}$$

Thus
$$f_{c.01} = h_{01}\frac{u}{2\pi} = \frac{2.405}{0.012}\left(\frac{3 \times 10^8}{2\pi}\right) = 9.57 \text{ GHz}$$

7.150 The waveguide of Problem 7.149 operates at 10 GHz. Determine the wavelength in the guide.

❙ The wavelength formula for rectangular guides (Problem 7.16) also holds for cylindrical guides:
$$\lambda_{np} = \frac{2\pi}{\beta_{np}} = \frac{u}{\sqrt{f^2 - f_{c.np}^2}} \tag{1}$$

Hence
$$\lambda_{01} = \frac{3 \times 10^8}{\sqrt{(10 \times 10^9)^2 - (9.57 \times 10^9)^2}} = 10.34 \text{ cm}$$

7.151 Sketch the electric and magnetic field lines in a transverse plane in the waveguide of Problem 7.143, for the TM$_{01}$ mode.

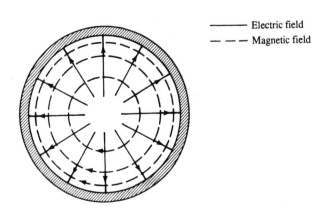

——— Electric field

— — — Magnetic field

Fig. 7-18

❙ The sketch is shown in Fig. 7-18. Notice that (2) of Problem 7.147 and Fig. 7-17 together imply that
$$|H_{\phi 0}| \propto |J_0'(h_{01}r)|$$
grows large as $r \to a$. Thus the magnetic field lines are drawn more and more densely as $r \to a$.

7.152 A cylindrical waveguide is operating in a TE mode. Write expressions for the magnetic field components.

❙ For TE modes, $E_z = 0$ and all field components are derived from $H_z(r, \phi, z) = H_{z0}(r, \phi)e^{-\gamma z}$. Now, H_{z0} obeys the Helmholtz equation (2) of Problem 7.144 and conditions (i) and (ii) of Problem 7.145. But condition (iii) is replaced by

(iii)′ $\left.\dfrac{\partial H_{z0}}{\partial r}\right|_{r=a} = 0$ [see (5) of Problem 7.143]

As a result, we have for TE$_{np}$:
$$H_{z0} = H_{np}J_n(h_{np}'r)\cos n\phi \tag{1}$$
where $n = 0, 1, 2, \ldots$ and $h_{np}'a$ is the pth positive root of $J_n'(v) = 0$.

In a process exactly similar to that of Problem 7.146, (1) and $E_{z0} = 0$ are substituted in the appropriate Maxwell's equations to give
$$H_{r0} = -\frac{\gamma_{np}'H_{np}}{h_{np}'}J_n'(h_{np}'r)\cos n\phi \qquad (\gamma_{np}' = j\beta_{np}') \tag{2}$$
$$H_{\phi 0} = \frac{\gamma_{np}'nH_{np}}{h_{np}'^2 r}J_n(h_{np}'r)\sin n\phi \qquad (\gamma_{np}' = j\beta_{np}') \tag{3}$$

7.153 Obtain the electric field components for the TE$_{np}$ modes.

❙ These are easiest derived from the orthogonality relation $\mathbf{E}_T \cdot \mathbf{H}_T = 0$:
$$E_{r0} = \frac{j\omega\mu nH_{np}}{h_{np}'^2 r}J_n(h_{np}'r)\sin n\phi \tag{1}$$
$$E_{\phi 0} = \frac{j\omega\mu H_{np}}{h_{np}'}J_n'(h_{np}'r)\cos n\phi \tag{2}$$

Of course, $E_{z0} = 0$.

7.154 Determine h_{11}' for a guide of 1-cm inner radius.

▮ From tables of Bessel functions we find that the first zero of $J_1'(v)$ occurs at 1.841; thus

$$h_{11}' = \frac{1.841}{1 \times 10^{-2}} = 184.1 \text{ m}^{-1}$$

7.155 Find the cutoff frequency for the TE_{11} mode in the waveguide of Problem 7.152, assuming that the guide is air-filled and has an inner radius of 1.2 cm.

▮ The cutoff frequency for TE_{np} is given by

$$f_{c.np}' = h_{np}' \frac{u}{2\pi} \tag{1}$$

Hence, from Problem 7.154,

$$f_{c.11}' = (184.1) \frac{3 \times 10^8}{2\pi} = 8.79 \text{ GHz}$$

7.156 What is the wavelength in the guide of Problem 7.155, if it operates at 10 GHz?

▮ Analogous to (1) of Problem 7.150,

$$\lambda_{11} = \frac{u}{\sqrt{f^2 - f_{c.11}'^2}} = \frac{30}{\sqrt{(10)^2 - (8.79)^2}} = 6.29 \text{ cm}$$

7.157 Sketch the electric and magnetic field lines in a transverse plane of the waveguide of Problem 7.152, for the TE_{11} mode.

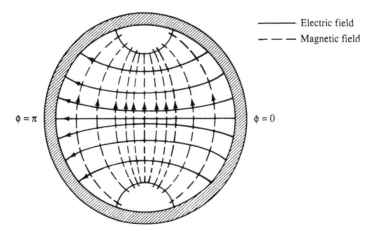

——— Electric field
– — – Magnetic field

$\phi = \pi$ $\phi = 0$

Fig. 7-19

▮ The sketch is shown in Fig. 7-19, which is in accordance with (2) and (3) of Problem 7.152 and (1) and (2) of Problem 7.153.

7.158 Show that TE_{11} is the dominant mode in any cylindrical waveguide.

▮ The cutoff frequency of a mode is proportional to the h- or h'-value for that mode, which in turn is proportional to a zero or a stationary point of a Bessel function. Of these, the smallest (positive) value is 1.841, the first maximum of J_1 (see Fig. 7-17). Thus, TE_{11} has the lowest cutoff frequency of any mode.

7.159 What modes can propagate in an air-filled guide of radius 10.89 mm, if the operating frequency is 12.5 GHz?

▮ By Problem 7.158 the guide cuts off at

$$f_{c.11}' = h_{11}' \frac{u}{2\pi} = \frac{1.841}{10.89 \times 10^{-3}} \frac{3 \times 10^8}{2\pi} = 8.075 \text{ GHz}$$

Moving to the right in Fig. 7-17, we have for the next two cutoff frequencies:

$$f_{c.01} = 8.075 \left(\frac{2.405}{1.841} \right) = 10.55 \text{ GHz}$$

$$f_{c.21}' = 8.075 \left(\frac{3.054}{1.841} \right) = 13.40 \text{ GHz}$$

Thus, at 12.5 GHz, only TE_{11} and TM_{01} propagate.

7.160 Determine the cutoff wavelength for the TE_{01} mode in a cylindrical waveguide having an inner radius of 5 cm.

▮ Analogous to (2) of Problem 7.67,

$$\lambda_{c.01} = \frac{u}{f'_{c.01}} = \frac{2\pi a}{h'_{01}a} = \frac{2\pi(5)}{3.832} = 8.2 \text{ cm}$$

since J_0 has its first minimum at 3.832.

7.161 Repeat Problem 7.160 for the dominant mode.

▮

$$\lambda_{c.11} = \frac{2\pi a}{h'_{11}a} = \frac{2\pi(5)}{1.841} = 17.1 \text{ cm}$$

7.162 A dielectric slab of thickness $2d$, shown in Fig. 7-20, extends to infinity in the x- and z-directions. This slab may be considered as a waveguide supporting propagation in the z-direction. For TM modes, $H_z = 0$ and $E_z = E_{z0}(y)e^{-\gamma z}$. Determine the function E_{z0}.

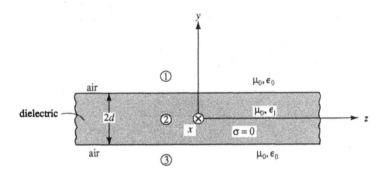

Fig. 7-20

▮ The Helmholtz equation for E_{z0} becomes

$$\frac{d^2 E_{z0}}{dy^2} + h^2 E_{z0} = 0 \qquad \text{where} \qquad h^2 = \gamma^2 + \omega^2 \mu_0 \epsilon \tag{1}$$

Since the dielectric is lossless, we set $\gamma = j\beta$ and write $h = k_y$ (k_y real and positive), or

$$k_y^2 = \omega^2 \mu_0 \epsilon_1 - \beta^2 \tag{2}$$

Then the general solution to (1) within the dielectric is:

$$E_{z0}(y) = C_1 \sin k_y y + C_2 \cos k_y y \qquad (-d < y < d) \tag{3}$$

In the (lossless) air surrounding the dielectric we require decreasing exponential solutions of (1), to make E_{z0} vanish at great distances. Therefore we set $h = j\kappa$ (κ real and positive), or

$$\kappa^2 = \beta^2 - \omega^2 \mu_0 \epsilon_0 \tag{4}$$

to obtain

$$E_{z0}(y) = \begin{cases} C_3 e^{-\kappa(y-d)} & y \geq d \\ C_4 e^{\kappa(y+d)} & y \leq -d \end{cases} \tag{5}$$

7.163 In the dielectric of Problem 7.162, for odd TM modes, display the components of the electric and magnetic fields.

▮ For odd TM modes, $E_{z0}(y)$ is an odd function:

$$E_{z0}(y) = C_1 \sin k_y y \qquad (-d \leq y \leq d) \tag{1}$$

$$E_{x0}(y) = 0 \tag{2}$$

$$E_{y0}(y) = -\frac{j\beta}{k_y^2} \frac{\partial E_{z0}}{\partial y} = -\frac{j\beta C_1}{k_y} \cos k_y y \tag{3}$$

From Maxwell's equations:

$$H_{x0}(y) = \frac{j\omega\epsilon}{k_y^2} \frac{\partial E_{z0}}{\partial y} = \frac{j\omega\epsilon_1 C_1}{k_y} \cos k_y y \tag{4}$$

$$H_{y0}(y) = 0 \tag{5}$$

$$H_{z0}(y) = 0 \tag{6}$$

7.164 Express the electric and magnetic field components in region 1 of Fig. 7-20, for odd TM modes.

▮

$$E_{x0}(y) = 0 \tag{1}$$

$$E_{y0}(y) = -\frac{j\beta}{\kappa} C_3 e^{-\kappa(y-d)} \tag{2}$$

$$E_{z0}(y) = C_3 e^{-\kappa(y-d)} \tag{3}$$

$$H_{x0}(y) = \frac{j\omega\epsilon_0}{\kappa} C_3 e^{-\kappa(y-d)} \tag{4}$$

$$H_{y0}(y) = H_{z0}(y) = 0 \tag{5}$$

7.165 Evaluate the constant C_3, appearing in (2), (3), and (4) of Problem 7.164 in terms of C_1, d, and k_y, for odd TM modes.

▌ The continuity of E_{z0} at the interface $y = d$ requires
$$C_3 = C_1 \sin k_y d \tag{1}$$

7.166 Let a signal of frequency ω excite an odd TM mode in the dielectric waveguide of Fig. 7-20. Determine the parameters k_y and κ for this frequency.

▌ Since H_{x0} is continuous at $y = d$, we must have—from (4) of Problem 7.163, (4) of Problem 7.164, and (1) of Problem 7.165—

$$\frac{j\omega\epsilon_1}{k_y}\cos k_y d = \frac{j\omega\epsilon_0 C_1}{\kappa}\sin k_y d \qquad \text{or} \qquad \frac{\kappa}{k_y} = \frac{\epsilon_0}{\epsilon_1}\tan k_y d \tag{1}$$

And (2) and (4) of Problem 7.162 yield:
$$\kappa^2 + k_y^2 = \omega^2\mu_0(\epsilon_1 - \epsilon_0) \tag{2}$$
(1) and (2) may be solved (by an iterative technique) for k_y and κ.

7.167 Verify that the characteristics of odd TM waves along a dielectric slab of thickness $2d$ are the same as those of odd TM modes supported by a dielectric slab of thickness d that is backed by a perfectly conductive plane.

▌ As a continuous odd function of y, E_{z0} necessarily vanishes at $y = 0$.

7.168 Repeat Problem 7.166 for even TM modes.

▌ For even TM modes, $E_{z0}(y)$ is an even function:
$$E_{z0}(y) = C_2 \cos k_y y \qquad (-d \le y \le d) \tag{1}$$
In air,
$$E_{z0}(y) = C_4 e^{-\kappa(y-d)} \qquad (y \ge d) \tag{2}$$
Continuity of E_{z0} at $y = d$ requires that
$$C_4 = C_2 \cos k_y d \tag{3}$$
From Maxwell's equations, we have, in the dielectric slab,
$$H_{x0}(y) = \frac{j\omega\epsilon_1 C_2}{k_y}\sin k_y y \qquad (-d \le y \le d) \tag{4}$$
and in the air above the dielectric,
$$H_{x0}(y) = -\frac{j\omega\epsilon_0}{\kappa} C_4 e^{-\kappa(y-d)} \qquad (y \ge d) \tag{5}$$
Continuity of H_{x0} at $y = d$ yields
$$\frac{\kappa}{k_y} = -\frac{\epsilon_0}{\epsilon_1}\cot k_y d \tag{6}$$
Equation (6) takes the place of (1) of Problem 7.166. Equation (2) of Problem 7.166, which holds for any sort of mode, is the requisite second equation.

7.169 Determine the range of the phase constant for TM waves in the dielectric waveguide of Problem 7.162.

▌ For real κ and k_y, (2) and (4) of Problem 7.162 require
$$\omega\sqrt{\mu_0\epsilon_0} < \beta < \omega\sqrt{\mu_0\epsilon_1}$$
Equivalently, the phase velocity ω/β in the guide must be intermediate between the plane-wave velocity in the dielectric material and that in air.

7.170 Find the cutoff frequencies of the odd TM modes of the dielectric waveguide of Problem 7.162.

▌ From Problem 7.169, at cutoff $\beta = \omega_c\sqrt{\mu_0\epsilon_0}$. Combine this with (2) of Problem 7.162 to obtain
$$(k_y^2)_c = \omega_c^2(\mu_0\epsilon_1 - \mu_0\epsilon_0) \tag{1}$$
This, together with (2) of Problem 7.166, implies that $\kappa_c = 0$; hence, by (1) of Problem 7.166,
$$\tan\left[\omega_c d\sqrt{\mu_0(\epsilon_1 - \epsilon_0)}\right] = 0 \qquad \text{or} \qquad \omega_c d\sqrt{\mu_0(\epsilon_1 - \epsilon_0)} = (n-1)\pi \quad (n = 1, 2, \ldots)$$

Finally,

$$f_c = \frac{n-1}{2d\sqrt{\mu_0(\epsilon_1 - \epsilon_0)}} \quad (n = 1, 2, \ldots) \tag{2}$$

The integer n figuring in (2) is the eigenvalue of the mode, which may henceforth be designated $TM_{odd.n}$.

7.171 Repeat Problem 7.170 for even TM modes

▮ Now $\kappa_c = 0$ and (6) of Problem 7.168 together yield

$$f_c = \frac{n - 1/2}{2d\sqrt{\mu_0(\epsilon_1 - \epsilon_0)}} \quad (n = 1, 2, \ldots) \tag{1}$$

as the cutoff frequency of $TM_{even.n}$.

7.172 Determine $H_{z0}(y)$ for TE waves in the dielectric slab of Fig. 7-20.

▮ All equations of Problem 7.162 remain valid when E_{z0} for a TM mode is replaced by H_{z0} for a TE mode.

7.173 Repeat Problem 7.163 for odd TE modes.

▮ This time, start with $H_{z0} = C_1 \sin k_y y$ and $E_{z0} \equiv 0$; you find, for $-d \leq y \leq d$,

$$H_{z0}(y) = C_1 \sin k_y y \tag{1}$$

$$H_{y0}(y) = -\frac{j\beta}{k_y} C_1 \cos k_y y \tag{2}$$

$$H_{x0}(y) = 0$$

The electric field components, obtained from Maxwell's equations, are

$$E_{z0}(y) = E_{y0}(y) = 0$$

$$E_{x0}(y) = -\frac{j\omega\mu_0}{k_y} C_1 \cos k_y y \tag{3}$$

Needless to say, the same symbol (C_1) stands for two unrelated arbitrary constants in Problems 7.163 and 7.173.

7.174 Express the electric and magnetic field components in region 1 of Fig. 7-20, for odd TE modes.

▮ With

$$H_{z0}(y) = \begin{cases} C_1 \sin k_y y & -d \leq y \leq d \\ C_3 e^{-\kappa(y-d)} & d \leq y \end{cases}$$

the continuity of tangential **H** at $y = d$ leads to

$$C_3 = C_1 \sin k_y d$$

Then, proceeding as in Problem 7.173, we obtain, for $y \geq d$,

$$E_{z0}(y) = C_1 \sin k_y d\, e^{-\kappa(y-d)} \tag{1}$$

$$H_{y0}(y) = -\frac{j\beta C_1}{\kappa} \sin k_y d\, e^{-\kappa(y-d)} \tag{2}$$

$$E_{x0}(y) = -\frac{j\omega\mu_0 C_1}{\kappa} \sin k_y d\, e^{-\kappa(y-d)} \tag{3}$$

7.175 Find an expression for the cutoff frequencies of odd TE modes in the dielectric waveguide of Fig. 7-20.

▮ Continuity of E_{x0} at $y = d$ requires, in view of (3) of Problem 7.173 and (3) of Problem 7.174,

$$\frac{\kappa}{k_y} = \tan k_y d \tag{1}$$

which is essentially the same as (1) of Problem 7.166. It follows that odd TE modes and odd TM modes have the same set of cutoff frequencies, as given by (2) of Problem 7.170.

7.176 Determine the thickness of a dielectric-slab waveguide, having $\epsilon_r = 5.0$, to support a $TM_{even.1}$ wave at a cutoff frequency of 10 GHz.

▮ From (1) of Problem 7.171 we have, with $n = 1$,

$$f_c = \frac{1/2}{2d\sqrt{\mu_0\epsilon_0(\epsilon_r - 1)}} = \frac{u_0/2}{2d\sqrt{\epsilon_r - 1}} = \frac{1.5 \times 10^8}{2d\sqrt{5 - 1}} = 10 \times 10^9$$

or $2d = 0.75$ cm.

7.177 Repeat Problem 7.176 for $TM_{odd.1}$ and $TM_{odd.2}$.

▮ From (2) of Problem 7.170 it follows that the $TM_{odd.1}$ wave will propagate regardless of the thickness of the dielectric. For $TM_{odd.2}$,

$$f_c = \frac{1}{2d\sqrt{\mu_0\epsilon_0(\epsilon_r - 1)}} = \frac{3 \times 10^8}{2d\sqrt{5-1}} = 10 \times 10^9 \quad \text{or} \quad 2d = 1.5 \text{ cm}$$

7.178 A surface wave propagates in the odd TM mode along a very thin dielectric having a relative permittivity ϵ_r. If $k_y d \ll 1$, obtain an approximate expression for the attenuation constant κ.

▌ From (1) of Problem 7.166,

$$\kappa = \frac{k_y}{\epsilon_r}\tan k_y d \approx \frac{k_y^2 d}{\epsilon_r} \quad \text{or} \quad k_y^2 \approx \epsilon_r \kappa/d$$

Substituting this in (2) of Problem 7.166 yields, after rearrangement,

$$\kappa d + \epsilon_r = \frac{d}{\kappa}[\omega^2\mu_0\epsilon_0(\epsilon_t - 1)]$$

If we now assume that $\kappa d \ll \epsilon_r$ and solve for κ, we obtain

$$\kappa \approx \mu_0\epsilon_0\omega^2(\epsilon_r - 1)\frac{d}{\epsilon_r} \quad (\text{Np/m}) \tag{1}$$

7.179 A 1-mm-thick dielectric sheet supports a surface wave in the odd TM mode at 10 GHz. Determine the relative permittivity of the dielectric, if the attenuation is not to exceed 16 Np/m.

▌ Since the dielectric is very thin, we may use (1) of Problem 7.178:

$$16 = \frac{(2\pi \times 10^{10})^2}{(3 \times 10^8)^2}(\epsilon_r - 1)\frac{0.5 \times 10^{-3}}{\epsilon_r} \quad \text{whence} \quad \epsilon_r = 3.7$$

7.180 Detemine the time-average power per unit width (x-direction) for a TM_{odd} wave propagative within the dielectric waveguide of Fig. 7-20.

▌ The power flows in the z-direction, and the average value of the linear power density is obtained from the average Poynting vector, $(S_z)_{av} = \frac{1}{2}E_{y0}H_{x0}$, where E_{y0} and H_{x0} are, respectively, given by (3) and (4) of Problem 7.163. Thus

$$P_{av} = \int_{-d}^{d}\frac{1}{2}E_{y0}H_{x0}\,dy = \int_0^d E_{y0}H_{x0}\,dy$$
$$= \int_0^d \frac{\beta\omega_1\epsilon_1 C_1^2}{k_y^2}\cos^2 k_y y\,dy = \frac{\omega\epsilon_1\beta C_1^2}{2k_y^2}\left(2d + \frac{1}{k_y}\sin 2k_y d\right)$$

where C_1 is the maximum amplitude of E_z.

7.181 Dielectric rods or fibers (Fig. 7-21), also known as *optical fibers*, can act as cylindrical waveguides. As indicated, the refractive indices of the core (or fiber) and cladding (or sheath) are n_1 and $n_2 < n_1$, respectively. Rays (plane waves) enter the core from an external medium of refractive index n_3. Let θ_e denote the largest entry angle such that the ray is propagated down the fiber by total internal reflection. (θ_e is thus the *aperture* of the system.) Express θ_e in terms of n_1, n_2, and n_3.

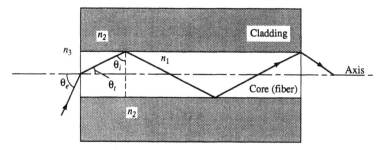

Fig. 7-21

▌ We seek θ_e such that $\theta_i = 90° - \theta_t = \theta_c$, the critical angle at the core-cladding interface. By Problem 5.140,

$$\sin \theta_c = \frac{n_2}{n_1} \tag{1}$$

Thus, from Snell's law,

$$\sin \theta_e = \frac{n_1}{n_3}\sin \theta_t = \frac{n_1}{n_3}\cos \theta_c$$
$$= \frac{n_1}{n_3}\sqrt{1 - \frac{n_2^2}{n_1^2}} = \frac{\sqrt{n_1^2 - n_2^2}}{n_3} \tag{2}$$

For nonmagnetic materials, $n = \sqrt{\epsilon_r}$, and (2) becomes

$$\sin \theta_e = \sqrt{\frac{\epsilon_1 - \epsilon_2}{\epsilon_3}} \tag{3}$$

In many applications the external medium is air ($n_3 = 1$), and the result becomes

$$\text{numerical aperture (NA)} \equiv \sin \theta_e = \sqrt{n_1^2 - n_2^2} = \sqrt{\epsilon_{r1} - \epsilon_{r2}} \tag{4}$$

7.182 For the dielectric-slab waveguide shown in Fig. 7-22 we have $d = 1 \, \mu$m, and refractive indices are $n_1 = 2.23$ and $n_2 = 2.2$. An argon-ion laser is used to focus radiation into this slab, at $\lambda_0 = 514.5$ nm. At cutoff, calculate the angle the propagation vector makes with the normal to the interface.

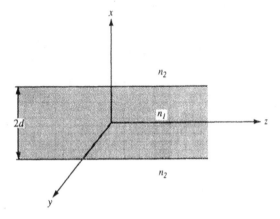

Fig. 7-22

▮ The required angle is the critical angle θ_c. By (1) of Problem 7.181,

$$\sin \theta_c = \frac{2.2}{2.23} = 0.9865 \qquad \text{or} \qquad \theta_c = 80.6°$$

7.183 For the data of Problem 7.182, determine the evanescent wave decay constant κ for an angle $\theta = 85°$ ($> 80.6°$), measured from the perpendicular to the interface.

▮ The decay constant is given by

$$\kappa = \beta_2 \sqrt{\left(\frac{n_1}{n_2}\right)^2 \sin \theta - 1}$$

However, by definition of the refractive index, $\beta_2 = 2\pi/\lambda_2 = 2\pi n_2/\lambda_0$; whence

$$\kappa = \frac{2\pi n_2}{\lambda_0} \sqrt{\left(\frac{n_1}{n_2}\right)^2 \sin \theta - 1} = \frac{2\pi \times 2.2}{514.5 \times 10^{-9}} \sqrt{\left(\frac{2.23}{2.2}\right)^2 \sin 85° - 1} = 4.12 \times 10^6 \, \text{m}^{-1}$$

7.184 What are the *mode numbers* of an optical fiber, and how are they related to the cutoff wavelengths of the modes?

▮ A mode number V is the product of the fiber radius and a radial eigenvalue; in the notation of previous problems, $V = h_{np}a$ (TM modes) or $V = h'_{np}a$ (TE modes).

For an ordinary cylindrical waveguide, the cutoff wavelength and the mode number are related by

$$\lambda_c = \frac{2\pi a}{V} \tag{1}$$

(see Problems 7.67 and 7.160). For an optical fiber in air, relation (1) is changed to

$$\lambda_c = \frac{2\pi a \sqrt{n_1^2 - n_2^2}}{V} = \frac{2\pi a (\text{NA})}{V} \tag{2}$$

7.185 Give a fairy-tale derivation of (2) of Problem 7.184.

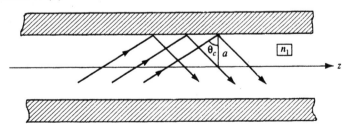

Fig. 7-23

❚ See Fig. 7-23, which shows a bundle of rays that have entered the fiber at the limiting angle θ_c (Fig. 7-21). Their wavelength, λ_c in air, becomes λ_c/n_1 in the core; therefore they "perceive" the core dimensions as magnified by the factor n_1. As against this, the effective radius of the core is not a, but its perpendicular projection

$$a \cos \theta_c = a \frac{\sqrt{n_1^2 - n_2^2}}{n_1}$$

With both effects taken into account, the bundle "sees itself" as propagating straight down the axis of a cylindrical waveguide of radius

$$n_1 \left(a \frac{\sqrt{n_1^2 - n_2^2}}{n_1} \right) = a\sqrt{n_1^2 - n_2^2}$$

and (1) of Problem 7.184 yields (2).

7.186 Design a multimode fiber with a step-function refractive-index profile. Let the mode number be 100 and the numerical aperture be 0.27. This fiber is to be used in a data link with a source wavelength of 0.85 μm. In particular, determine the core radius and the refractive indices n_1 and n_2.

❚ The governing equation is (2) of Problem 7.184. Choose a fused silica core with $n_1 = 1.458$; then

$$a = \frac{V\lambda_c}{2\pi(\mathrm{NA})} = \frac{100 \times 0.85 \times 10^{-6}}{2\pi \times 0.27} = 50.1 \ \mu\mathrm{m}$$

and, since $\mathrm{NA} = \sqrt{n_1^2 - n_2^2}$,

$$n_2 = \sqrt{n_1^2 - (\mathrm{NA})^2} = \sqrt{(1.458)^2 - (0.27)^2} = 1.433$$

(A cladding radius of 62.5 μm may be chosen.)

7.187 Can the fiber designed in Problem 7.186 operate in single (TE_{11}) mode?

❚ No: For single-mode operation the free-space wavelength must exceed the cutoff wavelength of the second-lowest mode, TM_{01}. Equivalently, the mode number must be smaller than that of TM_{01}, which is 2.405 (first zero of J_0).

7.188 A graded-index fiber has the refractive-index profile $n_1 = 3.3\,[1 - (r/a)^2]$ ($r < a$), where a is the core radius. Solve the paraxial-ray equation

$$-r\left(\frac{d\phi}{dz}\right)^2 = \frac{1}{n_a}\frac{dn_1}{dr} \tag{1}$$

for $\phi(z)$, given $\phi = \pi$ at $z = 0$. In (1), n_a is the average refractive index:

$$n_a = \frac{1}{a}\int_0^{\bullet} n_1 \, dr = 2.2$$

❚

$$-r\left(\frac{d\phi}{dz}\right)^2 = \frac{1}{2.2}(3.3)\left(-\frac{2r}{a^2}\right)$$

$$\frac{d\phi}{dz} = \pm \frac{\sqrt{3}}{a}$$

$$\int_\pi^\phi d\phi = \pm \frac{\sqrt{3}}{a}\int_0^z dz$$

$$\phi = \pi \pm \frac{\sqrt{3}}{a}z$$

7.189 An optical fiber has a numerical aperture of 0.2. Determine the acceptance angle for this fiber in water ($n_3 = 1.33$).

❚ By (2) of Problem 7.181,

$$\sin \theta_e = \frac{0.2}{1.33} \qquad \text{or} \qquad \theta_e = 8.65°$$

7.190 For the dielectric waveguide (fiber) shown in Fig. 7-24, calculate the time required for an electromagnetic wave to propagate through 1 km along the axis.

❚ The velocity along the axis is

$$v_{\mathrm{core}} = \frac{u_0}{n_1} = \frac{3 \times 10^8}{1.45} = 2.069 \times 10^8 \ \mathrm{m/s}$$

giving the required time as

$$t = \frac{1000}{2.069 \times 10^8} = 4.83 \ \mu\mathrm{s}$$

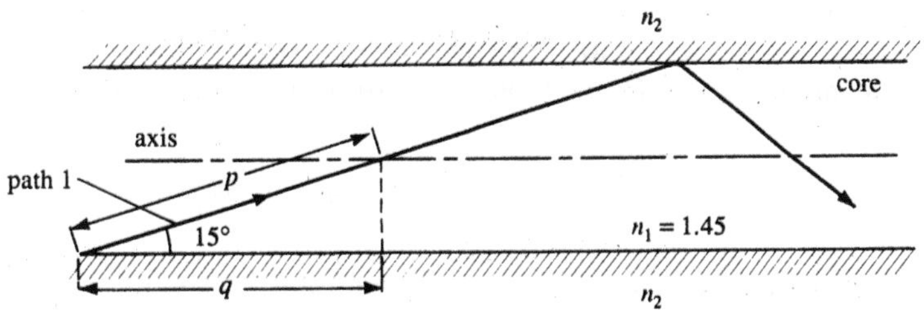

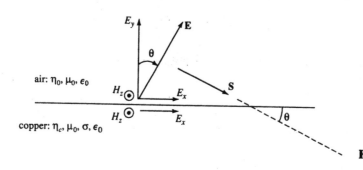

Fig. 7-24

7.191 For the guide of Problem 7.190, find the time of travel along path 1, shown in Fig. 7-24.

    / Path 1 is longer than the axial path by the factor $p/q = \sec 15° = 1.0353$. Thus, the required time is $t = (1.0353)(4.83) = 5\ \mu s$.

7.192 Electromagnetic power launched into an optical fiber link with lossless connectors is 1.5 mW. The fiber has a 0.5 dB/km attenuation. Calculate the fiber length such that a power of 2 μW is received at the other end of the fiber.

    / With L in km, the attenuation in decibels per kilometer is
$$\alpha = \frac{10}{L}\log_{10}\frac{P_{\text{in}}}{P_{\text{out}}} = \frac{10}{L}\log\left(\frac{1.5\times 10^{-3}}{2\times 10^{-6}}\right) = \frac{28.75}{L}$$
Thus $L = 28.75/0.5 = 57.50$ km.

7.193 A 15-km optical fiber link uses fibers with a loss of 1.5 dB/km. The fiber is joined every kilometer with connectors that contribute an attenuation of 0.8 dB each. Determine the minimum optical power that must be launched into the fiber in order to maintain an optical power level of 0.3 μW at the detector.

    /   Attenuation in dB $= 10\log_{10}\dfrac{P_{\text{near end}}}{P_{\text{far end}}}$

    Total contributed attenuation (14 connectors) $= 14 \times 0.8 = 11.2$ dB
    Attenuation over fiber $= 15 \times 1.5 = 22.5$ dB
    Total attenuation $= 11.2 + 22.5 = 33.7$ dB

Thus $33.7 = 10\log_{10}\dfrac{P_{\text{near end}}}{0.3\times 10^{-6}}$ whence $P_{\text{near end}} = 70.32$ mW

7.194 A single-mode fiber has mode number $V = 2.10$. For a core radius $a = 5\ \mu$m and refractive indices $n_1 = n_2 + 0.003 = 1.445$, calculate the cutoff wavelength.

    / Substitution of the data in (2) of Problem 7.184 yields $\lambda_c = 1.392\ \mu$m.

7.195 A multimode graded-index fiber has an acceptance angle in air of 8°. Determine $n_2 = n_{\text{cladding}}$, if $n_1 = n_{\text{core}} = 1.52$.

    / By (4) of Problem 7.181,
$$\sin 8° = \sqrt{(1.52)^2 - n_2^2} \quad \text{or} \quad n_2 = 1.514$$

7.196 Refer to the interface of Fig. 7-25. A surface wave travels along the copper surface. However, the Poynting vector **S** has a tilt, as shown, because copper has a finite conductivity ($E_x \neq 0$). Express the time-average Poynting vector in terms of the intrinsic impedances η_0 and η_c and the amplitude of the magnetic field H_{z0}.

Fig. 7-25

▌ By Problem 4.95,

$$\mathbf{S}_{av} = \tfrac{1}{2}\operatorname{Re}(\mathbf{E} \times \mathbf{H}^*) = \tfrac{1}{2}\operatorname{Re}(E_y H_z^*)\,\mathbf{a}_x + \tfrac{1}{2}\operatorname{Re}(-E_x H_z^*)\,\mathbf{a}_y \qquad (1)$$

However, by definition, $E_y = \eta_0 H_z$ and (both tangential components are continuous) $E_x = \eta_c H_z$. Substitution in (1) then gives

$$\mathbf{S}_{av} = \frac{1}{2}H_{z0}^2(\operatorname{Re}\eta_0\,\mathbf{a}_x - \operatorname{Re}\eta_c\,\mathbf{a}_y) \qquad (2)$$

7.197 Obtain an expression for the tilt angle θ (Fig. 7-25) for a 10-GHz surface wave.

▌ From (2) of Problem 7.196,

$$\tan\theta = \frac{\operatorname{Re}\eta_c}{\operatorname{Re}\eta_0} \qquad (1)$$

For air, $\eta_0 = 377\,\Omega$, and for copper (Problem 5.21),

$$\eta_c \approx \sqrt{\frac{\mu_0\omega}{\sigma}}\,\underline{/45^\circ} = \sqrt{\frac{4\pi \times 10^{-7} \times 2\pi \times 10 \times 10^9}{5.7 \times 10^7}}\,\underline{/45^\circ} = (2.63 + j2.63)10^{-2}\,\Omega$$

Thus

$$\tan\theta = \frac{2.63 \times 10^{-2}}{377} \quad \text{or} \quad \theta = 0.004^\circ$$

7.198 Repeat Problem 7.197 for an interface between air and seawater ($\epsilon_r = 81$, $\sigma = 4\,\text{S/m}$).

▌ Because seawater is a poor conductor, the exact impedance formula must be used:

$$\eta_c = \sqrt{\frac{j\omega\mu_0}{\sigma + j\omega\epsilon}} = \sqrt{\frac{j2\pi \times 10 \times 10^9 \times 4\pi \times 10^{-7}}{4 + j2\pi \times 10 \times 10^9 \times (10^{-9}/36\pi) \times 81}}$$

$$= 72.836\,\underline{/7.8^\circ} = 72.16 + j9.89\quad\Omega$$

Thus

$$\tan\theta = \frac{72.16}{377} \quad \text{or} \quad \theta = 10.8^\circ$$

7.199 A 10-GHz wave travels along the surface of the copper sheet shown in Fig. 7-25. Determine $(S_x)_{av}$ and $(S_y)_{av}$ if $(E_y)_{rms} = 100\,\text{V/m}$.

▌ From Problem 7.197, $\eta_c = (2.63 + j2.63)10^{-2}\,\Omega$ and $\eta_0 = 377\,\Omega$. The derivation of Problem 7.196 then yields

$$(S_x)_{av} = \frac{[(E_y)_{rms}]^2}{\eta_0} = \frac{100^2}{377} = 26.53\,\text{W/m}^2$$

$$(S_y)_{av} = [(H_z)_{rms}]^2\operatorname{Re}\eta_c = \frac{[(E_y)_{rms}]^2}{\eta_0^2}\operatorname{Re}\eta_c = \frac{100^2}{377^2} \times 2.63 \times 10^{-2} = 1.85\,\text{mW/m}^2$$

7.200 A wave travels parallel to an interface between air and a metallic sheet (Fig. 7-25). Given $\eta_c = (0.03 + j0.03)\,\Omega$ and $(E_y)_{rms} = 200\,\text{V/m}$, determine $|\mathbf{S}_{av}|$.

▌ Following Problem 7.199, we have

$$(S_x)_{av} = \frac{(200)^2}{377} = 106.1\,\text{W/m}^2 \qquad (S_y)_{av} = \left(\frac{200}{377}\right)^2 \times 0.03 = 8.44\,\text{mW/m}^2$$

whence $|\mathbf{S}_{av}| \approx 106.1\,\text{W/m}^2$.

7.201 For the interface shown in Fig. 7-25 we have $\sigma = 20\,\text{MS/m}$ and $(E_y)_{rms} = 200\,\text{mW/m}$. Calculate the average power dissipated in the conducting medium at 4 GHz.

▌ The dissipated power is given by $(S_x)_{av} = \tfrac{1}{2}H_{z0}^2\operatorname{Re}\eta_c$, where

$$\operatorname{Re}\eta_c = \sqrt{\frac{\mu_0\omega}{2\sigma}} = \sqrt{\frac{4\pi \times 10^{-7} \times 2\pi \times 4 \times 10^9}{20 \times 10^6}} = 0.0397\,\Omega$$

$$H_{z0} = \frac{\sqrt{2}\,(E_y)_{rms}}{\eta_0} = \frac{\sqrt{2} \times 200 \times 10^{-3}}{377} = 0.75 \times 10^{-3}\,\text{A/m}$$

Thus, $(S_x)_{av} = \tfrac{1}{2}(0.75 \times 10^{-3})^2 \times 0.0397 = 0.0112\,\mu\text{W/m}^2$.

CHAPTER 8
Cavity Resonators

8.1 A rectangular *cavity resonator* may be constructed by inserting conducting plates in a rectangular waveguide at $z = 0$ and $z = d$, as shown in Fig. 8-1. Write an expression for E_z within the cavity, assuming a TM mode.

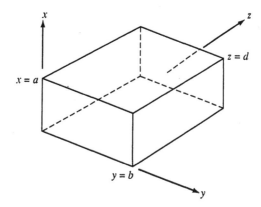

Fig. 8-1

▌ Refer to Problems 7.1–7.12. The propagation constant remains

$$\gamma^2 = M^2 + N^2 - \omega^2 \mu \epsilon \qquad (1)$$

where

$$M = \frac{m\pi}{a} \quad (m = 0, 1, 2, \ldots) \qquad N = \frac{n\pi}{b} \quad (n = 0, 1, 2, \ldots)$$

But now a backward-traveling wave must be included in the solution for E_z, which therefore is given by the superposition

$$E_z = E_z^+ + E_z^- = \sin Mx \sin Ny \, (C_E^+ e^{-\gamma z} + C_E^- e^{\gamma z}) \qquad (2)$$

In (2) the undetermined constants C_E^+ and C_E^- carry the units V/m.

8.2 Write the forms of field components E_x, E_y, H_x, H_y, and H_z for the TM mode in Problem 8.1.

▌ Use superposition. In Problem 7.21, first replace $j\beta_{mn}$ by $+\gamma$ and then by $-\gamma$, and add corresponding results:

$$E_x = \frac{M}{M^2 + N^2} \cos Mx \sin Ny \, (-\gamma C_E^+ e^{-\gamma z} + \gamma C_E^- e^{\gamma z}) \qquad (1)$$

$$E_y = \frac{N}{M^2 + N^2} \sin Mx \cos Ny \, (-\gamma C_E^+ e^{-\gamma z} + \gamma C_E^- e^{\gamma z}) \qquad (2)$$

$$H_x = \frac{j\omega\epsilon N}{M^2 + N^2} \sin Mx \cos Ny \, (C_E^+ e^{-\gamma z} + C_E^- e^{\gamma z}) \qquad (3)$$

$$H_y = \frac{-j\omega\epsilon M}{M^2 + N^2} \cos Mx \sin Ny \, (C_E^+ e^{-\gamma z} + C_E^- e^{\gamma z}) \qquad (4)$$

$$H_z = 0 \quad \text{(assumed)} \qquad (5)$$

8.3 Apply the boundary conditions at $z = 0$ and $z = d$ in the cavity of Fig. 8-1.

▌ E_x and E_y must vanish at both ends; therefore, by (1) or (2) of Problem 8.2,

$$C_E^+ = C_E^- \equiv C_E \qquad (1)$$

and

$$e^{2\gamma d} = 0 \quad \text{or} \quad \gamma = j\frac{p\pi}{d} \equiv jP \qquad (2)$$

where $p = 0, 1, 2, \ldots$.

8.4 Using Problems 8.1–8.3, write explicit expressions for all the field components in TM_{mnp}.

▌

$$E_x = -\frac{2MPC_E}{M^2 + N^2} \cos Mx \sin Ny \sin Pz \qquad (1)$$

$$E_y = -\frac{2NPC_E}{M^2 + N^2} \sin Mx \cos Ny \sin Pz \tag{2}$$

$$E_z = 2C_E \sin Mx \sin Ny \cos Pz \tag{3}$$

$$H_x = \frac{2j\omega\epsilon NC_E}{M^2 + N^2} \sin Mx \cos Ny \cos Pz \tag{4}$$

$$H_y = \frac{-2j\omega\epsilon MC_E}{M^2 + N^2} \cos Mx \sin Ny \cos Pz \tag{5}$$

$$H_z = 0 \tag{6}$$

8.5   Give the time-domain expressions for the field components of Problem 8.4.

▍ Multiply by $e^{j\omega t}$ and take the real part:

$$E_z = 2C_E \sin Mx \sin Ny \cos Pz \cos \omega t$$

$$E_x = -2\frac{MPC_E}{M^2 + N^2} \cos Mx \sin Ny \sin Pz \cos \omega t$$

$$E_y = -2\frac{NPC_E}{M^2 + N^2} \sin Mx \cos Ny \sin Pz \cos \omega t$$

$$H_x = -2\frac{\omega\epsilon NC_E}{M^2 + N^2} \sin Mx \cos Ny \cos Pz \sin \omega t$$

$$H_y = 2\frac{\omega\epsilon MC_E}{M^2 + N^2} \cos Mx \sin Ny \cos Pz \sin \omega t$$

Observe that the boundary conditions of Problem 8.3 force each component to be a standing wave along the z-axis, as well as along the x- or y-axis.

8.6   Referring to Problem 8.5, find the unique (resonance) frequency at which TM$_{mnp}$ "stands."

▍ From (1) of Problem 8.1 and (2) of Problem 8.3:

$$-\pi^2\left(\frac{p}{d}\right)^2 = \pi^2\left(\frac{m}{a}\right)^2 + \pi^2\left(\frac{n}{b}\right)^2 - \pi^2 4f_{mnp}^2 \mu\epsilon$$

or $$f_{mnp} = \frac{u}{2}\sqrt{\left(\frac{m}{a}\right)^2 + \left(\frac{n}{b}\right)^2 + \left(\frac{p}{d}\right)^2} \qquad (u = 1/\sqrt{\mu\epsilon}) \tag{1}$$

8.7   What is the lowest-order (lowest-resonant-frequency) TM mode that can exist in a cavity such as shown in Fig. 8-1?

▍ Problem 8.4 shows that neither m nor n can be zero; otherwise, all field components vanish. However, p can be zero. Thus, from Problem 8.6, the lowest-order TM mode is TM$_{110}$.

8.8   Obtain expressions for the field components for the TE modes in the cavity of Fig. 8-1.

▍ Proceeding as in Problems 8.1 and 8.2,

$$H_z = \cos Mx \cos Ny \,(C_H^+ e^{-\gamma z} + C_H^- e^{\gamma z}) \tag{1}$$

$$H_x = \frac{M}{M^2 + N^2} \sin Mx \cos Ny \,(\gamma C_H^+ e^{-\gamma z} - \gamma C_H^- e^{\gamma z}) \tag{2}$$

$$H_y = \frac{N}{M^2 + N^2} \cos Mx \sin Ny \,(\gamma C_H^+ e^{-\gamma z} - \gamma C_H^- e^{\gamma z}) \tag{3}$$

$$E_x = \frac{j\omega\mu N}{M^2 + N^2} \cos Mx \sin Ny \,(C_H^+ e^{-\gamma z} + C_H^- e^{\gamma z}) \tag{4}$$

$$E_y = \frac{-j\omega\mu M}{M^2 + N^2} \sin Mx \cos Ny \,(C_H^+ e^{-\gamma z} + C_H^- e^{\gamma z}) \tag{5}$$

$$E_z = 0 \quad \text{(assumed)} \tag{6}$$

Here, C_H^+ and C_H^- are in A/m.

8.9   It is obvious that the z-boundary conditions require $C_H^+ = C_H^- \equiv C_H$ and $\gamma = jP$ in Problem 8.8, just as with TM modes. Rewrite the field components for TE$_{mnp}$.

▍

$$H_z = j2C_H \cos Mx \cos Ny \sin Pz \tag{1}$$

$$H_x = -j\frac{2MP}{M^2 + N^2} C_H \sin Mx \cos Ny \cos Pz \tag{2}$$

$$H_y = -j\frac{2NP}{M^2 + N^2} C_H \cos Mx \sin Ny \cos Pz \qquad (3)$$

$$E_x = -\frac{2\omega\mu N}{M^2 + N^2} C_H \cos Mx \sin Ny \sin Pz \qquad (4)$$

$$E_y = \frac{2\omega\mu M}{M^2 + N^2} C_H \sin Mx \cos Ny \sin Pz \qquad (5)$$

$$E_z = 0 \qquad (6)$$

8.10 Write the time-domain expressions for the field components of Problem 8.9.

▮

$$H_z = -2C_H \cos Mx \cos Ny \sin Pz \sin \omega t$$

$$H_x = 2\frac{MPC_H}{M^2 + N^2} \sin Mx \cos Ny \cos Pz \sin \omega t$$

$$H_y = 2\frac{NPC_H}{M^2 + N^2} \cos Mx \sin Ny \cos Pz \sin \omega t$$

$$E_x = -2\frac{\omega\mu NC_H}{M^2 + N^2} \cos Mx \sin Ny \sin Pz \cos \omega t$$

$$E_y = 2\frac{\omega\mu MC_H}{M^2 + N^2} \sin Mx \cos Ny \sin Pz \cos \omega t$$

8.11 What is the resonant frequency of TE_{mnp}?

▮ See (1) of Problem 8.6.

8.12 What is the lowest-order TE mode in the cavity of Fig. 8-1?

▮ Problem 8.9 shows that either m or n (but not both) can be zero, but p cannot be zero. Hence, the lowest-order mode is TE_{011}, if $a < b$, or TE_{101}, if $a > b$.

8.13 Assuming that $a \geq d > b$ for the cavity of Fig. 8-1, exhibit the **H**- and **E**-fields of the lowest-order mode.

▮ In the TE_{101} mode, the nonzero components are

$$H_z = j2C_H \cos\frac{\pi x}{a} \sin\frac{\pi z}{d} \qquad (1)$$

$$H_x = -j2\frac{a}{d} C_H \sin\frac{\pi x}{a} \cos\frac{\pi z}{d} \qquad (2)$$

$$E_y = \frac{2\omega\mu a}{\pi} C_H \sin\frac{\pi x}{a} \sin\frac{\pi z}{d} \qquad (3)$$

8.14 Verify that in TE_{101} (as in any TE or TM mode) $\mathbf{E} \cdot \mathbf{H} = 0$.

▮ By Problem 8.13, $E_x H_x + E_y H_y + E_z H_z = (0)H_x + E_y(0) + (0)H_z = 0$.

8.15 For an air-filled, lossless, cavity resonator of dimensions $a = 60$ cm, $b = 50$ cm, and $d = 40$ cm, list, in order of ascending resonant frequencies, the ten lowest modes.

▮ The resonant frequencies are, from (1) of Problem 8.6,

$$f_{mnp} = (1.5 \times 10^8)\sqrt{\frac{m^2}{0.36} + \frac{n^2}{0.25} + \frac{p^2}{0.16}}$$

The required listing is: TM_{110} (390.5 MHz), TE_{101} (450.7 MHz), TE_{011} (480.2 MHz), TE_{111} and TM_{111} (541.4 MHz), TM_{210} (583.1 MHz), TE_{201} (625 MHz), TM_{120} (650 MHz), TE_{102} (790.6 MHz), TE_{012} (807.8 MHz), TE_{112} and TM_{112} (845.6 MHz).

8.16 Shielded rooms can be viewed as resonant cavities. Consequently, operation of equipment in such a room at a resonant frequency of the cavity should be avoided. A typical shielded room has dimensions (408 in) × (348 in) × (142 in). Determine its lowest resonant frequency.

▮ From (1) of Problem 8.6, converting to meters,

$$f_{011} = (1.5 \times 10^8)\sqrt{\frac{0^2}{13.01} + \frac{1^2}{78.13} + \frac{1^2}{107.4}} = 22.3 \text{ MHz}$$

8.17 Verify that the TM **E**- and **H**-fields given by (*1*) through (*6*) of Problem 8.4 indeed satisfy the boundary conditions $\mathbf{E}_{tang} = \mathbf{H}_{tang} = \mathbf{0}$ at each wall (Fig. 8-1).

❚ top wall: $E_z|_{x=a} = E_y|_{x=a} = 0, \quad H_x|_{x=a} = 0$
bottom wall: $E_z|_{x=0} = E_y|_{x=0} = 0, \quad H_x|_{x=0} = 0$
left wall: $E_z|_{y=0} = E_x|_{y=0} = 0, \quad H_y|_{y=0} = 0$
right wall: $E_z|_{y=b} = E_x|_{y=b} = 0, \quad H_y|_{y=b} = 0$
left end: $E_x|_{z=0} = E_y|_{z=0} = 0, \quad H_z|_{z=0} = 0$
right end: $E_x|_{z=d} = E_y|_{z=d} = 0, \quad H_z|_{z=d} = 0$

8.18 For TE_{101} plot the field lines at the instant $\omega t = \pi/2$.

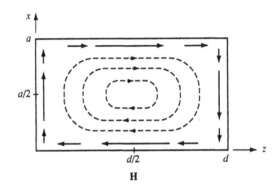

H   **Fig. 8-2**

❚ By Problem 8.10, H_z and H_x are the only nonzero field components, and
$$\frac{dx}{dz} = \frac{H_x}{H_z} = (\text{const.}) \frac{\tan(\pi x/a)}{\tan(\pi z/d)}$$
These curves have already been calculated [see (*1a*) and (*1b*) of Problem 7.37]; suitably relabeled, Fig. 7-3 becomes Fig. 8-2.

8.19 For a rectangular cavity resonator with perfectly conducting walls, calculate the surface current density on the top wall ($x = a$) for TE_{111}.

❚ Using Problem 8.10,
$$\mathbf{K} = (-\mathbf{a}_x \times \mathbf{H})_{x=a} = H_z|_{x=a}\,\mathbf{a}_y - H_y|_{x=a}\,\mathbf{a}_z$$
$$= 2C_H\left[\cos\frac{\pi y}{b}\sin\frac{\pi z}{d}\,\mathbf{a}_y + \frac{1/bd}{(1/a^2) + (1/b^2)}\sin\frac{\pi y}{b}\cos\frac{\pi z}{d}\,\mathbf{a}_z\right]\sin\omega t$$

8.20 Obtain an expression for the time-average energy stored in the electric field of the TE_{101} mode in the cavity of Fig. 8-1.

❚ Taking the electric field components from Problem 8.13,
$$W_e = \frac{\epsilon}{4}\int_V (|E_x|^2 + |E_y|^2 + |E_z|^2)\,dV = \frac{\epsilon}{4}\int_{x=0}^a\int_{y=0}^b\int_{z=0}^d |E_y|^2\,dz\,dy\,dx$$
$$= \frac{\epsilon}{4}(4f_{101}\mu a\,|C_H|)^2 b\int_{x=0}^a\int_{z=0}^d \sin^2\frac{\pi x}{a}\sin^2\frac{\pi z}{d}\,dx\,dz = \mu^2\epsilon a^3 bdf_{101}^2\,|C_H|^2$$
Substituting for f_{101} from (*1*) of Problem 8.6, we find
$$W_e = \frac{\mu abd}{4}\left(1 + \frac{a^2}{d^2}\right)|C_H|^2 \tag{1}$$

8.21 Show that the time-average energy stored in the magnetic field of the cavity of Problem 8.20 is the same as the time-average energy stored in the electric field.

❚ Taking the magnetic field components from Problem 8.13,
$$W_m = \frac{\mu}{4}\int_V (|H_x|^2 + |H_y|^2 + |H_z|^2)\,dV = \frac{\mu}{4}\int_{x=0}^a\int_{y=0}^b\int_{z=0}^d (|H_x|^2 + |H_z|^2)\,dz\,dy\,dx$$
$$= \frac{\mu b 4\,|C_H|^2}{4}\int_{x=0}^a\int_{z=0}^d \left(\frac{a^2}{d^2}\sin^2\frac{\pi x}{a}\cos^2\frac{\pi z}{d} + \cos^2\frac{\pi x}{a}\sin^2\frac{\pi z}{d}\right)dx\,dz$$
$$= \frac{\mu}{4}\left\{\frac{4a^2}{d^2}\,|C_H|^2\frac{a}{2}b\frac{d}{2} + 4\,|C_H|^2\frac{a}{2}b\frac{d}{2}\right\} = \frac{\mu abd}{4}\left(1 + \frac{a^2}{d^2}\right)|C_H|^2 \tag{1}$$
which is the same as (*1*) of Problem 8.20.

8.22 Check the units of $W_e = W_m$ as given by Problems 8.20 and 8.21.

▮

$$(W) = (H/m)(m^3)(A/m)^2 = (H \cdot A^2) = (J)$$

8.23 Determine the total time-average power loss in the walls of the cavity of Fig. 8-1, for the TE_{101} mode.

▮ As in Problem 7.80, we assume small losses; thus $\mathbf{H}_{tang}$ is the same as in the case of perfectly conducting walls. The average power-density vector interior to and directed into the walls is, from (2) of Problem 7.80,

$$(S_{av})_{loss} = \frac{1}{2\sigma_c\delta}|\mathbf{H}_{tang}|^2$$

The average power loss is obtained by integrating the above over the surfaces of the six walls:

$$
\begin{aligned}
(P_{av})_{loss} &= \int_S (S_{av})_{loss}\, dS \\
&= \frac{1}{2\sigma_c\delta}\left\{2\int_{x=0}^{a}\int_{y=0}^{b}(|H_x|^2)|_{z=0,d}\,dx\,dy + 2\int_{x=0}^{a}\int_{z=0}^{d}(|H_x|^2+|H_z|^2)|_{y=0,b}\,dx\,dz + 2\int_{y=0}^{b}\int_{z=0}^{d}(|H_z|^2)|_{z=0,a}\,dy\,dz\right\} \\
&= \frac{|C_H|^2}{\sigma_c\delta d^2}(2a^3b + a^3d + ad^3 + 2bd^3)
\end{aligned}
\tag{1}
$$

8.24 The *quality factor* (or Q) of a resonant cavity mode is defined in a manner similar to that for resonant lumped circuits:

$$Q \equiv 2\pi f \frac{\text{time-average energy stored}}{\text{time-average power loss}} = \omega\frac{W_e + W_m}{(P_{loss})_{av}} \tag{1}$$

where f is the resonant frequency of that mode. Evaluate the Q of the cavity of Fig. 8-1 for the TE_{101} mode.

▮ In the definition, substitute values from Problems 8.6, 8.20, 8.21, and 8.23, to obtain

$$Q_{101} = \sigma_c\delta\frac{\pi f_{101}\mu abd(a^2+d^2)}{2a^3b+a^3d+ad^3+2bd^3} = \frac{\pi\eta\sigma_c\delta}{2}\frac{b(a^2+d^2)^{3/2}}{2a^3b+a^3d+ad^3+2bd^3} \tag{2}$$

where $\eta = \sqrt{\mu/\epsilon}$ is the intrinsic impedance of the (lossless) cavity.

8.25 Determine Q of the TE_{101} mode in an air-filled brass cavity having dimensions $a = 4\,\text{cm}$, $d = 4\,\text{cm}$, and $b = 2\,\text{cm}$.

▮ The resonant frequency is obtained from (1) of Problem 8.6 as

$$f_{101} = (1.5 \times 10^8)\sqrt{\frac{1}{(0.04)^2}+\frac{1}{(0.04)^2}} = 5.3\,\text{GHz}$$

The skin depth of brass at the resonant frequency is ($\sigma_c = 15.7\,\text{MS/m}$)

$$\delta = \frac{1}{\sqrt{\pi f_{101}\mu\sigma_c}} = 1.744\,\mu\text{m}$$

and, from (2) of Problem 8.24, $Q_{101} = 5733$.

8.26 Design a cubic ($a = b = d$) cavity resonator to have a dominant resonant frequency of 7 GHz.

▮ From (1) of Problem 8.6,

$$f_{mnp} = \frac{1.5 \times 10^8}{a}\sqrt{m^2+n^2+p^2} \tag{1}$$

Lowest-order modes are TM_{110}, TE_{101}, TE_{011}; thus

$$7 \times 10^9 = \frac{1.5 \times 10^8}{a}\sqrt{2} \qquad\text{or}\qquad a = 3.03\,\text{cm}$$

8.27 Calculate the Q of the dominant mode of the cavity of Problem 8.26, assuming that the walls are brass.

▮ With $\sigma_c = 15.7\,\text{MS/m}$ and

$$\delta = \frac{1}{\sqrt{\pi f_{101}\mu\sigma_c}} = 1.518\,\mu\text{m}$$

(2) of Problem 8.24 (either form) yields $Q_{101} = 6654$.

8.28 Repeat Problem 8.26 if the cavity is filled with polyethylene having $\epsilon_r = 2.25$.

▮ Since $f \propto u \propto 1/\sqrt{\epsilon_r}$,

$$7 \times 10^9 = \frac{1.5 \times 10^8\sqrt{2}}{a\sqrt{2.25}} \qquad\text{or}\qquad a = 2.02\,\text{cm}$$

8.29   The dimensions of a rectangular cavity are each multiplied by $\rho > 0$. What is the effect on the Q_{101}?

▮ By Problem 8.6, $f_{101} \propto 1/\rho$. Now, Q_{101} is given by (2) of Problem 8.24. Noting that the geometrical factor in the second form is dimensionless, we have

$$Q_{101} \propto \delta \propto 1/\sqrt{f_{101}} \propto \sqrt{\rho}$$

8.30   In Fig. 8-1, $a = 4$ cm, $b = 5$ cm, $d = 3$ cm; the cavity is filled with a material having a relative dielectric constant of 100. What is the lowest resonant frequency of the cavity?

▮ The resonant frequency is given by

$$f = \frac{u_0}{\sqrt{\epsilon_r}} \sqrt{\left(\frac{m}{a}\right)^2 + \left(\frac{n}{b}\right)^2 + \left(\frac{p}{d}\right)^2} = (3 \times 10^9) \sqrt{\left(\frac{m}{4}\right)^2 + \left(\frac{n}{5}\right)^2 + \left(\frac{p}{3}\right)^2}$$

From Problems 8.7 and 8.12, the contest is limited to TM_{110} and TE_{011}; the former wins.

$$f_{110} = (3 \times 10^9)\sqrt{(\tfrac{1}{4})^2 + (\tfrac{1}{5})^2} = 480 \text{ MHz}$$

8.31   Prove that in the dominant mode of any rectangular (noncubic) cavity the electric field is polarized parallel to the shortest dimension.

▮ Choose rectangular coordinates such that the shortest dimension is the z-dimension. Then (see Problem 8.30) the dominant mode is TM_{110}, which has $E_x = E_y = 0$. [The rule remains valid for a cubic cavity, in the sense that TM_{110}, TE_{011}, and TE_{101} are all dominant and yield polarizations parallel to the z axis, x axis, and y axis, respectively.]

8.32   Refer to Problem 8.31. Show that if all dimensions of the cavity are doubled, the total energy content increases by the factor 8.

▮ In TM_{110} the electric field is of the form

$$E_x = E_y = 0 \qquad E_z = E_0 \sin \frac{\pi x}{a} \sin \frac{\pi y}{b} \cos \omega t$$

so that the total energy density is

$$w = 2w_e = \epsilon E_z^2 = \epsilon E_0^2 \sin^2 \frac{\pi x}{a} \sin^2 \frac{\pi y}{b} \cos^2 \omega t$$

It is seen that w has the same value at the point (x, y) of an original cross section and the homologous point $(2x, 2y)$ of the magnified cross section. Hence the integrated energy over a cross section increases by the factor 4. But the length is also doubled, so that the volume integral increases by the factor $4 \times 2 = 8$.

8.33   A cubical cavity measures 3 cm on each side. What is the lowest resonant frequency?

▮

$$f_{110} = f_{011} = f_{101} = \frac{3 \times 10^{10}}{2} \sqrt{2\left(\frac{1}{3}\right)^2} = 7.07 \text{ GHz}$$

8.34   A one-dimensional resonator is constructed of parallel perfectly-conducting plates, as shown in Fig. 8-3. The plates are separated by a distance w. A plane wave polarized in the y-direction is reflecting back and forth between the plates. Given E_0 as the amplitude of the electric field, write an equation for the electric field between the plates.

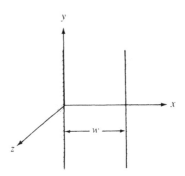

Fig. 8-3

▮ This is a standing wave along the x axis:

$$E_y = E_0 \sin \frac{n\pi x}{w} e^{j\omega t} \quad (n = 1, 2, \dots) \tag{1}$$

8.35 In the resonator of Problem 8.34, calculate the resonant frequency for the lowest mode, when $w = 6$ cm.

▌ In the lowest mode, the plate separation is a half-wavelength:

$$w = \frac{\lambda}{2} = \frac{u}{2f} \quad \text{or} \quad f = \frac{u}{2w} = \frac{3 \times 10^{10}}{2 \times 6} = 2.5 \text{ GHz}$$

8.36 Find the magnetic field in the resonator of Problem 8.34.

▌ Maxwell's equation

$$\text{curl } \hat{\mathbf{E}} = -j\omega\mu\hat{\mathbf{H}}$$

and (1) of Problem 8.34 yield

$$\frac{\partial E_y}{\partial x} = \frac{n\pi}{w} E_0 \cos \frac{n\pi x}{w} = -j\omega_n\mu H_z$$

Now, $\omega_n = n\omega_1 = n\pi u/w$ (cf. Problem 8.35), so that the above gives (restoring the time factor)

$$H_z = j\frac{E_0}{\eta} \cos \frac{n\pi x}{w} e^{j\omega t} \tag{1}$$

as expected for plane waves.

8.37 Derive an expression for the time-average energy density (electric plus magnetic) in the resonator of Problem 8.34.

▌

$$w_{av} = \frac{1}{4}\epsilon |\mathbf{E}|^2 + \frac{1}{4}\mu |\mathbf{H}|^2 = \frac{E_0^2}{4}\left(\epsilon \sin^2\frac{n\pi x}{w} + \frac{\mu}{\eta^2}\cos^2\frac{n\pi x}{w}\right) = \frac{\epsilon E_0^2}{4}$$

since $\eta^2 = \mu/\epsilon$. Note that while $(w_E)_{av}$ and $(w_H)_{av}$ vary across the gap, their sum is fixed (and is independent of the mode number).

8.38 In the resonator of Problem 8.34, determine where the average energy density in the electric field is maximum for the lowest mode.

▌ From Problem 8.37,

$$(w_E)_{av} = \frac{1}{4}\epsilon E_0^2 \sin^2\frac{\pi x}{w}$$

which is maximum at $x = w/2$, the center of the cavity.

8.39 A cubical resonant cavity, 10 cm on a side, operates in the TE_{011} mode (see Problem 8.31). If the amplitude of the electric field in the cavity is 100 kV/m, determine the time-average energy stored.

▌ The electric field in the cavity is of the form $E_x = E_0 \sin k_y y \sin k_z z\, e^{j\omega t}$, so that the stored electric energy is given by

$$W_e = \frac{1}{4}\epsilon_0 \int_{vol} |E_x|^2\, dx\, dy\, dz$$

$$= \frac{1}{4}\epsilon_0 E_0^2(0.1) \int_{z=0}^{0.1}\int_{y=0}^{0.1} \frac{\sin^2 k_y y \sin^2 k_z z}{k_y k_z}\, d(k_z z)\, d(k_y y)$$

$$= \frac{\epsilon_0 E_0^2}{4\pi^2}(0.1)^3\left(\frac{\pi^2}{4}\right) = \frac{10^{-9}}{36\pi \times 4\pi^2}(10^5)^2(0.1)^3\frac{\pi^2}{4} \approx 5.5\ \mu\text{J}$$

Since $W_m = W_e$, the total stored energy will be 11 μJ.

8.40 A cubical resonant cavity measures 12 cm on each side. What is the lowest resonance frequency?

▌ Since $f \propto 1/a$, Problem 8.33 gives $f = 7.07/4 = 1.77$ GHz.

8.41 The structure shown in Fig. 8-4 is a lossless coaxial cable short-circuited at the receiving end to form a *coaxial cavity*. Find the magnetic and electric fields at a point P within the cavity, assuming that a voltage wave propagates in the z-direction with an amplitude V_0.

▌ From transmission-line theory (Chapter 6) the voltage and current along the cavity are given by

$$V_s = V_0 e^{-j\beta z} - V_0 e^{j\beta z} = -j2V_0 \sin\beta z \tag{1}$$

$$I_s = \frac{V_0}{Z_0}e^{-j\beta z} + \frac{V_0}{Z_0}e^{j\beta z} = \frac{2V_0}{Z_0}\cos\beta z \tag{2}$$

where $Z_0 = \sqrt{l/c}$, with l and c respectively the per-unit-length inductance and capacitance of the cable. By Ampère's law, the magnetic field is

$$H_\phi = \frac{I_s}{2\pi r} = \frac{V_0}{\pi r Z_0}\cos\beta z \tag{3}$$

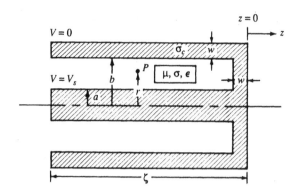

Fig. 8-4

and (Problem 2.192), the electric field is

$$E_r = \frac{V_s}{r \ln (b/a)} = \frac{-j2V_0}{r \ln (b/a)} \sin \beta z \tag{4}$$

8.42 Determine net average power loss ($\sigma > 0$) in the material filling the cavity of Problem 8.41.

   ■ By Ohm's law the current density in the material is (from Problem 8.41)

$$J_r = \sigma E_r = \frac{-j2\sigma V_0}{r \ln (b/a)} \sin \beta z$$

which may be written in the time domain as

$$J_r(r, z, t) = \frac{2\sigma V_0}{r \ln (b/a)} \sin \beta z \sin \omega t$$

Then

$$P_{\text{loss}} = \int_{\text{vol}} \frac{J_r^2}{\sigma} d\tau = \int_{-\zeta}^{0} \int_{0}^{2\pi} \int_{a}^{b} \frac{4\sigma V_0^2}{r[\ln (b/a)]^2} \sin^2 \beta z \sin^2 \omega t \, dr \, d\phi \, dz$$

$$= \frac{4\pi\sigma V_0^2}{\ln (b/a)} \left(\zeta - \frac{\sin^2 2\beta\zeta}{2\beta} \right) \sin^2 \omega t$$

The average power loss is thus

$$(P_{\text{loss}})_{\text{av}} = \frac{2\pi\sigma V_0^2}{\ln (b/a)} \left(\zeta - \frac{\sin 2\beta\zeta}{2\beta} \right) \tag{1}$$

8.43 Assuming that the cavity of Problem 8.41 is of length $\lambda/4$, where λ is the wavelength at the resonant frequency, find the power loss in dielectric material with which the cavity is filled.

   ■ Substitute $\beta\zeta = (2\pi/\lambda)(\lambda/4) = \pi/2$ in (1) of Problem 8.42 to obtain

$$(P_{\text{loss}})_{\text{av}} = \frac{2\pi\sigma V_0^2}{\ln (b/a)} \tag{1}$$

8.44 What is the resonant frequency of the cavity of Problem 8.43, if the cavity is 0.3 m long, $\epsilon = 100\epsilon_0$, and $\mu = \mu_0$?

   ■

$$f_r = \frac{u}{\lambda} = \frac{u_0\sqrt{\epsilon_r}}{4\zeta} = \frac{(3 \times 10^8)/10}{4(0.3)} = 25 \text{ MHz}$$

8.45 Define an effective resistance R_e that accounts for the dielectric loss in the cavity of Problem 8.43.

   ■ From (1) of Problem 8.41, the voltage at any instant within the cavity is

$$V(z, t) = 2V_0 \sin \beta z \sin \omega t$$

In particular, at $z = -\zeta$,

$$V_{\text{in}} = -2V_0 \sin \beta\zeta \sin \omega t = -2V_0 \sin \omega t \quad \text{(see Problem 8.43)}$$

If this voltage is applied across a resistor R_e, the time-average power loss is

$$(P_{\text{loss}})_{\text{av}} = \frac{1}{2} \frac{(2V_0)^2}{R_e} = \frac{2V_0^2}{R_e}$$

A comparison between this and (1) of Problem 8.43 gives

$$R_e = \frac{\ln (b/a)}{\pi\sigma\zeta} \tag{1}$$

8.46 Determine the time-average energy stored in the resonant cavity of Problem 8.43.

▋ Taking the field components from Problem 8.41, and using $\beta\zeta = \pi/2$, we obtain

$$W_e + W_m = \frac{1}{4}\int_{vol}(\epsilon|E_r|^2 + \mu|H_\phi|^2)\,d\tau$$

$$= \frac{V_0^2}{4}(2\pi)\int_{-\zeta}^{0}\left[\frac{4\epsilon}{\ln^2(b/a)}\sin^2\beta z + \frac{\mu}{\pi^2 Z_0^2}\cos^2\beta z\right]dz\int_a^b\frac{1}{r^2}r\,dr$$

$$= \frac{V_0^2\pi\zeta}{2}\left[\frac{2\epsilon}{\ln^2(b/a)} + \frac{\mu}{2\pi^2 Z_0^2}\right]\ln(b/a) \tag{1}$$

Now, from Problems 3.135 and 2.192,

$$Z_0 = \frac{1}{2\pi}\sqrt{\frac{\mu}{\epsilon}}\ln\left(\frac{b}{a}\right) \tag{2}$$

so that (1) becomes

$$W_e + W_m = \frac{2\pi\epsilon V_0^2\zeta}{\ln(b/a)} \tag{3}$$

8.47 Calculate the Q of the resonant cavity of Problem 8.43.

▋ In the definition, (1) of Problem 8.24, substitute (3) of Problem 8.46, (1) of Problem 8.43, and the resonance condition

$$\omega_r = \frac{\pi}{2\zeta\sqrt{\mu\epsilon}} \tag{1}$$

to find

$$Q = \omega_r\frac{\epsilon}{\sigma} = \frac{\pi}{2\zeta\sigma\sqrt{\mu/\epsilon}} \tag{2}$$

A comparison of (2) with Problem 5.34 shows that Q *is the loss cotangent at resonance*—a clear indication of a general relationship. Note that this parameter is independent of radial dimensions, but varies inversely with length.

8.48 Let the cavity of Problem 8.43 be represented by a parallel resonant circuit (Fig. 8-5). Assuming that R_e is as given by (1) of Problem 8.45, determine L_e and C_e.

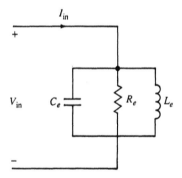

Fig. 8-5

▋ The Q of a parallel resonant circuit is given by $Q = \omega_r C_e R_e$. Thus, from (1) of Problem 8.45 and (2) of Problem 8.47,

$$C_e = \frac{Q/\omega_r}{R_e} = \frac{\pi\epsilon\zeta}{\ln(b/a)} \tag{1}$$

Similarly, from $Q = R_e/\omega_r L_e$ and (1) of Problem 8.47,

$$L_e = \frac{R_e}{\omega_r Q} = \frac{4\mu\zeta}{\pi^3}\ln\frac{b}{a} \tag{2}$$

8.49 Calculate the equivalent circuit parameters (Fig. 8-5) of a coaxial cavity having $b = 27.2$ mm, $a = 10$ mm, $f_r = 100$ MHz, $\sigma/\omega_r\epsilon = 0.001$, $\mu = \mu_0$, and $\epsilon_r = 4$.

▋ From (1) of Problem 8.47,

$$\zeta = \frac{1}{4f_r\sqrt{\mu\epsilon}} = \frac{u_0}{4f_r\sqrt{\epsilon_r}} = \frac{3\times10^8}{4\times100\times10^6\sqrt{4}} = 0.375\text{ m}$$

From (2) of Problem 8.48,

$$L_e = \frac{4\times4\pi\times10^{-7}\times0.375}{\pi^3}\ln2.72 = 60.8\text{ nH}$$

From (*1*) of Problem 8.48,

$$C_e = \frac{\pi \times 4 \times (10^{-9}/36\pi) \times 0.375}{\ln 2.72} = 41.64 \text{ pF}$$

Now

$$\sigma = \omega_r \epsilon (0.001) = 2\pi \times 100 \times 10^6 \times 4 \times \frac{10^{-9}}{36\pi} \times 10^{-3} = \frac{2}{9} \times 10^{-4} \text{ S/m}$$

Thus, from (*1*) of Problem 8.45,

$$R_e = \frac{\ln 2.72}{\pi \times (2/9)10^{-4} \times 0.375} = 38.22 \text{ k}\Omega$$

8.50 Determine the average power loss at resonance ($\beta\zeta = \pi/2$) in the (outer) cylindrical wall of the cavity of Problem 8.41.

▮ From (*3*) of Problem 8.41 at $r = b$ we have

$$H_\phi = \frac{V_0}{\pi b Z_0} \cos \beta z$$

Correspondingly, the linear current density at $r = b$ is $K_z = -H_{\phi b}$; to a good approximation, this current density is uniformly diffused through one skin depth δ, and the (uniform) current density within the wall is

$$J_z = \frac{K_z}{\delta} = \frac{-V_0}{\pi b \delta Z_0} \cos \beta z$$

Time-average power loss in the wall (annulus of thickness δ):

$$(P_{\text{loss}})_{\text{wall}} = \frac{1}{2} \int_{\text{vol}} \frac{J_z^2}{\sigma_c} \, d\tau = \frac{1}{2} \int_{-\zeta}^{0} \frac{J_z^2}{\sigma_c} \delta (2\pi b) \, dz = \frac{V_0^2 \zeta}{2\pi \sigma_c b \delta Z_0^2} \tag{1}$$

8.51 Repeat Problem 8.50 for the inner wall (of radius a).

▮ Replace b in (*1*) of Problem 8.50 by a.

8.52 Repeat Problem 8.50 for the end plate at $z = 0$.

▮ From Problem 8.41,

$$H_\phi \big|_{z=0} = \frac{V_0}{\pi r Z_0}$$

Reasoning as in Problem 8.50, we have in the end plate (thickness δ):

$$J_r = \frac{V_0}{\pi r \delta Z_0}$$

The corresponding average power loss is

$$(P_{\text{loss}})_{\text{end}} = \frac{1}{2} \int_a^b \int_0^{2\pi} \frac{V_0^2}{\sigma_c \pi^2 r^2 \delta^2 Z_0^2} r \, dr \, d\phi = \frac{V_0^2}{\pi \sigma_c \delta Z_0^2} \ln \left(\frac{b}{a} \right) \tag{1}$$

8.53 The cavity of Problem 8.41 is operating under resonance and is filled with a lossless dielectric having $\mu = \mu_0$; thus, only the walls have losses. Determine the Q of this cavity.

▮ From Problems 8.50 through 8.52 we have the total average wall losses

$$P_{\text{loss}} = \frac{V_0^2 \zeta}{2\pi \sigma_c \delta Z_0^2} \left[\frac{1}{a} + \frac{1}{b} + \frac{2}{\zeta} \ln \left(\frac{b}{a} \right) \right] = \frac{2\pi \epsilon \zeta V_0^2}{\sigma_c \delta \mu_0 \ln^2 (b/a)} \left[\frac{1}{a} + \frac{1}{b} + \frac{2}{\zeta} \ln \left(\frac{b}{a} \right) \right]$$

where (*2*) of Problem 8.46 has been used. Combining this with (*3*) of Problem 8.46 yields

$$Q_w = \omega_r \frac{W_e + W_m}{P_{\text{loss}}} = \frac{\omega_r \mu_0 \sigma_c \delta \ln (b/a)}{\dfrac{1}{a} + \dfrac{1}{b} + \dfrac{2}{\zeta} \ln \left(\dfrac{b}{a} \right)} \tag{1}$$

But

$$\delta = \frac{1}{\sqrt{\pi f_r \sigma_c \mu_0}} \quad \text{or} \quad \omega_r \mu_0 \sigma_c \delta = \frac{2}{\delta}$$

which when substituted in (*1*) yields

$$Q_w = \frac{2 \ln (b/a)}{\delta [(1/a) + (1/b) + (2/\zeta) \ln (b/a)]} \tag{2}$$

8.54 Obtain expressions for the equivalent-circuit (Fig. 8-5) parameters for the cavity of Problem 8.53.

▮ Refer to Problem 8.48. Since the total stored energy has not changed, C_e and L_e are unchanged; these are respectively given by (1) and (2) of Problem 8.48. The parameter R_e is recalculated as

$$R_{ew} = \frac{Q_w}{\omega_r C_e} = \frac{4\eta \, [\ln \, (b/a)]^2}{\pi^2 \delta [(1/a) + (1/b) + (2/\zeta) \ln \, (b/a)]} \tag{1}$$

where $\eta = \sqrt{\mu_0/\epsilon}$ is the intrinsic impedance of the lossless dielectric.

8.55 In the cavity of Problem 8.43, dielectric losses and wall losses are both present. What is the Q of this resonant cavity?

▌ The total power loss, being the sum of the wall losses and the dielectric loss, may be represented by an equivalent resistor which is a parallel combination of R_e, given by (1) of Problem 8.45, and R_{ew}, given by (1) of Problem 8.54. Consequently, the modified Q is given by

$$Q_T = \frac{1}{(1/Q) + (1/Q_w)} \tag{1}$$

where Q and Q_w are, respectively, given by (2) of Problem 8.47 and (2) of Problem 8.53.

8.56 Determine the Q of the cavity of Problem 8.49, if its walls are silver-plated ($\sigma_c = 61.8 \, \text{MS/m}$).

▌ From the data of Problem 8.49,

$$Q = \frac{\omega_r \epsilon}{\sigma} = \frac{1}{0.001} = 1000.$$

and

$$\delta = \frac{1}{\sqrt{\pi f_r \mu \sigma_c}} = \frac{1}{\sqrt{\pi \times 100 \times 10^6 \times 4\pi \times 10^{-7} \times 61.8 \times 10^6}} = 6.40 \, \mu\text{m}$$

Thus

$$Q_w = \frac{2 \ln 2.72}{(6.40 \times 10^{-6}) \left[\dfrac{1}{0.010} + \dfrac{1}{0.0272} + \dfrac{2}{0.375} \ln 2.72 \right]} = 2200$$

Finally, from (1) of Problem 8.55,

$$Q_T = \frac{1}{(1/1000) + (1/2200)} = 687.5$$

8.57 For the cavity shown in Fig. 8-4 we have $a = 0.8 \, \text{cm}$, $b = 4 \, \text{cm}$, and $\zeta = 12 \, \text{cm}$. The cavity is air-filled. Calculate the resonance frequency.

▌ At resonance,

$$\zeta = \frac{\lambda}{4} = 0.12 \, \text{m} \qquad \text{or} \qquad \lambda = 4 \times 0.12 = 0.48 \, \text{m}$$

Hence

$$f_r = \frac{u}{\lambda} = \frac{3 \times 10^8}{0.48} = 625 \, \text{MHz}$$

8.58 How does the resonance frequency change in the cavity of Problem 8.57 if both ends of the structure of Fig. 8-4 are short-circuited?

▌ In this case $\zeta = \lambda/2$, so f_r is doubled.

8.59 What is the resonance frequency of the cavity of Problem 8.57 if one-quarter of the length (from the short-circuited end) is filled with a dielectric of relative permittivity $\epsilon_r = 9$?

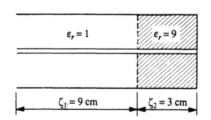

$\epsilon_r = 1$ $\epsilon_r = 9$

$\zeta_1 = 9 \, \text{cm}$ $\zeta_2 = 3 \, \text{cm}$ **Fig. 8-6**

▌ The configuration is shown in Fig. 8-6, from which

$$\beta_1 \zeta_1 + \beta_2 \zeta_2 = \frac{\pi}{2} \qquad \text{or} \qquad 2f_r \left(\frac{\zeta_1}{u_1} + \frac{\zeta_2}{u_2} \right) = \frac{1}{2}$$

Thus

$$2f_r \left(\frac{0.09}{3 \times 10^8} + \frac{0.03}{3 \times 10^8/\sqrt{9}} \right) = \frac{1}{2} \qquad \text{or} \qquad f_r = 416.67 \, \text{MHz}$$

8.60 For the cavity shown in Fig. 8-4, $a = 0.8$ cm, $b = 4$ cm, $\zeta = \lambda/4 = 15$ cm, and $\epsilon_r = 6.25$. Determine the equivalent-circuit (Fig. 8-5) parameters L_e and C_e.

▌ From (1) of Problem 8.48,

$$C_e = \frac{\pi \times 6.25 \times (10^{-9}/36\pi) \times 0.15}{\ln(4/0.8)} = 16.2 \text{ pF}$$

From (2) of Problem 8.48,

$$L_e = \frac{4 \times 4\pi \times 10^{-7} \times 0.15}{\pi^3} \ln\left(\frac{4}{0.8}\right) = 38.9 \text{ nH}$$

8.61 The cavity of Problem 8.60 is now filled with a lossy dielectric having $\epsilon_r = 6.25$ and $\sigma = 200 \ \mu\text{S/m}$. Determine the resistance R_e in the equivalent circuit of Fig. 8-5; also find the Q of the cavity.

▌ From (1) of Problem 8.45,

$$R_e = \frac{\ln(4/0.8)}{\pi \times 2 \times 10^{-4} \times 0.15} = 17.08 \text{ k}\Omega$$

The resonance frequency is

$$f_r = \frac{u}{\lambda_r} = \frac{u_0/\sqrt{\epsilon_r}}{4\zeta} = \frac{(3 \times 10^8)/2.5}{4 \times 0.15} = 200 \text{ MHz}$$

Thus

$$Q = \frac{\omega_r \epsilon}{\sigma} = \frac{2\pi(200 \times 10^6) \times (10^{-9}/36\pi) \times 6.25}{200 \times 10^{-6}} = 347.2$$

8.62 Repeat Problem 8.61, if now the dielectric is lossless and the cavity walls are coated with copper ($\sigma_c = 58$ MS/m).

▌ The skin depth at resonance is
$$\delta = 1/\sqrt{\pi f_r \mu \sigma_c} = 1/\sqrt{\pi \times 200 \times 10^6 \times 4\pi \times 10^{-7} \times 58 \times 10^6} = 4.67 \ \mu\text{m}$$
Thus, from (2) of Problem 8.53,

$$Q_w = \frac{2\ln(4/0.8)}{(4.67 \times 10^{-6})\left(\dfrac{10^2}{0.8} + \dfrac{10^2}{4} + \dfrac{2}{0.15}\ln\dfrac{4}{0.8}\right)} = 4020$$

and, taking C_e from Problem 8.60,

$$R_{ew} = \frac{Q_w}{\omega_r C_e} = \frac{4020}{2\pi(200 \times 10^6)(16.2 \times 10^{-12})} = 197.5 \text{ k}\Omega$$

8.63 Repeat Problem 8.61, if the dielectric in the cavity is lossy ($\epsilon_r = 6.25$ and $\sigma = 200 \ \mu\text{S/m}$) and the cavity walls are coated with copper.

▌ From Problems 8.61 and 8.62, $Q = 347.2$ and $Q_w = 4020$. Using these results in (1) of Problem 8.55 yields

$$Q_T = \frac{1}{(1/347.2) + (1/4020)} = 319.6$$

The corresponding resistors are in parallel. Hence, using the numerical values from Problems 8.61 and 8.62, we obtain

$$R_T = \frac{R_e R_{ew}}{R_e + R_{ew}} = \frac{(17.08 \times 197.5)10^6}{(17.08 + 197.5)10^3} = 15.72 \text{ k}\Omega$$

8.64 In the equivalent circuit of a resonant cavity (Fig. 8-5), $L_e = 0.2 \ \mu\text{H}$, $C_e = 20$ pF, and $R_e = 25$ kΩ. Determine the rms input current at resonance frequency for a 1 V (rms) input voltage.

▌ The admittance of the circuit is

$$Y = \frac{1}{R_e} + j\left(\omega C_e - \frac{1}{\omega L_e}\right) \tag{1}$$

Under resonance, $\omega C_e = 1/\omega L_e$; so that $Y = 1/R_e$ and the input current is given by

$$I_{\text{in}} = V_{\text{in}} Y = \frac{1}{25 \times 10^3} = 40 \ \mu\text{A (rms)}$$

8.65 Repeat Problem 8.64 if the input frequency is 1 percent greater than the resonance frequency.

▌ The resonance frequency is

$$\omega_r = \frac{1}{\sqrt{L_e C_e}} = \frac{1}{\sqrt{0.2 \times 20 \times 10^{-18}}} = 0.5 \times 10^9 \text{ rad/s}$$

Substituting $\omega = 10^9 (0.5 + 0.005)$ and other numerical values in (1) of Problem 8.64 yields

$$Y = \frac{1}{25 \times 10^3} + j\left(5.05 \times 10^8 \times 20 \times 10^{-12} - \frac{1}{5.05 \times 10^8 \times 0.2 \times 10^{-6}}\right) = 40 + j200 \ \mu S$$

Thus $|Y| = 204 \ \mu S$ and $I_{in} = 1 \times 204 = 204 \ \mu A$ (rms).

8.66 Determine the equivalent-circuit (Fig. 8-5) parameters for the cavity shown in Fig. 8-4, which is air-filled and has $a = 4$ mm, $b = 25$ mm, and $\zeta = 300$ mm $= \lambda/4$. The cavity walls are coated with a material having $\sigma_c = 50$ MS/m.

▌ From (2) of Problem 8.48,

$$L_e = \frac{4}{\pi^3} \times 4\pi \times 10^{-7} \times 0.300 \ln \frac{25}{4} = 89.1 \text{ nH}$$

From (1) of Problem 8.48,

$$C_e = \frac{\pi \times 10^{-9}}{36\pi} \times 0.300 \times \frac{1}{\ln (25/4)} = 4.55 \text{ pF}$$

Assuming that $\mu = \mu_0$, the skin depth is given by

$$\delta = 2\sqrt{\frac{\zeta}{\pi \sigma_c \eta}} = \frac{2}{\pi} \sqrt{\frac{0.300}{50 \times 10^6 \times 120}} = (\sqrt{2}/\pi) \times 10^{-5} \text{ m}$$

and (1) of Problem 8.54 gives

$$R_{ew} = \frac{4(120)}{\sqrt{2} \times 10^5} \left[\frac{10^{-3} \ln^2 (25/4)}{(1/4) + (1/25) + (1/150) \ln (25/4)}\right] = 377 \text{ k}\Omega$$

8.67 An air-filled coaxial resonant cavity 60 cm long is short-circuited at both ends. With reference to Fig. 8-4, the radial dimensions are $a = 0.8$ cm and $b = 4.0$ cm. The cavity walls are coated with copper ($\sigma_c = 58$ MS/m). External connections are made to the cavity at its midpoint. Determine C_e, L_e, and R_{ew} of the equivalent circuit.

▌ The cavity may be considered as two, semi-shorted, 30-cm-long cavities in parallel. Thus $0.30 = \lambda/4$ or $\lambda = 1.2$ m, giving:

$$f_r = \frac{u}{\lambda} = \frac{3 \times 10^8}{1.2} = 250 \text{ MHz}$$

$$\delta = 1/\sqrt{\pi f_r \mu \sigma_c} = 1/\sqrt{2\pi \times 250 \times 10^6 \times 4\pi \times 10^{-7} \times 58 \times 10^6} = 4.18 \ \mu m$$

$$C_e = 2\left[\frac{\pi \epsilon \zeta}{\ln (b/a)}\right] = \frac{2\pi \times (10^{-9}/36\pi) \times 0.30}{\ln (4/0.8)} = 10.36 \text{ pF}$$

$$L_e = \frac{1}{\omega_r^2 C_e} = \frac{1}{(2\pi \times 250 \times 10^6)^2 \times 10.36 \times 10^{-12}} = 39.12 \text{ nH}$$

$$Q_w = \frac{(2/\delta) \ln (b/a)}{(1/a) + (1/b) + (2/\zeta) \ln (b/a)} = \frac{[2/(4.18 \times 10^{-6})] \ln (4/0.8)}{\frac{10^2}{0.8} + \frac{10^2}{4} + \frac{2}{0.30} \ln \frac{4}{0.8}} = 4791$$

$$R_{ew} = \frac{Q_w}{\omega_r C_e} = \frac{4791}{2\pi \times 250 \times 10^6 \times 10.36 \times 10^{-12}} = 294.4 \text{ k}\Omega$$

8.68 For the air-filled coaxial cavity of Fig. 8-4 we have $\zeta = \lambda/4$ and $a = 0.5$ cm. The division of losses among the walls is as follows: center conductor, 70 percent; outer conductor, 20 percent; end plate, 10 percent. What is the resonance frequency?

▌ From Problems 8.50–8.52, the losses in the outer conductor, inner conductor, and end plate may be expressed as

$$P_b = \frac{k}{b} \qquad P_a = \frac{k}{a} \qquad P_e = \frac{k}{\zeta/2} \ln \left(\frac{b}{a}\right)$$

where $k = V_0^2 \zeta/2\pi\sigma_c \delta Z_0^2 = $ a constant. Thus

$$\frac{10}{70} = \frac{P_e}{P_a} = \frac{2a}{\zeta} \ln \frac{P_a}{P_b} = \frac{1.0}{\zeta} \ln \frac{70}{20}$$

or $\zeta = 8.78$ cm. Then

$$f_r = \frac{u}{4\zeta} = \frac{3 \times 10^8}{4 \times 0.0878} = 855 \text{ MHz}$$

8.69 Consider a rectangular cavity resonator of which the cross section perpendicular to the shortest dimension is square. Show that for the dominant mode of such a cavity,

$$Q = \frac{2}{\delta} \times \frac{\text{volume}}{\text{surface area}} \tag{1}$$

where δ is the skin depth, at resonance, in the cavity walls.

▮ To make available previous results, choose $a = d > b$ in Fig. 8-1; then TE_{101} is dominant, and (Problem 8.24)

$$Q_{101} = \frac{\sigma_c \delta \omega_r \mu}{2} \times \frac{(\text{volume})(2a^2)}{(\text{surface area})(a^2)} = \sigma_c \delta \omega_r \mu \times \frac{\text{volume}}{\text{surface area}}$$

But (see Problem 8.53) $\sigma_c \delta \omega_r \mu = 2/\delta$.

8.70 Show that if the square resonator of Problem 8.69 has unit volume—which can always be arranged for by a suitable choice of the length unit—then, in the dominant mode,

$$Q < \left(\frac{2}{9\pi}\right)^{1/3} \Big/ \delta \approx 0.414/\delta \tag{1}$$

▮ This follows immediately from (1) of Problem 8.69 and the fundamental geometrical fact that, among all solids of volume 1, a sphere has the smallest surface area; namely, $(36\pi)^{1/3}$.

8.71 Check the bound (1) of Problem 8.70 against the exact value for a cubical resonator of side 1 cm.

▮ The exact value is

$$Q = \frac{2}{[\delta \, (\text{cm})]} \times \frac{1 \, \text{cm}^3}{6 \, \text{cm}^2} = \frac{0.333}{\delta}$$

8.72 Argue that (1) of Problem 8.69, and with it (1) of Problem 8.70, should extend to cylindrical cavity resonators.

▮ The basic definition reads

$$Q = \omega_r \frac{W}{P_{\text{loss}}}$$

First, an argument exactly paralleling Problem 8.32 will establish that $W \propto (\text{volume})$ (the fact that a Bessel function takes the place of the sine-squared, changes nothing). Second, because the mode is symmetric, because the lines of **H** must hug the wall, and because the dissipation is presumed to occur in a layer of uniform thickness δ, it is most plausible that P_{loss} be proportional to the dissipating volume: $P_{\text{loss}} \propto (\text{surface area}) \times \delta$. Thus we are lead to

$$Q = \frac{C}{\delta}\left(\frac{\text{volume}}{\text{surface area}}\right)$$

and all that remains is to show that $C = 2$.

8.73 A cylindrical cavity resonator of length ζ and radius a supports the TE_{011} mode. (The scheme here is

$$TE_{\text{azimuthal no.} \,|\, \text{radial no.} \,|\, \text{axial no.}}$$

making this mode cylindrically symmetric.) Obtain the form of the axial magnetic field in the cavity.

▮ From the theory of cylindrical waveguides (see Problem 7.152, for instance), for a cylindrically symmetric TE mode,

$$H_z = C_1 J_0(hr)e^{\gamma z} + C_2 J_0(hr)e^{-\gamma z} \tag{1}$$

with

$$h^2 = \gamma^2 + \omega^2 \mu\epsilon \tag{2}$$

Now, by Maxwell's equations, $E_\phi \propto \partial H_z / \partial r$; and the boundary condition $\mathbf{E}_{\text{tang}} = \mathbf{0}$ forces

$$\left.\frac{\partial H_z}{\partial r}\right|_{z=0} = \left.\frac{\partial H_z}{\partial r}\right|_{z=\zeta} = \left.\frac{\partial H_z}{\partial r}\right|_{r=a} = 0$$

from which, respectively, $C_2 = -C_1$; $\gamma = j\beta = j\pi/\zeta$; $ha = 3.832$ (the first positive zero of J_0'). Hence,

$$H_z = j2C_1 J_0\left(3.832\frac{r}{a}\right) \sin\frac{\pi z}{\zeta} \tag{3}$$

8.74 Determine H_r and E_ϕ in the cavity of Problem 8.73 for the TE_{011} mode.

▮ The required field components are given by

$$H_r = \frac{\beta}{h^2}\frac{\partial H_z}{\partial r} \qquad \text{and} \qquad E_\phi = \frac{j\omega\mu}{h^2}\frac{\partial H_z}{\partial r}$$

Combining these with (1) of Problem 8.73 yields

$$H_r = -j\frac{2\pi a C_1}{3.832\zeta} J_1\left(3.832\frac{r}{a}\right) \cos\frac{\pi z}{\zeta} \tag{1}$$

$$E_\phi = \frac{2\omega\mu a C_1}{3.832} J_1\left(3.832\frac{r}{a}\right) \sin\frac{\pi z}{\zeta} \qquad (2)$$

8.75 Give the resonant frequency of the cavity of Problem 8.73 in the TE_{011} mode.

▮ By (2) of Problem 8.73 and the results $\gamma = j\pi/\zeta$, $ha = 3.832$,

$$\left(\frac{3.832}{a}\right)^2 = -\frac{\pi^2}{\zeta^2} + \frac{4\pi^2 f_r^2}{u^2}$$

where $u = 1/\sqrt{\mu\epsilon}$. Solving,

$$f_r = \frac{u}{2}\sqrt{\left(\frac{3.832}{\pi a}\right)^2 + \frac{1}{\zeta^2}}$$

8.76 Write expressions for the surface currents corresponding to the fields of Problems 8.73 and 8.74.

▮ Surface current density is given by $\mathbf{k} = \mathbf{a}_n \times \mathbf{H}$. Thus, (1) of Problem 8.73 and (1) of Problem 8.74 yield:
At the curved surface $(r = a)$:

$$k_\phi = j2C_1 J_0\left(3.832\frac{r}{a}\right)\sin\frac{\pi z}{\zeta}$$

At the end faces $(z = 0, z = \zeta)$:

$$k_\phi = \pm j\frac{2\pi a C_1}{3.832} J_1\left(3.832\frac{r}{a}\right)$$

8.77 The TM_{010} electric field in a cylindrical cavity (Fig. 8-7) is given by

$$E_z = \frac{h^2}{j\omega\epsilon} J_0\left(2.405\frac{r}{a}\right)$$

Determine the time-average stored energy in the cavity.

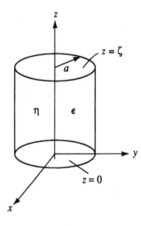

Fig. 8-7

▮
$$W = 2W_e = \frac{\epsilon}{2}\int_{vol} |\mathbf{E}|^2\, d\tau$$

$$= \frac{h^4}{2\omega^2\epsilon}(2\pi\zeta)\int_0^a J_0^2\left(2.405\frac{r}{a}\right) dr = \frac{\pi h^4 \zeta a^2}{2\omega^2\epsilon} J_1^2(2.405) \qquad (1)$$

8.78 The intrinsic wave impedance of the walls of cavity of Fig. 8-7 is $\eta_c + j\eta_c$ (Problem 5.45). The cavity is operating in the TM_{010} mode. Determine the time-average power dissipated in the walls of the cavity.

▮ For the TM_{010} mode the **H**-field is given by

$$H_\phi = \frac{2.405}{a} J_1\left(2.405\frac{r}{a}\right)$$

Thus
$$P_{loss} = \frac{\eta_c}{2}\int_{surface} |\mathbf{H}_{tang}|^2\, dS$$

$$= \frac{\eta_c}{2}\left(\frac{2.405}{a}\right)^2 (2\pi)\left[a\zeta J_1^2(2.405) + 2\int_0^a rJ_1^2\left(2.405\frac{r}{a}\right) dr\right]$$

$$= \eta_c\left(\frac{2.405}{a}\right)^2 \pi a(\zeta + a)J_1^2(2.405) \qquad (1)$$

8.79 Determine the Q of the TM_{010} mode in the cylindrical cavity of Problem 8.77.

❚ The Q of the cavity is given by

$$Q = \omega \frac{W}{P_{\text{loss}}}$$

Substituting the results of Problems 8.77 and 8.78 yields

$$Q = \frac{h^4 \zeta a^3}{2\omega\epsilon\eta_c(2.405)^2(\zeta + a)} \qquad (1)$$

8.80 Rewrite (1) of Problem 8.79 in terms of the intrinsic impedance η of the dielectric within the cavity, eliminating ω by use of the condition for resonance.

❚ In TM_{010} the field is independent of both ϕ and z. Hence $\gamma = 0$; so that

$$ha = 2.405 \qquad \text{and} \qquad h = \omega\sqrt{\mu\epsilon} = \omega\epsilon\eta$$

Therefore, $h^4 a^3 = h(ha)^3 = \omega\epsilon\eta(2.405)^3$, and

$$Q = \frac{1.2025\eta}{\eta_c(1 + a/\zeta)} \qquad (1)$$

8.81 The cylindrical cavity of Fig. 8-7 has $a = \zeta$. Determine ζ for the cavity to operate at a wavelength of 1 cm in the dominant mode.

❚ The dominant mode is TM_{010}, so that the resonance frequency is (Problem 8.80):

$$\omega_r = 2\pi f_r = hu = \frac{(ha)u}{\zeta} = \frac{(2.405)\lambda f_r}{\zeta}$$

or

$$\zeta = \frac{(2.405)\lambda}{2\pi} = 0.383 \text{ cm}$$

8.82 The resonant cavity shown in Fig. 8-7 is filled with a dielectric having $\epsilon_r = 4$. The cavity is operating in the TM_{010} mode. Calculate the time-average energy stored in the cavity under resonant conditions, assuming that the length of the cavity is 10 cm.

❚ Under resonant conditions, $ha = 2.405$ and ($\gamma = 0$) $h^2 = \omega^2\mu_0\epsilon$. Thus ($1$) of Problem 8.77 becomes

$$W = \frac{\pi(2.405)^2\omega^2\mu_0\epsilon\zeta}{2\omega^2\epsilon} J_1^2(2.405)$$

$$= 9.086 \times 4\pi \times 10^{-7} \times (0.10) \times (0.52)^2 = 0.308 \ \mu\text{J} \qquad (1)$$

8.83 Determine the radius of the cavity of Problem 8.82 so that the cavity may be resonant at 1.15 GHz in the TM_{010} mode.

❚ Under resonant conditions, $ha = 2.405$ and ($\gamma = 0$) $h = \omega_r/u = 4\pi f_r/u_0$. Thus,

$$a = \frac{2.405u_0}{4\pi f_r} = \frac{2.405 \times 3 \times 10^8}{4\pi \times 1.15 \times 10^9} = 5 \text{ cm}$$

8.84 The resonant cavity of Problem 8.83 is coated with gold ($\sigma_c = 41$ MS/m). How much (time-average) power is dissipated in the walls of the cavity, in the TM_{010} mode?

❚ $$\eta_c = \sqrt{\frac{\mu\omega}{2\sigma_c}} = \sqrt{\frac{4\pi \times 10^{-7} \times 2\pi \times 1.15 \times 10^9}{2 \times 41 \times 10^6}} = 0.010523 \ \Omega$$

Substituting this value of η_c along with $a = 0.05$ m, $\zeta = 0.10$ m and $J_1(2.405) = 0.52$ (from tables), in (1) of Problem 8.78 yields

$$P_{\text{loss}} = (0.010523)\left(\frac{2.405}{0.05}\right)^2 \pi \times 0.05(0.10 + 0.05)(0.52)^2 = 0.1551 \text{ W} \qquad (1)$$

8.85 Find the Q of the cavity of Problem 8.84.

❚ By (1) of Problem 8.82 and (1) of Problem 8.84,

$$Q = \frac{2\pi \times 1.15 \times 10^9 \times 0.308 \times 10^{-6}}{0.1551} = 14\,346$$

8.86 Verify that the results of Problems 8.82 through 8.85 are consistent with (1) of Problem 8.80.

∎ Substituting

$$\eta = \sqrt{\frac{\mu}{\epsilon}} = \sqrt{\frac{\mu_0}{4\epsilon_0}} = (377/2)\ \Omega$$

$\eta_c = 0.010523\ \Omega$, and $a/\zeta = 0.5$ in (1) of Problem 8.80 yields

$$Q = \frac{1.2025\eta}{\eta_c(1 + a/\zeta)} = \frac{1.2025 \times 377}{2 \times 1.0523 \times 10^{-2} \times 1.5} = 14\ 360$$

in excellent agreement with Problem 8.85.

8.87 A cubical cavity is to be resonant at 9 GHz in the TE_{101} mode. Determine the dimension of the cavity.

∎ By Problem 8.6,

$$f_r = \frac{u}{\sqrt{2}\ a} \qquad \text{or} \qquad a = \frac{u}{\sqrt{2}\ f_r}$$

Substituting numerical values yields

$$a = \frac{3 \times 10^{10}}{\sqrt{2} \times 9 \times 10^9} = 2.36\ \text{cm}$$

8.88 The walls of the cavity of Problem 8.87 are coated with silver ($\sigma = 61$ MS/m). What is the Q of the cavity for the TE_{101} mode?

∎ Since

$$Q = \frac{2\ (\text{volume})}{\delta\ (\text{area})} = \frac{2a^3}{6\delta a^2} = \frac{a}{3\delta}$$

first determine $1/\delta$ from

$$1/\delta = \sqrt{\pi f_r \mu \sigma} = \sqrt{\pi \times 9 \times 10^9 \times 4\pi \times 10^{-7} \times 61 \times 10^6} = 1.47 \times 10^6\ \text{m}^{-1}$$

Then

$$Q = \tfrac{1}{3} \times 2.36 \times 10^{-2} \times 1.47 \times 10^6 = 11\ 564$$

8.89 What is the resonance frequency of a cylindrical cavity (Fig. 8-7), of radius 1.15 cm, that operates in the TM_{010} mode?

∎ From Problem 8.81,

$$f_r = \frac{(ha)u}{2\pi a} = \frac{2.405 \times 3 \times 10^8}{2\pi \times 1.15 \times 10^{-2}} = 10\ \text{GHz}$$

8.90 Determine the Q of the cavity of Problem 8.89, assuming that its length is 2.3 cm and its walls are gold-plated ($\sigma = 41$ MS/m).

∎ Substituting $a/\zeta = 0.5$ and

$$\eta = \sqrt{\frac{\sigma}{\omega\epsilon}} = \sqrt{\frac{41 \times 10^6}{(2\pi \times 10 \times 10^9)(10^{-9}/36\pi)}} = 8.6\ \text{k}\Omega$$

in (1) of Problem 8.80 yields

$$Q = \frac{1.2025 \times 8.6 \times 10^3}{1.5} = 6894$$

8.91 We formulate Maxwell's equations in spherical coordinates to handle spherical cavity resonator problems; such problems generally have axial symmetry (see Problem 2.249). With this constraint, write the equations governing the fields for TM spherical modes.

∎ For axially symmetric TM modes, we have waves with components E_r, E_θ, and H_ϕ that depend on r and θ only. Maxwell's equations become

$$\frac{\partial}{\partial r}(rE_\theta) - \frac{\partial E_r}{\partial \theta} = -j\omega\mu(rH_\phi) \qquad (1)$$

$$\frac{1}{r\sin\theta}\frac{\partial}{\partial\theta}(H_\phi\sin\theta) = j\omega\epsilon E_r \qquad (2)$$

$$-\frac{\partial}{\partial r}(rH_\phi) = j\omega\epsilon(rE_\theta) \qquad (3)$$

8.92 Obtain an equation governing H_ϕ (actually, rH_ϕ) alone for TM spherical modes.

∎ First, differentiate (2) of Problem 8.91 with respect to θ, and multiply and divide by r, to obtain

$$-\frac{\partial E_r}{\partial\theta} = -\frac{1}{j\omega\epsilon r^2}\frac{\partial}{\partial\theta}\left[\frac{1}{\sin\theta}\frac{\partial}{\partial\theta}(rH_\phi\sin\theta)\right] \qquad (1)$$

Next, differentiate (3) of Problem 8.91 with respect to r:

$$\frac{\partial}{\partial r}(rE_\theta) = -\frac{1}{j\omega\epsilon}\frac{\partial^2}{\partial r^2}(rH_\phi) \tag{2}$$

Now substitute (1) and (2) into (1) of Problem 8.91, to obtain, after rearrangement,

$$\frac{\partial^2}{\partial r^2}(rH_\phi) + \frac{1}{r^2}\frac{\partial}{\partial\theta}\left[\frac{1}{\sin\theta}\frac{\partial}{\partial\theta}(rH_\phi\sin\theta)\right] + k^2(rH_\phi) = 0 \tag{3}$$

where $k \equiv \omega\sqrt{\mu\epsilon} = 2\pi/\lambda$

8.93   Attack (3) of Problem 8.92 by separation of variables.

∎   Assuming that $rH_\phi = R(r)\Theta(\theta)$, we find in the usual way:

$$r^2\frac{R''}{R} + k^2r^2 = -\frac{1}{\Theta}\frac{d}{d\theta}\left[\frac{1}{\sin\theta}\frac{d}{d\theta}(\Theta\sin\theta)\right] = n(n+1) \tag{1}$$

where the separation constant has been written in the form $n(n+1)$ for a reason that will become clear in Problems 8.94 and 8.95.

8.94   Solve (1) of Problem 8.93 for $R(r)$.

∎   Under the substitution $R_1 = R/\sqrt{r}$ the r-equation becomes

$$\frac{d^2R_1}{dr^2} + \frac{1}{r}\frac{dR_1}{dr} + \left[k^2 - \frac{(n+\frac{1}{2})^2}{r^2}\right]R_1 = 0 \tag{1}$$

which is Bessel's differential equation of order $n + \frac{1}{2}$. A complete solution may then be written as

$$R_1 = R/\sqrt{r} = A_nJ_{n+1/2}(kr) + B_nJ_{-n-1/2}(kr) \tag{2}$$

When the order is half an odd integer, Bessel functions reduce to algebraic combinations of sinusoids; for example,

$$J_{1/2}(x) = \sqrt{\frac{2}{\pi x}}\sin x \qquad\qquad J_{-1/2}(x) = \sqrt{\frac{2}{\pi x}}\cos x$$

$$J_{3/2}(x) = \sqrt{\frac{2}{\pi x}}\left[\frac{\sin x}{x} - \cos x\right] \qquad\qquad J_{-3/2}(x) = \sqrt{\frac{2}{\pi x}}\left[\sin x + \frac{\cos x}{x}\right] \tag{3}$$

$$J_{5/2}(x) = \sqrt{\frac{2}{\pi x}}\left[\left(\frac{3}{x^2} - 1\right)\sin x - \frac{3}{x}\cos x\right] \qquad J_{-5/2}(x) = \sqrt{\frac{2}{\pi x}}\left[\frac{3}{x}\sin x + \left(\frac{3}{x^2} - 1\right)\cos x\right]$$

8.95   Obtain a partial solution to (1) of Problem 8.93 for $\Theta(\theta)$.

∎   The transformation $u = \cos\theta$ takes the θ-equation into

$$(1 - u^2)\frac{d^2\Theta}{du^2} - 2u\frac{d\Theta}{du} + \left[n(n+1) - \frac{1}{1-u^2}\right]\Theta = 0 \tag{1}$$

which is a special case of *Legendre's equation*

$$(1 - x^2)\frac{d^2y}{dx^2} - 2x\frac{dy}{dx} + \left[n(n+1) - \frac{m^2}{1-x^2}\right]y = 0 \tag{2}$$

One solution of (2), denoted $P_n^m(x)$, is called an *associated Legendre function of the first kind, order n, degree m*. These are related to the Legendre *polynomials* through

$$P_n^m(x) = (1 - x^2)^{m/2}\frac{d^mP_n(x)}{dx^m} \tag{3}$$

(The second solution is of no importance here, because it is singular for $x = \pm1$, which maps into the entire z-axis.)

A solution to (1) may then be written as

$$\Theta = P_n^1(\cos\theta) = -\frac{d}{d\theta}P_n(\cos\theta) \tag{4}$$

Now, for integral values of n, $P_n(\cos\theta)$ is an nth-degree polynomial in $\cos\theta$; whence Θ will be an $(n-1)$st-degree polynomial in $\cos\theta$, times $\sin\theta$. Explicitly,

$$\begin{aligned}
P_0^1(\cos\theta) &= 0 \quad \text{(trivial)}\\
P_1^1(\cos\theta) &= \sin\theta\\
P_2^1(\cos\theta) &= 3\sin\theta\cos\theta\\
P_3^1(\cos\theta) &= \tfrac{3}{2}\sin\theta\,(5\cos^2\theta - 1)\\
P_4^1(\cos\theta) &= \tfrac{5}{2}\sin\theta\,(7\cos^3\theta - 3\cos\theta)
\end{aligned} \tag{5}$$

8.96   Obtain expressions for H_ϕ, E_θ, and E_r in spherical coordinates for axially symmetric TM modes.

▮ By Problems 8.93 and 8.94,

$$H_\phi = \frac{1}{\sqrt{r}}[A_n J_{n+1/2}(kr) + B_n J_{-n-1/2}(kr)] P_n^1(\cos\theta)$$

In this expression it is convenient to replace $J_{n+1/2}(kr)$ and $J_{-n-1/2}(kr)$ by two linear combinations of these Bessel functions; namely, the *Hankel function of the first kind*, $H_{n+1/2}^{(1)}(kr)$, which represents a *converging* wave; and the *Hankel function of the second kind*, $H_{n+1/2}^{(2)}(kr)$, a *diverging* wave. In general, both Hankel functions will be needed for a standing-wave solution; let us denote the correct linear combination of Hankel functions as $Z_{n+1/2}(kr)$. Then

$$H_\phi = \frac{A_n}{\sqrt{r}} P_n^1(\cos\theta) Z_{n+1/2}(kr) \tag{1}$$

Substituting (1) in (2) and (3) of Problem 8.91 gives (Hankel functions obey the same recursion relations as do Bessel functions)

$$E_\theta = \frac{A_n P_n^1(\cos\theta)}{j\omega\epsilon r^{3/2}}[n Z_{n+1/2}(kr) - kr Z_{n-1/2}(kr)] \tag{2}$$

$$E_r = -\frac{A_n n Z_{n+1/2}(kr)}{j\omega\epsilon r^{3/2}\sin\theta}[\cos\theta\, P_n^1(\cos\theta) - P_{n+1}^1(\cos\theta)] \tag{3}$$

8.97 Obtain expressions for E_ϕ, H_θ, and H_r for axially symmetric TE modes.

▮ The results are directly obtainable from (1)–(3) of Problem 8.96: replace E_r and E_θ by H_r and H_θ, respectively; H_ϕ by $-E_\phi$; and ϵ by μ (according to the principle of duality).

$$E_\phi = \frac{B_n}{\sqrt{r}} P_n^1(\cos\theta) Z_{n+1/2}(kr) \tag{1}$$

$$H_\theta = -\frac{B_n P_n^1(\cos\theta)}{j\omega\mu r^{3/2}}[n Z_{n+1/2}(kr) - kr Z_{n-1/2}(kr)] \tag{2}$$

$$H_r = \frac{B_n n Z_{n+1/2}(kr)}{j\omega\mu r^{3/2}\sin\theta}[\cos\theta\, P_n^1(\cos\theta) - P_{n+1}^1(\cos\theta)] \tag{3}$$

8.98 Write the expressions for the field components for the lowest-order, axially symmetric TM mode in a spherical cavity.

▮ For the lowest-order mode, which is $TM_{101[r\phi\theta]}$, it can be shown that $Z_{3/2}(kr) = J_{3/2}(kr)$. Hence, by (3) of Problem 8.94,

$$H_\phi = \frac{C\sin\theta}{kr}\left(\frac{\sin kr}{kr} - \cos kr\right) \tag{1}$$

$$E_r = -\frac{2j\eta C\cos\theta}{k^2 r^2}\left(\frac{\sin kr}{kr} - \cos kr\right) \tag{2}$$

$$E_\theta = \frac{j\eta C\sin\theta}{k^2 r^2}\left[\frac{(kr)^2-1}{kr}\sin kr + \cos kr\right] \tag{3}$$

where $\eta = \sqrt{\mu/\epsilon}$.

8.99 The spherical cavity of Problem 8.98 is of radius a. What is the resonant wavelength for the lowest-order TM mode?

▮ Apply the boundary condition $E_\theta = 0$ at $r = a$ to (3) of Problem 8.98, to obtain:

$$\tan ka = \frac{ka}{1-(ka)^2} \tag{1}$$

The first positive root of this transcendental equation is $ka \approx 2.74$; hence,

$$\lambda = \frac{2\pi a}{ka} \approx 2.293a \tag{2}$$

8.100 Determine the time-average energy stored at resonance in the cavity of Problems 8.98 and 8.99.

▮ Assume C real in (1) of Problem 8.98. Then the stored energy is given by

$$W = \int_0^a \int_0^\pi \frac{\mu}{2}|H_\phi|^2\, 2\pi r^2\sin\theta\, d\theta\, dr = \frac{4\pi\mu C^2}{3k^2}\int_0^a\left(\frac{\sin kr}{kr} - \cos kr\right)^2 dr$$

Evaluating this integral and using Problem 8.99 yields

$$W = \frac{2\pi\mu C^2 a^3}{3(ka)^3}\left[ka - \frac{1+(ka)^2}{ka}\sin^2 ka\right] \quad \text{where} \quad ka \approx 2.74 \tag{1}$$

8.101 The inner surface of the cavity of Problems 8.98 and 8.99 is coated with a good conductor, with an intrinsic impedance $\eta_c + j\eta_c$. Determine the time-average power loss for the TM_{101} mode under resonance.

▌ The power loss is given by

$$P_{\text{loss}} = \frac{1}{2}\eta_c \int_0^\pi |H_\phi|^2 \, 2\pi a^2 \sin\theta \, d\theta$$

Substituting (1) of Problem 8.98, and applying the condition (1) of Problem 8.99, yields

$$P_{\text{loss}} = \tfrac{4}{3}\pi\eta_c a^2 C^2 \sin^2 ka \qquad \text{where} \qquad ka \approx 2.74 \tag{1}$$

8.102 Find the Q of the resonant cavity of Problems 8.98 and 8.99.

▌ By (1) of Problem 8.100 and (1) of Problem 8.101,

$$Q = \omega \frac{W}{P_{\text{loss}}} = \frac{\mu\omega}{2(ka)^2 k\eta_c}\left[\frac{ka}{\sin^2 ka} - \frac{1+(ka)^2}{ka}\right]$$

But $\mu\omega = k\eta$, whence

$$Q = \frac{\eta}{2\eta_c(ka)^2}\left[\frac{ka}{\sin^2 ka} - \frac{1+(ka)^2}{ka}\right] \tag{1}$$

8.103 A spherical cavity of radius a is resonant in the TE_{101} mode. Give the field components.

▌ By duality with TM_{101}, we obtain the required components by substituting, in (1) through (3) of Problem 8.98, E_ϕ for H_ϕ, $-H_r$ for E_r, and $-H_\theta$ for E_θ:

$$E_\phi = \frac{D\sin\theta}{kr}\left(\frac{\sin kr}{kr} - \cos kr\right) \tag{1}$$

$$H_r = \frac{j2\eta D\cos\theta}{k^2 r^2}\left(\frac{\sin kr}{kr} - \cos kr\right) \tag{2}$$

$$H_\theta = \frac{-j\eta D\sin\theta}{k^2 r^2}\left[\frac{(kr)^2 - 1}{kr}\sin kr + \cos kr\right] \tag{3}$$

8.104 Find the resonant wavelength for the cavity of Problem 8.103.

▌ The resonance condition is obtained by setting $E_\phi = 0$ at $r = a$. Thus, (1) of Problem 8.103 gives

$$\tan ka = ka \tag{1}$$

The first positive root of (1) is $ka \approx 4.5$; hence

$$\lambda = \frac{2\pi a}{ka} \approx 1.396a \tag{2}$$

8.105 Determine the resonant frequency of a spherical cavity of 5-cm radius, filled with air and operating in the TM_{101} mode.

▌ From (2) of Problem 8.99,

$$f = \frac{u_0}{\lambda} = \frac{3\times 10^8}{2.293 \times 5\times 10^{-2}} = 2.617\ \text{GHz}$$

8.106 Repeat Problem 8.105 for the TE_{101} mode.

▌ From (2) of Problem 8.104,

$$f = \frac{u_0}{\lambda} = \frac{3\times 10^8}{1.396 \times 5\times 10^{-2}} = 4.298\ \text{GHz}$$

8.107 Find the Q of the cavity of Problem 8.105, assuming that the walls of the cavity are coated with gold ($\sigma_c = 41\ \text{MS/m}$).

▌ Use (1) of Problem 8.102, with $\eta = 377\ \Omega$ and

$$\eta_c = \sqrt{\frac{\mu\omega}{2\sigma_c}} = \sqrt{\frac{4\pi\times 10^{-7}\times 2\pi\times 2.617\times 10^9}{2\times 41\times 10^6}} = 0.01587\ \Omega$$

Thus

$$Q \approx (0.987)\frac{377}{0.01587} = 23\,447$$

9.1 A source current density $\mathbf{J}_s$, varying sinusoidally in time, results in a magnetic vector potential $\mathbf{A}$ in free space. Obtain a differential equation relating $\mathbf{A}$ to $\mathbf{J}_s$.

▮ Maxwell's equations for the phasor fields are
$$\text{curl } \mathbf{E} = -j\omega\mu_0\mathbf{H} \qquad \text{curl } \mathbf{H} = j\omega\epsilon_0\mathbf{E} + \mathbf{J}_s \qquad (1)$$
Introducing the vector potential, via $\mathbf{B} \equiv \text{curl } \mathbf{A}$, into the first equation (1), we obtain:
$$\text{curl } \mathbf{E} = -j\omega \text{ curl } \mathbf{A} \qquad \text{or} \qquad \text{curl } (\mathbf{E} + j\omega\mathbf{A}) = \mathbf{0}$$
which last implies (conservative field)
$$\mathbf{E} + j\omega\mathbf{A} = -\text{grad } V \qquad \text{or} \qquad \mathbf{E} = -j\omega\mathbf{A} - \text{grad } V \qquad (2)$$
where V is some scalar potential function. Substitution of (2) in the second equation (1) and use of the identity (cartesian coordinates implied)
$$\text{curl curl } \mathbf{A} = \text{grad div } \mathbf{A} - \nabla^2\mathbf{A}$$
yields
$$\nabla^2\mathbf{A} + \omega^2\mu_0\epsilon_0\mathbf{A} = -\mu_0\mathbf{J}_s + \text{grad } (\text{div } \mathbf{A} + j\omega\epsilon_0\mu_0 V) \qquad (3)$$
The divergence of $\mathbf{A}$ may be arbitrarily prescribed. Whereas in static problems one sets div $\mathbf{A} = 0$, the smart move here is clearly
$$\text{div } \mathbf{A} = -j\omega\mu_0\epsilon_0 V \qquad (4)$$
so that the required equation is
$$\nabla^2\mathbf{A} + \beta_0^2\mathbf{A} = -\mu_0\mathbf{J}_s \qquad (5)$$
where $\beta_0 \equiv \omega\sqrt{\mu_0\epsilon_0}$.

9.2 Express (5) of Problem 9.1 as a set of three scalar equations.

▮ $$\nabla^2 A_x + \beta_0^2 A_x = -\mu_0(J_s)_x \qquad \nabla^2 A_y + \beta_0^2 A_y = -\mu_0(J_s)_y \qquad \nabla^2 A_z + \beta_0^2 A_z = -\mu_0(J_s)_z$$

9.3 Give an integral (Green's function) solution to (5) of Problem 9.1.

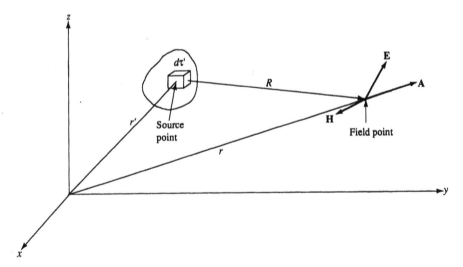

Fig. 9-1

▮ Analogous to the static solution (1) of Problem 3.49, we have
$$\mathbf{A} = \frac{\mu_0}{4\pi}\int_{\text{vol}} \frac{\mathbf{J}_s e^{-j\beta_0 R}}{R}\, d\tau' \qquad (1)$$
In (1), $R = [(x - x')^2 + (y - y')^2 + (z - z')^2]^{1/2}$ is the distance between the differential current element $\mathbf{J}_s\, d\tau'$ (the *source point* x', y', z') and the point at which we are computing $\mathbf{A}$ (the *field point* x, y, z), as shown in Fig. 9-1; the volume integration is extended over the entire source distribution. The x-, y-, and z-components of (1) respectively solve the three scalar equations of Problem 9.2.

9.4 A *Hertzian dipole* (*antenna*) consists of an infinitesimal current element, of length $d\zeta$, carrying a phasor current $\hat{I} = I\underline{/\theta_I}$. Determine **A** at some point (r, θ, ϕ).

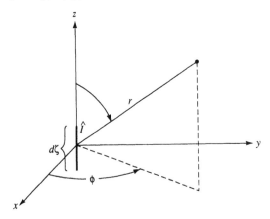

Fig. 9-2

▌ With the current element located as in Fig. 9-2, (*1*) of Problem 9.3 reduces to

$$\mathbf{A} = A_z\mathbf{a}_z = \frac{\mu_0}{4\pi}(\hat{I}\,d\zeta)\frac{e^{-j\beta_0 r}}{r}\mathbf{a}_z = \frac{\mu_0}{4\pi}(\hat{I}\,d\zeta)\frac{e^{-j\beta_0 r}}{r}(\cos\theta\,\mathbf{a}_r - \sin\theta\,\mathbf{a}_\theta) \qquad (1)$$

9.5 Explain how the electric and magnetic fields consistent with Maxwell's equations may be obtained from the magnetic vector potential in free space.

▌ In free space, away from the source, $\mathbf{J}_s = \mathbf{0}$; hence, by (*1*) of Problem 9.1,

$$\mathbf{E} = \frac{1}{j\omega\epsilon_0}\,\text{curl}\,\mathbf{H} = \frac{1}{j\omega\mu_0\epsilon_0}\,\text{curl curl}\,\mathbf{A} \qquad (1)$$

because

$$\mathbf{H} = \frac{1}{\mu_0}\,\text{curl}\,\mathbf{A} \qquad (2)$$

By use of (*5*) of Problem 9.1, (*1*) may be alternatively expressed as

$$\mathbf{E} = -j\omega\mathbf{A} + \frac{1}{j\omega\mu_0\epsilon_0}\,\text{grad div}\,\mathbf{A} \qquad (3)$$

9.6 Find the **H**-field due to the dipole antenna of Problem 9.4.

▌ In (*1*) of Problem 9.4, $A_\phi \equiv 0$ and A_r and A_θ are independent of ϕ; therefore

$$\mathbf{H} = \frac{1}{\mu_0}\,\text{curl}\,\mathbf{A} = \frac{\mathbf{a}_\phi}{\mu_0}\left\{\frac{1}{r}\left[\frac{\partial}{\partial r}(rA_\theta) - \frac{\partial A_r}{\partial\theta}\right]\right\} = \mathbf{a}_\phi\frac{\hat{I}\,d\zeta}{4\pi}\beta_0^2\sin\theta\left(j\frac{1}{\beta_0 r} + \frac{1}{\beta_0^2 r^2}\right)e^{-j\beta_0 r} \qquad (1)$$

Note that **H**, like **A**, is independent of ϕ.

9.7 Find the **E**-field due to the dipole antenna of Problem 9.4.

▌ Take the curl of (*1*) of Problem 9.6 and substitute in (*1*) of Problem 9.5, obtaining

$$E_r = 2\frac{\hat{I}\,d\zeta}{4\pi}\eta_0\beta_0^2\cos\theta\left(\frac{1}{\beta_0^2 r^2} - j\frac{1}{\beta_0^3 r^3}\right)e^{-j\beta_0 r} \qquad (1)$$

$$E_\theta = \frac{\hat{I}\,d\zeta}{4\pi}\eta_0\beta_0^2\sin\theta\left(j\frac{1}{\beta_0 r} + \frac{1}{\beta_0^2 r^2} - j\frac{1}{\beta_0^3 r^3}\right)e^{-j\beta_0 r} \qquad (2)$$

$$E_\phi = 0 \qquad (3)$$

where $\eta_0 = \sqrt{\mu_0/\epsilon_0}$ is the intrinsic impedance of free space.

9.8 Obtain expressions for the far field of the dipole antenna of Problem 9.4.

▌ In the expressions of Problems 9.6 and 9.7 retain only terms in $1/r$:

$$\mathbf{H}_{\text{farfield}} = \frac{j\beta_0(\hat{I}\,d\zeta)\sin\theta}{4\pi r}e^{-j\beta_0 r}\mathbf{a}_\phi \qquad \mathbf{E}_{\text{farfield}} = \frac{j\eta_0\beta_0(\hat{I}\,d\zeta)\sin\theta}{4\pi r}e^{-j\beta_0 r}\mathbf{a}_\theta \qquad (1)$$

9.9 Determine the intrinsic impedance for the far field of the dipole antenna of Problem 9.4.

▌

$$\frac{|\mathbf{E}_{\text{farfield}}|}{|\mathbf{H}_{\text{farfield}}|} = \eta_0 = 377.0\ \Omega \qquad (1)$$

9.10 Find the time-average power density at a distance r from the dipole antenna of Problem 9.4.

▌ With $H_r = H_\theta = E_\phi = 0$,
$$\mathbf{S}_{av} = \tfrac{1}{2} \operatorname{Re} (\mathbf{E} \times \mathbf{H}^*) = \tfrac{1}{2} \operatorname{Re} (-E_r H_\phi^* \mathbf{a}_\theta + E_\theta H_\phi^* \mathbf{a}_r)$$
But (*1*) of Problem 9.7 and (*1*) of Problem 9.6 show that
$$E_r \sim \left(1 - j \frac{1}{\beta_0 r}\right) \qquad H_\phi^* \sim -j\left(1 - j \frac{1}{\beta_0 r}\right)^*$$
so that $E_r H_\phi^*$ will be pure imaginary. Hence,
$$\mathbf{S}_{av} = \left(\frac{1}{2} \operatorname{Re} E_\theta H_\phi^*\right)\mathbf{a}_r = \frac{\eta_0 I^2 \, d\zeta^2 \, \beta_0^2}{32\pi^2 r^2} \sin^2\theta \, \mathbf{a}_r \tag{1}$$

9.11 Rewrite (*1*) of Problem 9.10 in terms of the wavelength.

▌ Substitution of
$$\eta_0 = \sqrt{\frac{\mu_0}{\epsilon_0}} = 120\pi \ \Omega \qquad \text{and} \qquad \beta_0 = \frac{2\pi}{\lambda_0} \ (\text{m}^{-1})$$
gives
$$\mathbf{S}_{av} = 15\pi I^2 \left(\frac{d\zeta}{\lambda_0}\right)^2 \frac{\sin^2\theta}{r^2}\mathbf{a}_r, \quad (\text{W/m}^2) \tag{1}$$

9.12 Consider the energy radiated by a Hertzian dipole through a fixed solid angle Ω (Fig. 9-3). By conservation,
$$|\mathbf{S}_{av}(\kappa r, \theta)| = \frac{1}{\kappa^2} |\mathbf{S}_{av}(r, \theta)| \tag{1}$$
Verify (*1*), supposing κ so large that the far field may be used to evaluate the left-hand side.

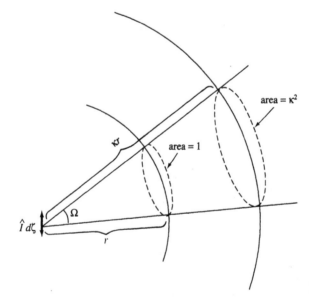

Fig. 9-3

▌ By Problem 9.8,
$$\text{L.H.S.} = \frac{1}{2} \frac{\eta_0 \beta_0^2 (I \, d\zeta)^2 \sin^2\theta}{(4\pi\kappa r)^2}$$

By Problem 9.10,
$$\text{R.H.S.} = \frac{1}{\kappa^2} \frac{\eta_0 \beta_0^2 (I \, d\zeta)^2 \sin^2\theta}{32\pi^2 r^2}$$

9.13 A 1-cm-long Hertzian dipole carries phasor current $\hat{I} = 10 \underline{/30°}$ A. Assuming that the wavelength is 3 m, determine the electric and magnetic fields at a distance of 10 cm from the dipole and at polar angle $\theta = 45°$. Compute the ratios $|E_\theta|/|E_r|$ and $|E_\theta|/|H_\phi|$ on this circle.

▌ From (*2*) of Problem 9.6,
$$\mathbf{H} = \frac{\hat{I} \, d\zeta}{4\pi}\left(\frac{2\pi}{\lambda_0}\right)^2 \sin\theta \left[\frac{j}{(2\pi r/\lambda_0)} + \frac{1}{(2\pi r/\lambda_0)^2}\right] e^{-j2\pi r/\lambda_0} \mathbf{a}_\phi$$
$$= (2.468 \times 10^{-2} \underline{/30°})(j4.77465 + 22.7973)e^{-j12°}\mathbf{a}_\phi = 0.57484 \underline{/29.83°} \, \mathbf{a}_\phi \quad \text{A/m}$$

From Problem 9.7,

$$E_r = (18.61 \underline{/30°})\left[\frac{1}{(2\pi r/\lambda_0)^2} - j\frac{1}{(2\pi r/\lambda_0)^3}\right]e^{-j12°} = 2069.67 \underline{/-60.17°} \quad \text{V/m}$$

$$E_\theta = 991.4 \underline{/-59.64} \quad \text{V/m}$$

Thus

$$\frac{|E_\theta|}{|E_r|} = 0.479 \quad \text{and} \quad \frac{|E_\theta|}{|H_\phi|} = 1724.6 \ \Omega$$

9.14 Check Problem 9.9 by repeating Problem 9.13 for $r = 10$ m, $\theta = 45°$.

❚ Proceeding as in Problem 9.13:

$$H_\phi = 1.18 \times 10^{-3} \underline{/-2.7°} \quad \text{A/m} \qquad E_r = 4.247 \times 10^{-2} \underline{/-92.73°} \quad \text{V/m} \qquad E_\theta = 0.444 \underline{/-2.74°} \quad \text{V/m}$$

whence

$$\frac{|E_\theta|}{|E_r|} = 10.45 \quad \text{and} \quad \frac{|E_\theta|}{|H_\phi|} = 376.08 \ \Omega \approx \eta_0$$

9.15 For the dipole antenna of Problem 9.4, plot $|\mathbf{S}_{av}|$ against θ for a fixed distance r. (This plot is called the *power pattern* or *radiation pattern*.)

❚ See Fig. 9-4. Note that no power is being radiated in the direction of the dipole current.

9.16 Determine the total average power radiated by the antenna of Problem 9.4.

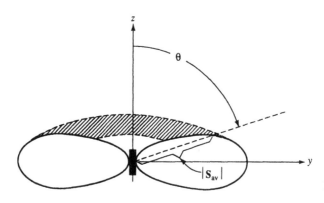

Fig. 9-4

❚ The required power may be obtained by integrating the normal component of $\mathbf{S}_{av}$, as given by (*1*) of Problem 9.11, over a spherical surface of radius R:

$$P_{\text{rad}} = \int_{\theta=0}^{\pi}\int_{\phi=0}^{2\pi}\left[15\pi\left(\frac{d\zeta}{\lambda_0}\right)^2 I^2 \frac{\sin^2\theta}{R^2}\right]R^2\sin\theta\,d\phi\,d\theta = 80\pi^2\left(\frac{d\zeta}{\lambda_0}\right)^2 I_{\text{rms}}^2 \quad \text{(W)} \tag{1}$$

where $I_{\text{rms}} = I/\sqrt{2}$.

9.17 Calculate the value of a (fictitious) *radiation resistance* which dissipates the same amount of power as that radiated by a Hertzian dipole, when both carry the same rms current.

❚ By (*1*) of Problem 9.16,

$$R_{\text{rad}} = \frac{P_{\text{rad}}}{I_{\text{rms}}^2} = 80\pi^2\left(\frac{d\zeta}{\lambda_0}\right)^2 \tag{1}$$

9.18 Calculate the rms current required in a 1-cm-long Hertzian dipole antenna to radiate 1 W of power at a frequency of 300 MHz.

❚ At 300 MHz, $\lambda_0 = 1$ m. Thus, from (*1*) of Problem 9.17,

$$R_{\text{rad}} = 80\pi^2\left(\frac{1 \times 10^{-2}}{1}\right)^2 = 78.9 \ \text{m}\Omega$$

Thus, for 1 W radiated power,

$$I_{\text{rms}}^2 = \frac{P_{\text{rad}}}{R_{\text{rad}}} = \frac{1}{78.9 \times 10^{-3}} \quad \text{or} \quad I_{\text{rms}} = 3.56 \ \text{A}$$

9.19 Repeat Problem 9.18 for a frequency of 3 MHz. What conclusion may be drawn from the result?

▮ Proceeding as in Problem 9.18, we have $\lambda_0 = 100$ m and $R_{\text{rad}} = 7.89\ \mu\Omega$; thus

$$I_{\text{rms}} = \sqrt{\frac{1}{7.89 \times 10^{-6}}} = 356\ \text{A}$$

This high current indicates that the Hertzian dipole is not a very effective radiator.

9.20 Compute the radiation resistance of, and the total average power radiated by, the dipole of Problem 9.13.

▮ From (1) of Problem 9.17,

$$R_{\text{rad}} = 80\pi^2 \left(\frac{d\zeta}{\lambda_0}\right)^2 = 80\pi^2 \left(\frac{10^{-2}}{3}\right)^2 = 7.018\ \text{m}\Omega$$

$$P_{\text{rad}} = |I_{\text{rms}}|^2 R_{\text{rad}} = \frac{100}{2} \times 8.773 \times 10^{-3} = 350.92\ \text{mW}$$

9.21 Suppose that an observer is in the far field of a wire-type antenna such as the Hertzian dipole, at a point 100 m away from the antenna. Assuming that the observer measures the magnitude of the electric field to be 1 V/m, determine the magnitudes of the electric and magnetic field intensity vectors at a distance of 1000 m along this same radial line.

▮ From Problems 9.8 and 9.9, $|\mathbf{E}_{\text{farfield}}| = \eta_0 |\mathbf{H}_{\text{farfield}}| \sim 1/r$; thus,

$$\frac{E_{1000\ \text{m}}}{E_{100\ \text{m}}} = \frac{100}{1000} \quad \text{or} \quad E_{1000\ \text{m}} = 0.1\ \text{V/m}$$

and

$$H_{1000\ \text{m}} = \frac{E_{1000\ \text{m}}}{\eta_0} = 0.265\ \text{mA/m}$$

9.22 Determine the average power densities at points 100 m and 1000 m away from the antenna of Problem 9.21.

▮ At any location, $|\mathbf{S}_{\text{av}}| = |\mathbf{E}|^2/2\eta_0$; thus

$$|\mathbf{S}_{\text{av}}|_{100\ \text{m}} = \frac{1}{2 \times 377}\, (1)^2 = 1.33\ \text{mW/m}^2$$

$$|\mathbf{S}_{\text{av}}|_{1000\ \text{m}} = \frac{1}{2 \times 377}\, (0.1)^2 = 13.3\ \mu\text{W/m}^2$$

9.23 Verify that the spherical waves emitted by a Hertzian dipole antenna resemble plane waves in the far field.

▮ From Problem 9.8,

$$\mathbf{H}_{\text{farfield}} = \frac{1}{\eta_0}\, (\mathbf{a}_r \times \mathbf{E}_{\text{farfield}})$$

Thus, $\mathbf{E}$ and $\mathbf{H}$ are mutually perpendicular and are related in magnitude via η_0, just as in a plane wave. They are not, however, uniform over the plane containing them (a tangent plane to the spherical wavefront).

9.24 Obtain the time-domain expressions for the far fields of the dipole antenna of Problem 9.4, for an input phasor current $I\underline{/0^\circ}$.

▮ In Problem 9.8 let $E_0 \equiv \eta_0\beta_0(I\,d\zeta)(\sin\theta)/4\pi$; then, since $\beta_0 = \omega/u_0$,

$$\mathbf{E}_{\text{farfield}} = \frac{E_0}{r}\cos\left[\omega\left(t - \frac{r}{u_0}\right) + 90^\circ\right]\mathbf{a}_\theta = -\frac{E_0}{r}\sin\omega\left(t - \frac{r}{u_0}\right)\mathbf{a}_\theta \tag{1}$$

$$\mathbf{H}_{\text{farfield}} = \frac{E_0}{\eta_0 r}\cos\left[\omega\left(t - \frac{r}{u_0}\right) + 90^\circ\right](\mathbf{a}_r \times \mathbf{a}_\theta) = -\frac{E_0}{\eta_0 r}\sin\omega\left(t - \frac{r}{u_0}\right)(\mathbf{a}_r \times \mathbf{a}_\theta) \tag{2}$$

9.25 Discuss the presence of the term $(t - r/u_0)$ in (1) and (2) of Problem 9.24.

▮ At any instant the field at an external point reflects the state of the source at an earlier time; the time lag, r/u_0, is precisely the transit time of the wave.

9.26 As shown in Fig. 9-5, a very small loop of radius b lying in the xy plane carries a phasor current $\hat{\mathbf{I}} = \hat{I}\mathbf{a}_{\phi'}$. Obtain an approximate expression for the magnetic vector potential at a distance r far away from the loop.

▮ From (1) of Problem 9.3, as applied to the torus,

$$\mathbf{A} = \frac{\mu_0}{4\pi}\oint_{\text{loop}} \hat{\mathbf{I}}\,\frac{e^{-j\beta_0 R}}{R}\,ds' \tag{1}$$

with $ds' = b\,d\phi'$. Under the assumption of an "electrically small" loop, $\beta_0(R - r) \ll 1$, or

$$e^{-j\beta_0 R} = e^{-j\beta_0 r}e^{-j\beta_0(R-r)} \approx e^{-j\beta_0 r}[1 - j\beta_0(R - r)] \tag{2}$$

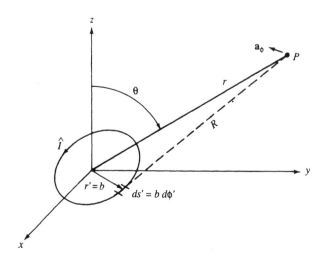

Fig. 9-5

and (1) becomes

$$\mathbf{A} \approx \frac{\mu_0}{4\pi} e^{-j\beta_0 r} \left[(1 + j\beta_0 r) \oint_{\text{loop}} \frac{\hat{\mathbf{I}} \, ds'}{R} - j\beta_0 \oint_{\text{loop}} \hat{\mathbf{I}} \, ds' \right] \tag{3}$$

The second integral in (3) is zero, since the vector current $\hat{\mathbf{I}}$ is assumed to be constant in magnitude and phase around the loop, and since $\mathbf{a}_{\phi'}$ integrates to the zero vector. Therefore, (3) becomes

$$\mathbf{A} \approx \frac{\mu_0 (1 + j\beta_0 r) e^{-j\beta_0 r}}{4\pi} \oint_{\text{loop}} \frac{\hat{\mathbf{I}} \, ds'}{R} \tag{4}$$

Now, in Problems 3.53 and 3.54 it was shown—under an assumption equivalent to "electrical smallness"—that the loop integral in (4) has the value $[(\hat{m} \sin \theta)/r^2] \mathbf{a}_\phi$, where $\hat{m} = \hat{I} \pi b^2 = \hat{I} A$ is the phasor magnetic dipole moment of the current loop. Consequently,

$$\mathbf{A} \approx \frac{\mu_0 \hat{m}}{4\pi r^2} (1 + j\beta_0 r) e^{-j\beta_0 r} \sin \theta \, \mathbf{a}_\phi = \frac{\mu_0}{4\pi r^3} (1 + j\beta_0 r) e^{-j\beta_0 r} (\hat{\mathbf{m}} \times \mathbf{r}) \tag{5}$$

9.27 Find the magnetic field, at the point P (Fig. 9-5), produced by the current loop of Problem 9.26.

▌ By (5) of Problem 9.26, $\mathbf{A} = A_\phi \mathbf{a}_\phi$; so that

$$\mu_0 \mathbf{H} = \text{curl } \mathbf{A} = \left(\frac{\cos \theta}{r \sin \theta} A_\phi + \frac{1}{r} \frac{\partial A_\phi}{\partial \theta} \right) \mathbf{a}_r + \left(-\frac{\partial A_\phi}{\partial r} - \frac{1}{r} A_\phi \right) \mathbf{a}_\theta$$

Going through the arithmetic we find:

$$H_r = j \frac{2\omega\mu_0}{4\pi\eta_0} \hat{m} \beta_0^2 \cos \theta \left(\frac{1}{\beta_0^2 r^2} - j \frac{1}{\beta_0^3 r^3} \right) e^{-j\beta_0 r} \tag{1}$$

$$H_\theta = \frac{\hat{m}}{4\pi} \sin \theta \left(-\frac{\beta_0^2}{r} + \frac{j\beta_0}{r^2} + \frac{1}{r^3} \right) e^{-j\beta_0 r}$$

$$= j \frac{\hat{m}\omega\mu_0}{4\pi\eta_0} \beta_0^2 \sin \theta \left(\frac{j}{\beta_0 r} + \frac{1}{\beta_0^2 r^2} - \frac{j}{\beta_0^3 r^3} \right) e^{-j\beta_0 r} \tag{2}$$

$$H_\phi = 0 \tag{3}$$

9.28 Find the electric field at P (Fig. 9-5) due to the loop of current of Problem 9.26.

▌ Since we know $\mathbf{H}$ from Problem 9.27, we determine $\mathbf{E}$ from

$$\text{curl } \mathbf{H} = j\omega\epsilon_0 \mathbf{E} = \frac{j\hat{m}\omega\mu_0}{4\pi\eta_0} \beta_0^2 \sin \theta \left(\frac{1}{r} - \frac{j}{\beta_0 r_2} \right) e^{-j\beta_0 r} \mathbf{a}_\phi$$

or

$$E_\phi = -j \frac{\hat{m}\omega\mu_0}{4\pi} \beta_0^2 \sin \theta \left(\frac{j}{\beta_0 r} + \frac{1}{\beta_0^2 r^2} \right) e^{-j\beta_0 r}$$

and $E_\theta = E_r = 0$.

9.29 What conclusions may be drawn by comparing the results of Problems 9.27 and 9.28 with those of Problems 9.6 and 9.7?

▌ Comparing the results of Problem 9.27 with those of Problem 9.7, we observe that the forms of solutions are identical, but the roles of $\mathbf{E}$ and $\mathbf{H}$ have been interchanged. The same conclusion holds for Problems 9.28 and 9.6. In other words, a duality exists between the structures of Figs. 9-2 and 9-5; the latter is often called a *magnetic dipole*.

9.30 Determine the far field due to the magnetic dipole of Problem 9.26.

▮ The far field is given by the terms in $1/r$ in the results of Problems 9.27 and 9.28:

$$\mathbf{E}_{\text{farfield}} = \frac{\omega\mu_0\hat{m}\beta_0}{4\pi r}\sin\theta\, e^{-j\beta_0 r}\mathbf{a}_\phi \qquad (1)$$

$$\mathbf{H}_{\text{farfield}} = -\frac{\omega\mu_0\hat{m}\beta_0}{4\pi\eta_0 r}\sin\theta\, e^{-j\beta_0 r}\mathbf{a}_\theta = \frac{1}{\eta_0}(\mathbf{a}_r \times \mathbf{E}_{\text{farfield}}) \qquad (2)$$

As was the case for the electric dipole (Problem 9.23), the far field of the magnetic dipole is such that $\mathbf{E}$ and $\mathbf{H}$ (i) decay as $1/r$, (ii) lie in a (local) plane perpendicular to the radial direction, and (iii) are related in magnitude through η_0.

9.31 Determine the time-average power density at P (Fig. 9-5) due to the current loop of Problem 9.26.

▮ The average power density is given by
$$\mathbf{S}_{\text{av}} = \tfrac{1}{2}\operatorname{Re}(\mathbf{E} \times \mathbf{H}^*) = \tfrac{1}{2}\operatorname{Re}(-E_\phi H_\theta^*\mathbf{a}_r + E_\phi H_r^*\mathbf{a}_\theta)$$
Using the results of Problems 9.27 and 9.28, it may be verified that $E_\phi H_r^*$ is pure imaginary, and that $[m = |\hat{m}|]$

$$-E_\phi H_\theta^* = \frac{(\omega\mu_0 m\beta_0^2)^2}{(4\pi)^2\eta_0}\sin^2\theta\left(\frac{1}{\beta_0^2 r^2} + \frac{j}{\beta_0^5 r^5}\right)$$

Hence
$$\mathbf{S}_{\text{av}} = \frac{\eta_0 m^2\beta_0^4}{32\pi^2}\frac{\sin^2\theta}{r^2}\mathbf{a}_r \qquad (1)$$

where the relation $\omega\mu_0 = \eta_0\beta_0$ was used.

9.32 Put (1) of Problem 9.31 in analogous form to (1) of Problem 9.11.

▮ $$\mathbf{S}_{\text{av}} = 60\pi^3 I^2\left(\frac{A}{\lambda_0^2}\right)^2\frac{\sin^2\theta}{r^2}\mathbf{a}_r \quad (\text{W/m}^2)$$

in which $A = \pi b^2$ is the area of the loop. Very nearly, $60\pi^3 = 1860$.

9.33 Find the total average power radiated by the magnetic dipole of Fig. 9-5.

▮ Integrate $|\mathbf{S}_{\text{av}}|$ over the surface of a sphere of radius R:

$$P_{\text{rad}} = 1860\left(\frac{A}{\lambda_0^2}\right)^2 I^2\int_{\theta=0}^{\pi}\int_{\phi=0}^{2\pi}\left(\frac{\sin^2\theta}{R^2}\right)R^2\sin\theta\, d\phi\, d\theta = 15\,585 I^2\left(\frac{A}{\lambda_0^2}\right)^2 \quad (\text{W}) \qquad (1)$$

9.34 Determine the value of a fictitious resistance that dissipates the same amount of power as is radiated by a magnetic dipole (Fig. 9-5) when both carry the same rms current.

▮ The radiation resistance is found from (1) of Problem 9.33:
$$R_{\text{rad}} = \frac{P_{\text{rad}}}{I_{\text{rms}}^2} = \frac{2P_{\text{rad}}}{I^2} = 31\,170\left(\frac{A}{\lambda_0^2}\right)^2 \quad (\Omega)$$

9.35 Find the rms current required in a magnetic dipole made of a loop 1 cm in radius, assuming that the loop radiates 1 W in power at 300 MHz.

▮ With $\lambda_0 = 1$ m and $A = \pi \times 10^{-4}$ m^2, (1) of Problem 9.34 yields
$$R_{\text{rad}} = 31\,170\left(\frac{\pi \times 10^{-4}}{1^2}\right)^2 = 3.08\,\text{m}\Omega \qquad \text{and} \qquad I_{\text{rms}} = \sqrt{\frac{1}{3.08 \times 10^{-3}}} = 18\,\text{A}$$

9.36 Repeat Problem 9.35 for a frequency of 3 MHz. What conclusion may be drawn from these results?

▮ $$R_{\text{rad}} = 31\,170\left(\frac{\pi \times 10^{-4}}{100^2}\right)^2 = 3.08\,\text{p}\Omega \qquad I_{\text{rms}} = \sqrt{\frac{1}{3.08 \times 10^{-11}}} = 0.18\,\text{MA}$$

Results of Problems 9.35 and 9.36 indicate that a magnetic dipole is not an effective radiator.

9.37 Consider a magnetic dipole antenna of radius 1 cm carrying phasor current $\hat{I} = 10\underline{/30°}$ A. If the frequency is 100 MHz, what are the electric and magnetic field intensities at $\theta = 45°$ for $r = 10$ cm, 1 m? Determine the ratios $|H_\theta|/|H_r|$ and $|E_\phi|/|H_\theta|$ on these circles.

▮ Comparison between the forms of the fields of the magnetic dipole and those of the electric dipole allows an easy calculation using the results of Problems 9.13 and 9.14.

At $r = 10$ cm:

$$E_\phi = 14.26 \, \underline{/-60.2°} \quad \text{V/m} \qquad H_r = 0.361 \, \underline{/29.83°} \quad \text{A/m} \qquad H_\theta = 122.96 \, \underline{/30.30°} \quad \text{A/m}$$

$$\frac{|H_\theta|}{|H_r|} = \frac{122.96}{0.361} = 340.61 \qquad \frac{|E_\phi|}{|H_\theta|} = \frac{14.26}{122.96} = 0.11597 \ \Omega$$

At $r = 1$ m:

$$E_\phi = 0.32395 \, \underline{/-115.52°} \quad \text{V/m} \qquad H_r = 8.205 \times 10^{-4} \, \underline{/-25.5°} \quad \text{A/m} \qquad H_\theta = 7.039 \times 10^{-4} \, \underline{/58.27°} \quad \text{A/m}$$

$$\frac{|H_\theta|}{|H_r|} = \frac{7.039}{8.205} = 0.8579 \qquad \frac{|E_\phi|}{|H_\theta|} = \frac{3239.5}{7.039} = 460.23 \ \Omega$$

9.38 For the dipole antenna of Problem 9.37, verify that in the far field, $|E_\phi|/|H_\theta|$ is essentially equal to the intrinsic impedance of free space. Also evaluate $|H_\theta|/|H_r|$.

▌ Represent the far field by $r = 10$ m. Then, proceeding as in Problem 9.37,

$$E_\phi = 2.927 \times 10^{-2} \, \underline{/-92.7°} \quad \text{V/m} \qquad H_r = 7.412 \times 10^{-6} \, \underline{/-2.73°} \quad \text{A/m} \qquad H_\theta = 7.745 \times 10^{-5} \, \underline{/87.26°} \quad \text{A/m}$$

$$\frac{|H_\theta|}{|H_r|} = \frac{7.745}{0.7412} = 10.45 \qquad \frac{|E_\phi|}{|H_\theta|} = \frac{2.927}{7.745 \times 10^{-3}} = 377.9 \ \Omega \approx \eta_0$$

9.39 Calculate the radiation resistance of the magnetic dipole antenna of Problem 9.37.

▌ The radiation resistance is given by (1) of Problem 9.34. With $A = 10^{-4}\pi$ m^2 and $\lambda_0 = 3$ m,

$$R_{\text{rad}} = 31\,170 \left(\frac{10^{-4}\pi}{3^2} \right)^2 = 3.85 \ \mu\Omega$$

9.40 How much average power is radiated by the magnetic dipole antenna of Problems 9.37 and 9.39?

▌
$$P_{\text{rad}} = I_{\text{rms}}^2 R_{\text{rad}} = \left(\frac{10}{\sqrt{2}} \right)^2 (38 \times 10^{-6}) = 0.193 \ \text{mW}$$

9.41 Consider a square loop carrying a phasor current $\hat{I}$, as shown in Fig. 9-6. Assume that the loop is electrically small at the frequency of interest, so that $\hat{I}$ is uniform (magnitude and phase) around the loop. Modeling this loop as four Hertzian dipoles, determine the far electric field at a distance $R \gg d$ away from each side in the plane of the loop.

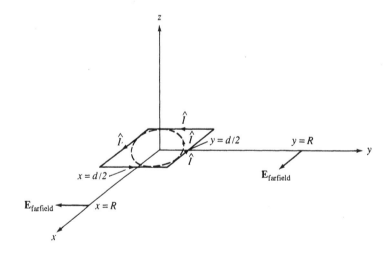

Fig. 9-6

▌ At $y = R$, the fields due to the sides at $x = d/2$ and $x = -d/2$ cancel because of symmetry and because the currents in these sides go in opposite directions. The fields due to the sides at $y = d/2$ and $y = -d/2$ are given by (1) of Problem 9.8, with $\theta = 90°$ and $\mathbf{a}_\theta = \mathbf{a}_x$ and $\mathbf{a}_\theta = -\mathbf{a}_x$, respectively:

$$\mathbf{E}_{\text{farfield}} = \frac{j\eta_0 \beta_0 \hat{I} d}{4\pi} \left(\frac{e^{-j\beta_0(R-d/2)}}{R - d/2} - \frac{e^{-j\beta_0(R+d/2)}}{R + d/2} \right) \mathbf{a}_x$$

The terms $(R - d/2)$ and $(R + d/2)$ in the denominators may be approximated by R, giving

$$\mathbf{E}_{\text{farfield}} = \frac{j\eta_0 \beta_0 \hat{I} d}{4\pi R} e^{-j\beta_0 R} (e^{j\beta_0 d/2} - e^{-j\beta_0 d/2}) = -\frac{2\eta_0 \beta_0 \hat{I} d}{4\pi R} e^{-j\beta_0 R} \sin(\beta_0 d/2) \mathbf{a}_x$$

Since the loop is assumed to be electrically small, $\sin(\beta_0 d/2) \approx \beta_0 d/2$; thus

$$\mathbf{E}_{\text{farfield}} = -\frac{\eta_0 \beta_0^2 \hat{I} d^2}{4\pi R} e^{-j\beta_0 R} \mathbf{a}_x = -\frac{\omega \mu_0 \beta_0 \hat{I} d^2}{4\pi R} e^{-j\beta_0 R} \mathbf{a}_x = -1184 \frac{\hat{I} d^2}{\lambda_0^2 R} e^{-j\beta_0 R} \mathbf{a}_x \quad \text{(V/m)} \quad (1)$$

By symmetry, the far field at $x = R$ is given by (1) with $\mathbf{a}_x$ replaced by $-\mathbf{a}_y$.

9.42 Compare (1) of Problem 9.41 with the far field of a (circular) magnetic dipole antenna (Problem 9.30). What conclusion may be drawn from this comparison?

▮ By (1) of Problem 9.30, at $r = R$, $\theta = 90°$ (from the z axis), $\phi = 90°$ (from the x axis),

$$\mathbf{E}_{\text{farfield}} = \frac{\omega \mu_0 \hat{m} \beta_0}{4\pi R} e^{-j\beta_0 R}(-\mathbf{a}_x)$$

in exact agreement with (1) of Problem 9.41, provided $\hat{m} = \hat{I} d^2$. The conclusion is that, at great distances, the shape of a loop is irrelevant and only its area matters.

9.43 A square loop, of side 1 cm, carries a current $1\,\underline{/0°}$ A at 300 MHz. Find the electric field 3 m away from the loop.

▮ In terms of relative distance, we have to find the far field. Thus, (1) of Problem 9.41 is applicable, with: $\omega = 2\pi \times 300 \times 10^6$ rad/s, $\mu_0 = 4\pi \times 10^{-7}$ H/m, $\beta_0 = 2\pi$ m^{-1}, $|\hat{I}| = 1$ A, $d = 1 \times 10^{-2}$ m, and $R = 3$ m.

$$|\mathbf{E}| = \frac{2\pi \times 300 \times 10^6 \times 4\pi \times 10^{-7} \times 2\pi \times 1 \times (10^{-2})^2}{4\pi \times 3} = 39.46 \text{ mV/m}$$

9.44 A dipole antenna of finite length ζ is driven by a voltage source at its input. Assume that the antenna is sufficiently short for the phasor current to decrease uniformly from its maximum at the center to zero at the endpoints. Derive an expression for the far electric field in terms of the input current.

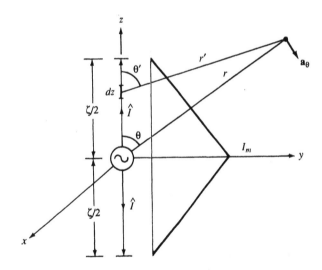

Fig. 9-7

▮ The given current distribution, shown in Fig. 9-7, may be written as

$$\hat{I}(z) = \frac{2I_m}{\zeta}\left(\frac{\zeta}{2} - |z|\right) \qquad \left(|z| \leq \frac{\zeta}{2}\right)$$

and, from (1) of Problem 9.8,

$$dE_\theta = j\eta_0 \beta_0 \frac{\sin \theta'\, e^{-j\beta_0 r'}}{4\pi r'} \hat{I}(z)\, dz \tag{1}$$

Since $r' \approx r - z \cos \theta$ and $\sin \theta' = (r/r') \sin \theta$, we may approximate (1) as

$$dE_\theta \approx j\eta_0 \beta_0 \frac{\sin \theta\, e^{-j\beta_0 r}}{4\pi r} \left[e^{j\beta_0 z \cos \theta} \hat{I}(z)\, dz\right] \tag{2}$$

In the farfield approximation, (2), the field contributions of all elements $\hat{I}\, dz$ of the upper half-antenna have the same direction $\mathbf{a}_\theta$; for the lower half-antenna (where the current direction is opposite) the common direction is $-\mathbf{a}_\theta$. Hence the inserted minus sign in the following superposition:

$$E_\theta \approx \frac{j\eta_0 \beta_0}{4\pi r} \sin \theta\, e^{-j\beta_0 r} \frac{2I_m}{\zeta}\left[\int_0^{\zeta/2}\left(\frac{\zeta}{2} - z\right)e^{j\beta_0 z \cos \theta}\, dz - \int_0^{-\zeta/2}\left(\frac{\zeta}{2} + z\right)e^{j\beta_0 z \cos \theta}\, dz\right]$$

In the second integral change the variable from z to $-z$:

$$E_\theta \approx \frac{j\eta_0\beta_0}{4\pi r}\sin\theta\, e^{-j\beta_0 r}\frac{2I_m}{\zeta}\left[\int_0^{\zeta/2}\cos(\beta_0 z\cos\theta)\,dz - 2\int_0^{\zeta/2} z\cos(\beta_0 z\cos\theta)\,dz\right]$$

$$=\frac{j\eta_0 I_m}{\pi\beta_0\zeta}\frac{e^{-j\beta_0 r}}{r}\left[\frac{1-\cos\left(\beta_0\frac{\zeta}{2}\cos\theta\right)}{\cos^2\theta}\right]\sin\theta \qquad (3)$$

For an electrically small antenna, $\beta_0\zeta \ll 1$ rad; thus

$$\cos\left(\beta_0\frac{\zeta}{2}\cos\theta\right)\approx 1-\frac{1}{2}\beta_0^2\frac{\zeta^2}{4}\cos^2\theta$$

and (3) becomes

$$E_\theta = j\eta_0\beta_0\frac{\sin\theta\, e^{-j\beta_0 r}}{4\pi r}\left(\frac{I_m}{2}\right)\zeta \qquad (4)$$

9.45 How might (4) of Problem 9.44 have been written at once, without any integration?

▌ We might have realized that, viewed from afar, a finite dipole looks infinitesimal; moreover, only the mean value of the current distribution can be of significance. We could then have gone straight from (2) to (4).

9.46 Repeat Problem 9.44 for the current distribution

$$\hat{I}(z) = I_m\sin\beta_0\left(\frac{\zeta}{2}-|z|\right) \qquad \left(|z|\le\frac{\zeta}{2}\right) \qquad (1)$$

Do not assume an electrically small antenna.

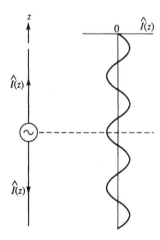

Fig. 9-8

▌ When the time-dependence $\cos\omega t$ is factored in, (1) is seen to represent a two-sided standing wave on the antenna, with an antinode at the central oscillator and a node at either end (Fig. 9-8). Points symmetric in the center are precisely in phase. This (and any similar) current distribution is handled exactly like the straight-line distribution of Problem 9.44. Substitute (1) in (2) of Problem 9.44 and carry out the integration in the same way, to obtain

$$E_\theta \approx j\frac{\eta_0 I_m e^{-j\beta_0 r}}{2\pi r}F(\theta) \qquad (2)$$

in which the *angle factor* or *pattern function* for this current distribution is given by

$$F(\theta) = \left[\cos\left(\beta_0\frac{\zeta}{2}\cos\theta\right)-\cos\beta_0\frac{\zeta}{2}\right]\Big/\sin\theta \qquad (3)$$

9.47 From Problems 9.44 and 9.46 it is clear that the far electric field set up by any distribution of antenna current must have the form

$$E_\theta \propto \frac{e^{-j\beta_0 r}}{r}F(\theta)$$

for a suitable angle factor. Characterize this factor in a general way, when the antenna is viewed from afar at $\theta = 90°$.

▌ From (2) of Problem 9.44,

$$F(90°) \propto \int_{-\zeta/2}^{\zeta/2} \hat{I}(z)\, dz \propto \text{(mean value of current distribution)}$$

(cf. Problem 9.45).

9.48 Find the far magnetic field due to the long dipole antenna of Problem 9.46.

▌ The point here is that, by Problems 9.23 and 9.44, the far field of a finite dipole antenna also is a quasi-plane wave. Hence

$$\mathbf{H}_{\text{farfield}} = H_\phi \mathbf{a}_\phi = \frac{E_\theta}{\eta_0} \mathbf{a}_\phi \qquad (1)$$

where E_θ is given by (2) of Problem 9.46.

9.49 Plot the electric-field pattern for a dipole antenna of electrical length $\zeta/\lambda_0 = 1/2$, known as a *half-wave dipole*.

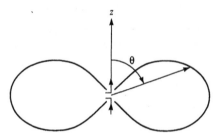

Fig. 9-9

▌ Refer to Problem 9.46. For $\beta_0 \zeta/2 = \pi/2$, (3) becomes

$$F(\theta) = \cos\left(\frac{\pi}{2} \cos\theta\right) \Big/ \sin\theta \qquad (1)$$

which is plotted in Fig. 9-9.

9.50 Repeat Problem 9.49 for a dipole antenna of electrical length 1.

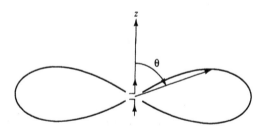

Fig. 9-10

▌ Now

$$F(\theta) = \frac{\cos(\pi \cos\theta) + 1}{\sin\theta} \qquad (1)$$

which is sketched in Fig. 9-10.

9.51 Repeat Problem 9.49 for $\zeta_e = 3/2$.

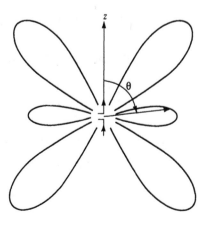

Fig. 9-11

■
$$F(\theta) = [\cos(\tfrac{3}{2}\pi \cos\theta)]/\sin\theta$$
See Fig. 9-11.

9.52 What is the maximum value of the electric field for the half-wave dipole antenna?

■ The angle function (*1*) of Problem 9.49 is a maximum for $\theta = 90°$ (broadside to the antenna).

9.53 Determine the time-average power density radiated by the dipole antenna of Problem 9.46.

■
$$\mathbf{S}_{av} = \frac{1}{2}\operatorname{Re}(\mathbf{E}\times\mathbf{H}^*) = \frac{1}{2}\operatorname{Re}(E_\theta H_\phi^*)\mathbf{a}_r = \frac{|E_\theta|^2}{2\eta_0}\mathbf{a}_r$$
By (*2*) of Problem 9.46,
$$\mathbf{S}_{av} = \eta_0\frac{I_m^2}{8\pi^2 r^2}F^2(\theta)\mathbf{a}_r = \frac{4.77 I_m^2}{r^2}F^2(\theta)\mathbf{a}_r \quad (\text{W/m}^2) \tag{1}$$

9.54 Find the total average power radiated by the dipole antenna of Problem 9.46.

■ The total average power radiated is obtained by integrating the normal component of (*1*) of Problem 9.53 over some closed surface that encloses the dipole. The obvious choice for the closed surface is a sphere:
$$P_{rad} = \int_{\theta=0}^{\pi}\int_{\phi=0}^{2\pi}\mathbf{S}_{av}\cdot\mathbf{a}_r\, r^2\sin\theta\, d\theta\, d\phi = \frac{\eta_0 I_m^2}{4\pi}\int_{\theta=0}^{\pi}F^2(\theta)\sin\theta\, d\theta \tag{1}$$

9.55 The current distribution in a half-wave dipole antenna is given by
$$I(z) = 0.25\sin\left(\frac{\pi}{2} - \beta_0|z|\right)\quad\text{A}$$
Calculate the power radiated by this antenna.

■ With $F(\theta)$ as given by (*1*) of Problem 9.49, a numerical integration yields
$$\int_{\theta=0}^{\pi}F^2(\theta)\sin\theta\, d\theta \approx 1.2186$$
Then, by (*1*) of Problem 9.54,
$$P_{rad} = \frac{120\pi}{4\pi}(0.25)^2(1.2186) = 2.28\text{ W}$$

9.56 Find the radiation resistance of the half-wave dipole antenna.

■ From Problems 9.54 and 9.55,
$$R_{rad} = \frac{P_{rad}}{I_{rms}^2} = \frac{2P_{rad}}{I_m^2} = 60\times 1.2186 = 73\ \Omega$$

9.57 Determine the magnitudes of the electric and magnetic fields of a half-wave dipole operated at a frequency of 300 MHz, at a distance of 100 m in the broadside plane ($\theta = 90°$). The input current to the terminals is $100\ \underline{/0°}$ mA.

■ Substitute $I_m = 100$ mA and $F(\theta) = 1$ in (*2*) of Problem 9.46:
$$|E_\theta| = \frac{377\times 100}{2\pi\times 100}\times 10^{-3}\times 1 = 60\text{ mV/m}$$
Also
$$|H_\phi| = \frac{1}{\eta_0}|E_\theta| = 159.15\ \mu\text{A/m}$$

9.58 How much average power is radiated by the antenna of Problem 9.57?

■ From the result of Problem 9.56,
$$P_{av} = I_{rms}^2 R_{rad} = \left(\frac{100}{\sqrt{2}}\times 10^{-3}\right)^2\times 73 = 0.365\text{ W}$$

9.59 A quarter-wave ($h = \lambda_0/4$) monopole antenna is shown in Fig. 9-12. Obtain its radiation resistance.

■ The infinite, perfectly conducting ground plane may be replaced with the image of the monopole, as indicated; this is possible because the (far) electric field of the resulting half-wave dipole is vertical in the broadside plane. Then, since the monopole radiates half the power radiated by the dipole for the same current,
$$R_{rad} = \tfrac{1}{2}\times 73 = 36.5\ \Omega$$

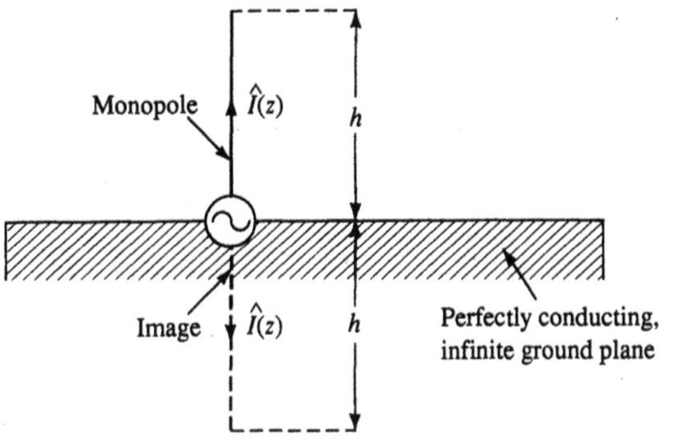

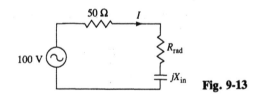

Fig. 9-12

9.60 The input impedance of a dipole antenna is defined by

$$Z_{in} = R_{in} + jX_{in} \tag{1}$$

Consider the antenna to be lossless, and express its radiation resistance R_{rad} in terms of its input resistance R_{in}.

▌ At the input terminals of a dipole antenna ($z = 0$), from (1) of Problem 9.46 we have

$$\hat{I}_{in} = I_m \sin \frac{\beta_0 \zeta}{2} \tag{2}$$

The average power delivered to the antenna is

$$P_{in} = \frac{|\hat{I}_{in}|^2}{2} R_{in} = \frac{I_m^2}{2} R_{in} \sin^2 \frac{\beta_0 \zeta}{2} \tag{3}$$

Since the antenna is lossless, the average power delivered to the antenna is simply the average power radiated: $P_{in} = P_{rad} = \frac{1}{2} I_m^2 R_{rad}$. Thus

$$R_{rad} = R_{in} \sin^2 \frac{\beta_0 \zeta}{2} \tag{4}$$

9.61 Express the radiation resistance of a dipole antenna in terms of $F(\theta)$, where $F(\theta)$ is defined by (2) of Problem 9.46.

▌ As in Problem 9.56,

$$R_{rad} = 60 \int_{\theta=0}^{\pi} F^2(\theta) \sin \theta \, d\theta \quad (\Omega) \tag{1}$$

9.62 A lossless quarter-wave monopole antenna (Problem 9.59) is situated above a perfectly conducting ground plane and is driven by a 100-V, 300-MHz source that has an internal impedance of 50 Ω. Compute the average power radiated by the antenna, given $X_{in} = 21.25$ Ω [in (1) of Problem 9.60].

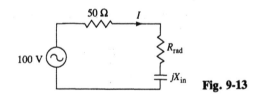

Fig. 9-13

▌ From the equivalent circuit of the antenna, shown in Fig. 9-13,

$$I = \frac{V}{R_{source} + R_{rad} + jX_{in}}$$

With $R_{rad} = 36.5$ Ω from Problem 9.59, we have

$$I = \frac{100 \underline{/0°}}{50 + 36.5 + j21.25} = 1.12 \underline{/-13.8°} \quad A$$

and $P_{rad} = \frac{1}{2} |I|^2 R_{rad} = 23$ W

9.63 Repeat Problem 9.62 for a $\lambda_0/5$, lossless monopole antenna having an input impedance of $(20 - j50)$ Ω, the input voltage remaining unchanged.

▌
$$I = \frac{100 \underline{/0°}}{50 + 20 - j50} = 1.163 \underline{/35.6°} \ A \qquad P_{rad} = P_{in} = \frac{1}{2}(1.163)^2(20) = 13.52 \ W$$

9.64   Repeat Problem 9.62 for a short $(\lambda_0/10)$ lossless monopole that has an input impedance of $(4 - j180)$ Ω.

I
$$I = \frac{100\,\underline{/0°}}{50 + 4 - j180} = 0.532\,\underline{/73.3°}\ \text{A} \qquad P_{\text{rad}} = P_{\text{in}} = \frac{1}{2}(0.532)^2(4) = 0.566\ \text{W}$$

9.65   A lossless dipole antenna, for which $Z_{\text{ant}} = (73 + j42.5)$ Ω, is attached to a source with a length of lossless 50-Ω coaxial cable. The source has an open-circuit voltage of 100 V (rms) and an internal impedance of 50 Ω. The frequency of the source is such that the dipole length is 0.5λ and the transmission-line length is $L = 1.3\lambda$; determine the average power radiated by the antenna.

I   The total input impedance is obtained from
$$Z_{\text{in}} = R_C\left[\frac{Z_{\text{ant}} + jR_C \tan \beta L}{R_C + jZ_{\text{ant}} \tan \beta L}\right] \qquad \text{where} \qquad \tan \beta L = \tan 2.6\pi = -3.0777$$
Thus, $Z_{\text{in}} = 23.09\,\underline{/-5.585°} = (22.98 - j2.247)$ Ω and
$$I_{\text{into line}} = \frac{100\,\underline{/0°}}{50 + Z_{\text{in}}} = \frac{100\,\underline{/0°}}{50 + 22.98 - j2.247} = 1.37\,\underline{/1.76°}\ \text{A (rms)}$$
$$(P_{\text{av}})_{\text{to line}} = (P_{\text{av}})_{\text{to ant}} = (P_{\text{av}})_{\text{rad}} = (1.37)^2 \times 22.98 = 43.1\ \text{W}$$

9.66   Calculate the voltage standing-wave ratio (VSWR) on the cable of Problem 9.65.

I   The reflection coefficient is given by
$$\Gamma_L = \frac{Z_{\text{ant}} - Z_{\text{source}}}{Z_{\text{ant}} + Z_{\text{source}}} = \frac{73 + j42.5 - 50}{73 + j42.5 + 50} = 0.371\,\underline{/42.52°}$$
and
$$\text{VSWR} = \frac{1 + |\Gamma_L|}{1 - |\Gamma_L|} = 2.18$$

9.67   Sketch the current distribution along a dipole for dipole lengths $3\lambda/4$, λ, $5\lambda/4$, $3\lambda/2$, and 2λ.

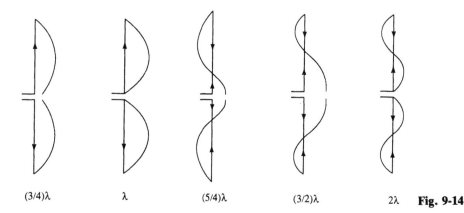

$(3/4)\lambda$ λ $(5/4)\lambda$ $(3/2)\lambda$ 2λ **Fig. 9-14**

I   The sketches are shown in Fig. 9-14.

9.68   Consider the radiation pattern function (3) of Problem 9.46, which has zeros at the endpoints $\theta = 0°$ and $\theta = 180°$ (use L'Hospital's rule). Prove that interior zeros are absent or present according as $\zeta_e \leq 1$ or $\zeta_e > 1$, where $\zeta_e \equiv \zeta/\lambda_0$ is the electrical length of the dipole.

I   Using standard trigonometric formulas the numerator of (3) may be written as
$$2\sin\left[\frac{\pi\zeta_e}{2}(1 - \cos\theta)\right]\sin\left[\frac{\pi\zeta_e}{2}(1 + \cos\theta)\right]$$
Hence, for an interior zero,
$$\frac{\zeta_e}{2}(1 \mp \cos\theta) = k \qquad (k = 1, 2, 3, \dots)$$
or
$$\cos\theta = \pm\left(1 - \frac{2k}{\zeta_e}\right) \qquad (k = 1, 2, 3, \dots) \tag{1}$$
When $\zeta_e \leq 1$, $1 - (2k/\zeta_e) \leq -1$, and (1) has no real solution. But when $\zeta_e > 1$, there is always at least one solution. For instance, if $\zeta_e = 2.5$, there are four solutions, symmetric with respect to $\theta = 90°$:
$$\theta = \cos^{-1}\left(\tfrac{3}{5}\right) \qquad \theta = \cos^{-1}\left(\tfrac{1}{5}\right) \qquad \theta = \cos^{-1}\left(-\tfrac{1}{5}\right) \qquad \theta = \cos^{-1}\left(-\tfrac{3}{5}\right)$$

9.69 Given a *lossy* dipole antenna, with per-unit-length resistance r_{wire}, obtain an expression for the net ohmic resistance of the antenna in terms of r_{wire}, ζ and λ_0.

▎ The total ohmic power loss is

$$P_{\text{loss}} = \int_{-\zeta/2}^{\zeta/2} r_{\text{wire}} \frac{|\hat{I}(z)|^2}{2}\, dz$$

whence

$$R_{\text{loss}} = \frac{P_{\text{loss}}}{|\hat{I}_{\text{in}}|^2/2} = \frac{r_{\text{wire}} \int_{-\zeta/2}^{\zeta/2} |\hat{I}(z)|^2\, dz}{|\hat{I}_{\text{in}}|^2} \qquad (1)$$

This expression can be evaluated by substituting the expressions for current distribution given in (1) of Problem 9.46 and the expression for input current given in (2) of Problem 9.60.

$$\int_{-\zeta/2}^{\zeta/2} |\hat{I}(z)|^2\, dz = 2\int_0^{\zeta/2} I_m^2 \sin^2 \beta_0\!\left(\frac{\zeta}{2}-z\right) dz = I_m^2 \frac{\zeta}{2}\left[1 - \frac{\sin 2\pi\zeta/\lambda_0}{2\pi\zeta/\lambda_0}\right] \qquad (2)$$

Substituting (2) in (1) yields

$$R_{\text{loss}} = \frac{r_{\text{wire}} \dfrac{\zeta}{2}\left[1 - \dfrac{\sin (2\pi\zeta/\lambda_0)}{2\pi\zeta/\lambda_0}\right]}{\sin^2 (\pi\zeta/\lambda_0)} \qquad (3)$$

9.70 A 1-m-long dipole antenna is driven by a 150-MHz source having a source resistance of 50 Ω and an open-circuit voltage of 100 V, as shown in Fig. 9-15. Determine the net ohmic resistance of the antenna. The antenna is constructed of 20-gauge (0.406 mm radius) copper wires. The skin depth is 5.4 μm.

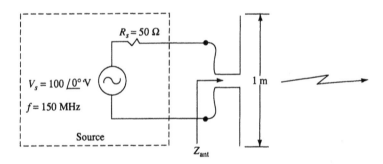

Fig. 9-15

▎ At 150 MHz, the 1-m dipole is exactly $\lambda_0/2$ in length, so that $R_{\text{rad}} = 73\ \Omega$ and $X_{\text{in}} = 42.5\ \Omega$. The 20-gauge wires have radii much larger than a skin depth at this frequency ($\delta = 5.4\ \mu$m). Thus, the high-frequency approximation for wire resistance yields

$$r_{\text{wire}} = \frac{R_s}{2\pi r_w} = \frac{1}{2\pi r_w}\sqrt{\frac{\omega\mu}{2\sigma}} = \frac{1}{2\pi r_w \sigma\delta}$$

$$= \frac{1}{2\pi \times 0.406 \times 10^{-3} \times 5.8 \times 10^7 \times 5.4 \times 10^{-6}} = 1.25\ \Omega/\text{m}$$

and (3) of Problem 9.69 gives

$$R_{\text{loss}} = \frac{(1.25)(0.5)(1-0)}{1^2} = 0.625\ \Omega$$

9.71 Evaluate the average power dissipated in the antenna of Problem 9.70.

▎ We first determine the input impedance to the antenna. Since the dipole is a half-wave dipole, the total input impedance to the antenna as seen by the source is

$$Z_{\text{ant}} = R_{\text{loss}} + R_{\text{rad}} + jX_{\text{in}} = (0.63 + 73 + j42.5)\ \Omega$$

Then

$$I_{\text{ant}} = \frac{V_S}{R_S + Z_{\text{ant}}} = \frac{100\ \underline{/0^\circ}}{50 + 73.63 + j42.5} = 0.765\ \underline{/-18.97^\circ}\ \ \text{A}$$

and

$$P_{\text{loss}} = \tfrac{1}{2}|I_{\text{ant}}|^2 R_{\text{loss}} = 184\ \text{mW}$$

9.72 Determine the *radiative efficiency* of the antenna of Problems 9.70 and 9.71. (The radiative efficiency is defined as the ratio of the radiated power to the power delivered to the antenna.)

▎ The average power radiated is given by

$$P_{\text{rad}} = \tfrac{1}{2}|I_{\text{ant}}|^2 R_{\text{rad}} = \tfrac{1}{2}(0.765)^2(73) = 21.36\ \text{W}$$

and $P_{\text{loss}} = 0.184$ W. Thus,

$$\text{radiative efficiency} = \frac{21.36}{21.36 + 0.184} = 0.991 \text{ or } 99.1\%$$

9.73 The antenna of Problems 9.70 and 9.71 is attached to the source via a $\lambda/4$ length of lossless 50-Ω transmission line. Recompute the average radiated power.

▮ From Problem 9.71, $Z_L = 50 + 73.63 + j42.5 = 123.63 + j42.5$ Ω; thus,

$$Z_{\text{in}} = \frac{R_C^2}{Z_L} = \frac{2500}{123.63 + j42.5} = 18.084 - j6.217 \quad \Omega$$

$$I_{\text{into line}} = \frac{100 \ /0^\circ}{50 + Z_{\text{in}}} = \frac{100 \ /0^\circ}{50 + 18.084 - j6.217} = 1.463 \ /5.217^\circ \quad \text{A}$$

$$P_{\text{into line}} = \tfrac{1}{2}(1.463)^2(18.084) = 19.353 \text{ W}$$

But $\qquad P_{\text{rad}} = \text{efficiency} \times P_{\text{ant}} = \text{efficiency} \times P_{\text{into line}} = 0.991 \times 19.353 = 19.179$ W

9.74 An *antenna array* is used to produce particular directional properties of the radiated field. Consider two vertical-wire antennas (Fig. 9-16). Obtain an expression for the **E**-field at a point P far away from the antennas. The respective currents in the antennas are $\hat{I}_1 = I \ /\alpha$ and $\hat{I}_2 = I \ /0$, where α (and 0) is in radians.

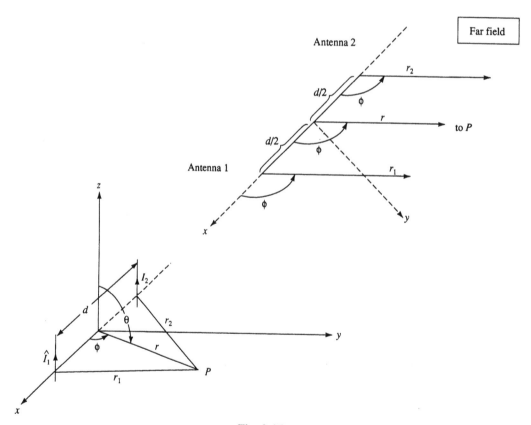

Fig. 9-16

▮ For the far fields use (2) of Problem 9.46:

$$E_{\theta 1} \approx \frac{KI \ /\alpha}{r_1} e^{-j\beta_0 r_1} \qquad E_{\theta 2} \approx \frac{KI \ /0}{r_2} e^{-j\beta_0 r_2}$$

where $K = K(\theta)$ gives the common dependence on the polar angle. Hence

$$E_\theta = E_{\theta 1} + E_{\theta 2} \approx KI\left(\frac{e^{-j\beta_0 r_1}}{r_1} e^{j\alpha} + \frac{e^{-j\beta_0 r_2}}{r_2}\right)$$

$$= KIe^{j\alpha/2}\left[\frac{e^{-j(\beta_0 r_1 - \alpha/2)}}{r_1} + \frac{e^{-j(\beta_0 r_2 + \alpha/2)}}{r_2}\right]$$

$$(1)$$

9.75 Complete the farfield approximation of E_θ begun in Problem 9.74.

▌ Whereas we may substitute $r_1 \approx r_2 \approx r$ in the denominators of (1) of Problem 9.74, in the numerators we must use

$$r_1 \approx r - \frac{d}{2}\cos\phi \qquad r_2 \approx r + \frac{d}{2}\cos\phi$$

Thus
$$E_\theta = \frac{KI}{r}e^{j\alpha/2}e^{-j\beta_0 r}\{e^{j[\beta_0(d/2)\cos\phi+\alpha/2]} + e^{-j[\beta_0(d/2)\cos\phi+\alpha/2]}\}$$

$$= \frac{2KI}{r}e^{j\alpha/2}e^{-j\beta_0 r}\cos\psi(\phi) \tag{1}$$

where
$$\psi(\phi) \equiv \beta_0\frac{d}{2}\cos\phi + \frac{\alpha}{2} \tag{2}$$

9.76 The antenna spacing in Fig. 9-16 is $d = \lambda_0/2$, and there is no phase difference between the currents. Sketch the electric field pattern for fixed r and fixed θ such that $K(\theta) \neq 0$.

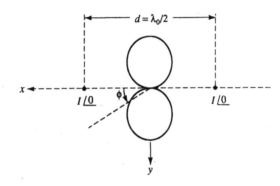

Fig. 9-17

▌ By (1) of Problem 9.75,
$$|E_\theta| \propto \left|\cos\left(\frac{\pi}{2}\cos\phi\right)\right|$$

See Fig. 9-17.

9.77 Repeat Problem 9.76 for $d = \lambda_0/4$ and $\alpha = \pi/2$.

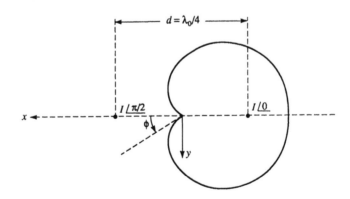

Fig. 9-18

▌ Now
$$|E_\theta| \propto \left|\cos\left[\frac{\pi}{4}(1+\cos\phi)\right]\right|$$

See Fig. 9-18.

9.78 Repeat Problem 9.76 for $d = \lambda_0/2$ and $\alpha = \pi$.

▌ Now
$$|E_\theta| \propto \left|\cos\left[\frac{\pi}{2}(1+\cos\phi)\right]\right| = \left|\sin\left(\frac{\pi}{2}\cos\phi\right)\right|$$

See Fig. 9-19.

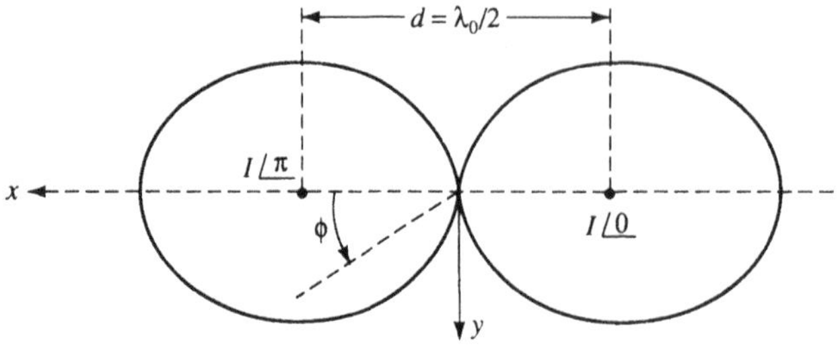

Fig. 9-19

9.79 For the array of Fig. 9-16, the far electric field pattern is governed by the function $|\cos \psi(\phi)|$. Because the pattern is symmetric about the axis through the two antennas, $\phi = 0°$ and $\phi = 180°$ are necessarily extreme points (sites of maxima or minima) of the pattern function. Show that at an off-axis extreme point the pattern function is either 0 or 1, and locate all such points.

▮ It is easier to deal with the differentiable function $\cos^2 \psi(\phi)$, which has the same extreme points as $|\cos \psi(\phi)|$. Equating the derivative to zero:

$$\cos \psi \sin \psi \sin \phi = 0$$

Axial extreme points: $\sin \phi = 0$
  At $\phi = 0°, 180°$ the extreme values are not necessarily 0 or 1.

Off-axis minima (nulls): $\cos \psi = |\cos \psi| = 0$
  Here the condition is

$$\psi = \frac{\pi d}{\lambda_0} \cos \phi + \frac{\alpha}{2} = \left(k + \frac{1}{2}\right)\pi$$

or $\qquad \cos \phi = \left(k - \frac{\alpha - \pi}{2\pi}\right)\frac{\lambda_0}{d} \qquad \left(\frac{\alpha - \pi}{2\pi} - \frac{d}{\lambda_0} < k < \frac{\alpha - \pi}{2\pi} + \frac{d}{\lambda_0}\right)$ $\qquad$ (1)

Note that the condition on the integers k precludes duplication of the axial extreme points. There may, of course, be no real ϕ for which (1) holds.

Off-axis maxima (unities): $\sin \psi = 0$, or $|\cos \psi| = 1$
  Here the condition is

$$\psi = \frac{\pi d}{\lambda_0} \cos \phi + \frac{\alpha}{2} = k\pi$$

or $\qquad \cos \phi = \left(k - \frac{\alpha}{2\pi}\right)\frac{\lambda_0}{d} \qquad \left(\frac{\alpha}{2\pi} - \frac{d}{\lambda_0} < k < \frac{\alpha}{2\pi} + \frac{d}{\lambda_0}\right)$ $\qquad$ (2)

The remarks on (1) also apply to (2).

9.80 Use Problem 9.79 to check Fig. 9-18.

▮ Minimum (null) at $\phi = 0°$; maximum (unity) at $\phi = 180°$.

Off-axis nulls given by

$$\cos \phi = 4k + 1 \qquad (k \neq 0)$$

No solution

Off-axis unities given by

$$\cos \phi = 4k - 1 \qquad (k \neq 0)$$

No solution

9.81 Use Problem 9.79 to check Fig. 9-19.

▮ Maxima (unities) at $\phi = 0°, 180°$.

Off-axis nulls given by

$$\cos \phi = 2k = 0 \qquad \text{or} \qquad \phi = 90°, 270°$$

Off-axis unities given by

$$\cos \phi = 2k - 1 \qquad (k \neq 0, 1)$$

No solution

9.82 Two identical monopole antennas are perpendicular to the earth. The antennas are separated by d and fed with currents of equal magnitude, as shown in Fig. 9-16. Sketch the pattern of the array around a large circle parallel to the earth, for $d = \lambda_0/2$ and $\alpha = 90°$.

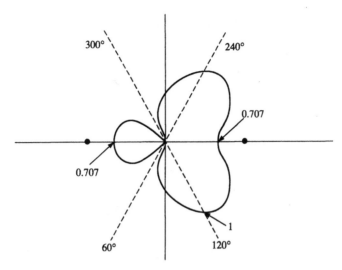

Fig. 9-20

▌ Use Problem 9.79 to locate the extreme points: maximum $(1/\sqrt{2})$ at $\phi = 0°$; minimum $(1/\sqrt{2})$ at $\phi = 180°$. Off-axis nulls at

$$\cos\phi = 2k + \frac{1}{2} = \frac{1}{2} \quad \text{or} \quad \phi = 60°, 300°$$

Off-axis unities at

$$\cos\phi = 2k - \frac{1}{2} = -\frac{1}{2} \quad \text{or} \quad \phi = 120°, 240°$$

Knowing these, and a few intermediate, values, we draw Fig. 9-20.

9.83   Repeat Problem 9.82 for $d = 5\lambda_0/8$ and $\alpha = \pi/4$.

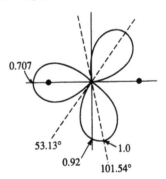

Fig. 9-21

▌ Maximum $(1/\sqrt{2})$ at $\phi = 0°$; minimum (null) at $\phi = 180°$. Off-axis nulls at

$$\cos\phi = \frac{8}{5}k + \frac{3}{5} = \frac{3}{5} \quad \text{or} \quad \phi = 53.13°, 306.87°$$

Off-axis unities at

$$\cos\phi = \frac{8}{5}k - \frac{1}{5} = -\frac{1}{5} \quad \text{or} \quad \phi = 101.54°, 258.46°$$

See Fig. 9-21.

9.84   Repeat Problem 9.82 for $d = \lambda_0$ and $\alpha = \pi$.

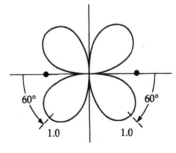

Fig. 9-22

▌ Minima (nulls) at $\phi = 0°$, $180°$. Off-axis nulls at
$$\cos \phi = k = 0 \quad \text{or} \quad \phi = 90°, 270°$$
Off-axis unities at
$$\cos \phi = k - \frac{1}{2} = -\frac{1}{2}, \frac{1}{2} \quad \text{or} \quad \phi = 60°, 120°, 240°, 300°$$
See Fig. 9-22.

9.85 Repeat Problem 9.82 for $d = \lambda_0/4$ and $\alpha = \pi$.

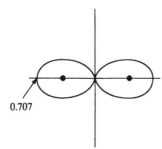

0.707

Fig. 9-23

▌ Maxima $(1/\sqrt{2})$ at $\phi = 0°$, $180°$. Off-axis nulls at
$$\cos \phi = 4k = 0 \quad \text{or} \quad \phi = 90°, 270°$$
Off-axis unities at
$$\cos \phi = 4k - 2 \quad \text{(no solution)}$$
See Fig. 9-23.

9.86 A four-element array is shown in Fig. 9-24(a) in the plane $\theta = \pi/2$. The magnitudes of the four currents are equal, but their respective phase displacements are as shown. Find the resultant far electric field.

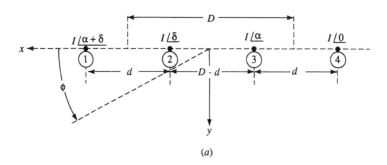

(a)

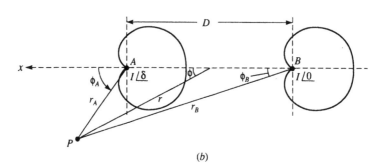

(b)

Fig. 9-24

▌ If antennas 1 and 2 are taken as a group, and antennas 3 and 4 as another group, then these two groups produce two identical far field patterns, A and B, separated by D, with phase difference δ, as shown in Fig. 9-24(b). From Problem 9.75,

$$E_{\theta A} = \frac{2KI}{r_A} e^{-j\beta_0 r_A} e^{j\alpha/2} e^{j\delta} \cos \left(\beta_0 \frac{d}{2} \cos \phi_A + \frac{\alpha}{2} \right) \qquad (1)$$

$$E_{\theta B} = \frac{2KI}{r_B} e^{-j\beta_0 r_B} e^{j\alpha/2} \cos \left(\beta_0 \frac{d}{2} \cos \phi_B + \frac{\alpha}{2} \right) \qquad (2)$$

If the field point P is sufficiently far removed from the center of the array, we may assume that $\phi_A \approx \phi_B \approx \phi$ in (1) and (2). Also, we may assume that $r_A \approx r_B \approx r$ in the denominators of (1) and (2); in the exponentials we use an approximation similar to that in the case of two elements. The result is

$$E_\theta = E_{\theta A} + E_{\theta B} = \frac{4KI}{r} e^{-j\beta_0 r} e^{j\alpha/2} e^{j\delta/2} \cos\left(\beta_0 \frac{d}{2}\cos\phi + \frac{\alpha}{2}\right) \cos\left(\beta_0 \frac{D}{2}\cos\phi + \frac{\delta}{2}\right) \tag{3}$$

Equation (3) shows that the overall pattern of the array may be found as the product of the array factor for either group A or group B,

$$\left|\cos\left(\beta_0 \frac{d}{2}\cos\phi + \frac{\alpha}{2}\right)\right|$$

and the array factor for the group of the two groups,

$$\left|\cos\left(\beta_0 \frac{D}{2}\cos\phi + \frac{\delta}{2}\right)\right|$$

It should be clear that this *principle of pattern multiplication* extends to any number of antennas.

9.87 For the array of Fig. 9-24(a), let $d = \lambda_0/4$, $\alpha = \pi/2$, $D = \lambda_0/2$, and $\delta = \pi$. Sketch the electric field pattern, using the principle of pattern multiplication.

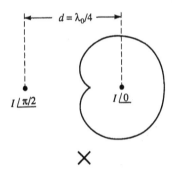

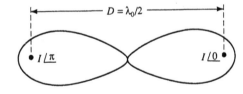

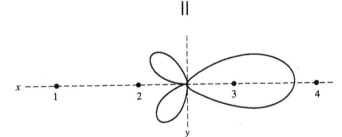

Fig. 9-25

▮ Notice that either pair generates the pattern of Fig. 9-18, while the pair of pairs generates the pattern of Fig. 9-19 [with d replaced by D]. The pattern multiplication is illustrated in Fig. 9-25.

9.88 For the array of Fig. 9-24(a), let $D = \lambda_0/2$, $d = \lambda_0/4$, $\alpha = 180°$, and $\delta = 0°$. Sketch the electric field pattern at the surface of the earth.

▮ The pattern of either pair was developed in Problem 9.85; it is reproduced in Fig. 9-26(a). The pattern of the pair of pairs [Fig. 9-26(b)] comes from Problem 9.76. Figure 9-26(c) shows the product pattern.

9.89 Show that if pattern A, with n_A lobes, is multiplied by pattern B, with n_B lobes—and if the two patterns have no null in common—then the product pattern has exactly $n_A + n_B$ lobes.

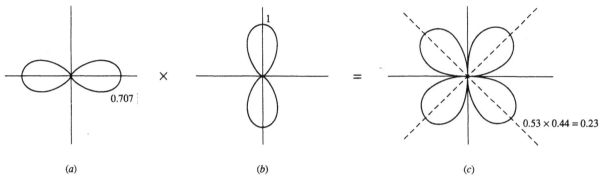

Fig. 9-26

▮ In any pattern the number of lobes equals the number of nulls. Thus, pattern A contributes n_A nulls to the product pattern and pattern B contributes an additional n_B nulls. Since the product pattern has $n_A + n_B$ nulls, and only these, it must have exactly $n_A + n_B$ lobes.

9.90 Two dipoles are separated in space by one wavelength. The terminal currents are of equal magnitudes but are out of phase by 90°. Sketch the electric field pattern in a plane perpendicular to the dipoles.

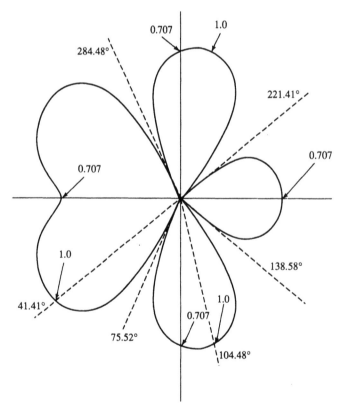

Fig. 9-27

▮ Use Problem 9.79. Minimum $(1/\sqrt{2})$ at $\phi = 0°$; maximum $(1/\sqrt{2})$ at $\phi = 180°$. Off-axis nulls:

$$\cos \phi = k + \frac{1}{4} = -\frac{3}{4}, \frac{1}{4} \quad \text{or} \quad \phi = 138.58°, 221.41°, 75.52°, 284.48°$$

Off-axis unities:

$$\cos \phi = k - \frac{1}{4} = -\frac{1}{4}, \frac{3}{4} \quad \text{or} \quad \phi = 104.48°, 255.52°, 41.41°, 318.59°$$

See Fig. 9-27.

9.91 Four identical monopole antennas have adjacent spacings of $\lambda/2$. The antenna currents are in phase and of equal magnitudes. Sketch the resulting electric field pattern along the surface of the earth.

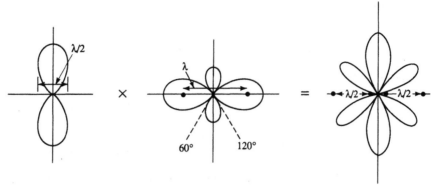

Fig. 9-28

❚ Figure 9-28 shows the pattern multiplication.

9.92 The *radiation intensity* $U(\theta, \phi)$ for an antenna is defined by
$$U(\theta, \phi) \equiv r^2 S_{av} \qquad (1)$$
where $S_{av} \sim 1/r^2$ (Problem 9.12) is the radiated average-power density. Show that the total radiated power is the integral of the radiation intensity over a solid angle of 4π sr (steradian).

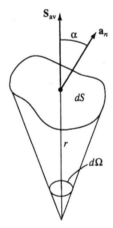

Fig. 9-29

❚ On a surface enclosing the antenna, let α denote the local angle between the time-average Poynting vector and the surface normal. Then
$$P_{rad} = \iint \mathbf{S}_{av} \cdot \mathbf{a}_n \, dS = \iint S_{av} \cos \alpha \, dS$$
$$= \iint r^2 S_{av} \frac{dS \cos \alpha}{r^2} = \iint U \, d\Omega$$
The last integral is extended over the solid angle subtended by the closed surface, which is 4π sr. See Fig. 9-29.

9.93 The *directive gain* of an antenna in a particular direction, $D(\theta, \phi)$, is the ratio of the radiation intensity in that direction to the average (with respect to solid angle) radiation intensity:
$$D(\theta, \phi) \equiv \frac{U(\theta, \phi)}{U_{av}} = \frac{4\pi U(\theta, \phi)}{P_{rad}} = \frac{4\pi r^2 S_{av}}{P_{rad}} \qquad (1)$$
The *directivity* of the antenna is the directive gain in the direction that yields a maximum:
$$D_{max} \equiv \frac{U_{max}}{U_{av}} \qquad (2)$$
Determine the directive gain and directivity of a Hertzian dipole.

❚ The radiation intensity, as found from (1) of Problem 9.11, is independent of ϕ:
$$U(\theta) = r^2 S_{av} = 15\pi \left(\frac{d\zeta}{\lambda_0}\right)^2 I^2 \sin^2 \theta$$
The radiated power is given by (1) of Problem 9.16:
$$P_{rad} = 40\pi^2 \left(\frac{d\zeta}{\lambda_0}\right)^2 I^2$$

Thus, the directive gain is

$$D(\theta) = \frac{4\pi U(\theta)}{P_{rad}} = 1.5 \sin^2 \theta$$

The directivity is therefore the directive gain at $\theta = \pi/2$: $D_{max} = 1.5$.

9.94 Repeat Problem 9.93 for a half-wave dipole.

❚ For the half-wave dipole, we obtain ($\eta_0 = 120\pi$ Ω, $R_{rad} = 73$ Ω)

$$D(\theta) = \frac{\eta_0}{\pi R_{rad}} F^2(\theta) = 1.64 F^2(\theta) \qquad (1)$$

where $F(\theta)$ is given by (1) of Problem 9.49. It follows that $D_{max} = 1.64$, which occurs for $\theta = \pi/2$.

9.95 Show that the elementary magnetic dipole (Problem 9.26) has the same directive gain and directivity as the Hertzian electric dipole (Problem 9.93).

❚ For the magnetic dipole, Problem 9.32 gives

$$U(\theta) = r^2 S_{av} = 1860 \left(\frac{A}{\lambda_0^2}\right)^2 I^2 \sin^2 \theta \quad (W)$$

and Problem 9.33 gives

$$P_{rad} = 15\,585 I^2 \left(\frac{A}{\lambda_0^2}\right)^2 \quad (W)$$

Thus

$$D(\theta) = \frac{4\pi U(\theta)}{P_{rad}} = 1.5 \sin^2 \theta \qquad \text{and} \qquad D_{max} = 1.5$$

9.96 Two antennas, A and B (Fig. 9.30), are said to be *coupled* if some power applied to antenna A results in some power being absorbed by the load Z_L connected to antenna B. Relate the terminal voltages and currents at the two antennas via a pair of coupled equations.

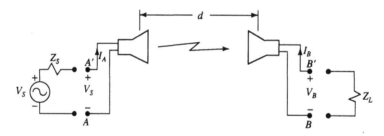

Fig. 9-30

❚ Denoting as Z_{AA} and Z_{BB} the self-impedances, and as Z_{BA} and Z_{AB} the mutual impedances, of the antennas, from the circuit theory the required equations are

$$V_A = Z_{AA} I_A + Z_{AB} I_B \qquad (1)$$
$$V_B = Z_{BA} I_A + Z_{BB} I_B \qquad (2)$$

9.97 According to the *reciprocity theorem*, if a current I_A is applied to the terminals of antenna A (Fig. 9-30) and the terminals of antenna B are opened ($I_B = 0$), a voltage V_B will appear at the terminals of antenna B. If, on the other hand, a current I_B is applied to the terminals of antenna B and the terminals of antenna A are opened ($I_A = 0$), a voltage V_A will appear across the terminals of antenna A. Apply this theorem to verify that $Z_{AB} = Z_{BA}$ in Problem 9.96.

❚ The reciprocity theorem implies that the ratio of driving current to resulting open-circuit voltage is the same in A as in B:

$$\left.\frac{V_B}{I_A}\right|_{I_B=0} = \left.\frac{V_A}{I_B}\right|_{I_A=0} \qquad (1)$$

But then, from (2) and (1) of Problem 9.96, $Z_{BA} = Z_{AB} = Z_M$.

9.98 Using reciprocity, rewrite (1) and (2) of Problem 9.96, and represent the resulting equations by an equivalent circuit.

❚ The equations

$$V_A = Z_{AA} I_A + Z_M I_B \qquad (1)$$
$$V_B = Z_M I_A + Z_{BB} I_B \qquad (2)$$

may be represented as a lumped, two-port equivalent circuit, as shown in Fig. 9-31.

ANTENNAS ∅ 351

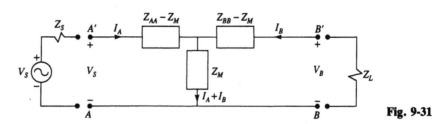

Fig. 9-31

9.99 An alternative equivalent circuit for the coupled antennas of Fig. 9-30 is shown in Fig. 9-32. Compare this circuit with that shown in Fig. 9-31 and express Z_A, Z_B and $V_{OC.BA}$ in terms of the circuit parameters of Fig. 9-31.

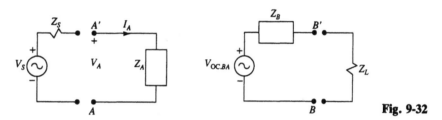

Fig. 9-32

▌ The input impedance to antenna A, Z_A, in Fig. 9-32 becomes
$$Z_A = (Z_{AA} - Z_M) + \{Z_M \parallel [(Z_{BB} - Z_M) + Z_L]\} \qquad (1)$$
where $\parallel$ means "in parallel." The open-circuit voltage source $V_{OC.BA}$ in Fig. 9-32 is given by
$$V_{OC.BA} = I_A Z_M \qquad (2)$$
and the equivalent source impedance is
$$Z_B = (Z_{BB} - Z_M) + \{Z_M \parallel [(Z_{AA} - Z_M) + Z_S]\} \qquad (3)$$

9.100 Suppose that the load is matched to the receiving antenna of Fig. 9-30. Assuming the antennas to be lossless, find the ratio of the power delivered to the load to the power delivered by the source to the transmitting antenna.

▌ Since the load is matched, $Z_L = Z_B^*$. Now, referring to the equivalent circuit of Fig. 9-32, if Z_A is written as $R_A + jX_A$, the power delivered by the source to the transmitting antenna (the radiated power for a lossless antenna) becomes
$$P_T = \tfrac{1}{2} |I_A|^2 R_A \qquad (1)$$
Similarly, if $Z_B = R_B + jX_B$, the power delivered to the matched load is
$$P_R = \frac{|I_A Z_M|^2}{8 R_B} \qquad (2)$$
Taking the ratio of (1) and (2), we obtain
$$\frac{P_R}{P_T} = \frac{|Z_M|^2}{4 R_A R_B} \qquad (3)$$

9.101 Modify the circuit of Fig. 9-32 by placing the voltage source V_S in the terminal circuit of antenna B, while retaining the impedances Z_S and Z_L in their previous locations. What conclusions may be derived from this exercise?

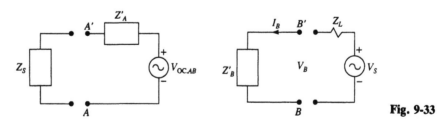

Fig. 9-33

▌ The modified circuit is shown in Fig. 9-33, where $V_{OC.AB} = I_B Z_M$. We can see that $Z_A' = Z_A$ and $Z_B' = Z_B$; in other words, the input impedance to an antenna when it is used for transmission is equal to the equivalent source impedance when it is used for reception. Note that for this statement to be true, the terminal impedances for each antenna must remain unchanged. If the antennas are widely separated, Z_M will be small in comparison with Z_{AA} and Z_{BB}. For this situation, (1) and (3) of Problem 9.99 imply $Z_A \approx Z'_A$ and $Z_B \approx Z'_B$, independent of Z_S and Z_L.

9.102 Verify the statement that the transmission pattern of an antenna is the same as its reception pattern.

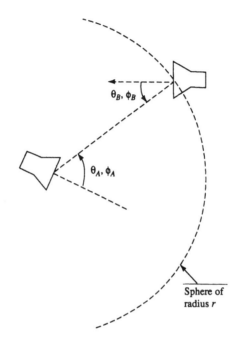

Fig. 9-34

▌ Consider the configuration of Fig. 9-34. Antenna A is held in a fixed position and antenna B is moved about on a sphere of constant radius that is centered at antenna A. Angles θ_B and ϕ_B give the orientation of the broadside axis of antenna B with respect to a radial line between the two antennas. Similarly, angles θ_A and ϕ_A give the orientation of this radial line with respect to the broadside axis of antenna A. Now let θ_B and ϕ_B be fixed, but vary θ_A and ϕ_A by moving antenna B on the surface of the sphere. If antenna B is receiving, the received power at antenna B at various positions on the sphere gives the transmission pattern of antenna A and, according to (3) of Problem 9.100, is related to θ_A and ϕ_A by $Z_M(\theta_A, \phi_A)$. Now let antenna B be the transmitter and antenna A be the receiver. Once again, moving antenna B about on the sphere, the received power at antenna A is given by (3) of Problem 9.100; this received power is again related to θ_A and ϕ_A via $Z_M(\theta_A, \phi_A)$. [Simply interchange R_A and R_B in (3) of Problem 9.100 when the roles of the antennas are reversed.] By reciprocity, $Z_M(\theta_A, \phi_A)$ is the same whether antenna A is transmitting and antenna B is receiving, or vice versa; and this yields the desired result.

9.103 Modify the parameters of the circuit of Fig. 9-31, assuming that the represented antennas (Fig. 9-30) are widely separated.

▌ For large d, Z_M will be negligible in comparison with Z_{AA} or Z_{BB}.

9.104 Consider two lossless, widely separated, half-wave dipoles. Given 10 W delivered to the transmitting antenna and 1 mW received in the other antenna (in a matched load), determine the mutual impedance $|Z_M|$ for this transmission path, and the current $|I_A|$.

▌ Substituting $R_A = R_B = 73\ \Omega$, $P_T = 10$ W and $P_R = 10^{-3}$ W in (3) of Problem 9.100 yields
$$|Z_M|^2 = \frac{4 \times 73 \times 73 \times 10^{-3}}{10} \qquad \text{or} \qquad |Z_M| = 1.46\ \Omega$$
From (2) of Problem 9.100,
$$10^{-3} = \frac{|I_A Z_M|^2}{8 \times 73} = |I_A|^2 \frac{73 \times 10^{-3}}{2 \times 10} \qquad \text{or} \qquad |I_A| = 0.52\ \text{A}$$

9.105 For the antennas of Problem 9.104, determine the received power if the receiving antenna's load, instead of being matched, is $(10 + j0)\ \Omega$.

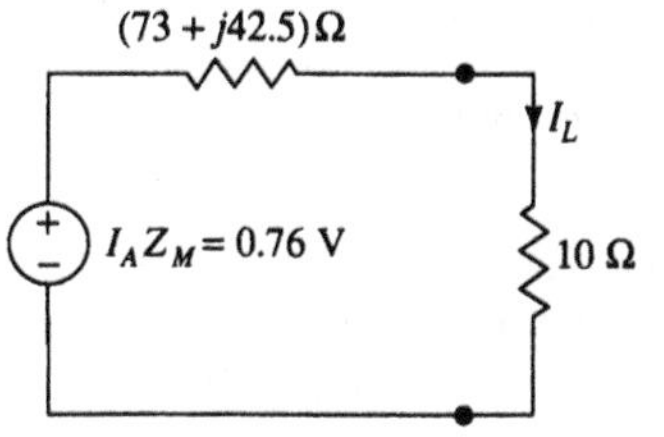

Fig. 9-35

❚ The equivalent circuit for the unmatched load is shown in Fig. 9-35, from which

$$|I_L| = \frac{|0.76|}{|73 + j42.5 + 10|} = 8.15 \times 10^{-3}\,\text{A}$$

$$P_R = \tfrac{1}{2}|I_L|^2 R_L = \tfrac{1}{2}(8.15 \times 10^{-3})^2 \times 10 = 0.332\,\text{mW}$$

9.106 The *effective aperture* of an antenna, $A_e(\theta, \phi)$, is the ratio of the power received (in its load), P_R, to the power density of the incident wave, S_{av}, when the incident wave arrives in the direction (θ, ϕ) and is polarized in the same plane as a wave *emitted* in that direction:

$$A_e(\theta, \phi) \equiv \frac{P_R}{S_{\text{av}}} \quad (\text{m}^2) \tag{1}$$

The maximum effective aperture, $A_{em}(\theta, \phi)$, is achieved when the load impedance is the complex conjugate of the antenna impedance (Problem 9.100). Determine the maximum effective aperture of the Hertzian dipole antenna for an incident, linearly polarized, plane wave.

❚ Assume the matched dipole load $Z_L = R_{\text{rad}} - jX$, where the input impedance to the dipole is $R_{\text{rad}} + jX$ and the dipole is assumed lossless. Suppose that the incident wave arrives at an angle θ with respect to the dipole axis, with the electric field polarized in the θ-direction (as an emitted wave would be). The power density in the incident wave is $S_{\text{av}} = |\mathbf{E}|^2/2\eta_0 = |\mathbf{E}|^2/240\pi$ (W/m²). The open-circuit voltage induced in the antenna is proportional to the parallel component of $\mathbf{E}$:

$$|V_{\text{OC}}| = |\mathbf{E}|\sin\theta\,d\zeta \tag{2}$$

Since the load is matched for maximum power transfer, the power received is

$$P_R = \frac{|V_{\text{OC}}|^2}{8R_{\text{rad}}} = \frac{|\mathbf{E}|^2\sin^2\theta\,\lambda_0^2}{640\pi^2} \quad (\text{W}) \tag{3}$$

wherein R_{rad} was evaluated from Problem 9.17. Therefore

$$A_{em}(\theta) = \frac{P_R}{S_{\text{av}}} = \frac{3\lambda_0^2\sin^2\theta}{8\pi} = \frac{\lambda_0^2}{4\pi}(1.5\sin^2\theta) \tag{4}$$

The presence in (4) of the directive gain of the (lossless) dipole, $D(\theta) = 1.5\sin^2\theta$ (see Problem 9.93), is no accident: the relation

$$A_{em}(\theta, \phi) = \frac{\lambda_0^2}{4\pi}D(\theta, \phi) \tag{5}$$

is valid for any lossless antenna.

9.107 Determine the maximum effective aperture of an elemental magnetic-dipole antenna.

❚ Same as for electric dipole— see Problem 9.95 and (2) of Problem 9.106.

9.108 Find A_{em} for a lossless half-wave dipole.

❚ From (5) of Problem 9.106,

$$A_{em}(\theta, \phi) = \frac{\lambda_0^2}{4\pi}D(\theta, \phi)$$

in which we substitute $D(\theta, \phi) = 1.64F^2(\theta)$, from (1) of Problem 9.94, to obtain

$$A_{em}(\theta, \phi) = 0.131F^2(\theta)\lambda_0^2$$

Substitution for $F(\theta)$ from (1) of Problem 9.49 yields

$$A_{em} = \frac{0.131\lambda_0^2\cos^2\left(\frac{\pi}{2}\cos\theta\right)}{\sin^2\theta}$$

9.109 An aircraft transmitter is designed to communicate with a ground station; the ground antenna must receive at least 1 μW. Assume that both antennas are omnidirectional. When the airplane flies directly over the station at an altitude of 5000 ft, a signal of 500 mW is received by the station. Determine the maximum communication range of the airplane.

∎ Power decreases as $1/r^2$; therefore

$$r_{\text{max}} = (5000 \text{ ft}) \sqrt{\frac{500 \text{ mW}}{10^{-3} \text{ mW}}} = 3.535 \times 10^6 \text{ ft} = 670 \text{ mi}$$

9.110 The *power gain* (or simply *gain*) of an antenna in a particular direction, $G(\theta, \phi)$, is defined as the ratio of the power required from an imaginary isotropic source to produce the given intensity at (θ, ϕ), to the power *supplied to* the actual antenna. Relate the power gain to the directive gain.

∎ By definition,

$$G(\theta, \phi) = \frac{4\pi U(\theta, \phi)}{P_{\text{in}}} \tag{1}$$

which, when compared to (1) of Problem 9.93, gives

$$G(\theta, \phi) = \frac{P_{\text{rad}}}{P_{\text{in}}} D(\theta, \phi) \equiv \epsilon_{\text{rad}} D(\theta, \phi) \tag{2}$$

where ϵ_{rad} is the radiative efficiency of the antenna (Problem 9.72).

9.111 Argue that (5) of Problem 9.106 holds for any antenna, if $D(\theta, \phi)$ is replaced by $G(\theta, \phi)$.

∎ Suppose the dipole in Problem 9.106 to be lossy. Then, on the assumption (which could be supported by a reciprocity argument) that its radiative and absorptive efficiencies are equal, the power received in the load would be decreased by the factor ϵ_{rad}; and so (5) would become

$$A_{em}(\theta, \phi) = \frac{\lambda_0^2}{4\pi} [\epsilon_{\text{rad}} D(\theta, \phi)] = \frac{\lambda_0^2}{4\pi} G(\theta, \phi) \tag{1}$$

9.112 Consider two antennas in free space (Fig. 9-36). One antenna is being fed P_T watts of power, and the other is receiving P_R watts of power in its terminal impedance. The transmitting antenna has gain $G_T(\theta_T, \phi_T)$ and effective aperture $A_{eT}(\theta_T, \phi_T)$ in the direction of transmission, (θ_T, ϕ_T); the receiving antenna has gain and effective aperture $G_R(\theta_R, \phi_R)$ and $A_{eR}(\theta_R, \phi_R)$ in the direction of reception, (θ_R, ϕ_R). Express P_R/P_T in terms of G's, λ_0, and d.

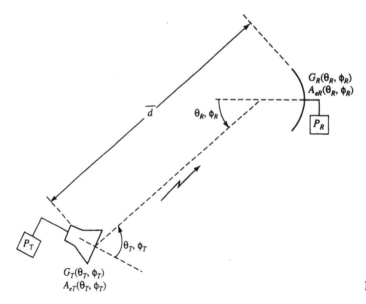

Fig. 9-36

∎ From (1) of Problem 9.110, the power density at the receiving antenna is

$$S_{\text{av}} = \frac{P_T}{4\pi d^2} G_T(\theta_T, \phi_T) \tag{1}$$

and the received power is, by (1) of Problem 9.106,

$$P_R = S_{\text{av}} A_{eR}(\theta_R, \phi_R) \tag{2}$$

By division,

$$\frac{P_R}{P_T} = \frac{G_T(\theta_T, \phi_T) A_{eR}(\theta_R, \phi_R)}{4\pi d^2} \tag{3}$$

Now replace the effective aperture of the receiving antenna with its gain via (1) of Problem 9.111 (assuming a matched load and matched polarization so that the effective apertures are the maximum effective apertures), to obtain the *Friis transmission equation*:

$$\frac{P_R}{P_T} = G_T(\theta_T, \phi_T) G_R(\theta_R, \phi_R)\left(\frac{\lambda_0}{4\pi d}\right)^2 \tag{4}$$

Because antenna efficiencies run very high (see Problem 9.72), the term P_T in the above equations may be identified with the *output* of the transmitting antenna.

9.113 In Problem 9.112, what is the magnitude of the electric field at the receiver?

❚ From (1) of Problem 9.112,

$$S_{av} = \frac{|\mathbf{E}|^2}{2\eta_0} = \frac{P_T G_T(\theta_T, \phi_T)}{4\pi d^2} \quad \text{or} \quad |\mathbf{E}| = \frac{\sqrt{60 P_T G_T(\theta_T, \phi_T)}}{d} \quad (\text{V/m}) \tag{1}$$

since $\eta_0 = 120\pi \ \Omega$.

9.114 Rewrite the Friis transmission equation, (4) of Problem 9.112, assuming that the gains of the antennas are given in decibels.

❚ $$10 \log \frac{P_R}{P_T} = G_{T,dB} + G_{R,dB} - 20 \log f - 20 \log d + 148 \tag{1}$$

in which $f = (3 \times 10^8)/\lambda_0$ is the operating frequency.

9.115 In the Friis formula, it is implicit that either antenna be in the far field of the other. Obtain a quantitative criterion for the receiving antenna to be in the far field of the transmitting antenna.

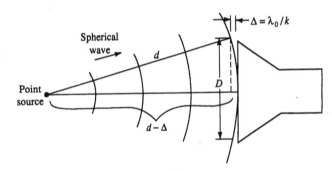

Fig. 9-37

❚ Refer to Fig. 9-37. If the maximum dimension of the receiving antenna is D, we require that the incident spherical wave deviate from planarity by a fraction of a wavelength, $\Delta = \lambda_0/k$ $(k \gg 1)$. Then, by geometry,

$$d^2 = (d - \Delta)^2 + \left(\frac{D}{2}\right)^2 = d^2 - 2d\Delta + \Delta^2 + \frac{D^2}{4} \approx d^2 - 2d\Delta + \frac{D^2}{4}$$

whence $$d \approx \frac{D^2}{8\Delta} = \frac{kD^2}{8\lambda_0} \tag{1}$$

The usual choice is $k = 16$, whereby (1) becomes the *phase-curvature criterion*

$$d_{\text{farfield}} \approx \frac{2D^2}{\lambda_0} \tag{2}$$

9.116 A telemetry transmitter placed on the moon is to transmit data to the earth. The transmitter power is 100 mW and the gain of the transmitting antenna in the direction of transmission is 12 dB. Determine the minimum gain of the receiving antenna in order to receive 1 nW. The distance from the moon to earth is 238,857 mi and the transmitter frequency is 100 MHz.

❚ In (1) of Problem 9.114 substitute the data ($d = 3.844 \times 10^8$ m), to find $G_{R,dB} = 92.14$.

9.117 A microwave relay link is to be designed. The transmitting and receiving antennas are separated by 30 mi, and the power gain in the direction of transmission for both antennas is 45 dB. Assuming that both antennas are lossless and matched, and the frequency is 3 GHz, determine the minimum transmitter power for a received power of 1 mW.

❚ Substitute the data ($d = 4.828 \times 10^4$ m) in (1) of Problem 9.114, to find
$$\log P_T = 1.5660 \quad \text{or} \quad P_T = 36.81 \text{ W}$$

9.118   An antenna on an aircraft is being used to jam an enemy radar. Assuming that the antenna has a gain of 12 dB in the direction of transmission and the transmitted power is 5 kW, determine the electric field intensity in the vicinity of the radar, which is 2 mi away. The frequency of transmission is 7 GHz.

▮ Substitute $P_T = 5 \times 10^3$ W, $G_T = 12$ dB $= 15.85$, and $d = 2$ mi $= 3218.7$ m in (*1*) of Problem 9.113, obtaining $|\mathbf{E}| = 0.68$ V/m.

9.119   Two identical horn antennas are separated by a distance of 100 m. Both antennas have directive gains of 15 dB in the direction of transmission, and their dimensions are 12 cm by 6 cm. The transmitting antenna sends out 5 W at 3 GHz; determine the received power.

▮ By Problem 9.115,

$$d_{\text{farfield}} \approx \frac{2D^2}{\lambda_0} = \frac{2(0.12)^2}{0.1} = 0.29 \text{ m} \ll 100 \text{ m}$$

so that the Friis formula of Problem 9.114 applies. Substituting the data, we get

$$\log P_R = -4.499 \qquad \text{or} \qquad P_R = 31.7 \, \mu\text{W}$$

9.120   Calculate the electric field intensity at the receiving antenna of Problem 9.119.

▮ By (*1*) of Problem 9.113, with $G_T = 15$ dB $= 31.62$,

$$|\mathbf{E}| = \frac{1}{100}\sqrt{60 \times 5 \times 31.62} = 0.974 \text{ V/m}$$

9.121   A lossless half-wave dipole, having a gain of 2.15 dB, is being driven by a 10-V, 50-Ω generator. Determine the electric intensity in the far field at a distance of 10 km in a plane perpendicular to the antenna.

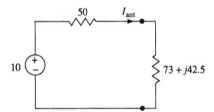

Fig. 9-38

▮ In the equivalent circuit in Fig. 9-38, the antenna impedance was taken from Problem 9.70.

$$I_{\text{ant}} = \frac{10}{123 + j42.5} = 7.68 \times 10^{-2} \underline{/-19.1^\circ} \text{ A}$$

$$P_{\text{rad}} = \tfrac{1}{2}|I_{\text{ant}}|^2 (73) = \tfrac{1}{2}(7.68 \times 10^{-2})^2(73) = 0.215 \text{ W} = P_T$$

and $G_T = 2.15$ dB $= 1.64$. Substituting these values and $d = 10^4$ m in (*1*) of Problem 9.113 yields $|\mathbf{E}| = 461 \, \mu$V/m.

9.122   Antennas are often located above the surface of some conducting (ground) plane, as shown in Fig. 9-39. For the antenna system shown, obtain a correction factor for the Friis equation, (*4*) of Problem 9.112, to account for the effect of ground reflections on signal transmission.

▮ The received signal consists of a direct wave and a wave reflected at the ground plane. If the antennas are in the far fields of each other, the received voltage (as well as the incident electric field) at the receiving antenna due to the direct wave may be written as

$$V_d = V_0 F_T(\theta_{Td}) F_R(\theta_{Rd}) \frac{e^{-j\beta_0 d}}{d} \tag{1}$$

where F_T and F_R denote the pattern functions (Problem 9.46) of the transmitting and receiving antennas, respectively, and where θ_{Td} and θ_{Rd} are the angles shown in Fig. 9-39. The reflected wave will contribute

$$V_r = V_0 F_T(\theta_{Tr}) F_R(\theta_{Rr}) \Gamma \frac{e^{-j\beta_0 d_r}}{d_r} \tag{2}$$

where $\Gamma = |\Gamma| \underline{/\theta_\Gamma}$ is the complex reflection coefficient at the ground plane, and d_r is the total path length of the reflected wave. The total received voltage is the sum of (*1*) and (*2*):

$$V_R = V_d + V_r = V_d\left(1 + \frac{V_r}{V_d}\right) \equiv V_d F$$

Thus, the ground reflection modifies the free-space propagation by the multiplicative factor

$$F = 1 + \frac{F_T(\theta_{Tr}) F_R(\theta_{Rr}) \Gamma}{F_T(\theta_{Td}) F_R(\theta_{Rd})} \frac{d}{d_r} e^{-j\beta_0(d_r - d)} \tag{3}$$

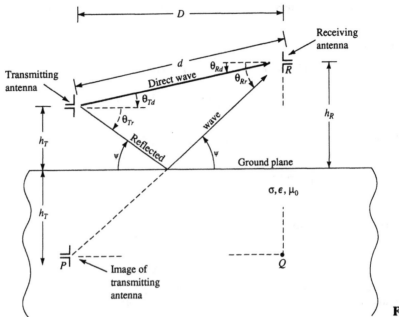

Fig. 9-39

Consequently, the Friis transmission equation can be modified to account for ground reflection by multiplying the right-hand side by $|F|^2$ [the Friis equation involves P_R, whereas F pertains to V_R].

Warning: If an actual evaluation of F is required, the angles θ_T and θ_R must be converted to polar angles, to allow the use of previously developed pattern functions [e.g., (3) of Problem 9.46]. Thus $\theta_T = 10°$ becomes $\theta = 80°$ or $100°$, depending on the sense.

9.123 In practice, the antenna separation, D, is very large compared to the antenna heights h_T and h_R shown in Fig. 9-39. Subject to this restriction, obtain an approximate expression corresponding to (3) of Problem 9.122.

▮ With $D \gg h_R, h_T$, we may assume the radiation patterns in the directions of the direct and reflected waves to be the same:

$$F_T(\theta_{Td}) \approx F_T(\theta_{Tr}) \qquad F_R(\theta_{Rd}) \approx F_R(\theta_{Rr})$$

Similarly, we may take $d \approx d_r$, except in the exponential phase-shift term. Thus, the correction factor becomes

$$F \approx 1 + \Gamma e^{-j\beta_0(d_r - d)} \tag{1}$$

9.124 Depending on the orientation of the transmitting antenna in Fig. 9-39, radiation may arrive at the ground plane horizontally polarized; that is, with the electric field parallel to the ground plane. Obtain an expression for the corresponding reflection coefficient Γ_h.

▮ This is "perpendicular" (to the vertical) polarization in the sense of Chapter 5. The reflection coefficient—between two nonconducting media—is given by (1) and (2) of Problem, 5.139, in which $n = \sqrt{\epsilon_r}$. The expressions remain valid when medium 2 is a lossy dielectric, provided its complex relative permittivity,

$$\hat{\epsilon}_{r2} = \frac{\epsilon}{\epsilon_0} - j\frac{\sigma}{\omega\epsilon_0} = \epsilon_{r2}\left(1 - j\frac{\sigma}{\omega\epsilon}\right) \tag{1}$$

is understood (see Problem 5.197). Thus, with $\epsilon_{r1} = 1$ and $\theta_i = 90° - \psi$, one finds

$$\Gamma_h = \frac{\sin\psi - \sqrt{\hat{\epsilon}_{r2} - \cos^2\psi}}{\sin\psi + \sqrt{\hat{\epsilon}_{r2} - \cos^2\psi}} \tag{2}$$

9.125 Repeat Problem 9.124 for vertical polarization.

▮ This is "parallel" (to the vertical) polarization. Equation (1) of Problem 5.150, with the same substitutions as in Problem 9.124, yields

$$\Gamma_v = \frac{\hat{\epsilon}_{r2}\sin\psi - \sqrt{\hat{\epsilon}_{r2} - \cos^2\psi}}{\hat{\epsilon}_{r2}\sin\psi + \sqrt{\hat{\epsilon}_{r2} - \cos^2\psi}} \tag{1}$$

9.126 Show that for grazing incidence on the ground plane, $\Gamma_h = \Gamma_v = -1$.

▮ Let $\psi \to 0$ in (2) of Problem 9.124 and in (1) of Problem 9.125.

9.127 Refer to Problem 9.123. Further approximate $|F|$, under the condition of grazing incidence on the ground plane.

▌ Writing $\Delta \equiv d_r - d$ for the path-length difference,

$$|F|^2 = FF^* \approx (1 - e^{-j\beta_0\Delta})(1 - e^{j\beta_0\Delta}) = 2(1 - \cos\beta_0\Delta) = 4\sin^2\frac{\beta_0\Delta}{2}$$

or

$$|F| \approx 2\left|\sin\frac{\beta_0\Delta}{2}\right| = 2\left|\sin\frac{\pi\Delta}{\lambda_0}\right| \qquad (1)$$

9.128 With the magnitude of the correction factor as given by (1) of Problem 9.127, state conditions for constructive and for destructive interference between direct and reflected waves at the receiving antenna.

▌ For constructive interference:

$$|F| = 2 \quad\text{or}\quad \frac{\Delta}{\lambda_0} = \frac{1}{2}, \frac{3}{2}, \frac{5}{2}, \ldots$$

For destructive interference:

$$|F| = 0 \quad\text{or}\quad \frac{\Delta}{\lambda_0} = 1, 2, 3, \ldots$$

9.129 The path-length difference, as a difference of very nearly equal numbers, invites serious computational errors. Obtain an approximate expression for Δ to avoid such errors.

▌ In Fig. 9-39, right triangle PQR gives

$$d_r = \sqrt{D^2 + (h_T + h_R)^2} \qquad (1)$$

But clearly, $D^2 = d^2 - (h_T - h_R^2)$, and (1) becomes

$$d_r = \sqrt{d^2 - (h_T - h_R)^2 + (h_T + h_R)^2} = d\sqrt{1 + (4h_Th_R/d^2)} \approx d[1 + (2h_Th_R/d^2)]$$

by a binomial expansion. Hence

$$\Delta = d_r - d \approx \frac{2h_Th_R}{d} \qquad (2)$$

9.130 In view of the result of Problem 9.129, extend the approximation (1) of Problem 9.127.

▌

$$|F| = 2\left|\sin\frac{2\pi h_Th_R}{\lambda_0 d}\right| \qquad (1)$$

The argument of the sine function may or may not be small; if it is,

$$|F| \approx \frac{4\pi h_Th_R}{\lambda_0 d} \qquad (2)$$

9.131 Consider the problem of communication between an airport traffic control tower and an airplane approaching to land. The tower frequency is 119.1 MHz, and the antenna is vertically polarized and located at a height of 100 ft above the airport surface, An aircraft approaching the airport is at a distance D from the tower and has altitude 1000 ft. Calculate the radii of circles centered on the tower at which communication would be theoretically impossible.

▌ The correction factor will be equal to zero when the path-length difference is a multiple of λ_0 (Problem 9.128). The wavelength at the tower frequency is $\lambda_0 = 2.52$ m $= 8.26$ ft. In terms of the horizontal separation $D \approx d$, (2) of Problem 9.129 gives as the noncommunication criterion:

$$\frac{2h_Th_R}{D} = n\lambda_0$$

or

$$D = \frac{2h_Th_R}{n\lambda_0} = \frac{2 \times 100 \times 1000}{n \times 8.26} = \frac{24,213}{n}\text{ ft} = \frac{4.6}{n}\text{ mi}$$

for $n = 1, 2, 3, \ldots$. The outermost null circle is thus of radius 4.6 mi.

9.132 Refer to Fig. 9-39. Show that if the transmitted wavelength is not too small compared to the *height* (not the *length*) of the transmitting antenna, then the received electric field is proportional to the heights of the transmitting and receiving antennas and inversely proportional to the square of the distance between the antennas.

▌ By Problem 9.47.

$$|\mathbf{E}_R| \propto \frac{|F|}{d} \approx \frac{|F|}{D} \qquad (1)$$

Now look at (*1*) of Problem 9.130. A basic assumption is that $h_R/d \approx h_R/D \ll 1$. If, in addition, $h_T/\lambda_0 \approx 100$ (say), then (*2*) of Problem 9.130 may be presumed valid; together with (*1*) above, it gives

$$|\mathbf{E}_R| \propto \frac{h_T h_R}{D^2}$$

9.133 A half-wave dipole is oriented perpendicular to the earth, which is characterized by $\epsilon_r = 10$ and $\sigma = 10^{-2}\,\text{S/m}$. The antenna is at a height of 100 ft and the operating frequency is 100 MHz. Evaluate the vertical component of the electric field at point *P*, at a height of 100 ft and a distance of 1000 ft from the antenna.

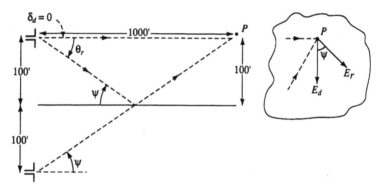

Fig. 9-40

▌ Figure 9-40 is the diagram for this problem; note that only the directional properties of the dipole transmitter matter, there being no receiving antenna. In the notation of Fig. 9-39, we have, at *P*:

$$E_d = K\frac{e^{-j\beta_0 d}}{d} F_T(\theta_d) = K\frac{e^{-j\beta_0 d}}{d} \quad \text{[heeding the \textit{Warning}, Problem 9.122]}$$

$$E_r = \Gamma K\frac{e^{-j\beta_0 d_r}}{d_r} F_T(\theta_r)$$

$$|E_{\text{vert}}| = |E_d + E_r \cos\psi| = |E_d|\left|1 + \frac{E_r}{E_d}\cos\psi\right|$$

$$= \frac{|K|}{d}\left|1 + \Gamma F_T(\theta_r)\frac{d}{d_r}e^{-j\beta_0(d_r-d)}\cos\psi\right|$$

Now to evaluate: $\psi = \theta_r = \tan^{-1}(200/1000) = 11.3°$, which is small enough for vertical polarization. By Problem 9.125, with $\hat{\epsilon}_{r2} = 10 - j1.8$,

$$\Gamma_v = 0.213\,\underline{/-170°}$$

Also: $d = 304.8$ m; $d/d_r = \cos\psi = 0.9806$;

$$F_T(11.3°) = \left[\cos\left(\frac{\pi}{2}\cos 101.3°\right)\right]\Big/(\sin 101.3°) = 0.972$$

$\beta_0 = (2\pi/3)$ rad/m; and, by fiat, $|K| = 1$ mV. Substitution of these values yields $|E_{\text{vert}}| = 2.633\,\mu\text{V/m}$.

9.134 Two half-wave dipoles are oriented parallel to the earth, which is characterized by $\epsilon_r = 10$ and $\sigma = 10^{-2}\,\text{S/m}$. Both antennas are at a height of 100 ft and are separated by 300 ft. If one antenna is driven by a 100-MHz, 1-V sinusoidal signal, determine the power received at the base of the other antenna into a matched load.

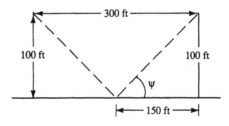

Fig. 9-41

▌ From Fig. 9-41,

$$\tan\psi = \frac{100}{150} \quad \text{or} \quad \psi = 33.69°$$

For this angle and $\sigma/\omega\epsilon = 0.18$, $\Gamma_h = 0.696\,\underline{/177.966°}$; gain in the direction of direct and reflected waves is the same ($G = 1.64$). Thus

$$I_{\text{in}} = \frac{1}{73 + j42.5} = 11.84 \underline{/-30.21°} \quad \text{mA}$$

Now, $d = 300' = 91.44$ m and $d_r = 360.6' = 109.9$ m; thus the received electric field is

$$E_T = 60 |I_{\text{in}}| \left(\frac{e^{-j\beta_0 d}}{d} + \Gamma_h \frac{e^{-j\beta_0 d_r}}{d_r} \right)$$

$$= 0.7103 \left(\frac{1}{91.44} e^{-j(2\pi/3) \times 91.44} + 0.696 \underline{/177.966°} \frac{e^{-j(2\pi/3) \times 109.9}}{109.9} \right)$$

$$= (6.513 \times 10^{-3} \underline{/35.36°}) e^{-j(2\pi/3) \times 91.44} \quad \text{V/m}$$

It follows that

$$S_{\text{av}} = \frac{|E_T|^2}{2\eta_0} = 56.24 \text{ nW/m}^2$$

$$A_{em} = \frac{\lambda_0^2}{4\pi} G = \frac{9}{4\pi} \times 1.64 = 1.175 \text{ m}^2$$

$$P_R = A_{em} S_{\text{av}} = 66 \text{ nW}$$

9.135 Check Problem 9.134, using the Friis transmission equation.

   ▮ The correction factor is $F = 0.8383 \underline{/35.3°}$ [from (1) of Problem 9.123]; hence
$$P_T = \tfrac{1}{2} |I_{\text{in}}|^2 \times 73 = 5.117 \times 10^{-3} \text{ W}$$

$$P_R = G_T G_R |F|^2 \left(\frac{\lambda_0}{4\pi d} \right)^2 P_T$$

$$= 1.64 \times 1.64 \times 0.8383^2 \left(\frac{3}{4\pi \times 91.44} \right)^2 \times 5.117 \times 10^{-3} = 65.9 \text{ nW}$$

9.136 A one-fifth-wave dipole antenna is used to radiate a 200-MHz signal to a satellite in space. What is the radiation resistance of the antenna?

   ▮
$$R_{\text{rad}} = 80\pi^2 \left(\frac{\zeta}{\lambda} \right)^2 = 80\pi^2 \times \frac{1}{25} = 31.58 \text{ } \Omega$$

9.137 A dipole antenna radiating at 200 MHz is fed from a 60-Ω transmission line matched to the source. Determine the length of the dipole to match the line impedance at the signal frequency.

   ▮ For matching, $60 = R_{\text{rad}} = 80\pi^2 (\zeta/\lambda)^2$. But
$$\lambda = \frac{u}{f} = \frac{3 \times 10^8}{200 \times 10^6} = 1.5 \text{ m}$$

Thus
$$60 = 80\pi^2 \left(\frac{\zeta}{1.5} \right)^2 \quad \text{or} \quad \zeta = 0.413 \text{ m}$$

9.138 In the computation of Problem 9.133, could the approximation (2) of Problem 9.129 have been applied? If so, could the problem have been solved by use of the Friis formula?

   ▮ The exact path-length difference, as used in the problem solution, is
$$d_r - d = d \left(\frac{1}{\cos \psi} - 1 \right) = (304.8) \left(\frac{1}{0.9806} - 1 \right) = 6.035 \text{ m} = 2.012\lambda_0$$

while the approximation yields
$$d_r - d \approx \frac{2h^2}{d} = 20 \text{ ft} = 6.096 \text{ m} = 2.032\lambda_0$$

By itself, an error in the phase of $\frac{1}{50}$ wavelength would be acceptable. However, use of the full Friis formula would be tantamount to taking $F_T(\theta_r) = 1$, instead of 0.972. This error would combine with the small phase error to produce an unacceptable result.

9.139 Obtain a general expression for the ratio E_θ / H_ϕ for a Hertzian dipole. Evaluate the magnitude of this ratio at a distance of one wavelength from the dipole.

   ▮ From (1) of Problem 9.6 and (2) of Problem 9.7,

$$\frac{E_\theta}{H_\phi} = \eta_0 \frac{\dfrac{j}{\beta_0 r} + \dfrac{1}{\beta_0^2 r^2} - \dfrac{j}{\beta_0^3 r^3}}{\dfrac{j}{\beta_0 r} + \dfrac{1}{\beta_0^2 r^2}} \tag{1}$$

Substituting $\eta_0 = 377\ \Omega$, and $\beta_0 r = 2\pi$ in (1),

$$\left|\frac{E_\theta}{H_\phi}\right| = 377 \left|\frac{\dfrac{j}{2\pi} + \dfrac{1}{4\pi^2} - \dfrac{j}{8\pi^3}}{\dfrac{j}{2\pi} + \dfrac{1}{4\pi^2}}\right| = 367\ \Omega \qquad (2)$$

9.140 The far field due to a certain Hertzian dipole is $E_\theta = (100/r) \sin^2 \theta\, e^{-j\beta_0 r}$ (V/m). Evaluate the corresponding average radiated power.

▮ Integrating $S_{av} = |E_\theta|^2/2\eta_0$ over the surface of a large sphere, we obtain

$$P_{av} = \frac{1}{2\eta_0} \times 2\pi \times 10^4 \int_0^\pi \sin^5 \theta\, d\theta = \frac{2\pi \times 10^4}{377} \left(-\frac{5}{8}\cos\theta + \frac{5}{48}\cos 3\theta - \frac{1}{80}\cos 5\theta\right)_0^\pi = 88.9\ \text{W}$$

9.141 An antenna has a uniform radiation intensity in all directions. What is its directivity?

▮ By the definition, (1) of Problem 9.893,

$$D(\theta,\,\phi) = \frac{U_{av}}{U_{av}} = 1 \qquad \text{whence} \qquad D_{max} = 1$$

9.142 A receiving antenna connected to a load Z_L may be represented by the equivalent circuit shown in Fig. 9-42. Express the effective aperture (Problem 9.106) in the direction of reception in terms of incident power density and circuit parameters.

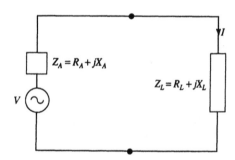

Fig. 9-42

▮ In terms of peak current and voltage,

$$A_e = \frac{P_R}{S_{av}} = \frac{|I_m|^2 R_L}{2S_{av}}$$

But

$$I = \frac{V}{(R_A + R_L) + j(X_A + X_L)}$$

Thus

$$A_e = \frac{|V_m|^2 R_L}{(R_A + R_L)^2 + (X_A + X_L)^2}\left(\frac{1}{2S_{av}}\right) \qquad (1)$$

9.143 Obtain an expression for the maximum effective aperture (in the given direction) of the antenna of Problem 9.142. Assume the antenna to be lossless.

▮ The conditions for maximum aperture are that the antenna be oriented for a maximum induced $|V|$ and that the power transfer to the load from the antenna be maximum. Since the antenna is lossless, $R_A = R_{rad}$, the radiation resistance. Thus for maximum power transfer we must have $R_A = R_{rad} = R_L$ and $X_A = -X_L$. Substituting these in (1) of Problem 9.142 yields

$$A_{em} = \frac{|V_m|^2}{8R_{rad}S_{av}}$$

in agreement with (3) of Problem 9.106.

9.144 Certain antennas (e.g., horns) emit essentially all their energy in a narrow beam, and the radiation intensity is reasonably uniform through the beam solid angle $\Omega_A < 4\pi$. Show that if such an antenna is lossless, the radiated wavelength is given by

$$\lambda_0 \approx \sqrt{A_{em}\Omega_A}$$

where A_{em} is the maximum effective aperture for any direction within the beam.

■ For any direction (θ, ϕ) within the beam,

$$D(\theta, \phi) \equiv \frac{4\pi U(\theta, \phi)}{P_{\text{rad}}} \approx \frac{4\pi U}{U\Omega_A} = \frac{4\pi}{\Omega_A} = D_{\max} \qquad (1)$$

since, by assumption, the radiation intensity is constant within the beam and vanishes outside the beam. Then (5) of Problem 9.106 gives, within the beam,

$$A_{em}(\theta, \phi) \approx \frac{\lambda_0^2}{\Omega_A} \equiv A_{em} \qquad \text{. or} \qquad \lambda_0 \approx \sqrt{A_{em}\Omega_A} \qquad (2)$$

Observe that (1), but not (2), is also valid for a lossy antenna.

9.145 Show that (1) of Problem 9.144, $D_{\max}\Omega_A = 4\pi$, becomes valid for an arbitrary (nonuniform) radiation pattern under a suitable definition of *beam solid angle* Ω_A.

■ Let $\Omega_A \equiv P_{\text{rad}}/U_{\max}$; i.e., an isotropic source of intensity $U_{\max}$ would send out through a (geometrical) solid angle Ω_A the same power that the actual antenna emits through a solid angle 4π sr. From Problem 9.93, $D_{\max} = U_{\max}/U_{\text{av}} = 4\pi U_{\max}/P_{\text{rad}}$. Therefore, by multiplication,

$$D_{\max}\Omega_A = \frac{4\pi U_{\max}}{P_{\text{rad}}} \frac{P_{\text{rad}}}{U_{\max}} = 4\pi$$

9.146 A Hertzian dipole, of length ζ, is made of round wire of radius a and conductivity σ. If this antenna operates at a frequency f, what is its *surface resistance*?

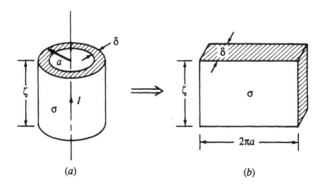

(a)   (b)   **Fig. 9-43**

■ From Fig. 9-43(a) we notice that the current is confined to the skin depth δ. Thus, the effective cylindrical shell is equivalent to the rectangular block shown in Fig. 9-43(b), the resistance of which is given by

$$R_A = \frac{\zeta}{\sigma\delta(2\pi a)}$$

Substituting $1/\delta = \sqrt{\pi f \mu_0 \sigma}$ yields

$$R_A = \frac{\zeta \sqrt{\pi f \mu_0 \sigma}}{\sigma(2\pi a)} \equiv R_s\left(\frac{\zeta}{2\pi a}\right) \qquad (1)$$

where the surface resistance is defined by

$$R_s \equiv \sqrt{\pi f \mu_0 / \sigma} \qquad (2)$$

9.147 Determine the radiative efficiency of the Hertzian dipole antenna of Problem 9.146.

■ By definition,

$$\epsilon_{\text{rad}} \equiv \frac{\text{radiated power}}{\text{input power}} = \frac{P_{\text{rad}}}{P_{\text{loss}} + P_{\text{rad}}} \qquad (1)$$

But $P_{\text{loss}} = I^2 R_A$ and $P_{\text{rad}} = I^2 R_{\text{rad}}$; so that (1) becomes

$$\epsilon_{\text{rad}} = \frac{R_{\text{rad}}}{R_A + R_{\text{rad}}} \qquad (2)$$

From Problems 9.17 and 9.146,

$$R_{\text{rad}} = 80\pi^2\left(\frac{\zeta}{\lambda_0}\right)^2 \qquad \text{and} \qquad R_A = R_s\left(\frac{\zeta}{2\pi a}\right) \qquad (3)$$

Substituting these in (2) yields

$$\epsilon_{\text{rad}} = \frac{1}{1 + \dfrac{R_s \lambda_0^2}{160\pi^3 a \zeta}} \qquad (4)$$

9.148 A 1-m-long dipole antenna, made of copper wire of 2 mm radius, radiates at 2 MHz. Calculate the radiative efficiency of the antenna.

▮ From (2) of Problem 9.146,
$$R_s = \sqrt{\pi \times 2 \times 10^6 \times 4\pi \times 10^{-7}/5.7 \times 10^7} = 372 \ \mu\Omega$$

Now
$$\lambda_0 = \frac{u_0}{f} = \frac{3 \times 10^8}{2 \times 10^6} = 150 \text{ m}$$

Then, from (3) of Problem 9.147,
$$R_{rad} = 80\pi^2 \left(\frac{1}{150}\right)^2 = 35.1 \text{ m}\Omega \quad \text{and} \quad R_A = (372 \times 10^{-6}) \left(\frac{1}{2\pi \times 2 \times 10^{-3}}\right) = 29.6 \text{ m}\Omega$$

Then (2) of Problem 9.147 gives
$$\epsilon_{rad} = \frac{35.1}{29.6 + 35.1} = 54.25\%$$

9.149 An antenna that radiates uniformly in all directions is called *isotropic* or *omnidirectional*. Find the radiation resistance of an isotropic antenna, if its field pattern is given by $E = KI/r$, where K is a constant.

▮ At radius r the time-average power density is
$$S_{av} = \frac{E^2}{2\eta_0} = \frac{K^2 I^2}{2r^2 \eta_0}$$

Then, by the definition of radiation resistance,
$$R_{rad} = \frac{P_{rad}}{I_{rms}^2} = \frac{2P_{rad}}{I^2} = \frac{2S_{av}(4\pi r^2)}{I^2} = \frac{4\pi K^2}{\eta_0} = \frac{K^2}{30} \quad (\Omega) \tag{1}$$

9.150 Calculate the radiation resistance of the isotropic antenna of Problem 9.149, assuming that the field distribution is given by $E = 25I/r$ (V/m).

▮ For $K = 25 \ \Omega$, (1) of Problem 9.149 gives
$$R_{rad} = \frac{(25)^2}{30} = 20.83 \ \Omega$$

9.151 What is the beam solid angle of a Hertzian dipole?

▮ By Problem 9.93, $\Omega_A = 4\pi/D_{max} = 4\pi/1.5 = 8\pi/3$.

9.152 In Problem 9.44, the length of the antenna is $\lambda_0/10$. What is its radiation resistance?

▮ For a nonsinusoidal current, (1) of Problem 9.17 becomes
$$R_{rad} = 80\pi^2 \left(\frac{\zeta}{\lambda_0}\right)^2 \left(\frac{I_{av}}{I_m}\right)^2 \quad \text{(see Problem 9.172)} \tag{1}$$

where I_{av} = (average current on the dipole) and I_m = [amplitude (maximum value) of the current at the terminals of the dipole]. Now, for the given current distribution, $I_{av} = I_m/2$. Substituting this and $\zeta = \lambda_0/10$ in (1) yields
$$R_{rad} = 80\pi^2 (\tfrac{1}{10})^2 (\tfrac{1}{2})^2 = 1.974 \ \Omega$$

9.153 If the loss resistance of the antenna of Problem 9.152 is $0.9 \ \Omega$, what is the radiative efficiency of the antenna?

▮ From (2) of Problem 9.147, with $R_A = 0.9 \ \Omega$ and $R_{rad} = 1.974 \ \Omega$ (from Problem 9.152),
$$\epsilon_{rad} = \frac{1.974}{0.9 + 1.974} = 68.7\%$$

9.154 Assuming that the antenna of Problem 9.152 radiates at 2 MHz, determine its length.

▮
$$\zeta = \frac{\lambda_0}{10} = \frac{u_0}{10f} = \frac{3 \times 10^8}{10 \times 2 \times 10^6} = 15 \text{ m}$$

9.155 A certain half-wave dipole antenna, transmitting at 3 MHz, has efficiency 95%. Calculate its maximum effective aperture.

▮ By Problem 9.94, $D_{max} = D(90°) = 1.64$; hence, by Problem 9.111,
$$A_{em}(90°) = \frac{\lambda_0^2}{4\pi} \epsilon_{rad} D_{max} = \frac{10^4}{4\pi} (0.95)(1.64) = 1240 \text{ m}^2$$

9.156 Determine the directivity of an antenna which has a conical field pattern that is uniform for $0 < \theta < 60°$ and zero for $60° < \theta < 180°$, and is independent of ϕ.

    ▮ The beam solid angle is

$$\Omega_A = 2\pi \int_0^{\pi/3} \sin\theta \, d\theta = \pi \text{ sr}$$

Thus, by (1) of Problem 9.144,

$$D_{\max} = \frac{4\pi}{\Omega_A} = 4$$

9.157 Find the gain of the antenna of Problem 9.156, presuming it lossless.

    ▮ Given that $\epsilon_{\text{rad}} = 1$,

$$G(\theta) = D(\theta) = \begin{cases} 4 & 0 < \theta < 60° \\ 0 & 60° < \theta < 180° \end{cases}$$

9.158 Find the directivity of the antenna of Problem 9.156 if the conical field pattern is uniform for $0 < \theta < 30°$ and zero outside.

    ▮

$$\Omega_A = 2\pi \int_0^{\pi/6} \sin\theta \, d\theta = 0.842 \text{ sr} \qquad D_{\max} = \frac{4\pi}{0.842} = 14.93$$

9.159 The antenna of Problem 9.158 radiates at 5 MHz. Calculate its maximum effective aperture, assuming no losses.

    ▮ For a lossless antenna, we have, within the cone,

$$A_{em} = \frac{\lambda_0^2}{4\pi} D_{\max} = \frac{\lambda_0^2}{\Omega_A} \qquad (1)$$

and, of course, $A_{em} = 0$ outside the cone. At 5 MHz, $\lambda_0 = 60$ m; and (Problem 9.158) $\Omega_A = 0.842$ sr. Thus

$$A_{em} = \frac{(60)^2}{0.842} = 4276.6 \text{ m}^2$$

9.160 For a 1.5 A current, the antenna of Problem 9.159 produces a 3-V/m field at a distance of 60 m. Determine the radiation resistance.

    ▮ For a lossless antenna, we may write

$$\frac{1}{2}|I|^2 R_{\text{rad}} = P_R = P_{\text{rad}} = \frac{|E|^2}{2\eta_0}(r^2\Omega_A)$$

or

$$R_{\text{rad}} = \frac{|E/I|^2 r^2\Omega_A}{\eta_0} = \frac{4(60)^2(0.842)}{377} = 32.16 \ \Omega$$

9.161 The field of an antenna is uniform over $0 < \theta < 60°$ and $0 < \phi < 120°$. Determine the beam solid angle.

    ▮

$$\Omega_A = \int_{\theta=0}^{\pi/3} \int_{\phi=0}^{2\pi/3} \sin\theta \, d\theta \, d\phi = \frac{\pi}{3} \text{ sr}$$

9.162 For the antenna of Problem 9.161 the field in the specified region is 3 V/m at points 60 m from the antenna, for a 1.5 A current. Calculate the radiation resistance.

    ▮ See Problem 9.160; by proportion,

$$R_{\text{rad}} = \frac{\pi/3}{0.842}(32.16) = 40.0 \ \Omega$$

9.163 What is the directivity of the antenna of Problem 9.161?

    ▮

$$D_{\max} = \frac{4\pi}{\Omega_A} = \frac{4\pi}{\pi/3} = 12$$

9.164 The lossless antenna of Problem 9.161 radiates at 9 MHz; determine its maximum effective aperture.

    ▮ By (1) of Problem 9.159,

$$A_{em} = \frac{\lambda_0^2}{\Omega_A} = \left(\frac{100}{3}\right)^2 \Big/ \frac{\pi}{3} = 1061.03 \text{ m}^2$$

9.165 Obtain an expression for the power radiated by an electrically short dipole antenna, of physical length ζ and carrying a uniform current $\hat{I} = I \underline{/0°}$.

    ■ It is clear from Problems 9.44 and 9.45 that (1) of Problem 9.16 remains valid for a short dipole if $d\zeta$ is replaced by ζ; thus

$$P_{\text{rad}} = 40 \left(\frac{\pi I \zeta}{\lambda_0} \right)^2 \quad \text{(W)} \tag{1}$$

9.166 Calculate the radiation resistance of the antenna of Problem 9.165.

    ■
$$R_{\text{rad}} = \frac{2 P_{\text{rad}}}{I^2} = 80 \left(\frac{\pi \zeta}{\lambda_0} \right)^2 \quad (\Omega)$$

(Compare Problem 9.17.)

9.167 Calculate the average radiated power from a 6-MHz, 2-m-long dipole that carries 2 A in current.

    ■ At 6 MHz, $\lambda_0 = 50$ m; the antenna is electrically short. Thus, (1) of Problem 9.165 gives

$$P_{\text{rad}} = 40 \left(\frac{\pi \times 2 \times 2}{50} \right)^2 = 2.53 \text{ W}$$

9.168 For the antenna of Problem 9.167, find the angle θ_1, such that one-half of the radiated power is in the region defined by $0 < \phi < 2\pi$ and $90° - \theta_1 < \theta < 90° + \theta_1$.

    ■ With the time-average Poynting vector as given by (1) of Problem 9.11, we obtain the equation

$$20 \pi^2 I^2 \left(\frac{\zeta}{\lambda_0} \right)^2 = 15 \pi I^2 \left(\frac{\zeta}{\lambda_0} \right)^2 \int_{\pi/2 - \theta_1}^{\pi/2 + \theta_1} \int_0^{2\pi} \sin^3 \theta \, d\phi \, d\theta$$

or
$$\frac{1}{3} = \sin \theta_1 - \frac{1}{3} \sin^3 \theta_1$$

Solving this cubic equation (by Newton's method, say) we get $\theta_1 \approx 21°$. Note that the result is valid for any (small) ζ_e and any I.

9.169 The far electric field of a certain antenna is given by

$$|\mathbf{E}| = \frac{150}{r} \sin^2 \theta \quad \text{(V/m)}$$

Determine the average power radiated.

    ■
$$P_{\text{rad}} = \int_{\phi=0}^{2\pi} \int_{\theta=0}^{\pi} \frac{1}{2\eta_0} |\mathbf{E}|^2 r^2 \sin \theta \, d\theta \, d\phi = \left(\frac{150}{r} \right)^2 \frac{r^2}{2\eta_0} (2\pi) \int_0^{\pi} \sin^5 \theta \, d\theta$$
$$= \frac{2\pi}{2 \times 377} (150)^2 \left(-\frac{5}{8} \cos \theta + \frac{5}{48} \cos 3\theta - \frac{1}{80} \cos 5\theta \right)_0^{\pi} = 200 \text{ W}$$

9.170 The radiation resistance of a lossless quarter-wave monopole antenna is 36.56 Ω. The electric field 1600 m away from the antenna is 0.293 V/m. Determine the power delivered to the antenna.

    ■ From Problem 9.46,

$$|E_\theta| = \frac{\eta_0 I_m}{2\pi r} \frac{\cos \left(\frac{\pi}{4} \cos \theta \right) - \frac{\sqrt{2}}{2}}{\sin \theta}$$

Substituting numerical values ($\theta = 90°$) yields
$$0.293 = \frac{120 \pi I_m}{2\pi \times 1600} \left(1 - \frac{\sqrt{2}}{2} \right) \quad \text{or} \quad I_m = 2.67 \text{ A}$$
Then
$$P_{\text{in}} = P_{\text{rad}} = \tfrac{1}{2} I_m^2 R_{\text{rad}} = \tfrac{1}{2} (2.67)^2 \times 36.56 = 130.32 \text{ W}$$

9.171 Repeat Problem 9.170 for a half-wave dipole, for the same distance and the same magnitude of the electric field.

    ■ R_{rad} will be twice as large, so that $P_{\text{in}} = 260.64$ W.

9.172 The *effective length* L_{eff} of an antenna carrying peak current I_m is defined as the length of an imaginary linear antenna with a uniform current I_m such that both antennas have the same far field in the $\theta = \pi/2$ plane. Obtain a general expression for the effective length.

▮ Superposition of the far-field elements, (2) of Problem 9.44, yields for the actual antenna field at $\theta = \pi/2$:

$$(E_\theta)_{\text{act}} = \frac{j\eta_0\beta_0}{4\pi r}e^{-j\beta_0 r}\left[\int_{-\zeta/2}^{\zeta/2}\hat{I}(z)\,dz\right] \tag{1}$$

(cf. Problem 9.47). On the other hand,

$$(E_\theta)_{\text{imag}} = \frac{j\eta_0\beta_0}{4\pi r}e^{-j\beta_0 r}\int_{-L_{\text{eff}}/2}^{L_{\text{eff}}/2}I_m\,dz = \frac{j\eta_0\beta_0}{4\pi r}e^{-j\beta_0 r}(I_m L_{\text{eff}}) \tag{2}$$

The two fields will be identical if

$$L_{\text{eff}} = \frac{1}{I_m}\int_{-\zeta/2}^{\zeta/2}\hat{I}(z)\,dz = \frac{\hat{I}_{\text{av}}}{I_m}\zeta \tag{3}$$

9.173 A center-fed half-wave dipole antenna carries a sinusoidal current distribution of amplitude I_0. What is the effective length of the dipole?

▮ By (3) of Problem 9.172,

$$L_{\text{eff}} = \frac{1}{I_m}\int_{-\lambda_0/4}^{\lambda_0/4}I_m\sin\left(\frac{\pi}{2} - \frac{2\pi}{\lambda_0}|z|\right)dz = \frac{\lambda_0}{\pi}$$

9.174 From Problem 9.76, the far field of two identical in-phase point sources (actually, line sources) located $\lambda/2$ apart along the z axis is given by

$$|\mathbf{E}| \propto \left|\cos\left(\frac{\pi}{2}\cos\theta\right)\right|$$

A second identical array of two point sources is placed a distance $\lambda/2$ from the first. Determine the resulting field pattern. Continue this process to obtain the pattern of an array of three pairs of sources $\lambda/2$ apart.

▮ By pattern multiplication the resultant pattern due to a two-pair array is given by

$$|\mathbf{E}| \propto \cos^2\left(\frac{\pi}{2}\cos\theta\right)$$

Similarly, for a three-pair array,

$$|\mathbf{E}| \propto \left|\cos^3\left(\frac{\pi}{2}\cos\theta\right)\right|$$

9.175 Find the far field of n isotropic point sources of equal amplitudes and uniform spacing.

▮ If E_0 is the amplitude of the far field due to each source, d is the spacing between sources, and α is the progressive phase difference between sources, then, with the reference at source 1, the resultant far field is given by

$$E = E_0[1 + e^{j\psi} + e^{j2\psi} + \cdots + e^{j(n-1)\psi}] = E_0\frac{1 - e^{jn\psi}}{1 - e^{j\psi}}$$
$$= E_0 e^{j(n-1)\psi/2}\frac{\sin(n\psi/2)}{\sin(\psi/2)} \tag{1}$$

where $\psi \equiv \beta d\cos\theta + \alpha$. (In Problem 9.75, ff., this quantity was notated as 2ψ.)

9.176 When n is odd ($n = 2m + 1$) in Problem 9.175, (1) may be simplified by taking the middle source as the reference for phase. Carry this out.

▮
$$E = E_0[e^{-jm\psi} + e^{-j(m-1)\psi} + \cdots + e^{-j\psi} + 1 + e^{j\psi} + \cdots + e^{j(m-1)\psi} + e^{jm\psi}]$$
$$= E_0 e^{-jm\psi}[1 + e^{j\psi} + \cdots + e^{j2m\psi}] = E_0 e^{-j(n-1)\psi/2}[1 + e^{j\psi} + \cdots + e^{j(n-1)\psi}]$$
$$= E_0\frac{\sin(n\psi/2)}{\sin(\psi/2)} \qquad (n \text{ odd}) \tag{1}$$

where the last step followed from (1) of Problem 9.175.

9.177 In the *diffraction pattern* of Problem 9.175 or 9.176, what is the height of the principal maximum?

▮ As $\psi \to 0$,

$$|E| \to |E_0|\frac{n\psi/2}{\psi/2} = n\,|E_0|$$

i.e., constructive interference among all n sources.

9.178 Locate the nulls of the diffraction pattern of Problem 9.175 or 9.176.

▮ For a null, $n\psi/2$ must be a nonzero integral multiple of π, but $\psi/2$ must not be an integral multiple of π.

The first condition is satisfied if
$$n\psi = n\beta d \cos\theta_0 + n\alpha = \pm 2k\pi$$
or
$$\cos\theta_0 = \frac{\pm(k/n)2\pi - \alpha}{\beta d} \quad (k = 1, 2, 3, \ldots) \tag{1}$$

In (1) we must rule out all values of k that make $|\cos\theta_0| > 1$; and, additionally, all k that are divisible by n (such k violate the second condition).

9.179 A 25-dB-gain, right-circularly polarized antenna in a radio link radiates 2 W of power at 1.5 GHz. The receiving antenna presents a voltage standing-wave ratio of 2 due to an impedance mismatch at its terminals. The receiving antenna is 90 percent efficient and has a field pattern near the beam maximum given by $\mathbf{E} = (2\mathbf{a}_x + j\mathbf{a}_y)F(\theta, \phi)$, where F gives the angular dependence of the pattern. The distance between the two antennas is 3000 km, and the receiving antenna is required to deliver 10^{-14} W to the receiver. Determine the maximum effective aperture of the receiving antenna.

∎ The normalized transmitting and receiving antenna field patterns are:
$$\mathbf{E}_T = \frac{1}{\sqrt{2}}(\mathbf{a}_x - j\mathbf{a}_y)F_T(\theta, \phi) \equiv \mathbf{P}_T F_T(\theta, \phi)$$
$$\mathbf{E}_R = \frac{1}{\sqrt{5}}(2\mathbf{a}_x + j\mathbf{a}_y)\sqrt{5}\,F_R(\theta, \phi) \equiv \mathbf{P}_R \sqrt{5}\,F_R(\theta, \phi)$$

Polarization loss factor: $\quad \text{PLF} = |\mathbf{P}_T \cdot \mathbf{P}_R^*|^2 = \frac{1}{10}(2 - 1)^2 = 0.1$

Reflection coefficient: $\quad |\Gamma_R| = \dfrac{\text{VSWR} - 1}{\text{VSWR} + 1} = \dfrac{2 - 1}{2 + 1} = \dfrac{1}{3}$

Wavelength: $\quad \lambda = \dfrac{3 \times 10^8}{1.5 \times 10^9} = 0.2$ m

Separation between the antennas: $\quad r = 3 \times 10^6$ m (given)
Gain of transmitting antenna: $\quad G_T = 25$ dB $= 10^{2.5} = 316.23$ (given)
The Friis formula gives the received power as
$$P_R = P_T \frac{G_R G_T \lambda^2}{(4\pi r)^2}(1 - |\Gamma_R|^2)(\text{PLF}) \tag{1}$$

Substituting numerical values in (1) yields
$$10^{-14} = 2 \times \frac{G_R(316.23)(0.2)^2(1 - \frac{1}{9})^2 0.1}{(4\pi \times 3 \times 10^6)^2} \quad \text{or} \quad G_R = 6.32$$

From this, $D_R = G_R/\epsilon_{\text{rad}} = 6.32/0.90 = 7.022$, and, finally,
$$A_{em,R} = \frac{\lambda^2}{4\pi}D_R = \frac{(0.2)^2}{4\pi} \times 7.022 = 0.02235 \text{ m}^2$$

9.180 The radiation intensity of an antenna is given by
$$U = U_m \sin\theta \sin^2\phi \quad (0 < \theta < \pi/2; 0 < \phi < 2\pi)$$
Determine the directivity and the beam solid angle.

∎ The maximum intensity is at $\theta = \pi/2$ and $\phi = \pi/2$, so that $U_{\max} = U_m$; the directivity is thus
$$D_{\max} = \frac{4\pi U_{\max}}{P_{\text{rad}}} = \frac{4\pi U_m}{U_m \int_0^{2\pi}\int_0^{\pi/2}\sin^2\theta \sin^2\phi\, d\theta\, d\phi} = \frac{16}{\pi} = 5.093$$
and the beam solid angle is $\Omega_A = 4\pi/D_{\max} = \pi^2/4 = 2.467$ sr.

9.181 The maximum current and maximum radiation intensity for the antenna of Problem 9.180 are related via $I_m = 0.1 U_m$. Find the radiation resistance of the antenna, in terms of U_m.

∎
$$\frac{1}{2}I_m^2 R_{\text{rad}} = P_{\text{rad}} \quad \text{or} \quad \frac{1}{2}(0.1U_m)^2 R_{\text{rad}} = \frac{4\pi U_m}{D_{\max}}$$
from which $R_{\text{rad}} = (50\pi^2/U_m)$ (Ω).

9.182 Determine the elevational (θ) and azimuthal (ϕ) band widths for the antenna of Problem 9.180.

∎ The required half-power beam widths, from Fig. 9-44, are
$$\text{HPBW}_\theta = \theta_2 - \theta_1 = 150° - 30° = 120°$$
$$\text{HPBW}_\phi = \phi_2 - \phi_1 = 135° - 45° = 90°$$

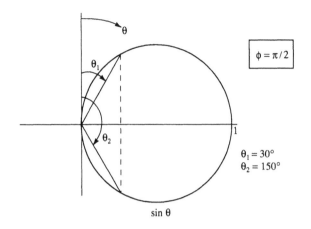

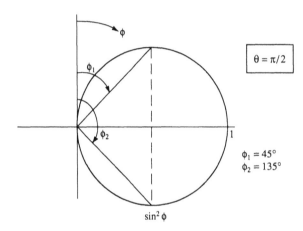

Fig. 9-44

9.183 A thin linear antenna of length L supports the current distribution
$$I(z) = I_0 \cos{(\pi z/L)} \qquad (-L/2 \leq z \leq L/2)$$
Determine the far-field vector potential and obtain the normalized pattern factor.

❚ For the given current distribution, (1) of Problem 9.3 becomes, in the far field,
$$A_z = \frac{\mu_0}{4\pi} \frac{e^{-j\beta_0 r}}{r} I_0 \int_{-L/2}^{L/2} \cos{\frac{\pi z'}{L}} e^{j\beta_0 z' \cos{\theta}} dz'$$
so that the unnormalized pattern factor is
$$F(\theta) = \int_{-L/2}^{L/2} \cos{\frac{\pi z'}{L}} e^{j\beta_0 z' \cos{\theta}} dz' = \frac{1}{2} \int_{-L/2}^{L/2} [e^{j(\pi/L + \beta_0 \cos{\theta})z'} + e^{-j(\pi/L - \beta_0 \cos{\theta})z'}] dz'$$
$$= \frac{2L}{\pi} \frac{\cos{\left(\frac{\beta_0 L}{2} \cos{\theta}\right)}}{1 - \left(\frac{\beta_0 L}{2} \cos{\theta}\right)^2}$$
The normalized pattern factor is $f(\theta) = (\pi/2L)F(\theta)$.

9.184 For a large broadside array of $n \gg 1$ sources of equal amplitudes and uniform spacing, find the beam width between first nulls (BWFN).

❚ Refer to Problems 9.175 and 9.178. The sources must be in phase ($\alpha = 0$) to put the principal maximum at $\theta = 90°$. In terms of the angle γ (Fig. 9-45), the null angles are [see (1) of Problem 9.178],
$$\gamma_0 = \sin^{-1}\left(\pm\frac{k\lambda}{nd}\right) \qquad (1)$$
For large n and small k, (1) reduces to
$$\gamma_0 = \pm\frac{k}{nd/\lambda} \approx \pm\frac{k}{L/\lambda} \quad \text{(rad)} \qquad (2)$$

where $L = (n-1)d$ = length of the array. Since the first nulls correspond to $k = 1$,

$$\text{BWFN} = \frac{1}{L/\lambda} - \left(-\frac{1}{L/\lambda}\right) = \frac{2\,\text{rad}}{L/\lambda} = \frac{114.6°}{L/\lambda} \tag{3}$$

9.185 Find the half-power beam width (HPBW) of the array of Problem 9.184.

$$\text{HPBW} \approx \tfrac{1}{2}\,\text{BWFN} = \frac{57.3°}{L/\lambda} \tag{1}$$

9.186 Obtain an expression for the BWFN for an end-fire array of n sources of equal amplitudes and uniform spacing (having a maximum at $\theta = 0$, Fig. 9-45).

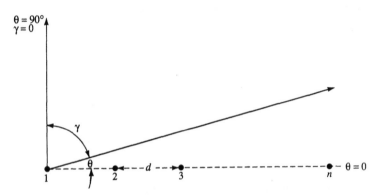

Fig. 9-45

For $\psi = \beta d \cos\theta + \alpha$ to vanish at $\theta = 0$, we require that $\alpha = -\beta d$. Then (1) of Problem 9.178 gives

$$\cos\theta_0 - 1 = \pm\frac{k}{nd/\lambda} \qquad \text{or} \qquad \frac{\theta_0}{2} = \sin^{-1}\left(\pm\sqrt{\frac{k}{2nd/\lambda}}\right) \tag{1}$$

Applying the condition $nd \gg k\lambda$ to (1) yields

$$\theta_0 = \pm\sqrt{\frac{2k}{nd/\lambda}} \approx \pm\sqrt{\frac{2k}{L/\lambda}} \tag{2}$$

For the first nulls, $k = 1$, and so

$$\text{BWFN} = \sqrt{\frac{2}{L/\lambda}} - \left(-\sqrt{\frac{2}{L/\lambda}}\right) = (2\,\text{rad})\sqrt{\frac{2}{L/\lambda}} = (114.6°)\sqrt{\frac{2}{L/\lambda}} \tag{3}$$

9.187 A broadside array has 20 sources of equal amplitudes that are spaced $\lambda/2$ apart. Calculate the BWFN and the HPBW.

From (3) of Problem 9.184,
$$\text{BWFN} = \frac{114.6°}{L/\lambda} = \frac{114.6°}{10\lambda/\lambda} = 11.46° \qquad \text{and} \qquad \text{HPBW} = \frac{1}{2}(11.46°) = 5.73°$$

9.188 Repeat Problem 9.187 for an end-fire array of 20 sources of equal amplitudes spaced $\lambda/2$ apart.

From (3) of Problem 9.186,
$$\text{BWFN} = (114.6°)\sqrt{\frac{2}{L/\lambda}} = (114.6°)\sqrt{\frac{2}{10}} = 51.25° \qquad \text{and} \qquad \text{HPBW} = \frac{1}{2}(51.25°) = 25.63°$$

9.189 The directivity of an antenna may be approximately expressed as
$$D_{\max} \approx \frac{41\,253}{\theta_{\text{HP}}^0 \phi_{\text{HP}}^0} \tag{1}$$
where θ_{HP}^0 and ϕ_{HP}^0 are, respectively, HPBW in θ- and ϕ-planes, in degrees. Find the directivity (in dB) of the antenna of Problem 9.187.

From Problem 9.187, $\theta_{\text{HP}}^0 = 5.73°$ and $\phi_{\text{HP}}^0 = 360°$. Thus, substitution in (1) yields
$$D_{\max} = \frac{41\,253}{5.73 \times 360} = 20 = 13\,\text{dB}$$

9.190 Repeat Problem 9.189 for the antenna of Problem 9.188.

▌ In this case, from Problem 9.188, we have $\theta_{HP}^0 = \phi_{HP}^0 = 25.63°$; hence,

$$D_{max} = \frac{41\,253}{25.63 \times 25.63} = 62.80 = 17.98 \text{ dB}$$

9.191 Two point sources, 2λ apart, are of equal amplitudes and are in phase. Find the angles for maxima and nulls.

▌ Here $\psi = 4\pi \cos \theta$, giving the normalized field

$$E_n = \cos \frac{\psi}{2} = \cos (2\pi \cos \theta)$$

From this we locate the maxima at $0°$, $180°$, $\pm 60°$, $\pm 90°$, and $\pm 120°$. Similarly, the nulls are at $\pm 41.4°$, $\pm 75.5°$, $\pm 104.5°$, and $\pm 138.6°$.

9.192 Repeat Problem 9.191 for point sources in opposite phase and 1.5λ apart.

▌ Now, $\psi = 3\pi \cos \theta - \pi$. For nulls,

$$\frac{3\pi}{2} \cos \theta_0 - \frac{\pi}{2} = \pm \frac{\pi}{2}, \pm \frac{3\pi}{2}, \ldots \quad \text{or} \quad \theta_0 = \pm 48.2°, \pm 90°, \pm 131.8°$$

For maxima,

$$\frac{3\pi}{2} \cos \theta - \frac{\pi}{2} = 0, \pm \pi, \pm 2\pi, \ldots \quad \text{or} \quad \theta = 0°, 180°, \pm 70.5°, \pm 109.5°$$

9.193 The *Hansen–Woodyard criteria* for increased directivity of an array are

$$(i) \quad \alpha < \pi \quad \text{and} \quad (ii) \quad d < \frac{\lambda}{2}\left(1 - \frac{1}{n}\right)$$

Verify that the criteria are satisfied for an end-fire array (principal maximum at $\theta = 180°$) with $n = 3$ and $d = 0.3\lambda$.

▌ The condition here is $\psi(180°) = \pi/n$, or

$$\alpha = \beta d + \frac{\pi}{n} = 2\pi(0.3) + \frac{\pi}{3} = 0.933\pi < \pi$$

Also
$$\frac{1}{2}\left(1 - \frac{1}{n}\right) = \frac{1}{3} > \frac{3}{10} = \frac{d}{\lambda}$$

9.194 Determine the normalized array factor for the array of Problem 9.193.

▌ Here, $\psi = \beta d \cos \theta + \alpha = 0.6\pi \cos \theta + 0.933\pi$ and $n = 3$; thus

$$E_n = \frac{1}{n} \frac{\sin (n\psi/2)}{\sin (\psi/2)} = \frac{\sin (0.9\pi \cos \theta + 1.4\pi)}{3 \sin (0.3\pi \cos \theta + 0.467\pi)}$$

9.195 The far field of a magnetic dipole (a loop of radius a carrying a current I_0) may be written as

$$E_\phi \approx \eta(\beta a)^2 \frac{I_0 \sin \theta}{4r} e^{-j\beta r} \quad \text{and} \quad H_\theta = -\frac{E_\phi}{\eta}$$

Determine the radiation intensity distribution.

▌ The radiation intensity is defined as

$$U(\theta) \equiv r^2 S_{av}(\theta) = r^2 \frac{|E_\phi|^2}{2\eta} = \frac{1}{32} \eta(\beta a)^4 |I_0|^2 \sin^2 \theta \qquad (1)$$

9.196 What is the radiation resistance of the antenna of Problem 9.195?

▌ The radiation resistance is given by

$$R_{rad} = \frac{2P_{rad}}{|I_0|^2} \quad \text{where} \quad P_{rad} = \int_0^{2\pi} \int_0^\pi U(\theta) \sin \theta \, d\theta \, d\phi$$

By (1) of Problem 9.195,

$$P_{rad} = \frac{2\pi}{32} \eta(\beta a)^4 |I_0|^2 \int_0^\pi \sin^3 \theta \, d\theta = \frac{\pi}{12} \eta(\beta a)^4 |I_0|^2$$

whence $R_{rad} = \pi\eta(\beta a)^4/6$.

9.197 A magnetic dipole antenna operates at 100 MHz. If the radius of the loop is 20 cm, what is the radiation resistance of the antenna?

$$\beta = \frac{2\pi f}{u} = \frac{2\pi \times 100 \times 10^6}{3 \times 10^8} = 2.094 \text{ rad/m}$$

and so, by Problem 9.196,

$$R_{\text{rad}} = \frac{\pi}{6} \times 377 \times (2.094 \times 0.2)^4 = 6.07 \ \Omega$$

9.198 Sketch the power pattern of the antenna of Problem 9.195 and find the half-power beam width (HPBW).

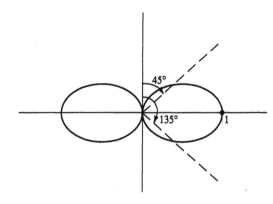

Fig. 9-46

▮ From Problem 9.195, $S_{\text{av}} \propto \sin^2 \theta$. Thus we obtain the power pattern shown in Fig. 9-46, from which HPBW $= 135° - 45° = 90°$.

9.199 A circularly polarized wave with $\mathbf{E}_w = (1/\sqrt{2})(\mathbf{a}_x - j\mathbf{a}_y)e^{-j\beta z}$ is incident on an antenna whose polarization is given by $\mathbf{E}_A = (1/\sqrt{2})(\mathbf{a}_x + j\mathbf{a}_y)e^{-j\beta z}$. Determine the polarization loss factor (PLF).

▮ Proceed as in Problem 9.179:

$$\text{PLF} = |\mathbf{P}_w \cdot \mathbf{P}_A^*|^2 = (\tfrac{1}{2})^2(1-1)^2 = 0$$

9.200 Two spacecraft are separated by 3000 km. Each has an antenna with $D_{\text{max}} = 200$, operating at 2 GHz. If craft A's receiver requires 20 dB over 1 pW, what transmitter power is required on craft B to achieve this signal level?

▮ We are given that

$$20 = 10 \log_{10} \frac{P_R}{1 \text{ pW}} \qquad \text{or} \qquad P_R = 100 \text{ pW}$$

Substituting in

$$P_R = P_T \frac{D_T D_R \lambda^2}{(4\pi r)^2}$$

yields

$$10^{-10} = P_T \frac{200 \times 200 \times (3/20)^2}{(4\pi \times 3 \times 10^6)^2} \qquad \text{or} \qquad P_T = 157.91 \text{ W}$$

9.201 A system consisting of a transmitting antenna and a receiving antenna may be represented by a two-port network having the admittance matrix shown in Fig. 9-47. By applying voltage sources at the terminals and using reciprocity, show that $Y_{12} = Y_{21}$.

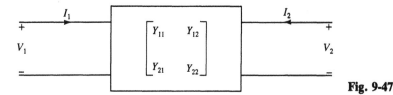

Fig. 9-47

▮ Using the admittance matrix representation, we have

$$Y_{12} = \frac{I_1}{V_2} \bigg|_{V_1=0} = \frac{I_1}{V_{21}} \tag{1}$$

where V_{21} is the voltage at port 2 due to the current I_1. Similarly,

$$Y_{21} = \frac{I_2}{V_1}\bigg|_{V_2=0} = \frac{I_2}{V_{12}} \tag{2}$$

where V_{12} is the voltage at port 1 due to the current I_2. From reciprocity,

$$V_{21}I_2 = V_{12}I_1 \tag{3}$$

Combining (1) through (3) yields $Y_{12} = Y_{21}$.

9.202 A parabolic reflector antenna is circular in cross section, with a diameter of 1.22 m. If the maximum effective aperture equals 55 percent of the physical aperture, compute the gain of the antenna, in decibels, at 20 GHz ($\lambda = 15$ mm).

$\blacksquare$

      Physical aperture: $A_p = \frac{\pi}{4}d^2 = \frac{\pi}{4}(1.22)^2 = 1.169\ \text{m}^2$

      Effective aperture: $A_{em} = 0.55\,A_p = 0.55 \times 1.169 = 0.643\ \text{m}^2$

      Gain: $G = \frac{4\pi}{\lambda^2}A_{em} = \frac{4\pi}{(0.015)^2} \times 0.643 = 35\,912 = 45.55\ \text{dB}$

9.203 A radio link from the moon to the earth has a moon-based antenna with $D_{\max} = 1000$, operating at 100 MHz. The transmitter power is 2000 W. What should be the maximum effective aperture of the earth-based antenna in order to deliver 20 dB above 1 pW? Take the earth-moon distance as 381 000 km.

$\blacksquare$ The wavelength is 3 m, and, from Problem 9.200, the needed power is $P_R = 100$ pW. Then, from

$$P_R = P_T \frac{D_R D_T \lambda^2}{(4\pi r)^2}$$

we have

$$D_R = P_R \frac{(4\pi r)^2}{P_T D_T \lambda^2} = \frac{10^{-10}(4\pi \times 3.81 \times 10^8)^2}{2000 \times 1000 \times (3)^2} = 127.35$$

and

$$A_{emR} = \frac{\lambda^2}{4\pi}D_R = \frac{9 \times 127.35}{4\pi} = 91.2\ \text{m}^2$$

9.204 For a linear antenna which is an integral number of half-wavelengths long ($L = n\lambda/2$), the current distribution will be of the form

$$I(z) = I_m \sin\left[\beta\left(z + \frac{L}{2}\right)\right]$$

Show that the radiation (far) field of the antenna is

$$E_\theta = \begin{cases} \dfrac{j\eta I_m}{2\pi r}e^{-j\beta r}\dfrac{\cos\left(\dfrac{n\pi}{2}\cos\theta\right)}{\sin\theta} & (n\ \text{odd}) \\[4mm] \dfrac{\eta I_m}{2\pi r}e^{-j\beta r}\dfrac{\sin\left(\dfrac{n\pi}{2}\cos\theta\right)}{\sin\theta} & (n\ \text{even}) \end{cases} \tag{1}$$

$\blacksquare$ For the given current, the vector potential may be written as

$$A_z = \frac{I_m e^{-j\beta r}}{4\pi r}\int_{-L/2}^{L/2}\sin\left[\beta\left(\frac{L}{2} + z'\right)\right]e^{j\beta z'\cos\theta}\,dz' \qquad (\beta L = n\pi) \tag{2}$$

The integral in (2) may be evaluated by expressing the sine function in terms of complex exponentials; the procedure is elementary, but so tedious that only the final result will be given:

$$\int_{-L/2}^{L/2}\cdots = \begin{cases} \dfrac{2}{\beta\sin^2\theta}\cos\left(\dfrac{n\pi}{2}\cos\theta\right) & (n\ \text{odd}) \\[4mm] \dfrac{2}{j\beta\sin^2\theta}\sin\left(\dfrac{n\pi}{2}\cos\theta\right) & (n\ \text{even}) \end{cases} \tag{3}$$

With A_z given by (2) and (3), the electric field,

$$E_\theta = j\omega\mu\sin\theta\,A_z$$

is as written in (1).

9.205 Specialize the result of Problem 9.204 to the case of a half-wave dipole.

❚
$$E_\theta = \frac{j\eta I_m}{2\pi r} e^{-j\beta r} \frac{\cos\left(\frac{\pi}{2}\cos\theta\right)}{\sin\theta}$$

which agrees precisely with (2) of Problem 9.46 and (1) of Problem 9.49.

9.206 For the antenna of Problem 9.204, with n odd, show that the radiation resistance referred to I_m is

$$R_{\text{rad}} = \frac{\eta}{4\pi}[\gamma + \ln(2n\pi) - \text{Ci}(2n\pi)] \tag{1}$$

where $\gamma = 0.5772\cdots$ is Euler's constant and Ci is the tabulated *cosine integral*.

❚ First, we evaluate the radiated power:

$$P_{\text{rad}} = \int_0^{2\pi}\int_0^\pi \frac{|E_\theta|^2}{2\eta} r^2 \sin\theta\, d\theta\, d\phi = \frac{\pi}{\eta}\frac{\eta^2 |I_m|^2}{(2\pi)^2}\int_0^\pi \frac{\cos^2\left(\frac{n\pi}{2}\cos\theta\right)}{\sin\theta}\, d\theta \tag{2}$$

With $u = \cos\theta$ and using the relation $\cos^2 x = \frac{1}{2}(1 + \cos 2x)$, the integral in (2) becomes

$$I_{\text{odd}} \equiv \int_0^\pi \frac{\cos^2\left(\frac{n\pi}{2}\cos\theta\right)}{\sin\theta}\, d\theta = \frac{1}{2}\int_{-1}^1 \frac{1 + \cos n\pi u}{1 - u^2}\, du$$

$$= \frac{1}{4}\int_{-1}^1 \frac{1 + \cos n\pi u}{1 + u}\, du + \frac{1}{4}\int_{-1}^1 \frac{1 + \cos n\pi u}{1 - u}\, du = \frac{1}{2}\int_{-1}^1 \frac{1 + \cos n\pi u}{1 + u}\, du$$

Now let $v = n\pi(1 + u)$ [n odd], to obtain

$$I_{\text{odd}} = \frac{1}{2}\int_0^{2n\pi} \frac{1 - \cos v}{v}\, dv = \frac{1}{2}[\gamma + \ln(2n\pi) - \text{Ci}(2n\pi)] \tag{3}$$

From (2) and (3) it follows at once that $R_{\text{rad}} = 2P_{\text{rad}}/|I_m|^2$ is as given in (1).

9.207 Show that (1) of Problem 9.206 is also valid for n even.

❚ In
$$I_{\text{even}} \equiv \int_0^\pi \frac{\sin^2\left(\frac{n\pi}{2}\cos\theta\right)}{\sin\theta}\, d\theta$$

first make the transformation $u = \cos\theta$ and then the transformation $v = n\pi(1 - u)$, to obtain

$$I_{\text{even}} = \frac{1}{2}\int_0^{2n\pi} \frac{1 - \cos v}{v}\, dv$$

But this is formally identical to I_{odd}.

9.208 If the feed point of the antenna of Problem 9.204 is $z = a\lambda$, determine the input resistance. Assume a lossless antenna.

❚ The input current may be written as

$$I_{\text{in}} = I_m \sin\left[\beta\left(a\lambda + \frac{n\lambda}{4}\right)\right] = I_m \sin\left[2\pi\left(a + \frac{n}{4}\right)\right]$$

If the antenna is lossless, all the input power is radiated, $\frac{1}{2}|I_{\text{in}}|^2 R_{\text{in}} = \frac{1}{2}|I_m|^2 R_{\text{rad}}$, whence

$$R_{\text{in}} = \frac{R_{\text{rad}}}{\sin^2\left[2\pi\left(a + \frac{n}{4}\right)\right]} \tag{1}$$

in which R_{rad} is given by (1) of Problem 9.206.

9.209 Using Problem 9.208, confirm that the input resistance of a center-fed half-wave dipole is 73 Ω.

❚ For $n = 1$ and $a = 0$, (1) of Problem 9.208 becomes $R_{\text{in}} = R_{\text{rad}}$. Consequently, from (1) of Problem 9.206,

$$R_{\text{in}} = \frac{\eta}{4\pi}[0.5772 + \ln(2\pi) - \text{Ci}(2\pi)]$$
$$= 30[0.5772 + 1.8379 - (-0.02)] \approx 73\ \Omega$$

9.210 For a certain current distribution, the far electric field from a dipole antenna is given by

$$E_\theta = \frac{Ke^{-j\beta r}}{r}\sin\theta\, [2\cos(\beta h \cos\theta)] \tag{1}$$

Find the total radiated power over a large upper hemisphere.

▮ The radiated power is obtained from

$$P_{rad} = \frac{2\pi}{2\eta} \int_0^{\pi/2} |E_\theta|^2 r^2 \sin\theta \, d\theta = \frac{4\pi |K|^2}{\eta} \int_0^{\pi/2} \cos^2(\beta h \cos\theta) \sin^3\theta \, d\theta$$

Let $u = \cos\theta$; then

$$P_{rad} = \frac{4\pi |K|^2}{\eta} \int_0^1 (1 - u^2) \cos^2(\beta hu) \, du$$

$$= \frac{4\pi |K|^2}{\eta} \left[\frac{u}{2} + \frac{\sin 2\beta hu}{4\beta h} - \frac{u \cos 2\beta hu}{4(\beta h)^2} - \left(\frac{u^2}{4\beta h} - \frac{1}{8(\beta h)^3} \right) \sin 2\beta hu - \frac{u^3}{6} \right]_0^1$$

$$= \frac{4\pi |K|^2}{\eta} \left[\frac{1}{3} - \frac{\cos 2\beta h}{(2\beta h)^2} + \frac{\sin 2\beta h}{(2\beta h)^3} \right]$$

9.211 Obtain the vector potential governing the far field of the finite current sheet shown in Fig. 9-48.

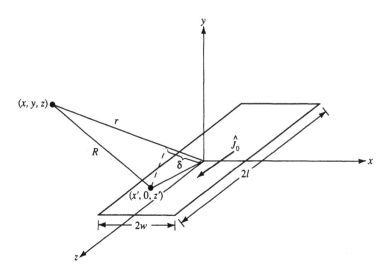

Fig. 9-48

▮ By Problem 9.3, $\mathbf{A} = A_z \mathbf{a}_z$, where

$$A_z(x, y, z) = \frac{\mu_0 \hat{J}_0}{4\pi} \int_{x'=-w}^w \int_{z'=-l}^l \frac{e^{-j\beta_0 R}}{R} \, dx' \, dz'$$

Under the far-field approximation, with $r^2 = x^2 + y^2 + z^2$,

$$\frac{1}{R} \approx \frac{1}{r} \qquad \text{and} \qquad e^{-j\beta_0 R} \approx e^{-j\beta_0(r-\delta)}$$

where $\delta = (x/r)x' + (z/r)z'$. Hence,

$$A_z(x, y, z) \approx \frac{\mu_0 \hat{J}_0}{4\pi r} e^{-j\beta_0 r} \int_{x'=-w}^w \exp[j\beta_0(x/r)x'] \, dx' \int_{z'=-l}^l \exp[j\beta_0(z/r)z'] \, dz'$$

$$= \frac{\mu_0 \hat{J}_0}{\pi r} e^{-j\beta_0 r} \frac{\sin(\beta_0 wx/r)}{\beta_0 x/r} \frac{\sin(\beta_0 lz/r)}{\beta_0 z/r} \tag{1}$$

9.212 From Problem 9.211 derive the far-field vector potential of a dipole antenna of length ζ that carries uniform current $\hat{I}_0$.

▮ Rewrite (1) of Problem 9.211 as

$$A_z(x, y, z) = \frac{\mu_0[\hat{J}_0(2w)]}{2\pi r} e^{-j\beta_0 r} \frac{\sin(\beta_0 wx/r)}{\beta_0 wx/r} \frac{\sin[(\beta_0 \zeta/2)(z/r)]}{\beta_0 z/r}$$

where l has been replaced by $\zeta/2$. Now, holding $\hat{J}_0(2w) = \hat{I}_0$ constant, let $w \to 0$, obtaining in the limit

$$A_z(x, y, z) = \frac{\mu_0 \hat{I}_0}{2\pi \beta_0 z} e^{-j\beta_0 r} \sin\left(\frac{\beta_0 \zeta}{2} \frac{z}{r} \right)$$

or, in the usual spherical coordinates,

$$A_z(r, \theta) = \frac{\mu_0 \hat{I}_0}{2\pi \beta_0 r} e^{-j\beta_0 r} \frac{\sin\left(\frac{\beta_0 \zeta}{2} \cos\theta \right)}{\cos\theta} \tag{1}$$

9.213 Continue Problem 9.212 to retrieve the far-field vector potential for a Hertzian dipole.

▌ For $\beta_0 \zeta \to \beta_0 \, d\zeta \ll 1$, the sine function in (1) of Problem 9.212 may be replaced by its argument, giving

$$A_z(r, \theta) \to \frac{\mu_0(\hat{I}_0 \, d\zeta)}{4\pi r} e^{-j\beta_0 r} \ .$$

as in Problem 9.4.

9.214 Give the radiation pattern function for the finite dipole antenna of Problem 9.212.

▌ Since $E_\theta \propto (\sin \theta) A_z$, the required function is

$$F(\theta) = \sin\left(\frac{\beta_0 \zeta}{2} \cos \theta\right) \tan \theta$$

9.215 A transmitting antenna produces a maximum far field (in a certain direction) given by

$$E_m = 90 I_m \frac{e^{-j\beta r}}{r} \quad \text{(V/m)}$$

The input impedance of this lossless antenna is 50 Ω. Express the maximum effective aperture in terms of the wavelength of the radiation.

▌ The maximum aperture is given by

$$A_{em} = \frac{\lambda^2}{4\pi} D_{max} = \lambda^2 \frac{U_{max}}{P_{rad}}$$

Now

$$U_{max} = \frac{|E_m|^2 r^2}{2\eta} = \frac{90^2 |I_m|^2}{240\pi} \quad \text{(W)}$$

and, since $P_{rad} = P_{in}$ for a lossless antenna,

$$P_{rad} = \tfrac{1}{2} |I_m|^2 R_{in} = 25 |I_m|^2 \quad \text{(W)}$$

Thus

$$A_{em} = \frac{90^2}{(240\pi)(25)} \lambda^2 = 0.4297 \lambda^2 \quad \text{(m}^2)$$

CHAPTER 10
Review Problems

Note The problems are not arranged in any specific order, and solutions are abbreviated.

10.1 Two parallel wires, located in air, carry equal and opposite currents and are separated by a distance of 0.5 m. Determine the maximum current they may carry such that the force per unit length of the conductors does not exceed 5 N/m.

▮ From the result of Problem 3.2,

$$\text{force/unit length} = \frac{\mu_0 I^2}{2\pi b} \quad \text{or} \quad 5 = \frac{(4\pi \times 10^{-7})I^2}{2\pi(0.5)}$$

Hence $I \le 3535.5$ A.

10.2 The **B**-field in a certain region in space is given by $\mathbf{B} = 0.03\mathbf{a}_x + 0.2\mathbf{a}_y$ T. In this field a conductor, carrying 20 A of current, is located in the xz plane. The coordinates of the ends of the conductor are $(0, 0, 0)$ and $(1, 0, 1)$ m. What is the force on the conductor?

▮ $\mathbf{F} = I\boldsymbol{\ell} \times \mathbf{B}$. Here, $I\boldsymbol{\ell} = 20\mathbf{a}_x + 20\mathbf{a}_z$ and $\mathbf{B} = 0.03\mathbf{a}_x + 0.2\mathbf{a}_y$. Hence, $\mathbf{F} = -4\mathbf{a}_x + 0.6\mathbf{a}_y + 4\mathbf{a}_z$ N.

10.3 The magnetic vector potential in a certain region is given by $\mathbf{A} = a\mathbf{a}_r + br\mathbf{a}_\phi + c\mathbf{a}_z$. Determine the corresponding **B**.

▮ Since $\mathbf{B} = \text{curl } \mathbf{A}$, in cylindrical coordinates we have

$$\mathbf{B} = \frac{1}{r}\frac{\partial}{\partial r}(rA_\phi)\mathbf{a}_z = \frac{1}{r}\frac{\partial}{\partial r}(br^2)\mathbf{a}_z = 2b\,\mathbf{a}_z$$

10.4 A 2-m-long straight conductor carries 5 A in current. Locate the origin of a cylindrical coordinate system at the midpoint of the conductor and find the magnetic vector potential at $(0.5, \phi, 0)$.

▮ From (1) of Problem 3.49,

$$\mathbf{A} = \left(\frac{\mu_0 I}{4\pi}\int_{-L/2}^{L/2}\frac{dz}{r}\right)\mathbf{a}_z$$

Substituting numerical values yields $\mathbf{A} = (2 \times 10^{-6})\mathbf{a}_z$.

10.5 In a certain conducting region, $\mathbf{H} = -4y^3\mathbf{a}_x$ (A/m). Calculate the current through a square of side 1 m, with one corner at the origin and the sides coinciding with the positive x and y axes.

▮
$$\mathbf{J} = \nabla \times \mathbf{H} = -\mathbf{a}_z\frac{\partial H_x}{\partial y} = 12y^2\mathbf{a}_z \quad (\text{A/m}^2)$$

Hence
$$I = \int_0^1\int_0^1 (12y^2)\,dx\,dy\,(\mathbf{a}_z \cdot \mathbf{a}_z) = 12\left(\frac{y^3}{3}\right)\Big|_0^1 = 4\text{ A}$$

10.6 Find the maximum energy-product of a magnet having the characteristic (2) of Fig. 3-57.

▮ The energy product is $-BH = -(10^{-5}H + 1.0)H$ (J/m³), which is maximized for $H = -50$ kA/m. Correspondingly, $B = 0.5$ T, for a maximum of $(-0.5)(-50) = 25$ kJ/m³.

10.7 A two-winding electromechanical system has its inductances given by

$$L_{11} = \frac{k_1}{x} = L_{22} \qquad L_{12} = L_{21} = \frac{k_2}{x}$$

where x is the separation, and k_1 and k_2 are constants. Neglecting the winding resistances, derive an expression for the force between the windings when both windings are connected to the same voltage source, $v = V_m \sin \omega t$.

▮ Because $L_{11} = L_{22}$ and $L_{12} = L_{21}$, $i_1 = i_2 \equiv i$. Hence the stored magnetic energy is
$$W_m = \tfrac{1}{2}L_{11}i_1^2 + \tfrac{1}{2}L_{22}i_2^2 + L_{12}i_1i_2 = (L_{11} + L_{12})i^2$$
and the electrical force is

$$F_e = \frac{\partial W_m(i, x)}{\partial x} = i^2\frac{\partial}{\partial x}(L_{11} + L_{12}) = -\frac{(k_1 + k_2)i^2}{x^2} \tag{1}$$

The current i is related to the voltage $v_1 = v_2 \equiv v$ through

$$v = \frac{d}{dt}[(L_{11} + L_{12})i] \quad \text{or} \quad i = \frac{1}{L_{11} + L_{12}} \int v \, dt = \frac{x}{k_1 + k_2}\left(-\frac{V_m}{\omega}\cos \omega t\right) \quad (2)$$

Then (1) and (2) give

$$F_e = -\frac{V_m^2 \cos^2 \omega t}{(k_1 + k_2)\omega^2}$$

10.8 Given the following complex form of the electric field of a wave traveling along the z axis

$$E_x(z) = E_0[e^{-jkz} + j\tfrac{1}{10}e^{j(kz+\phi)}]$$

find the instantaneous value $E_x(t, z)$.

❚ $$E_x(t, z) = \text{Re}\,[(E_x(z)e^{j\omega t}] = E_0[\cos{(\omega t - kz)} - \tfrac{1}{10}\sin{(\omega t + kz + \phi)}]$$

10.9 For the **E**-field of Problem 10.8, find the corresponding **H**-field in complex form.

❚ Maxwell's equation

$$\nabla \times \mathbf{E} = -\frac{\partial \mathbf{B}}{\partial t} = -j\omega\mu_0\mathbf{H}$$

yields

$$\frac{\partial E_x}{\partial z} = -j\omega\mu_0 H_y \quad \text{or} \quad H_y(z) = \frac{kE_0}{\mu_0\omega}\left(e^{-jkz} - j\frac{1}{10}e^{j(kz+\phi)}\right)$$

10.10 For the **E**-field given in Problem 10.8 and the corresponding **H**-field determined in Problem 10.9, what is the average electromagnetic power transmitted through a unit transverse area?

❚ The power density is given by the Poynting vector

$$\mathbf{S}_{av} = \frac{1}{2}\text{Re}\,(\mathbf{E} \times \mathbf{H}^*) = \frac{1}{2}\text{Re}\,(E_x H_y^*)$$
$$= \frac{1}{2}\text{Re}\left[E_0\left\{e^{j(\omega t - kz)} + \frac{j}{10}e^{j(\omega t + kz + \phi)}\right\} \times \frac{kE_0}{\mu_0\omega}\left\{e^{-j(\omega t - kz)} + \frac{j}{10}e^{-j(\omega t + kz + \phi)}\right\}\right]$$
$$= \frac{1}{2}\text{Re}\left[\frac{kE_0^2}{\mu_0\omega}\left(1 - \frac{1}{100}\right)\right] = \frac{99}{200}\frac{kE_0^2}{\mu_0\omega}$$

10.11 The electric field in the space-charge wave for a slab beam of electrons has components in rectangular coordinates as follows:

$$E_x = -j\beta aC \cos{(ax)}\, e^{-j\beta z} \qquad E_z = C(\omega^2\mu\epsilon - \beta^2)\sin{(ax)}\, e^{-j\beta z} \qquad E_y = 0$$

(where C, a, and β are constants). What is the corresponding charge density ρ?

❚ By Gauss' law,

$$\rho = \nabla \cdot \mathbf{D} = \frac{\partial D_x}{\partial x} + \frac{\partial D_y}{\partial y} + \frac{\partial D_z}{\partial z} = \epsilon\left(\frac{\partial E_x}{\partial x} + \frac{\partial E_z}{\partial z}\right)$$
$$= j\beta\epsilon C[a^2 - (\mu\epsilon\omega^2 - \beta^2)]\sin{(ax)}\, e^{-j\beta z}$$

10.12 Determine the corresponding **H**-field for the **E**-field given in Problem 10.11.

❚ For sinusoidal time variation, $\nabla \times \mathbf{E} = -j\omega\mu\mathbf{H}$; consequently, $H_x = H_z = 0$ and
$$-j\omega\mu H_y = -\beta^2 aC \cos{(ax)}\, e^{-j\beta z} - C(\mu\epsilon\omega^2 - \beta^2)a \cos{(ax)}\, e^{-j\beta z}$$

or $$H_y = \frac{aC}{j\omega\mu}\cos{(ax)}\, e^{-j\beta z}(\beta^2 + \mu\epsilon\omega^2 - \beta^2) = \frac{a\omega C\epsilon}{j}\cos{(ax)}\, e^{-j\beta z}$$

10.13 For a coaxial cylindrical capacitor of radii a and b and length L, evaluate the total displacement current flowing across a cylindrical surface of radius $a < r < b$, taking the voltage variation as sinusoidal in time, and the variation of electric field with radius the same as in statics. Show that the result is independent of r and equal to the charging current for the capacitor. The capacitance of the coaxial cylinder is given by $C = 2\pi\epsilon L/[\ln{(b/a)}]$.

❚ By Gauss' law, from Fig. 10-1, we have $\epsilon E_r 2\pi r = q$ (C/m), so that
$$\int_a^b E_r\, dr = \frac{q}{2\pi\epsilon}\int_a^b \frac{dr}{r} = \frac{q\ln{(b/a)}}{2\pi\epsilon} = V_0 \sin \omega t$$

Thus

$$E_r = \frac{V_0 \sin \omega t}{r \ln{(b/a)}} \quad \text{and} \quad i_D = \epsilon\frac{\partial E_r}{\partial t} = \frac{\epsilon V_0 \omega \cos \omega t}{r \ln{(b/a)}}$$

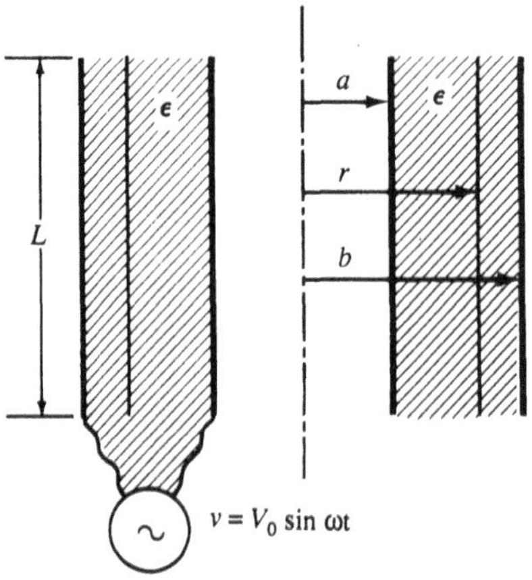

Fig. 10-1

Total displacement current:

$$I_D = i_D(2\pi rL) = \frac{2\pi\epsilon LV_0\omega}{\ln(b/a)}\cos\omega t \qquad (1)$$

Total conduction current:

$$I_c = C\frac{dv}{dt} = \frac{2\pi\epsilon L}{\ln(b/a)}V_0\omega\cos\omega t \qquad (2)$$

Notice that (1) and (2) are identical.

10.14 A capacitor formed by two parallel circular plates has an essentially uniform axial electric field produced by a voltage $V_0\sin\omega t$ across the plates. Utilize the symmetry to find the magnetic field at radius r between the plates. Assume that the plates are separated by a distance d and that ϵ is the permittivity of the dielectric.

▌ Since the electric field is

$$E(t) = \frac{V_0}{d}\sin\omega t = \frac{D(t)}{\epsilon}$$

we have

$$i_D = \frac{\partial D}{\partial t} = \frac{\epsilon V_0\omega}{d}\cos\omega t \qquad \text{and} \qquad I_D = i_D\pi r^2 = \pi r^2\frac{\epsilon V_0}{d}\omega\cos\omega t$$

By Ampère's law, at a radius r,

$$2\pi rH_\phi = \frac{\pi r^2\epsilon V_0}{d}\omega\cos\omega t \qquad \text{or} \qquad H_\phi = \frac{r\epsilon V_0}{2d}\omega\cos\omega t$$

10.15 Given $\mathbf{A} = 2\mathbf{a}_x + 3\mathbf{a}_y$ and $\mathbf{B} = 3\mathbf{a}_x - 4\mathbf{a}_y$, determine the (smaller) angle between $\mathbf{A}$ and $\mathbf{B}$.

▌

$$\cos\theta = \frac{\mathbf{A}\cdot\mathbf{B}}{|\mathbf{A}|\,|\mathbf{B}|} = \frac{-6}{\sqrt{13}\sqrt{25}} = -0.3328$$

or $\theta = 109.44°$. (The other possibility, $\theta = -109.44°$, represents the same angular separation.)

10.16 Find a unit vector perpendicular to the plane containing vectors $\mathbf{A}$ and $\mathbf{B}$ of Problem 10.15.

▌ $\pm\mathbf{a}_z$ (perpendicular to the xy plane).

10.17 A certain scalar field is given by $V = (-Q\cos\theta)/r^2$ $(r \neq 0)$, where Q is a constant. Find the gradient of this field. At what value of θ are the r- and θ-components of this vector field equal?

▌ In spherical coordinates (with no ϕ-dependence) the gradient is given by

$$\nabla V = \mathbf{a}_r\frac{\partial V}{\partial r} + \mathbf{a}_\theta\frac{1}{r}\frac{\partial V}{\partial\theta} = \frac{2Q}{r^3}\cos\theta\,\mathbf{a}_r + \frac{Q}{r^3}\sin\theta\,\mathbf{a}_\theta$$

For the two components to be equal, we must have

$$2\cos\theta = \sin\theta \qquad \text{or} \qquad \tan\theta = 2 \qquad \text{or} \qquad \theta = 63.43°$$

10.18 Water flows in the z-direction through a cylindrical pipe of radius R; the flow vector over the cross section is

$$\mathbf{F} = \frac{R-r}{1+r}\mathbf{a}_z$$

Show that the amount of water entering any arbitrary closed surface is the same as the amount leaving the surface.

▮ For the given **F**, div **F** = 0. Consequently, by the divergence theorem,

$$\iint\limits_{\text{closed surface}} \mathbf{F} \cdot d\mathbf{S} = 0 \qquad [\text{flow in} = \text{flow out}]$$

10.19 In Problem 10.18, find curl **F** at the axis and at the inner surface of the pipe.

▮ In cylindrical coordinates,

$$\text{curl } \mathbf{F} = -\frac{\partial F_z}{\partial r}\mathbf{a}_\phi = \frac{R+1}{(1+r)^2}\mathbf{a}_\phi = \begin{cases} (R+1)\mathbf{a}_\phi & r = 0 \\ \dfrac{1}{R+1}\mathbf{a}_\phi & r = R \end{cases}$$

10.20 Find the **H**-field at point P due to the current I flowing in the bent conductor shown in Fig. 10-2(a).

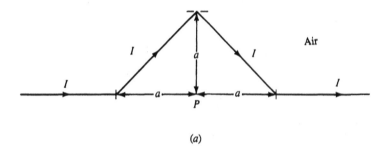

(a)

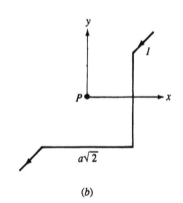

(b)   **Fig. 10-2**

▮ Redraw the configuration as in Fig. 10-2(b). From the result of Problem 3.8, with $L = a\sqrt{2}$, and for four half-sides, we obtain

$$\mathbf{B} = \mu_0\mathbf{H} = \frac{\mu_0\sqrt{2}\,I}{\pi(\sqrt{2}\,a)}(-\mathbf{a}_z) \qquad \text{or} \qquad \mathbf{H} = -\frac{I}{\pi a}\mathbf{a}_z$$

10.21 The interface shown in Fig. 10-3 has no sources. The magnetic flux density in region 1, at point P, is $\mathbf{B}_1 = 0.8\mathbf{a}_x + 0.6\mathbf{a}_y$ T. Find the flux density in region 2, at P.

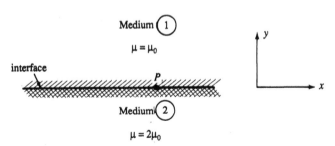

Fig. 10-3

▮ Since $B_{n1} = B_{n2}$, we have $B_{y1} = B_{y2} = 0.6$ T. Similarly, $H_{t1} = H_{t2}$ implies that

$$\frac{B_{x1}}{\mu_0} = \frac{B_{x2}}{2\mu_0} \quad \text{or} \quad B_{x1} = \frac{1}{2}B_{x2} = 0.8 \text{ T}$$

Thus $B_{x2} = 1.6$ T, and $\mathbf{B}_2 = 1.6\mathbf{a}_x + 0.6\mathbf{a}_y$ T.

10.22 A magnetic circuit is 25 cm long and has a 10-cm² uniform cross section. Assuming that the circuit is excited by a 500-turn coil carrying 10 A in current, calculate the flux in the circuit. The relative permeability of the circuit is 500.

▮

$$\psi_m = \frac{NI}{R} = \frac{NI}{L/\mu A} = \frac{\mu ANI}{L}$$

$$= \frac{(500 \times 4\pi \times 10^{-7})(10 \times 10^{-4})(500)(10)}{25 \times 10^{-2}} = 12.566 \text{ mWb}$$

10.23 An infinite slab of dielectric, of thickness t, has a uniform volume charge density ρ (C/m³). Find the magnitude of the electric field (*a*) inside and (*b*) outside the slab. From symmetry the field is purely x-directed (Fig. 10-4).

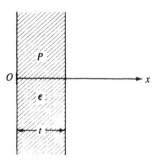

Fig. 10-4

▮ (*a*) By Gauss' law,

$$\text{div } \mathbf{D} = \frac{\partial D_x}{\partial x} = \rho \quad \text{or} \quad D_x = \rho x \quad \text{or} \quad E_x = \frac{\rho x}{\epsilon}$$

(*b*) $\iint \mathbf{D} \cdot d\mathbf{S} = Q$. Thus, $2D_x \times$ area $= \rho \times$ area $\times t$, or $E_x = \rho t / 2\epsilon$.

10.24 An electric field impinges at 45° on the interface of two perfect dielectrics, exiting at 60°. Determine the ratio of the permittivities of the dielectrics.

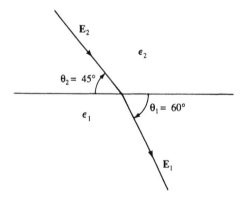

Fig. 10-5

▮ Since $E_{t1} = E_{t2}$ and $D_{n1} = D_{n2}$, we obtain $\epsilon_1 \tan \theta_1 = \epsilon_2 \tan \theta_2$, where θ_1 and θ_2 are as shown in Fig. 10-5. Hence,

$$\frac{\epsilon_1}{\epsilon_2} = \frac{\tan 45°}{\tan 60°} = \frac{1}{\sqrt{3}}$$

10.25 Given: $\hat{\mathbf{H}} = 10e^{j4\pi/5}\mathbf{a}_z$ at 10 MHz. Express the field in the time domain.

▮ With $\omega = 2\pi \times 10^7$ rad/s,

$$\mathbf{H} = \text{Re}\left[10e^{j4\pi/5}e^{j\omega t}\right]\mathbf{a}_z = 10 \text{ Re}\left[e^{j(\omega t + 4\pi/5)}\right]\mathbf{a}_z = 10\cos\left(2\pi \times 10^7 t + \frac{4\pi}{5}\right)\mathbf{a}_z$$

10.26 Given $\hat{\rho} = j4e^{-j\pi/3}$ at 10 MHz. Obtain an expression for ρ in the time domain.

▌ With $\omega = 2\pi \times 10^7$ rad/s,
$$\rho = \mathrm{Re}\,[j4e^{-j\pi/3}e^{j\omega t}] = 4\,\mathrm{Re}\,[e^{j(\omega t - \pi/3 + \pi/2)}] = 4\cos(2\pi \times 10^7 t + \pi/6)$$

10.27 Express as complex exponentials:
$$\mathbf{E} = 10e^{-200x}\cos(\omega t - 200x)\,\mathbf{a}_z \qquad \mathbf{H} = -\tfrac{1}{2}e^{-200x}\cos\left(\omega t - 200x - \frac{\pi}{4}\right)\mathbf{a}_y$$

▌
$$\hat{\mathbf{E}} = 10e^{-200x}e^{-j200x}\mathbf{a}_z \qquad \hat{\mathbf{H}} = -\tfrac{1}{2}e^{-200x}e^{-j(200x + \pi/4)}\mathbf{a}_y$$

10.28 Rework Problem 4.54 using phasors.

▌ The phasor fields $\hat{\mathbf{E}} = E_m \sin \alpha x\, e^{-j\beta z}\,\mathbf{a}_y$ and
$$\hat{\mathbf{H}} = -\frac{\beta}{\omega\mu_0}E_m\sin\alpha x\,e^{-j\beta z}\,\mathbf{a}_x - \frac{\alpha}{j\omega\mu_0}E_m\cos\alpha x\,e^{-j\beta z}\,\mathbf{a}_z$$
must satisfy Ampère's law $\nabla \times \hat{\mathbf{H}} = j\omega\epsilon_0\hat{\mathbf{E}}$; thus
$$\left(\frac{j\beta^2}{\omega\mu_0}E_m\sin\alpha x\,e^{-j\beta z} - \frac{\alpha^2}{j\omega\mu_0}E_m\sin\alpha x\,e^{-j\beta z}\right)\mathbf{a}_y = j\omega\epsilon_0 E_m\sin\alpha x\,e^{-j\beta z}\,\mathbf{a}_y$$
or $\alpha^2 + \beta^2 = \mu_0\epsilon_0\omega^2$, as before.

10.29 The magnetic field due to a pole of a magnet may be assumed to be approximately spherically symmetrical, of the form $B_r = c/r^2$, where c is a constant. A loop of wire is located at a distance x from the pole, as shown in Fig. 10-6. Determine the magnetic flux through the loop, assuming that its radius is b.

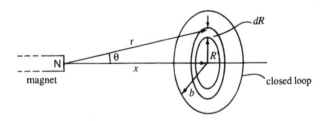

Fig. 10-6

▌
$$\psi_m = \int \mathbf{B}\cdot d\mathbf{S} = \int_{R=0}^{b} B_r\cos\theta\,(2\pi R\,dR)$$
$$= \int_0^b \frac{c}{r^2}\frac{x}{r}(2\pi R\,dR) = 2\pi cx\int_0^b \frac{R\,dR}{\sqrt{(x^2 + R^2)^3}}$$
$$= 2\pi cx\left[\frac{-1}{\sqrt{x^2 + R^2}}\right]_0^b = 2\pi c\left(1 - \frac{x}{\sqrt{x^2 + b^2}}\right)$$

10.30 Write Maxwell's equations for electromagnetic fields $\mathbf{E}$ and $\mathbf{H}$ varying sinusoidally in time.

▌ Equations (1) through (4) of Problem 4.41 become
$$\mathrm{curl}\,\mathbf{E} = -j\omega\mu\mathbf{H} \qquad (1)$$
$$\mathrm{curl}\,\mathbf{H} = \sigma\mathbf{E} + j\omega\epsilon\mathbf{E} \qquad (2)$$
$$\mathrm{div}\,\mathbf{E} = \rho/\epsilon \qquad (3)$$
$$\mathrm{div}\,\mathbf{H} = 0 \qquad (4)$$

10.31 A perfectly conducting sphere of radius R in free space has a charge Q uniformly distributed over its surface. Determine the electric field at the surface of the sphere.

▌ By Problem 2.50, $\mathbf{E} = (Q/4\pi\epsilon_0 R^2)\mathbf{a}_r$.

10.32 A straight conductor of length L moves with velocity $\mathbf{u}$ in a uniform magnetic field $\mathbf{B}$. Obtain an expression for the potential difference between the two ends.

▌ The emf in the moving conductor is induced by charge separation, which gives rise to an $\mathbf{E}$-field. Under equilibrium, the net force on the charge carriers is zero:
$$-e(\mathbf{E} + \mathbf{u}\times\mathbf{B}) = 0 \qquad \text{or} \qquad \mathbf{E} = -\mathbf{u}\times\mathbf{B}$$

The potential difference between the two ends of the conductor is given by

$$V = -\int_0^L \mathbf{E} \cdot d\mathbf{s} = \int_0^L (\mathbf{u} \times \mathbf{B}) \cdot d\mathbf{s} = (\mathbf{u} \times \mathbf{B}) \cdot \int_0^L d\mathbf{s} = (\mathbf{u} \times \mathbf{B}) \cdot \mathbf{L} = \mathbf{B} \cdot (\mathbf{L} \times \mathbf{u})$$

where $\mathbf{L}$ is the vector from the negative to the positive end.

10.33 Two current-carrying loops, C_i and C_j, are shown in Fig. 10-7. Defining the mutual inductance

$$M_{ij} = \frac{\mu_0}{4\pi} \oint_{C_i} \oint_{C_j} \frac{d\mathbf{s}_i \cdot d\mathbf{s}_j}{|\mathbf{r}_i - \mathbf{r}_j|} \qquad (1)$$

show that the flux linking loop C_i due to a current I_j in loop C_j is given by $\psi_{mij} = M_{ij}I_j$.

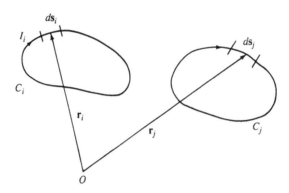

Fig. 10-7

❙ The flux of $\mathbf{B}_j$ through the loop C_i is

$$\psi_{mij} = \iint \mathbf{B}_j(\mathbf{r}_i) \cdot d\mathbf{S}_i \qquad (2)$$

where $\mathbf{r}_i$ is the vector integration variable, and the integration is extended over the area enclosed by C_i. Now, since $\mathbf{B} = \text{curl } \mathbf{A}$, (2) may be written as

$$\psi_{mij} = \int \text{curl}_j \mathbf{A}_j(\mathbf{r}_i) \cdot d\mathbf{S}_i = \oint_{C_i} \mathbf{A}_j(\mathbf{r}_i) \cdot d\mathbf{s}_i \qquad (3)$$

where Stokes' theorem was used. But the magnetic vector potential $\mathbf{A}_j$ at the location $\mathbf{r}_i$ is given by the Poisson integral

$$\mathbf{A}_j(\mathbf{r}_i) = \frac{\mu_0 I_j}{4\pi} \oint_{C_j} \frac{d\mathbf{s}_j}{|\mathbf{r}_i - \mathbf{r}_j|} \qquad (4)$$

Together, (3) and (4) yield

$$\psi_{mij} = \frac{\mu_0 I_j}{4\pi} \oint_{C_i} \oint_{C_j} \frac{d\mathbf{s}_i \cdot d\mathbf{s}_j}{|\mathbf{r}_i - \mathbf{r}_j|} = M_{ij}I_j$$

10.34 Show that the H-field within a conducting material medium satisfies a diffusion equation,

$$\nabla^2 u = a^2 \frac{\partial u}{\partial t}$$

❙ From Maxwell's equation, $\nabla \times \mathbf{H} = \mathbf{J} = \sigma\mathbf{E}$, since displacement current is negligible within a conductor. Taking the curl of both sides of this equation yields

$$\nabla \times \nabla \times \mathbf{H} \equiv \nabla(\nabla \cdot \mathbf{H}) - \nabla^2\mathbf{H} = -\nabla^2\mathbf{H} = \sigma\nabla \times \mathbf{E}$$

since $\nabla \cdot \mathbf{H} = 0$. But

$$\nabla \times \mathbf{E} = -\frac{\partial \mathbf{B}}{\partial t} = -\mu\frac{\partial \mathbf{H}}{\partial t}$$

Hence

$$\nabla^2\mathbf{H} = \mu\sigma\frac{\partial \mathbf{H}}{\partial t}$$

10.35 Graphite has a conductivity of 0.12 S/m and a relative permittivity of 5. Determine the frequency range over which graphite may be considered a good conductor.

❙ For a good conductor, the displacement current must be much less than the conduction current; i.e., $\omega \ll \sigma/\epsilon$. Substituting the given numerical values gives

$$2\pi f = \omega \ll \frac{0.12}{5 \times (10^{-9}/36\pi)} \qquad \text{or} \qquad f \ll 432\,\text{MHz}$$

10.36 Give the most general expression for the emf set up in a conductor which is moving through a time-varying A-field.

$$\mathbf{E} = -\frac{\partial \mathbf{A}}{\partial t} + (\mathbf{u} \times \text{curl } \mathbf{A}) - \nabla V$$

10.37 The conducting bar of length L shown in Fig. 10-8 oscillates across the rails with the velocity $\mathbf{u} = U_m \cos \omega t \, \mathbf{a}_x$ in a magnetic field $\mathbf{B} = B_m \cos \omega t \, \mathbf{a}_z$. Determine the instantaneous induced emf in the loop.

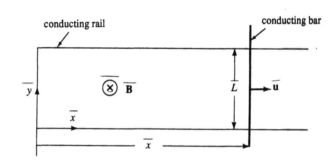

Fig. 10-8

Let $x = x_0$ at $t = 0$; then

$$x(t) = x_0 + \int_0^t U_m \cos \omega t \, dt = x_0 + \frac{U_m}{\omega} \sin \omega t$$

and the induced emf is given by

$$\begin{aligned}
\text{emf} &= -\frac{d}{dt}[B(t)Lx(t)] \\
&= -L\frac{d}{dt}\left[(B_m \cos \omega t)\left(x_0 + \frac{U_m}{\omega} \sin \omega t\right)\right] \\
&= -LB_m U_m \cos 2\omega t + Lx_0 B_m \omega \sin \omega t
\end{aligned}$$

10.38 If the electric field of a plane wave is given by

$$\mathbf{E} = 10 \cos\left(3\pi \times 10^8 t - \frac{4\pi}{3} z\right) \mathbf{a}_x + 5 \cos\left(3\pi \times 10^8 t - \frac{4\pi}{3} z\right) \mathbf{a}_y \quad \text{(V/m)}$$

write an expression for the associated magnetic field. The medium is lossless, with $\mu_r = 1$.

From $\omega\mu = \eta\beta$,

$$\eta = \frac{\omega\mu}{\beta} = \frac{(3\pi \times 10^8)(4\pi \times 10^{-7})}{4\pi/3} = 90\pi \ \Omega$$

Hence

$$\mathbf{H} = \frac{10}{90\pi} \cos\left(3\pi \times 10^8 t - \frac{4\pi}{3} z\right) \mathbf{a}_y - \frac{5}{90\pi} \cos\left(3\pi \times 10^8 t - \frac{4\pi}{3} z\right) \mathbf{a}_x \quad \text{(A/m)}$$

10.39 What can be said about the polarization of the wave of Problem 10.38?

Because $E_y/E_x = 5/10$ for all t, the wave is plane polarized.

10.40 A plane wave traveling in a lossless medium characterized by (ϵ_1, μ_1) is normally incident on a sheet of lossless material characterized by (ϵ_2, μ_2), of thickness d. Show that if $d = n\lambda_2/2$ $(n = 1, 2, \ldots)$, then the incident wave is not attenuated by the sheet.

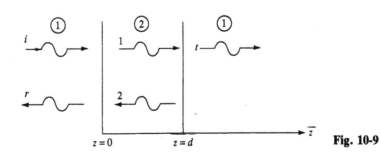

Fig. 10-9

▮ The incident, reflected and transmitted waves are shown in Fig. 10-9. The interface conditions read:

$$E_i + E_r = E_1 + E_2$$

$$\frac{E_i}{\eta_1} - \frac{E_r}{\eta_1} = \frac{E_1}{\eta_2} - \frac{E_2}{\eta_2}$$

$$E_1 e^{-jn\pi} + E_2 e^{jn\pi} = E_t e^{-j\beta_1 d}$$

$$\frac{E_1}{\eta_2} e^{-jn\pi} - \frac{E_2}{\eta_2} e^{jn\pi} = \frac{E_t}{\eta_1} e^{-j\beta_1 d}$$

Writing the four equations in matrix form yields

$$\begin{bmatrix} -1 & 1 & 1 & 0 \\ 1 & \eta_1/\eta_2 & -\eta_1/\eta_2 & 0 \\ 0 & -e^{-jn\pi} & -e^{jn\pi} & e^{-j\beta_1 d} \\ 0 & -e^{-jn\pi} & e^{jn\pi} & \frac{\eta_2}{\eta_1} e^{-j\beta_1 d} \end{bmatrix} \begin{bmatrix} E_r \\ E_1 \\ E_2 \\ E_t \end{bmatrix} = \begin{bmatrix} E_i \\ E_i \\ 0 \\ 0 \end{bmatrix}$$

Solving for E_t yields

$$E_t = \frac{4E_i}{(4\cos n\pi)e^{-j\beta_1 d}} \qquad \text{whence} \qquad |E_t| = |E_i|$$

10.41 A plane wave is incident normal to the surface of seawater ($\mu_r = 1$, $\epsilon_r = 79$, $\sigma = 3$ S/m). The electric field is parallel to the surface and is 1 V/m just inside the water. At what depth would it be possible for a submarine to receive a signal, if the receiver required a field intensity of 10 μV/m at 20 kHz?

▮ At 20 kHz,

$$\frac{\sigma}{\omega\epsilon} = \frac{3 \times 36\pi \times 10^9}{2\pi \times 10^3 \times 20 \times 79} \gg 1$$

Thus we may neglect $\partial D/\partial t$ and use the skin-depth approach:

$$E(x) = E_0 e^{-x/\delta} \qquad \text{or} \qquad x = \delta \ln\frac{E_0}{E} = 11.5\,\delta$$

But $\delta = 1/\sqrt{\pi f \mu \sigma} = 2.06$. Thus, $x = 23.7$ m.

10.42 Repeat Problem 10.41 for a frequency of 20 GHz.

▮ At 20 GHz: $\sigma/\omega\epsilon = 108/3160 \ll 1$, which implies an imperfect dielectric. The attenuation constant is now

$$\alpha \approx \frac{\sigma}{2}\sqrt{\frac{\mu}{\epsilon}} \approx 63.6\ \text{m}^{-1}$$

and

$$E = E_0 e^{-\alpha x} \qquad \text{or} \qquad x = \frac{1}{\alpha}\ln\frac{E_0}{E} \approx 0.181\ \text{m}$$

10.43 A traveling wave is given by $E_y = 10\sin(\beta z - \omega t)$. Sketch the wave at $t = 0$ and at $t = t_1$, when it has advanced $\lambda/8$, if the velocity is 3×10^8 m/s and the angular frequency $\omega = 10^6$ rad/s.

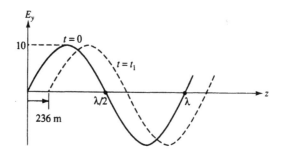

Fig. 10-10

▮

$$\frac{\lambda}{8} = \frac{1}{8}\left(\frac{3 \times 10^8}{10^6/2\pi}\right) = 236\ \text{m}$$

See Fig. 10-10.

10.44 In free space, $\mathbf{E}(z, t) = 1200\sin(\omega t - \beta z)\,\mathbf{a}_y$ (V/m). Obtain $\mathbf{H}(z, t)$.

∎ Notice that the direction of propagation is $+z$. Since $\mathbf{E} \times \mathbf{H}$ must be in $+z$, $\mathbf{H}$ must have the direction $-\mathbf{a}_x$. Consequently, in free space,

$$\mathbf{H}(z,\,t) = \frac{E_y}{120\pi}(-\mathbf{a}_x) = -3.18 \sin(\omega t - \beta z)\,\mathbf{a}_x \quad (\text{A/m})$$

10.45 Calculate the propagation constant for the wave of Problem 10.44, for operating frequency 100 MHz.

∎ Since $\sigma = 0$, the propagation constant is given by

$$\gamma = j\omega\sqrt{\mu_0\epsilon_0} = \frac{j2\pi f}{u_0} = \frac{j2\pi \times 100 \times 10^6}{3 \times 10^8} = j2.09 \text{ m}^{-1}$$

10.46 Show that the field given by $\mathbf{E}(z,\,t) = 10 \sin(\omega t + \beta z)\,\mathbf{a}_x + 10 \cos(\omega t + \beta z)\,\mathbf{a}_y$ is circularly polarized.

∎ In a fixed xy plane—which may always be taken as $z = 0$—$\hat{\mathbf{E}} = 10\,\underline{/90° - \omega t}$; i.e., the tip of the $\mathbf{E}$-vector traces a circle of radius 10.

10.47 A backward-traveling (in the $-z$-direction) $\mathbf{H}$-field, having an amplitude of 0.1 A/m and a wavelength of 0.05 m, is y-directed. Assuming that the wave exists in free space, write an expression for it in the time domain.

∎ The form must be: $\mathbf{H}(z,\,t) = H_m \cos(\omega t + \beta z)\,\mathbf{a}_y$. Substitute $\beta = 2\pi/\lambda = 40\pi\,\text{rad/m}$, $\omega = u_0\beta = 1.2\pi \times 10^{10}\,\text{rad/s}$, and $H_m = 0.1$ A/m to find

$$\mathbf{H}(z,\,t) = 0.1 \cos(1.2\pi \times 10^{10}t + 40\pi z)\,\mathbf{a}_y \quad (\text{A/m})$$

10.48 Assuming that the wave of Problem 10.47 is a plane wave, write an expression for the $\mathbf{E}$-field.

∎ For propagation along $-\mathbf{a}_z$, $\mathbf{E}$ must be along $-\mathbf{a}_x$:
$$\mathbf{E} = \eta_0 H_y(-\mathbf{a}_x) = -12\pi \cos(1.2\pi \times 10^{10}t + 40\pi z)\,\mathbf{a}_x \quad (\text{V/m})$$

10.49 A material has $\mu_r = 1$, $\epsilon_r = 8$, and $\sigma = 0.01$ S/m. Calculate the propagation constant for the material at 900 MHz.

∎ From the data,

$$\frac{\sigma}{\omega\epsilon} = \frac{0.01}{2\pi \times 900 \times 10^6 \times 8 \times 10^{-9}/36\pi} = 0.025 \ll 1$$

Thus $\alpha \approx 0$ and

$$\beta \approx \omega\sqrt{\mu\epsilon} = 2\pi \times 900 \times 10^6 \frac{\sqrt{1 \times 8}}{3 \times 10^8} = 53.3 \text{ rad/m}$$

Consequently, $\gamma = \alpha + j\beta = j53.3 \text{ m}^{-1}$.

10.50 Determine the conversion factor between the neper (Np) and the decibel (dB).

∎ Consider a plane wave traveling in the $+z$-direction, whose amplitude decays according to $E = E_0 e^{-\alpha z}$. The power carried by the wave is proportional to E^2, so that $P = P_0 e^{-2\alpha z}$. Now, by definition of the decibel, the power drop over the distance z is $10 \log_{10}(P_0/P)$ dB. But

$$10 \log_{10}\frac{P_0}{P} = \frac{10}{2.3026}\ln\frac{P_0}{P} = \frac{20}{2.3026}(\alpha z) = 8.686(\alpha z)$$

Thus, αz nepers is equivalent to $8.686(\alpha z)$ decibels; i.e., 1 Np = 8.686 dB.

10.51 For a certain material, $\mu_r = 1$, $\sigma = 5 \times 10^{-3}$ S/m, and $\epsilon_r = 8$. At what frequencies may the material be considered a perfect dielectric?

∎ We assume arbitrarily that

$$\frac{\sigma}{\omega\epsilon} \le \frac{1}{100}$$

marks the cutoff. Then

$$f = \frac{\omega}{2\pi} \ge \frac{100\sigma}{2\pi\epsilon} = 1.13 \text{ GHz}$$

10.52 Show that the attenuation constant, α, in the material of Problem 10.51 is approximately independent of frequency.

▌ For $\sigma/\omega\epsilon \leq 1/100$,

$$\alpha = \omega\sqrt{\frac{\mu\epsilon}{2}\left(\sqrt{1+\left(\frac{\sigma}{\omega\epsilon}\right)^2}-1\right)} \approx \omega\sqrt{\frac{\mu\epsilon}{2}\left[\frac{1}{2}\left(\frac{\sigma}{\omega\epsilon}\right)^2\right]} = \frac{\sigma}{2}\sqrt{\frac{\mu}{\epsilon}}$$

10.53 Find the skin depth δ at a frequency of 1.6 MHz in aluminum, for which $\sigma = 38.2\,\text{MS/m}$ and $\mu_r = 1$.

▌
$$\delta = \frac{1}{\sqrt{\pi f \mu \sigma}} = 6.44 \times 10^{-5}\,\text{m} = 64.4\,\mu\text{m}$$

10.54 Calculate the intrinsic impedance η and the propagation constant γ for a conducting medium in which $\sigma = 58\,\text{MS/m}$ and $\mu_r = 1$, at a frequency $f = 100\,\text{MHz}$.

▌
$$\gamma = \sqrt{\omega\mu\sigma}\,\underline{/45°} = 2.14 \times 10^5\,\underline{/45°}\;\;\text{m}^{-1} \qquad \eta = \sqrt{\frac{\omega\mu}{\sigma}}\,\underline{/45°} = 3.69 \times 10^{-3}\,\underline{/45°}\;\;\Omega$$

10.55 A plane wave traveling in the $+z$-direction in free space ($z < 0$) is normally incident at $z = 0$ on a conductor ($z > 0$) for which $\sigma = 61.7\,\text{MS/m}$, $\mu_r = 1$. The free-space **E**-wave has a frequency $f = 1.5\,\text{MHz}$ and an amplitude 1.0 V/m; at the interface it is given by

$$\mathbf{E}(0, t) = 1.0\sin 2\pi ft\,\mathbf{a}_y \quad (\text{V/m})$$

Find $\mathbf{H}(z, t)$ for $z > 0$.

▌ For $z > 0$, and in complex form,

$$\mathbf{E}(z, t) = 1.0e^{-\alpha z}e^{j(2\pi ft - \beta z)}\mathbf{a}_y \quad (\text{V/m})$$

In the conductor,

$$\alpha = \beta = \sqrt{\pi f \mu \sigma} = \sqrt{\pi(1.5 \times 10^6)(4\pi \times 10^{-7})(61.7 \times 10^6)} = 1.91 \times 10^4\,\text{m}^{-1}$$

$$\eta = \sqrt{\frac{\omega\mu}{\sigma}}\,\underline{/45°} = 4.38 \times 10^{-4}e^{j\pi/4}\;\Omega$$

Then, since $E_y/(-H_x) = \eta$,

$$\mathbf{H}(z, t) = -2.28 \times 10^3 e^{-\alpha z}e^{j(2\pi ft - \beta z - \pi/4)}\mathbf{a}_x \rightarrow -2.28 \times 10^3 e^{-\alpha z}\sin(2\pi ft - \beta z - \pi/4)\,\mathbf{a}_x \quad (\text{A/m})$$

where f, α, and β are as given above.

10.56 In free space, $\mathbf{E}(z, t) = 50\cos(\omega t - \beta z)\,\mathbf{a}_x$ (V/m). Find the average power crossing a circular area of radius 2.5 m in the plane $z = \text{const.}$

▌
$$P_{\text{av}} = S_{\text{av}} \times \text{area} = \frac{E_{\text{max}}^2}{2\eta_0}(\pi r^2) = \frac{50^2}{2(120\pi)}\pi(2.5)^2 = 65.1\,\text{W}$$

10.57 A wave from free space strikes a coated dielectric region, as shown in Fig. 10-11. The coating material has $\mu_1 = \mu_0$ and $\epsilon_1 = 4\epsilon_0$, and is a quarter-wavelength in thickness relative to the velocity of propagation in that material. The coated material has constants $\mu_2 = \mu_0$ and $\epsilon_2 = \epsilon_2'\epsilon_0$, and may be considered infinite in extent. It is observed that in the space in front of the system, 1 percent of the incident plane-wave energy is reflected, and that an electric field maximum occurs at the surface of reflection. Determine ϵ_2'.

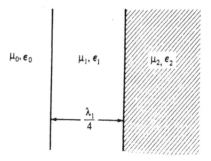

Fig. 10-11

▌ Defining the ratio of the powers by

$$\frac{P_z^-}{P_z^+} = |\rho|^2 = 0.01 \qquad \text{or} \qquad |\rho| = 0.1$$

we have

$$\rho = \frac{Z_{L1} - \eta_0}{Z_{L1} + \eta_0} \qquad Z_{L1} = \eta_1\left[\frac{\eta_2\cos(\beta\lambda/4) + j\eta_1\beta\sin(\lambda/4)}{\eta_1\cos(\beta\lambda/4) + j\eta_2\beta\sin(\lambda/4)}\right] = \eta_1\frac{j\eta_1}{j\eta_2} = \frac{\eta_1^2}{\eta_2} \quad (\text{real})$$

Thus, $\rho = 0.1 \underline{/0°}$; $\eta_1 = \sqrt{\mu_0/4\epsilon_0} = 2\eta_0$. Now,

$$0.1 = \frac{\eta_1^2/\eta_2 - \eta_0}{\eta_1^2/\eta_2 + \eta_0} = \frac{\eta_0 - 4\eta_2}{\eta_0 + 4\eta_2}$$

from which $44\eta_2 = 9\eta_0$, or

$$44\sqrt{\mu_0/(\epsilon_0\epsilon_2')} = 9\sqrt{\mu_0/\epsilon_0} \qquad \text{or} \qquad \sqrt{\epsilon_2'} = \frac{44}{9} \qquad \text{or} \qquad \epsilon_2' \approx 24$$

10.58 The electric field for a linearly polarized wave in a nonmagnetic dielectric is given by
$$\mathbf{E} = 150e^{j(\omega t - \beta z)}\mathbf{a_x} \quad \text{(V/m)}$$
For this material the wave impedance is $\eta = 100\,\Omega$. What is the maximum amplitude of the magnetic field vector, and what is the direction?

▌ $H_{max} = 150/100 = 1.5$ A/m; $\mathbf{H}$ in $+y$-direction.

10.59 Calculate the relative permittivity of the material medium of Problem 10.58.

▌ Since $\eta = \eta_0/\sqrt{\epsilon_r}$,

$$\epsilon_r = \left(\frac{\eta_0}{\eta}\right)^2 = \left(\frac{377}{100}\right)^2 = 14.2$$

10.60 Determine the time-average power density in the wave of Problem 10.58.

▌
$$S_{av} = \frac{E_{max}^2}{2\eta} = \frac{150^2}{200} = 112.5 \text{ W/m}^2$$

10.61 A copper conductor, of conductivity 5.7×10^7 S/m, carries current at 60 Hz. Calculate the ratio of conduction current to displacement current.

▌ Since the conduction and displacement currents are, respectively, given by $I_C = \sigma AE$ and $I_D = \epsilon A\omega E$, the ratio is

$$\frac{I_C}{I_D} = \frac{\sigma}{\omega\epsilon} = \frac{5.7 \times 10^7}{2\pi \times 60 \times (10^{-9}/36\pi)} = 1.71 \times 10^{16}$$

10.62 A silver conductor has a circular cross section 4 cm in diameter. Calculate the resistance per meter length of the conductor, at 10 kHz. Conductivity of silver is 6.1×10^7 S/m.

▌ At 10 kHz, we assume that the current is confined to one skin depth δ. Thus the area of cross section seen by the current is $2\pi r\delta$. The resistance per unit length of the conductor becomes

$$R_{ac} = \frac{1}{\sigma A} = \frac{1}{\sigma(2\pi r\delta)} \tag{1}$$

But $\delta = \sqrt{2/\omega\mu\sigma}$, so that (1) becomes

$$R_{ac} = \frac{1}{2r}\sqrt{\frac{\mu f}{\pi\sigma}} \tag{2}$$

Substituting numerical values yields
$$R_{ac} = \frac{1}{2 \times 2 \times 10^{-2}}\sqrt{\frac{4\pi \times 10^{-7} \times 10 \times 10^3}{\pi \times 6.1 \times 10^7}} = 2.02 \times 10^{-4} \text{ }\Omega/\text{m}$$

10.63 Assuming that the conductor of Problem 10.62 transmits power at 100 A (rms), calculate the ratio of the power loss per unit length of the conductor, at 10 kHz, to that at dc.

▌ From Problem 10.62, $(P_{loss})_{ac} = I^2 R_{ac} = (100)^2(2.02 \times 10^{-4}) = 2.02$ W/m. Now,
$$R_{dc} = \frac{1}{\sigma A} = \frac{1}{\sigma(\pi r^2)} = \frac{1}{6.1 \times 10^7 \times \pi(2)^2 \times 10^{-4}} = 1.3 \times 10^{-5} \text{ }\Omega/\text{m}$$
so that $(P_{loss})_{dc} = (100)^2(1.3 \times 10^{-5}) = 0.13$ W/m, and the required ratio is $2.02/0.13 = 15.54$ (independent of the current).

10.64 A meter conductor of square cross section and of conductivity 10^7 S/m is 100 m long and measures 0.5 cm on each side. Calculate the skin depth if the conductor carries current at 160 kHz.

▌
$$\delta = \frac{1}{\sqrt{\pi f\mu\sigma}} = \frac{1}{\sqrt{\pi \times 160 \times 10^3 \times 4\pi \times 10^{-7} \times 10^7}} = 3.98 \times 10^{-4} \text{ m}$$

10.65 For the conductor of Problem 10.64, assuming that almost the entire current is confined to the skin depth, obtain an expression for the ratio of ac resistance to dc resistance.

▌ Let A_{ac} be the effective area of the conductor through which the current flows at 160 kHz. The required ratio is

$$\frac{R_{ac}}{R_{dc}} = \frac{l/\sigma A_{ac}}{l/\sigma A} = \frac{A}{A_{ac}} \tag{1}$$

where A is the actual cross-sectional area of the conductor. Now, if the side of the square is a, $A = a^2$ and $A_{ac} = 4a\delta$. Thus, (1) becomes

$$\frac{R_{ac}}{R_{dc}} = \frac{a}{4\delta} = \frac{0.5}{4 \times 3.98 \times 10^{-2}} = 3.14$$

10.66 An **E**-field, $\mathbf{E} = (E_0\mathbf{a}_x + E_1\mathbf{a}_y)\sin(\omega t - \beta z)$, propagates in free space. Obtain an expression for the Poynting vector.

▌ From Maxwell's equation $\nabla \times \mathbf{E} = -\partial \mathbf{B}/\partial t$,

$$-\frac{\partial E_y}{\partial z} = -\frac{\partial B_x}{\partial t} \quad \text{and} \quad \frac{\partial E_x}{\partial z} = -\frac{\partial B_y}{\partial t}$$

Consequently,

$$B_x = -\beta E_1 \int \cos(\omega t - \beta z)\, dt = -\frac{\beta E_1}{\omega}\sin(\omega t - \beta z)$$

$$B_y = \beta E_0 \int \cos(\omega t - \beta z)\, dt = \frac{\beta E_0}{\omega}\sin(\omega t - \beta z)$$

Thus, the Poynting vector is given by

$$\mathbf{S} = \mathbf{E} \times \mathbf{H} = (E_x H_y - E_y H_x)\mathbf{a}_z$$
$$= \frac{\beta}{\mu_0 \omega}(E_0^2 + E_1^2)\sin^2(\omega t - \beta z)\,\mathbf{a}_z = \frac{\mathbf{E} \cdot \mathbf{E}}{\eta_0}\mathbf{a}_z$$

10.67 The phase velocity in a certain medium is given by

$$u_p = U_0\left(\frac{\omega}{\omega_0}\right)^{1/3} \tag{1}$$

where U_0 and ω_0 are constants. Obtain an expression for the group velocity u_g in terms of the phase velocity.

▌ By definition and from (1),

$$u_p \equiv \frac{\omega}{\beta} = U_0\left(\frac{\omega}{\omega_0}\right)^{1/3}$$

Thus
$$\omega^{2/3} = \beta U_0\left(\frac{1}{\omega_0}\right)^{1/3} \quad \text{or} \quad \frac{2}{3}\omega^{-1/3}\frac{d\omega}{d\beta} = U_0\left(\frac{1}{\omega_0}\right)^{1/3} \tag{2}$$

Since the group velocity $u_g \equiv d\omega/d\beta$, (2) yields

$$u_g = \frac{3}{2}U_0\left(\frac{\omega}{\omega_0}\right)^{1/3} = \frac{3}{2}u_p \tag{3}$$

10.68 Verify (3) of Problem 10.67, using (2) of Problem 5.183.

▌ With

$$u_p = U_0\left(\frac{\omega}{\omega_0}\right)^{1/3} \quad \text{and} \quad \frac{du_p}{d\omega} = \frac{1}{3}U_0\frac{\omega^{-2/3}}{\omega_0^{1/3}}$$

(2) of Problem 5.183 yields

$$u_g = \frac{u_p}{1 - \dfrac{\omega}{u_p}\dfrac{U_0}{3}\left(\dfrac{1}{\omega_0^{1/3}\omega^{2/3}}\right)} = \frac{u_p}{1 - \dfrac{1}{3}\dfrac{U_0}{u_p}\left(\dfrac{\omega}{\omega_0}\right)^{1/3}} = \frac{u_p}{1 - \dfrac{1}{3}} = \frac{3}{2}u_p$$

10.69 Calculate the Brewster angle for a parallel-polarized plane wave incident from air on a glass plate having a relative permittivity $\epsilon_r = 10$.

▌ From (1) of Problem 5.154, $\theta_B = \tan^{-1}\sqrt{10} = 72.45°$.

10.70 The transmission line of Fig. 10-12(a) transmits trapezoidal pulses with rise-fall times T [Fig. 10-12(b)], where T is the one-way transit time of the line. Sketch the source voltage for $0 \le t < 8T$, given $R_L = 2R_C$.

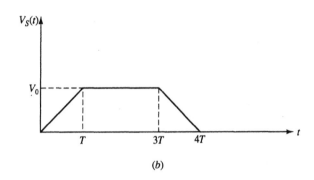

(a)

(b)

(c)

Fig. 10-12

∎ Since $R_L = 2R_C$, the voltage reflection coefficients are $\Gamma_S = 0$ and $\Gamma_L = 1/3$. With the initial voltage $V_1^+ = \frac{1}{2}V_S$, the source voltage is sketched in Fig. 10-12(c).

10.71 Sketch the load voltage for the line of Problem 10.70.

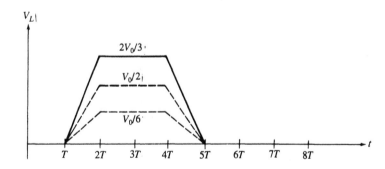

Fig. 10-13

∎ See Fig. 10-13.

10.72 Repeat Problem 10.70 for $R_L = \frac{1}{2}R_C$.

∎ For this case, $\Gamma_S = 0$, $\Gamma_L = -\frac{1}{3}$, and $V_1^+ = \frac{1}{2}V_S$. Consequently we obtain the sketch shown in Fig. 10-14.

10.73 Repeat Problem 10.71 for $R_L = \frac{1}{2}R_C$.

∎ See Fig. 10-15.

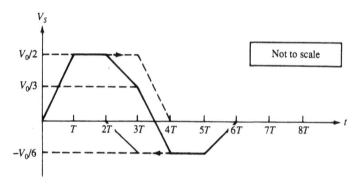

Fig. 10-14

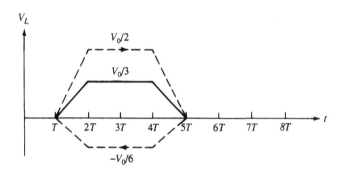

Fig. 10-15

10.74 Sketch the line voltage as a function of position z at $t = T/2$, $3T/2$, and $2T$, for the line of Problem 10.70. The length of the line is ζ.

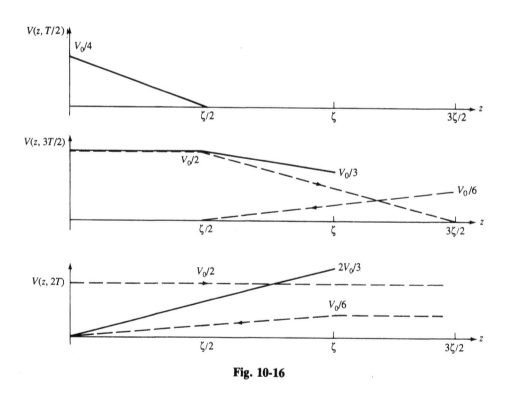

Fig. 10-16

▮ See Fig. 10-16.

10.75 Repeat Problem 10.74 for $R_L = \frac{1}{2}R_C$.

▮ See Fig. 10-17.

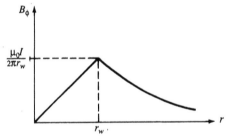

Fig. 10-17

10.76 A round solid conductor of radius r_w carries direct current I. Sketch the resulting **B**-field within and outside the conductor.

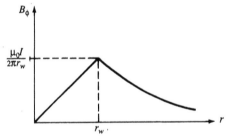

Fig. 10-18

I For dc excitation, the current distribution is uniform over the conductor cross section; hence, in a cylinder of radius $r < r_w$,

$$I_r = \frac{\pi r^2}{\pi r_w^2} I$$

By symmetry, the magnetic field internal to the wire is circumferentially directed. Using Ampère's law, we obtain the magnetic field along a contour of radius $r < r_w$ as

$$H_\phi = \frac{I_r}{2\pi r} \quad \text{or} \quad B_\phi = \mu_0 H_\phi = \frac{\mu_0 r}{2\pi r_w^2} I \tag{1}$$

For $r \geq r_w$,

$$B_\phi = \frac{\mu_0 I}{2\pi r} \tag{2}$$

Equations (1) and (2) are plotted in Fig. 10-18.

10.77 Determine the electrical length $\zeta_e = \zeta/\lambda$ of the following transmission lines: **(a)** $f = 3$ MHz, $\zeta = 130$ m, $\mu_r = 1$, $\epsilon_r = 1$; **(b)** $f = 2.5$ GHz, $\zeta = 5$ cm, $\mu_r = 1$, $\epsilon_r = 2.5$; **(c)** $f = 480$ MHz, $\zeta = 0.5$ m, $l = 5$ μH/m, $c = 20$ pF/m.

I
$$\zeta_e = \frac{\zeta f}{u} = \frac{\zeta f \sqrt{\mu_r \epsilon_r}}{u_0}$$

Thus: **(a)** $\zeta_e = 1.3$; **(b)** $\zeta_e = 0.659$; **(c)** $u = 1/\sqrt{lc} = 1 \times 10^8$ m/s and $\zeta_e = 2.4$.

10.78 Convert the electrical lengths of the lines in Problem 10.77 to degrees.

I Multiply ζ_e-value by 360°: **(a)** 468° = 108°; **(b)** 237.17°; **(c)** 864° = 144°.

10.79 Determine, using the Smith chart, the input impedance, VSWR, and voltage reflection coefficient at the load, for a transmission line having $Z_L = (100 - j80)$ Ω; $R_C = 75$ Ω; $\zeta_e = 0.6$.

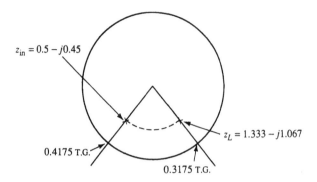

$z_{in} = 0.5 - j0.45$

0.4175 T.G.

0.3175 T.G.

$z_L = 1.333 - j1.067$

Fig. 10-19

▮ Since $z_L = 1.333 - j1.067$, Fig. 10-19 gives

$$z_{in} = 0.5 - j0.45 \qquad Z_{in} = (37.5 - j33.75)\ \Omega \qquad \text{VSWR} = 2.5 \qquad \Gamma_L = 0.45\ \underline{/-120.5°}$$

10.80 Repeat Problem 10.79 for a line with $Z_L = -j375\ \Omega$; $R_C = 300\ \Omega$; $\zeta_e = 0.8$.

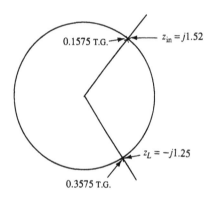

0.1575 T.G.

$z_{in} = j1.52$

$z_L = -j1.25$

0.3575 T.G.

Fig. 10-20

▮ Proceeding as in Problem 10.79, the normalized input impedance is $z_L = -j1.25$, and from Fig. 10-20 we have

$$z_{in} = 1.52 \qquad Z_{in} = j456\ \Omega \qquad \text{VSWR} = \infty \qquad \Gamma_L = 1\ \underline{/-77.5°}$$

10.81 Find the load impedance, VSWR, and load reflection coefficient for a transmission line with $Z_{in} = (150 + j230)\ \Omega$; $R_C = 100\ \Omega$; $\zeta_e = 0.6$. Use the Smith chart.

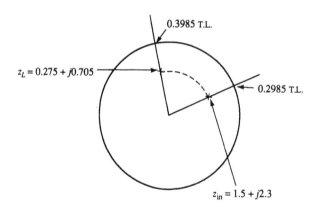

0.3985 T.L.

$z_L = 0.275 + j0.705$

0.2985 T.L.

$z_{in} = 1.5 + j2.3$

Fig. 10-21

▮ The normalized impedance is $z_{in} = 1.5 + j2.3$; then, from Fig. 10-21,

$$z_L = 0.275 + j0.705 \qquad Z_L = (27.5 + j70.5)\ \Omega \qquad \text{VSWR} = 5.5 \qquad \Gamma_L = 0.695\ \underline{/106.5°}$$

10.82 Repeat Problem 10.81 for a line with $Z_{in} = j250\ \Omega$; $R_C = 100\ \Omega$; $\zeta_e = 0.8$.

▮ The normalized input impedance is $z_{in} = j2.5$; thus, from Fig. 10-22,

$$z_L = -j0.83 \qquad Z_L = -j83\ \Omega \qquad \text{VSWR} = \infty \qquad \Gamma_L = 1\ \underline{/-100°}$$

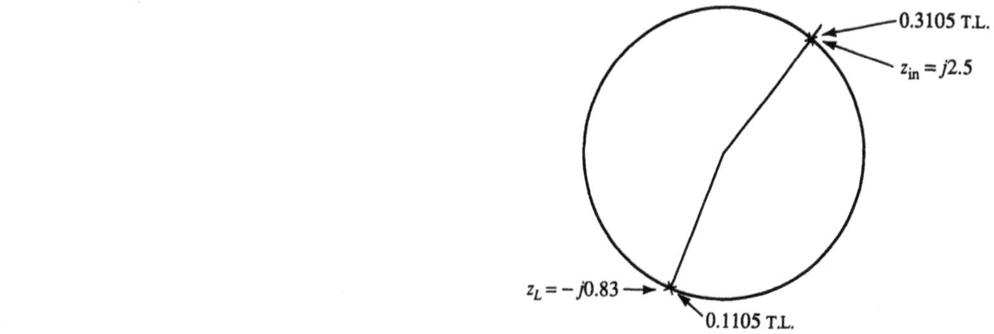

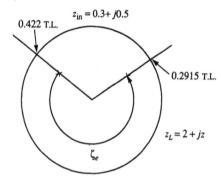

Fig. 10-22

10.83   From the Smith chart, find the shortest length of a transmission line having $Z_{in} = (30 + j50)\ \Omega$; $Z_L = (200 + j200)\ \Omega$; $R_C = 100\ \Omega$. Also determine the VSWR and Γ_L.

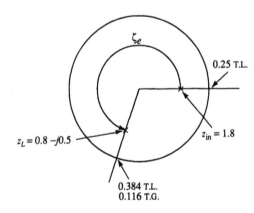

Fig. 10-23

▮   The normalized input impedance is $z_{in} = 0.3 + j0.5$. Thus, from Fig. 10-23, we get $z_L = 2 + j2$,
$$\zeta_e = 0.5 - 0.422 + 0.2915 = 0.370 \qquad \text{or} \qquad \zeta = 0.370\lambda$$
VSWR = 4.2, and $\Gamma_L = 0.615\ \underline{/123.5°}$.

10.84   Repeat Problem 10.83 for a line with $Z_{in} = (135 + j0)\ \Omega$; $Z_L = (60 - j37.5)\ \Omega$; $R_C = 75\ \Omega$.

Fig. 10-24

▮   With $z_{in} = 1.8$, Fig. 10-24 yields
$$z_L = 0.8 - j0.5 \qquad \zeta_e = 0.5 - 0.25 + 0.116 = 0.366 \qquad \text{VSWR} = 1.8 \qquad \Gamma_L = 0.285\ \underline{/96°}$$

10.85   By the Smith chart find the inverse of $C = 50 - j100$.

▮   Choose $R_C = 50$. Thus $c = 1 - j2$ (Fig. 10-25); $1/c = 0.2 + j0.4$; $1/C = (4 + j8) \times 10^{-3}$.

10.86   Repeat Problem 10.85 for $C = 5 - j10$.

▮   Choose $R_C = 5$. Then $c = 1 - j2$ (Fig. 10-25); $1/c = 0.2 + j0.4$; $1/C = (4 + j8) \times 10^{-2}$.

10.87   An air-filled rectangular waveguide measuring $a = 2.29$ cm and $b = 1.02$ cm operates in the dominant TE_{10} mode. It is required that the operating frequency be at least 30 percent above the cutoff frequency of the TE_{10}

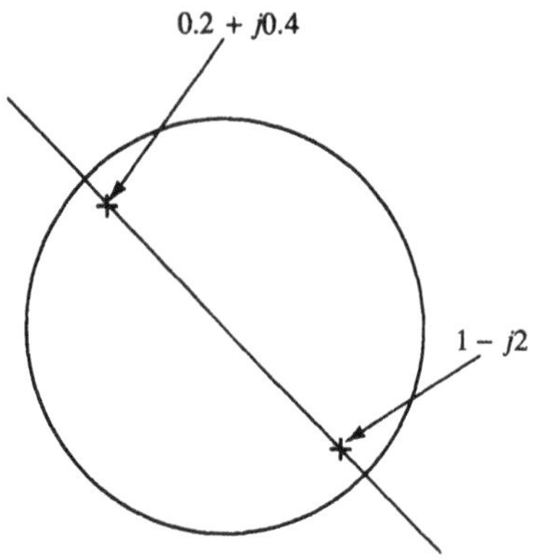

Fig. 10-25

mode but less than 90 percent of the next-higher cutoff frequency. Determine the range of the operating frequency.

❚ Proceed as in Problem 7.47. For the TE$_{10}$ mode,

$$f_{c.10} = \frac{u_0}{2a} = \frac{3 \times 10^8}{2 \times 2.29 \times 10^{-2}} = 6.55 \text{ GHz}$$

For the next (TE$_{20}$) mode,

$$f_{c.20} = \frac{u_0}{a} = 13.1 \text{ GHz}$$

The required operating frequency range is

$$1.3f_{c.10} \leq f \leq 0.9f_{c.20} \quad \text{or} \quad 8.515 \leq f \leq 11.79 \text{ GHz}$$

10.88 Describe the surface charge distribution on the top wall of the cavity of Problem 8.19.

❚ It is evident that the current density **K** is zero at the centerpoint of the wall; hence (negative) charge accumulates there.

10.89 Determine the **H**-field produced by a Hertzian dipole antenna fed with a dc current I_{dc}.

❚ In (1) of Problem 9.6, let $\beta_0 = \omega/u_0 \to 0$ and $\hat{I} \to I_{\text{dc}}$, to obtain

$$\mathbf{H} = \mathbf{a}_\phi \frac{I_{\text{dc}} \, d\zeta}{4\pi r^2} \sin \theta$$

10.90 Find the **E**-field very near the dipole antenna of Problem 9.4, under dc conditions.

❚ For near fields we retain only $(1/r^3)$-terms in (1) and (2) of Problem 9.7:

$$E_r = \frac{-j2\hat{I} \, d\zeta}{4\pi\epsilon_0 \omega r^3} \cos \theta \, e^{-j\beta_0 r} \qquad E_\theta = \frac{-j\hat{I} \, d\zeta}{4\pi\epsilon_0 \omega r^3} \sin \theta \, e^{-j\beta_0 r}$$

in which η_0/β_0 has been replaced with $1/\epsilon_0\omega$. Now, since $I = dq/dt$, $\hat{I} = j\omega\hat{q}$. Substituting for $\hat{I}$ in the above components and allowing $\omega \to 0$, we obtain

$$\mathbf{E} = \frac{\hat{q} \, d\zeta}{4\pi\epsilon_0 r^3} (2 \cos \theta \, \mathbf{a}_r + \sin \theta \, \mathbf{a}_\theta)$$

which is the same as (1) of Problem 2.156 $[p = \hat{q} \, d\zeta]$.

10.91 A 50-Ω transmission line is terminated in a load $Z_L = 90 + j60$ Ω. Determine the reflection coefficient and the VSWR due to this load. The line operates at 60 MHz.

❚

$$\Gamma_L = \frac{Z_L - R_C}{Z_L + R_C} = \frac{90 + j60 - 50}{90 + j60 + 50} = 0.473 \underline{/33.1°}$$

$$\text{VSWR} = \frac{1 + |\Gamma_L|}{1 - |\Gamma_L|} = \frac{1 + 0.473}{1 - 0.473} = 2.8$$

10.92 For the line of Problem 10.91, calculate the shortest length from the input to the load in order to have a purely resistive impedance. What is the value of the input resistance?

❚ If ζ is the required length, then from the result of Problem 10.91.

$$2\beta\zeta = 33.1° \quad \text{or} \quad 2\left(\frac{2\pi}{\lambda}\right)\zeta = 33.1°$$

Substituting $\lambda = (3 \times 10^8)/(60 \times 10^6) = 5$ m yields

$$\zeta = \frac{33.1 \times 5}{2 \times 360} = 0.23 \text{ m} \quad \text{and} \quad R = 2.8 \times 50 = 140 \text{ }\Omega$$

10.93 From the Smith chart, find the input impedance when the line of Problem 10.91 is $\lambda/6$ long.

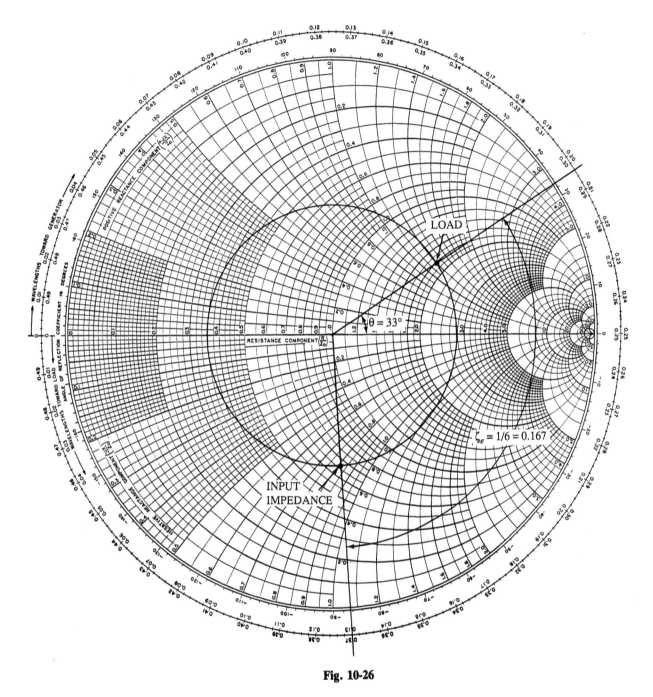

Fig. 10-26

❙ From Fig. 10-26, new input location $= (0.202 + 0.167) = 0.369$. Now, $4\pi\zeta_e = 2\pi/3 = 120°$; $z_{in} = 0.67 - j0.8$; $Z_{in} = (0.67 - j0.8)(50) = 33 - j40$ Ω.

10.94 A 100-Ω line is terminated in a load $Z_L = 60 + j40$. It is desired to match this load to the line with a parallel single-stub shorted tuner. Using the Smith chart, find the location of the stub, in wavelengths from the load.

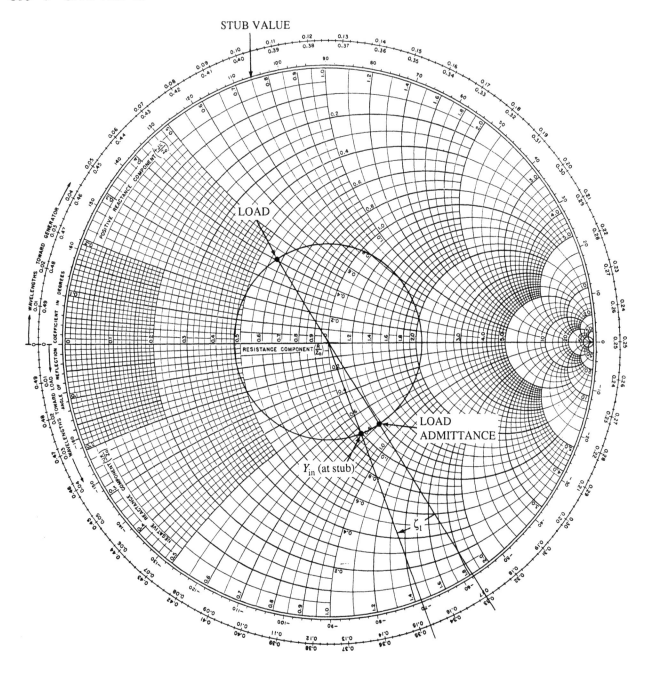

STUB VALUE

LOAD

LOAD
ADMITTANCE

Y_{in} (at stub)

ζ_1

Fig. 10-27

∎ $z_L = 0.6 + j0.4$ $y_L = 1.18 - j0.8$

See Fig. 10-27, where ζ_1 is the location of the shorted stub; thus

$$\zeta_1 = 0.346 - 0.330 = 0.016 \qquad \text{and} \qquad Y_{in} \text{ (at stub)} = 1 - j0.75$$

10.95   From the Smith chart determine the electrical length of the shorted stub of Problem 10.94.

∎   From Fig. 10-27, $Y_s = $ stub admittance $= j0.75$ and

$$\zeta_{es} = \text{electrical length of stub} = 0.250 + 0.102 = 0.352$$

10.96   A pure resistive load of 200 Ω terminates a 100-Ω transmission line. The input impedance of the line is capacitive, with a 60-Ω resistive component. Determine the input impedance of the line, using the Smith chart.

▌ Normalized impedance $z_{in} = 0.6 - jx$. From Fig. 10-28, $x = 0.37$; hence $Z_{in} = (0.6 - j0.37)(100) = (60 - j37)\ \Omega$.

10.97 From the Smith chart, determine the minimum electrical length of the line of Problem 10.96.

▌ From Fig. 10-28, $\zeta_e = 0.421 - 0.250 = 0.171$.

10.98 A short-circuited stub is placed across the load of the line of Problem 10.96, resulting in a standing-wave ratio of 4. From the Smith chart, determine the normalized susceptance added to the load by the stub.

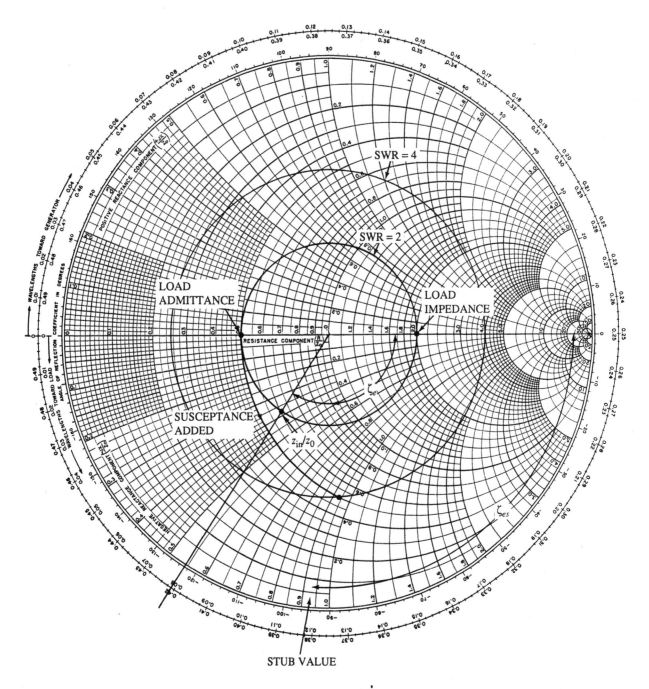

Fig. 10-28

▌ From Fig. 10-28, susceptance added $= -0.93$.

10.99 What is the minimum electrical length of the short-circuited stub in Problem 10.98?

❚ See Fig. 10-28. Stub admittance $= Y_s = j0.75$; $\zeta_{es} = 0.250 + 0.102 = 0.352$.

10.100 A plane electromagnetic wave propagates through a nonmagnetic dielectric material with a magnetic field amplitude of 6 A/m at a sinusoidal frequency of 200 MHz. The relative dielectric constant is 9. Determine the amplitude of the associated electric field.

❚
$$|E| = \eta\,|H| = \frac{\eta_0}{\sqrt{\epsilon_r}}|H| = \frac{377}{\sqrt{9}}(6) = 754 \text{ V/m}$$

10.101 What is the time-average power density in the wave of Problem 10.100?

❚
$$\langle S \rangle = \frac{1}{2}\,\eta\,|H|^2 = \frac{1}{2}\left(\frac{377}{3}\right)(6)^2 = 2.26 \text{ kW/m}^2$$

10.102 Determine the maximum energy density in the wave of Problem 10.100.

❚
$$W_{\text{max}} = \tfrac{1}{2}\epsilon_0\,|E|^2 + \tfrac{1}{2}\mu_0\,|H|^2 = \mu_0\,|H|^2 = (4\pi \times 10^{-7})(6)^2 = 4.5 \times 10^{-5} \text{ J/m}^3$$

10.103 A plane electromagnetic wave propagating through free space encounters the surface of a material with a relative dielectric constant of 16. The incident wave has an electric-field amplitude of $E_0 = 2 \times 10^4$ V/m. Assuming that the wave is traveling normal to the dielectric surface, derive an expression for the reflection coefficient.

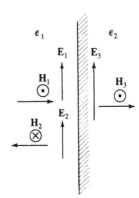

Fig. 10-29

❚ From Fig. 10-29:
$$E_1 + E_2 = E_3$$

and
$$H_1 - H_2 = H_3 \qquad \text{or} \qquad \frac{E_1}{\eta_1} - \frac{E_2}{\eta_1} = \frac{E_3}{\eta_2}$$

·Eliminate E_3 between these two equations, to find
$$\Gamma \equiv \frac{E_2}{E_1} = \frac{1 - (\eta_1/\eta_2)}{1 + (\eta_1/\eta_2)} = \frac{1 - \sqrt{\epsilon_2/\epsilon_1}}{1 + \sqrt{\epsilon_2/\epsilon_1}}$$

Substituting numerical values gives
$$\Gamma = \frac{1 - \sqrt{16}}{1 + \sqrt{16}} = -0.6$$

10.104 For the data of Problem 10.103, calculate the amplitude of the magnetic field in the reflected wave.

❚ With $\Gamma = -0.6$,
$$E_r = E_2 = \Gamma E_1 = -0.6 \times 2 \times 10^4 = -1.2 \times 10^4 \text{ V/m}$$
$$H_r = H_2 = \frac{E_2}{\eta_1} = \frac{-1.2 \times 10^4}{377} = -31.83 \text{ A/m}$$

10.105 What is the standing-wave ratio in the space outside the dielectric of Problem 10.103?

❚
$$\text{SWR} = \frac{1 + |\Gamma|}{1 - |\Gamma|} = \frac{1 + 0.6}{1 - 0.6} = 4.0$$

10.106 Verify the result of Problem 10.105 by computing the ratio of the maximum to the minimum electric field.

$$\text{SWR} = \frac{E_{\max}}{E_{\min}} = \frac{(2+1.2)10^4}{(2-1.2)10^4} = 4.0$$

10.107 Determine the time-average power/area transmitted into the dielectric of Problem 10.103.

■ Incident power density: $\langle P_1 \rangle = \frac{1}{2}|\mathbf{E}_1 \times \mathbf{H}_1| = \frac{1}{2\eta_1} E_1^2$

$$= \frac{1}{2 \times 377}(2 \times 10^4)^2 = 0.5305 \text{ MW/m}^2$$

Reflected power density: $\langle P_2 \rangle = \frac{1}{2}|\mathbf{E}_2 \times \mathbf{H}_2| = \frac{1}{2\eta_2} E_2^2$

$$= \frac{1}{2 \times 377}(1.2 \times 10^4)^2 = 0.191 \text{ MW/m}^2$$

Transmitted power density: $\langle P_3 \rangle = \langle P_1 \rangle - \langle P_2 \rangle = 0.3395 \text{ MW/m}^2$

10.108 Verify the result of Problem 10.107 by evaluating $\langle P_3 \rangle = \frac{1}{2}|\mathbf{E}_3 \times \mathbf{H}_3|$.

■ From Fig. 10-29,
$$|\mathbf{E}_3| = |\mathbf{E}_1| - |\mathbf{E}_2| = (2-1.2)10^4 = 8 \times 10^3 \text{ V/m}$$
$$|\mathbf{H}_3| = |\mathbf{H}_1| + |\mathbf{H}_2| = \frac{1}{377}(2+1.2)10^4 = 84.88 \text{ A/m}$$
$$\langle P_3 \rangle = \frac{1}{2}|\mathbf{E}_3 \times \mathbf{H}_3| = \frac{1}{2}E_3 H_3 = \frac{1}{2} \times 8 \times 10^3 \times 84.88 = 0.3395 \text{ MW/m}^2$$

10.109 For a plane wave in free space, show that the magnitude of the average Poynting vector is given by the product of the average energy density and the phase velocity of the wave.

■ We have $W_{av} = \frac{1}{2}\epsilon_0 E_m^2$ and $S_{av} = E_m^2/2\eta_0$; therefore
$$\frac{S_{av}}{W_{av}} = \frac{1}{\epsilon_0\eta_0} = \frac{1}{\epsilon_0\sqrt{\mu_0/\epsilon_0}} = \frac{1}{\sqrt{\mu_0\epsilon_0}} = u_0$$

10.110 Verify that the result of Problem 10.109 is dimensionally correct.

$$\frac{\text{W/m}^2}{\text{J/m}^3} = \frac{\text{W/m}^2}{\text{W} \cdot \text{s/m}^3} = \frac{\text{m}}{\text{s}}$$

10.111 A wave propagates (in the z-direction) between two parallel plates of infinite extent that are separated by a distance w. The electric field for the TE mode is
$$E_y = E_0 \sin \alpha_e x \sin (\omega t - \beta z) \quad \text{(V/m)} \tag{1}$$
and the magnetic field for the TM mode is
$$H_y = H_0 \cos \alpha_m x \sin (\omega t - \beta z) \quad \text{(A/m)} \tag{2}$$
Determine the time-average Poynting vector for the TM mode.

■ Since the power flow is in the z-direction, we must find E_x for H_y as given by (2). For the TM mode, $H_z = 0$, and from Maxwell's equation we have
$$-\frac{\partial H_y}{\partial z} = \epsilon \frac{\partial E_x}{\partial t}$$
which gives
$$E_x = \frac{\beta}{\omega\epsilon} H_0 \cos \alpha_m x \sin (\omega t - \beta z) = \frac{\beta}{\omega\epsilon} H_y \tag{3}$$
The instantaneous Poynting vector is then given by
$$S_z = E_x H_y = \frac{\beta H_0^2}{\omega\epsilon} \cos^2 \alpha_m x \sin^2 (\omega t - \beta z) \quad \text{(W/m}^2) $$
the average value of which is
$$(S_z)_{av} = \frac{\beta H_0^2}{2\omega\epsilon} \cos^2 \alpha_m x \quad \text{(W/m}^2) \tag{4}$$

10.112 For the structure of Problem 10.111, write the electric and magnetic field equations for the TEM mode.

▌ For the TEM mode, $E_z = H_z = 0$; also, $\alpha_m = 0$. Thus, from (2) and (3) of Problem 10.111,

$$H_y = H_0 \sin(\omega t - \beta z) \quad \text{(A/m)} \qquad E_x = \frac{\beta}{\omega \epsilon} H_0 \sin(\omega t - \beta z) \quad \text{(V/m)}$$

10.113 Show that the time-average Poynting vector for the TEM wave is independent of the frequency.

▌ For $\alpha_m = 0$, (4) of Problem 10.111 yields

$$(S_z)_{av} = \frac{\beta H_0^2}{2\omega \epsilon}$$

But

$$\beta = \sqrt{\frac{\omega^2}{u^2} - 0^2} = \frac{\omega}{u} \qquad \text{whence} \qquad (S_z)_{av} = \frac{H_0^2}{2u\epsilon}$$

which is independent of ω.

10.114 The input impedance of a transmission line of length ζ is given by

$$Z_{in} = Z_0 \frac{Z_L \cos \beta \zeta + j Z_0 \sin \beta \zeta}{Z_0 \cos \beta \zeta + j Z_L \sin \beta \zeta} \tag{1}$$

A line of 150 Ω characteristic impedance is terminated in a capacitive reactance $X_c = j50$ Ω. Calculate the input impedance if $\zeta = 3\lambda/8$.

▌ For $\beta \zeta = 3\pi/4$,

$$Z_{in} = 150 \left[\frac{(-j50)(-1/\sqrt{2}) + j(150)(1/\sqrt{2})}{(150)(-1/\sqrt{2}) + j(-j50)(1/\sqrt{2})} \right] = -j300 \ \Omega$$

10.115 Determine the magnitude of the reflection coefficient of the line of Problem 10.114.

▌ The reflection coefficient Γ is obtained from

$$Z_L = \left(\frac{1+\Gamma}{1-\Gamma}\right) Z_0 \qquad \text{or} \qquad -j50 = \left(\frac{1+\Gamma}{1-\Gamma}\right)150 \qquad \text{or} \qquad \Gamma = \frac{j+3}{j-3}$$

Hence

$$|\Gamma| = \frac{|j+3|}{|j-3|} = 1$$

10.116 The line of Problem 10.114 is now terminated in a resistance of 15 Ω. A signal is applied to the line and a maximum voltage of 30 V appears across the load. Determine the voltage of the incident wave.

▌ Let subscripts + and − respectively denote incident and reflected quantities; then

$$Z_L = \left(\frac{V_+ + V_-}{V_+ - V_-}\right) Z_0 \tag{1}$$

But the load voltage is $V_L = V_+ + V_- = 30$ V, which when substituted in (1) gives, in conjunction with the data,

$$15 = \left(\frac{30}{V_+ - V_-}\right)150 \qquad \text{or} \qquad V_+ - V_- = 300 \text{ V}$$

This, together with $V_+ + V_- = 30$ V, gives $V_+ = 165$ V.

10.117 Calculate the current in the reflected wave on the transmission line of Problem 10.116.

▌ From the results of Problem 10.116, $V_- = 30 - V_+ = 30 - 165 = -135$ V; hence

$$I_- = \frac{V_-}{-Z_0} = \frac{-135}{-150} = 0.9 \text{ A}$$

10.118 Verify that the results of Problems 10.116 and 10.117 are consistent with a load impedance of 15 Ω.

▌ From Problem 10.116,

$$I_+ = \frac{V_+}{Z_0} = \frac{165}{150} = 1.1 \text{ A} \qquad I_L = I_+ + I_- = 1.1 + 0.9 = 2.0 \text{ A}$$

whence

$$Z_L = \frac{V_L}{I_L} = \frac{30}{2} = 15 \ \Omega$$

10.119 Rework Problem 10.114 using the Smith chart.

▌

$$z_L = \frac{Z_L}{Z_0} = \frac{-j50}{150} = -j0.333 \quad \text{(capacitive)}$$

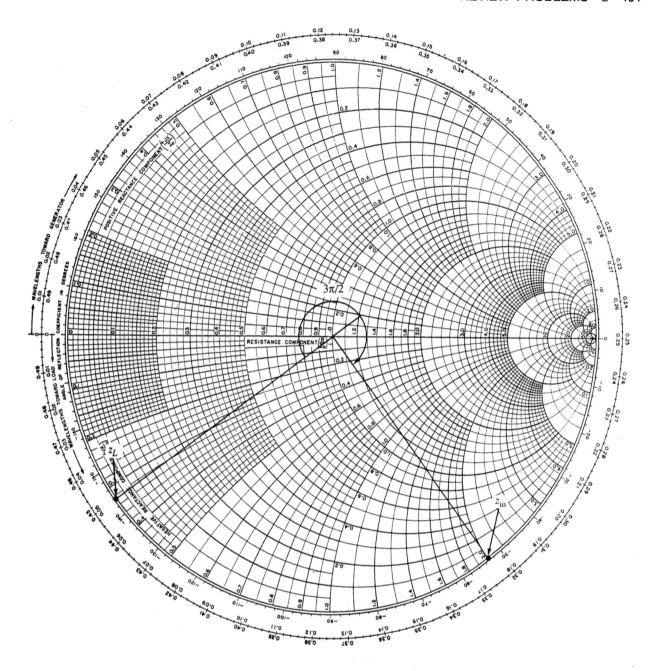

Fig. 10-30

and the angle of rotation is $2\beta\zeta = 3\pi/2$. Thus, from Fig. 10-30, $Z_{in} = (-j2)150 = -j300 \ \Omega$.

10.120 Rework Problem 10.116 using the Smith chart.

▌ In this case,

$$z_L = \frac{Z_L}{Z_0} = \frac{15}{150} = 0.1$$

From Fig. 10-30 (outer scale),

$$\text{SWR} = 10 = \frac{1 + |\Gamma|}{1 - |\Gamma|} \quad \text{or} \quad |\Gamma| = 0.818 = \frac{V_+}{V_-}$$

But $V_+ + V_- = 30 = V_+(1 - 0.818)$, and so

$$V_+ = \frac{30}{0.182} = 164.8 \text{ V}$$

10.121 A transmission line of characteristic impedance $Z_0 = 70\ \Omega$ is terminated in a resistance of $120\ \Omega$. The amplitude of the voltage in the incident wave is 25 V. What is the reflected-wave voltage?

$\blacksquare$
$$Z_L = \left(\frac{1+\Gamma}{1-\Gamma}\right)Z_0 \quad \text{or} \quad 120 = \left(\frac{1+\Gamma}{1-\Gamma}\right)(70) \quad \text{or} \quad \Gamma = 0.263$$

Then the voltage in the reflected wave is $V_- = \Gamma V_+ = 0.263 \times 25 = 6.575$ V.

10.122 For the transmission line of Problem 10.121, calculate (*a*) the current in the incident wave, and (*b*) the voltage standing-wave ratio.

$\blacksquare$ Using the results of Problem 10.121, we have

(*a*) $\quad I_+ = \dfrac{V_+}{Z_0} = \dfrac{25}{70} = 0.357$ A $\qquad$ (*b*) $\quad$ VSWR $= \dfrac{1+|\Gamma|}{1-|\Gamma|} = \dfrac{1+0.263}{1-0.263} = 1.714$

10.123 What is the input impedance of the line of Problem 10.121, if the line is $3\lambda/4$ long?

$\blacksquare$ With $\beta\zeta = 3\pi/2$, (*1*) of Problem 10.114 gives
$$Z_{\text{in}} = \frac{Z_0^2}{Z_L} = \frac{70^2}{120} = 40.83\ \Omega$$

10.124 A 300-Ω coaxial cable is terminated in a load $Z_L = (300 + j300)\ \Omega$. Using the Smith chart, find the reflection coefficient and the VSWR.

$\blacksquare$
$$z_L = \frac{300 + j300}{300} = 1 + j1$$

and Fig. 10-31 gives $\Gamma = 0.45\ \underline{/63°}$ and VSWR = 2.6.

10.125 The load on the cable of Problem 10.124 is changed to $j120\ \Omega$ and a 200-kHz signal is applied to the line input. Calculate, from the Smith chart, the length of the line to make it appear as a short circuit.

$\blacksquare$ The normalized impedance is
$$z_L = \frac{j120}{300} = j0.4$$

From Fig. 10-31, $\zeta = (0.5 - 0.06)\lambda = 0.44\lambda$. But
$$\lambda = \frac{u}{f} = \frac{3 \times 10^8}{2 \times 10^5} = 1.5 \times 10^3 \text{ m}$$

and so $\zeta = 0.44 \times 1.5 \times 10^3 = 660$ m.

10.126 An electric field in free space is given by
$$\mathbf{E} = E_0 e^{j(\omega t - \beta_1 y - \beta_2 z)}\mathbf{a}_x \tag{1}$$
Find the corresponding magnetic field.

$\blacksquare$ Applying Maxwell's equation to (*1*) yields
$$(\nabla \times \mathbf{E})_x = \frac{\partial E_z}{\partial y} - \frac{\partial E_y}{\partial z} = -\mu\frac{\partial H_x}{\partial t} \quad \text{whence} \quad H_x = 0$$

$$(\nabla \times \mathbf{E})_y = \frac{\partial E_x}{\partial z} - \frac{\partial E_z}{\partial x} = \frac{\partial E_x}{\partial z} = -\mu\frac{\partial H_y}{\partial t} \quad \text{or} \quad -j\beta_2 E_x = -j\omega\mu H_y \quad \text{or} \quad H_y = \frac{\beta_2}{\mu\omega}E_x$$

and
$$(\nabla \times \mathbf{E})_z = \frac{\partial E_y}{\partial x} - \frac{\partial E_x}{\partial y} = -\mu\frac{\partial H_z}{\partial t} \quad \text{or} \quad +j\beta_1 E_x = -j\omega\mu H_z \quad \text{or} \quad H_z = -\frac{\beta_1}{\mu\omega}E_x$$

Hence
$$\mathbf{H} = \frac{\beta_2}{\omega\mu}E_x\mathbf{a}_y + \left(\frac{-\beta_1}{\omega\mu}E_x\right)\mathbf{a}_z = \frac{E_0}{\omega\mu}[\beta_2\mathbf{a}_y - \beta_1\mathbf{a}_z]e^{j(\omega t - \beta_1 y - \beta_2 z)}$$

10.127 Obtain an equation which must be satisfied by β_1 and β_2 of (*1*) of Problem 10.126.

$\blacksquare$ E_x must satisfy the free-space wave equation:
$$\frac{\partial^2 E_x}{\partial x^2} + \frac{\partial^2 E_x}{\partial y^2} + \frac{\partial^2 E_x}{\partial z^2} = \frac{1}{u^2}\frac{\partial^2 E_x}{\partial t^2}$$

For the function (*1*) of Problem 10.126, this becomes
$$-\beta_1^2 E_x - \beta_2^2 E_x = -\frac{\omega^2}{u^2}E_x$$

Hence, the required constraint is $u^2(\beta_1^2 + \beta_2^2) = \omega^2$.

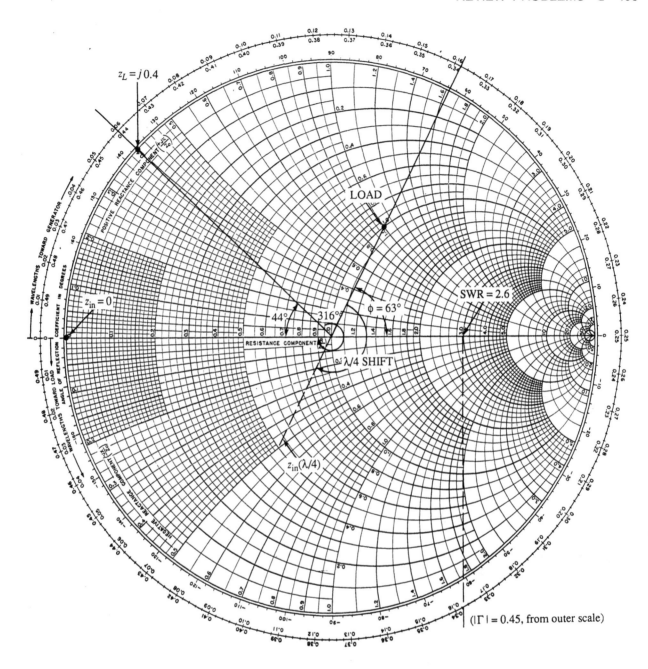

Fig. 10-31

10.128 A sinusoidal magnetic field propagating in the y-direction in a dielectric is polarized in the z-direction and has an amplitude of 5 A/m at 50 MHz. The dielectric has $\epsilon_r = 100$. Determine the polarization and amplitude of the corresponding electric field.

▌ $\mathbf{E}$ must have the direction $-\mathbf{a}_x$, and

$$|\mathbf{E}| = \eta \, |\mathbf{H}| = \frac{377}{\sqrt{100}} (5) = 188.5 \text{ V/m}$$

10.129 Evaluate the velocity of propagation and β for the wave of Problem 10.128.

▌
$$u = \frac{u_0}{\sqrt{\epsilon_r}} = \frac{3 \times 10^8}{\sqrt{100}} = 3 \times 10^7 \text{ m/s} \qquad \beta = \frac{\omega}{u} = \frac{2\pi \times 50 \times 10^6}{3 \times 10^7} = 10.47 \text{ m}^{-1}$$

10.130   A plane wave diffuses into a nonmagnetic conductor with $\sigma = 10^5$ S/m. The wavelength of the signal is 300 m. Calculate the skin depth.

$$\delta = \frac{1}{\sqrt{\pi f \mu \sigma}} = \sqrt{\frac{\lambda}{\pi u \mu \sigma}} = \sqrt{\frac{300}{\pi \times 3 \times 10^8 \times 4\pi \times 10^{-7} \times 10^5}} = 1.6 \text{ mm}$$

10.131   Calculate the absolute value of the intrinsic impedance of the material of Problem 10.130.

Solving the diffusion equation in a conducting material yields

$$H_y = \left(\frac{\sigma\delta}{1+j}\right) E_x$$

where δ is the skin depth. Then,

$$|\eta| = \left|\frac{E_x}{H_y}\right| = \frac{\sqrt{2}}{\sigma\delta}$$

From Problem 10.130, $\sigma = 10^5$ S/m and $\delta = 1.6$ mm, so that

$$|\eta| = \frac{\sqrt{2}}{10^5 \times 1.6 \times 10^{-3}} = 8.84 \text{ m}\Omega$$

10.132   A 3-MHz plane wave propagates in free space. The electric field has an amplitude of 1 kV/m. What is the average power density in the wave?

$$S_{av} = \frac{E_0^2}{2\eta} = \frac{(1000)^2}{2 \times 377} = 1.326 \text{ kW/m}^2$$

10.133   For the plane wave of Problem 10.132, find the time-average energy stored in a cube of space measuring one wavelength on each side.

The time-average energy density is

$$w = \frac{1}{4}(\mu_0 H_0^2 + \epsilon_0 E_0^2) = \frac{1}{2}\epsilon_0 E_0^2$$

Now

$$\lambda = \frac{u}{f} = \frac{3 \times 10^8}{3 \times 10^6} = 100 \text{ m}$$

Hence

$$W = w\lambda^3 = \frac{1}{2} \times \frac{10^{-9}}{36\pi}(1000)^2(100)^3 = 4.42 \text{ J}$$

10.134   A coaxial line carries a current $i = I_0 \cos(\beta x - \omega t)$ in the central conductor. The inner and outer radii of the line conductors are a and b, respectively. Derive an expression for the time-average energy stored in the magnetic field of the line, if the line is two wavelengths long.

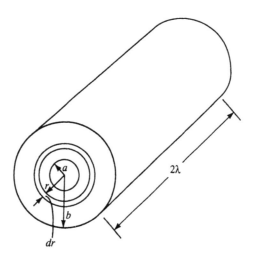

Fig. 10-32

Referring to Fig. 10-32, from Ampère's law,

$$H_\phi = \frac{i}{2\pi r} = \frac{I_0 \cos(\beta x - \omega t)}{2\pi r}$$

The time-average magnetic-energy density is given by

$$w_m = \frac{1}{4} \mu |H_\phi|^2 = \frac{\mu I_0^2}{16\pi^2 r^2}$$

Thus, the average stored energy in two wavelengths becomes

$$W_m = \frac{\mu I_0^2}{16\pi^2} \int_{x=0}^{2\lambda} \int_{r=a}^{b} \frac{1}{r^2} (2\pi r \, dr) \, dx = \frac{\mu I_0^2}{4\pi} \lambda \ln \frac{b}{a}$$

10.135 For the coaxial line of Problem 10.134, verify that the time-average energy density in the electric field is equal to that in the magnetic field.

▌ Because $E_r = \eta H_\phi$,

$$w_e = \frac{1}{4} \epsilon |E_r|^2 = \frac{1}{4} \epsilon \eta^2 |H_\phi|^2 = \frac{1}{4} \epsilon \frac{\mu}{\epsilon} |H_\phi|^2 = w_m$$

10.136 A coaxial cable is filled with a nonmagnetic material having relative permittivity $\epsilon_r = 25$. Find the wavelength on the line for a signal of 2 MHz.

▌

$$\lambda = \frac{u_0}{f \sqrt{\epsilon_r}} = \frac{3 \times 10^8}{2 \times 10^6 \times 5} = 30 \, \text{m}$$

10.137 A plane wave of amplitude $E_0 = 10^4$ V/m traveling in air strikes normally a dielectric material of permittivity $\epsilon = 12\epsilon_0$. Determine the amplitude and sign of the reflection coefficient.

▌ Here, $\epsilon_2/\epsilon_1 = 12$, and Problem 10.103 yields

$$\Gamma = \frac{1 - \sqrt{12}}{1 + \sqrt{12}} = -0.552$$

10.138 Calculate the standing-wave ratio in the air outside the dielectric described in Problem 10.137: Also determine the amplitude of the wave transmitted into the dielectric.

▌

$$\text{SWR} = \frac{1 + |\Gamma|}{1 - |\Gamma|} = \frac{1 + 0.552}{1 - 0.552} = 3.46$$

$$E_t = E_i + E_r = E_i(1 + \Gamma) = 10^4(1 - 0.552) = 4.48 \, \text{kV/m}$$

10.139 A load at the end of a transmission line 1.2 m in length has the value $Z_L = 75 + j60$ Ω when a signal of 50 MHz is applied to the line. Use the Smith chart to find the standing-wave ratio, given the characteristic impedance of the line $Z_0 = 50$ Ω.

▌

$$z_L = \frac{Z_L}{Z_0} = \frac{75 + j60}{50} = 1.5 + j1.2$$

and from the Smith chart (Fig. 10-33), SWR = 2.8.

10.140 Determine the input impedance of the line of Problem 10.139 from the Smith chart.

▌ Electrical length of line:

$$\zeta_e = \frac{\zeta f}{u} = \frac{(1.2)(50 \times 10^6)}{3 \times 10^8} = 0.2$$

From Fig. 10-33, location of input = 0.192 + 0.2 = 0.392; thus

$$z_{in} = \frac{Z_{in}}{Z_0} = 0.55 - j0.65 \quad \text{or} \quad Z_{in} = (0.55 - j0.65)(50) = 27.5 - j32.5 \quad \Omega$$

10.141 A standard load assembly for a laboratory experiment is constructed of a 150-Ω resistor at the end of a cable of length 0.1λ. It is desired to admittance-match this assembly to a longer cable by using a single parallel-shorted stub tuner. (*a*) Where should the stub be located? (*b*) What is the length of the shorted stub? Use the Smith chart.

▌ (*a*) The normalized load impedance is

$$z_L = \frac{Z_L}{Z_0} = \frac{150}{50} = 3.0$$

The normalized impedance of the load assembly including the 0.1λ cable is $0.8 - j1.0$. Then, from Fig. 10-34, the distance from load assembly to input is $\zeta_1 = 0.166\lambda - 0.10\lambda = 0.066\lambda$. (*b*) The input admittance at the stub

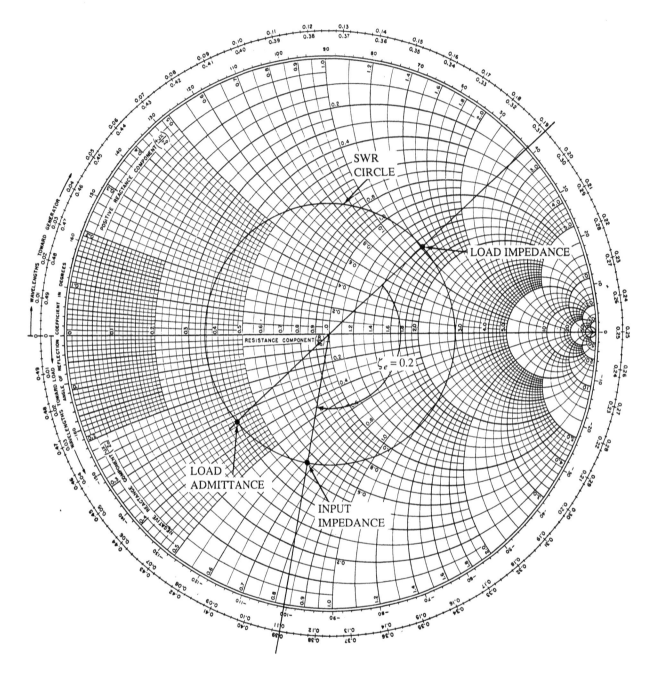

Fig. 10-33

location is $Y_{in}/Y_0 = 1.0 + j1.16$; hence the admittance due to the shorted stub is $-j1.16$. Thus the length of the stub is

$$\zeta_2 = (0.364 - 0.250)\lambda = 0.114\lambda$$

10.142 A plane wave travels in free space. The amplitude of the electric field is 0.1 MV/m at 500 kHz. Determine **(a)** the wavelength, **(b)** the propagation constant, and **(c)** the amplitude of the magnetic flux density.

▮ (a) $\quad \lambda = \dfrac{u}{f} = \dfrac{3 \times 10^8}{500 \times 10^3} = 600 \text{ m}$

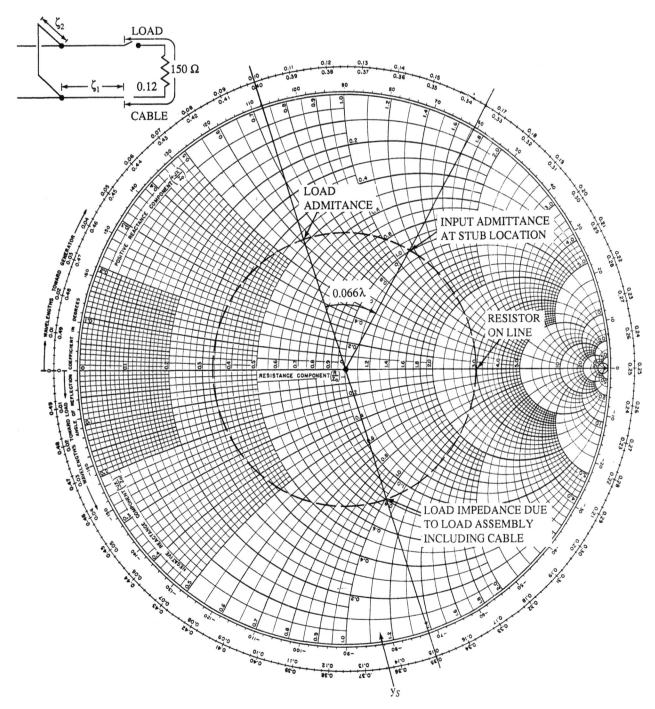

Fig. 10-34

(b) $\beta = \dfrac{2\pi}{\lambda} = \dfrac{2\pi}{600} = 10.47$ mm

(c) $B_m = \mu_0 H_m = \mu_0 \dfrac{E_m}{\eta_0} = \dfrac{E_m}{u_0} = \dfrac{10^5}{3 \times 10^8} = 0.333$ mT

10.143 A plane wave traveling in air strikes a dielectric, of relative permittivity ϵ_r, normally and is reflected. The incident magnetic field is 3.0 A/m and the magnetic field in the reflected wave is 1.8 A/m. Determine the reflection coefficient of the dielectric and the standing-wave ratio in the air.

$$\Gamma = \frac{E_r}{E_i} = -\frac{H_r}{H_i} = -\frac{1.8}{3.0} = -0.6 \qquad \text{SWR} = \frac{1+|\Gamma|}{1-|\Gamma|} = \frac{1+0.6}{1-0.6} = 4.0$$

10.144 For the wave of Problem 10.143, find the electric field in the incident wave and the magnetic field just inside the surface of the dielectric.

▮ $$E_i = u\mu H_i = 3 \times 10^8 \times 4\pi \times 10^{-7} \times 3.0 = 1131 \text{ V/m} \qquad H_t = H_i + H_r = 4.8 \text{ A/m}$$

10.145 What is the relative permittivity of the dielectric of Problem 10.143?

▮ The relative permittivity and the reflection coefficient are related to each other by

$$\Gamma = \frac{\sqrt{\epsilon_0} - \sqrt{\epsilon_0 \epsilon_r}}{\sqrt{\epsilon_0} + \sqrt{\epsilon_0 \epsilon_r}} = \frac{1 - \sqrt{\epsilon_r}}{1 + \sqrt{\epsilon_r}}$$

Since $\Gamma = -0.6$, from Problem 10.143, we solve to find $\epsilon_r = 16$.

10.146 A material is characterized by $\sigma = 3 \times 10^{-4}$ S/m and $\epsilon_r = 60$. A thin (1-mm) sheet of this material is sandwiched between two parallel conducting plates, and a 30-kHz signal of 200-V peak value is applied to these plates. Calculate the peak values of the displacement current density and the conduction current density.

▮ $$|J_d| = \epsilon \left| \frac{\partial E}{\partial t} \right| = \omega \epsilon |E| = (2\pi \times 30 \times 10^3)\left(60 \times \frac{1}{36\pi} \times 10^{-9} \right) \frac{200}{10^{-3}} = 20 \text{ A/m}^2$$

$$|J_c| = \sigma |E| = 3 \times 10^{-4} \times \frac{200}{10^{-3}} = 60 \text{ A/m}^2$$

10.147 In the material of Problem 10.146, calculate the skin depth when the ratio of the conduction current to displacement current is 10^{-10}.

▮ The operating frequency ω is obtained from $\sigma/\omega\epsilon = 10^{-10}$. The skin depth is then given by

$$\delta = \sqrt{\frac{2}{\omega\mu\sigma}} = \frac{10^{-5}\sqrt{2}}{\sigma\eta} = \frac{10^{-5}\sqrt{2}}{3 \times 10^{-4} \times 377} = 0.125 \text{ mm}$$

10.148 Calculate the frequency and the phase velocity of a plane wave when its magnetic field is
$$H_z = 120 \sin (6 \times 10^8 t + 12x) \quad (\text{A/m})$$
The wave is propagating through a dielectric.

▮ $$f = \frac{\omega}{2\pi} = \frac{6 \times 10^8}{2\pi} = 95.5 \text{ MHz} \qquad u = \frac{\omega}{\beta} = \frac{6 \times 10^8}{12} = 5 \times 10^7 \text{ m/s}$$

10.149 What is the intrinsic impedance of the dielectric of Problem 10.148?

▮ $$\eta = \frac{\eta_0 u}{u_0} = \frac{377 \times 5 \times 10^7}{3 \times 10^8} = 62.83 \text{ }\Omega$$

10.150 Determine the time-average value of the Poynting vector in the wave of Problem 10.148.

▮ $$S_{\text{av}} = \tfrac{1}{2}\eta H_0^2 = \tfrac{1}{2}(62.83)(120)^2 = 0.452 \text{ MW/m}^2$$

10.151 The electric-field components of an electromagnetic wave are
$$E_x = 0 \qquad E_y = 2E_0 \cos (\omega t - \beta x) \qquad E_z = \tfrac{1}{2}E_0 \sin (\omega t - \beta x)$$
How is the wave polarized?

▮ $$\frac{E_y^2}{(2E_0)^2} + \frac{E_z^2}{(E_0/2)^2} = 1 \quad (\text{elliptical polarization})$$

10.152 Determine the magnetic field corresponding to the electric field given in Problem 10.151.

▮ From Maxwell's equations we have

$$(\boldsymbol{\nabla} \times \mathbf{E})_x = \frac{\partial E_z}{\partial y} - \frac{\partial E_y}{\partial z} = 0 = -\mu \frac{\partial H_x}{\partial t} \qquad \text{or} \qquad H_x = 0$$

$$(\boldsymbol{\nabla} \times \mathbf{E})_y = \frac{\partial E_x}{\partial z} - \frac{\partial E_z}{\partial x} = 0 - \left(\frac{1}{2}E_0 \right)(-\beta) \cos (\omega t - \beta x) = -\mu \frac{\partial H_y}{\partial t}$$

or

$$H_y = -\frac{\beta E_0}{2\mu} \int \cos (\omega t - \beta x) \, dt = -\frac{\beta E_0}{2\mu\omega} \sin (\omega t - \beta x)$$

$$(\boldsymbol{\nabla} \times \mathbf{E})_z = \frac{\partial E_y}{\partial x} - \frac{\partial E_x}{\partial y} = 2E_0(-\beta)[-\sin (\omega t - \beta x)] - 0 = -\mu \frac{\partial H_z}{\partial t}$$

or
$$H_z = -\frac{2\beta E_0}{\mu}\int \sin(\omega t - \beta x)\,dt = \frac{2\beta E_0}{\mu\omega}\cos(\omega t - \beta x)$$

10.153 Obtain an expression for the instantaneous Poynting vector in the wave of Problem 10.151.

■ The power flow is in the x-direction and $S_x = E_y H_z - E_z H_y$. Substituting the field components from Problems 10.151 and 10.152 yields

$$S_x = [(2E_0\cos(\omega t - \beta x)]\left[\frac{2\beta E_0}{\omega\mu}\cos(\omega t - \beta x)\right] - \left[\frac{E_0}{2}\sin(\omega t - \beta x)\right]\left[-\frac{\beta E_0}{2\omega\mu}\sin(\omega t - \beta x)\right]$$

$$= \frac{\beta E_0^2}{\omega\mu}\left[4\cos^2(\omega t - \beta x) + \frac{1}{4}\sin^2(\omega t - \beta x)\right]$$

10.154 A wave in free space is incident normally on a dielectric material whose relative permittivity is 12. What percentage of incident power is reflected?

■
$$\Gamma \equiv \frac{E_r}{E_i} = \frac{1 - \sqrt{\epsilon/\epsilon_0}}{1 + \sqrt{\epsilon/\epsilon_0}} = \frac{1 - \sqrt{12}}{1 + \sqrt{12}} = -0.552$$
$$\frac{\text{reflected power}}{\text{incident power}} = \Gamma^2 = (0.552)^2 = 30.47\%$$

10.155 From the data of Problem 10.154, calculate the ratio of the magnitude of the electric field just inside the dielectric to that in the incident wave.

■
$$\frac{E_t}{E_i} = \frac{E_i + E_r}{E_i} = 1 + \Gamma = 0.448$$

10.156 What is the velocity of the wave in the dielectric of Problem 10.154?

■
$$u = \frac{u_0}{\sqrt{\epsilon_r}} = \frac{3\times 10^8}{\sqrt{12}} = 8.66\times 10^7 \text{ m/s}$$

10.157 The electric field for a 100-MHz, plane-polarized wave in a conducting material ($\mu = \mu_0$) is given by

$$E_x = 0 \qquad \frac{d^2 E_y}{dx^2} = -(2\times 10^4 - j7\times 10^5)E_y \qquad E_z = 0 \tag{1}$$

Determine the relative permittivity of the material.

■ The electric field must satisfy the vector Helmholtz equation
$$\nabla^2 \mathbf{E} = j\omega\mu(\sigma + j\omega\epsilon)\mathbf{E} \tag{2}$$
Comparing (1) and (2) yields
$$\omega^2\mu\epsilon = 2\times 10^4 = \omega^2\epsilon_r/u_0^2 \qquad \text{and} \qquad \omega\mu\sigma = 7\times 10^5 \tag{3}$$
From the first of these,
$$\epsilon_r = \frac{2\times 10^4 u_0^2}{4\pi^2 f^2} = \frac{2\times 10^4 \times 9\times 10^{16}}{4\pi^2 \times 10^{16}} = 4559$$

10.158 Calculate the conductivity of the material of Problem 10.157.

■ From the second equation (3) of Problem 10.157,
$$\sigma = \frac{7\times 10^5}{\omega\mu} = \frac{7\times 10^5}{2\pi \times 10^8 \times 4\pi \times 10^{-7}} = 886.6 \text{ S/m}$$

10.159 Is the material of Problem 10.157 a reasonably good conductor at the given frequency?

■ Yes: $\dfrac{\sigma}{\omega\epsilon} = \dfrac{886.6}{2\pi \times 10^8 \times 4559 \times 10^{-9}/36\pi} \approx 35.$

10.160 The electric field of a plane wave in a dielectric is given by
$$E_x(y, t) = 500 e^{j(2\times 10^6 t + 4\times 10^{-2}y)} \quad \text{(V/m)}$$
Find the (**a**) direction of propagation; (**b**) amplitude of the electric field; (**c**) frequency of the wave; (**d**) phase velocity.

■ (**a**) $-y$-direction; (**b**) 500 V/m; (**c**) $(2\times 10^6)/2\pi = 318.3$ kHz; (**d**) $(2\times 10^6)/(4\times 10^{-2}) = 5\times 10^7$ m/s.

10.161 Determine the relative permittivity of the material medium specified in Problem 10.160.

❚ From Problem 10.160(*d*),

$$\epsilon_r = \left(\frac{u_0}{u}\right)^2 = \left(\frac{3 \times 10^8}{5 \times 10^7}\right)^2 = 36$$

10.162 What is the intrinsic impedance of the material of Problem 10.160?

❚

$$\eta = \frac{\eta_0}{\sqrt{\epsilon_r}} = \frac{377}{\sqrt{36}} = 62.83 \ \Omega$$

10.163 Determine the amplitude of the magnetic field corresponding to the electric field of Problem 10.160.

❚ Using the result of Problem 10.162, we obtain

$$|H_z| = \frac{|E_x|}{\eta} = \frac{500}{62.83} = 7.96 \ \text{A/m}$$

10.164 A conductor of circular cross section 2.5 cm in diameter, and having a conductivity 10^9 S/m, is 10 m long. Find the resistance of the conductor at 60 Hz and at 400 Hz.

❚ Under ac conditions, the effective cross-sectional area is $\pi D \delta$, where $\delta = \sqrt{2/\omega\mu\sigma}$ is the skin depth. Thus

$$R_{\text{ac}} = \frac{l}{\sigma(\pi D \delta)} = \frac{l}{\pi D} \sqrt{\frac{\omega\mu}{2\sigma}}$$

Substituting numerical data gives $R_{60\text{Hz}} = 62.1 \ \mu\Omega$ and $R_{400\text{Hz}} = 160 \ \mu\Omega$.

10.165 For the air-dielectric-air system shown in Fig. 10-35, determine the ratio E_{t3}/E_i. The frequency of the incident wave is 1 GHz.

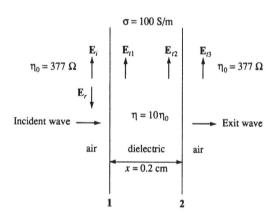

Fig. 10-35

❚ Applying the boundary conditions at the interface 1, we have

$$E_{t1} = E_i \left(\frac{2\eta}{\eta_0 + \eta}\right) \qquad (1)$$

Within the imperfect dielectric the field decays as

$$E_{t2} = E_{t1} e^{-\alpha x} \qquad (2)$$

where $\alpha = 1/\delta = \sqrt{\pi f \mu_0 \sigma}$. At the interface 2:

$$E_{t3} = E_{t2} \left(\frac{2\eta_0}{\eta_0 + \eta}\right) \qquad (3)$$

Combining (1) through (3) yields

$$\frac{E_{t3}}{E_i} = \frac{4\eta_0\eta}{(\eta_0 + \eta)^2} e^{-\alpha x} \qquad (4)$$

Substituting $\eta_0 = 377 \ \Omega$, $\eta = 3770 \ \Omega$, $x = 2 \times 10^{-3}$ m, and $\alpha = 200\pi$ m^{-1} in (4) yields

$$\frac{E_{t3}}{E_i} = 9.4 \times 10^{-2}$$

10.166 A pure standing wave results from the total reflection of a perpendicularly polarized radio wave from a glass skyscraper (at $z = 0$). If the incident electric field is given by
$$E_i = E_0 \cos (\omega t - \beta z)$$
express the total electric field.

▮ $E_r = E_0 \cos (\omega t + \beta z)$, and so
$$E_T = E_i + E_r = E_0[\cos (\omega t - \beta z) + \cos (\omega t + \beta z)] = 2E_0 \cos \beta z \cos \omega t$$

10.167 Determine the power density at a point on the negative z axis in Problem 10.166.

▮ Choose coordinates in Problem 10.166 that make $E_T = E_y$. Then, from curl $\mathbf{E} = -\mu_0(\partial \mathbf{H}/\partial t)$,
$$H_x = -\frac{2E_0\beta}{\mu_0\omega} \sin \beta z \sin \omega t$$
and
$$S_z = -E_y H_x = \frac{\beta E_0^2}{\mu_0\omega} \sin 2\beta z \sin 2\omega t$$

10.168 In Problem 10.167, the time-average $\langle S_z \rangle$ is zero. What does this mean?

▮ There is no net transport of energy by a standing wave. (Or, the incident and reflected waves transport equal amounts in opposite directions.)

10.169 A transmission line, operating at 300 MHz and having a 200 Ω characteristic impedance, is connected to a 100-Ω, purely resistive load. Calculate the input impedance, if the length of the line is 15 cm.

▮ The input impedance is given by
$$Z_{\text{in}} = Z_0 \frac{Z_L + jZ_0 \tan \beta \zeta}{Z_0 + jZ_L \tan \beta \zeta} \tag{1}$$
Now
$$\beta \zeta = \frac{\omega \zeta}{u} = \frac{2\pi(3 \times 10^8)(0.15)}{3 \times 10^8} = 0.30\pi$$
Substituting this and other given numerical values in (1) yields
$$Z_{\text{in}} = 200\left(\frac{100 + j200 \times 1.3764}{200 + j100 \times 1.3764}\right) = 241 \underline{/35.50°} \ \ \Omega$$

10.170 For the standing wave of Problem 10.166, evaluate the instantaneous energy density in space.

▮ Because the medium (space) is nondissipative, the energy densities of the incident and reflected waves simply add:
$$w = \epsilon_0 E_i^2 + \epsilon_0 E_r^2 = \epsilon_0 E_0^2[\cos^2 (\omega t - \beta z) + \cos^2 (\omega t + \beta z)]$$
$$= 2\epsilon_0 E_0^2[\sin^2 \beta z \sin^2 \omega t + \cos^2 \beta z \cos^2 \omega t] \tag{1}$$

10.171 A nonmagnetic metallic conductor has a conductivity of $\sigma = 10^7$ S/m. It is 100 m long and has a square cross section of 0.5 cm on each side. Assuming that the conductor operates at 160 kHz, obtain the ratio of the ac resistance to the dc resistance.

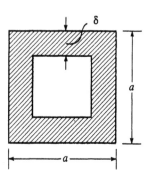

Fig. 10-36

▮ Under the usual assumptions, we have (Fig. 10-36)
$$\frac{R_{\text{ac}}}{R_{\text{dc}}} = \frac{A_{\text{dc}}}{A_{\text{ac}}} = \frac{a^2}{4a\delta} = \frac{a}{4\delta}$$

Now $\qquad \delta = 1/\sqrt{\pi f \mu \sigma} = 1/\sqrt{\pi(160 \times 10^3)4\pi \times 10^{-7} \times 10^7} = 0.398 \text{ mm}$

so that $\qquad \dfrac{R_{ac}}{R_{dc}} = \dfrac{5 \times 10^{-3}}{4 \times 0.398 \times 10^{-3}} = 3.14$

10.172 The average power density in a wave incident on a building is 3.2 kW/m². If the reflection coefficient is 30 percent, what is (a) the resulting standing-wave ratio, and (b) the power density in the reflected wave?

▮ (a)
$$\text{SWR} = \frac{1 + |\Gamma|}{1 - |\Gamma|} = \frac{1 + 0.3}{1 - 0.3} = 1.857$$

(b) Since power $\propto E^2$,

$$\text{(reflected power density)} = \Gamma^2 \times \text{(incident power density)}$$
$$= (0.3)^2 \times 3.2 = 0.288 \text{ kW/m}^2$$

10.173 Show that at a distance of $\lambda/8$ from the building of Problem 10.166 the stored electromagnetic energy has a permanent value.

▮ Substitute $\beta z = -\pi/4$ in (1) of Problem 10.170.

10.174 A radar transmitter sends out a 20-kW pulse which lasts for 0.5 μs. At sufficiently large distances the wave is plane, with a transverse area of 0.25 m². Calculate the (a) length of the pulse in space and (b) average energy density in the pulse.

▮ (a) The leading edge has been traveling (at speed u_0) for 0.5 μs longer than the trailing edge. Thus, the pulse length is

$$l = (3 \times 10^8)(0.5 \times 10^{-6}) = 150 \text{ m}$$

(b) Energy in pulse: $\quad W = 20 \times 10^3 \times 0.5 \times 10^{-6} = 10^{-2} \text{ J}$

Energy density: $\quad \dfrac{W}{lA} = \dfrac{10^{-2}}{(1.5 \times 10^2)(0.25)} = 266.7 \text{ μJ/m}^3$

10.175 For the wave of Problem 10.174, determine the (a) maximum values of the electric and magnetic fields and (b) the time-average value of the Poynting vector.

▮ (a) The *time*-average energy density is given by $\frac{1}{2}\epsilon_0 E_m^2$; the *space*-average energy density is, from Problem 10.174(b), 266.7 μJ/m³. Equating these two averages leads to $E_m = 7766$ V/m. Then

$$H_m = \frac{E_m}{\mu_0 u} = \frac{7766}{4\pi \times 10^{-7} \times 3 \times 10^8} = 20.6 \text{ A/m}$$

(b) $\qquad S_{av} = \frac{1}{2}E_m H_m = \frac{1}{2}(7766)(20.6) \approx 80 \text{ kW/m}^2$

10.176 A 100-MHz plane wave travels in free space with a magnetic field amplitude of $H_0 = 20$ A/m. Determine the (a) phase constant, (b) amplitude of the corresponding electric field, and (c) wavelength.

▮ (a) $\qquad \beta = \dfrac{\omega}{u} = \dfrac{2\pi \times 100 \times 10^6}{3 \times 10^8} = 2.094 \text{ m}^{-1}$

(b) $\qquad E_0 = \mu_0 u H_0 = 4\pi \times 10^{-7} \times 3 \times 10^8 \times 20 = 7.54 \text{ kV/m}$

(c) $\qquad \lambda = \dfrac{2\pi}{\beta} = \dfrac{2\pi}{2.094} = 3 \text{ m}$

10.177 The current density distribution in a certain region (μ, ϵ) is given by $\mathbf{J} = \mathbf{a}_x J_0 \sin(\omega t - kz)$; find the magnetic vector potential in that region.

▮ In the given region, the magnetic vector potential satisfies

$$\nabla^2 \mathbf{A} - \mu\epsilon \frac{\partial^2 \mathbf{A}}{\partial t^2} = -\mu \mathbf{J} \qquad (1)$$

Clearly, (1) will have a solution of the form $\mathbf{A} = K\mathbf{J}$, for some scalar constant K. By substitution,

$$K(-k^2\mathbf{J}) + \omega^2\mu\epsilon K\mathbf{J} = -\mu\mathbf{J} \qquad \text{or} \qquad K = \frac{\mu}{k^2 - \omega^2\mu\epsilon} \qquad (2)$$

10.178 Find the magnetic and electric fields corresponding to the current density given in Problem 10.177.

▮ $$\mathbf{H} = \frac{1}{\mu}\text{curl }\mathbf{A} = \frac{K}{\mu}\text{curl }\mathbf{J} = -\frac{kJ_0}{k^2 - \omega^2\mu\epsilon}\cos(\omega t - kz)\,\mathbf{a}_y$$

and, from the plane wave relation $E_x = (\mu\omega/k)H_y$,

$$\mathbf{E} = -\frac{\mu\omega J_0}{k^2 - \omega^2\mu\epsilon}\cos(\omega t - kz)\,\mathbf{a}_x$$

10.179 A plane wave traveling in air is normally incident on a metal surface. The standing-wave ratio outside the metal is 99 and the amplitude of the electric field in the incident wave is 10^4 V/m. What is the amplitude of the electric field just inside the metal?

■ $$\text{SWR} = \frac{1 + |\Gamma|}{1 - |\Gamma|} \quad \text{or} \quad |\Gamma| = \frac{\text{SWR} - 1}{\text{SWR} + 1} = \frac{99 - 1}{99 + 1} = 0.98$$

Thus, transmission coefficient $|T| = 1 - |\Gamma| = 0.02$. The transmitted field E_t is related to the incident field E_i by $E_t = |T| E_i$; thus $E_t = 0.02 \times 10^4 = 200$ V/m.

10.180 A 500-MHz wave is incident on a metallic sheet of conductivity $\sigma = 3 \times 10^7$ S/m. Determine the wave impedance of the metal at this frequency.

■ $$\eta = \sqrt{\omega\mu_0/2\sigma}\,(1 + j) = 8.11(1 + j)\text{ m}\Omega$$

10.181 A plane wave incident on a metal has a SWR in the air outside the metal of 98. What percentage of the incident energy is transmitted into the metal?

■ $$\text{reflection coefficient } \Gamma = \frac{\text{SWR} - 1}{\text{SWR} + 1} = \frac{98 - 1}{98 + 1} = 0.98$$

Then $\Gamma^2 = 96$ percent of the incident energy is reflected, and so 4 percent is transmitted.

10.182 A 1-m-long lossless coaxial cable, having a characteristic impedance of 50 Ω, is filled with a dielectric of relative permittivity $\epsilon_r = 3$. The cable is short-circuited at one end. What is its equivalent inductance at 100 MHz?

■ The input impedance, in general, is given by

$$Z_{\text{in}} = Z_0\left[\frac{Z_L + jZ_0\tan\beta\zeta}{Z_0 + jZ_L\tan\beta\zeta}\right] \tag{1}$$

For a short-circuited condition, $Z_L = 0$ and (1) becomes

$$Z_{\text{in}} = jZ_0\tan\beta\zeta = jZ_0\tan(2\pi f\sqrt{\mu\epsilon}\,\zeta) \equiv j2\pi fL_e$$

Thus $$L_e = \frac{Z_0}{2\pi f}\tan(2\pi f\sqrt{\mu\epsilon}\,\zeta) = \frac{50}{2\pi \times 100 \times 10^6}\tan\left[\frac{2\pi \times 10^8}{3 \times 10^8}(\sqrt{3})(1)\right] = 42.04\text{ nH}$$

10.183 An electric field $E_z = 1000\sin 10\pi y \sin 10\pi x$ (V/m) exists in an imperfect dielectric characterized by $\mu_r = 1$, $\epsilon_r = 2$, and $\sigma = 5 \times 10^{-6}$ S/m. How much power will be dissipated in a 10-cm cube of this dielectric at 750 MHz?

■ $$P_{\text{av}} = \frac{1}{2}\int_{\text{vol}} \sigma E_z^2\,d\tau = \frac{1}{2} \times 5 \times 10^{-6} \times 10^6 \times 0.1\int_0^{0.1}\int_0^{0.1}\sin^2 10\pi x \sin^2 10\pi y\,dx\,dy = 625\ \mu\text{W}$$

10.184 A plane wave having an electric field $E_x = E_0\sin(10^8 t - \beta z)$ (V/m) is normally incident on an iron surface characterized by $\mu_r = 250$ and $\sigma_c = 2 \times 10^6$ S/m. The average power density in the incident wave is 2 W/m^2. Determine the average power-loss density in the iron.

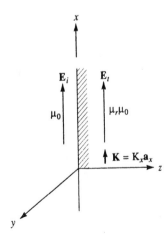

Fig. 10-37

▮ The incident power is

$$\frac{E_0^2}{2\eta_0} = \frac{E_0^2}{2(377)} = 2 \text{ W/m}^2 \qquad \text{whence} \qquad E_0 = 38.8 \text{ V/m}$$

The amplitude of the corresponding **H**-field is

$$H_0 = \frac{E_0}{\eta_0} = \frac{38.8}{377} = 0.103 \text{ A/m}$$

At the interface (Fig. 10-37), $K_x = 2H_0 = 0.206$ A/m. The surface resistance (the resistance of a $1 \times 1 \times \delta$ block of iron) is

$$R_s = \frac{(1 \text{ m})}{\sigma_c \times \delta \times (1 \text{ m})} \quad (\Omega)$$

Now $\qquad \delta = 1/\sqrt{\pi f \mu \sigma} = 1/\sqrt{\pi \times 10^8 \times 4\pi \times 10^{-7} \times 250 \times 2 \times 10^6} = 2.25 \times 10^{-6}$ m

whence $R_s = 0.222$ Ω. Since the current in the block is $I = K_x \times (1 \text{ m}) = 0.206$ A, the average power loss is $\frac{1}{2}I^2 R_s = 4.71$ mW. Since this is per 1 m² of surface area, the desired density is 4.71 mW/m².

10.185 A 10-cm-long Hertzian dipole is excited by a current of 20-A amplitude at 300 MHz. Determine H_ϕ along the broadside axis of the antenna when $r = 10$ cm.

▮ By Problem 9.6,

$$H_\phi = \frac{I\zeta}{4\pi} \sin\theta \, e^{-j2\pi r/\lambda} \left(\frac{j2\pi}{\lambda r} + \frac{1}{r^2}\right)$$

Substituting $\sin\theta = \sin \pi/2 = 1$, $\lambda = u/f = (3 \times 10^8)/(300 \times 10^6) = 1$ m, $r = \zeta = 0.1$ m, and $I = 20$ A, we obtain

$$H_\phi = \frac{1}{2\pi} e^{-j2\pi \times 0.1} \left(\frac{j2\pi}{0.1} + \frac{1}{0.1^2}\right) = 18.8 \, \underline{/-3.86°} \quad \text{A/m}$$

10.186 An air-filled rectangular waveguide measuring 1 by 10 cm operates at 2 GHz. What is the phase velocity in the guide for the lowest-order mode?

▮ Cutoff and free-space wavelengths are

$$\lambda_{0c} = \frac{2 \times 10}{1} = 20 \text{ cm} \qquad \lambda_0 = \frac{u_0}{f} = \frac{3 \times 10^{10}}{2 \times 10^9} = 15 \text{ cm}$$

Thus, the phase velocity is given by

$$v_p = \frac{u_0}{\sqrt{1 - (\lambda_0/\lambda_{0c})}} = \frac{3 \times 10^8}{\sqrt{1 - (15/20)^2}} = 4.54 \times 10^8 \text{ m/s}$$

10.187 What are the propagating modes at wavelengths greater than 8 cm in a waveguide of square cross section measuring 10 cm on a side?

▮ The cutoff wavelengths are given by

$$\lambda_{0c} = \frac{2}{\sqrt{(m/a)^2 + (n/a)^2}} = \frac{2a}{\sqrt{m^2 + n^2}}$$

TE$_{10}$ mode: $\quad \lambda_{0c} = 2a = 20$ cm
TE$_{01}$ mode: $\quad \lambda_{0c} = 2a = 20$ cm
TE$_{11}$ mode: $\quad \lambda_{0c} = \sqrt{2}\, a = 14.14$ cm
TE$_{21}$ mode: $\quad \lambda_{0c} = \frac{2}{\sqrt{5}} a = 8.94$ cm
TE$_{22}$ mode: $\quad \lambda_{0c} = \frac{a}{\sqrt{2}} = 7.07$ cm < 8 cm

So, the propagating modes are: TE$_{10}$, TE$_{01}$, TE$_{02}$, TE$_{20}$, TE$_{11}$, TM$_{11}$, TE$_{12}$, TE$_{21}$, TM$_{12}$, and TM$_{21}$.

10.188 A cavity resonator of square cross section a^2 (mm²) operates in the TE$_{110}$ mode at a wavelength of 12 mm. The cavity is $a/4$ (mm) deep. The cavity walls are silver-plated. Determine the Q of the cavity.

▮ The Q is given by

$$Q = \frac{2 \text{ (volume)}}{\delta \text{ (area)}} \qquad \text{where} \qquad \frac{1}{\delta} = \sqrt{\pi f \mu \sigma} \tag{1}$$

For silver, $\sigma = 6.1 \times 10^7$ S/m, and $f = u/\lambda = (3 \times 10^8)/(12 \times 10^{-3}) = 25$ GHz; thus

$$\frac{1}{\delta} = \sqrt{\pi \times 25 \times 10^9 \times 4\pi \times 10^{-7} \times 6.1 \times 10^7} = 245.37 \times 10^4 \text{ m}^{-1}$$

Now, volume $= a^2(a/4) = a^3/4$, and area $= 2a^2 + 4a(a/4) = 3a^2$. Thus, (1) yields

$$Q = \frac{2(a^3/4)}{\delta(3a^2)} = \frac{a}{6\delta} = \frac{\lambda/\sqrt{2}}{6\delta} = \frac{12 \times 10^{-3}}{\sqrt{2}} \times \frac{245.37 \times 10^4}{6} = 3470$$

10.189 A cavity of square cross section, measuring 8 by 8 cm, is 4 cm deep. Find the wavelength and the operating frequency for the TE_{110} mode.

▌ For the square cross section (a^2), $\lambda = \sqrt{2}\,a = \sqrt{2} \times 8 = 11.314$ cm and

$$f = \frac{u}{\lambda} = \frac{3 \times 10^8}{11.314 \times 10^{-2}} = 2.652 \text{ GHz}$$

10.190 If the walls of the cavity of Problem 10.189 are gold-plated, what is the Q in the TE_{110} mode?

▌ Proceeding as in Problem 10.188, we have

$$Q = \frac{2a^2(a/2)}{\delta[2a^2 + 4a(a/2)]} = \frac{a}{4\delta} = \frac{a}{4}\sqrt{\pi f \mu \sigma}$$
$$= \frac{8 \times 10^{-2}}{4}\sqrt{\pi \times 2.652 \times 10^9 \times 4\pi \times 10^{-7} \times 4.1 \times 10^7} = 13\,100$$

10.191 A 1-ns pulse is sent into a 1000-m, multimode, dielectric-slab waveguide whose refractive indices are $n_1 = 1.48$ and $n_2 = n_3 = 1.47$ (Fig. 10-38). (a) What is the bandwidth of this pulse before it enters the slab? (b) Assuming no dispersion, determine the differential group delay of the pulse when it reaches the far end of the slab. (c) What is the bandwidth of the delayed pulse?

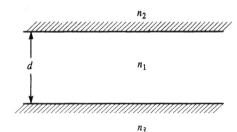

Fig. 10-38

▌ (a) $\quad BW = \dfrac{1}{2\,\Delta t} = \dfrac{1}{2 \times 10^{-9}} = 500 \text{ MHz}$

   (b) $\quad \Delta t = \dfrac{l}{u}(n_1 - n_2) = \dfrac{1000}{3 \times 10^8}(1.48 - 1.47) = 33 \text{ ns}$

   (c) $\quad BW = \dfrac{1}{2 \times 33 \times 10^{-9}} = 15.15 \text{ MHz}$

10.192 A symmetric dielectric-slab waveguide (Fig. 10-38) has $d = 1\ \mu m$, $n_1 = 2.24$, $n_2 = n_3 = 2.2$, and $\lambda_0 = 514.5$ nm. (a) At cutoff, obtain the angle the propagation vector makes with the normal to the core/cladding interface. (b) Calculate the number of even TE modes that can propagate.

▌ (a) The required angle is given by

$$\tan \alpha_c = \frac{n_2}{\sqrt{n_1^2 - n_2^2}} = \frac{2.2}{\sqrt{(2.24)^2 - (2.2)^2}} = 5.22 \qquad \text{or} \qquad \alpha_c = 79.16°$$

   (b) $\quad R = \sqrt{n_1^2 - n_2^2}\,(\beta_0 d) = \dfrac{\sqrt{(2.24)^2 - (2.2)^2}\,(2\pi \times 10^{-6})}{514.5 \times 10^{-9}} \approx 0.819(2\pi)$

which corresponds to two even modes.

10.193 The components of an elliptically polarized wave in air are
$$E_x = 4 \sin(\omega t - \beta z) \qquad \text{and} \qquad E_y = 8 \sin(\omega t - \beta z + 60°)$$
Calculate the average power density carried by the wave.

▌
$$S_{av} = \frac{|E|^2}{2\eta_0} = \frac{|E_x|^2 + |E_y|^2}{2\eta_0} = \frac{4^2 + 8^2}{2(377)} = 106.1 \text{ mW/m}^2$$

10.194 In Problem 8.102 it was shown that the Q of the TM_{101} in a spherical cavity of radius a is given by

$$Q = \frac{\eta}{2\eta_c(ka)^2}\left[\frac{ka}{\sin^2 ka} - \frac{1+(ka)^2}{ka}\right] \tag{1}$$

Show that $Q \approx \eta/\eta_c$.

❚ Rewrite (1) as

$$Q = \frac{\eta}{\eta_c}\frac{1}{2(ka)^2}\left[ka\left(\frac{1-\sin^2 ka}{\sin^2 ka}\right) - \frac{1}{ka}\right] = \frac{\eta}{\eta_c}\frac{1}{2(ka)^2}\left(\frac{ka}{\tan^2 ka} - \frac{1}{ka}\right) \tag{2}$$

Now use the resonance condition (see Problem 8.99), $\tan ka = ka/[1-(ka)^2]$, in (2) to obtain

$$Q = \frac{\eta}{\eta_c}\frac{1}{2(ka)^2}\left[\frac{[1-(ka)^2]^2 - 1}{ka}\right] = \frac{\eta}{\eta_c}\frac{1}{2ka}[-2+(ka)^2] \tag{3}$$

But, again from Problem 8.99, at resonance $ka \approx 2.74$. Substituting this in (3) yields

$$Q = \frac{\eta}{\eta_c}\frac{1}{2(2.74)}[-2+(2.74)^2] = \frac{\eta}{\eta_c}(1.005) \approx \frac{\eta}{\eta_c}$$

10.195 A wave travels in the z-direction. In the plane $z = 0$, the electric field in the wave is given by

$$\mathbf{E} = \mathbf{a}_x\,2\cos\omega t - \mathbf{a}_y\,3\sin\omega t$$

Express this field in phasor form.

❚
$$\mathbf{E} = \mathbf{a}_x\,2\cos\omega t - \mathbf{a}_y\,3\sin\omega t = \mathbf{a}_x\,2\cos\omega t + \mathbf{a}_y\,3\cos(\omega t + 90°)$$
$$= \mathrm{Re}\,[(2\mathbf{a}_x + 3e^{j\pi/2}\mathbf{a}_y)e^{j\omega t}]$$

or $\hat{\mathbf{E}} = 2\mathbf{a}_x + 3\mathbf{a}_y$.

10.196 The field of Problem 10.195 is elliptically polarized. Determine (*a*) the axial ratio; (*b*) the tilt angle of the major axis of the polarization ellipse with respect to the x axis; (*c*) the sense of polarization. (*d*) Draw the polarization ellipse in the xy plane.

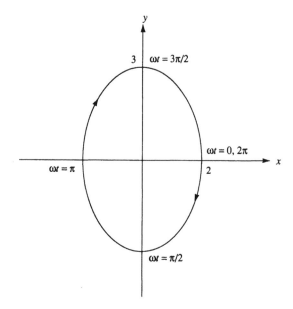

Fig. 10-39

❚ The ellipse is given by

$$\frac{E_x^2}{4} + \frac{E_y^2}{9} = 1$$

(*a*) $\mathrm{AR} \equiv \dfrac{\text{major axis}}{\text{minor axis}} = \dfrac{3}{2} = 1.5$

(*b*)   tilt angle $= \pi/2 = 90°$

(*c*)   left-handed polarization

(*d*)   See Fig. 10-39.

10.197 Determine the power gain (in decibels) of an antenna having a directivity of 20 and a radiation efficiency of 90 percent.

❚   gain = efficiency × directivity = $0.9 \times 20 = 18 = 12.55$ dB

10.198 The radiation field of a particular antenna is given by

$$\mathbf{E} = j\omega\mu \sin\theta \frac{I_0 L e^{-jkr}}{4\pi r} \mathbf{a}_\theta + \omega\mu k \sin\theta \, I_0 S \frac{e^{-jkr}}{4\pi r} \mathbf{a}_\phi \qquad (k \equiv 2\pi/\lambda)$$

where I_0 is a constant current along the antenna, and where L and S depend on the antenna geometry. Obtain an expression for the radiation resistance.

▮ We obtain R_{rad} from $R_{\text{rad}} = 2P_{\text{rad}}/|I_0|^2$, where

$$P_{\text{rad}} = \int_{r=\text{const.}} \mathbf{S}_{\text{av}} \cdot \mathbf{a}_r \, r^2 \, d\Omega$$

Now
$$\mathbf{S}_{\text{av}} = \frac{1}{2\eta}(|E_\theta|^2 + |E_\phi|^2)\mathbf{a}_r = \frac{1}{2\eta} \frac{\omega^2\mu^2 |I_0|^2}{16\pi^2 r^2} \sin^2\theta \, (L^2 + k^2 S^2) \, \mathbf{a}_r \qquad (1)$$

Hence
$$P_{\text{rad}} = \frac{1}{2\eta} \frac{\omega^2\mu^2 |I_0|^2}{16\pi^2} (L^2 + k^2 S^2) \int_0^{2\pi} \int_0^\pi \sin^3\theta \, d\theta \, d\phi$$

$$= 40\pi^2 |I_0|^2 \left[\left(\frac{L}{\lambda}\right)^2 + \left(\frac{kS}{\lambda}\right)^2 \right] \qquad (2)$$

and
$$R_{\text{rad}} = 80\pi^2 \left[\left(\frac{L}{\lambda}\right)^2 + \left(\frac{kS}{\lambda}\right)^2 \right] \qquad (3)$$

10.199 What is the direction of maximum radiation intensity of the antenna of Problem 10.198?

▮ $U(\theta) = r^2 S_{\text{av}} \propto \sin^2\theta$, which is maximized at $\theta = 90°$.

10.200 Calculate the maximum directivity of the antenna of Problem 10.198.

▮ Defining $B_0 \equiv \omega^2\mu^2 |I_0|^2 (L^2 + k^2 S^2)/32\pi^2\eta$, we have, from (1) and (2) of Problem 10.198,

$$D_{\text{max}} = \frac{4\pi U_{\text{max}}}{P_{\text{rad}}} = \frac{4\pi B_0}{B_0 \displaystyle\int_0^{2\pi} \int_0^\pi \sin^3\theta \, d\theta \, d\phi} = \frac{4\pi}{8\pi/3} = 1.5$$

10.201 What is the polarization of the antenna of Problem 10.198?

▮ In the time domain,

$$\mathbf{E}(t) = -\frac{\omega\mu \sin\theta}{4\pi r} I_0 L \sin(\omega t - kr) \mathbf{a}_\theta + \frac{\omega\mu k \sin\theta}{4\pi r} I_0 S \cos(\omega t - kr) \mathbf{a}_\phi \qquad (1)$$

Clearly, the amplitudes of the θ- and ϕ-components are unequal and are 90° out of phase in time. Thus, the antenna is elliptically polarized for all values of θ.

10.202 Determine the relationship between L and S of the antenna of Problem 10.198 such that the antenna is circularly polarized.

▮ For equal amplitudes in (1) of Problem 10.201 we must have $L = kS$.

10.203 Define the *polarization loss factor*. Show that the power loss due to a linearly polarized wave being incident on a circularly polarized antenna is 3 dB.

▮ The power loss that occurs when the receiving antenna does not have the same polarization state as the incoming wave is measured by the polarization loss factor,
$$\text{PLF} \equiv |\mathbf{p}_w \cdot \mathbf{p}_a^*|^2$$
where $\mathbf{p}_w \equiv$ unit vector representing the incident electric field
$\mathbf{p}_a \equiv$ unit vector representing the polarization state of the receiving antenna
Note that $\mathbf{p}_a$ also represents the polarization of the field radiated by the same antenna in the transmitting mode.
For the interaction between a linearly polarized wave and a circularly polarized antenna, we may write

$$\mathbf{p}_w = \mathbf{a}_x \qquad \mathbf{p}_a = \frac{1}{\sqrt{2}}(\mathbf{a}_x + j\mathbf{a}_y)$$

whence
$$\text{PLF} = \left| \mathbf{a}_x \cdot \frac{1}{\sqrt{2}}(\mathbf{a}_x - j\mathbf{a}_y) \right|^2 = \frac{1}{2} = -3 \text{ dB}$$

10.204 Consider a y-directed dipole, of moment $I_0 l$, located at $(a, b, 0)$ in front of a pair of intersecting ground planes [Fig. 10-40(a)]. Using image theory, find the pattern function $F(\theta, \phi)$ for this antenna.

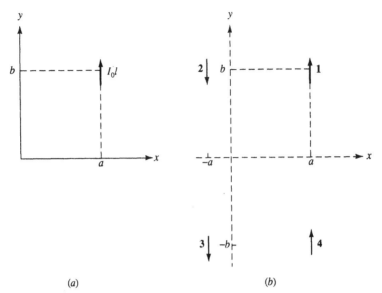

(a)   (b)   **Fig. 10-40**

❚ Three, symmetrically disposed, image sources are required to satisfy the boundary conditions [Fig. 10-40(a)]. Consider the subarray composed of sources **1** and **2**; its pattern factor is

$$F_x = 1e^{j(ka\cos\gamma_x + 0)} + 1e^{-j(ka\cos\gamma_x + \pi)} = 2j\sin(ka\cos\gamma_x)$$

where γ_x is the polar angle with respect to the x axis. We have $\cos\gamma_x = \sin\theta\cos\phi$; thus,

$$F_x(\theta, \phi) = 2j\sin(ka\sin\theta\cos\phi)$$

Now, for the array of pairs **1–2** and **4–3**, we have the pattern function (equal amplitudes, identical phases)

$$F_y(\theta, \phi) = 1e^{jkb\cos\gamma_y} + 1e^{jkb\cos\gamma_y} = 2\cos(kb\sin\theta\sin\phi)$$

Then, by pattern multiplication,

$$F(\theta, \phi) = F_x(\theta, \phi)F_y(\theta, \phi) = 4j\sin(ka\sin\theta\cos\phi)\cos(kb\sin\theta\sin\phi) \tag{1}$$

10.205 Determine the far electric field of the antenna of Problem 10.204.

❚ From Problem 9.8 and the pattern function (1) of Problem 10.204:

$$E_\theta = -j\omega\mu I_0 l\frac{e^{-jkr}}{4\pi r}\cos\theta\sin\phi\,[4j\sin(ka\sin\theta\cos\phi)\cos(kb\sin\theta\sin\phi)]$$

$$E_\phi = -j\omega\mu I_0 l\frac{e^{-jkr}}{4\pi r}\cos\phi[4j\sin(ka\sin\theta\cos\phi)\cos(kb\sin\theta\sin\phi)]$$

10.206 For an unsymmetric, infinite biconical antenna with half-angles θ_1 and θ_2 (Fig. 10-41), derive an expression for the characteristic impedance.

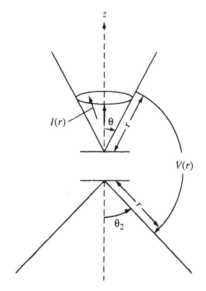

Fig. 10-41

▮ For the TEM mode, the far fields are

$$\mathbf{E} = \mathbf{a}_\theta \eta H_0 \frac{e^{-jkr}}{r \sin \theta} \qquad \mathbf{H} = \mathbf{a}_\phi H_0 \frac{e^{-jkr}}{r \sin \theta}$$

Thus

$$V(r) = \int_{\theta_1}^{\pi-\theta_2} E_\theta \, r \, d\theta = \eta H_0 e^{-jkr} \int_{\theta_1}^{\pi-\theta_2} \frac{d\theta}{\sin \theta} = \eta H_0 e^{-jkr} \ln \left[\cot \frac{\theta_1}{2} \cot \frac{\theta_2}{2} \right]$$

and

$$I(r) = \int_0^{2\pi} H_\phi \, r \sin \theta \, d\phi = 2\pi H_0 e^{-jkr}$$

Then

$$Z_c = \frac{V(r)}{I(r)} = \frac{\eta}{2\pi} \ln \left[\cot \frac{\theta_1}{2} \cot \frac{\theta_2}{2} \right]$$

10.207 Four half-wave dipoles, each parallel to the y axis, are arranged with their centers along the z axis, as shown in Fig. 10-42. Verify that the spacing $d = 0.35\lambda$ satisfies the Hansen–Woodyard criterion for increased directivity of an end-fire array.

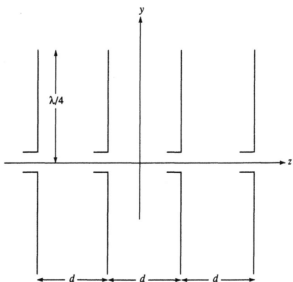

Fig. 10-42

▮ The criterion is

$$d < \frac{\lambda}{2} \left(1 - \frac{1}{n} \right) = \frac{\lambda}{2} \left(1 - \frac{1}{4} \right) = 0.375\lambda$$

and $d = 0.35\lambda$ meets it.

10.208 For the antenna of Problem 10.207, determine the interelement phase shift required to produce an increased-directivity, end-fire beam at $\theta_0 = 180°$.

▮ As in Problem 9.193,

$$\alpha = kd + \frac{\pi}{n} = \frac{2\pi}{\lambda}(0.35\lambda) + \frac{\pi}{4} = 0.95\pi \text{ rad}$$

10.209 The radiation resistance R_{rad} of center-fed dipole antennas is usually referred to the peak of the antenna current. For a lossless dipole, it is possible to change the input resistance and obtain better matching with the transmission line at the feed, by putting the feed off-center (Fig. 10-43). If the feed point is at a distance $z_0 > 0$ from the center of a dipole, with the current distribution

$$I(z) = I_0 \sin \left[\beta \left(\frac{\zeta}{2} - |z| \right) \right] \qquad (|z| \leq \zeta/2)$$

obtain a relationship between R_{in} and R_{rad} in terms of β, ζ, and z_0.

▮ Since $z_0 > 0$,

$$I(z_0) = I_0 \sin \left[\beta \left(\frac{\zeta}{2} - z_0 \right) \right] \qquad (1)$$

For a lossless antenna, $P_{\text{in}} = P_{\text{rad}}$, or

$$\frac{1}{2} |I(z_0)|^2 R_{\text{in}} = \frac{1}{2} |I_0|^2 R_{\text{rad}} \qquad (2)$$

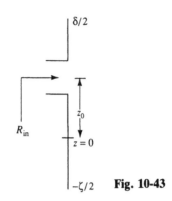

Fig. 10-43

Together, (1) and (2) imply:

$$R_{in} = \frac{R_{rad}}{\sin^2\left[\beta\left(\frac{\zeta}{2} - z_0\right)\right]} \qquad (3)$$

10.210 Apply Problem 10.209 to a half-wave dipole.

▮
$$R_{in} = \frac{R_{rad}}{\cos^2 \beta z_0}$$

10.211 Find the distance between the points $P(r = 4,\ \theta = 60°,\ \phi = 0°)$ and $Q(6, 120°, 120°)$ in the spherical coordinate system.

▮ The cartesian coordinates of P are
$$x = r \sin\theta\cos\phi = 3.464 \qquad y = r\sin\theta\sin\phi = 0 \qquad z = r\cos\theta = 2$$
Similarly, for Q: $x = -2.598$, $y = 4.5$, and $z = -3$. Thus
$$D_{PQ} = \sqrt{(3.464 + 2.598)^2 + 4.5^2 + (2 + 3)^2} = 9.055$$

10.212 A field is given in cartesian coordinates as
$$\mathbf{G} = \frac{10}{\sqrt{x^2 + y^2}}[x\mathbf{a}_x + y\mathbf{a}_y + z\sqrt{2(x^2 + y^2)}\,\mathbf{a}_z]$$
Give the direction of the field at $P(3, 4, -2)$.

▮
$$\mathbf{G}(P) = \frac{10}{\sqrt{3^2 + 4^2}}[3\mathbf{a}_x + 4\mathbf{a}_y + (-2\sqrt{2(3^2 + 4^2)}\,\mathbf{a}_z)] = 6\mathbf{a}_x + 8\mathbf{a}_y - 4\sqrt{50}\,\mathbf{a}_z$$

$$|\mathbf{G}(P)| = \sqrt{6^2 + 8^2 + (-4\sqrt{50})^2} = 30$$
The required unit vector is
$$\frac{\mathbf{G}(P)}{|\mathbf{G}(P)|} = \frac{6}{30}\mathbf{a}_x + \frac{8}{30}\mathbf{a}_y - \frac{4}{30}\sqrt{50}\,\mathbf{a}_z$$

10.213 Repeat Problem 10.212 in cylindrical coordinates.

▮ In cylindrical coordinates,
$$\mathbf{G} = \frac{10}{r}(r\mathbf{a}_r + zr\sqrt{2}\,\mathbf{a}_z) = 10(\mathbf{a}_r + z\sqrt{2}\,\mathbf{a}_z)$$
and so
$$\mathbf{G}(P) = 10(\mathbf{a}_r - 2\sqrt{2}\,\mathbf{a}_z)$$
$$|\mathbf{G}(P)| = 10\sqrt{1 + 8} = 30 \quad [check]$$
$$\frac{\mathbf{G}(P)}{|\mathbf{G}(P)|} = \frac{1}{3}(\mathbf{a}_r - 2\sqrt{2}\,\mathbf{a}_z)$$
where, of course, $\mathbf{a}_r$ is to be evaluated at the point $P(5, \tan^{-1}(4/3), -2)$.

10.214 Find the scalar projection of the vector $\mathbf{A} = -3\mathbf{a}_y - 2\mathbf{a}_z$ on the line joining the points $P_1(0, -2, 3)$ and $P_2(\sqrt{3}/2, -3/2, 1)$.

▮ The unit vector along the line is
$$\mathbf{u} = \frac{(\sqrt{3}/2)\mathbf{a}_x + (1/2)\mathbf{a}_y - 2\mathbf{a}_z}{\sqrt{(3/4) + (1/4) + 4}} = \frac{1}{\sqrt{5}}\left(\frac{\sqrt{3}}{2}\mathbf{a}_x + \frac{1}{2}\mathbf{a}_y - 2\mathbf{a}_z\right)$$

The desired projection is therefore

$$\mathbf{A} \cdot \mathbf{u} = \frac{1}{\sqrt{5}}\left(0 - \frac{3}{2} + 4\right) = \frac{\sqrt{5}}{2}$$

10.215 Given a scalar field

$$V = \sin\left(\frac{\pi}{2}x\right)\sin\left(\frac{\pi}{3}y\right)e^{-z}$$

determine the magnitude and the direction of the maximum rate of increase of V at point $P(1, 2, 3)$.

$\blacksquare$ The maximum rate of change is obtained from the gradient.

$$\nabla V = \mathbf{a}_x \sin\left(\frac{\pi y}{3}\right)e^{-z}\left[\frac{\pi}{2}\cos\left(\frac{\pi x}{2}\right)\right] + \mathbf{a}_y \sin\left(\frac{\pi x}{3}\right)e^{-z}\left[\frac{\pi}{3}\cos\left(\frac{\pi y}{3}\right)\right] + \mathbf{a}_z \sin\left(\frac{\pi x}{2}\right)(-e^{-z})\sin\left(\frac{\pi y}{3}\right)$$

Evaluating at $P(1, 2, 3)$ yields

$$\nabla V(P) = e^{-3}\left[\frac{\pi}{3}\cos\left(\frac{2\pi}{3}\right)\mathbf{a}_y - \sin\left(\frac{2\pi}{3}\right)\mathbf{a}_z\right] = 10^{-3}(-22.58\mathbf{a}_y - 43.12\mathbf{a}_z) \qquad (1)$$

From (1), $|\nabla V(P)| = 48.67 \times 10^{-3}$, and the direction vector of $\nabla V(P)$ is $\mathbf{a} = -0.4639\mathbf{a}_y - 0.886\mathbf{a}_z$.

10.216 For the scalar field of Problem 10.215, find the rate of change of V at $P(1, 2, 3)$ in the direction of the origin.

$\blacksquare$ The unit vector from P to the origin is

$$\mathbf{u} = \frac{-\mathbf{a}_x - 2\mathbf{a}_y - 3\mathbf{a}_z}{\sqrt{1+4+9}} = -\frac{1}{\sqrt{14}}(\mathbf{a}_x + 2\mathbf{a}_y + 3\mathbf{a}_z)$$

and the desired rate of change is

$$\mathbf{u} \cdot \nabla V(P) = -\frac{10^{-3}}{\sqrt{14}}[0 + 2(-22.58) + 3(-43.12)] = 0.0466$$

10.217 A field is expressed in spherical coordinates by $\mathbf{E} = \mathbf{a}_r(25/r^2)$. Find $|\mathbf{E}|$ at point $P(-3, 4, -5)$.

$\blacksquare$
$$\frac{25}{3^2 + 4^2 + 5^2} = \frac{1}{2}$$

10.218 For the field given in Problem 10.217, find the angle α which $\mathbf{E}(P)$ makes with the vector $\mathbf{B} = 2\mathbf{a}_x - 2\mathbf{a}_y + 1\mathbf{a}_z$.

$\blacksquare$ Since $\mathbf{E}$ is radially directed, it has the direction of the position vector $\mathbf{R}_p = -3\mathbf{a}_x + 4\mathbf{a}_y - 5\mathbf{a}_z$; thus

$$\alpha = \cos^{-1}\left(\frac{\mathbf{R}_p \cdot \mathbf{B}}{|\mathbf{R}_p|\,|\mathbf{B}|}\right) = \cos^{-1}\left[\frac{-3(2) + 4(-2) + (-5)(1)}{(\sqrt{50})(\sqrt{9})}\right] = 153.6°$$

10.219 Given a vector field $\mathbf{E} = \mathbf{a}_x y + \mathbf{a}_y x$, evaluate the line integral of $\mathbf{E}$ from $P_1(2, 1, -1)$ to $P_2(8, 2, -1)$ along the parabola $x = 2y^2$.

$\blacksquare$
$$\int_{P_1}^{P_2} \mathbf{E} \cdot d\mathbf{s} = \int_{P_1}^{P_2} (y\mathbf{a}_x + x\mathbf{a}_y) \cdot (dx\, \mathbf{a}_x + dy\, \mathbf{a}_y)$$
$$= \int_2^8 y\, dx + \int_1^2 x\, dy = \frac{1}{\sqrt{2}}\int_2^8 \sqrt{x}\, dx + 2\int_1^2 y^2\, dy$$
$$= \frac{28}{3} + \frac{14}{3} = 14$$

10.220 Repeat Problem 10.219 by evaluating the integral along the line segment joining P_1 and P_2.

$\blacksquare$ The segment is given parametrically by

$$x = 2 + t(8 - 2) = 2 + 6t$$
$$y = 1 + t(2 - 1) = 1 + t$$

where $0 \leq t \leq 1$. Hence

$$\int_{P_1}^{P_2} \mathbf{E} \cdot d\mathbf{s} = \int_0^1 \left(y\frac{dx}{dt} + x\frac{dy}{dt}\right) dt$$
$$= \int_0^1 [6(1 + t) + 1(2 + 6t)]\, dt$$
$$= \int_0^1 (8 + 12t)\, dt = 14$$

10.221 Explain the uniform result of Problems 10.219 and 10.220.

    ▌ $\mathbf{E} = \mathrm{grad}\, xy$.

10.222 A vector field is given by $\mathbf{A} = \mathbf{a}_x 3x^2 y^2 - \mathbf{a}_y x^3 y^2$. Calculate $\oint \mathbf{A} \cdot d\mathbf{s}$ around the triangular contour shown in Fig. 10-44.

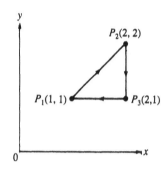

Fig. 10-44

    ▌

$$\oint \mathbf{A} \cdot d\mathbf{s} = \int_{P_1}^{P_2} 3x^2 y^2\, dx + \int_{P_2}^{P_3} 3x^2 y^2\, dx + \int_{P_3}^{P_1} 3x^2 y^2\, dx$$

$$- \int_{P_1}^{P_2} x^3 y^2\, dy - \int_{P_2}^{P_3} x^3 y^2\, dy - \int_{P_3}^{P_1} x^3 y^2\, dy$$

$$= 3 \int_1^2 x^2 (x^2)\, dx + 0 + 3 \int_2^1 x^2 (1)\, dx - \int_1^2 (y)^3 y^2\, dy - \int_2^1 (2)^3 y^2\, dy - 0$$

$$= 3 \left[\frac{1}{5} x^5 \right]_1^2 + 3 \left[\frac{1}{3} x^3 \right]_2^1 - \left[\frac{1}{6} y^6 \right]_1^2 - 8 \left[\frac{1}{3} y^3 \right]_2^1 = \frac{593}{30}$$

10.223 For the **A** given in Problem 10.222, evaluate $\int_S (\nabla \times \mathbf{A}) \cdot d\mathbf{S}$ over the triangular area shown in Fig. 10-44 and verify Stokes' theorem.

    ▌

$$\nabla \times \mathbf{A} = \left[-\frac{\partial}{\partial x}(x^3 y^2) - \frac{\partial}{\partial y}(3x^2 y^2) \right] \mathbf{a}_z = -(3x^2 y^2 + 6x^2 y)\mathbf{a}_z$$

and, consistent with the direction of the line integral, $d\mathbf{S} = dx\, dy\, (-\mathbf{a}_z)$.

Hence

$$\int_S (\nabla \times \mathbf{A}) \cdot d\mathbf{S} = \int_1^2 \int_1^x 3x^2 y^2\, dy\, dx + \int_1^2 \int_1^x 6x^2 y\, dy\, dx$$

$$= 3 \int_1^2 x^2 \left[\frac{1}{3} y^3 \right]_1^x dx + 6 \int_1^2 x^2 \left[\frac{1}{2} y^2 \right]_1^x dx = \int_1^2 x^2 (x^3 - 1)\, dx + 3 \int_1^2 x^2 (x^2 - 1)\, dx$$

$$= \frac{1}{6} [x^6]_1^2 - \frac{1}{3} [x^3]_1^2 + \frac{3}{5} [x^5]_1^2 - [x^3]_1^2 = \frac{593}{30}$$

10.224 A material medium is characterized by $\mu = \mu_0$, $\sigma = 1\ \text{S/m}$, and $\epsilon_r = 5$. The material is exposed to an electric field $\mathbf{E} = 200 \cos \omega t\, \mathbf{a}_z$ (V/m). At what frequency will the displacement and conduction current densities be equal?

    ▌ For equality $\sigma / 2\pi f \epsilon = 1$, or

$$f = \frac{\sigma}{2\pi\epsilon} = \frac{1}{2\pi \times 5 \times (10^{-9}/36\pi)} = 3.6\ \text{GHz}$$

10.225 Find the E-field corresponding to $\mathbf{B} = 3e^{-(x+at)}\mathbf{a}_z$ (T) in free space.

    ▌ Maxwell's equation gives

$$\nabla \times \mathbf{B} = \mu_0 \epsilon_0 \frac{\partial \mathbf{E}}{\partial t} \qquad \text{or} \qquad 3e^{-(x+at)}\mathbf{a}_y = \mu_0 \epsilon_0 \frac{\partial \mathbf{E}}{\partial t}$$

Thus

$$\mathbf{E} = -\frac{3}{\mu_0 \epsilon_0 a} e^{-(x+at)}\mathbf{a}_y \equiv C e^{-(x+at)}\mathbf{a}_y \qquad (1)$$

Now $\nabla \times \mathbf{E} = -\partial \mathbf{B}/\partial t$ requires

$$-C e^{-(x+at)}\mathbf{a}_z = -\frac{\partial \mathbf{B}}{\partial t} \qquad \text{or} \qquad \mathbf{B} = -\frac{C}{a} e^{-(x+at)}\mathbf{a}_z$$

Comparing with the given **B**, we must have

$$-\frac{C}{a} = \frac{3}{\mu_0\epsilon_0 a^2} = 3 \quad \text{or} \quad a = \frac{1}{\sqrt{\mu_0\epsilon_0}} = 3 \times 10^8 \text{ m/s}$$

Consequently, the required **E**-field is $\mathbf{E} = -3ae^{-(x+at)}\mathbf{a}_y$ (V/m).

10.226 If the **E**-field in free space is given by $\mathbf{E} = (A/r)\sin\theta\cos(\omega t - \beta r)\mathbf{a}_\theta$, where $\beta = \omega\sqrt{\mu_0\epsilon_0}$, show that the corresponding **H**-field is

$$\mathbf{H} = \sqrt{\frac{\epsilon_0}{\mu_0}} E_\theta \mathbf{a}_\phi$$

▌ By Faraday's law,

$$\text{curl } \mathbf{E} = \frac{1}{r}\frac{\partial}{\partial r}(rE_\theta)\mathbf{a}_\phi = \frac{\beta A \sin\theta}{r}\sin(\omega t - \beta r)\mathbf{a}_\phi = -\mu_0\frac{\partial \mathbf{H}}{\partial t}$$

or

$$\mathbf{H} = \frac{\beta}{\mu_0\omega}\left[\frac{A}{r}\sin\theta\cos(\omega t - \beta r)\right]\mathbf{a}_\phi = \sqrt{\frac{\epsilon_0}{\mu_0}} E_\theta \mathbf{a}_\phi$$

10.227 A lossless dielectric is backed by a perfect conductor, as shown in Fig. 10-45. Within the dielectric, $\mathbf{E}_1 = 2\cos 500t\,\mathbf{a}_x$ (V/m) and $\mathbf{H}_1 = \cos 500t\,(0.005\mathbf{a}_y + 0.006\mathbf{a}_z)$ (A/m). Find the surface current density at the interface.

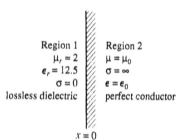

Region 1
$\mu_r = 2$
$\epsilon_r = 12.5$
$\sigma = 0$
lossless dielectric

Region 2
$\mu = \mu_0$
$\sigma = \infty$
$\epsilon = \epsilon_0$
perfect conductor

$x = 0$ **Fig. 10-45**

▌ From Ampère's law, at $x = 0$, $\mathbf{J}_s = \mathbf{a}_x \times (\mathbf{H}_2 - \mathbf{H}_1)$. But $\mathbf{H}_2 = 0$ in a perfect conductor (for a time-varying field), whence

$$\mathbf{J}_s = \mathbf{a}_x \times (-\mathbf{H}_1) = \cos 500t\,(0.006\,\mathbf{a}_y - 0.005\mathbf{a}_z) \quad (\text{A/m}^2)$$

10.228 Find the surface charge density at the interface of Problem 10.227.

▌ Within the perfect conductor, $\mathbf{D}_2 = 0$. By Gauss' law, at $x = 0$,

$$\rho_s = D_{x1} = \epsilon_1 E_{x1} = \epsilon_0\epsilon_r E_{x1}$$
$$= \frac{10^{-9}}{36\pi} \times 12.5 \times 2\cos 500t = 221\cos 500t \quad (\text{pC/m}^2)$$

10.229 The circuit of Fig. 10-46 is located in a field $\mathbf{B} = 3 \times 10^{-6}\cos(5\pi \times 10^7 t - 2\pi x/3)\,\mathbf{a}_z$ (T). Find i, given $R = 15\,\Omega$.

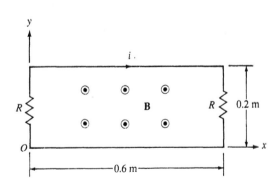

Fig. 10-46

❚ The flux linkage is given by ($\omega = 5\pi \times 10^7$ rad/s)

$$\psi_m = \int_{\text{area}} \mathbf{B} \cdot d\mathbf{S} = 3 \times 10^{-6} \times 0.2 \int_{x=0}^{0.6} \cos(\omega t - 2\pi x/3)\, dx$$

$$= 3 \times 10^{-6} \times 0.2 \times \left(\frac{-3}{2\pi}\right) \sin(\omega t - 2\pi x/3)]_0^{0.6}$$

$$= \left(\frac{0.9}{\pi} \times 10^{-6}\right)(0.691 \sin \omega t + 0.951 \cos \omega t) \quad \text{(Wb)}$$

Induced voltage and current:

$$e = -\frac{d\psi_m}{dt} = -31.095 \cos \omega t + 42.795 \sin \omega t \quad \text{(V)}$$

$$i = \frac{e}{2R} = \frac{-1}{30}(31.095 \cos \omega t - 42.795 \sin \omega t)$$

$$= -1.0365 \cos 5\pi \times 10^7 t + 1.4265 \sin 5\pi \times 10^7 t \quad \text{(A)}$$

10.230 Verify the current direction indicated in Fig. 10-46.

❚ Clockwise current produces magnetic flux through the loop in the $-z$-direction; this flux counters the flux of **B**, as required by Lenz's law.

10.231 A conducting circular loop of radius b is situated in the neighborhood of a straight conductor that carries a sinusoidal current $I_0 \sin \omega t$, as shown in Fig. 10-47. Calculate the magnetic flux that penetrates the loop.

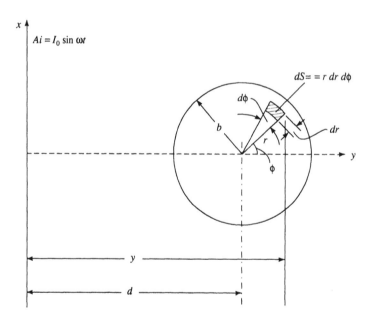

Fig. 10-47

❚ By Ampère's law, at a point (r, ϕ) within the loop,

$$i = I_0 \sin \omega t = H_z(2\pi y) = B_z \frac{2\pi y}{\mu_0}$$

or

$$B_z = \frac{\mu_0 I_0 \sin \omega t}{2\pi y} = \frac{\mu_0 I_0 \sin \omega t}{2\pi (d + r \cos \phi)}$$

Thus

$$\psi_m = \iint B_z\, dS = \frac{\mu_0 I_0}{2\pi} \sin \omega t \int_0^b \int_0^{2\pi} \frac{r\, dr\, d\phi}{d + r \cos \phi}$$

$$= \mu_0 I_0 \sin \omega t \int_0^b \frac{r\, dr}{\sqrt{d^2 - r^2}} = \mu_0 I_0 (d - \sqrt{d^2 - b^2}) \sin \omega t \tag{1}$$

10.232 For the configuration of Problem 10.231 (Fig. 10-47): $b = 0.1$ m, $f = 60$ Hz, $d = 0.15$ m. An ac milliammeter inserted into a small gap in the loop reads 0.3 mA. If the total impedance of the loop, including the milliammeter, is 0.01 Ω, determine I_0.

$\blacksquare$ The voltage induced in the loop is, from (1) of Problem 10.231,

$$e = \frac{d\psi_m}{dt} = \mu_0 I_0 (d - \sqrt{d^2 - b^2})\omega \cos \omega t$$

For 0.3 mA (rms) current, Ohm's law gives

$$\frac{\mu_0 I_0 (d - \sqrt{d^2 - b^2})\omega}{\sqrt{2}} = (0.3 \times 10^{-3})Z \tag{1}$$

Substitute data in (1) and solve for I_0, obtaining $I_0 = 0.234$ A.

10.233 For a plane wave, propagating in the z-direction in air,

$$\hat{\mathbf{E}}(z) = (30 - j40)e^{-j\pi z}\mathbf{a}_x + (20 + j10)e^{-j\pi z}\mathbf{a}_y$$

Calculate $\mathbf{E}(\frac{1}{3}$ m, 5 ns).

$\blacksquare$ The given phasor may also be written as

$$\hat{\mathbf{E}} = 50\underline{/-53.1°}\, e^{-j\pi z}\mathbf{a}_x + 22.36\underline{/26.6°}\, e^{-j\pi z}\mathbf{a}_y$$

For this wave, $\omega = \beta u = \pi(3 \times 10^8)$ rad/s. Thus, in the time domain we have

$$\mathbf{E}(z, t) = 50 \cos (\omega t - \pi z - 53.1°)\, \mathbf{a}_x + 22.36 \cos (\omega t - \pi z + 26.6°)\, \mathbf{a}_y$$

Substituting $z = \frac{1}{3}$ m and $\omega t = 1.5\pi$ rad yields $\mathbf{E} = -46\mathbf{a}_x - 12.32\mathbf{a}_y$.

10.234 The electric field of a plane wave in air is given by

$$\hat{\mathbf{E}} = 100e^{-j\pi z/3}\mathbf{a}_x + j100e^{-j\pi z/3}\mathbf{a}_y \quad \text{(V/m)} \tag{1}$$

Express this field in the time domain and determine the frequency.

$\blacksquare$
$$\mathbf{E}(z, t) = 100 \cos \left(\omega t - \frac{\pi}{3}z \right) \mathbf{a}_x + 100 \cos \left(\omega t - \frac{\pi}{3}z + \frac{\pi}{2} \right) \mathbf{a}_y$$

$$= 100 \cos \left(\omega t - \frac{\pi}{3}z \right) \mathbf{a}_x - 100 \sin \left(\omega t - \frac{\pi}{3}z \right) \mathbf{a}_y \quad \text{(V/m)}$$

$$\omega = \beta u = \frac{\pi}{3} \times 3 \times 10^8 = 10^8\pi \text{ rad/s}$$

10.235 For Problem 10.234, determine the $\mathbf{E}$-field in the $z = 0$ plane at $t = 0, 5, 10, 15$ ns.

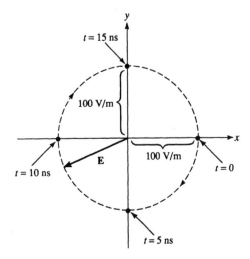

Fig. 10-48

$\blacksquare$ See Fig. 10-48.

10.236 Refer to Fig. 10-48. What is the sense of the circular polarization?

$\blacksquare$ The wave is propagating out of the paper, in the direction $+\mathbf{a}_z$. Looking in that direction, the rotation of $\mathbf{E}$ is counterclockwise; i.e. exhibits left-hand polarization.

10.237 A 3-GHz plane wave, with $\mathbf{E}$ along the y-direction, travels along the x-direction in a lossy medium characterized by $\mu = \mu_0$, $\epsilon_r = 2.5$, and $\sigma/\omega\epsilon = 0.01$. Calculate the propagation constant γ.

▌ The propagation constant is given by
$$\gamma = \sqrt{j\omega\mu(\sigma + j\omega\epsilon)} = j\omega\sqrt{\mu\epsilon}\ \sqrt{1 - j(\sigma/\omega\epsilon)}$$
Substituting $\omega = 2\pi \times 3 \times 10^9$ rad/s, $\mu = 4\pi \times 10^{-7}$ H/m, $\epsilon = 2.5 \times (10^{-9}/36\pi)$ F/m, and $\sigma/\omega\epsilon = 0.01$ yields
$$\gamma = (j6\pi \times 10^9)\sqrt{(4\pi \times 10^{-7} \times 2.5 \times (10^{-9}/36\pi)}\ \sqrt{1 - j0.01} = 0.496 + j99.345 \quad \text{m}^{-1}$$

10.238 Determine the distance over which the amplitude of the wave of Problem 10.237 is reduced by 50 percent.

▌ From Problem 10.237, the attenuation constant $\alpha = 0.496$ Np/m, and the required distance, d, is given by
$$e^{-\alpha d} = 0.5 \quad \text{or} \quad d = \frac{1}{\alpha}\ln 2 = \frac{1}{0.496}(0.693) = 1.397 \text{ m}$$

10.239 Determine the intrinsic impedance of the medium in Problem 10.237.

▌
$$\eta = \frac{j\omega\mu}{\gamma} = \frac{j2400\pi^2}{0.496 + j99.345} = 238.4\underline{/0.286°}\ \ \Omega$$

10.240 If the E-field at $x = 0$ in Problem 10.237 is
$$\mathbf{E}(0, t) = 50\cos(6\pi \times 10^9 t + \pi/3)\,\mathbf{a}_y \quad (\text{V/m})$$
what is the corresponding $\mathbf{H}(x, t)$?

▌ From Problem 10.237, $\alpha = 0.496$ m^{-1} and $\beta = 99.345 = 31.6\pi$ rad/m. Thus, at any point within the material medium, the E-field may be written as
$$\mathbf{E}(x, t) = 50e^{-0.496x}\cos\left(6\pi \times 10^9 t + \frac{\pi}{3} - 31.6\pi x\right)\mathbf{a}_y \quad (\text{V/m})$$
Since $\eta = 238.4\ \underline{/0.286°}\ \Omega$, from Problem 10.239, and since $\hat{\mathbf{H}} = \mathbf{a}_z(\hat{\mathbf{E}} \cdot \mathbf{a}_y/\eta)$,
$$\mathbf{H}(x, t) = \frac{50}{238.4}e^{-0.496x}\cos\left(6\pi \times 10^9 t + \frac{\pi}{3} - 31.6\pi x - 0.286°\right)\mathbf{a}_z$$
$$= 0.21e^{-0.496x}\cos(6\pi \times 10^9 t - 31.6\pi x + 59.71°)\,\mathbf{a}_z \quad (\text{A/m})$$

10.241 A plane wave, traveling in the y-direction in seawater ($\mu_r = 1$, $\epsilon_r = 80$, $\sigma = 4$ S/m), has, at $y = 0$, the magnetic vector
$$\mathbf{H}(0, t) = 0.1\sin(10^{10}\pi t - \pi/3)\,\mathbf{a}_x \quad (\text{A/m})$$
Determine the attenuation constant, phase constant, and intrinsic impedance.

▌
$$\gamma = \sqrt{j\omega\mu(\sigma + j\omega\epsilon)} = \sqrt{j(10^{10}\pi)(4\pi \times 10^{-7})[4 + j(10^{10}\pi)(80 \times 10^{-9}/36\pi)]} = 84 + j300\pi \quad \text{m}^{-1}$$
Thus $\alpha = 84$ Np/m; $\beta = 300\pi$ rad/m; and
$$\eta = \frac{j\omega\mu}{\gamma} = 41.8\ \underline{/5.1°}\ \ \Omega$$

10.242 Determine $\mathbf{E}(y, t)$ corresponding to the H-field of Problem 10.241.

▌ For all y, $\mathbf{H}(y, t) = 0.1e^{-84y}\sin(10^{10}\pi t - \pi/3 - 300\pi y)\,\mathbf{a}_x$ (A/m). Then, with η as found in Problem 10.241, we have $\hat{\mathbf{E}} = \mathbf{a}_z(\eta\hat{\mathbf{H}} \cdot \mathbf{a}_x)$, or
$$\mathbf{E}(y, t) = 4.18e^{-84y}\sin\left(10^{10}\pi t - \frac{\pi}{3} - 300\pi y + 5.1°\right)\mathbf{a}_z = 4.18e^{-84y}\sin(10^{10}\pi t - 300\pi y - 54.9°)\,\mathbf{a}_z \quad (\text{V/m})$$

10.243 At 100 MHz, the skin depth of graphite is 0.16 mm. What is the conductivity of graphite?

▌ Since $\delta = 1/\sqrt{\pi f\mu\sigma}$,
$$\sigma = 1/\pi f\mu_0\delta^2 = 1/(\pi \times 100 \times 10^6 \times 4\pi \times 10^{-7} \times 0.16 \times 0.16 \times 10^{-6}) \approx 0.1 \text{ MS/m}$$

10.244 For the graphite of Problem 10.243, determine the distance a 1-GHz wave travels before its field intensity is reduced by 50 percent.

▌ Since graphite is a good conductor, its attenuation constant is $\alpha \approx 1/\delta$. Now, at 1 GHz = 1000 MHz,
$$\delta = \sqrt{\frac{100}{1000}}(0.16 \text{ mm}) = 50\ \mu\text{m}$$
The required distance d is given by
$$e^{-\alpha d} = 0.5 \quad \text{or} \quad d = \delta\ln 2 = (50)(0.693) = 35\ \mu\text{m}$$

10.245 Describe the polarization of the electric wave
$$\mathbf{E} = E_0\sin(\omega t - \beta z)\,\mathbf{a}_x - \frac{1}{2}E_0\sin(\omega t - \beta z - 45°)\,\mathbf{a}_y$$

▌ The two components are not in phase and have different amplitudes; hence, the wave exhibits elliptical polarization. Since E_y leads E_x (by 135°) and propagation is in the z-direction, the polarization is left-handed.

10.246 A right-hand circularly polarized plane wave in air, with
$$\hat{\mathbf{E}}_i(z) = E_0(\mathbf{a}_x - j\mathbf{a}_y)e^{-j\beta z}$$
impinges normally on a perfectly conducting wall at $z = 0$. Determine the polarization of the reflected wave.

▌ $\hat{\mathbf{E}}_r(0) = -\hat{\mathbf{E}}_i(0) = -E_0(\mathbf{a}_x - j\mathbf{a}_y)$; thus
$$\hat{\mathbf{E}}_r(z) = -E_0(\mathbf{a}_x - j\mathbf{a}_y)e^{j\beta z} = E_0(\mathbf{a}_x - j\mathbf{a}_y)e^{j\beta(z+\lambda/2)}$$
Only the direction of propagation has been reversed: left-hand circular polarization.

10.247 Find the linear current density $\hat{\mathbf{K}}$ on the conducting wall of Problem 10.246.

▌ The usual plane wave relations give
$$\hat{\mathbf{H}}_i(0) = \frac{E_0}{\eta}(j\mathbf{a}_x + \mathbf{a}_y) = \hat{\mathbf{H}}_r(0)$$
whence
$$\hat{\mathbf{K}} = -\mathbf{a}_z \times [\hat{\mathbf{H}}_i(0) + \hat{\mathbf{H}}_r(0)] = \frac{2E_0}{\eta}(\mathbf{a}_x - j\mathbf{a}_y)$$

10.248 Obtain an expression for the instantaneous total electric field intensity in Problem 10.246. Assume E_0 real.

▌
$$\hat{E}_T(z) = \hat{E}_i(z) + \hat{E}_r(z) = -2E_0 \sin \beta z\, (\mathbf{a}_y + j\mathbf{a}_x)$$
$$E_T(z, t) = \text{Re}\,[\hat{E}_T(z)(\cos \omega t + j \sin \omega t]$$
$$= 2E_0 \sin \beta z\, (\mathbf{a}_x \sin \omega t - \mathbf{a}_y \cos \omega t)$$
As predicted in Problem 10.246, this is a *standing*, right-hand circularly polarized wave.

10.249 A plane wave, whose electric field is given by $\mathbf{E}_i = \mathbf{a}_x\, 10 \cos (6\pi \times 10^9 t - 30\pi z)$ (V/m), is incident normally on a dielectric interface ($z = 0$). Given $\mu_r = 1$ everywhere, calculate ϵ_r for $z < 0$.

▌ The required ϵ_{r1} (for $z < 0$) is obtained from the relationship $\beta_1 = \omega\sqrt{\mu_0\epsilon_0\epsilon_{r1}}$. Substituting $\beta_1 = 30\pi$, $\omega = 6\pi \times 10^9$, and $\mu_0 = 4\pi \times 10^{-7}$, we obtain
$$\epsilon_{r1} = \left(\frac{30\pi \times 3 \times 10^8}{6\pi \times 10^9}\right)^2 = 2.25$$

10.250 In Problem 10.249 both regions are lossless, and ϵ_{r2} (for $z > 0$) is 4. Determine the reflection coefficient Γ and the transmission coefficient T.

▌ For lossless regions,
$$\eta_1 = \sqrt{\frac{\mu_1}{\epsilon_1}} = \sqrt{\frac{\mu_0}{\epsilon_0\epsilon_{r1}}} = 80\pi\ \Omega \qquad \eta_2 = \sqrt{\frac{\mu_2}{\epsilon_2}} = \sqrt{\frac{\mu_0}{\epsilon_0\epsilon_{r2}}} = 60\pi\ \Omega$$
Then
$$\Gamma = \frac{\eta_2 - \eta_1}{\eta_2 + \eta_1} = \frac{60 - 80}{60 + 80} = -\frac{1}{7} \qquad \text{and} \qquad T = 1 + \Gamma = \frac{6}{7}$$

10.251 Determine the instantaneous vector expressions for the reflected and transmitted magnetic fields in Problem 10.249.

▌ The phase constant of the transmitted wave is
$$\beta_2 = \beta_1\sqrt{\epsilon_{r2}/\epsilon_{r1}} = 30\pi\sqrt{16/9} = 40\pi \text{ rad/m}$$
Taking Γ and T from Problem 10.250, we obtain the reflected fields:
$$\mathbf{E}_r = -\frac{10}{7} \cos (6\pi \times 10^9 t + 30\pi z)\, \mathbf{a}_x \quad \text{(V/m)}$$
$$\mathbf{H}_r = \frac{10}{7} \times \frac{1}{80\pi} \cos (6\pi \times 10^9 t + 30\pi z)\, \mathbf{a}_y = 5.68 \times 10^{-3} \cos (6\pi \times 10^9 t + 30\pi z)\, \mathbf{a}_y \quad \text{(A/m)}$$
The transmitted fields are
$$\mathbf{E}_t = \frac{6}{7} \times 10 \cos (6\pi \times 10^9 t - 40\pi z)\, \mathbf{a}_x \quad \text{(V/m)}$$
$$\mathbf{H}_t = \frac{6}{7} \times \frac{10}{60\pi} \cos (6\pi \times 10^9 t - 40\pi z)\, \mathbf{a}_y = 45.5 \times 10^{-3} \cos (6\pi \times 10^9 t - 40\pi z)\, \mathbf{a}_y \quad \text{(A/m)}$$

10.252 A 10-MHz plane wave propagating in free space in the direction $\mathbf{a}_y$ is normally incident on a perfectly conducting surface, $y = 0$. The incident electric field has a maximum amplitude E_0 and is directed along the unit vector $(0.6\mathbf{a}_x - 0.8\mathbf{a}_z)$. Write the complete vector expressions for the incident and reflected electric and magnetic fields, using a cosine reference; i.e., $\mathbf{E} = \text{Re}\,(\hat{\mathbf{E}}e^{j\omega t})$.

∎ At 10 MHz,

$$\omega = 2\pi \times 10^7 \text{ rad/s} \qquad \beta = \frac{\omega}{u} = \frac{\pi}{15} \text{ rad/m}$$

and $\eta = 120\pi \,\Omega$. The incident fields are then

$$\mathbf{E}_i(y, t) = E_0 \cos\left(2\pi \times 10^7 t - \frac{\pi}{15}y\right)(0.6\mathbf{a}_x - 0.8\mathbf{a}_z) \quad \text{(V/m)}$$

$$\mathbf{H}_i(y, t) = \frac{1}{\eta}\,\mathbf{a}_y \times \mathbf{E}_i = \frac{E_0}{120\pi} \cos\left(2\pi \times 10^7 t - \frac{\pi}{15}y\right)[-(0.8\mathbf{a}_x + 0.6\mathbf{a}_z)]$$

$$= -E_0 \cos\left(2\pi \times 10^7 t - \frac{\pi}{15}y\right)(2.12 \times 10^{-3}\mathbf{a}_x + 1.59 \times 10^{-3}\mathbf{a}_y) \quad \text{(A/m)}$$

Since the reflection coefficient $\Gamma = -1$ at a perfect conductor, the reflected fields are

$$\mathbf{E}_r(y, t) = -E_0 \cos\left(2\pi \times 10^7 t + \frac{\pi}{15}y\right)(0.6\mathbf{a}_x - 0.8\mathbf{a}_z) \quad \text{(V/m)}$$

$$\mathbf{H}_r(y, t) = \frac{1}{\eta}(-\mathbf{a}_y) \times \mathbf{E}_r = -E_0 \cos\left(2\pi \times 10^7 t + \frac{\pi}{15}y\right)(2.12 \times 10^{-3}\mathbf{a}_x + 1.59 \times 10^{-3}\mathbf{a}_y) \quad \text{(A/m)}$$

10.253 Evaluate $\rho_s(t)$, the surface charge density on the conducting surface of Problem 10.252.

∎ Because the incident and reflected waves are perpendicularly polarized, $\mathbf{D}_n = \epsilon_0 \mathbf{E}_n = 0$ at $y = 0$, and so $\rho_s(t) \equiv 0$.

10.254 A plane wave in air (medium 1), with $\hat{\mathbf{E}}_i(z) = 10e^{-j6z}\mathbf{a}_x$, is incident normally at an interface (at $z = 0$) with a lossy dielectric (medium 2) characterized by $\epsilon_r = 2.5$ and $\sigma/\omega\epsilon = 0.5$. Determine the reflection and transmission coefficients and the propagation constant for medium 2.

∎ The intrinsic impedances are $\eta_1 = 377\,\Omega = \eta_0$ and

$$\eta_2 = \sqrt{\frac{j\omega\mu_0}{\sigma + j\omega\epsilon_0\epsilon_r}} = \sqrt{\frac{j\omega\mu_0}{j\omega\epsilon_0\epsilon_r(1 - j\sigma/\omega\epsilon)}}$$

$$= \frac{\eta_0}{\sqrt{2.5}}\sqrt{\frac{1}{1 - j\sigma/\omega\epsilon}} = \frac{377}{\sqrt{2.5}}\sqrt{\frac{1 + j0.5}{1.25}} = 225.5\,\underline{/13.28°}\quad \Omega$$

Thus, required coefficients are

$$\Gamma = \frac{\eta_2 - \eta_1}{\eta_2 + \eta_1} = \frac{(219.5 - 377) + j51.81}{(219.5 + 377) + j51.81} = 0.277\,\underline{/156.8°} \qquad T = 1 + \Gamma = 0.753\,\underline{/8.3°}$$

Finally, since $\omega\mu_0 = \beta_0\eta_0$,

$$\gamma_2 = \frac{j\omega\mu_0}{\eta_2} = \frac{j\beta_0\eta_0}{\eta_2} = \frac{j6(377)}{225.5\,\underline{/13.28°}}$$

$$= 10.0\,\underline{/76.72°} = 2.30 + j9.76 \quad \text{m}^{-1}$$

10.255 Write time-domain expressions for the incident, reflected, and transmitted fields in the wave of Problem 10.254.

∎ Since $\beta_0 = 6 \text{ rad/m}$ and $\beta_0 u_0 = 6 \times 3 \times 10^8 = 1.8 \times 10^9 \text{ rad/s}$, the incident fields are

$$\mathbf{E}_i = 10 \cos(1.8 \times 10^9 t - 6z)\,\mathbf{a}_x \quad \text{(V/m)} \qquad \mathbf{H}_i = \frac{10}{377} \cos(1.8 \times 10^9 t - 6z)\,\mathbf{a}_y \quad \text{(A/m)}$$

With Γ as in Problem 10.254 [$156.8° = 2.737$ rad], the reflected fields are:

$$\hat{\mathbf{E}}_r = \Gamma\hat{\mathbf{E}}_i^* = 2.77e^{j6z}e^{j2.737}\,\mathbf{a}_x \quad \text{or} \quad \mathbf{E}_r = 2.77 \cos(1.8 \times 10^9 t + 6z + 2.737)\,\mathbf{a}_x \quad \text{(V/m)}$$

$$\hat{\mathbf{H}}_r = -\Gamma\hat{\mathbf{H}}_i^* = -\frac{0.277\,\underline{/2.737}}{377}(10e^{j6z})\mathbf{a}_y \quad \text{or} \quad \mathbf{H}_r = -7.347 \times 10^{-3} \cos(1.8 \times 10^9 t + 6z + 2.737)\,\mathbf{a}_y \quad \text{(A/m)}$$

Knowing the transmission coefficient and the propagation constant from Problem 10.254, we may write the transmitted fields as

$$\hat{\mathbf{E}}_t = T\hat{\mathbf{E}}_i(0)e^{-\gamma_2 z} \quad \text{or} \quad \mathbf{E}_t = 7.23e^{-2.30z} \cos(1.8 \times 10^9 t - 9.76z + 8.3°)\,\mathbf{a}_x \quad \text{(V/m)}$$

$$\hat{\mathbf{H}}_t = \frac{|\hat{\mathbf{E}}_t|}{\eta_2}\,\mathbf{a}_y \quad \text{or} \quad \mathbf{H}_t = 0.033e^{-2.30z} \cos(1.8 \times 10^9 t - 9.76z - 5°)\,\mathbf{a}_y \quad \text{(A/m)}$$

10.256 For the wave of Problem 10.254, find the time-average Poynting vectors in the air and in the lossy medium.

▌ In medium 1 (air),

$$\mathbf{S}_{av} = (\mathbf{S}_{av})_i + (\mathbf{S}_{av})_r = \frac{1}{2}\left[\frac{10^2}{377} - \frac{(10 \times 0.277)^2}{377}\right]\mathbf{a}_z = 0.122\mathbf{a}_z \ \text{W/m}^2$$

In medium 2 (lossy medium; see Problem 5.88),

$$\mathbf{S}_{av} = \frac{E_m^2 |T|^2}{2|\eta_2|}e^{-2\alpha_2 z}\cos\theta_{\eta 2}\,\mathbf{a}_z = \frac{(10 \times 0.753)^2}{2(225.5)}e^{-4.60z}\cos 13.3°\,\mathbf{a}_z = 0.122e^{-4.60z}\mathbf{a}_z \quad (\text{W/m}^2)$$

10.257 Consider a wave propagating in air in the $\mathbf{a}_z$ direction, with
$$\mathbf{E}(x, z) = \mathbf{a}_y \sin(10\pi x)\,e^{-j5\pi z} \quad (\text{V/m})$$
Find the frequency and the magnetic field intensity $\mathbf{H}$.

▌ This is a nonuniform wave along the z axis, since the amplitude varies over a (xy) plane perpendicular to the direction of travel. Identify $\beta_x = 10\pi$ and $\beta_z = 5\pi$; then

$$\beta_0 = \omega\sqrt{\mu_0\epsilon_0} = \frac{\omega}{u_0} = \sqrt{\beta_x^2 + \beta_z^2} \quad \text{or} \quad \omega = (3 \times 10^8)\sqrt{(10\pi)^2 + (5\pi)^2} = 105.4 \times 10^8 \ \text{rad/s}$$

From Maxwell's equation,

$$\mathbf{H} = -\frac{1}{j\omega\mu_0}\nabla \times \mathbf{E} = -\mathbf{a}_x\frac{5\pi}{\omega\mu_0}\sin 10\pi x\, e^{-j5\pi z} + \mathbf{a}_z j\frac{10\pi}{\omega\mu_0}\cos 10\pi x\, e^{-j5\pi z}$$

Substituting the numerical values for ω and μ_0 finally yields
$$\mathbf{H} = (-\mathbf{a}_x 1.186 \sin 10\pi x + j\mathbf{a}_z 2.372 \cos 10\pi x)\,10^{-3}\,e^{-j5\pi z} \quad (\text{A/m})$$

10.258 Show that the wave of Problem 10.257 is the superposition of two *plane* waves. Determine unit vectors $\mathbf{a}_{n1}$ and $\mathbf{a}_{n2}$ in the directions of propagation of these plane waves.

▌ The given $\mathbf{E}$-field may be written as

$$\mathbf{E} = \mathbf{a}_y\frac{1}{2j}(e^{j\beta_x x} - e^{-j\beta_x x})e^{-j\beta_z z} = \mathbf{a}_y\frac{1}{2j}e^{-j(-\beta_x x + \beta_z z)} - \mathbf{a}_y\frac{1}{2j}e^{-j(\beta_x x + \beta_z z)}$$

The two waves on the right are clearly plane waves; indeed, waves of constant amplitude. Looking at the phase of the first wave, we see that $\mathbf{a}_{n1}$ is obtained by normalizing the vector $-\beta_x\mathbf{a}_x + \beta_z\mathbf{a}_z$:

$$\mathbf{a}_{n1} = -\frac{\beta_x}{\beta_0}\mathbf{a}_x + \frac{\beta_z}{\beta_0}\mathbf{a}_z = \frac{1}{\sqrt{5}}(-2\mathbf{a}_x + \mathbf{a}_z)$$

Similarly,

$$\mathbf{a}_{n2} = \frac{1}{\sqrt{5}}(2\mathbf{a}_x + \mathbf{a}_z)$$

10.259 Determine the $\mathbf{H}$-fields of the two plane waves of Problem 10.258. Add the two vectors and show that the result is consistent with that obtained in Problem 10.257.

▌

$$\mathbf{H}_1 = \frac{1}{\eta_0}\mathbf{a}_{n1} \times \mathbf{E}_1 = \frac{1}{120\pi}\left[\frac{1}{2j\sqrt{5}}(-2\mathbf{a}_z - \mathbf{a}_x)\right]e^{j(10\pi x - 5\pi z)}$$

$$\mathbf{H}_2 = \frac{1}{\eta_0}\mathbf{a}_{n2} \times \mathbf{E}_2 = \frac{1}{120\pi}\left[\frac{1}{2j\sqrt{5}}(-2\mathbf{a}_z + \mathbf{a}_x)\right]e^{-j(10\pi x + 5\pi z)}$$

Adding the two vectors yields

$$\mathbf{H} = \mathbf{H}_1 + \mathbf{H}_2 = \mathbf{a}_z\frac{j2/\sqrt{5}}{120\pi}\left[\frac{1}{2}(e^{j10\pi x} + e^{-j10\pi x})\right]e^{-j5\pi z} - \mathbf{a}_x\frac{1/\sqrt{5}}{120\pi}\left[\frac{1}{2j}(e^{j10\pi x} - e^{-j10\pi x})\right]e^{-j5\pi z}$$

$$= (-\mathbf{a}_x 1.186 \sin 10\pi x + j\mathbf{a}_z 2.372 \cos 10\pi x)\,10^{-3}\,e^{-j5\pi z} \quad (\text{A/m})$$

10.260 A plane wave in air, with electric field
$$\mathbf{E}(y, z) = 5(\mathbf{a}_y + \mathbf{a}_z\sqrt{3})e^{j6(\sqrt{3}y - z)} \quad (\text{V/m})$$
is incident (obliquely) on a perfectly conducting plane located at $z = 0$. Determine the instantaneous surface charge density on the plane. (Compare Problem 10.253.)

▌ The wave approaches from $z < 0$; at $z = 0^-$ the normal component of $\mathbf{E}$ is
$$E_n = -E_z(y, 0) = -5\sqrt{3}\,e^{j6\sqrt{3}y} \quad \text{V/m}$$
and Gauss' law gives
$$\rho_s = D_n = \epsilon_0 E_n = -5\sqrt{3}\,\epsilon_0\cos(\omega t + 6\sqrt{3}\,y) \quad \text{C/m}^2$$
where $\omega = u_0\sqrt{\beta_y^2 + \beta_z^2} = 3.6 \times 10^9 \ \text{rad/s}$.

10.261 Consider oblique incidence on a perfect conductor (perpendicular polarization), as shown in Fig. 10-49. Write the expressions for the instantaneous incident fields, given that the incident $\mathbf{E}$-field has an amplitude $E_{i0} = 1 \ \text{V/m}$ at 150 MHz.

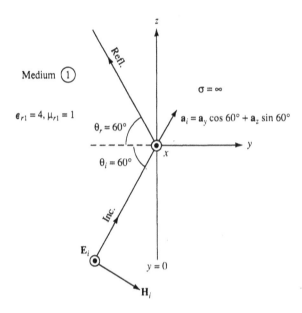

Fig. 10-49

$\blacksquare$ First we calculate

$$\beta_1 = \frac{\omega}{u_1} = \frac{\omega\sqrt{\epsilon_{r1}}}{u_0} = 2\pi \text{ rad/m} \qquad \eta_1 = \frac{\eta_0}{\sqrt{\epsilon_{r1}}} = 60\pi \ \Omega$$

$$\hat{\mathbf{E}}_i = \mathbf{a}_x \, 1 \, e^{-j2\pi(y\cos 60° + z\sin 60°)} = \mathbf{a}_x e^{-j(\pi y + \pi\sqrt{3}\,z)}$$

or $\mathbf{E}_i(y, z, t) = \mathbf{a}_x \cos(3\pi \times 10^8 t - \pi y - \pi\sqrt{3}\,z)$ (V/m). From this,

$$\hat{\mathbf{H}}_i = \left(-\mathbf{a}_z\frac{1}{2} + \mathbf{a}_y\frac{\sqrt{3}}{2}\right)\frac{1}{\eta_1} e^{-j(\pi y + \pi\sqrt{3}\,z)}$$

or $$\mathbf{H}_i(y, z, t) = \frac{1}{60\pi}\left(-\frac{1}{2}\mathbf{a}_z + \frac{\sqrt{3}}{2}\mathbf{a}_y\right)\cos(3\pi \times 10^8 t - \pi y - \pi\sqrt{3}\,z) \quad \text{(A/m)}$$

10.262 Write the reflected E-field for Problem 10.261.

$\blacksquare$ The boundary condition is $\hat{\mathbf{E}}_r(0, 0) = -\hat{\mathbf{E}}_i(0, 0) = (-1)\mathbf{a}_x$, and the direction of the reflected wave is
$$\mathbf{a}_r = -\mathbf{a}_y \cos 60° + \mathbf{a}_z \sin 60°$$
Then, $\mathbf{E}_r(y, z, t) = -\mathbf{a}_x \cos(3\pi \times 10^8 t + \pi y - \pi\sqrt{3}\,z)$ (V/m).

10.263 A light ray is incident on a $\frac{1}{8}$-in-thick glass pane at 85° to the normal, as shown in Fig. 10-50. If $\epsilon_r = 2.25$ for glass, determine the displacement S.

$\blacksquare$ By Snell's law,

$$\frac{\sin\theta_t}{\sin 85°} = \frac{\eta_1}{\eta_2} = \frac{1}{\sqrt{2.25}} \qquad \text{or} \qquad \theta_t = 41.6°$$

Now $$\cos 41.6° = \frac{1/8''}{d} \qquad \text{or} \qquad d = 0.1671 \text{ in}$$

Finally, $S = d\sin 43.4° = 0.1148$ in.

10.264 Light can be guided along dielectric fibers of transparent material. Determine the minimum dielectric constant of the fiber so that a wave incident on one end at any angle will be confined to the fiber until it emerges at the other end (Fig. 10-51).

$\blacksquare$ For total internal reflection, the angle θ_{i2} must be such that

$$\sin\theta_{i2} \geq \frac{1}{\sqrt{\epsilon_r}}$$

Since $\theta_{i2} = 90° - \theta_{t1}$, we also have

$$\sin\theta_{i2} = \cos\theta_{t1} \geq \frac{1}{\sqrt{\epsilon_r}} \qquad \text{or} \qquad 1 - \sin^2\theta_{t1} \geq \frac{1}{\epsilon_r} \tag{1}$$

But, by Snell's law,

$$\frac{\sin\theta_{t1}}{\sin\theta_{i1}} = \frac{1}{\sqrt{\epsilon_r}} \qquad \text{or} \qquad \sin^2\theta_{t1} = \frac{\sin^2\theta_{i1}}{\epsilon_r} \tag{2}$$

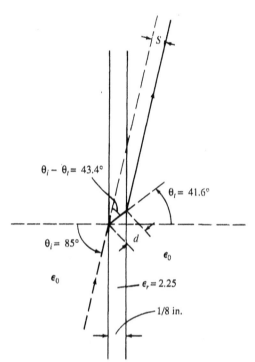

Fig. 10-50

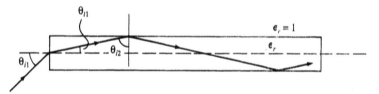

Fig. 10-51

Substituting (2) in (1) yields

$$\epsilon_r \geq 1 + \sin^2 \theta_{i1} \qquad \text{whence} \qquad \epsilon_{r\min} = 2$$

10.265 For the glass prism shown in Fig. 10-52, find the percent of incident light power reflected back by the prism.

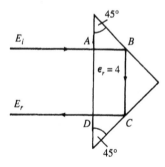

Fig. 10-52

▌ Transmission coefficient at A:

$$T_A = \frac{2\eta_2}{\eta_2 + \eta_1} = \frac{2(\eta_0/2)}{(\eta_0/2) + \eta_0} = \frac{2}{3}$$

At B and C there is total internal reflection:

$$\Gamma_B = \Gamma_C = 1$$

Transmission coefficient at D:

$$T_D = \frac{2\eta_0}{\eta_0 + (\eta_0/2)} = \frac{4}{3}$$

Hence $$\frac{E_r}{E_i} = \frac{2}{3} \times 1 \times 1 \times \frac{4}{3} = \frac{8}{9} \quad \text{and} \quad \frac{P_r}{P_i} = \left(\frac{8}{9}\right)^2 = 79\%$$

10.266 An airplane uses a radar altimeter to determine its height above ground. If the airplane is flying over the ocean ($\epsilon_r = 81$, $\mu_r = 1$, $\sigma = 4$ S/m) and the radar frequency is 7 GHz, determine the percent power reduction, $(P_{\text{trans}} - P_{\text{recd}})/P_{\text{trans}}$.

▮ With the air as medium 1 and the ocean as medium 2,

$$\eta_1 = \eta_0 = 377 \ \Omega$$
$$\eta_2 = \sqrt{j\omega\mu/(\sigma + j\omega\epsilon)} = 41.721 \ \underline{/3.619°} \ \ \Omega$$
$$\Gamma = \frac{\eta_2 - \eta_1}{\eta_2 + \eta_1} = 0.8011 \ \underline{/179.192°}$$

Then

$$\text{power reduction} = 1 - \frac{P_{\text{recd}}}{P_{\text{trans}}} = 1 - \left(\frac{|\mathbf{E}_r|}{|\mathbf{E}_i|}\right)^2 = 1 - |\Gamma|^2 = 0.3583$$

or 35.83 percent.

10.267 A lossy transmission line, 60 m long, is connected between a sinusoidal source and a load. Given that $Z_C = 703 \ \underline{/-13.2°} \ \ \Omega$, $\gamma = \alpha + j\beta = 0.0083 + j0.035 \ \ \text{m}^{-1}$, $V_S = 20 \ \underline{/0°} \ \ \text{V}$, $Z_S = 700 \ \Omega$, $f = 1$ kHz, and $Z_L = 1000 \ \Omega$, determine the per-unit-length resistance, inductance, capacitance, and conductance of the line.

▮ With $Z_C = 703 \ \underline{/-13.2°} = 684.43 - j160.53 \ \ \Omega$, we have
$$Z = r + j\omega l = \gamma Z_C = (0.0083 + j0.035)(684.43 - j160.53) = 11.3 + j22.62 \ \ \Omega/\text{m}$$
Thus, $r = 11.3 \ \Omega/\text{m}$, and $\omega l = 22.62 \ \Omega/\text{m}$ or $l = 3.60$ mH/m. Finally,
$$g + j\omega c = Y = \frac{\gamma}{Z_C} = \frac{0.0083 + j0.035}{684.43 - j160.53} = 0.0125 \times 10^{-5} + j5.120 \times 10^{-5} \ \ \text{S/m}$$
or $g = 0.125 \ \mu$S/m and $c = (5.120 \times 10^{-5})/(2\pi \times 10^3) = 8.15$ nF/m.

10.268 For the line of Problem 10.267, calculate the input reflection coefficient. Take $z = 0$ as the location of the input terminals.

▮ The input reflection coefficient is given by $\Gamma(0) = \Gamma_L e^{-2\gamma\zeta}$, where
$$\Gamma_L = \frac{Z_L - Z_C}{Z_L + Z_C} = \frac{1000 - 703 \ \underline{/-13.2°}}{1000 + 703 \ \underline{/-13.2°}} = 0.209 \ \underline{/32.5°} = 0.209 e^{j0.567}$$

Thus $$\Gamma(0) = 0.209 e^{j0.567} e^{-2(0.0083 \times 60)} e^{-j2(0.035 \times 60)} = 0.0675 + j0.0371$$

10.269 Determine the input impedance of the line of Problems 10.267 and 10.268.

▮
$$Z_{\text{in}}(0) = Z_C \frac{1 + \Gamma(0)}{1 - \Gamma(0)} = (703 \ \underline{/-13.20°}) \left[\frac{1 + (0.0675 + j0.0371)}{1 - (0.0675 + j0.0371)}\right]$$
$$= 804 \ \underline{/-13.49°} = 782 - j187 \ \ \Omega$$

10.270 For the line of Problems 10.267 and 10.268 determine the VSWR at the load and at the input.

▮ At any location z,

$$\text{VSWR}(z) = \frac{1 + |\Gamma(z)|}{1 - |\Gamma(z)|}$$

Thus

$$\text{VSWR}(\zeta) = \frac{1 + |\Gamma_L|}{1 - |\Gamma_L|} = \frac{1 + 0.209}{1 - 0.209} = 1.528$$

$$\text{VSWR}(0) = \frac{1 + |\Gamma(0)|}{1 - |\Gamma(0)|} = \frac{1 + \sqrt{(0.0675)^2 + (0.0371)^2}}{1 - \sqrt{(0.0675)^2 + (0.0371)^2}} = 1.167$$

10.271 Calculate the average power delivered to the line of Problems 10.267 to 10.269.

▮ The equivalent circuit for the system is shown in Fig. 10-53, from which

$$I(0) = \frac{20 \ \underline{/0°}}{700 + 804 \ \underline{/-13.49°}} = 0.0134 e^{j7.19°} \ \text{A}$$

The average power delivered to the line is
$$P_{\text{av}}(0) = \frac{1}{2} \text{Re} \left[V(0)I^*(0)\right] = \frac{1}{2} \text{Re} \left[|I(0)|^2 Z_{\text{in}}\right]$$
$$= \frac{1}{2} |I(0)|^2 |Z_{\text{in}}| \cos \theta_{\text{zin}} = \frac{1}{2}(0.0134)^2(804) \cos 13.49° = 70.2 \ \text{mW}$$

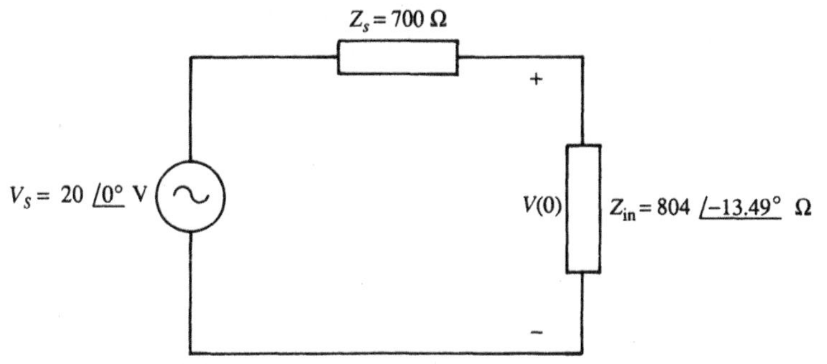

Fig. 10-53

10.272 For the line of Problems 10.267 to 10.271, calculate the load current and the average power delivered to the load.

    ❚ The current along the line is

$$I(z) = \frac{V_m^+}{Z_C} e^{-\gamma z}[1 - \Gamma(z)]$$

At $z = 0$, with $V_m^+/Z_C \equiv I_m^+$, we have $I(0) = I_m^+[1 - \Gamma(0)]$, or

$$I_m^+ = \frac{I(0)}{1 - \Gamma(0)} = \frac{0.0134\,\underline{/7.19°}}{1 - (0.0675 + j0.0371)} = 0.0125\,\underline{/5.2°} = 0.0125 e^{j0.091} \quad \text{A}$$

The load current is given by

$$I_L = I_m^+ e^{-\gamma\zeta}(1 - \Gamma_L) = 0.0125 e^{j0.091} e^{-(0.0083 + j0.035)60}(1 - 0.209 e^{j0.567}) = 6.32\,\underline{/-4.64°} \quad \text{mA}$$

and

$$P_{av}(\text{load}) = \tfrac{1}{2}|I_L|^2 R_L = \tfrac{1}{2}(6.32 \times 10^{-3})^2 1000 = 19.97\,\text{mW}$$

10.273 Given a rectangular waveguide with dimensions $a/b = 9/4$, list, in ascending order, the cutoff frequencies of the TE_{01}, TE_{10}, TE_{11}, TM_{11}, TE_{20}, TE_{21}, TE_{12}, TM_{21}, TM_{12}, TE_{22}, and TM_{22} modes.

    ❚ The cutoff frequency of an (m, n) mode is given by

$$f_{c.mn}^2 = \frac{u^2}{4}\left[\left(\frac{m}{a}\right)^2 + \left(\frac{n}{b}\right)^2\right] = \frac{u^2}{4a^2}\left[m^2 + \left(\frac{9}{4}n\right)^2\right] \equiv \frac{u^2}{4a^2} F(m, n)$$

Since the order of f is the order of f^2, is the order of F, the required list is as follows:

(m, n)	$(1, 0)$	$(2, 0)$	$(0, 1)$	$(1, 1)$	$(2, 1)$	$(0, 2)$	$(1, 2)$	$(2, 2)$
$F(m, n)$	1	4	5.0625	6.0625	9.0625	20.25	21.25	24.25

10.274 A rectangular waveguide is air-filled and operated at 30 GHz. Given that the cutoff frequency of the TM_{21} mode is 18 GHz, determine the wavelength, phase constant, phase velocity, and intrinsic impedance of this mode.

    ❚ First calculate $F_{21} = \sqrt{1 - (f_{c.21}/f)^2} = \sqrt{[1 - (18/30)^2]} = 0.8$. Then:

$$\lambda_{21} = \frac{\lambda_0}{F_{21}} = \frac{u}{fF_{21}} = \frac{3 \times 10^8}{30 \times 10^9 \times 0.8} = 12.5\,\text{mm}$$

$$\beta_{21} = \beta_0 F_{21} = \frac{\omega}{u} F_{21} = \frac{2\pi \times 30 \times 10^9}{3 \times 10^8} \times 0.8 = 502.7\,\text{rad/m}$$

$$u_{21} = \frac{u}{F_{21}} = \frac{3 \times 10^8}{0.8} = 3.75 \times 10^8\,\text{m/s}$$

$$\eta_{21} = \eta_0 F_{21} = 377 \times 0.8 = 301.6\,\Omega$$

10.275 For the guide of Problem 10.274, assuming that the operating frequency in TM_{21} is reduced to 15 GHz, determine the distance down the guide such that the input power is reduced in magnitude by 20 dB (a factor of 10).

    ❚ Since 15 GHz is below cutoff (18 GHz), there is no propagation; so that

$$\alpha_{21} = \beta_0\sqrt{(f_{c.21}/f)^2 - 1} = \frac{2\pi \times 15 \times 10^9}{3 \times 10^8} \sqrt{(18/15)^2 - 1} = 208.39\,\text{Np/m}$$

Then

$$e^{-\alpha_{21}d} = \frac{1}{10} \quad \text{or} \quad d = \frac{1}{\alpha_{21}} \ln 10 = \frac{1}{208.39}(2.3026) = 1.1\,\text{cm}$$

10.276 Will a 20-kHz antenna buried 2 ft in the ground be effective for communication with a point on the surface, if soil is characterized by $\epsilon_r = 2.8$, $\sigma = 2 \times 10^{-4}$ S/m, and $\mu_r = 1$? Take amplitude reduction to 5 percent of the maximum value as the criterion.

$\blacksquare$ Since

$$\frac{\sigma}{\omega\epsilon} = \frac{2 \times 10^{-4}}{2\pi \times 20 \times 10^3 \times 2.8 \times (10^{-9}/36\pi)} = 64.286 \gg 1$$

we can write (good conductor):

$$\alpha \approx \sqrt{\omega\mu\sigma/2} = \sqrt{2\pi \times 20 \times 10^3 \times 4\pi \times 10^{-7} \times 2 \times 10^{-4}/2} = 0.00397$$

For a 5 percent amplitude reduction,

$$e^{-\alpha d} = e^{-0.00397d} = 0.05 \qquad \text{or} \qquad d = -\frac{1}{0.00397}\ln 0.05 = 754.6 \text{ m}$$

which is much greater than 2 ft. Therefore, the antenna will be effective.

10.277 Can a 10-GHz radar system be used to look for metal ores 100 m below the surface of earth, for which $\epsilon_r = 2.5$, $\sigma = 8 \times 10^{-3}$ S/m, and $\mu_r = 1$?

$\blacksquare$

$$\frac{\sigma}{\omega\epsilon} = \frac{8 \times 10^{-3}}{2\pi \times 10 \times 10^9 \times 2.5 \times (10^{-9}/36\pi)} = 5.76 \times 10^{-3} \ll 1$$

For this good dielectric,

$$\alpha \approx \frac{\sigma}{2}\sqrt{\frac{\mu}{\epsilon}} = \frac{8 \times 10^{-3}}{2} \times \frac{120\pi}{\sqrt{2.5}} = 0.9537 \text{ m}^{-1}$$

For a 5 percent amplitude reduction,

$$e^{-\alpha d} = e^{-0.9537d} = 0.05 \qquad \text{or} \qquad d = -\frac{1}{0.9537}\ln 0.05 = 3.14 \text{ m} \ll 100 \text{ m}$$

implying that the radar system is ineffective.

10.278 The wave $\mathbf{E}_i = \mathbf{a}_x\, 10 \cos(3 \times 10^9 t - 30\pi z)$ (V/m) is normally incident on the planar boundary, $z = 0$, between two lossless dielectrics. Given $\mu_r = 1$ everywhere and $\epsilon_{r2} = 4$ for $z > 0$, find the reflection and transmission coefficients.

$\blacksquare$ Since $\beta_1 = 30\pi = \omega\sqrt{\mu_0\epsilon_0}\sqrt{\epsilon_{r1}}$,

$$\sqrt{\epsilon_{r1}} = \frac{30\pi \times 3 \times 10^8}{3 \times 10^9} = 3\pi$$

Then, since $\mu = \mu_0$ everywhere,

$$\Gamma = \frac{\sqrt{\epsilon_{r1}} - \sqrt{\epsilon_{r2}}}{\sqrt{\epsilon_{r1}} + \sqrt{\epsilon_{r2}}} = \frac{3\pi - 2}{3\pi + 2} = 0.65 \qquad \text{and} \qquad T = 1 + \Gamma = 1.65$$

10.279 From the data of Problem 10.278, find the instantaneous fields everywhere.

$\blacksquare$ The wave impedances are $\eta_1 = \eta_0/\sqrt{\epsilon_{r1}} = 40\ \Omega$ and $\eta_2 = \eta_0/\sqrt{\epsilon_{r2}} = 60\pi\ \Omega$, and $\beta_2 = (\sqrt{\epsilon_{r2}}/\sqrt{\epsilon_{r1}})\beta_1 = 20$ rad/m. Thus,

$$\mathbf{H}_i = \frac{|\mathbf{E}_i|}{\eta_1}\mathbf{a}_y = \mathbf{a}_y\, 0.25 \cos(3 \times 10^9 t - 30\pi z) \quad \text{(A/m)}$$

$$\mathbf{E}_r = \mathbf{a}_x\Gamma\, 10 \cos(3 \times 10^9 t + 30\pi z) = \mathbf{a}_x\, 6.5 \cos(3 \times 10^9 t + 30\pi z) \quad \text{(V/m)}$$

$$\mathbf{H}_r = \frac{|\mathbf{E}_r|}{\eta_1}(-\mathbf{a}_y) = -\mathbf{a}_y\, 0.1625 \cos(3 \times 10^9 t + 30\pi z) \quad \text{(A/m)}$$

$$\mathbf{E}_t = \mathbf{a}_x T\, 10 \cos(3 \times 10^9 t - 20z) = \mathbf{a}_x\, 16.5 \cos(3 \times 10^9 t - 20z) \quad \text{(V/m)}$$

$$\mathbf{H}_t = \frac{|\mathbf{E}_t|}{\eta_2}\mathbf{a}_y = \mathbf{a}_y\, 0.0875 \cos(3 \times 10^9 t - 20z) \quad \text{(A/m)}$$

10.280 Derive Snell's law from *Fermat's principle*.

$\blacksquare$ See Fig. 10-54. According to Fermat's principle, of all paths joining (x_1, y_1) and (x_2, y_2), the path of least time is the actual light path. Obviously, this quickest path must be a broken-line path. For that shown in Fig. 10-54 the traversal time may be taken as $(n \propto 1/u)$:

$$T = n_1 r_1 + n_2 r_2 = n_1\sqrt{(x - x_1)^2 + y_1^2} + n_2\sqrt{(x - x_2)^2 + y_2^2}$$

Differentiating for a minimum,

$$0 = \frac{dT}{dx} = n_1\frac{x - x_1}{r_1} + n_2\frac{x - x_2}{r_2} = \pm n_1 \sin\theta_i \mp n_2 \sin\theta_r$$

which is Snell's law.

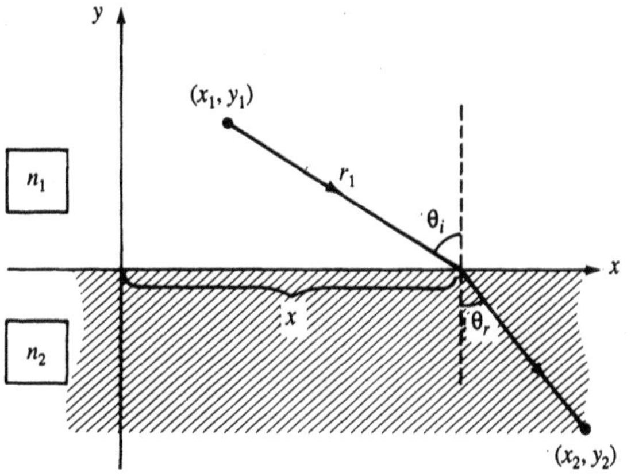

Fig. 10-54

10.281 Find the surface charge density at the surface of the wall of Problem 10.246.

▮ The **E**-field (and therefore the **D**-field) is tangential to the wall, so that $\rho_s \equiv 0$.

10.282 What is the Brewster angle (for parallel-polarized waves) at the interface of Problem 10.278?

▮
$$\tan \theta_B = \sqrt{\epsilon_{r2}}/\sqrt{\epsilon_{r1}} = 2/3\pi = 0.2122 \quad \text{or} \quad \theta_B = 12.0°$$

10.283 Two lossless nonmagnetic materials have a plane interface. A plane wave is incident on the boundary from region 1. Calculate the ratio $r \equiv \epsilon_{r2}/\epsilon_{r1}$, assuming that one-third of the incident power is transmitted into region 2.

▮ For these media,

$$\Gamma = \frac{1 - \sqrt{r}}{1 + \sqrt{r}} \tag{1}$$

and, by conservation of energy,

$$\frac{P_{\text{trans}}}{P_{\text{inc}}} = \frac{1}{3} = 1 - \Gamma^2 \tag{2}$$

Eliminate Γ between (1) and (2) to find

$$r = \left(\frac{1 - \sqrt{2/3}}{1 + \sqrt{2/3}}\right)^2 = 0.0102$$

10.284 The semi-infinite region $z < 0$ is free space, while $\epsilon = 20$ pF/m, $\mu = 5$ μH/m, and $\sigma = 0.004$ S/m for $z > 0$. A plane wave with
$$\mathbf{E}_i = \mathbf{a}_x 100 \cos (10^8 t - \beta_1 z) \quad \text{(V/m)}$$
is incident on the boundary from medium 1. Find the propagation and transmission coefficients.

▮
$$\frac{\sigma_2}{\omega\epsilon_2} = \frac{0.004}{10^8 \times 20 \times 10^{-12}} = 2$$

$$\beta_1 = \frac{\omega}{u} = \frac{10^8}{3 \times 10^8} = 0.333 \text{ rad/m}$$

$$\gamma_2 = \sqrt{j\omega\mu_2(\sigma_2 + j\omega\epsilon_2)} = j\omega\sqrt{\mu_2\epsilon_2}\sqrt{1 - (j\sigma_2/\omega\epsilon_2)}$$
$$= j10^8\sqrt{5 \times 10^{-6} \times 20 \times 10^{-12}}\sqrt{1 - j2} = 0.7862 + j1.272 \text{ m}^{-1} \equiv \alpha_2 + j\beta_2$$

$$\eta_2 = \sqrt{j\omega\mu_2/(\sigma_2 + j\omega\epsilon_2)} = \sqrt{\mu_2/\epsilon_2}/\sqrt{1 - (j\sigma_2/\omega\epsilon_2)}$$
$$= \sqrt{\left(\frac{5 \times 10^{-6}}{20 \times 10^{-12}}\right)}\bigg/\sqrt{1 - j2} = 284.43 + j175.79 \quad \Omega$$

Therefore,

$$\Gamma = \frac{\eta_2 - \eta_1}{\eta_2 + \eta_1} = \frac{284.43 + j175.79 - 120\pi}{284.43 + j175.79 + 120\pi} = 0.2903e^{j1.796}$$

$$T = \frac{2\eta_2}{\eta_2 + \eta_1} = \frac{2(284.43 + j175.79)}{284.43 + j175.79 + 120\pi} = 0.9771e^{j0.294}$$

10.285 Find the total instantaneous fields in media 1 and 2 of Problem 10.284.

❚ Proceeding as in Problem 10.279, but using Γ and T from Problem 10.284,

$\mathbf{E}_i = \mathbf{a}_x \, 100 \cos (10^8 t - 0.333z)$ (V/m)

$\mathbf{E}_r = \mathbf{a}_x \, 29.03 \cos (10^8 t + 0.333z + 1.796)$ (V/m)

$\mathbf{H}_i = \mathbf{a}_y \dfrac{100}{120\pi} \cos (10^8 t - 0.333z) = \mathbf{a}_y \, 0.2653 \cos (10^8 t - 0.333z)$ (A/m)

$\mathbf{H}_r = -\mathbf{a}_y \dfrac{29.03}{120\pi} \cos (10^8 t + 0.333z + 1.796) = -\mathbf{a}_y \, 0.077 \cos (10^8 t + 0.333z + 1.796)$ (A/m)

$\mathbf{E}_t = \mathbf{a}_x \, 97.71 \, e^{-0.7862z} \cos (10^8 t - 1.272z + 0.294)$ (V/m)

$\mathbf{H}_t = \mathbf{a}_y \left(\dfrac{97.71}{334.37} \right) e^{-0.7862z} \cos (10^8 t - 1.272z + 0.294 - 0.554) = \mathbf{a}_y \, 0.292 \, e^{-0.7862z} \cos (10^8 t - 1.272z - 0.26)$ (A/m)

Printed in the USA
CPSIA information can be obtained
at www.ICGtesting.com
JSHW051458221024
72172JS00012B/104